T0327561

APPLIED STATISTICS FOR CIVIL AND ENVIRONMENTAL ENGINEERS

APPLIED STATISTICS FOR CIVIL AND ENVIRONMENTAL ENGINEERS

Second Edition

Nathabandu T. Kottegoda
Department of Hydraulic, Environmental, and Surveying Engineering
Politecnico di Milano, Italy

Renzo Rosso
Department of Hydraulic, Environmental, and Surveying Engineering
Politecnico di Milano, Italy

Blackwell
Publishing

This edition first published 2008

Blackwell Publishing was acquired by John Wiley & Sons in February 2007. Blackwell's publishing programme has been merged with Wiley's global Scientific, Technical, and Medical business to form Wiley-Blackwell.

Registered office
John Wiley & Sons Ltd, The Atrium, Southern Gate, Chichester, West Sussex, PO19 8SQ, United Kingdom

Editorial office
9600 Garsington Road, Oxford, OX4 2DQ, United Kingdom

For details of our global editorial offices, for customer services and for information about how to apply for permission to reuse the copyright material in this book please see our website at www.wiley.com/wiley-blackwell.

ISBN: 978-1-4051-7917-1

Library of Congress Cataloging-in-Publication Data

Kottegoda, N. T.
Applied statistics for civil and environmental engineers / Nathabandu T. Kottegoda, Renzo Rosso. – 2nd ed.
p. cm.
Prev. ed. published as: Statistics, probability, and reliability for civil and environmental engineers. New York : McGraw-Hill, c1997.
Includes bibliographical references and index.
ISBN-13: 978-1-4051-7917-1 (hardback : alk. paper)
ISBN-10: 1-4051-7917-1 (hardback : alk. paper) 1. Civil engineering–Statistical methods. 2. Environmental engineering–Statistical methods. 3. Probabilities. I. Rosso, Renzo. II. Kottegoda, N. T. Statistics, probability, and reliability for civil and environmental engineers. III. Title.
TA340.K67 2008
519.502′4624–dc22 2007047496

A catalogue record for this book is available from the British Library.

Set in 10/12pt Times by Aptara Inc., New Delhi, India

1 2008

Contents

Dedication

To my parents. To estimate the debt I owe them requires a lifespan of nibbanic extent. To Mali, Shani, Siraj, and Natasha. N.T.K.

A mamma Aria, a Donatella, ai due Riccardi della mia vita e al nostro indimenticabile Rufus. R.R.

Preface to the First Edition

Statistics, probability, and reliability are subject areas that are not commonly easy for students of civil and environmental engineering. Such difficulties notwithstanding, a greater emphasis is currently being made on the teaching of these methods throughout institutions of higher learning. Many professors with whom we have spoken have expressed the need for a single textbook of sufficient breadth and clarity to cover these topics.

One might ask why it is necessary to write a new book specifically for civil and environmental engineers. Firstly, we see a particular importance of statistical and associated methods in our disciplines. For example, some modes of failure, interactions, probability distributions, outliers, and spatial relationships that one encounters are unique and require different approaches. Secondly, colleagues have said that existing books are either old and outdated or omit particularly important engineering problems, emphasizing instead areas that may not be directly relevant to the practitioner.

We set ourselves several objectives in writing this book. First, it was necessary to update much of the older material, which have rightly stood for decades, even centuries. Indeed. Second, we had to look at the engineer's structures, waterways, and the like and bring in as much material as possible for the tasks at hand. We felt an urgent need to modernize, incorporate new concepts throughout, and reduce or eliminate the impact of some topics. We aimed to order the material in a logical sequence. In particular we tried to adopt a writing style and method of presentation that are lively and without overrigorous drudgery. These had to be accomplished without compromising a deep and thorough treatment of fundamentals.

The layout of the book is sraightforward, so it can be used to suit one's personal needs. We apologize to any readers who think we have strayed from the path of simplicity in certain parts, such as the associated variables and contagious distributions of Chapter 3 and the order statistics of Chapter 7. One might wish to omit these sections on a first reading. The introductions to the chapters will be helpful for this purpose.

The explanation of the theory is accompanied by the assumptions made. Definitions are separately highlighted. In many places we point out the limitations and pitfalls or violations. There are warnings of possible misuses, misunderstandings, and misinterpretations. We provide guidance to the proper interpretation of statistical results.

The numerous examples, for which we have for the most part used recorded observations, will be helpful to beginners as well as to mature students who will consult the text as a reference. We hope these examples will lead to a better understanding of the material and design variabilities, a prelude to the making of sound decisions.

Each chapter concludes with extensive homework problems. In many instances, as in Chapter 1, they are based on real data not used elsewhere in the text. We have not used cards or dice or coins or black and red balls in any of the problems and examples. Answers to selected problems are summarized in Appendix D. A detailed manual of solutions is available.

Computers are continuously becoming cheaper and more powerful. Newer ways of handling data are being devised. At the inception, we seriously considered the use of commercial software packages to enhance the scope of the book. However, the problem of choosing one, from the many suitable packages acted as a deterrent. Our concern was the serious limitations imposed by utilizing a source that necessitates corresponding purchase by an adopting school or by individual engineers. Besides, the calculations illustrated in the book can be made using worksheets available as standard software for personal computers. As an aid, the data in Appendix E will be placed on the Internet.

We have utilized the space saved (from jargon and notation of a particular software, output, graphs, and tables) to widen the scope, make our explanations more thorough, and insert additional illustrations and problems. Readers also have an almost all-inclusive index, a comprehensive glossary of notation, additional mathematical explanations, and other material in the appendixes. Furthermore, we hope that the extensive, annotated bibliographies at the end of each chapter, numerous citations and tables, will make this a useful reference source.

The book is written for use by students, practicing engineers, teachers, and researchers in civil and environmental engineering and applied statistics; female readers will find no hint of male chauvinism here. It is designed for a one- or two-semester course and is suitable for final-year undergraduate and first-year graduate students. The text is self-contained for study by engineers. A background of elementary calculus and matrix algebra is assumed.

ACKNOWLEDGMENTS

We acknowledge with thanks the work of the staff at Publication Services, Inc., in Champaign, IL. Gianfausto Salvadori gave his time generously in reviewing the manuscript and providing solutions to some homework problems. Thanks are due again to Adri Buishand for his elaborate and painstaking reviews. Our publisher solicited other reviewers whose reports were useful. Howard Tillotson and colleagues at the University of Birmingham, England, provided data and some student problems. Discussions with Tony Lawrance at lunch in the University Staff House and the example problem he solved at Helsinki Airport are appreciated. Valuable assistance was provided by Giovanni Solari and Giulio Ballio in wind and steel engineering, respectively. In addition, Giovanni Vannuchi was consulted on geotechnical engineering. Research staff and doctoral students at the Politecnico di Milano helped with the homework problems and the preparation of the index. Dora Tartaglia worked diligently on revisions to the manuscript. We thank the publishers, companies, and individuals who gave us permission to use their material, data, and tables; some of the tables were obtained through our own resources We shall be pleased to have any omissions brought to our notice. The support and hospitality provided at the Università degli Studi di Pavia by Luigi Natale and others are acknowledged with thanks. Most importantly, without the patience and tolerance of our families this book could not have been completed.

N. T. Kottegoda
R. Rosso
Milano, Italy
1 July 1996

Preface to the Second Edition

Last year a senior European professor, who uses our book, was visiting us in Milano. When told of the revisions underway he expressed some surprise. "There is nothing to revise," he said. But all books need revision sooner or later, especially a multidimensional one. The equations, examples, problems, figures, tables, references, and footnotes are all subject to inevitable human fallibilities: typographical errors and errors of fact. Our first objective was to bring the text as close to the ideal state as possible. The second priority was to modernize.

In Chapter 10, a new section is added on Markov chain Monte Carlo modeling; this has popularized Bayesian methods in recent years; there is a full description and case study on Gibbs sampling. In Chapter 8 on simulation, we include a new section on sensitivity analysis and uncertainty analysis; a clear and detailed distinction is made between epistemic and aleatory uncertainties; their implications in decision-making are discussed. In Chapter 7 on Frequency Analysis of Extreme Events, natural hazards and flood hydrology are updated. In Chapter 6 on regression analysis, further considerations have been made on the diagnostics of regression; there are new discussions on general and generalized linear models. In Chapter 5 on Model Estimation and Testing we give special importance to the Anderson-Darling goodness-of-fit test because of its sensitivity to departures in the tail areas of a probability density function; we make applications to nonnormal distributions using the same data as in the estimation of parameters. In Chapter 3 a section is added on the novel method of copulas with particular emphasis on bivariate distributions. We have revised the problems following Basic Probability Concepts in Chapter 2. Other chapters are also revised and modernized and the annotated references are updated.

As before, we have kept in mind the scientific method of Claude Bernard, the French medical researcher of the nineteenth century. This had three essential parts: observation of phenomena in nature (seen in Appendix E, and in the examples and problems), observation of experiments (as reported in each chapter), and the theoretical part (clear enough for the audience in mind, but without over-simplification).

"Nobody trusts a model except the one who originated it; everybody trusts data except those who record it." Models and data are subject to uncertainty. There is still a gap between models and data. We attempt to bridge this gap.

The title of the book has been abridged from *Statistics, Probability, and Reliability for Civil and Environmental Engineers* to *Applied Statistics for Civil and Environmental Engineers.* The applications and problems pertain almost equally to both disciplines and all areas are included.

Another aspect we emphasized before was that the calculations illustrated in the book can be made using worksheets available as standard software for personal computers. Alternatively, R which is now commonplace can be downloaded free of charge and adopted to run some of the homework problems, if one so prefers. Our decision not to recommend the use of particular commercial software packages, by giving details of jargon, notation, and so on, seems to be justified. We find that a specific version soon become obsolete with the advent of a new version.

A limited access solutions manual is available with the data from Appendix E on the Wiley-Blackwell website [www.blackwellpublishing.com/kottegoda].

We are grateful for the encouragement given by many users of the first edition, and to the few who pointed out some discrepancies. We thank the anonymous reviewers for their useful comments. Gianfausto Salvadori, Carlo De Michele, Adri Buishand, and Tony Lawrance assisted us again in the revisions. Julia Burden and Lucy Alexander of Blackwell Publishing supported us throughout the project. Università degli Studi di Pavia is thanked for continued hospitality. The help provided by Fabrizio Borsa and Enrico Raiteri in the preparation of some figures is acknowledged.

N. T. Kottegoda
R. Rosso
Milano, Italy
14 September 2007

Introduction

As a wide-ranging discipline, statistics concerns numerous procedures for deriving information from data that have been affected by chance variations. On the basis of scientific experiments, one may record and make summaries of observations, quantify variations, or other changes of significance, and compare data sequences by means of some numbers or characteristics. The use of statistics in this way is for descriptive purposes. At a more sophisticated level of analysis and interpretation, one can, for instance, test hypotheses using the inferential approach developed during the twentieth century. Thus it may be ascertained, for instance, whether the change of an ingredient affects the properties of a concrete or whether a particular method of surfacing produces a longer-lasting road; this approach often includes the estimation by means of observations of the parameters of a statistical model. Then inferences can be drawn from data and predictions made or decisions taken. When faced with uncertainty, this last phase is the principal aim of a civil or environmental engineer acting as an applied statistician.

In all activities, engineers have to cope with possible uncertainties. Observations of soil pressures, tensile strengths of concrete, yield strengths of steel, traffic densities, rainfalls, river flows, and pollution loads in streams vary from one case to the next for apparently unknown reasons or on account of factors that cannot be assessed to any degree of accuracy. However, designs need to be completed and structures, highways, water supply, and sewerage schemes constructed. Sound engineering judgment, in fact, springs from physical and mathematical theories, but it goes far beyond that. Randomness in nature must be taken into account. Thus the onus of dealing with the uncertainties lies with the engineer.

The appropriate methods of tackling the uncertainty vary with different circumstances. The key is often the dispersion that is commonly evidenced in available data sets. Some phenomena may have negligible or low variability. In such a case, the mean of past observations may be used as a descriptor, for example, the elastic constant of a steel. Nevertheless, the consequences of a possible change in the mean should also be considered. Frequently, the variability in observations is found to be quite substantial. In such situations, an engineer sometimes uses, rather conservatively, a design value such as the peak storm runoff or the compressive strength of a concrete. Alternatively, it has been the practice to express the ability of a component in a structure to withstand a specified loading without failure or a permissible deflection by a so-called factor of safety; this is in effect a blanket to cover all possible contingencies. However, we envisage some problems here in following a purely deterministic approach because there are doubts concerning the consistency of specified strengths, flows, loads, or factors from one case to another. These cannot be lightly dismissed or easily compounded when the consequences of ignoring variability are detrimental or, in general, if the decision is sensitive to a particular uncertainty. (Often there are crucial economic considerations in these matters.) This obstacle strongly suggests that the way forward is by treating statistics and probability as necessary aids in decision making, thus coping with uncertainty through the engineering process.

Note that statistical methods are in no way intended to replace the physical knowledge and experience of the engineer and his or her skills in experimentation. The engineer should know how the measurements are made and recorded and how errors may arise from possible limitations in the equipment. There should be readiness to make changes and improvements so that the data-gathering process is as reliable and representative as possible.

On this basis, statistics can be a complementary and a valuable aid to technology. In prudent hands it can lead to the best practical assessment of what is partially known or uncertain.

The quantification of uncertainty and the assessment of its effects on design and implementation must include concepts and methods of probability, because statistics is built on the foundation of probability theory. In addition, decision making under risk involves the use of applied probability. Historically, probability theory arose as a branch of mathematics concerned with the analysis of certain games of chance; it consequently found applications in the measurement and understanding of uncertainty in innumerable natural phenomena and human activities. The fundamental interrelationship between statistics and probability is clearly evident in practice. As seen in past decades, there has been an irreversible change in emphasis from descriptive to inferential statistics. In this respect we must note that statistical inferences and the risk and reliability of decision making under uncertainty are evaluated through applied probability, using frequentist or Bayesian estimation. This applies to the most widely used methods. Alternatives that come under generalized information theory are now available.

The reliability of a system, structure, or component is the complement of its probability of failure. Risk and reliability analysis, however, entail many activities. The survival probability of a system is usually stated in terms of the reliabilities of its components. The modeling process is an essential part of the analysis, and time can be an important factor. Also, the risk factor that one computes may be inherent, additional, or composite. All these points show that reliability design deserves special emphasis.

Methods of reducing data, reviewed in Chapter 1, begin with tabulation and graphical representation, which are necessary first steps in understanding the uncertainty in data and the inherent variability. Numerical summaries provide descriptions for further analysis. Exploratory methods are followed by relationships between data observed in pairs. Thus the investigation begins. The route is long and diverse, because *statistics is the science and art of experimenting, collecting, analyzing, and making inferences from data.* This opening chapter provides a route map of what is to follow so that one can gain insight into the numerous tools statistics offers and realizes the variety of problems that can be tackled. In Chapters 2 and 3, we develop a background in probability theory for coping with uncertainty in engineering. Using basic concepts, we then discuss the total probability and Bayes' theorems and define statistical properties of distributions used for estimation purposes. Chapter 4 examines various mathematical models of random processes. There is a wide-ranging discussion of discrete and continuous distributions; joint and derived types are also given in Chapters 3 and 4; we introduce copulas that can effectively model joint distributions. Model estimation and testing methods, such as confidence intervals, hypothesis testing, analysis of variance, probability plotting, and identification of outliers, are treated in Chapter 5. The estimation and testing are based on the principle that all suppositions need to be carefully examined in light of experimentation and observation. Details of regression and multivariate statistical methods are provided in Chapter 6, along with principal component analysis and associated methods and spatial correlation. Extreme value analysis applied to floods, droughts, winds, earthquakes, and other natural hazards is found in Chapter 7; some special types of models are included. Simulation is the subject of Chapter 8, which comprises the use of simulation in design and for other practical purposes; also, we discuss sensitivity analysis and uncertainty analysis of the aleatory and epistemic types. In Chapter 9, risk and reliability analysis and reliability design are developed in detail. Chapter 10 is devoted to Bayesian and other types of economic decision making, used when the engineer faces uncertainty; we include here Markov chain Monte Carlo methods that have recently popularized the Bayesian approach.

Chapter 1
Preliminary Data Analysis

All natural processes, as well as those devised by humans, are subject to variability. Civil engineers are aware, for example, that crushing strengths of concrete, soil pressures, strengths of welds, traffic flow, floods, and pollution loads in streams have wide variations. These may arise on account of natural changes in properties, differences in interactions between the ingredients of a material, environmental factors, or other causes. To cope with uncertainty, the engineer must first obtain and investigate a *sample* of data, such as a set of flow data or triaxial test results. The sample is used in applying statistics and probability at the descriptive stage. For inferential purposes, however, one needs to make decisions regarding the *population* from which the sample is drawn. By this we mean the total or aggregate, which, for most physical processes, is the virtually unlimited universe of all possible measurements. The main interest of the statistician is in the aggregation; the individual items provide the hints, clues, and evidence.

A data set comprises a number of measurements of a phenomenon such as the failure load of a structural component. The quantities measured are termed *variables*, each of which may take any one of a specified set of values. Because of its inherent randomness and hence unpredictability, a phenomenon that an engineer or scientist usually encounters is referred to as a *random variable*, a name given to any quantity whose value depends on chance.[1] Random variables are usually denoted by capital letters. These are classified by the form that their values can possibly take (or are assumed to take). The pattern of variability is called a *distribution*. A *continuous* variable can have any value on a continuous scale between two limits, such as the volume of water flowing in a river per second or the amount of daily rainfall measured in some city. A *discrete* variable, on the contrary, can only assume countable isolated numbers like integers, such as the number of vehicles turning left at an intersection, or other distinct values.

Having obtained a sample of data, the first step is its presentation. Consider, for example, the modulus of rupture data for a certain type of timber shown in Table E.1.1, in Appendix E. The initial problem facing the civil engineer is that such an array of data by itself does not give a clear idea of the underlying characteristics of the stress values in this natural type of construction material. To extract the salient features and the particular types of information one needs, one must summarize the data and present them in some readily comprehensible forms. There are several methods of presentation, organization, and reduction of data. Graphical methods constitute the first approach.

1.1 GRAPHICAL REPRESENTATION

If "a picture is worth a thousand words," then graphical techniques provide an excellent method to visualize the variability and other properties of a set of data. To the powerful interactive system of one's brain and eyes, graphical displays provide insight into the form

[1] The term will be formally defined in Section 3.1.

and shape of the data and lead to a preliminary concept of the generating process. We proceed by assembling the data into graphs, scanning the details, and noting the important characteristics. There are numerous types of graphs. Line and dot diagrams, histograms, relative frequency polygons, and cumulative frequency curves are given in this section. Subsequently, exploratory methods, such as stem-and-leaf plots and box diagrams and graphs depicting a possible association between two variables, are presented in Sections 1.3 and 1.4. We begin with the simple task of counting.

1.1.1 Line diagram or bar chart

The occurrences of a discrete variable can be classified on a line diagram or bar chart. In this type of graph, the horizontal axis gives the values of the discrete variable and the occurrences are represented by the heights of vertical lines. The horizontal spread of these lines and their relative heights indicate the variability and other characteristics of the data.

Example 1.1. Flood occurrences. Consider the annual number of floods of the Magra River at Calamazza, situated between Pisa and Genoa in northwestern Italy, over a 34-year period, as shown in Table 1.1.1.

A flood in the river at the point of measurement means the river has risen above a specified level, beyond which the river poses a threat to lives and property. The data are plotted in Fig. 1.1.1 as a line diagram.

The data suggest a symmetrical distribution with a midlocation of four floods per year. In some other river basins, there is a nonlinear decrease in the occurrences for increasing numbers of floods in a year commencing at zero, showing a negative exponential type of variation.

1.1.2 Dot diagram

A different type of graph is required to present continuous data. If the data are few (say, less than 25 items) a dot diagram is a useful visual aid. Consider the possibility that only

Table 1.1.1 Number of flood occurrences per year from 1939 to 1972 at the gauging station of Calamazza on the Magra River, between Pisa and Genoa in northwestern Italy[a]

Number of floods in a year	Number of occurrences
0	0
1	2
2	6
3	7
4	9
5	4
6	1
7	4
8	1
9	0
Total	34

[a] A flood occurrence is defined as river discharge exceeding 300 m^3/s.

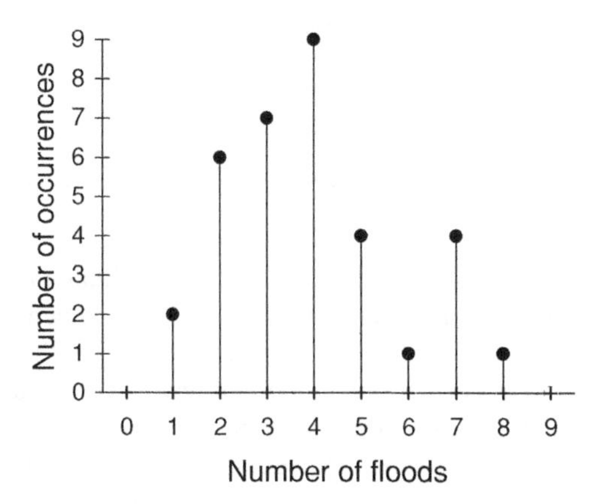

Fig. 1.1.1 Line diagram for flood occurrences in the Magra River at Calamazza between Genoa and Pisa in northwestern Italy.

the first 15 items of data in Table E.1.1—which shows the modulus of rupture in N/mm^2 for 50 mm $\times$ 150 mm Swedish redwood and whitewood—are available. The abridged data are ranked in ascending order and are given in Table 1.1.2 and plotted in Fig. 1.1.2.

The reader can see that the midlocation is close to 40 N/mm^2 but the wide spread makes this location difficult to discern. A larger sample should certainly be helpful.

1.1.3 Histogram

If there are at least, say, 25 observations, one of the most common graphical forms is a block diagram called the *histogram*. For this purpose, the data are divided into groups according to their magnitudes. The horizontal axis of the graph gives the magnitudes. Blocks are drawn to represent the groups, each of which has a distinct upper and lower limit. The area of a block is proportional to the number of occurrences in the group. The variability of the data is shown by the horizontal spread of the blocks, and the most common values are found in blocks with the largest areas. Other features such as the symmetry of the data or lack of it are also shown.

The first step is to take into account the *range* r of the observations, that is, the difference between the largest and smallest values.

Example 1.2. Timber strength. We go back to the timber strength data given in Table E.1.1. They are arranged in order of magnitude in Table 1.1.3.

There are $n = 165$ observations with somewhat high variability, as expected, because timber is a naturally variable material. Here the range $r = 70.22 - 0.00 = 70.22$ N/mm^2.

To draw a histogram, one divides the range into a number of *classes* or *cells* n_c. The number of occurrences in each class is counted and tabulated. These are called *frequencies*.

Table 1.1.2 The first 15 items of modulus of rupture data measuring timber strengths in N/mm^2, from Table E.1.1 (commencing with the top row), ranked in increasing order

29.11	29.93	32.02	32.40	33.06	34.12	35.58	39.34
40.53	41.64	45.54	48.37	48.78	50.98	65.35	

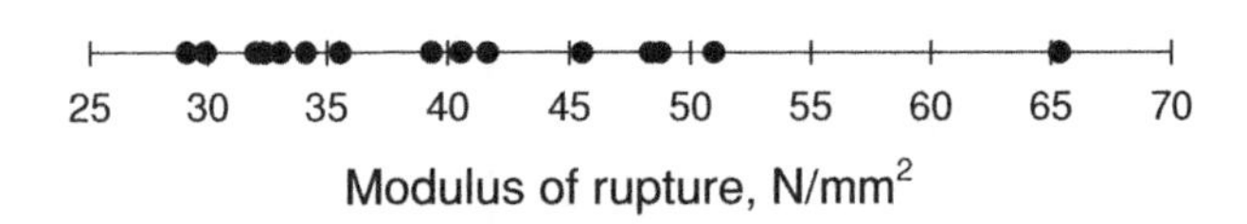

Fig. 1.1.2 Dot diagram for a short sample of timber strengths from Table 1.1.3.

The width of the classes is usually made equal to facilitate interpretation. For some work such as the fitting of a theoretical function to observed frequencies, however, unequal class widths are used. Care should be exercised in the choice of the number of classes, n_c. Too few will cause an omission of some important features of the data; too many will not give a clear overall picture because there may be high fluctuations in the frequencies. A rule of thumb is to make $n_c = \sqrt{n}$ or an integer close to this, but it should be at least 5 and not greater than 25. Thus, histograms based on fewer than 25 items may not be meaningful. Sturges (1926) suggested the approximation

$$n_c = 1 + 3.3 \log_{10} n. \tag{1.1.1}$$

A more theoretically based alternative follows the work of Freedman and Diaconis (1981):[2]

$$n_c = \frac{r\, n^{1/3}}{2 \text{ iqr}}. \tag{1.1.2}$$

Here iqr is the *interquartile range*. To clarify this term, we must define Q_2, or the *median*. This denotes the middle term of a set of data when the values are arranged in ascending order, or the average of the two middle terms if n is an even number. The first or lower quartile, Q_1, is the median of the lower half of the data, and likewise the third

Table 1.1.3 Ranked modulus of rupture data for timber strengths in N/mm^2, in ascending order[a]

0.00	28.00	31.60	34.44	36.84	39.21	41.75	44.30	47.25	53.99
17.98	28.13	32.02	34.49	36.85	39.33	41.78	44.36	47.42	54.04
22.67	28.46	32.03	34.56	36.88	39.34	41.85	44.36	47.61	54.71
22.74	28.69	32.40	34.63	36.92	39.60	42.31	44.51	47.74	55.23
22.75	28.71	32.48	35.03	37.51	39.62	42.47	44.54	47.83	56.60
23.14	28.76	32.68	35.17	37.65	39.77	43.07	44.59	48.37	56.80
23.16	28.83	32.76	35.30	37.69	39.93	43.12	44.78	48.39	57.99
23.19	28.97	33.06	35.43	37.78	39.97	43.26	44.78	48.78	58.34
24.09	28.98	33.14	35.58	38.00	40.20	43.33	45.19	49.57	65.35
24.25	29.11	33.18	35.67	38.05	40.27	43.33	45.54	49.59	65.61
24.84	29.90	33.19	35.88	38.16	40.39	43.41	45.92	49.65	69.07
25.39	29.93	33.47	35.89	38.64	40.53	43.48	45.97	50.91	70.22
25.98	30.02	33.61	36.00	38.71	40.71	43.48	46.01	50.98	
26.63	30.05	33.71	36.38	38.81	40.85	43.64	46.33	51.39	
27.31	30.33	33.92	36.47	39.05	40.85	43.99	46.50	51.90	
27.90	30.53	34.12	36.53	39.15	41.64	44.00	46.86	53.00	
27.93	31.33	34.40	36.81	39.20	41.72	44.07	46.99	53.63	

[a] The original data set is given in Table E.1.1; $n = 165$. The median is underlined.

[2] See also Scott (1979).

Table 1.1.4 Frequency computations for the modulus of rupture data ranked in Table 1.1.3[a]

Class upper limit (N/mm^2)	Class center (N/mm^2)	Absolute frequency	Relative frequency	Cumulative relative frequency (%)
5	2.5	1	0.006	0.61
10	7.5	0	0.000	0.61
15	12.5	0	0.000	0.61
20	17.5	1	0.006	1.21
25	22.5	9	0.055	6.67
30	27.5	18	0.109	17.58
35	32.5	26	0.158	33.33
40	37.5	38	0.230	56.36
45	42.5	34	0.206	76.97
50	47.5	20	0.121	89.09
55	52.5	9	0.055	94.55
60	57.5	5	0.030	97.58
65	62.5	0	0.000	97.58
70	67.5	3	0.018	99.39
75	72.5	1	0.006	100.00

[a] The width of each class is 5 N/mm^2 in this example.

or upper quartile, Q_3, is the median of the upper half of the data. This definition will be used throughout.[3] Thus,

$$\text{iqr} = Q_3 - Q_1. \tag{1.1.3}$$

Example 1.3. Timber strength. For the timber strength data of Table E.1.1, the median, that is, Q_2, is 39.05 N/mm^2. Also Q_3 and Q_1 are 44.57 and 32.91 N/mm^2, respectively, and hence iqr = 11.66 N/mm^2. From the simple square-root rule, the number of classes, n_c = 12.84. However, by using Eqs. (1.1.1) and (1.1.2), the number of classes are 8.32 and 16.52, respectively. If these are rounded to 9 and 15 and the range is extended to 72 and 75 N/mm^2 for graphical purposes, the equal class widths become 8 and 5 N/mm^2, respectively. Let us use these widths. It is important to specify the class boundaries without ambiguity for the counting of frequencies; for example, in the first case, these should be from 0 to 7.99, 8.00 to 15.99, and so on. As already mentioned, the vertical axis of a histogram is made to represent the frequency and the horizontal axis is used as a measurement scale on which the class boundaries are marked. For each of these class widths, 8 and 5 N/mm^2, class boundaries are made and counting of frequencies is completed using Table 1.1.3; the lowest boundary is at 0 and the highest boundaries are at 72 and 75 N/mm^2, respectively. Table 1.1.4 gives the absolute and relative frequencies for class widths of 5 N/mm^2.

Rectangles are then erected over each of the classes, proportional in area to the class frequencies. When equal class widths are used, as shown here, the heights of the rectangles represent the frequencies. Thus, Figs. 1.1.3 and 1.1.4 are obtained.

The information conveyed by the two histograms seems to be similar. The diagrams are almost symmetrical with a peak in the class below 40 N/mm^2 and a steady decrease on either side. This type of diagram usually brings out any possible imperfections in the data, such as

[3] There are alternatives, such as rounding $(n + 1)/4$ and $(n + 1) \times (3/4)$ to the nearest integers to calculate the locations of Q_1 and Q_3, respectively. The rounding is upward or downward, respectively, when the numbers fall exactly between two integers.

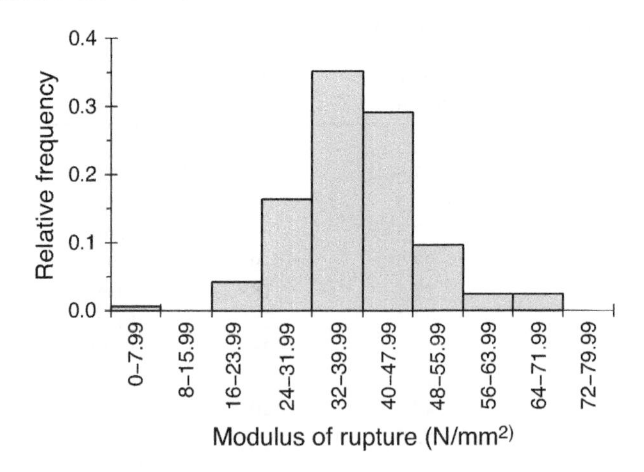

Fig. 1.1.3 Histogram for timber strength data with class width of 8 N/mm^2.

the gaps at the ends. Further investigations are required to understand the true nature of the population. More on these aspects will follow in this and subsequent chapters.

1.1.4 Frequency polygon

A frequency polygon is a useful diagnostic tool to determine the distribution of a variable. It can be drawn by joining the midpoints of the tops of the rectangles of a histogram after extending the diagram by one class on both sides. We assume that equal class widths are used. If the ordinates of a histogram are divided by the total number of observations, then a relative frequency histogram is obtained. Thus, the ordinates for each class denote the *probabilities* bounded by 0 and 1, by which we simply mean the chances of occurrence. The resulting diagram is called the relative frequency polygon.

Example 1.4. Timber strength. Corresponding to the histogram of Fig. 1.1.4, the values of class center are computed and a relative frequency polygon is obtained; this is shown in Fig. 1.1.5.

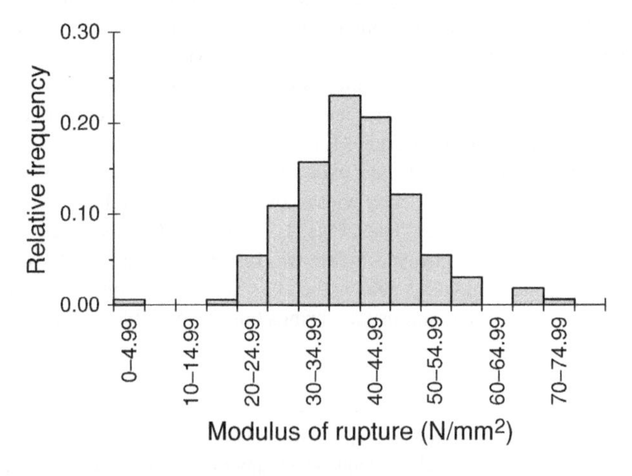

Fig. 1.1.4 Histogram for timber strength data with class width of 5 N/mm^2.

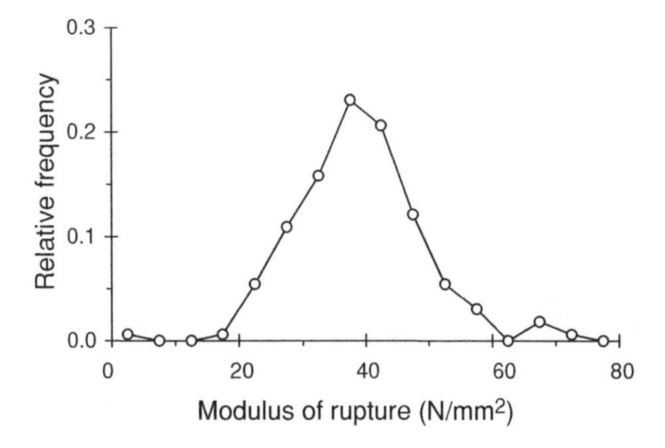

Fig. 1.1.5 Relative frequency polygon for timber strength data with class width of 5 N/mm^2.

As the number of observations becomes large, the class widths theoretically tend to decrease and, in the limiting case of an infinite sample, a relative frequency polygon becomes a frequency curve. This is in fact a probability curve, which represents a mathematical *probability density function*, abbreviated as pdf, of the population.[4]

1.1.5 Cumulative relative frequency diagram

If a cumulative sum is taken of the relative frequencies step by step from the smallest class to the largest, then the line joining the ordinates (cumulative relative frequencies) at the ends of the class boundaries forms a cumulative relative frequency or probability diagram. On the vertical axis of the graph, this line gives the probabilities of nonexceedance of values shown on the horizontal axis. In practice, this plot is made by utilizing and displaying every item of data distinctly, without the necessity of proceeding via a histogram and the restrictive categories that it entails. For this purpose, one may simply determine (e.g., from the ranked data of Table 1.1.3) the number of observations less than or equal to each value and divide these numbers by the total number of observations. This procedure is adopted here.[5]

Thus, the probability diagram, as represented by the cumulative relative frequency diagram, becomes an important practical tool. This diagram yields the median and other quartiles directly. Also, one can find the 9 values that divide the total frequency into 10 equal parts called *deciles* and the so-called *percentiles*, where the pth percentile is the value that is greater than p percent of the observations. In general, it is possible to obtain the $(n - 1)$ values that divide the total frequency into n equal parts called the *quantiles*. Hence a cumulative frequency polygon is also called a *quantile* or *Q-plot*; a Q-plot though has quantiles on the vertical axis unlike a cumulative frequency diagram.

Example 1.5. Timber strength. Figure 1.1.6 is the cumulative frequency diagram obtained from the ranked timber strength data of Table 1.1.3 using each item of data as just described.

---

[4] This function is discussed in Chapter 3. One of the first tasks in applying inferential statistics, as presented in Chapters 4 and 5, will be to estimate the mathematical function from a finite sample and examine its closeness to the histogram.

[5] Further aspects of this subject, as related to probability plots, are described in Chapter 5.

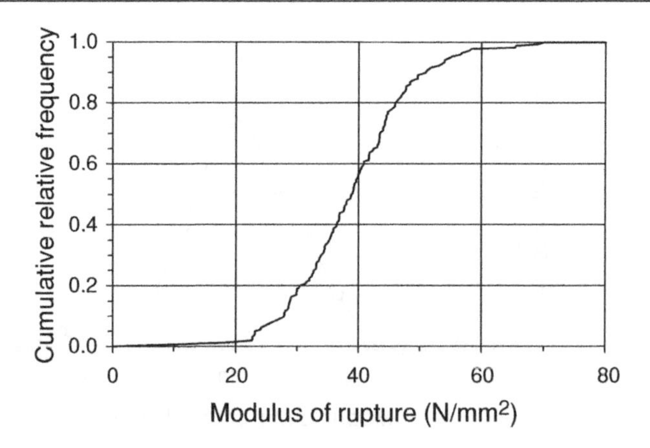

Fig. 1.1.6 Cumulative relative frequency diagram for timber strength data.

The deciles and percentiles can be abstracted. By convention a vertical probability or proportionality scale is used rather than one giving percentages (except in duration curves, discussed shortly). The 90th percentile, for instance, is 51 N/mm^2 approximately and the value 40 N/mm^2 has a probability of nonexceedance of approximately 0.56.

If the sample size increases indefinitely, the cumulative relative frequency diagram will become a *distribution curve* in the limit. This represents the population by means of a (mathematical) distribution function, usually called a *cumulative distribution function*, abbreviated to cdf, just as a relative frequency polygon leads to a probability density function.

As a graphical method of ascertaining the distribution of the population, the quantile plot can be drawn using a modified nonlinear scale for the probabilities, which represents one of several types of theoretical distributions.[6] Also, as shown in Section 1.4, two distributions can be compared using a Q-Q plot.

1.1.6 Duration curves

For the assessment of water resources and for associated design and planning purposes, engineers find it useful to draw *duration curves*. When dealing with flows in rivers, this type of graph is known as a *flow duration curve.* It is in effect a cumulative frequency diagram with specific time scales. The vertical axis can represent, for example, the percentage of the time a flow is exceeded; and in addition, the number of days per year or season during which the flow is exceeded (or not) may be given. The volume of flow per day is given on the horizontal axis. For some purposes, the vertical and horizontal axes are interchanged as in a Q-plot. One example of a practical use is the scaled area enclosed by the curve, a horizontal line representing 100% of the time, and a vertical line drawn at a minimum value of flow, which is desirable to be maintained in the river. This area represents the estimated supplementary volume of water that should be diverted to the river on an annual basis to meet such an objective.

Example 1.6. Streamflow duration. Figure 1.1.7 gives the flow duration curve of the Dora Riparia River in the Alpine region of northern Italy, calculated over a period of 47 years from the records at Salbertrand gauging station. This figure is drawn using the same procedure

[6] This method is demonstrated in Section 5.8.

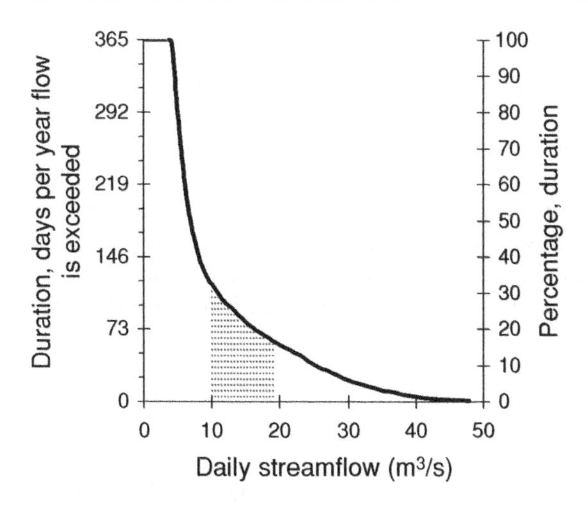

Fig. 1.1.7 Flow duration curve of Dora Riparia River at Salbertrand in the Alpine region of Italy.

adopted for a cumulative relative frequency diagram, such as Fig. 1.1.6. For instance, suppose it is decided to divert a proportion of the discharges above 10 m^3/s and below 20 m^3/s from the river. Then the area bounded by the curve and the vertical lines drawn at these discharges, using the vertical scale on the left-hand side, will give the estimated maximum amount available for diversion during the year in m^3 after multiplication by the number of seconds in a day. This area is hatched in Fig. 1.1.7. If such a decision were to be implemented over a long-term basis, it should be essential to use a long series of data and to estimate the distribution function.

1.1.7 Summary of Section 1.1

In this section we have introduced some of the basic graphical methods. Other procedures such as stem-and-leaf plots and scatter diagrams are presented in Sections 1.3 and 1.4, respectively. More advanced plots are introduced in Chapters 5 and 6. In the next section we discuss associated numerical methods.

1.2 NUMERICAL SUMMARIES OF DATA

Useful graphical procedures for presenting data and extracting knowledge on variability and other properties were shown in Section 1.1. There is a complementary method through which much of the information contained in a data set can be represented economically and conveyed or transmitted with greater precision. This method utilizes a set of characteristic numbers to summarize the data and highlight their main features. These numerical summaries represent several important properties of the histogram and the relative frequency polygon. The most important purpose of these descriptive measures is for statistical inference, a role that graphs cannot fulfill. Basically, there are three distinctive types: measures of central tendency, of dispersion, and of asymmetry, all of which can be visualized through the histogram as discussed in Section 1.1. The additional measure of "peakedness," that is, the relative height of the peak, requires a large sample for its estimation and is mainly relevant in the case of symmetric distributions.

1.2.1 Measures of central tendency

Generally data from many natural systems, as well as those devised by humans, tend to cluster around some values of variables. A particular value, known as the central value, can be taken as a representative of the sample. This feature is called central tendency because the spread seems to take place about a center. The definition of the central value is flexible, and its magnitude is obtained through one of the measures of its location. There are three such well-known measures: the mean, the mode, and the median. The choice depends on the use or application of the central value.

The *sample arithmetic mean* is estimated from a sample of observations: $x_1, x_2, \ldots, x_n$, as

$$\bar{x} = \frac{1}{n}\sum_{i=1}^{n} x_i. \tag{1.2.1}$$

If one uses a single number to represent the data, the sample mean seems ideal for the purpose. After counting, this calculation is the next basic step in statistics. For theoretical purposes the mean is the most important numerical measure of location. As stated in Section 1.1, if the sample size increases indefinitely a curve is obtained from a frequency polygon; the mean is the centroid of the area between this curve and the horizontal axis and it is thus the *balance point* of the frequency curve.

The population value of the mean is denoted by μ. We reiterate our definition of population with reference to a phenomenon such as that represented by the timber strength data of Table E.1.1. A population is the aggregate of observations that might result by making an experiment in a particular manner.

The sample mean has a disadvantage because it may sometimes be affected by unexpectedly high or low values, called *outliers*. Such values do not seem to conform to the distribution of the rest of the data. There may be physical reasons for outliers. Their presence may be attributed to conditions that have perhaps changed from what were assumed, or because the data are generated by more than one process. On the other hand, they may arise on account of errors of faulty instrumentation, measurement, observation, or recording. The engineer must examine any visible outliers and ascertain whether they are erroneous or whether their inclusion is justifiable. The occurrence of any improbable value requires careful scrutiny in practice, and this should be followed by rectification or elimination if there are valid reasons for doing so.

Example 1.7. Timber strength. A case in point is the value of zero in the timber strength data of Table E.1.1 This value is retained here for comparative purposes. The mean of the 165 items, which is 39.09 N/mm^2, becomes 39.33 N/mm^2 without the value of zero.

Example 1.8. Concrete test Table E.1.2 is a list of the densities and compressive strengths at 28 days from the results of 40 concrete cube test records conducted in Barton-on-Trent, England, during the period 8 July 1991 to 21 September 1992, and arranged in reverse chronological order.

These have sample means of 2445 kg/m^3 and 60.14 N/mm^2, respectively. The two numbers are measures of location representing the density and compressive strength of concrete.

With many discordant values at the extremes, a *trimmed mean*, such as a 5% trimmed mean, may be calculated. For this purpose, the data are ranked and the mean is obtained after ignoring 5% of the observations from each of the two extremities (see Problem 1.16).

The technique of *coding* is sometimes used to facilitate calculations when the data are given to several significant figures but the digits are constant except for the last few. For example, the densities in Table E.1.2 are higher than 2400 N/mm^2 and less than 2500 N/mm^2, so that the number 2400 can be subtracted from the densities. The remainders will retain the essential characteristics of the original set (apart from the enforced shift in the mean), thus simplifying the arithmetic.

In considering the entire data set, a *weighted mean* is obtained if the variables of a sample are multiplied by numbers called weights and then divided by the sum of the weights. It is used if some variables should contribute more (or less) to the average than others.

The *median* is the central value in an ordered set or the average of the two central values if the number of values, n, is even, as specified in Section 1.1.

Example 1.9. Concrete test. The calculation of the median and other measures of location will be greatly facilitated if the data are arranged in order of magnitude. For example, the compressive strengths of concrete given in Table E.1.2 are rewritten in ascending order in Table 1.2.1.

The median of these data is 60.1 N/mm^2, which is the average of 60.0 and 60.2 N/mm^2.

The median of the timber strength data of Table 1.1.3 is 39.05 N/mm^2, as noted in the table. The median has an advantage over the mean. It is relatively unaffected by outliers and is thus often referred to as a *resistant* measure. For instance, the exclusion of the zero value in Table 1.1.3 results only in a minor change of the median from 39.05 to 39.10 N/mm^2.

One of the countless practical uses of the median is the application of a disinfectant to many samples of bacteria. Here, one seeks an association between the proportion of bacteria destroyed and the strength of the disinfectant. The concentration that kills 50% of the bacteria is the *median dose*. This is termed *LD50* (lethal dose for 50%) and provides an excellent measure.

The *mode* is the value that occurs most frequently. Quite often the mode is not unique because two or more sets of values have equal status. For this reason and for convenience, the mode is often taken from the histogram or frequency polygon.

Example 1.10. Concrete test. For the ranked compressive strengths of concrete in Table 1.2.1, the mode is 60.5 N/mm^2.

Example 1.11. Timber strength. From Fig. 1.1.4, for example, the mode of the timber strength data is 37.5 N/mm^2, which corresponds to the midpoint of the class with the highest frequency. However, there is ambiguity in the choice of the class widths as already noted. On the other hand, in Table 1.1.3 there are nine values in the range 38.64–39.34 N/mm^2, and thus 39 N/mm^2 seems a more representative value, but this problem can only be resolved theoretically.

As the sample size becomes indefinitely large, the modal value will correspond to the peak of the relative frequency curve on a theoretical basis. The mode may often have greater practical significance than the mean and the median. It becomes more useful as the asymmetry of the distribution increases. For instance, if an engineer were to ask a person who sits habitually on the banks of a river fishing to indicate the mean level of the river, he or she is inclined to point out the modal level. It is the value most likely to occur and it

Table 1.2.1 Ordered data of density and compressive strength of concrete[a]

Order	Density (kg/m^3)	Compressive strength (N/mm^2)
1	2411	49.9
2	2415	50.7
3	2425	52.5
4	2427	53.2
5	2427	53.4
6	2428	54.4
7	2429	54.6
8	2433	55.8
9	2435	56.3
10	2435	56.7
11	2436	56.9
12	2436	57.8
13	2436	57.9
14	2436	58.8
15	2437	58.9
16	2437	59.0
17	2441	59.6
18	2441	59.8
19	2444	59.8
20	2445	60.0
21	2445	60.2
22	2446	60.5
23	2447	60.5
24	2447	60.5
25	2448	60.9
26	2448	60.9
27	2449	61.1
28	2450	61.5
29	2454	61.9
30	2454	63.3
31	2455	63.4
32	2456	64.9
33	2456	64.9
34	2457	65.7
35	2458	67.2
36	2469	67.3
37	2471	68.1
38	2472	68.3
39	2473	68.9
40	2488	69.5

[a] The original data sets are given in Table E.1.2.

is not affected by exceptionally high or low values. Clearly, the deletion of the zero value from Table 1.1.3 does not alter the mode, as we have also seen in the case of the median.

These positive attributes of the mode and median notwithstanding, the mean is indispensable for many theoretical purposes. Also in the same class as the sample arithmetic mean, there are two other measures of location that are used in special situations. These are the harmonic and geometric means.

The *harmonic mean* is the reciprocal of the mean of the reciprocals. Thus the harmonic mean for a sample of observations, $x_1, x_2, \ldots, x_n$, is defined as

$$\bar{x}_h = \frac{1}{1/n[(1/x_1) + (1/x_2) + \cdots + 1/x_n)]}. \tag{1.2.2}$$

It is applied in situations where the reciprocal of a variable is averaged.

Example 1.12. Stream flow velocity. A practical example of the harmonic mean is the determination of the mean velocity of a stream based on measurements of travel times over a given reach of the stream using a floating device. For instance, if three velocities are calculated as 0.20, 0.24, and 0.16 m/s, then the sample harmonic mean is

$$\bar{x}_h = \frac{1}{(1/3)[(1/0.20) + (1/0.24) + (1/0.16)]} = 0.19 \text{ m/s}.$$

The *geometric mean* is used in averaging values that represent a rate of change. Here the variable follows an exponential, that is, a logarithmic law. For a sample of observations, $x_1, x_2, \ldots, x_n$, the geometric mean is the positive nth root of the product of the n values. This is the same as the antilog of the mean of the logarithms:

$$\bar{x}_g = (x_1 x_2 \ldots x_n)^{1/n} = \exp\left(\frac{1}{n}\sum_{i=1}^{n} \ln x_i\right) = \left(\prod_{i=1}^{n} x_i^{1/n}\right). \tag{1.2.3}$$

Example 1.13. Population growth. Consider the case of populations of towns and cities that increase geometrically, which means that a future increase is expected that is proportional to the current population. Such information is invaluable for planning and designing urban water supplies and sewerage systems. Suppose, for example, that according to a census conducted in 1970 and again in 1990 the population of a city had increased from 230,000 to 310,000. An engineer needs to verify, for purposes of design, the per capita consumption of water in the intermediate period and hence tries to estimate the population in 1980. The central value to use in this situation is the geometric mean of the two numbers which is

$$\bar{x}_g = (230,000 \times 310,000)^{1/2} = 267,021.$$

(Note that the sample arithmetic mean $\bar{x} = 270,000$.)

As we see in Example 1.13, the geometric mean is less than the arithmetic mean.[7]

1.2.2 Measures of dispersion

Whereas a measure of central tendency is obtained by locating a central or representative value, a measure of dispersion represents the degree of scatter shown by observations or the inherent variability in a phenomenon under observation. Dispersion also indicates the precision of the data. One method of quantification is through an order statistic, that is, one of ranked data.[8] The simplest in the category is the range, which is the difference between the largest and smallest values, as defined in Section 1.1.

---

[7] This theoretical property is demonstrated in Example 3.10.

[8] We shall discuss order statistics formally in Chapter 7; see also Chapter 5.

Example 1.14. Timber strength. As noted before, the range of the timber strength data of Table 1.1.3 is 70.22 – 0.00 = 70.22 N/mm^2.

Example 1.15. Concrete test. For the compressive strengths of concrete given in Table E.1.2 and ranked in Table 1.2.1, the range is $r = 69.5 - 49.9 = 19.6$ N/mm^2; the range of the concrete densities is $2488 - 2411 = 77$ kg/m^3. These numbers provide a measure of the spread of the data in each case.

The range, however, is a nondecreasing function of the sample size and thus characterizes the population poorly. Moreover, the range is unduly affected by high and low values that may be somewhat incompatible with the rest of the data even though they may not always be classified as outliers. For this reason, the interquartile range, iqr, which is relatively a resistant measure, is preferable. As defined in Section 1.1, in a ranked set of data this is the difference between the median of the top half and the median of the bottom half.

Example 1.16. Concrete test. For the compressive strengths of concrete, the iqr is 6.55 N/mm^2.

Example 1.17. Timber strength. The timber strength data in Table 1.1.3 have an iqr of 11.66 and 11.47 N/mm^2, respectively, with or without the zero value. A similar and more general measure is given by the interval between two symmetrical percentiles. For example, the 90–10 percentile range for the timber strength data is approximately $52 - 28 = 24$ N/mm^2 from Fig. 1.1.6.

The aforementioned measures of dispersion can be easily obtained. However, their shortcoming is that, apart from two values or numbers equivalent to them, the vast information usually found in a sample of data is ignored. This criticism is not applicable if one determines the average deviation about some central value, thus including all the observations. For example, the *mean absolute deviation*, denoted by d, measures the average absolute deviation from the sample mean. For a sample of observations, $x_1, x_2, \ldots, x_n$, it is defined as

$$d = \frac{|x_1 - \bar{x}| + |x_2 - \bar{x}| + \cdots + |x_n - \bar{x}|}{n} = \sum_{i=1}^{n} \frac{|x_i - \bar{x}|}{n}. \tag{1.2.4}$$

Example 1.18. Annual rainfall. If the annual rainfalls in a city are 50, 56, 42, 53, and 49 cm over a 5-year period, the absolute deviation with respect to the sample mean of 50 cm is given by

$$d = \frac{1}{5}(|50 - 50| + |56 - 50| + |42 - 50| + |53 - 50| + |49 - 50|) = 3.6\,\text{cm}.$$

This measure of dispersion is easily understood and practically useful. However, it is valid only if the large and small deviations are as significant as the average deviations. There are strong theoretical reasons (as seen in Chapters 3, 4, and 5), on the other hand, for using the sample *standard deviation*, denoted by s, which is the root mean square deviation about the mean. Indeed, this is the principal measure of dispersion (although the interquartile range is meaningful and expedient). For a sample of observations, $x_1, x_2, \ldots, x_n$ it is defined by

$$s = \sqrt{\frac{1}{n}[(x_1 - \bar{x})^2 + (x_2 - \bar{x})^2 + \cdots + (x_n - \bar{x})^2]} = \sqrt{\frac{1}{n}\sum_{i=1}^{n}(x_i - \bar{x})^2}. \quad (1.2.5)$$

By expanding and summarizing the terms on the extreme right-hand side,

$$s = \sqrt{\frac{1}{n}\left(\sum_{i=1}^{n} x_i^2 - 2\bar{x}\sum_{i=1}^{n} x_i + n\bar{x}^2\right)} = \sqrt{\frac{1}{n}\sum_{i=1}^{n} x_i^2 - \bar{x}^2}. \quad (1.2.6)$$

Engineers will recognize that this measure is analogous to the radius of gyration of a structural cross section. In contrast to the mean absolute deviation, it is highly influenced by the largest and smallest values. The standard deviation of the population is denoted by σ. It is common practice to replace the divisor n of Eq. (1.2.5) by $(n-1)$ and denote the left-hand side by $\hat{s}$. Consequently, the estimate of the standard deviation is, *on average*, closer to the population value because it is said to have smaller *bias*. Therefore, Eq. (1.2.5) will, on average, give an underestimate of σ except in the rare case in which μ is known.[9] The required modification to Eq. (1.2.6) is as follows:

$$\hat{s} = \sqrt{\frac{1}{n-1}\sum_{i=1}^{n} x_i^2 - \frac{n}{n-1}\bar{x}^2}. \quad (1.2.7)$$

This reduction in n can be justified by means of the concept of *degrees of freedom.* It is a consequence of the fact that the sum of the n deviations $(x_1 - \bar{x}), (x_2 - \bar{x}), \ldots, (x_n - \bar{x})$ is zero, which follows from Eq. (1.2.1) for the mean. Hence, regardless of the arrangement of the data, if any $(n - 1)$ terms are specified the remaining term is fixed or known, because

$$x_n - \bar{x} = -\sum_{i=1}^{n-1}(x_i - \bar{x}).$$

It follows from this equation that one degree of freedom is lost in defining the sample standard deviation. The concept of degrees of freedom was introduced by the English statistician R. A. Fisher on the analogy of a dynamical system in which the term denotes the number of independent coordinate values necessary to determine the system.

Example 1.19. Annual rainfall. From the annual rainfall data in Example 1.18 (50, 56, 42, 53, and 49 cm), one can estimate the standard deviation σ by using Eq. (1.2.5), as follows:

$$\hat{s} = \sqrt{\frac{1}{5}[(50-50)^2 + (56-50)^2 + (42-50)^2 + (53-50)^2 + (49-50)^2]}$$
$$= \sqrt{\frac{1}{5}(0^2 + 6^2 + 8^2 + 3^2 + 1^2)} = \sqrt{\frac{110}{5}} = 4.69 \text{ cm}.$$

An alternative estimate of σ (which is, on average, less biased) is obtained using Eq. (1.2.7) as follows:

$$\hat{s} = \sqrt{\frac{110}{4}} = 5.24 \text{ cm}.$$

[9] Terms such as bias are discussed formally in Section 5.2. It is shown in Example 5.1 that $\hat{s}^2$ is unbiased; however, $\hat{s}$ is known to have bias, though less than s on average.

Example 1.20. Timber strength. By using Eq. (1.2.7), the sample standard deviation of the timber strength data of Table E.1.1 is 9.92 N/mm^2 (or 9.46 N/mm^2 if the zero value is excluded).

Example 1.21. Concrete test. By using Eq. (1.2.7), the sample standard deviation for the density and compressive strength of concrete in Table E.1.2 are 15.99 kg/m^3 and 5.02 N/mm^2, respectively.

Dividing the standard deviation by the mean gives the dimensionless measure of dispersion called the *sample coefficient of variation, v*:

$$v = \frac{\hat{s}}{\bar{x}} \tag{1.2.8}$$

This is usually expressed as a percentage. The coefficient of variation is useful in comparing different data sets with respect to central location and dispersion.

Example 1.22. Comparison of timber and concrete strength data. From the values of mean and standard deviation in Examples 1.7 and 1.20, the sample coefficient of variation of the timber strength data is 25.3% (or 24.0% without the value of zero). Similarly, from Examples 1.8 and 1.21 the density and compressive strength of concrete data have sample coefficients of 0.65 and 8.24%, respectively. The higher variation in the timber strength data is a reflection of the variability of the natural material, whereas the low variation in the density of the concrete is evidence of a uniform quality in the constituents and a high standard of workmanship, including care taken in mixing. The variation in the compressive strength of concrete is higher than that of its density. This can be attributed to random factors that influence strength, such as some subtle changes in the effectiveness of the concrete that do not alter its density.

From the square of the sample standard deviation one obtains the *sample variance*, $\hat{s}^2$, which is the mean of the squared deviations from the mean. The population variance is denoted by σ^2. The variance, like the mean, is important in theoretical distributions.

By squaring Eqs. (1.2.6) and (1.2.7), two estimators of the population variance are found. Here *estimator* refers to a method of estimating a constant in a parent population. As in all the foregoing equations, this term means the random variable of which the estimate is a realization. An *unbiased estimator* is obtained from Eq. (1.2.7) because on average (that is by repeated sampling) the estimator tends to the population variance σ^2. In other words, the *expectation* E, which is in effect the average from an infinite number of observations, of the square of the right-hand side of Eq. (1.2.7) is equal to σ^2.

There are also measures of dispersion pertaining to the mean of the deviations between the observations. *Gini's mean difference*, for example, is a long-standing method.[10] This is given by

$$g = \frac{2}{n\,(n-1)} \sum_{i>j} \sum_{j=1}^{n} [x_{(i)} - x_{(j)}], \tag{1.2.9}$$

in which the observations $x_1, x_2, \ldots, x_n$ are arranged in ascending order.

[10] See, for example, Stuart and Ord (1994, p. 58) for more details of this method originated by the Italian mathematician, Gini. See also Problem 1.7.

1.2.3 Measure of asymmetry

Another important property of the histogram or frequency polygon is its shape with respect to symmetry (on either side of the mode). The *sample coefficient of skewness* measures the asymmetry of a set of data about its mean. For a sample of observations, $x_1, x_2, \ldots, x_n$, it is defined as

$$g_1 = \frac{\sum_{i=1}^{n}(x_i - \overline{x})^3}{ns^3}. \tag{1.2.10}$$

Division by the cube of the sample standard deviation gives a dimensionless measure.

A histogram is said to have positive skewness if it has a longer tail on the right, which is toward increasing values, than on the left. In this case the number of values less than the mean is greater than the number that exceeds the mean. Many natural phenomena tend to have this property. For a positively skewed histogram,

$$\text{mode} < \text{median} < \text{mean}.$$

This inequality is reversed if skewness is negative. A symmetrical histogram suggests zero skewness.

Example 1.23. Comparison of timber and concrete strength data. The coefficient of skewness of the timber strength data of Table E.1.1 and the compressive strength data of Table E.1.2 are 0.15 (or 0.53 after excluding the zero value) and 0.03, respectively. These indicate a small skewness in the first case and a symmetrical distribution in the second case.

The example indicates that this measure of skewness is sensitive to the tails of the distribution.

1.2.4 Measure of peakedness

The extent of the relative steepness of ascent in the vicinity and on either side of the mode in a histogram or frequency polygon is said to be a measure of its *peakedness* or *tail weight*. This is quantified by the dimensionless *sample coefficient of kurtosis*, which is defined for a sample of observations, $x_1, x_2, \ldots, x_n$ by

$$g_2 = \frac{\sum_{i=1}^{n}(x_i - \overline{x})^4}{ns^4}. \tag{1.2.11}$$

Example 1.24. Comparison of timber and concrete strength data. The kurtosis of the timber strength data of Table E.1.1 is 4.46 (or 3.57 without the zero value) and that of the compressive strengths of Table E.1.2 is 2.33. One can easily see from Eq. (1.2.11) that even a small variation in one of the items of data may influence the kurtosis significantly. This observation warrants a large sample size, perhaps 200 or greater, for the estimation of the kurtosis. Small sample sizes, particularly in the second case with $n = 40$, preclude the attachment of any special significance to these estimates.

1.2.5 Summary of Section 1.2

Of the numerical summaries listed here, the mean, standard deviation, and coefficient of skewness are the best representative measures of the histogram or frequency polygon, from both visual and theoretical aspects. These provide economical measures for summarizing the information in a data set. Sample estimates for the data we have been discussing here, including the coefficients of variation and kurtosis, are given in Table 1.2.2.

Table 1.2.2 Sample estimates of numerical summaries of the timber strength data of Table 1.1.3 and the concrete strength and density data of Table 1.2.1

Data set	Sample size	Mean[a]	Standard deviation[a]	Coefficient of variation (%)	Coefficient of skewness	Coefficient of kurtosis
Estimated by equation		1.2.1	1.2.7	1.2.8	1.2.10	1.2.11
Timber strength—full sample	165	39.09	9.92	25.3	0.15	4.46
Timber strength without the zero value	164	39.33	9.46	24.0	0.53	3.57
Compressive strength of concrete	40	60.14	5.02	8.35	0.03	2.33
Density of concrete	40	2445	15.99	0.65	0.38	3.15

[a] Units for strength are N/mm^2; units for density are kg/m^3.

1.3 EXPLORATORY METHODS

Some graphical displays are used when one does not have any specific questions in mind before examining a data set. These methods were appropriately called *exploratory data analysis* by Tukey (1977). Among such procedures the *box plot* is advantageous, and the *stem-and-leaf plot* is also a valuable tool.

1.3.1 Stem-and-leaf plot

The histogram is a highly effective graphical procedure for showing various characteristics of data as seen in Section 1.1. However, for smaller samples, less than, say, 40 in size, it may not give a clear indication of the variability and other properties of the data. The *stem-and-leaf plot*, which resembles a histogram turned through a right angle, is a useful procedure in such cases. Its advantage is that the data are grouped without loss of information because the magnitudes of all the values are presented. Furthermore, its intrinsic tabular form highlights extreme values and other characteristics that a histogram may obscure. As in a histogram, the data are initially ranked in ascending order but a different approach is adopted in finding the number of classes. The class widths are almost invariably equal. For the increments or class intervals (and hence class widths) one uses 0.5, 1, or 2 multiplied by a power of 10, which means that the intervals are in units such as 0.1 or 200 or 10,000, which are more tractable than, say, 0.13 or 140 or 12,000. The terminology is best explained through the following worked example.

Example 1.25. Concrete test. For the concrete strength data of Table E.1.2, the maximum and minimum values are 69.5 and 49.9 N/mm^2, respectively. As a first choice, the data can be divided into 21 classes in intervals of 1 N/mm^2 with lower boundaries at 49, 50, 51 N/mm^2, and so on, up to 69 N/mm^2. For the *ordered stem-and-leaf plot* of Fig. 1.3.1, a vertical line is drawn with the class boundaries marked in increasing order immediately to its left.

The boundary values are called the *leading digits* and, together with the vertical line, constitute the *stem*. The *trailing digits* on the right represent the items of data in increasing order when read jointly with the leading digits using the indicated units. They are termed *leaves*, and their counts are the class frequencies. Thus the digits 49 (stem) and 9 (leaf) constitute 49.9. It is useful to provide an additional column at the extreme left, as shown here, giving the cumulative frequencies—called *depths*—up to each class. This is completed

1	49	9
2	50	7
2	51	
3	52	5
5	53	2 4
7	54	4 6
8	55	8
11	56	3 7 9
13	57	8 9
15	58	8 9
19	59	0 6 8 8
(7)	60	0 2 5 5 5 9 9
14	61	1 5 9
11	62	
11	63	3 4
9	64	9 9
7	65	7
6	66	
6	67	2 3
4	68	1 3 9
1	69	5

Fig. 1.3.1 Stem-and-leaf plot for compressive strengths of concrete in Table E.1.2; units for stem: 1 N/mm^2; units for leaves: 0.1 N/mm^2.

firstly by starting at the top and totaling downward to the line containing the median for which the individual frequency is given in parentheses, and secondly by starting at the bottom and totaling upward to the line containing the median.

The diagram gives all the information in the data, which is its main advantage. Furthermore, the range, median, symmetry, or gaps in the data, frequently occurring values, and any possible outliers can be highlighted. In this example, a symmetrical distribution is indicated. The plot may be redrawn with a smaller number of classes, perhaps for greater clarity, using the guidelines for choosing the intervals stipulated previously. The units of data in a plot can be rounded to any number of significant figures as necessary. Also, the number of stems in a plot can be doubled by dividing each stem into two lines. When 1 multiplied by a power of 10 is used as an interval, for example, the first line, which is denoted by an asterisk (*), will thus have leaves 0 to 4, and the leaves of the second, represented by a period (.), will be from 5 to 9. Likewise, one may divide a stem into five lines. The stem-and-leaf plot is best suited for small to moderate sample sizes, say, less than 200.

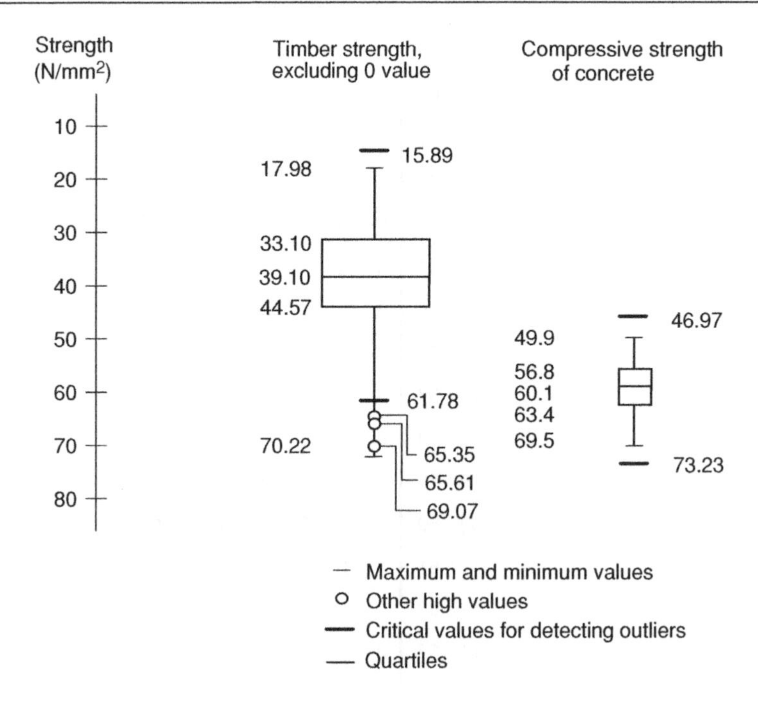

Fig. 1.3.2 Box plots for timber strength and compressive strength of concrete data from Tables 1.1.3 to 1.2.1.

1.3.2 Box plot

Another plot that is highly useful in data presentation is the *box plot*, which displays the three quartiles, Q_1, Q_2, Q_3, on a rectangular box aligned either horizontally or vertically. The box, together with the minimum and maximum values, which are shown at the ends of lines extended at either side from the box from the midpoints of its extremities, constitute the *box-and-whiskers plot*, as it is sometimes called. The numerical signposts are arranged as follows from top to bottom: minimum, Q_1, Q_2, Q_3, and maximum. Together they constitute a five-number summary. The minimum and maximum values may be replaced by the 5th and 95th (or other extreme) percentiles or supplemented by these and additional extreme values. These plots play an important role in comparing two or more samples. The width of the box is made proportional to the sample size in such cases, if they are different.

Example 1.26. Comparison of timber and concrete strength data. Let us use a box plot to compare the strengths of two representative materials used by civil engineers. Figure 1.3.2 shows the timber strength data ranked in Table 1.1.3, with the zero value excluded, and the compressive strength of concrete data that were ranked in Table 1.2.1. The box plot of compressive strengths of concrete shown on the right strongly indicates symmetry in their distribution. In the case of the timber strength data, the box is less symmetrical. However, there are clear signs of positive skewness; because the length of the line connecting the highest value to the box is longer than that connecting the lowest value to the box.

Empirical rules have been devised to detect outliers by means of box plots. As previously stated, this term signifies an excessive discordance with reference to an assumed distribution to which the majority of observations belong. One such procedure identifies outliers as those values located at distances greater than 1.5 iqr above the third quartile or less than 1.5 iqr below the first quartile.

Example 1.27. Comparison of timber and concrete strength data. The iqrs for the timber strength and compressive strength of concrete data are 11.47 and 6.55 N/mm^2 and thus the two critical distances for detecting outliers are 17.21 and 9.83 N/mm^2, respectively. These distances are set out on either from the extremities of the boxes and are shown by thick horizontal lines in Fig. 1.3.2. By this rule, the concrete data do not have any outliers, whereas there are four outliers beyond the demarcating line for high outliers in the timber strength data of Table 1.1.2. These are the values 65.35, 65.61, 69.07, and 70.22 N/mm^2. At the other extremity, there is the zero value that was discarded before the diagram was drawn. When such an observation is recorded one should verify whether it stems from a faulty calibration or other source of error; it is clearly an outlier by the method described here.[11]

1.3.3 Summary of Section 1.3

In general, box plots are helpful in highlighting distributional features, including the range and many of the properties of a histogram. They provide a valuable means of comparing data measuring related or similar characteristics. The stem-and-leaf plot is also clearly useful in presenting a set of data as an alternative to the histogram. Both diagrams can be easily drawn. These are two of the commonly used exploratory graphical methods. Other methods presented in subsequent chapters include the hanging histogram of Subsection 5.8.5.1.

1.4 DATA OBSERVED IN PAIRS

In the preceding sections, the behavior of one variable was considered. Let us extend this discussion to the case where simultaneous observations are made of two variables and a study is made to find an association between the variables. In this section the simple bivariate case of paired samples is examined, and the types of association between them are briefly assessed.

1.4.1 Correlation and graphical plots

A specific type of association that is frequently examined is known as *correlation* (from co-relation). In usual practice, graphical methods are initially applied; subsequently, numerical summaries provide a quantification and a means of assessment. For example, if there are n pairs of observations, $(x_1, y_1), (x_2, y_2), \ldots, (x_n, y_n)$, of two variables X and Y, a preliminary indication of the correlation is obtained through a *scatter diagram*. In this plot the coordinates denote the observed pairs of values.

Example 1.28. Concrete test. The scatter diagram of Fig. 1.4.1 represents the concrete data of Table E.1.2, with the density and compressive strength at 28 days given by the horizontal and vertical axes, respectively.

At first sight, there is no well-defined relationship between the two sets of observations although one would expect a density that is higher or lower than average to be associated with a compressive strength of concrete that is correspondingly higher or lower than its average.

---

[11] More precise methods of systematically detecting outliers (such as those investigated by Kottegoda, 1984) are discussed in Chapter 5.

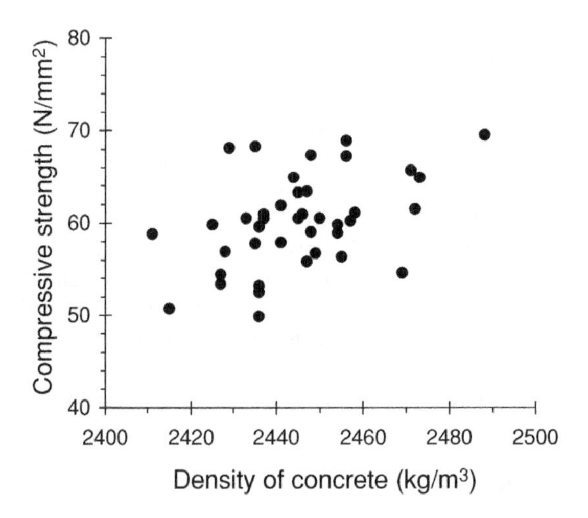

Fig. 1.4.1 Scatter diagram of concrete test data from Table E.1.2.

1.4.2 Covariance and the correlation coefficient

The *sample covariance*, $s_{X,Y}$, gives a numerical summary of the *linear* association between two quantitative variables X and Y. It is the average of the product of their deviations about the respective means. Thus,

$$s_{X,Y} = \frac{1}{n}\sum_{i=1}^{n}(x_i - \bar{x})(y_i - \bar{y}). \tag{1.4.1}$$

The covariance will be greater when there is a greater direct association between X and Y with respect to higher than average values and similarly for lower than average values. If the sample covariance is divided by the sample standard deviations of the two variables, s_X and s_Y [as in Eq. (1.2.6)], one obtains a dimensionless measure of *linear* association called the *sample coefficient of correlation*,

$$r_{X,Y} = \frac{1}{n s_X s_Y}\sum_{i=1}^{n}(x_i - \bar{x})(y_i - \bar{y}). \tag{1.4.2}$$

Substituting for s_X and s_Y, we find

$$r_{X,Y} = \frac{\sum_{i=1}^{n}(x_i - \overline{x})(y_i - \overline{y})}{\sqrt{\sum_{i=1}^{n}(x_i - \overline{x})^2 \sum_{i=1}^{n}(y_i - \overline{y})^2}}. \tag{1.4.3}$$

The correlation coefficient is constrained by $-1 \leq r_{X,Y} \leq 1$. Because the association measured here is defined by Eqs. (1.4.2) and (1.4.3), this result is called the *linear coefficient of correlation* or the *product-moment correlation coefficient.*[12]

The two limiting values in the preceding constraint are of theoretical interest and are applicable if all the points of a scatter diagram lie on a straight line of the type

$$Y_i = \beta_0 + \beta_i x_i, \tag{1.4.4}$$

[12] Another measure, Spearman's rank correlation coefficient, is discussed in Chapter 5.

where β_0 and β_1 are constants. The constant β_1 will be positive for all positive correlations including the maximum value, $r_{X,Y} = 1$. In the opposite case, β_1 will be negative, indicating negative correlation. That is, a high value of one variable tends to be associated with a low value of the other; the minimum value, $r_{X,Y} = -1$, is in this category.

In some cases the scatter diagram may indicate that there is an exponential or other nonlinear type of relationship between the two variables. In such cases, special procedures are necessary. For example, one may apply a logarithmic, square root, negative reciprocal, or other appropriate transformation to one or both variables prior to analysis (as discussed in Chapter 6).

Example 1.29. Concrete test. The scatter diagram of Fig. 1.4.1 does not show a strong relationship between the density and the compressive strength. This fact is confirmed by the correlation coefficient of +0.44 obtained from Eq. (1.4.3). It is possible that the inclusion of additional variables, such as the results of slump tests, will lead to an improved relationship for predictive purposes in a multiple regression analysis.

Note that a zero correlation does not show that the variables are independent. For variables that have no dependence, however, the correlation will not be of any significance.[13]

Note that one is only seeking an association between two variables through the correlation coefficient, not a cause and effect relationship. In some cases there are clear reasons for dependency, as in the case of a force exerted on a steel wire and the consequent increase in its length, or as in rainfall resulting in runoff. Often, however, one cannot reach such a conclusion when there is strong positive or negative correlation. One may find, for instance, that two variables are correlated because they are both associated with a third variable and not because there is a physical relationship between the first two.[14]

Equations of regression such as Eq. (1.4.4) are generally used to predict Y for a given value of X without invoking a causal relationship. Accordingly, the given value x is called the *explanatory* (nonrandom) variable and Y is the *response* (random) variable.

Example 1.30. Water quality. Another example of positive or negative correlation is the association between variables measuring water quality. A case study is taken from the Blackwater River in central England, which is constantly monitored for the control of pollution. The variables that are measured, among others, are the amounts of *dissolved oxygen*, DO, and the *biochemical oxygen demand*, BOD, in the water. Dissolved oxygen is required for the respiration of aerobic life forms such as fish. The BOD denotes the amount of oxygen used in meeting the metabolic needs of aerobic microorganisms in water, whether naturally occurring or resulting from sewage outflows and other discharges; thus, high values of BOD generally indicate high levels of pollution. Usually determined in a laboratory after a 5-day incubation of samples taken from the water, BOD is the most widely used indicator of pollution despite some shortcomings. Sampling at 38 stations along the river gives the data presented in Table E.1.3.

---

[13] The significance of small values of correlation and whether they probably indicate zero correlation are discussed in Chapter 6, in addition to other aspects of regression including the particular notation of Eq. (1.4.4). The concept of independence is discussed in Chapter 2.

[14] An absurdity cited in early literature is the apparent relationship between horse kicks suffered by cavalrymen and wheat production in Europe. Also, Yule (1926) correlates concurrent time series of the proportion of Church of England marriages and the standardized mortality rates per 1000 persons with a "nonsense" correlation coefficient of 0.95; he explains that both variables are highly influenced by a common factor; we now call this behavior spurious correlation.

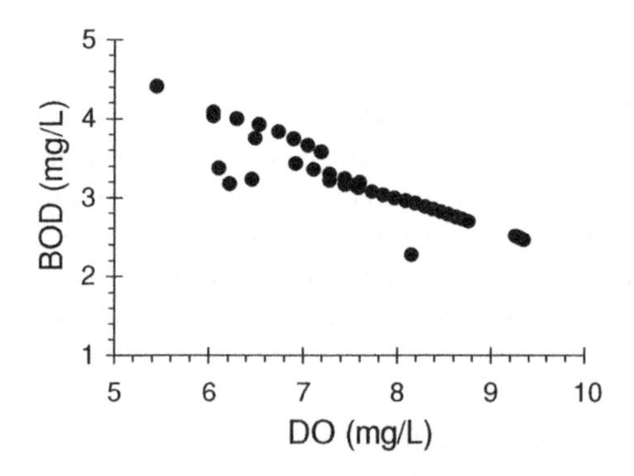

Fig. 1.4.2 Scatter diagram of water quality data from Table E.1.3.

The scatter diagram of the two indicators of water quality data is shown in Fig. 1.4.2. As expected, it strongly indicates a negative type of correlation with high values of DO associated with low values of BOD and vice versa. The coefficient of correlation from Eq. (1.4.3) is -0.90. It suggests that the value of BOD can be estimated from a measurement of the DO. The scatter in the diagram may be partly attributed to some inadequacies of the BOD test and partly to factors such as temperature and rate of flow, which affect the DO.

The presence of outliers tends to have a significant effect on the coefficient of correlation. Consider, for example, the lowest BOD in Fig. 1.4.2, which corresponds to the first pair of values in Table E.1.3. This may not warrant consideration as an outlier. It can, however, be due to an incorrect observation or an error in recording. With reference to Example 1.30, it is interesting to note that if one changes the first BOD value of Table E.1.3, from 2.27 to 2.77, the correlation coefficient decreases from -0.90 to -0.92.

1.4.3 Q-Q plots

Quantiles representing two attributes or phenomena that are considered to be associated may be compared using a *Q-Q plot*. Here one plots the quantiles of one data set against the corresponding quantiles of another set as a means of comparing their probability distributions. One proceeds initially with the ranking and calculation of cumulative relative frequencies for a quantile plot for each set of data (as a prerequisite to drawing Fig. 1.1.6, for example). The two quantile plots are then associated graphically by plotting values of data with equal cumulative relative frequencies. In this type of diagram the limiting case, in which the distributions differ only with respect to location and scale, is represented by a straight line. The manner in which the plot departs from linearity indicates other types of difference between the two distributions.

When one quantile function represents a theoretical distribution, the Q-Q plot becomes a probability plot. This is a very useful diagram adopted in practice initially by a civil engineer, R. W. Powell in 1943. The probability plot may be considered to be an extension of the box plot, because all the quantiles are used in this method of comparing empirical and theoretical distributions.[15]

---

[15] Details of this method are given in Section 5.8.

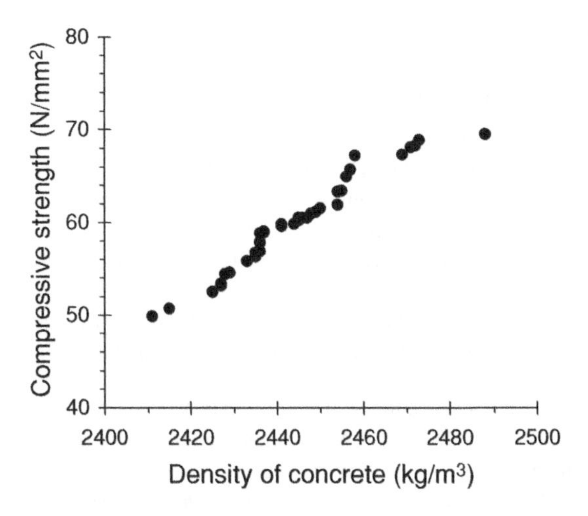

Fig. 1.4.3 Q-Q plot of concrete test data from Table E.1.2.

Example 1.31. Concrete test. The distributions of the concrete strengths and densities listed in Table E.1.2 are to be compared using a Q-Q plot. For this purpose the ranked data of Table 1.2.1 are used to obtain the cumulative relative frequencies for each item of data in the sample of concrete strengths and the sample of concrete densities. Then a Q-Q plot is drawn by associating data of equal cumulative frequencies. When sample sizes are the same, such as in the case of the data used here, one can proceed directly to the Q-Q plot; in other cases one calculates the quantiles of the smaller sample and then interpolates, correspondingly, the quantiles for the larger sample.

There are apparent similarities in the distributions of strengths and densities, as shown in Fig. 1.4.3. Although the distributions are not close, they do not seem to be divergent.

1.4.4 Summary of Section 1.4

A brief preliminary introduction is provided here on methods of investigating data observed in pairs. This is a prelude to the formal presentations in Chapters 3 and 5 and particularly in Chapter 6 on regression and multivariate analysis.

1.5 SUMMARY FOR CHAPTER 1

In this chapter numerous graphical methods for presenting data sets are introduced. These include line diagrams, histograms, relative frequency polygons, cumulative relative frequency diagrams, and scatter diagrams. Details of exploratory methods such as stem-and-leaf plots and box plots are also given.

Many of the numerical summaries for reducing data in this chapter are essential for the application of statistics and probability in engineering. Among the most important of these statistics are the mean, standard deviation, and the coefficient of correlation. Several sets of data are provided here as examples of random variables which engineers encounter. One needs to interpret these and draw sensible conclusions. The graphical and numerical methods here are a necessary first step and lead into the probabilistic methods of Chapters 2 and 3 and the verification of mathematical models in subsequent chapters.

REFERENCES

General. The following references are given for further reading as required:

Ang, A. M. S., and W. H. Tang (1975). *Probability Concepts in Engineering Planning and Design, Vol. 1: Basic Principles*, John Wiley and Sons, New York. A blend of theory and practice with wide appeal for practicing civil engineers.

Benjamin, J. R., and C. A. Cornell (1970). *Probability, Statistics and Decision for Civil Engineers*, McGraw-Hill, New York. A classic for civil engineers with examples, extensive case studies, and innumerable problems to solve.

Blank, L. T. (1980). *Statistical Procedures for Engineering, Management and Science*, McGraw-Hill, New York. Well-explained theory and practical examples, commendable as an introductory text.

Chambers, J. M., W. S. Cleveland, B. Kleiner, and P. A. Tukey (1983). *Graphical Methods for Data Analysis*, Wadsworth, Belmont, CA. A standard reference for those seeking further knowledge of graphical methods.

Groeneveld, R. A. (1979). *Introductory Statistical Methods—An Integrated Approach Using Minitab*, Marcel Dekker, New York. An ideal statistical guide with computer applications suitable for beginners.

Hahn, G. J., and S. S. Shapiro (1967). *Statistical Models for Engineering*, John Wiley and Sons, New York. Reprinted in 1994 as a Wiley Classic in the engineering series. Recommended as a reference book for understanding the basics of statistical applications in engineering.

Hand, D. J., F. Daly, A. D. Lunn, K. J. McConway, and E. Ostrowski (1994). *A Handbook of Small Data Sets*, Chapman and Hall, London. Diverse data sets.

Hines, W. H., and D. C. Montgomery (1990). *Probability and Statistics in Engineering and Management Science*, 3rd ed., John Wiley and Sons, New York. Comprehensive book of 700 pages, 300 examples, and 626 problems.

Hoaglin, D. C., F. Mosteller, and J. W. Tukey (eds.) (1983). *Understanding Robust and Exploratory Data Analysis*, John Wiley and Sons, New York. This authoritative book will further enhance one's knowledge of exploratory data analysis.

Johnson, R., and G. K. Bhattacharyya (1992). *Statistics—Principles and Methods*, 2nd ed., John Wiley and Sons, New York. Basic principles well explained.

Mendenhall, W., and R. J. Beaver (1994). *Introduction to Probability and Statistics*, 9th ed., Duxbury Press, Boston. Statistics and probability at beginners' level.

Mendenhall, W., and T. Sincich (1995). *Statistics for Engineering and the Sciences*, 4th ed., Prentice Hall, Englewood Cliffs, NJ. Introduction with many applications.

Moore, D. S., and G. P. McCabe (2003). *Introduction to the Practice of Statistics*, 4th ed., W. H. Freeman and Co., New York. A useful primer in statistics.

Moroney, M. J. (1975). *Facts from Figures*, reprinted 1990, Penguin Books, London. The best book written for an absolute beginner in statistics.

Scheaffer, R. L., and J. T. McClave (1995). *Probability and Statistics for Engineers*, 4th ed., Duxbury Press, Belmont, CA. A variety of charts and preliminary calculations in Chapter 1. In general, low emphasis in mathematics throughout. Highly commendable as an introduction.

Wackerly, D. D., W. Mendenhall, and R. L. Scheaffer (2002). *Mathematical Statistics with Applications*, 6th ed., Duxbury, Pacific Grove, CA. Comprehensive introduction.

Additional references quoted in text

Freedman, D., and P. Diaconis (1981). "On the histogram as a density estimator: L_2 theory," *Zeitschrift fur Wahrscheinlich keitstheorie und verwandte Gebiete.*, Vol. 57, pp. 453–476, Chap. 2. Related to the class intervals of a histogram.

Kottegoda, N. T. (1984). "Investigation of outliers in annual maximum flow series," *J. Hydrol.*, Vol. 72, No. 1, pp. 105–137. Methods of detecting outliers.

Scott, D. W. (1979). "On optimal and data-based histograms," *Biometrika*, Vol. 66, pp. 605–610. Number of classes for histograms.

Stuart, A., and J. K. Ord (1994). *Kendall's Advanced Theory of Statistics*, Vol. 1, 6th ed., Charles, Edward Arnold, London. Advanced reference. See Gini's mean difference in Chapter 2.

Sturges, H. A. (1926). "The choice of a class interval," *J. Am. Stat. Assoc.*, Vol. 21, pp. 65–66. Historical work on the histogram.

Tukey, P. A. (1977). *Exploratory Data Analysis*, Addison-Wesley, Reading, MA. Original reference on exploratory methods.

Yule, G. U. (1926). "Why do we sometimes get nonsense correlation between time series," *J. R. Stat. Soc.*, Vol. 89, pp. 1–69. Shows how two unrelated variables can have a high coefficient of correlation because they are influenced by a common factor.

PROBLEMS

1.1. Earthquake records. Measurements of engineering interest have been recorded during earthquakes in Japan and in other parts of the world since 1800. One of the critical recordings is of apparent relative density, RDEN. After the commencement of a strong earthquake, a saturated fine, loose sand undergoes vibratory motion and consequently the sand may liquefy without retaining any shear strength, thus behaving like a dense liquid. This will lead to failures in structures supported by the liquefied sand. These are often catastrophic. The standard penetration test is used to measure RDEN. Another measurement taken to estimate the prospect of liquefaction is that of the intensity at which the ground shakes. This is the peak surface acceleration of the soil during the earthquake, ACCEL. The data are from J. T. Christian and W. F. Swiger (1975), *J. Geotech. Eng. Div.*, Proc. ASCE, 101, GT111, 1135–1150, and are reproduced by permission of the publisher (ASCE):

RDEN (%)	ACCEL (units of g)	RDEN (%)	ACCEL (units of g)	RDEN (%)	ACCEL (units of g)
53	0.219	30	0.138	50	0.313
64	0.219	72	0.422	44	0.224
53	0.146	90	0.556	100	0.231
64	0.146	40	0.447	65	0.334
65	0.684	50	0.547	68	0.419
55	0.611	55	0.204	78	0.352
75	0.591	50	0.170	58	0.363
72	0.522	55	0.170	80	0.291
40	0.258	75	0.192	55	0.314
58	0.250	53	0.292	100	0.377
43	0.283	70	0.299	100	0.434
32	0.419	64	0.292	52	0.350
40	0.123	53	0.225	58	0.334

Note: g denotes acceleration due to gravity (9.81 m/s^2).

Compute the sample mean $\bar{x}$, standard deviation $\hat{s}$, and the coefficient of skewness, g_1, for RDEN and ACCEL. Construct stem-and-leaf plots for each set. Comment on the distributions. Plot the scatter diagram and calculate the correlation coefficient r. What conclusions can be reached?

1.2. Flood discharge. Annual maximum flood flows in the Po River at Pontelagoscuro, Italy, over a 61-year period from 1918 to 1978 are given in the second column of Table E.7.2. Compute the sample mean $\bar{x}$ and standard deviation $\hat{s}$. Sketch a histogram and the cumulative relative frequency diagram. Compute the quartiles and draw a box-and-whiskers plot. Comment on the distribution. Flood embankments along the banks of the river can withstand a flow of 5000 m^3/s. What is the probability that this will be exceeded during a 12-month period?

1.3. Flood discharge. The following are the annual maximum flows in m^3/s in the Colorado River at Black Canyon for the 52-year period from 1878 to 1929:

1980	1130	3120	2120	1700	2550	8500	3260	3960	2270
1700	1570	2830	2120	2410	2550	1980	2120	2410	2410
1420	1980	2690	3260	1840	2410	1840	3120	3290	3170
1980	4960	2120	2550	4250	1980	4670	1700	2410	4550
2690	2270	5660	5950	3400	3120	2070	1470	2410	3310
3230	3090								

[Adapted from E. J. Gumbel (1954), "Statistical theory of extreme values and some practical applications," *National Bureau of Standards, Applied Mathematics Series 33*, U.S. Govt. Printing Office, Washington, DC.]
Compute the mean $\bar{x}$ and standard deviation $\hat{s}$. Sketch a histogram and the relative frequency diagram. Compute the quartiles and draw a box-and-whiskers plot. How does this distribution differ from that of Problem 1.2?

1.4. Welding joints for steel. At the University of Birmingham, England, laboratory measurements were taken of the horizontal legs x and vertical legs y of numerous welding joints for steel buildings. The main objective was to make the legs equal to 6 mm. A part of the results is listed below in millimeters.

$x =$ 5.5, 5.0, 5.0, 6.0, 7.0, 5.2, 5.5, 5.5, 6.0, 6.0, 4.5, 6.0, 5.5, 7.7, 7.5, 6.0, 5.6, 5.0, 5.5, 5.5, 6.0, 6.5, 5.5, 5.0, 5.5, 5.5, 6.5, 6.5, 7.0, 5.5, 6.5, 5.5, 6.0, 6.5, 8.5, 5.0, 6.0, 6.5, 5.0, 7.0, 5.0, 5.0, 6.5, 6.5, 6.0, 4.7, 8.0, 7.0, 5.5, 7.0, 6.6, 6.5, 7.0, 6.0, 6.5, 5.0, 7.0, 7.5, 7.0, 7.0

$y =$ 6.5, 6.5, 5.5, 7.5, 6.0, 7.0, 5.0, 8.0, 6.7, 7.8, 5.7, 6.5, 5.5, 8.0, 8.0, 6.3, 6.0, 6.0, 6.0, 5.5, 6.5, 6.0, 6.0, 6.0, 6.0, 6.5, 6.5, 6.0, 6.0, 6.5, 7.5, 7.5, 6.0, 4.5, 7.0, 7.0, 6.0, 4.0, 4.0, 7.0, 7.0, 6.5, 7.0, 5.0, 5.0, 5.7, 5.0, 5.0, 6.0, 7.0, 6.0, 7.0, 6.0, 5.5, 6.0, 4.0, 5.5, 8.0, 7.5, 6.5

The data were provided by Dr A. G. Kamtekar.
Draw a scatter diagram for these data. Draw a line through the ideal point ($x = y = 6$ mm) and the origin. Draw two lines through the origin that are symmetrical about the first line and envelope all of the points. Comment on the results. Draw the cumulative sum (cusum) plots,

$$Cx_n = \sum_i^n (x_i - \mu_x) \quad \text{and} \quad Cy_n = \sum_i^n (y_i - \mu_y)$$

for $n = 1, 2, \ldots, 60$ and $\mu_x = \mu_y = 6$. Let

$$dx_n = Cx_n - \min_{i=1}^{n-1}[Cx_i]$$

and the critical limit be $\max(dx_n) = 12$ mm. Is the critical limit reached? Repeat for the vertical legs y. [Further details of cusum plots are given by W. H. Woodalland B. M. Adams (1993), "The statistical design of cusum charts," *Qual. Eng.*, Vol. 5, No. 4, pp. 550–570; the associated control chart is the subject of Problem 5.11.]

1.5. Frost frequency. Excessive frost can be harmful to roads. Frequencies of the number of days of frost during April in Greenwich, England, over a 65-year period are given by C. E. Brooks and N. Carruthers (1953), *Handbook of Statistical Methods in Hydrology*, H. M. Stationary Office, London, and are listed below:

Days of frost	0	1	2	3	4	5	6	7	8	9	10
Frequency	15	11	5	11	7	6	2	3	2	1	2

Draw a line diagram of the data. Comment on the results. Compute the mean number of days of frost in April. What is the probability of a frostfree April in a given year? What change would you expect in the frequency distribution for a month in midwinter?

1.6. Concrete cube test. From 28-day concrete cube tests made in England in 1990, the following results of maximum load at failure in kilonewtons and compressive strength in newtons per square millimeter were obtained:

Maximum load: 950, 972, 981, 895, 908, 995, 646, 987, 940, 937, 846, 947, 827, 961, 935, 956
Compressive strength: 42.25, 43.25, 43.50, 39.25, 40.25, 44.25, 28.75, 44.25, 41.75, 41.75, 38.00, 42.50, 36.75, 42.75, 42.00, 33.50

The data were supplied by Dr J. E. Ash, University of Birmingham, England.

Calculate the means $\bar{x}$, standard deviations $\hat{s}$, mean absolute deviations d, and the coefficients of skewness g_1. Draw two stem-and-leaf plots of the data. Draw a scatter diagram and calculate the coefficient of correlation. What conclusions can be drawn?

1.7. Timber strength. For the timber strength data of Table E.1.1 determine the following measures of dispersion:
(a) Interquantile range, iqr
(b) Mean absolute deviation, d
(c) Gini's mean difference, g
Compare results with the standard deviation $\hat{s}$ of Table 1.2.2. Repeat these determinations after deleting the zero value. Rank the measures of dispersion in increasing order of susceptibility to the exclusion of the zero value on the basis of percentage change.

1.8. Population growth. From past records, the population of an urban area has doubled every 10 years. Currently, it has a population of 200,000. An engineer needs to make an estimate of the requirements for water supply during the next 23 years. What maximum population does one assume for the period?

1.9. Traffic speed. The following is the frequency distribution of travel times of motorcars on the M1 motorway from Coventry, England, to M10, St Albans, according to a survey conducted in England (see Ph.D. thesis of A. W. Evans, University of Birmingham, England, 1967):

Mean times (min): 53, 58, 63, 68, 73, 78, 83, 88, 93, 98, 103, 108,113, 118, 123, 128, 133, 138, 143, 148, 153, 158, 163, 168

Corresponding frequencies: 10, 24, 109, 127, 122, 119, 97, 102, 104, 92, 68, 72, 66, 61, 36, 33, 17, 15, 10, 8, 9, 6, 7, 3

Draw the histogram. Describe the salient features. What is the likely reason for the twin peaks? What inference can be made from the mean time interval between the two peaks?

1.10. Average speed. On a certain country road that runs from a coastal town to a village in the mountains, the average speed of motorcars is 80 km/h uphill and 100 km/h downhill. What is the average speed for a journey from the town to the village and back?

1.11. Annual rainfall. Catchment-averaged annual rainfall in the Po River basin of Italy for the 61-year period from 1918 to 1978 are given in the penultimate column of Table E.7.2. Draw a stem-and-leaf plot and a box plot of the data. Comment on the type of distribution.

1.12. Rock test. A contractor engaged in building part of a sewer tunnel claimed that the rock was harder than described in his contract with a District Council in the United Kingdom and thus more work was required to construct the tunnel than anticipated. An independent company made tests to verify the contractor's claim. Among these were uniaxial compressive strengths, of which 123 specimens are listed here, in meganewtons per square meter.

2.40, 22.08, 16.80, 4.80, 21.36, 9.12, 9.36, 3.60, 15.36, 15.60, 6.24, 9.84, 16.08, 30.00, 20.40, 12.96, 19.20, 10.32, 15.84, 62.40, 40.80, 4.80, 7.20, 8.88, 14.40, 14.88, 5.76, 18.72, 12.48, 11.04, 8.64, 19.20, 8.16, 18.96, 8.64, 12.00, 14.88, 17.52, 12.48, 13.44, 9.36, 11.28, 8.88, 15.12, 9.36, 17.28, 26.40, 4.32, 11.28, 7.92, 13.92, 11.76, 9.60, 8.40, 9.84, 27.60, 6.00, 14.40, 8.88, 17.04, 12.48, 9.84, 10.80, 12.24, 12.00, 13.20, 11.28, 11.76, 11.76, 8.00, 9.36, 15.12, 11.52, 16.08, 10.80, 14.64, 8.40, 13.44, 10.56, 9.12, 13.44, 12.72, 13.68, 11.28, 5.52, 11.04, 12.00, 7.20, 8.64, 11.76, 8.64, 7.68, 7.68, 13.92, 6.48, 7.20, 7.92, 9.60, 8.64, 9.12, 12.96, 9.36, 14.64, 9.12, 8.88, 20.40, 17.28, 8.64, 11.76, 7.92, 7.68, 11.04, 12.48, 14.40, 9.84, 9.12, 8.40, 12.00, 4.80, 12.72, 9.60, 8.64, 9.84

Draw histograms using Eqs. (1.1.1) and (1.1.2) for the class widths. What do you notice about the histograms in general? Draw a box-and-whiskers plot. What evidence is there to support the contractor's claim?

1.13. Soil erosion. Measurements taken on farmlands of the amounts of soil washed away by erosion suggest a relationship with flow rates. The following results are taken from G. R. Foster, W. R. Ostercamp, and L. J. Lane (1982), "Effect of discharge rate on rill erosion," Winter 1982 Meeting of the American Society of Agricultural Engineers:

Flow (L/s)	0.31	0.85	1.26	2.47	3.75
Soil eroded (kg)	0.82	1.95	2.18	3.01	6.07

Draw a plot of the data. Comment on the results.

1.14. Concrete cube test. The following 28-day compressive strengths, in newtons per square millimeter, were obtained from test results on concrete cubes in England:

50.5, 45.8, 49.6, 47.7, 54.0, 49.4, 54.1, 53.1, 56.5, 55.2, 52.7, 52.0, 54.2, 55.2, 53.4, 51.0, 53.1, 48.5, 51.0, 58.6, 52.5, 49.5, 51.1, 48.1, 50.2, 49.3, 47.3, 52.9, 52.8, 49.5, 48.8, 53.8, 47.3, 47.7, 52.2, 45.7, 53.4, 48.5, 49.1, 43.3

The data were supplied by Dr J. E. Ash, University of Birmingham, England.

Compare these results with the compressive strengths in Table E.1.2 by drawing back-to-back stem-and-leaf plots. For this purpose, plot the foregoing results on the left of the stem with reference to Fig. 1.3.1 and omit the cumulative frequencies. Comment on the differences in the distributions.

1.15. Water quality. Water quality measurements are taken daily on the River Ouse at Clapham, England. The concentrations of chlorides and phosphates in solution, given below in milligrams per liter, are determined over a 30-day period.

Chloride: 64.0, 66.0, 64.0, 62.0, 65.0, 64.0, 64.0, 65.0, 65.0, 67.0, 67.0, 74.0, 69.0, 68.0, 68.0, 69.0, 63.0, 68.0, 66.0, 66.0, 65.0, 64.0, 63.0, 66.0, 55.0, 69.0, 65.0, 61.0, 62.0, 62.0
Phosphate: 1.31, 1.39, 1.59, 1.68, 1.89, 1.98, 1.97, 1.99, 1.98, 2.15, 2.12, 1.90 1.92, 2.00, 1.90, 1.74, 1.81, 1.86, 1.86, 1.65, 1.58, 1.74, 1.89, 1.94, 2.07, 1.58, 1.93, 1.72, 1.73, 1.82

Compare the coefficients of variation v. Draw a scatter diagram and compute the correlation coefficient r. Comment on the results. Do you see any role in this association for predictive purposes?

1.16. Timber strength. From the timber strength data of Table 1.1.3, compute the 3% trimmed mean by omitting 3% of the observations from the highest and the lowest extremities of the ranked data. Compute the standard deviation $\hat{s}$ and the coefficients of skewness g_1 and kurtosis g_2. Compare with the results for the full sample (as given in Table 1.2.2).

1.17. Concrete beam. Joist-hanger tests carried out at the University of Birmingham, England, on concrete beams gave observations of deflections in millimeters and failure load in kilograms. The following results pertain to 75 mm × 150 mm hangers on which timber joists rest:

Failure load: 1903, 1665, 1903, 1991, 2229, 1910, 2025, 1991, 1882, 2032, 1896, 1346
Deflection: 0.69, 0.67, 0.80, 0.50, 0.74, 0.78, 0.57, 0.91, 0.54, 0.50, 0.97, 0.62

Determine by drawing a scatter diagram and computing the correlation coefficients whether there is any association between the two variables. Discuss your results.

1.18. Hurricane frequency. Hurricane damage is of great concern to civil engineers. The frequencies of hurricanes affecting the east coast of the United States each year during a period of 69 years are given as follows by H. C. S. Thom (1966), *Some Methods of Climatological Analysis*, World Meteorological Organisation, Geneva:

Number of hurricanes	0	1	2	3	4	5	6	7	8	9
Frequency	1	6	10	16	19	5	7	3	1	1

Draw a line diagram and comment on its form. Discuss differences or similarities between this diagram and Fig. 1.1.1.

1.19. Air pollution. On 13 April 1994, the following concentration of pollutants were recorded at eight stations of the monitoring system for pollution control located in the downtown area of Milan, Italy:

	Station							
	Aquileia	Cenisio	Juvara	Liguria	Marche	Senato	Verziere	Zavattari
NO_2 (μg/m$^{3)}$	130	130	115	120	135	142	90	116
CO (mg/m$^{3)}$	2.9	4.4	3.6	4.1	3.3	5.7	4.8	7.3

Compare the coefficients of variation v of the pollutants and determine their correlation r.

1.20. Storm rainfall. The analysis of storm data is essential for predicting flood hazards in urban areas. Annual maximum rainfall depths (in millimeters) recorded at Genoa University in Italy, for durations varying from 5 minutes to 3 hours, are presented here for the years 1974–1987.

	Duration (min)								
Year	5	10	20	30	40	50	60	120	180
1974	12.1	19.5	28.8	30.5	32.4	35.5	38.7	48.0	51.6
1975	10.1	14.9	26.7	31.2	34.7	38.2	40.2	55.0	56.0
1976	17.9	20.0	31.1	37.2	41.1	51.0	55.7	67.1	80.6
1977	20.0	32.6	52.6	72.4	90.1	108.8	118.9	146.5	157.3
1978	5.1	13.6	16.0	21.3	24.1	24.6	25.0	40.7	49.9
1979	20.5	26.1	36.3	46.1	49.3	50.3	55.6	65.2	90.1
1980	10.0	15.7	20.9	25.0	30.5	38.0	40.1	58.0	63.8
1981	12.0	27.9	47.9	56.0	70.0	80.0	89.4	106.9	114.2
1982	10.0	14.4	20.0	23.3	25.1	26.4	27.2	34.3	41.2
1983	10.0	12.1	17.3	19.2	22.1	27.3	32.7	54.4	66.5
1984	20.1	32.8	60.0	65.7	76.1	92.8	105.7	122.3	122.3
1985	7.6	8.1	13.0	16.5	21.6	25.3	25.3	27.0	32.3
1986	8.7	11.7	20.0	22.9	26.1	26.3	27.6	41.1	56.7
1987	24.6	36.7	56.7	73.9	93.9	110.1	128.5	180.8	188.7

Compute the mean $\bar{x}$ and standard deviation $\hat{s}$ and coefficient of skewness g_1 for each duration. Are there some regularities in the growth of these statistics with increasing duration? Comment on the results and the physical relevance to storm characteristics.

1.21. Carbon dioxide. The records of atmospheric trace gases are used in the study of global climatic changes. Monthly carbon dioxide concentrations (in parts per million in volume) recorded at Mount Cimone, Italy, from 1980 to 1988 are given here.

	Month											
Year	Jan	Feb	Mar	Apr	May	Jun	Jul	Aug	Sep	Oct	Nov	Dec
1980	340.87	339.83	342.27	342.51	338.27	335.52	330.14	328.81	331.17	335.03	339.05	340.43
1981	341.47	343.11	342.39	342.51	339.49	335.28	330.77	330.30	333.55	336.80	339.41	343.18
1982	341.70	344.38	345.68	345.70	340.80	336.66	334.65	332.40	335.15	339.26	341.19	345.18
1983	342.38	346.18	345.00	344.24	342.32	338.34	336.03	335.00	336.57	339.86	343.97	345.61
1984	346.32	349.44	351.33	350.50	346.43	344.35	346.29	335.19	337.59	342.26	344.88	346.91
1985	349.92	348.17	350.62	350.61	345.93	341.43	337.67	337.16	339.40	344.07	349.49	347.40
1986	349.41	351.41	352.29	350.75	348.37	342.96	337.22	338.53	340.90	346.28	348.95	350.52
1987	351.94	353.75	354.79	352.61	350.39	347.38	341.64	341.64	342.19	345.60	350.39	352.36
1988	353.13	355.02	354.96	354.51	352.20	346.71	342.60	344.60	343.66	348.99	352.42	353.27

Compute the mean $\bar{x}$ and standard deviation $\hat{s}$ for each year (by rows) and for each month (by columns). Because the temporal evolution of the annual mean indicates that carbon dioxide increases (probably resulting in global warming), compute the annual rate of increase. Comment on the results.

1.22. Historical records of earthquake intensity. *Catalogo dei terremoti italiani dall'anno 1000 al 1980* ("Catalog of Italian earthquakes from year 1000 to 1980") was edited by D. Postpischl in 1985, and is available through the National Research Council of Italy. This directory contains all of the available historical information on earthquakes that occurred in Italy during the past (nearly) 1000 years. It also includes values of earthquake intensity in terms of the Mercalli–Canconi–Sieber (MCS) index. The following table gives the values of MCS intensity for the city of Rome:

	MCS intensity						
Century	2	3	4	5	6	7	Total
XI				2			2
XII				1			1
XIII				1			1
XIV							0
XV				1	1	1	3
XVI							0
XVII				1			1
XVIII		7	4	2	2		15
XIX	110	125	50	14	1	1	301
XX	3		2				5
Total	113	132	56	22	4	2	329

Draw the line diagram for the whole data and for those recorded in each century. Compare the data recorded in the eighteenth century with those recorded in the other centuries.

1.23. Sea waves. Because of scarcity of records, the characteristics of sea waves are often derived from other climatological data. For the purpose, the SMB method (named after Sverdrup, Munk, and Bretschneider) is widely used in engineering practice [see U.S. Army Corps of Engineers (1977), *Shore Protection Manual*,

Vol. 1, Coastal Engineering Research Center, Washington, DC]. Liberatore and Rosso used this model to simulate sea waves in the upper Adriatic Sea [Liberatore, G., and R. Rosso (1983). "Sulla valutazione stocastica dell'onda di progetto in base alla ricostruzione dello stato del mare: un esempio di applicazione per l'Adriatico centro-meridionale," *Giornale del Genio Civile*, Vol. 1–3, pp. 3–25]. They investigated two different strategies for model calibration, called "no. 1" and "no. 2" in the table presented here. The table also includes the observed and the simulated values of the height of the highest sea wave and of its period for measurements taken from August 1977 to September 1978.

Measured values		Simulated values: Calibration strategy no. 1		Simulated values: Calibration strategy no. 2	
Height (m)	Period (s)	Height (m)	Period (s)	Height (m)	Period (s)
2.26	6.1	1.81	5.4	1.54	5.8
3.10	4.3	2.93	6.8	2.54	6.4
3.22	5.7	3.24	7.2	2.80	6.7
3.84	7.7	3.18	7.1	2.69	6.6
2.56	5.3	2.74	6.6	2.32	6.1
2.74	5.7	3.49	7.4	3.00	6.9
2.28	4.9	2.12	5.8	1.80	5.4
3.88	6.7	5.10	9.0	4.43	8.4
2.49	5.0	2.14	5.8	1.81	5.4
4.22	6.9	4.45	8.8	3.77	7.7
2.01	5.0	2.57	6.4	2.19	5.9
2.77	5.9	2.68	6.5	2.27	6.0
3.61	6.5	3.86	7.8	3.36	7.3
3.51	7.4	4.02	8.0	3.51	7.5
2.52	5.0	3.39	7.3	2.95	6.9
2.12	5.1	2.61	6.5	2.21	6.0
2.73	6.5	2.22	6.0	1.88	5.5
3.30	5.4	4.05	8.0	3.49	7.5

Draw a scatter diagram to compare the observed and simulated values of wave heights and periods. Compute the correlation coefficients r. Compute the deviations of the simulated data from the observed data, and find the mean $\bar{x}_1$, standard deviation $\hat{s}_1$, and coefficient of variation v of these deviations. Do these results indicate which of the two investigated strategies provides the better representation of sea waves from climatological data?

1.24. Surveying. A triangulated network is used to determine the position of three points in space, denoted by $\mathbf{u}_1 \equiv (x_1, y_1)$, $\mathbf{u}_2 \equiv (x_2, y_2)$, and $\mathbf{u}_3 \equiv (x_3, y_3)$, by measuring their mutual distances and their distances from two reference points, $\mathbf{u}_A \equiv (x_A, y_A)$ and $\mathbf{u}_B \equiv (x_B, y_B)$, as shown in Fig. 1.P1.

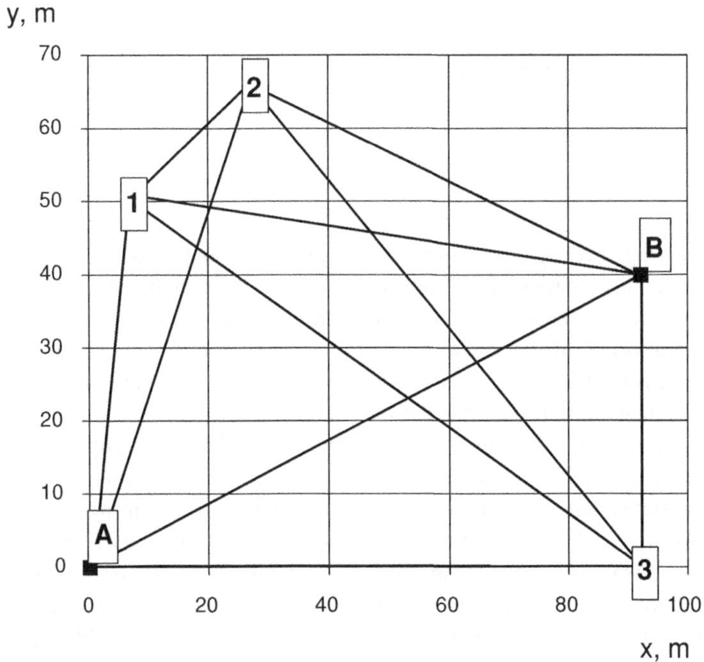

Fig. 1.P1 Survey configuration.

The Cartesian coordinates of the reference points are $x_A = y_A = 0$, $x_B = 92$, and $y_B = 40$ m. The table of the measured distances is given next.

	$\mathbf{u}_A$	$\mathbf{u}_B$	$\mathbf{u}_1$	$\mathbf{u}_2$	$\mathbf{u}_3$
$\mathbf{u}_A$	**0**	**100**	50	71	92
$\mathbf{u}_B$	**100**	**0**	86	70	40
$\mathbf{u}_1$	50	86	**0**	26	99
$\mathbf{u}_2$	71	70	26	**0**	93
$\mathbf{u}_3$	92	40	99	93	**0**

Using appropriate trigonometric methods, find the average location and coefficients of variation of the coordinates of point $\mathbf{u}_1 \equiv (x_1, y_1)$.

Chapter 2

Basic Probability Concepts

Engineering investigations involving natural phenomena as well as systems devised by humans exhibit scatter and variability as illustrated in Chapter 1. The resulting uncertainty that the engineer encounters is a major problem. By using the *theory of probability*, one can incorporate this uncertainty into the analysis and thus make rational decisions. The main focus of this chapter is to define the concept of probability and to discuss some of the associated axioms and basic properties.

Describing and predicting events in the real world can be approached by constructing mathematical models to describe practical events. For example, Newton's second law of motion states that when an object is subject to a force F it moves with an acceleration a proportional to F. The mathematical model can be written as $F = ma$, which expresses the relationship between the variables F, a, and the mass m. To use this equation to predict F, the force of a moving object in the gravitational field, as a function of the acceleration due to gravity, g, one must know the mass m of the object. Although this law (and its mathematical model) describes the motion of an object, it cannot be proved in a logical sense, but it can be verified experimentally.

Engineering problems sometimes require more complex models involving differential equations, the solution of which gives a prediction of specific values of some variables for known values of other variables. In all such cases, the outcome of an experiment is completely and precisely determined from the equations of the model, given some initial configuration of the variables. However, the hypothesis adopted and the set of differential equations can be affected by uncertainty arising from the natural variability of the independent variables or the inadequacy of model equations caused by incomplete knowledge or intractable mathematics.[1]

The engineer's use of probability theory is aimed at constructing mathematical models to describe events in the real world, as in the above case. The engineer must therefore consider the possibility of the occurrence of events that may influence experimental outcomes and estimate their likelihood. It might be desirable, for instance, to find a precise law to describe flood discharges at a given river site. The resulting mathematical model would be very complex, if it could be formulated at all, for it is not feasible to write a set of equations capable of predicting the number of times the river flow will exceed a specified threshold level in a given year. On the other hand, we can construct a *probability* model that, although not helpful in predicting the occurrence and magnitude of an individual flood, is quite useful in dealing with the flood regime characterizing that site. Specifically, one can postulate a number p, in the range zero to one, which represents the probability that a flood exceeding a given level occurs in a given year.[2] Like Newton's second law

[1] In some cases, the prototype is described by nonlinear differential equations yielding unpredictable dependent variables because of chaotic behavior. In a chaotic system, given some configuration of elements, the outcome of any experiment is unpredictable from the equations of the model owing to its sensitivity to the initial conditions; see Gleick (1987) and Lorenz (1993).

[2] A parallel scheme of modeling uncertainty is possible through the fuzzy set theory of Zadeh (1965). This arises from the generalization of the mathematical concept of an ordinary set (as defined shortly). The application of

of motion, this model is not provable from first principles; but if one observes river flows over a large number of years, one can ascertain the accuracy of the postulated p.

2.1 RANDOM EVENTS

Before introducing probability theory, let us define the mathematical concept of a random event and some related concepts such as sample space and event space. We shall also consider three of the ways in which events in a sample space are related and the use of Venn diagrams to illustrate relationships.

2.1.1 Sample space and events

Random events can be best described by assuming that an experiment has been performed and a series of observations taken under uniform conditions, so that there is no bias toward any particular result. Consider the determination of the strengths of concrete given in Table E.1.2. Because an individual outcome of this experiment cannot be predicted, the collection of all possible outcomes must initially be considered. It is convenient to represent this collection as a set, called the *sample space* or *value space* or *population* or *universe*. In a statistical sense, the set consists of *events* (as formally defined shortly) representing the phenomenon studied, and the outcomes are the actual values taken. The sample space is denoted by Ω and comprises a set of points, called *sample points*; each of the points is associated with one or more distinguishable events.

The term *space* is used to define the total collection of elements that are the results of an experiment but the term has much wider connotations than outcomes of a physical experiment. In set theory, *space* means the collection of all objects of interest in a given discussion. Use of the term *sample* arises from the uncertainty associated with the results of the experiment.

Definition: Sample space. The *sample space*, denoted by Ω, is the collection of all possible events arising from a conceptual experiment or from an operation that involves chance.

Example 2.1. Reservoir storage. The amount of water S stored in a reservoir varies in time from 0 to c, the active reservoir capacity, owing to the combined effect of inflows and outflows (see Fig. 2.1.1).

The sample space of the experiment measured as the volume of water in the reservoir at a given time can be defined as $\Omega \equiv \{S : 0 \leq S < c\}$. This is a set of sample points in the interval $[0, c)$.

Although Ω signifies a continuous sample space with an infinite number of points, one can also use a discrete representation of Ω by considering a finite number of states.[3] How one defines the discrete sample space depends on the judgement of the engineer. It is

fuzzy logic is based on a membership function with the same range as p. See the exposition by McNeill and Freiberger (1993). It has, however, aroused controversy among some statisticians and control engineers. Besides, a subjectively chosen optimization function is sometimes used in calibration; see, for instance, Chang and Chen (1998). Nevertheless, it has had wide acceptance and many highly successful engineering applications, as in automobile and other vehicle subsystems, washing machines, and cameras; see also Ross (2004).

Alternatively, Pawlak (1991) proposed a mathematical rough set theory to deal with a specific type of uncertainty; this leads to decision tables that expresse information in the data.

[3] A discrete sample space can have an infinite (countable) number of points in a theoretical representation.

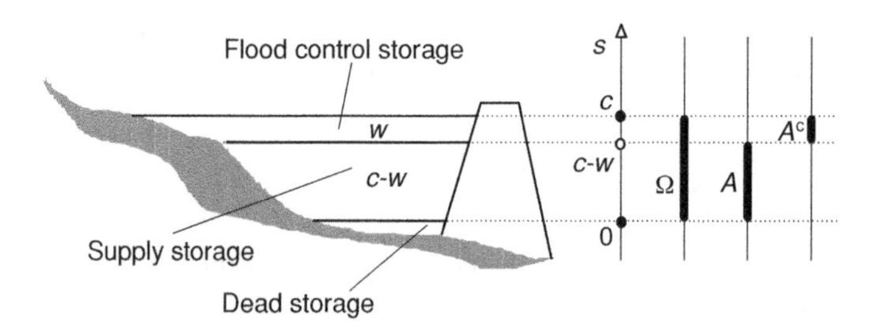

Fig. 2.1.1 Storage in a multipurpose reservoir.

mainly related to the specific problem and the use made of the model and is constrained by the resolution of the instrument, such as a water level indicator in a reservoir, used in measurements.

An *event* is a collection of sample points in the sample space Ω of an experiment. An event can consist of a single sample point called a *simple* or *elementary* event, or it can be made up of two or more sample points known as a *compound* event.

Definition: Event. An *event* (denoted by a capital letter A) is a subset of the sample space Ω.

Example 2.2. Reservoir storage. It is convenient to define reservoir storage S by a sequence of k states $\omega_1, \omega_2, \ldots, \omega_k$. The sample space is correspondingly given by the set

$$\Omega \equiv \{A_i, \quad \text{with } i = 1, 2, \ldots, k\},$$

where $A_i \equiv \{S\colon (i-1)c/k \leq S < ic/k\},\ i = 1, 2, \ldots, k$ is a set of events.

Consider four states of a reservoir: $\omega_i \equiv \{S\colon (i-1)c/4 \leq S < ic/4\},\ i = 1, \ldots, 4$, as shown in Fig. 2.1.2.

The event A defined as $A = \omega_4 \equiv \{S\colon 3c/4 \leq S < c\}$ is a simple event, because it corresponds with a single sample point (for this discretization). On the other hand, the event B defined as $B = \omega_1 + \omega_2 \equiv \{S\colon 0 \leq S < c/2\}$ is a compound event, because it comprises the collection of two simple events, namely $A_1 = \omega_1 \equiv \{S\colon 0 \leq S < c/4\}$ and $A_2 = \omega_2 \equiv \{S\colon c/4 \leq S < 2c/4\}$. Other possible events are shown by the pie charts.

Further examples of sample space and events are shown in Figs. 2.1.3 to 2.1.5.

Because an event A is defined as a subset of the sample space Ω, this subset is contained in Ω, that is, $A \subset \Omega$. The *complement* A^c of an event A consists of all those outcomes of Ω that are not included in the event A. This event can be also interpreted as the nonoccurrence of the event A.

Example 2.3. Reservoir storage. Let $\Omega \equiv \{S\colon 0 \leq S < c\}$ be the continuous sample space associated with the volume of water stored in a multipurpose reservoir at a certain time. Because mitigation of the downstream flood hazard is usually one of the objectives for construction of a reservoir, a portion of its capacity must be left empty at the beginning of the flood season. Let $w < c$ denote the residual reservoir capacity available for flood control storage. At the beginning of the flood season, the reservoir manager must investigate the event $A \equiv \{S\colon 0 \leq S \leq c - w\}$, which corresponds to the availability of sufficient flood storage in the reservoir (see Fig. 2.1.1). The complement of A is the event $A^c \equiv \{S\colon c - w < S < c\}$ which signifies that the reservoir has insufficient residual capacity to meet the flood control reservation. Both A and A^c are compound events in relation to Fig. 2.1.2.

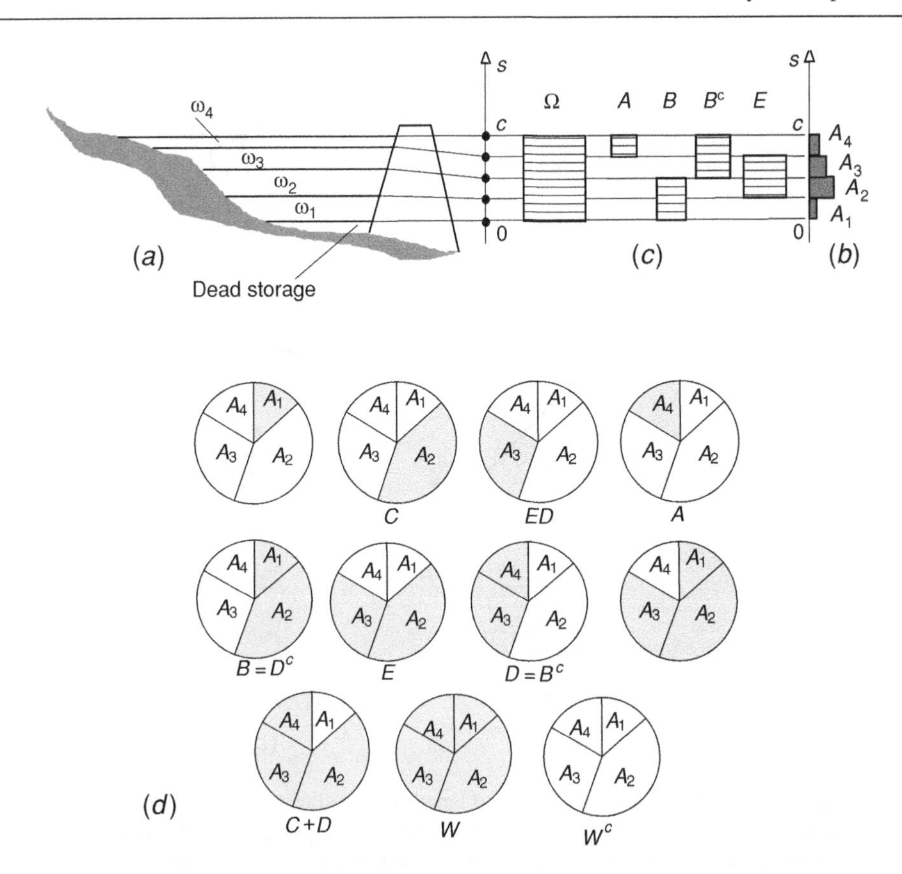

Fig. 2.1.2 (*a*) Reservoir storage as represented by four states, ω_1, ω_2, ω_3, and ω_4. (*b*) The widths of the rectangles on the extreme right are proportional to the relative frequencies of these states. (*c*) The events indicated in the text are represented by rectangles at the center on the right with areas proportional to the relative frequencies of these events. (*d*) The pie charts display all possible events (shaded) and also the empty state (unshaded). For example, $A = A_4$ means that $3c/4 \leq S < c$ in Example 2.2.

2.1.2 The null event, intersection, and union

There are many ways in which events in a sample space are related. Firstly, two events A_1 and A_2 are *mutually exclusive* or *disjoint* if the occurrence of one event excludes the occurrence of the other. This means that *none* of the points contained in A_1 is contained in A_2 and vice versa. Together A_1 and A_2 constitute the *null event*, denoted by $A_1A_2 = A_1 \cap A_2 = \emptyset$. For example, the events $A \equiv \{S\colon 3c/4 \leq S < c\}$ and $B \equiv \{S\colon 0 \leq S < c/2\}$ in Example 2.2 are mutually exclusive. One can extend the notion to more than two events. It is also clear that all simple events are mutually exclusive.

Secondly, two events A_1 and A_2 that are not mutually exclusive have some common sample points that constitute their *intersection*. This is denoted by $A_1 \cap A_2$ or A_1A_2 (but $A_1 \cap A_2 \neq \emptyset$). We note, for instance, from Example 2.2 and Fig. 2.1.2 that the intersection $A \cap B^c$ is the event $\{S\colon 3c/4 \leq S < c\}$, that is A_4, corresponding to state ω_4.

Thirdly, the *union* of two events A_1 and A_2 represent their joint occurrence, and it comprises the event containing all the sample points in A_1 and A_2. This is denoted by a $\cup$ sign between A_1 and A_2, that is, $A_1 \cup A_2$ or simply $A_1 + A_2$. For instance, from

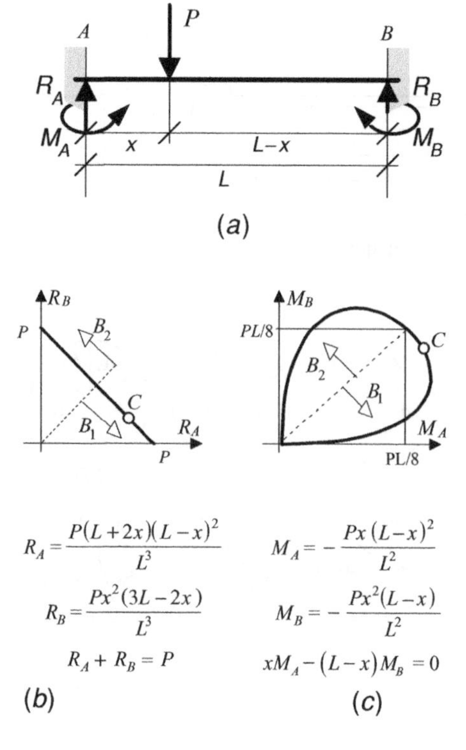

Fig. 2.1.3 (*a*) Sketch of a rigid beam of length L loaded with a concentrated load P which is located at a random distance x from the left joint. (*b*) Sample space of reactions R_A and R_B. (*c*) Sample space of moments M_A and M_B. The simple event C corresponds to the location of P shown in (*a*). The compound events B_1 corresponding to location $0 \le x \le L/2$, and B_2, with $L/2 < x \le L$, are mutually exclusive and collectively exhaustive.

Example 2.2 and Fig. 2.1.2 the union $A_1 \cup A_2$ is the event $\{S: 0 \le S < c/2\}$, that is, B. The null event, intersection, and union are three important relationships between events. Note that the intersection is equivalent to the "and" logical statement or function, whereas the union is equivalent to "and/or."

Definition: Mutually exclusive events, intersection, and union. Let A_1 and A_2 denote two events of a sample space Ω.

(1) A_1 and A_2 are *mutually exclusive* if the occurrence of one event excludes the other; A_1 and A_2 comprise the null set, that is, $A_1A_2 = A_1 \cap A_2 = \emptyset$.

(2) The common points, if any, of two events A_1 and A_2 constitute the *intersection*, denoted by $A_1 \cap A_2$ or A_1A_2.

(3) The *union* of two events A_1 and A_2 represent their joint or common occurrence. The combined event is denoted by $A_1 \cup A_2$ or $A_1 + A_2$.

The sample space Ω can be described by using a set of *mutually exclusive and collectively exhaustive* events, say, $B_1, B_2, \ldots, B_i, \ldots$. For example, if $B = A_1 + A_2$ and $D = A_3 + A_4$, the sample space of the discrete representation of reservoir storage (Example 2.2) can be given by $\Omega = B + D$. Since $BD = \emptyset$, we can define B and D as a set of mutually exclusive and collectively exhaustive events.

Further examples of mutually exclusive and collectively exhaustive events are shown in Figs. 2.1.3 and 2.1.5.

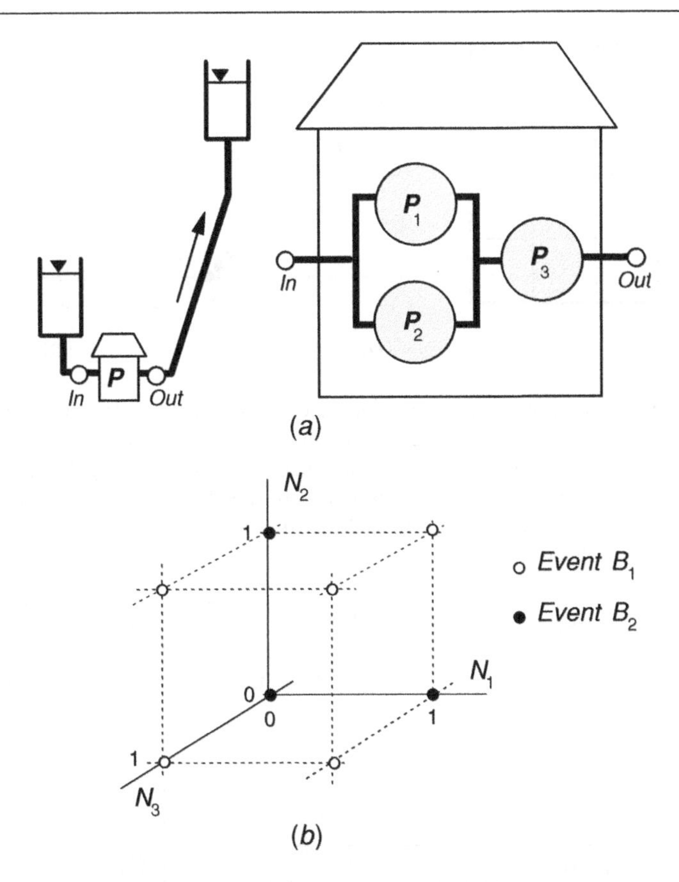

Fig. 2.1.4 (*a*) Pumping station and pipelines connecting two tanks. The station has two parallel pumps P_1 and P_2 serially connected to a third pump P_3. (*b*) Let $N_i = 1$ if the ith pump fails, and $N_i = 0$ otherwise. The sample space of the experiment to verify system operation has the eight simple events shown by points in the space (N_1, N_2, N_3). System failure is given by the occurrence of event $B_1 \equiv \{(0, 0, 1), (0, 1, 1), (1, 0, 1), (1, 1, 0), (1, 1, 1)\}$. The system is in operation if the event $B_2 \equiv \{(0, 0, 0), (0, 1, 0), (1, 0, 0)\}$ occurs. The events B_1 and B_2 are mutually exclusive and collectively exhaustive.

2.1.3 Venn diagram and event space

A collection of points in a sample space Ω as shown, for example, in Fig. 2.1.6 is called a set. Figure 2.1.6 shows a sequence of Venn diagrams (named after an English cleric and mathematician of the nineteenth century). Such diagrams provide a very useful visual representation of sets and set operations such as the complement, union, intersection, and other combinations. Because sample points, sample spaces, and events are sets, one can use this type of illustration to show the events in a sample space and important relationships among events.

The collection of all the events associated with an experiment and possible combinations of the events is defined as the *event space*, and it is a set denoted by A. This set contains all possible outcomes of the experiment included in the sample space Ω, as shown, for example, by the pie charts of Fig. 2.1.2. If A is an event that has some (nonzero) chance of occurrence, then A^c, the nonoccurrence of A, has also some chance of occurrence; therefore, A^c is also a subset of A. Further subsets are denoted by A_1 and A_2 in the

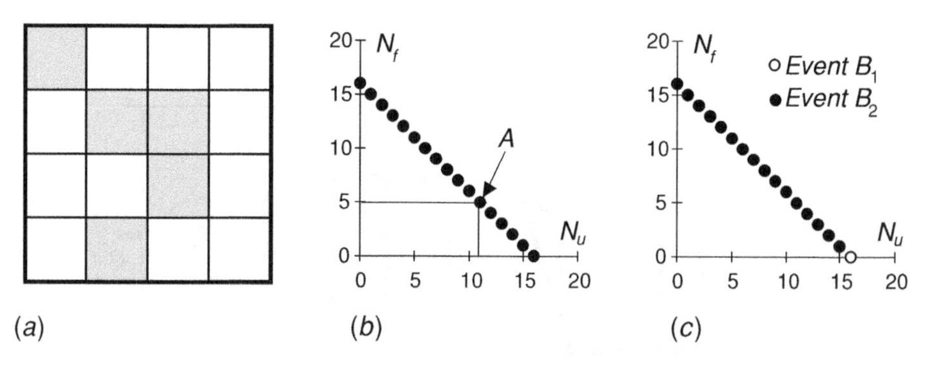

Fig. 2.1.5 (*a*) Two-phase pattern mosaic indicating forested (shaded) and unforested (unshaded) pixels from a remotely sensed image. This results from areal surveys used in the determination of land use and cover. (*b*) Sample space of the experiment of determining the numbers of forested N_f and unforested N_u pixels. The simple event A corresponds to the outcome displayed in (*a*). (*c*) The events $B_1 \equiv \{N_f = 0 \text{ and } N_u = 16\}$ and $B_2 \equiv \{0 < N_f \leq 16 \text{ and } N_u < 16\}$ are mutually exclusive and collectively exhaustive.

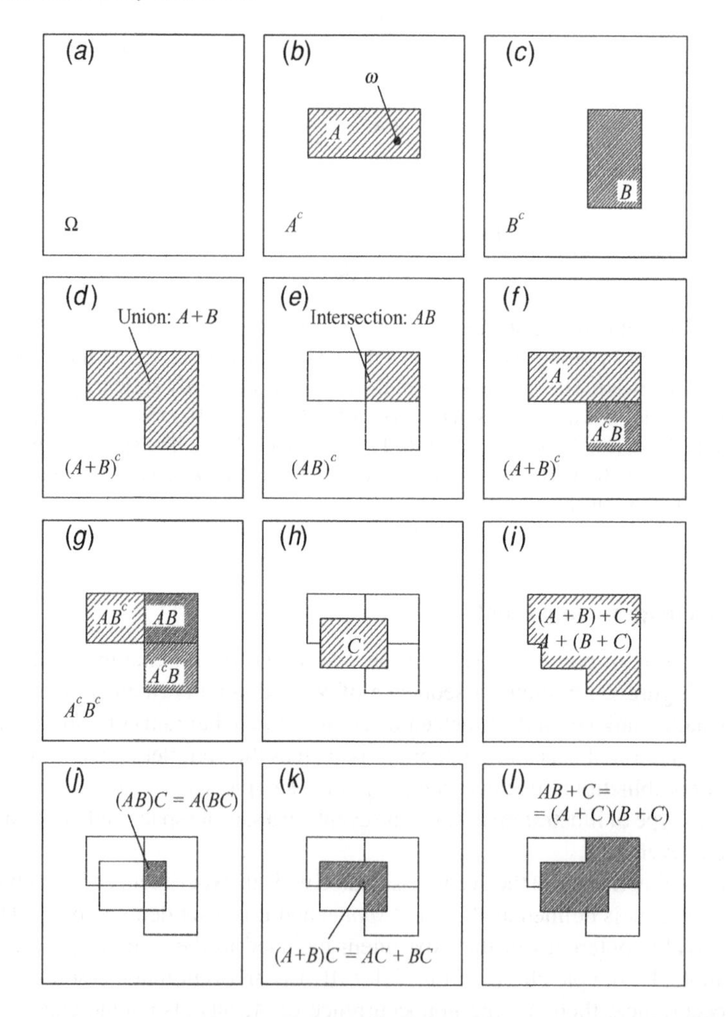

Fig. 2.1.6 Venn diagrams representing the sample space and different random events.

following definition. One can formally define *any* collection of events A with Properties 1 to 3, given in the following *definition*, as an *algebra of events*, which is characterized by the further property that A also includes the empty set, Ø (Property 5). Moreover, the *intersection* of two events, A_1 and A_2 (which, as noted, is the event consisting of those outcomes of Ω that are contained in both A_1 and A_2) is also an event of A (Property 4).

Definition: Event space. The collection of all possible events associated with a given experiment is defined as the *event space* and is denoted by A. This space is characterized by the following properties:

(1) $\Omega \in \mathsf{A}$.
(2) If $A \in \mathsf{A}$, then $A^c \in \mathsf{A}$.
(3) If $A_1 \in \mathsf{A}$ and $A_2 \in \mathsf{A}$, then $A_1 + A_2 \in \mathsf{A}$.

Also, it follows from the foregoing that

(4) If $A_1 \in \mathsf{A}$ and $A_2 \in \mathsf{A}$, then $A_1 A_2 \in \mathsf{A}$.
(5) If $\emptyset \in \mathsf{A}$, then $A^c \in \mathsf{A}$.

One can see that the concept of event space is more complex than that of sample space. Whereas Ω is the set representing all possible outcomes of an experiment, the event space A is a special set that contains Ω, as well as the events that can be defined by combining the outcomes of that experiment. In Example 2.2, for instance, the event space is given by

$$\mathsf{A} \equiv \{A_1, A_2, A_3, A_4, A_1 + A_2, A_2 + A_3, A_3 + A_4, A_1 + A_2 + A_3, A_2 + A_3 + A_4, A_1 + A_2 + A_3 + A_4, \emptyset\},$$

as shown in Fig. 2.1.2*d* using a pie chart representation. Because the algebra of events is analogous to areas on a plane, Venn diagrams such as those of Fig. 2.1.6 are most helpful for interpretation.

Example 2.4. Tipping bucket rain gauge. Consider the tipping bucket rain gauge shown in Fig. 2.1.7.

The experiment has only two possible outcomes: $A \equiv$ {the bucket is connected with switch a} and $B \equiv$ {the bucket is connected with switch b}. Therefore, the sample space of this experiment is $\Omega \equiv \{A, B\}$. *This constitutes the population for this sample space.* The events which can be defined from Ω are the elementary events A and B, their complements $A^c = B$ and $B^c = A$, their union $A + B$, and its complement $(A + B)^c$. Therefore, one has

$$\mathsf{A} \equiv \{A, B, A + B, (A + B)^c\}.$$

Both the sample space $\Omega = A + B$ and its complement $\Omega^c = (A + B)^c$ are events contained in A.

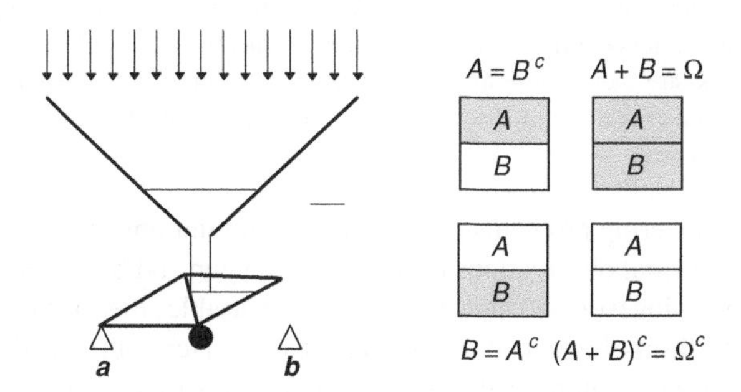

Fig. 2.1.7 Conceptual diagram of a tipping bucket rain gauge (left) and Venn diagrams (right) showing the events constituting the event space.

The union and intersection of events are said to be *associative* or *distributive*. The interiors of rectangles A and B shown in Fig. 2.1.6b and 2.1.6c, respectively, represent two simple events. The point ω represents an outcome of the event A. The union $A+B$ and intersection AB are seen in Fig. 2.1.6d and 2.1.6e, respectively. Figure 2.1.6f and 2.1.6g show events such as A^cB and AB^c. Event C, which intersects A and B, is shown in Fig. 2.1.6h. The compound events of Fig. 2.1.6i and 2.1.6j are, respectively,

$$(A+B)+C = A+(B+C), \quad \text{that is,} \quad (A \cup B) \cup C = A \cup (B \cup C)$$

and

$$(AB)C = A(BC), \quad \text{that is,} \quad (A \cap B) \cap C = A \cap (B \cap C).$$

These events are said to be associative, because they involve, respectively, *either* an addition *or* a multiplication of simple events. On the other hand, the compound events

$$(A+B)C = AC+BC, \quad \text{that is,} \quad (A \cup B) \cap C = (A \cap C) \cup (B \cap C)$$

and

$$AB+C = (A+C)(B+C), \quad \text{that is,} \quad (A \cap B) \cup C = (A \cup C) \cap (B \cup C),$$

presented in Fig. 2.1.6k and 2.1.6l, respectively, are distributive. This is because they arise from the addition *and* multiplication of simple events. The rules are equivalent to those governing the addition and multiplication of numbers, so that the axioms of conventional algebra apply to operations of sets and events. By using these properties, one can demonstrate that all the events that can be obtained by these operational rules are sets of the event space A.

Example 2.5. Reservoir storage. Let $\Omega \equiv \{S: 0 \le S < c\}$ denote the continuous sample space associated with the volume stored in a reservoir at the end of the dry season. Let A and B be two events defined as $A \equiv \{S: 0 \le S < c/3\}$ and $B \equiv \{S: c/4 \le S < c/2\}$. Both $AB \equiv \{S: c/4 \le S < c/3\}$ and $A+B \equiv \{S: 0 \le S < c/2\}$ are in the event space A, which is defined as $\mathsf{A} \equiv \{S: x \le S < y\}$ with $0 \le x \le y < c$ for any pair of x and y in $[0, c)$. Consider another event of A, say $C \equiv \{S: c/5 \le S < 3c/5\}$. From the algebra of events it will be noted that the event $(A+B)C = AC+BC$ corresponds with $\{S: c/5 \le S < c/2\}$. Also, the event defined as $AB+C = (A+C)(B+C)$ corresponds with $\{S: c/5 \le S < 3c/5\}$, which is equivalent to event C.

Example 2.6. Flood occurrence. Consider the number of floods, N, occurring in a year at a gauging station given in Table 1.1.1. Since an upper bound to annual flood occurrences cannot be established on a physical basis, one must assume that the possible outcomes of this experiment are the positive integer numbers including zero, that is, $\Omega \equiv \{N: 0 \le N < \infty$, with N an integer value$\}$. Let $A \equiv \{N: N = 0\}$ and $B \equiv \{N: N > 0\}$. These two events describe the nonoccurrence and occurrence, respectively, of at least one destructive flood in a year and are therefore mutually exclusive and exhaustive. To determine the likelihood with which the events A and B can occur is a fundamental problem in the assessment of flood risk at a river site.

Many engineering problems involve data from joint observations. For instance, in Section 1.4 concurrent observations of dissolved oxygen and biochemical oxygen demand in the water were investigated, because these two variables measuring water quality are typically observed simultaneously when investigating river pollution. Other examples of joint observations are compressive strength and density of concrete, as seen in Chapter 1; intensity and duration of storm rainfall at a recording rain gauge; wind speed and direction; period and height of sea waves; strength and settlement of a loaded beam; hourly numbers of cars and trucks in a highway; and the numbers of cars and their average velocities. All these experiments require a *two-dimensional sample space*. Depending on the experiment, this sample space can be either continuous (for example, wind speed and direction) or discrete (for example, paired numbers of cars and trucks), or it can also be discrete-continuous (for example, number of cars and the average velocity). Because this argument can be extended to any dimension, one can introduce a *multidimensional sample space* to describe experiments involving several variables.

Example 2.7. Number of rainy days and total rainfall. In a given location storm occurrences are random, because meteorological phenomena are such that one cannot predict the occurrence and intensity over a future time horizon. For irrigation purposes, one is interested in predicting the number of rainy days in a period of, say, 10 days, and the total amount of rainfall delivered during that period. The problem can be represented as a random experiment with sample space

$$\Omega \equiv \{(i, x): i = 0, 1, 2, 3, \ldots, 10; \quad \text{and} \quad 0 \leq x\},$$

where i denotes the number of rainy days. In this example x represents the total rainfall in millimeters in Milan, Italy, during the period September 11–20, in a year as shown in Fig. 2.1.8.

A sample point of Ω is given by two numerical values, the first of which indicates the integer number of rainy days i and the second number represents the total rainfall depth x. For instance, the farmers of Castle Park, Milan, know that if the number of rainy days is not less than four and the total rainfall during the period is in excess of 20 mm, irrigation of grassland is not required. Accordingly, the random event

$$A \equiv \{(i, x): i > 3, \quad \text{and} \quad x > 20\}$$

is of interest to them.

The set of all possible events is the event space A, which contains both Ω, the sure (certain) event, and $\emptyset = \Omega^c$, the null (impossible) event. It also contains the complement of A, that is, the set

$$A^c \equiv \{(i, x): i < 4, \quad \text{and} \quad x \leq 20\},$$

which corresponds to the event that irrigating is needed. Two disjoint or mutually exclusive events describe two different circumstances, which cannot occur simultaneously. For instance, it is clear that events

$$B \equiv \{(i, x): 3 \leq i < 5, \quad \text{and} \quad x \geq 10\},$$

and

$$C \equiv \{(i, x): 1 \leq i < 3, \quad \text{and} \quad 2 \leq x < 10\}$$

are mutually exclusive. In this case, $BC = \emptyset$. Because of Property 3 in the definition of event space, events such as

$$A + B \equiv \{(i, x): i > 2, \quad \text{and} \quad x \geq 10\}$$

and

$$B + C \equiv \{(i, x): 1 \leq i < 5, \quad \text{and} \quad 2 \leq x\}$$

are also part of the algebra. Because of Property 4 the same reasoning applies to the set

$$AB \equiv \{(i, x): i = 4, \quad \text{and} \quad x > 20\},$$

which describes the occurrence of 4 rainy days with more than 20 mm of rainfall over Milan during this period.

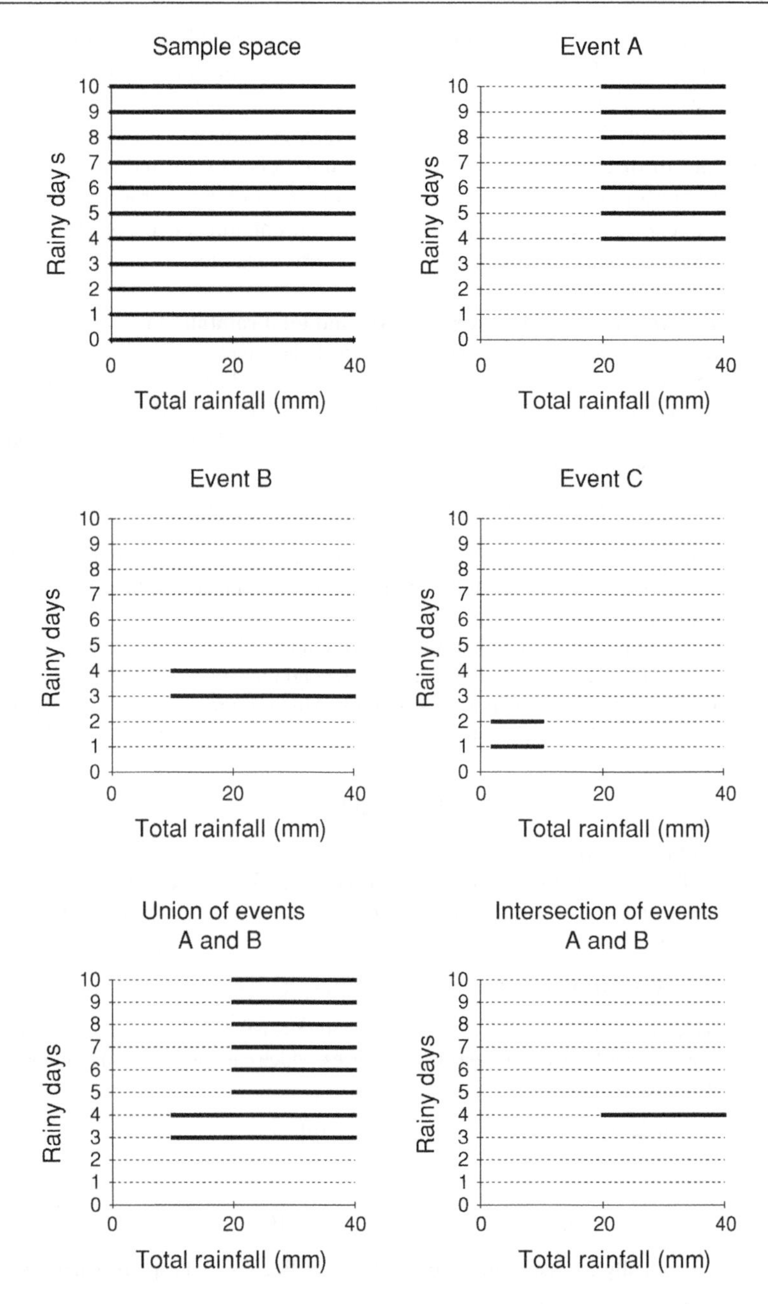

Fig. 2.1.8 Sample space and events representing rainy days and total rainfall in Milan during the period September 11–20. Event B, for example, denotes 3 or 4 rainy days with a total rainfall in the range 10–40 mm.

Quite often an engineer is interested in the possible outcomes of an experiment given that some event has occurred; the set of events associated with an event, say, A, can be considered as a new, reduced sample space. That is, by invoking the condition that event A has occurred one automatically restricts the sample points to the set representing A. For example, if an offshore engineer is interested in sea waves exceeding 2 m in height and

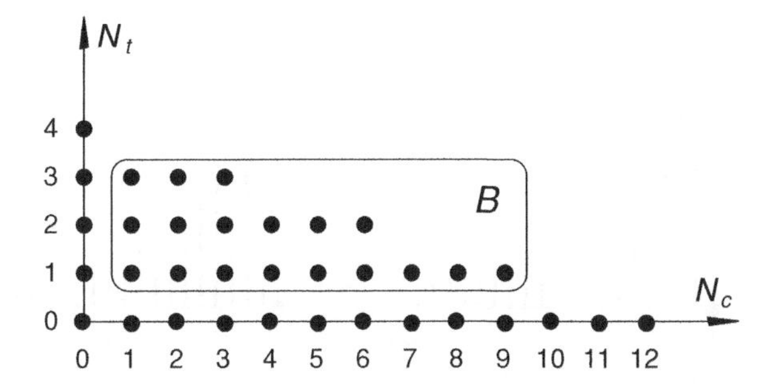

Fig. 2.1.9 Two-dimensional sample space of the paired numbers of cars N_c and trucks N_t that can be accommodated in a small ferry with a maximum loading capacity of six cars and two trucks, where one truck is equivalent to three cars. The event $B \equiv \{(N_c, N_t): N_c > 0, N_t > 0\}$ corresponds to the conditional sample space when the ferry carries both cars and trucks. The maximum load the ferry can take is the equivalent of twelve cars.

direction ranging from 25 to 120°, the original two-dimensional sample space given by $\Omega \equiv \{0^\circ \leq \theta \leq 360^\circ, h > 0\}$ is reduced to the subset $\{25^\circ \leq \theta \leq 120^\circ, h > 2\}$. Although the definition of a primary or a conditional sample space is often a matter of convenience, the notion of conditional sample space is essential in many applications. This concept will be applied when dealing with conditional probabilities in the next section.

Some examples of two-dimensional and conditional sample space are shown in Figs. 2.1.9 to 2.1.11.

2.1.4 Summary of Section 2.1

The sample space and event space, introduced here, are two of the basic concepts in probability theory. We also discuss the null event, intersection, and union and show how Venn diagrams are used.

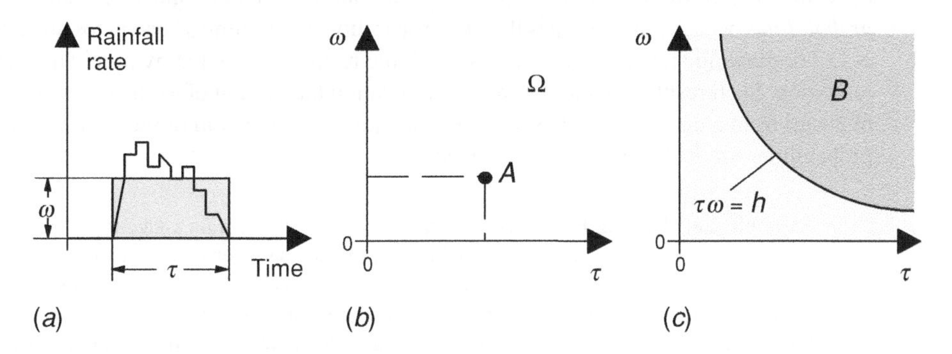

Fig. 2.1.10 (*a*) Average storm intensity, ϖ, and duration, τ, from a record of rainfall rate at a point in space. (*b*) Two-dimensional sample space Ω of the duration and intensity of a storm. The simple event A corresponds to the outcome displayed in (*a*). (*c*) The event $B \equiv \{(\tau, \varpi): \tau\varpi > h\}$ corresponds to the conditional sample space representing the storms with rainfall depth exceeding a value of h.

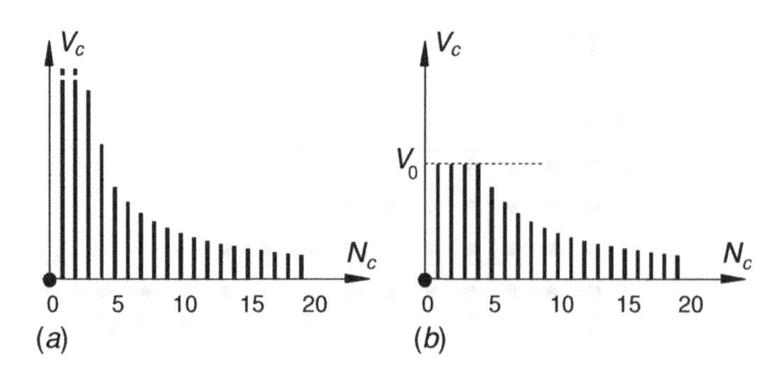

Fig. 2.1.11 (*a*) Two-dimensional sample space of the number N_c of cars per minute and the average velocity V_c on a road, where $N_c V_c \leq$ a constant. (*b*) Conditional sample space with the constraint that the speed limit V_0 is not exceeded, that is, $V_c \leq V_0$.

2.2 MEASURES OF PROBABILITY

In this section we discuss different aspects of probability. Axioms, the rules of addition and multiplication and independence are presented. The total probability and Bayes' theorems follow.

2.2.1 Interpretations of probability

Probability theory arose as a branch of mathematics dealing with the analysis of certain games of chance. The classical definition of probability is based on the results of a random experiment (such as tossing a coin or drawing a card from a pack). If this experiment can result in n mutually exclusive and equally likely outcomes and if n_A of these outcomes have an attribute A, then the probability of A is the fraction n_A/n. For example, if a card is drawn from an ordinary deck of playing cards, there are 52 possible outcomes that are *mutually exclusive* since two or more cards are not drawn simultaneously. If the deck and game are fair (that is, the pack is well shuffled), these 52 outcomes are *equally likely*, which means that the chance of drawing a particular card is the same as that of any other. For instance, the probability of drawing a diamond is 13/52, or 1/4, because there are 13 cards of this suit in the pack. The probabilities so determined are called *prior probabilities* because, if one states that the probability of obtaining a head in tossing a coin is 1/2 or drawing an ace from a pack is 1/13, the result is arrived at by purely deductive reasoning. Such results do not require for verification the tossing of a coin or the drawing of a card from a deck. On the basis of the assumption that the coin or the pack is fair, the probabilities are 1/2 and 1/13, respectively.

Example 2.8. Tipping bucket rain gauge. A tipping bucket rain gauge type operates by means of a pair of buckets (see Fig. 2.1.7). After the rain has been collected through a funnel at the top, it fills the first bucket, which then overbalances and empties, thereby directing the flow of water into the second bucket. The alternating motion of the tipping buckets is transmitted to a recording time device, which provides a measure of the rainfall intensity by counting the rate of tilting. Because the two buckets are equal in volume, one knows from prior reasoning that the probability that the water flows into either the left or the right bucket is 1/2 in any instant. Accordingly, from Example 2.4 one can assume that events A and B are mutually exclusive, collectively exhaustive, and equally likely, so that prior probabilities of 1/2 can be assigned to these events.

Nothing is said in the foregoing discussion regarding the determination of whether or not a particular coin or deck is fair. If the coin is biased in favor of heads, for example, the two possible outcomes of tossing that coin are not equally likely. One encounters other difficulties with the classical approach when trying to answer questions such as the following: What is the probability that, say, tomorrow will be a rainy day? What is the probability that a beam will collapse under a given load? What is the probability that a hurricane will occur in the next year? Notions such as *equally likely* cannot be used in this context, as they can be in games of chance. One needs to extend the definition to bring problems such as these into the framework of the theory.

The most widely applicable notion of probability is called *posterior* probability or *frequency*. We need again to refer to an experiment. Let us assume that a series of observations of a physical process can be made under uniform conditions. An observation is made; the experiment is repeated under similar conditions, and another observation is taken. After many repetitions, one comes to realize that there is an uncontrollable haphazard or random variation. For example, a large number of concrete specimens from the same source are tested until a state of rupture is reached. As usually happens, the critical load varies from one test to another, and one inevitably comes to the conclusion that the individual results are unpredictable; however, the engineer seeks some overall measure of loading capability to evaluate the safety of a structure built with that concrete. In many cases the observations fall into certain classes wherein the relative frequencies are quite stable. Therefore, one can postulate a number p, called the probability of a specified event, and approximate p by the *relative frequency* with which the repeated observations correspond with the event. This can be the relative frequency, for instance, with which the rupture load falls in a certain range after a large number of tests.

Example 2.9. Timber strength. In the previous chapter 165 timber strength values are ranked into different classes; in one example these are characterized by a width of 5 N/mm^2 (see Fig. 1.1.4 and Table 1.1.4). For example, the likelihood that the modulus of rupture of this material ranges from 20 to 24.99 N/mm^2 can be measured as the relative frequency of the specimens that were ruptured when loads within this range were applied. We see in Table 1.1.4 that 9 items of data out of 165 lie in the aforementioned range. Hence, we can infer that $p = 9/165$, or 0.055, say, 5.5%. If one is to use this material under a maximum design strength of 25 N/mm^2, it is possible to estimate the reliability of a structure using this timber as the probability that its modulus of rupture exceeds a value of 25 N/mm^2. Because 154 values equal or exceed 25 N/mm^2, the reliability $r = 154/165$, or 0.933, say, 93.3%.

Both general types of probability (prior and posterior) require an experiment in which the various outcomes can occur under somewhat uniform conditions. Examples of this are repeated card drawing for the prior case and repeated load testing for the posterior case. However, one should bring into the realm of probability theory situations that cannot conceivably fit into the framework of repeated outcomes under somewhat similar conditions. For example, what is the probability that a freeway or highway system will meet the demands for the next 25 years? Alternatively, what are the probabilities of having different types of soil and rock below the foundation of a proposed structure? These types of problems are also a legitimate part of general probability theory and are included in what is referred to as *subjective probability* (in contrast to objective methods based on theory or observation). Here, assigned probabilities are based on experience and personal judgement, and comprise a set of weights. Subjective factors are accounted for to some extent by the Bayesian approach (which is discussed in Subsection 2.2.7). Although problems tend mainly to be solved by the classical approach, there are many situations in which subjectivity can play a very useful role. In conclusion note that the following axioms of probability, from which probability theory is developed, can be applied to prior, posterior, or subjective probabilities.

2.2.2 Probability axioms

The classical axiomatic definition of probability requires the concept of a function, say, $f(\cdot)$. This is a rule that associates each point from one set of points with one and only one point from another set. The first collection of points is the *domain* and the second collection is the *counterdomain*. In the case of reservoir storage as given by Examples 2.1, 2.2, and 2.5, for example, the volume of water is usually obtained from measurements of the water surface elevation h and the computation of surface areas, once the bathymetry of the reservoir has been established. The values taken by h vary from the dead level ($h = h_{\min}$) to that representing the spillway crest ($h = h_{\sup}$). Correspondingly, the effective storage s varies from 0 to c, the active reservoir capacity. The stage-volume relationship can be represented as a function $s = f(h)$ where the domain is the set H given by the interval $[h_{\min}, h_{\sup})$, and the counterdomain is the set S given by the interval $[0, c)$. This is equivalent to stating that $f(\cdot)$ maps H into S. The mathematical form of this function will depend on the topography of the basin impounded by the dam.

As defined in Section 2.1, let Ω denote the sample space of a random experiment and A the collection of events assumed to be an event space for that experiment. A *probability function* $\Pr[\cdot]$ is a function with domain A and counterdomain in the interval [0, 1] that follows the three axioms given shortly. The uncertainty associated with the event space is measured by mapping any event, say, A, into the interval [0, 1]. Here, $\Pr[A]$ is defined as the chance or probability that event A occurs or the probability of event A. With reference to Ω, $\Pr[A]$ is the sum of the probabilities of the sample points that constitute A. In practice, when Ω contains an infinite number of sample points, probabilities are assigned to areas (or lengths). Thus A can represent a collection of points associated with a particular event, as noted in Section 2.1.

Definition: Probability function. A *probability function* $\Pr[\cdot]$ is a function mapping the event space A of a random experiment into the interval [0, 1] according to the following axioms:

(*i*) $\Pr[A] \geq 0$, for every $A \in \mathsf{A}$. (2.2.1)

(*ii*) $\Pr[\Omega] = 1$. (2.2.2)

(*iii*) If $A_1 \in \mathsf{A}$, $A_2 \in \mathsf{A}$, and $A_1A_2 = \emptyset$, then $\Pr[A_1 + A_2] = \Pr[A_1] + \Pr[A_2]$, (2.2.3)

where $\Omega \in \mathsf{A}$ denotes the sample space of the experiment, and A, A_1, and A_2 denote events that belong to that sample space.

Example 2.10. Reservoir storage. In Example 2.2 the storage S in a reservoir was discretized into four states, ω_1, ω_2, ω_3, and ω_4 (see Fig. 2.1.2). The sample space is therefore given by the set

$$\Omega \equiv \{A_1, A_2, A_3, A_4\},$$

where $A_i \equiv \omega_i \equiv \{S: (i-1)c/4 \leq S < ic/4\}$ for $i = 1, 2, 3, 4$. Observations of reservoir storage are made at the end of each period of operation, say, one year. Suppose that after 36 years of reservoir operation, the following frequencies have been observed: 5, 15, 10, and 6, for simple events A_1, A_2, A_3, and A_4, respectively. The probabilities, assigned on the basis of relative frequencies, are

$$\Pr[A_1] = 5/36, \quad \Pr[A_2] = 15/36, \quad \Pr[A_3] = 10/36, \quad \Pr[A_4] = 6/36,$$

which satisfy Axiom i. Since

$$5/36 + 15/36 + 10/36 + 6/36 = (5 + 15 + 10 + 6)/36 = 1,$$

$\Pr[\Omega] = \Pr[A_1 + A_2 + A_3 + A_4] = 1$, so that Axiom ii is also satisfied.

Consider two mutually exclusive events, say, $C = A_2 \equiv \{S: c/4 \leq S < c/2\}$ and $D = A_3 + A_4 \equiv \{S: c/2 \leq S < c\}$. By combining the foregoing frequencies, one can see that event C occurred 15 times in 36 years, whereas event D occurred $10 + 6 = 16$ times during that period. The associated probabilities are thus $\Pr[C] = 15/36$ and $\Pr[D] = 16/36$, respectively. Since $C + D = A_2 + A_3 + A_4$,

$$\begin{aligned}\Pr[C + D] = \Pr[A_2 + A_3 + A_4] &= (15 + 10 + 6)/36 = 15/36 + 16/36 \\ &= \Pr[C] + \Pr[D],\end{aligned}$$

which satisfies Axiom iii.

The theory of probability deals logically with the relationships among probability measures. Because of the deductive character of the theory, one can develop all such relationships entirely from the three axioms described by Eqs. (2.2.1) to (2.2.3).

2.2.3 Addition rule

The third axiom states that the basic addition property of probability can be extended to any sequence of mutually exclusive events. If $A_1, A_2, \ldots, A_k \in \mathsf{A}$, and $A_i A_j = \emptyset$ for any $i \neq j$, with $i, j = 1, 2, \ldots, k$, then

$$\Pr[A_1 + A_2 + \cdots + A_k] = \Pr[A_1] + \Pr[A_2] + \cdots + \Pr[A_k]. \quad (2.2.4)$$

From this rule one can derive a number of further properties of probability that can be used to perform additive operations in the event space, such as union and intersection of events.

Axiom ii can be applied to an event A and its complement, A^c; A and A^c jointly satisfy the conditions for exclusive events, thus obtaining $\Pr[A + A^c] = \Pr[A] + \Pr[A^c]$. But since $A + A^c = \Omega$, we also have $\Pr[A + A^c] = \Pr[\Omega] = 1$ from Eq. (2.2.2). By combining these results,

$$\Pr[A^c] = 1 - \Pr[A]. \quad (2.2.5)$$

This states that the probability of the complement of an event is given by the difference from unity of the probability of that event.

Property: Probability of complement. The probability of the complement A^c of an event A equals the difference from unity of $\Pr[A]$.

This property is useful in evaluating the probability of occurrence of a complementary event of practical interest from that estimated through relative frequencies or by developing physical or operational considerations.

Example 2.11. Flood occurrence. Consider the floods that exceed the previously established design flood in the outlet reach of the Bisagno River at Genoa, Italy, observed from 1931 to 1995. Records indicate that six floods occurred in the period, namely, in 1945, 1951 (twice),

1953, 1970, and 1992. Let N denote the number of flood occurrences per year. The engineer is interested in evaluating the likelihood of the occurrence of such a flood in any year. Define $\Omega \equiv \{N: N \geq 0\}$, and let $A \equiv \{N: N = 0\}$. The event representing the occurrence of at least one flood in a year is $A^c \equiv \{N: N > 0\}$. The records indicate that $\Pr[A^c] = 5/65 = 0.077$ and $\Pr[A] = 1 - 0.077 = 0.923$ is the probability that no flood occurs in any year. Thus $\Pr[A^c] = 0.077$ measures the hydrological risk affecting the river site.

Another important relationship among probabilities is the one linking the union and intersection of two events A and B that are not necessarily mutually exclusive. As shown in Fig. 2.1.6f, $A + B = A + A^c B$, where A and $A^c B$ are two events that are mutually exclusive. By using Axiom *iii* one obtains

$$\Pr[A + B] = \Pr[A + A^c B] = \Pr[A] + \Pr[A^c B].$$

Because event B is the union of two events, namely, $A^c B$ and AB, Axiom *iii* also yields

$$\Pr[B] = \Pr[A^c B] + \Pr[AB].$$

From this equation we can determine $\Pr[A^c B]$ as $(\Pr[B] - \Pr[AB])$, which can be substituted for $\Pr[A^c B]$ in the previous equation to give

$$\Pr[A \cup B] \equiv \Pr[A + B] = \Pr[A] + \Pr[B] - \Pr[AB], \tag{2.2.6}$$

which provides the general *addition rule* of probability theory. This rule states that the probability of occurrence of at least one of the events A and B equals the sum of their individual probabilities reduced by the probability of the joint occurrence of A and B. Intuitively, the addition rule is correct, because one can see from Fig. 2.1.6d and 2.1.6e that, to obtain the sample space of the event $A + B$, one adds the sample spaces of events A and B and subtracts that of AB.

If A and B are mutually *exclusive* events, that is, $AB = \emptyset$, then $\Pr[AB] = 0$. Hence, $\Pr[A \cup B] \equiv \Pr[A + B] = \Pr[A] + \Pr[B]$.

Property: Addition rule of probability theory. The probability of the union $A + B$ (that is, at least one) of two events A and B equals the difference between the sum of the probabilities of these events and the probability of their intersection. In the case of exclusive events the probability of the intersection is zero.

Note that Eq. (2.2.6) yields Eq. (2.2.4) for two (or more) events that are mutually exclusive according to Axiom *iii*.

Example 2.12. Reservoir storage. Following the material in Example 2.10, let us consider the complement of event D, that is, the event $D^c = \{S: 0 \leq S < c/2\}$ which has probability

$$\Pr[D^c] = \Pr[A_1 + A_2] = (5 + 15)/36 = 20/36.$$

Thus $\Pr[D^c] = 1 - \Pr[D]$, since $\Pr[D^c] = 1 - 16/36 = 20/36$.

Let $E = A_2 + A_3 \equiv \{\{S: c/4 \leq S < 3c/4\}$, which is not mutually exclusive with events C and D of Example 2.10. One obtains

$$\Pr[ED] = \Pr[E] + \Pr[D] - \Pr[E + D] = 25/36 + 16/36 - 31/36 = 10/36.$$

This can be easily verified because the intersection ED is the simple event A_3 which has a relative frequency of 10/36.

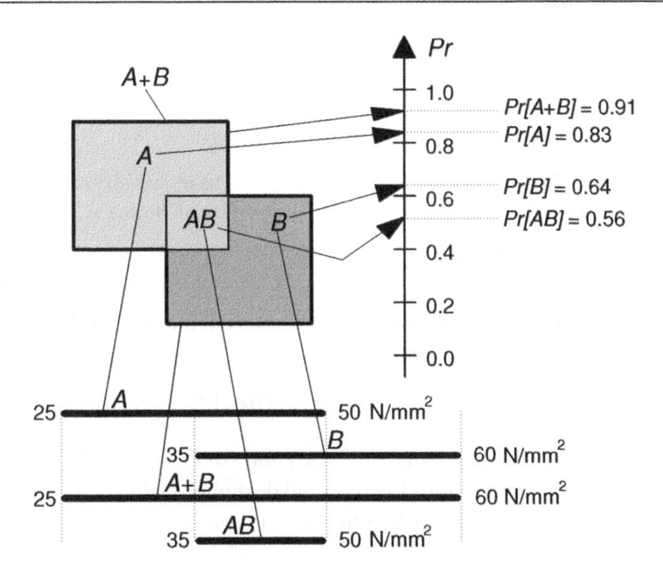

Fig. 2.2.1 Venn diagram showing the events indicated in the text and probabilities of these events as obtained from relative frequency of timber strength data of Table 1.1.4.

Example 2.13. Timber strength. Consider the timber strength data of Table 1.1.3. Define the following events (see Fig. 2.2.1):

$$A \equiv \{25 < \eta_t < 50 \text{ N/mm}^2\} \quad \text{and} \quad B \equiv \{35 < \eta_t < 60 \text{ N/mm}^2\},$$

where η_t denotes the modulus of rupture of a test sample. Using the relative frequencies to estimate probabilities, from Table 1.1.4, we write

$$\Pr[A] = (18 + 26 + 38 + 34 + 20)/165 = 136/165 = 0.824,$$
$$\Pr[B] = (38 + 34 + 20 + 9 + 5)/165 = 106/165 = 0.642.$$

The intersection event, AB, is given by the common outcomes of A and B, so that $AB \equiv \{35 < \eta_t < 50 \text{ N/mm}^2\}$. Accordingly,

$$\Pr[AB] = (38 + 34 + 20)/165 = 92/165 = 0.557.$$

The probability of the union event $A + B$ can be computed by Eq. (2.2.6). Thus,

$$\Pr[A + B] = \Pr[A] + \Pr[B] - \Pr[AB] = 0.824 + 0.642 - 0.557 = 0.909.$$

Since the event $A + B$ is given by the collection of the outcomes of both A and B, that is, $A + B \equiv \{25 < \eta_t < 60 \text{ N/mm}^2\}$, one can verify this result again from the relative frequencies:

$$\Pr[A + B] = (18 + 26 + 38 + 34 + 20 + 9 + 5)/165 = 150/165 = 0.909.$$

Therefore the probability that the strengths of the Swedish redwood and whitewood are between 25 and 60 N/mm^2 is more than 90%.

2.2.4 Further properties of probability functions

Some other useful properties can be demonstrated by using the three preceding axioms with basic aspects of set theory. These properties hold for any probability space.[4]

[4] A probability space is formally defined as the triplet (Pr[·], A, Ω), where Ω is the sample space, A is the event space (defined as an algebra of events), and Pr[·] is a probability function with domain A.

Property: Probability of null event. The probability of the null (impossible) event is zero; that is,

$$\Pr[\emptyset] = 0. \tag{2.2.7}$$

Property: Probability of a contained event. The probability of an event A that is contained in another event B does not exceed the probability of B; that is,

$$\Pr[A] \leq \Pr[B], \text{ if } A \subset B. \tag{2.2.8}$$

Property: Boole's inequality. The probability of the union of n events does not exceed the sum of their probabilities; that is,

$$\Pr[A_1 + A_2 + \ldots + A_n] \leq \Pr[A_1] + \Pr[A_2] + \ldots + \Pr[A_n]. \tag{2.2.9}$$

In certain cases, the inequality of (2.2.8) and Boole's inequality of (2.2.9) can provide conservative approximations of the required design probability in the absence of sufficient knowledge to determine the probability of a design event.

Example 2.14. Dam failure. Two natural events can result in the failure of a dam in an earthquake-prone area. Firstly, a very high flood, exceeding the design capability of its spillway, say, event A, may destroy it. Secondly, a destructive earthquake can cause a structural collapse, say, event B. Hydrological and seismological consultants estimate that the probability measures characterizing flood exceedance and earthquake occurrence on a yearly basis are $\Pr[A] = a$ and $\Pr[B] = b$, respectively. The occurrence of one or both events can result in the failure of the dam. Thus, the probability of failure of the dam is given by

$$\Pr[A + B] = \Pr[A] + \Pr[B] - \Pr[AB].$$

Only the first two probabilities on the right are known. However, the engineer can assume that the joint event AB has an extremely low probability, so that Boole's inequality can be used to obtain a conservative estimate of the probability of the union, $\Pr[A + B]$. Since

$$\Pr[A + B] \leq \Pr[A] + \Pr[B],$$

the engineer assumes

$$\Pr[A + B] \approx \Pr[A] + \Pr[B] = a + b.$$

If one takes, for example, values of a and b for small dams in seismic areas as 0.02 and 0.01, respectively,

$$\Pr[A + B] \cong 0.02 + 0.01 = 0.03.$$

This indicates a probability of not more than 3% that the dam will collapse in a given year.

2.2.5 Conditional probability and multiplication rule

As already mentioned, many engineering problems require answers to questions concerning the likelihood of the occurrence of some event that is conditional on the occurrence of one or more other events. For example, a dam has not failed in 100 years and one is interested in the probability that it will survive for another 100 years. As another example, one examines the pollutant concentration in 100 bottles of sea water (collected randomly at a popular resort) of which no more than 5 should exceed an allowable threshold: then one may need to evaluate the probability that the fourth sample is polluted given that the first three samples are not polluted. Alternatively, an engineer may wish to know the probability that monthly streamflow at a river outlet will exceed 2000 m^3/s during April given that higher values have been observed in each of the last three months. Such questions can be answered only in the context of conditional probability, a very useful concept in engineering because all events encountered in practice are conditioned. Another reason for using conditional probability is that engineers often assess the strength of a structure through its constituents by measuring a variable associated with the characteristic that is important. For instance, the density of a compacted soil is tested because it gives an indication of the strength and stability of an embankment. From the axioms given above, conditional probability is defined as follows:

Definition: Conditional probability. Let A and B be two events in the sample space, Ω, of a random experiment, and let $\Pr[B] > 0$. The *conditional probability* of event A given that event B has occurred, denoted by $\Pr[A|B]$, is defined by

$$\Pr[A|B] = \Pr[AB]/\Pr[B], \tag{2.2.10}$$

and is undefined for $\Pr[B] = 0$.

Note that the sample space is reduced from Ω to B and thus one needs to normalize using divisor $\Pr[B]$ [see Axiom *ii*, Eq. (2.2.2)]. Also, $\Pr[B]$ is termed the *marginal probability* of event B.

Example 2.15. Concrete test. Consider the $n = 40$ paired data of densities and compressive strengths of concrete given in Table E.1.2, as rearranged in Table 2.2.1.

Define the following events:

$$A \equiv \{2440 < \lambda_c < 2460 \text{ kg/m}^3\} \quad \text{and} \quad B \equiv \{55 < \eta_c < 65 \text{ N/mm}^2\},$$

where λ_c denotes the density of a concrete cube under test, measured in kg/m^3, and η_c denotes the compressive strength of that cube, measured in N/mm^2 (see Fig. 2.2.2).

Using relative frequencies, we write

$$\Pr[A] = n_A/n = 19/40, \quad \text{and} \quad \Pr[B] = n_B/n = 26/40.$$

Those experimental outcomes with both density and strength values lying in the ranges just defined are represented by the intersection event, $AB \equiv \{2440 < \lambda_c < 2460 \text{ kg/m}^3, 55 < \eta_c < 65 \text{ N/mm}^2\}$. Accordingly,

$$\Pr[AB] = n_{AB}/n = 16/40.$$

The probability that a concrete cube with density from 2440 to 2460 kg/m^3 yields a value of compressive strength in the range 55–65 N/mm^2 is

$$\Pr[A|B] = \Pr[AB]/\Pr[B] = (16/40)/(26/40) = 16/26.$$

The same result is obtained directly from the relative frequency approach by observing that 16 of the 26 outcomes with compressive strengths in the range 55–65 N/mm^2 have a density from 2440 to 2460 kg/m^3.

Example 2.16. Water distribution. Consider a pipeline for the distribution of a water supply of an urban area of 200 km^2. The city plan is approximately rectangular with dimensions of 10 by 20 km, and it is uniformly covered by the network shown in Fig. 2.2.3. Pressures and flow rates are uniform throughout the whole network, so that losses are equally likely to occur within it. Define the events

$$A \equiv \{\text{a severe water loss occurs in location } \mathbf{u} \equiv (u_1, u_2)$$

Table 2.2.1 Density, λ_c, and compressive strength, η_c, at 28 days from examination of 40 concrete cube test records (see Table E.1.2)

Density (kg/m^3)	Compressive strength (N/mm^2)	A^a	B^a	AB^a
2437	60.5		•	
2437	60.9		•	
2425	59.8		•	
2427	53.4			
2428	56.9		•	
2448	67.3	•		
2456	68.9	•		
2436	49.9			
2435	57.8		•	
2446	60.9	•	•	•
2441	61.9	•	•	•
2456	67.2	•		
2444	64.9	•	•	•
2447	63.4	•	•	•
2433	60.5		•	
2429	68.1			
2435	68.3			
2471	65.7			
2472	61.5		•	
2445	60.0	•	•	•
2436	59.6		•	
2450	60.5	•	•	•
2454	59.8	•	•	•
2449	56.7	•	•	•
2441	57.9	•	•	•
2457	60.2	•	•	•
2447	55.8	•	•	•
2436	53.2			
2458	61.1	•	•	•
2415	50.7			
2448	59.0	•	•	•
2445	63.3	•	•	•
2436	52.5			
2469	54.6			
2455	56.3	•	•	•
2473	64.9		•	
2488	69.5			
2454	58.9	•	•	•
2427	54.4			
2411	58.8		•	
$n = 40$	$n = 40$	$n_A = 19$	$n_B = 26$	$n_{AB} = 16$

[a] The absolute frequencies of the following events are computed: $A = \{2440 < \lambda_c < 2460 \text{ kg/m}^3\}$, $B = \{55 < \eta_c < 65 \text{ N/mm}^2\}$, and $AB = \{2440 < \lambda_c < 2460 \text{ kg/m}^3, 55 < \eta_c < 65 \text{ N/mm}^2\}$.

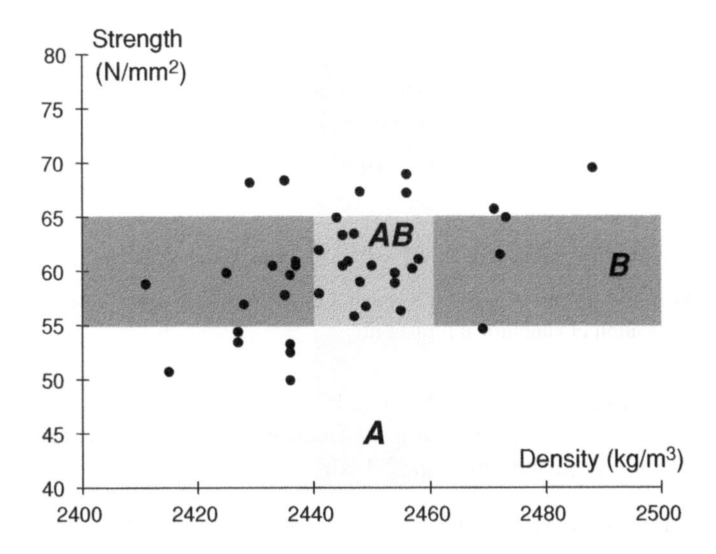

Fig. 2.2.2 Scatter diagram of concrete test data and events for application of the conditional probability concept.

where $0 < u_1 \leq 6$ km, $\quad 0 < u_2 \leq 3$ km} and

$$B \equiv \{\text{a severe water loss occurs in location } \mathbf{v} \equiv (v_1, v_2)$$
$$\text{where } 4 < v_1 \leq 12 \text{ km}, 2 < v_2 \leq 6 \text{ km}\}.$$

Assume that the probability of a loss in a given subarea is proportional to the area. Therefore, if a loss occurs in the pipe network,

$$\Pr[A] = (6 \times 3)/200 = 0.09,$$
$$\Pr[B] = (12 - 4) \times (6 - 2)/200 = (8 \times 4)/200 = 0.16.$$

Suppose that one seeks an answer to the following question: "If a loss occurs in the area affected by event B, what is the probability of event A?". It is necessary to know the proportion of the city area affected by event B within which A also occurs. From Fig. 2.2.3 one observes that this area is $(6 - 4) \times (3 - 2) = 2$ km^2, which must be divided by 32, that is, the area associated with B, to obtain the conditional probability. Therefore,

$$\Pr[A|B] = 2/32 = 1/16 = 0.0625.$$

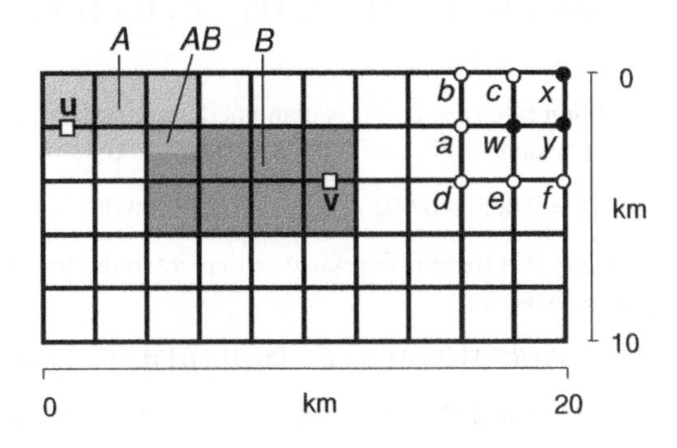

Fig. 2.2.3 Pipeline network for urban water supply.

This result can also be obtained by using Eq. (2.2.10) which requires the evaluation of $\Pr[AB]$. If one observes that the total area corresponding to both events taken separately is 48 km^2, the probability of occurrence of either A or B, or both A and B, is $\Pr[A + B] = 48/200 = 0.24$. By using the addition rule one gets

$$\Pr[AB] = \Pr[A] + \Pr[B] - \Pr[A + B] = 0.09 + 0.16 - 0.24 = 0.01.$$

Hence

$$\Pr[A|B] = \Pr[AB]/\Pr[B] = 0.01/0.16 = 0.0625.$$

As seen here, the evaluation through a reduced sample space gives the same result as the application of conditional probability.

We have seen Examples 2.15 and 2.16 that the definition of conditional probability by Eq. (2.2.10) is compatible with the frequency approach. This result occurs because $\Pr[AB] = n_{AB}/n_B$, and from Eq. (2.2.10)

$$\Pr[A|B] = \Pr[AB]/\Pr[B] = (n_{AB}/n)/(n_B/n) = n_{AB}/n_B,$$

where n_B and n_{AB} denote the numbers of occurrences of events B and AB, respectively, and n is the total number of events.

In some applications, the probabilities $\Pr[A]$ and $\Pr[A|B]$, for example, can be estimated directly, whereas the joint probability $\Pr[AB]$ may not be known. This can be obtained from Eq. (2.2.10) as follows:

$$\Pr[AB] = \Pr[A|B]\Pr[B] = \Pr[B|A]\Pr[A]. \tag{2.2.11}$$

Note that Eq. (2.2.11) can be also used to obtain the conditional probability $\Pr[A|B]$ from $\Pr[B|A]$ when the marginal probabilities of events A and B are known, and vice versa. Because $\Pr[A|B] \leq 1$ and $\Pr[B|A] \leq 1$ from Axiom *ii*, the following inequalities also hold:

$$\Pr[A|B] \leq \Pr[A]/\Pr[B], \quad \Pr[B|A] \leq \Pr[B]/\Pr[A].$$

Example 2.17. Wall foundation. The foundation of a wall can fail either by excessive settlement or from bearing capacity. The respective failures are represented by events A and B, with probabilities $\Pr[A] = a$, and $\Pr[B] = b$. The probability of failure in bearing capacity given that the foundation displays excessive settlement is $\Pr[B|A] = \beta_a$, say. The probability of failure of the wall foundation can be evaluated from

$$\begin{aligned}\Pr[A + B] &= \Pr[A] + \Pr[B] - \Pr[AB] = \Pr[A] + \Pr[B] - \Pr[B|A]\Pr[A] \\ &= a + b - \beta_a a,\end{aligned}$$

where the addition rule (Eq. 2.2.6) is combined with the concept of conditional probability (Eq. 2.2.11). For example, if $a = 0.005$, $b = 0.002$, and $\beta_a = 0.2$,

$$\Pr[A + B] = 0.005 + 0.002 - 0.2 \times 0.005 = 0.006.$$

The probability that there is excessive settlement in the foundation but there is no failure in bearing capacity is

$$\Pr[AB^c] = \Pr[B^c|A]\Pr[A] = (1 - \Pr[B|A])\Pr[A] = (1 - \beta_a)a$$

by using Eqs. (2.2.5) and (2.2.11). For the above values of a and β_a,

$$\Pr[AB^c] = (1 - 0.2) \times 0.005 = 0.004.$$

The conditional probability that the foundation has excessive settlement given that it fails in bearing capacity is obtained from Eq. (2.2.11) as follows:

$$\Pr[A|B] = \Pr[B|A]\Pr[A]/\Pr[B] = \beta_a a/b = 0.2 \times 0.005/0.002 = 0.5.$$

Also, $\beta_a = \Pr[B|A] \leq b/a = 0.4$.

The concept of conditional probability can be extended to any number of events. For example, for three events, A, B, and C,

$$\Pr[ABC] = \Pr[A|BC]\Pr[BC] = \Pr[A|BC]\Pr[B|C]\Pr[C].$$

In general, the following expansion holds for m events:

$$\Pr[A_1A_2A_3\ldots A_m] = \Pr[A_1|A_2A_3\ldots A_m]\Pr[A_2|A_3\ldots A_m]\ldots\Pr[A_m], \quad (2.2.12)$$

which can be interpreted as the *multiplication rule* of probability theory. This rule is primarily useful for experiments defined in terms of stages. Suppose an experiment has m stages and A_j is an event defined at stage j of the experiment; then $\Pr[A_j|A_1\ldots A_{j-1}]$ is the conditional probability of the event described in terms of the conditional probabilities of events occurring in stages $1, 2, \ldots, j-1$. The multiplication rule becomes simpler when the events at each stage of the experiment are independent from each other, as shown next.

2.2.6 Stochastic independence

When the probability of occurrence of an event is not affected by the occurrence of another event, these two event are (*statistically* or *stochastically*) *independent*. Thus, two events A and B are said to be independent if either

$$\Pr[A|B] = \Pr[A], \quad \text{if} \quad \Pr[B] > 0, \quad (2.2.13a)$$

or

$$\Pr[B|A] = \Pr[B], \quad \text{if} \quad \Pr[A] > 0. \quad (2.2.13b)$$

According to the definition of conditional probability, the *independence* of A and B as stated by Eq. (2.2.13) also implies that Eq. (2.2.11) yields

$$\Pr[A \cap B] \equiv \Pr[AB] = \Pr[A]\Pr[B]. \quad (2.2.14)$$

The preceding relationship is applicable only if Eq. (2.2.13) holds, and thus it can be used as an alternative definition of independence. Equation (2.2.14) states that the probability of the joint occurrence of two independent events equals the product of their marginal probabilities.

Definition: Independence. Two events defined in a given probability space A are *independent* if either the conditional probability of one event equals its marginal probability, or their joint probability equals the product of the marginals.

This definition applies to two physical events that are not related in any way, so that the probability measure of one event is not altered by the occurrence of the other.

Example 2.18. Concrete test. Consider the concurrent data of density λ_c and compressive strength η_c of concrete given in Table 2.2.1. For the two events A and B shown in Fig. 2.2.2,

$$A \equiv \{2440 < \lambda_c < 2460\ \text{kg/m}^3\} \quad \text{and} \quad B \equiv \{55 < \eta_c < 65\ \text{N/mm}^2\},$$

one obtains $\Pr[A] = 0.475$ using relative frequencies, and $\Pr[B] = 0.65$ using relative frequencies. Also, the probability of the joint occurrence,

$$AB \equiv \{2440 < \lambda_c < 2460 \text{ kg/m}^3, 55 < \eta_c < 65 \text{ N/mm}^2\}$$

is $\Pr[AB] = 0.40$. From Eq. (2.2.11)

$$\Pr[A|B] = \Pr[AB]/\Pr[B] = 0.615.$$

Clearly, $\Pr[A|B]$ differs significantly from $\Pr[A]$, so independence between events A and B does not hold. Therefore, the engineer cannot assume that density and compressive strength are independent. The absence of independence is also evident when comparing the value of the product of the marginal probabilities, $\Pr[A]\Pr[B] = .309$, with that of the joint probability, $\Pr[AB] = .40$.

For *independent events*, the joint probability in the addition rule of Eq. (2.2.6) is substituted by the product of the marginal probabilities, giving the following important rule:

$$\Pr[A + B] = \Pr[A] + \Pr[B] - \Pr[A]\Pr[B].$$

From this one can determine the probability of the union (sum) of two (or more) events from the knowledge of the marginal (individual) probabilities of these events.

One can prove independence of two events only by obtaining $\Pr[A]$, $\Pr[B]$, and $\Pr[AB]$ and then by demonstrating that either Eq. (2.2.13) or Eq. (2.2.14) holds. Engineers, who normally rely on their knowledge of physical situations to determine whether two particular events are independent, should adopt this procedure wherever possible to enhance the scientific arguments.

Example 2.19. Dam reliability. Consider again the problem of the failure of a dam (see Example 2.14) caused by the occurrence of either a flood exceeding the design capacity of the spillway (event A) or a destructive earthquake producing the structural collapse of the dam (event B). Let $\Pr[A] = a$ and $\Pr[B] = b$ denote the respective probabilities of occurrence in a year. If the two events are statistically independent, their joint probability equals the product of their marginals, that is,

$$\Pr[AB] = \Pr[A]\Pr[B] = ab.$$

Accordingly, the risk of failure of the dam in a year is given by

$$\Pr[A + B] = \Pr[A] + \Pr[B] - \Pr[AB] = a + b - ab.$$

If we assume typical values for a and b for small dams of 0.02 and 0.01, respectively, the risk of failure in a year is

$$\Pr[A + B] = a + b - ab = 0.02 + 0.01 - 0.02 \times 0.01 = 0.0298.$$

Note that this result is very close to that (.03) obtained by using Boole's inequality (see Example 2.14). When rare events with small marginal probabilities are considered, their joint probability generally plays a minor role, so that its effect can often be neglected in risk assessment.

One can see from Fig. 2.1.6g that the sample space for this experiment is given by four elementary events, namely,

$$\Omega \equiv \{AB, AB^c, A^cB, A^cB^c\},$$

whereas the failure event is represented by $A + B = (AB) \cup (AB^c) \cup (A^cB)$. Therefore, the probability of survival of the dam in any year can be measured as

$$\Pr[A^cB^c] = 1 - \Pr[A + B] = 1 - (a + b - ab),$$

which is defined as the reliability in a year. For $a = 0.02$ and $b = 0.01$,

$$\Pr[A^cB^c] = 1 - (a + b - ab) = 1 - (.02 + 0.01 - 0.01 \times 0.02) = 0.9702,$$

which means a chance of more than 97% that the dam will survive in a year. However, the designer is mainly interested in evaluating the system reliability during its lifetime, say, m years, and must therefore consider failure probability after $i = 1, 2, \ldots, m$ years from the construction of the dam. The survival probability after the first year is again $\Pr[A^cB^c]$. The experiment is repeated in subsequent years. The survival probability after the second year is given by $\Pr[(A^cB^c)_1 \cap (A^cB^c)_2]$ which can be written as

$$\Pr[(A^cB^c)_1(A^cB^c)_2] = \Pr[(A^cB^c)_1]\Pr[(A^cB^c)_2|(A^cB^c)_1].$$

where subscripts denote years. Because floods and earthquakes occurring in a year can be assumed to be independent of those occurring in another year, the conditional probability on the right-hand side can be simply written as $\Pr[(A^cB^c)_2|(A^cB^c)_1] = \Pr[(A^cB^c)_2]$. We also assume that the survival probability in any one year is the same as that in any other year. The survival probability after the second year is therefore

$$\begin{aligned}\Pr[(A^cB^c)_1(A^cB^c)_2] &= \Pr[(A^cB^c)_1]\ \Pr[(A^cB^c)_2] = \{\Pr[A^cB^c]\}^2\\ &= [1 - (a + b - ab)]^2.\end{aligned}$$

By using the same procedure, the m-year survival probability can be evaluated as

$$\Pr[(A^cB^c)_1(A^cB^c)_2 \ldots (A^cB^c)_m] = \{\Pr[A^cB^c]\}^m = [1 - (a + b - ab)]^m,$$

which can be taken as a reliability measure of the system, assuming constant probabilities of failure. For a design lifetime of $m = 50$ years,

$$\Pr[(A^cB^c)_1(A^cB^c)_2 \ldots (A^cB^c)_{50}] = 0.9702^{50} = 0.2203,$$

which means a design reliability of 22%. The design risk will be given by the complementary probability, which means that there is a design risk of 78% of dam failure within the design lifetime of 50 years.

The risk of failure in the ith year of dam operation is given by the probability that either an overtopping flood or a destructive earthquake or both will occur exactly in that year without any previous occurrence. This is given by

$$\begin{aligned}&\Pr[(A^cB^c)_1(A^cB^c)_2 \ldots (A^cB^c)_{i-1}(A + B)_i]\\ &\quad = (\Pr[A^cB^c])^{i-1}\Pr[(A + B)_i|(A^cB^c)_{i-1}] = (\Pr[A^cB^c])^{i-1}\Pr[A + B]\\ &\quad = [1 - (a + b - ab)]^{i-1}(a + b - ab),\end{aligned}$$

which is simply the design reliability rescaled by the elementary risk of failure. (Details of this geometric distribution with parameter $(a + b - ab)$ are given in Section 4.1.) For instance, the risk of dam failure during the tenth year of operation will be

$$\Pr[(A^cB^c)_1(A^cB^c)_2 \ldots (A^cB^c)_9(AB)_{10}] = 0.9702^9 \times 0.0298 = 0.0227.$$

Figure 2.2.4 gives reliabilities for varying design periods, say, m, and probabilities of first-time failure during the mth year of dam operation.

Equation (2.2.14) can be extended to any number of stochastically independent random events, so that it can be viewed as the multiplication rule in the independent case.

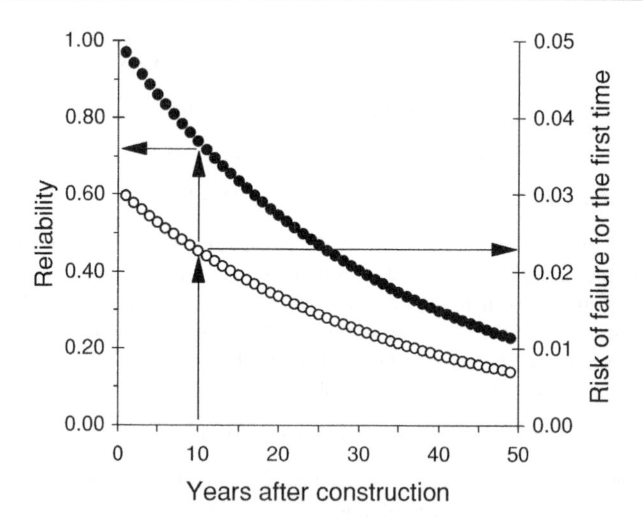

Fig. 2.2.4 Dam survival probability (m-year design reliability) and probability of failure in the mth year of operation (risk of failure for the first time).

Example 2.20. Water distribution. Consider again the pipe network for water distribution shown in Fig. 2.2.3. Since takeoff occurs at grid nodes, what is the probability that a node remains isolated, that is, all pipes connected to it become ineffective? Assuming that pipe failures are independent events, one obtains

$$\Pr[\text{node}k \text{ remains isolated}] = \prod p_{ik},$$

where the product is extended to all the ith adjacent nodes that are joined with any node k. Nodes w, y, and x, for example, are connected to four, three, and two pipes, respectively. The pipes have been fabricated from one material by a single manufacturer and installed during the same period. Therefore, one can assume that the probability of an individual pipe failure is constant, say, $p_{ik} = p$, for any pair of connected joint nodes i and k. Therefore,

$$\Pr[\text{node } w \text{ remains isolated}] = p_{aw}\,p_{cw}\,p_{ew}\,p_{yw} = p^4,$$
$$\Pr[\text{node } y \text{ remains isolated}] = p_{xy}\,p_{fy}\,p_{wy} = p^3,$$

and

$$\Pr[\text{node } x \text{ remains isolated}] = p_{cx}\,p_{yx} = p^2.$$

More generally, the probability that a node remains isolated is p^l, with l denoting the number of pipes connected with that node. Because $p \leq 1$, the risk that a node remains isolated decreases geometrically with increasing l.

Probability models dealing with independent events are very frequent in engineering practice, where they are applied to civil and environmental problems. Destructive floods, catastrophic earthquakes, hurricanes, and other geophysical hazards occur as a temporal sequence of independent events. The failure of power plants, building structures, water treatment facilities, and other systems constructed by humans can often be studied by analyzing the failures of different elementary components that are independent of each other. The multiplication rule for independent events allows one to determine the probability of a complex event (such as the occurrence of a flood exceeding the design capability of a dam spillway in a time horizon of 10 years, or the failure of a redundant pumping station) by multiplying the individual probabilities of simple events.

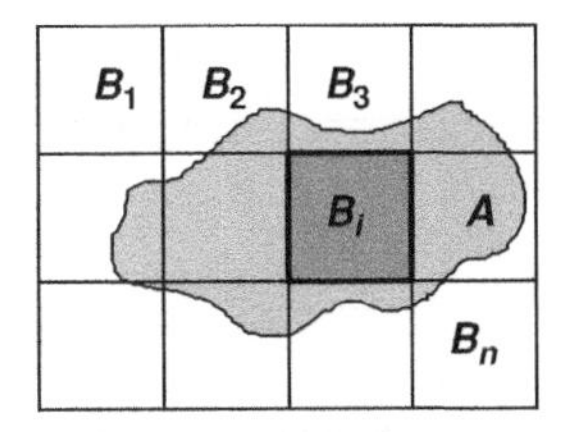

Fig. 2.2.5 Venn diagram for the theorem of total probability. Events $B_i, i = 1, \ldots, n$, are mutually exclusive and exhaustive, and some of them intersect A.

2.2.7 Total probability and Bayes' theorems

Sometimes the probability of an event A cannot be determined directly. However, its occurrence is accompanied by the occurrence of other events $B_i, i = 1, 2, \ldots, n$, such that the probability of A will depend on which of the events B_i has occurred. In such a case, the probability of A will be an expected probability, that is, the average probability weighted by those of B_i. This problem can be approached by using the theorem of total probability, which can be derived by the definition of conditional probability.

Consider a set of mutually exclusive, collectively exhaustive events, B_i, where $i = 1, 2, \ldots, n$. This statement means that $B_i B_j = \emptyset$, the null event, for any $i \neq j$ with $i, j = 1, 2, \ldots, n$, and that $B_1 + \cdots + B_n = \Omega$, the sample space. The probability of another event A can be given by using this set as follows (see Fig. 2.2.5):

$$\Pr[A] = \Pr[AB_1] + \Pr[AB_2] + \cdots + \Pr[AB_n] = \sum_{i=1}^{n} \Pr[AB_i].$$

Expanding each term in the sum using Eq. (2.2.11), we write

$$\Pr[A] = \sum_{i=1}^{n} \Pr[A|B_i] \Pr[B_i]. \tag{2.2.15}$$

This expression is known as the *theorem of total probability*.

Property: Total probability theorem. If B_i, where $i = 1, 2, \ldots, n$ is a set of mutually exclusive, collectively exhaustive events, the probability of an event A that occurs concurrently with the B_i, equals the sum of the products of the conditional probability of A given B_i, and the marginal probability of B_i.

Example 2.21. Timber strength. Consider the timber strength data of Table 1.1.3 and the frequency distribution given in Table 1.1.4. The sample space can be represented by the foregoing mutually exclusive, collectively exhaustive events (see Fig. 2.2.6).

$$B_1 \equiv \{0 \leq \eta_t < 25 \text{ N/mm}^2\},$$
$$B_2 \equiv \{25 \leq \eta_t < 45 \text{ N/mm}^2\},$$
$$B_3 \equiv \{45 \leq \eta_t < 65 \text{ N/mm}^2\}$$

and

$$B_4 \equiv \{\eta_t \geq 65 \text{ N/mm}^2\},$$

with η_t denoting the modulus of rupture in N/mm^2. Probabilities of these events are computed from relative frequencies as follows:

$$\Pr[B_1] = (nB_1/n) = 11/165,$$

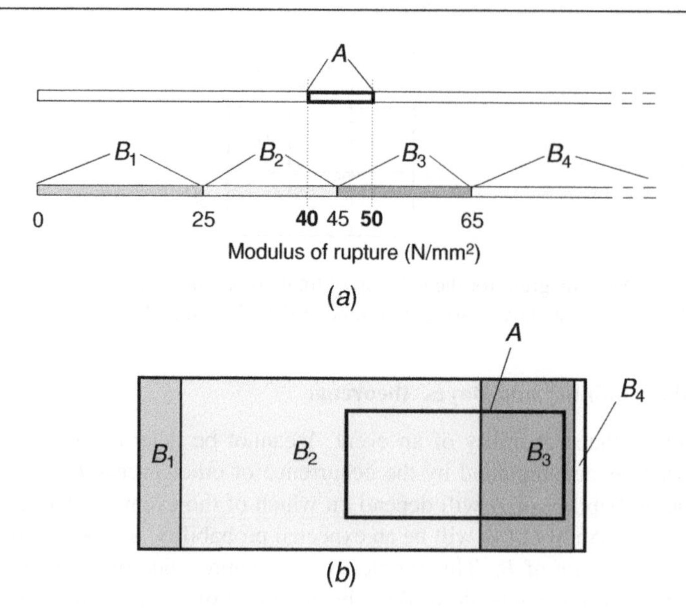

Fig. 2.2.6 (*a*) Events for application of the theorem of total probability to timber strengths. (*b*) Venn diagram in which areas are proportional to probabilities.

$$\Pr[B_2] = (nB_2/n) = 116/165,$$
$$\Pr[B_3] = (nB_3/n) = 34/165,$$

and

$$\Pr[B_4] = (nB_4/n) = 4/165,$$

where

$$\Pr[B_1 + B_2 + B_3 + B_4] = (11 + 116 + 34 + 4)/165 = 1.$$

Suppose an engineer is interested in estimating the probability that the modulus of rupture ranges from 40 to 50 N/mm^2, that is, the event

$$A \equiv \{40 \le \eta_t < 50 \text{ N/mm}^2\},$$

Since

$$\Pr[A|B_1] = \Pr[AB_1]/\Pr[B_1] = (n_{AB1}/n)/(n_{B1}/n) = (0/165)/(11/165) = 0,$$
$$\Pr[A|B_2] = \Pr[AB_2]/\Pr[B_2] = (n_{AB2}/n)/(n_{B2}/n) = (34/165)/(116/165) = 34/116,$$
$$\Pr[A|B_3] = \Pr[AB_3]/\Pr[B_3] = (n_{AB3}/n)/(n_{B3}/n) = (20/165)/(34/165) = 20/34,$$

and

$$\Pr[A|B_4] = \Pr[AB_4]/\Pr[B_4] = (n_{AB4}/n)/(n_{B4}/n) = (0/165)/(4/165) = 0,$$

the theorem of total probability gives

$$\Pr[A] = \Pr[A|B_1]\Pr[B_1] + \Pr[A|B_2]\Pr[B_2] + \Pr[A|B_3]\Pr[B_3] + \Pr[A|B_4]\Pr[B_4]$$
$$= 0 + (34/116) \times (116/165) + (20/34) \times (34/165) + 0 = (34 + 20)/165 = 54/165.$$

This can be verified, in this particular example, because the same result is found by considering the relative frequency in the range 40–50 N/mm^2.

This theorem is attributed to Thomas Bayes, an English cleric and philosopher of the eighteenth century.[5]

Example 2.23. Water quality. Consider concurrent data of DO and BOD recorded at 38 sites on the Blackwater River, England, in Table E.1.3. Owing to similarities in water uses, one can assume that the observations are from the same population. The means of the data are 7.5 and 3.2 mg/L, respectively. Define the following mutually exclusive and collectively exhaustive events:

$$B_1 \equiv \{\text{DO} \leq 7.5 \text{ mg/L}, \text{BOD} > 3.2 \text{ mg/L}\},$$
$$B_2 \equiv \{\text{DO} > 7.5 \text{ mg/L}, \text{BOD} > 3.2 \text{ mg/L}\},$$
$$B_3 \equiv \{\text{DO} > 7.5 \text{ mg/L}, \text{BOD} \leq 3.2 \text{ mg/L}\},$$
$$B_4 \equiv \{\text{DO} \leq 7.5 \text{ mg/L}, \text{BOD} \leq 3.2 \text{ mg/L}\}.$$

These are given in Fig. 2.2.8. By using relative frequencies,

$$\Pr[B_1] = 17/38 = 0.447,$$
$$\Pr[B_2] = 0/38 = 0.00,$$
$$\Pr[B_3] = 19/38 = 0.50,$$
$$\Pr[B_4] = 2/38 = 0.05.$$

The standard deviations are 1.0 and 0.5 mg/L, respectively. Let A be the event defined by concurrent values of DO and BOD within the range (mean – standard deviation) to (mean + standard deviation), that is,

$$A \equiv \{6.5 < \text{DO} < 8.5 \text{ mg/L}; 2.7 < \text{BOD} < 3.7 \text{ mg/L}\}.$$

The conditional probabilities of event A given that B_i occurs are

$$\Pr[A|B_1] = 7/17 = 0.41,$$
$$\Pr[A|B_2] \text{ is undefined because } \Pr[B_2] = 0,$$
$$\Pr[A|B_3] = 11/19 = 0.58,$$
$$\Pr[A|B_4] = 1/2 = 0.50.$$

From the theorem of total probability [Eq. (2.2.15)],

$$\Pr[A] = \Pr[A|B_1]\Pr[B_1] + \Pr[A|B_2]\Pr[B_2] + \Pr[A|B_3]\Pr[B_3] + \Pr[A|B_4]\Pr[B_4],$$

that is,

$$\Pr[A] = (7/17)(17/38) + (\text{undefined})(0) + (11/19)(19/38) + (1/2)(2/38)$$
$$= (7/38) + (11/38) + (1/38) = 19/38 = 0.50,$$

which means that the monitored values of DO and BOD have a 50% chance of lying in the previously defined range.

From Bayes' theorem [(Eq. 2.2.16)],

$$\Pr[B_1|A] = \frac{\Pr[A|B_1]\Pr[B_1]}{\sum_{i=1}^{n}\Pr[A|B_i]\Pr[B_i]} = \frac{(7/17)(17/38}{19/38} = \frac{7}{19} = 0.37,$$

[5] However, Stigler (1983), writing in the style of Agatha Christie, states that the odds in favor of Nicholas Saunderson (the blind professor who succeeded Newton's successor at Cambridge University) as the originator of the theorem are 3 to 1, which touches on Damon Runyan's limit to probability in life. So where should Hercule Poirot's finger finally point in this whodunit? Readers of Stigler (1983) will note the Laplacian indifference (vagueness) in the author's prior opinion.

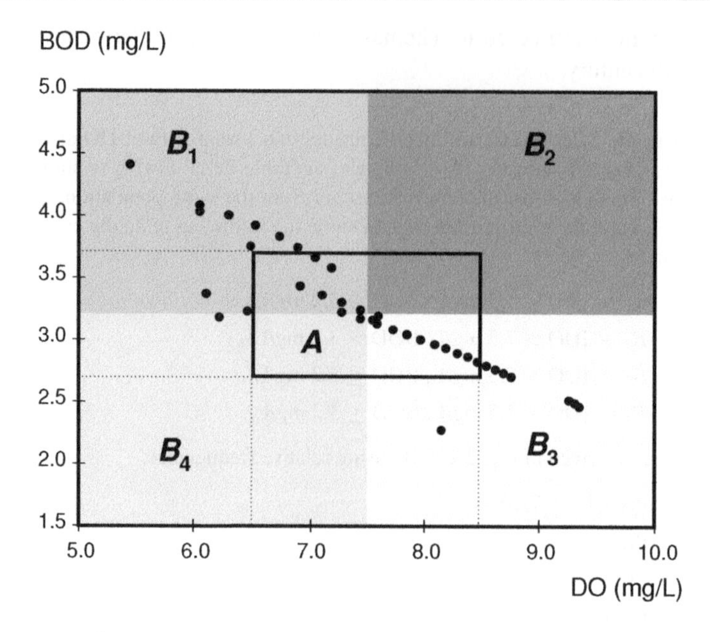

Fig. 2.2.8 Scatter diagram of water quality data and events in application of Bayes' theorem.

which means that, if the monitored values of DO and BOD lie in the previously defined range, there is a 37% chance that DO does not exceed its sample mean and BOD does exceed its sample mean. Using Bayes' theorem, one also obtains

$$\Pr[B_2|A] = 0,$$
$$\Pr[B_3|A] = 0.58,$$
$$\Pr[B_4|A] = 0.05.$$

As in the case of the theorem of total probability, Bayes' theorem is particularly useful for experiments carried out in stages. This has a potentially important role in engineering applications because it provides a method for continuously incorporating new information with previous data. The additional data may come, for example, from borehole results or tests on concrete cubes. By updating the prior probabilities, the engineer can assess the likelihood of design events by incorporating the additional information given by conditioned posterior probabilities.

If one defines as *state* the unknown quantification of the *population* and considering that some *sample* of observations is available, Bayes' theorem can be written as

$$\Pr[\text{state}|\text{sample}] = \frac{\Pr[\text{sample}|\text{state}]\,\Pr[\text{state}]}{\sum_{\text{all states}} \Pr[\text{sample}|\text{state}]\,\Pr[\text{state}]}.$$

In practice, an engineer often has prior knowledge of the occurrences of different states of a population (the so-called factors of information). In addition, there is frequently access to data from which one can estimate the likelihood of a measurable quantity or sample of data, given the true state of the population. By means of Bayes' theorem, one can then estimate the conditional probability of a given state of that population after a sample has been observed.

Table 2.2.2 Likelihood that the seismic recorder (showing the conditional probability of depths to bedrock) indicates state B_i given that the true state is B_j

	True state, B_j			
Measured state, B_i	$j = 1$ $h \leq 5$ m	$j = 2$ 5 m $< h \leq 10$ m	$j = 3$ 10 m $< h \leq 15$ m	$j = 4$ $h > 15$ m
$i = 1$ $h \leq 5$ m	0.90	0.05	0.03	0.02
$i = 2$ 5 m $< h \leq 10$ m	0.07	0.88	0.10	0.06
$i = 3$ 10 m $< h \leq 15$ m	0.03	0.05	0.81	0.12
$i = 4$ $h > 15$ m	0.00	0.02	0.06	0.80
Sum	1.00	1.00	1.00	1.00

Example 2.24. Imperfect testing of bedrock depth. An engineer designing a foundation for a tall structure needs to know the depth h of soil above bedrock at the site. For preliminary design purposes, the depth is divided into four states: $B_1 = \{h \leq 5m\}$, $B_2 = \{5\text{ m} < h \leq 10\text{ m}\}$, $B_3 = \{10\text{ m} < h \leq 15\text{ m}\}$, and $B_4 = \{h > 15\text{ m}\}$. The engineer then consults a local geologist who, from a knowledge of the geology of that area, assigns prior probabilities to the four states as follows:

$$\Pr[B_1] = 0.60, \quad \Pr[B_2] = 0.20, \quad \Pr[B_3] = 0.15, \quad \text{and} \quad \Pr[B_4] = 0.05.$$

For measuring the depth to bedrock, a seismic recorder is used which is subject to some error. From previous experience the geologist estimates the conditional probabilities that the instrument indicates a particular state out of four states (the sum of the probabilities of which is 1.0) for each of four actual states of nature. The likelihoods are given in Table 2.2.2.

The reading of the instrument is $h = 7$ m, which is henceforth referred to as sample no. 1. This corresponds with state B_2. The posterior probabilities of the actual states of nature are evaluated from Eq. (2.2.16) as follows:

$$\Pr[B_k|\text{sample no.1}] = \frac{\Pr[\text{sample no. }1|B_k]\Pr[B_k]}{\sum_{i=1}^{4}\Pr[\text{sample no. }1|B_i]\Pr[B_i]},$$

where

$$\sum_{i=1}^{4}\Pr[\{5\text{ m} < h \leq 10\text{ m}\}|B_i]\Pr[B_i] = .07\times.060 + .88\times.20 + .10\times.15 + .06\times.05$$
$$= .236.$$

Accordingly,

$$\Pr[B_1|\text{sample no. }1] = \frac{.07\times.60}{.236} = .178,$$

$$\Pr[B_2|\text{sample no. }1] = \frac{.88\times.20}{.236} = .746,$$

$$\Pr[B_3|\text{sample no. }1] = \frac{.10\times.15}{.236} = .063,$$

and

$$\Pr[B_4|\text{sample no. }1] = \frac{.06\times.05}{.236} = .013.$$

The sum of the preceding quantities is unity.

Because there is still a chance of about 25% that the true state may not be B_2, a second test is made and the reading is 8 m. Thus sample no. 2 also indicates state B_2.

Using the foregoing posterior probabilities (after sample no. 1) as revised prior probabilities for the site, we find again posterior probabilities (after sample no. 2), as follows:

$$\sum_{i=1}^{4} \Pr[\{5 \text{ m} < h < 10 \text{ m}\} | B_i] \Pr[B_i]$$
$$= 0.07 \times 0.178 + 0.88 \times 0.746 + 0.10 \times 0.063 + 0.06 \times 0.013 = 0.675.$$

With S.1,2 denoting samples 1 and 2, therefore,

$$\Pr[B_1|\text{S.1,2}] = \frac{0.07 \times 0.178}{0.675} = 0.018,$$
$$\Pr[B_2|\text{S.1,2}] = \frac{0.88 \times 0.746}{0.675} = 0.972,$$
$$\Pr[B_3|\text{S.1,2}] = \frac{0.10 \times 0.063}{0.675} = 0.009,$$

and

$$\Pr[B_4|\text{S.1,2}] = \frac{.06 \times .013}{0.675} = .001.$$

The sum of these probabilities is unity. It is now evident that the chance that the true state is not B_2 is very small. The engineer may therefore proceed on the assumption that the depth to rock is in the range 5–10 m.

We note that if the likelihoods of indicating correctly a state of nature by the instrument are smaller than the values of 0.90, 0.88, 0.81, and 0.80 given in Table 2.2.2, then the posterior probabilities will also be low. Such results are usually applicable to a cheaper type of testing procedure or experimentation. Furthermore, if sample no. 2 is quite different from sample no. 1 then some ambiguity will arise and tests will need to be repeated.

We are not sure, of course, what the true state of nature is; but good instrumentation with the help of Bayes' theorem, as shown here, should provide more realistic probabilities of individual states.

2.2.8 Summary of Section 2.2

Prior, posterior, and subjective probabilities were discussed here, and the axioms of probability introduced. Rules were specified for addition and multiplication of probabilities, and conditional probability was defined. Total probability and Bayes' theorem were presented with applications.

2.3 SUMMARY FOR CHAPTER 2

An event is a collection of outcomes or sample points in the sample space of an experiment that results in a set of observations. The sample space comprises an exclusive and exhaustive set of events, and all possible combinations are given by the event space. The axioms presented here governing complementary, null, union, and intersection events are a prelude to the introduction of probability. The theory of probability provides a deductive framework for evaluating the probabilities of different types of events. Probabilities can be interpreted as prior or posterior. A prior probability can be estimated from relative frequencies of observed events; but if it is assigned from one's experience or judgement it is called subjective. The operational rules of probability theory, which provide the basis for the relationships among probabilities of different events, are derived from three simple axioms and the notion of a probability function. These have led to the laws governing conditional probability, stochastic independence, the addition rule, the multiplication rule, the theorem of total probability, and Bayes' theorem.

Many engineers follow the probabilistic approach, perhaps intuitively at times. The purpose of this chapter which concluded with Bayes' theorem is to provide a rigorous foundation to the methods of sampling and experimentation. As stated in Chapter 1, however, we must emphasize the importance of the quality and extent of the available data because an engineer must appreciate that the significance of the results depends on the data from which the probabilities are estimated.

REFERENCES

General. The following references are given for further reading as required:

Ang, A. H. S., and W. H. Tang (1975). *Probability Concepts in Engineering Planning and Design, Vol. 1: Basic Principles*, John Wiley and Sons, New York. A blend of theory and practice with wide appeal for practicing civil engineers.

Benjamin, J. R., and C. A. Cornell (1970). *Probability, Statistics and Decision for Civil Engineers*, McGraw-Hill, New York. A classic for civil engineers with examples, extensive case studies, and innumerable problems to solve.

Blake, I. F. (1979). *An Introduction to Applied Probability*. John Wiley and Sons, New York. Provides an excellent introduction.

de Finetti, B. (1972). *Probability, Introduction and Statistics*, John Wiley and Sons, New York. A classical introduction to nonclassical statistical theory.

Evans, D. H. (1992). *Probability and Its Applications for Engineers*, Marcel Dekker, New York. Suitable for advanced undergraduate students.

Feller, W. (1968). *An Introduction to Probability Theory and Its Applications*, Vol. 1, 3rd ed., Chapter 1, John Wiley and Sons, New York. A classic for scientists with an extended overview of issues from elementary to advanced probability.

Lindgren, B. W., G. W. McElrath, and D. A. Berry (1978). *Introduction to Probability and Statistics*, 4th ed., Macmillan, New York. A general course book with examples from everyday life.

Mood, A. M., F. A. Graybill, and D. C. Boes (1974). *Introduction to the Theory of Statistics*, 3rd ed., McGraw-Hill, New York. A self-contained intermediate-level introduction to classical or mainstream statistical theory.

Pitman, J. (1993). *Probability*, Springer-Verlag, New York. Introductory text on probability recommended for a wide spectrum of readers.

Rozanov, Y. A. (1977). *Probability Theory. A Concise Course*, revised English ed., Dover, New York. A useful primer in probability.

Stirzaker, D. (1994). *Elementary Probability*, Cambridge University Press, Cambridge, England. An advanced introduction highly suitable for engineers.

Additional references quoted in text

Chang, F., and L. Chen (1998). "Real-coded genetic algorithm for rule-based flood control reservoir management," *Water Resour. Manage.*, Vol. 12, pp. 185–198. Application of fuzzy logic.

Gleick, J. (1987). *Chaos: Making a Science*, Viking Penguin, New York. Gives a clear account of the theory.

Lorenz, E. N. (1993). *The Essence of Chaos*, University of Washington Press, Seattle. The mathematical treatment will appeal to engineers. Gives widely quoted examples of scientific modeling.

McNeill, D., and P. Freiberger (1993). *Fuzzy Logic*, Simon and Schuster, New York. Justification for Zadeh's Fuzzy logic.

Pawlak, Z. (1991). *Rough Set: Theoretical Aspects of Reasoning About Data*, Kluwer Academic, Dordrecht. New practical approach to chance and uncertainty by a mathematician.

Ross, T. (2004). *Fuzzy Logic with Engineering Applications*, John Wiley and Sons, New York. Four chapters present specific case studies.

Soil Conservation Service (1983). *National Engineering Handbook*, "Section 3: Sedimentation" and "Section 4: Hydrology," 2nd ed., U.S. Department of Agriculture, Washington, DC. Used in Problems 2.25 and 2.26.

Stigler, S. M. (1983). "Who discovered Bayes's theorem?," *Am. Stat.*, Vol. 37, pp. 290–296. Entertaining research.

Zadeh, L. A. (1965). "Fuzzy sets," *Information and Control*, Vol. 8, 338–353. Original article on fuzzy logic, in which a fuzzy set is a generalization of the mathematical concept of a classical set; he suggests an alternative method of quantifying uncertainty.

PROBLEMS

2.1. Football stadium balcony. A civil engineer is asked to assess the reliability of a balcony overlooking a football stadium. The maximum number of people who can be accommodated in the balcony is 20. The weight of an individual can be approximately 50, 75, or 100 kg.

(a) Sketch the sample space.

(b) Show the following events involving numbers of people and their weights at any time:

$A \equiv$ {there are more than 16 people in the balcony},
$B \equiv$ {the total weight of people in the balcony is 1500 kg},
$C \equiv$ {there are more than 15 people of the maximum weight}.

2.2. Reservoir inflows. A reservoir impounds water from a stream X and receives water Y deviated via a tunnel from an adjoining catchment. The annual inflow from source X can be approximated to 1 or 2 or 3 units of 10^6 m^3, and that from source Y is 2 or 3 or 4 units of 10^6 m^3. On appropriate Venn diagrams show the following events:

(a) $A \equiv$ {source X is less than 3 units}.

(b) $B \equiv$ {source Y is more than 2 units}.

(c) $A + B$.

(d) AB.

2.3. Sequential construction. The sequence of construction of a structure involves two phases. Initially, the foundation is built, then work commences on the superstructure. The completion of the foundation can take 4 or 5 months, which are equally likely to be needed. The superstructure requires 5, 6, or 7 months to be completed, with equal likelihood for each period. The time of completion of the superstructure is independent of that taken to complete the foundation. List the possible combinations of times for the completion of the project and determine the associated probabilities.

2.4. Dam spillway. An engineer is designing a spillway for a dam. The evaluation of maximum flow data is based on a short period of recordkeeping. The critical flow rates and their probabilities are estimated from, A, discharge measurements, B, rainfall observations, and C, combination of flow discharge and rainfall data, as follows:

Event A from flow data: 8,000 to 12,000 m^3/s, $\Pr[A] = 0.5$.
Event B from rainfall data: 10,000 to 15,000 m^3/s, $\Pr[B] = 0.6$.
Event $C = A + B$: 8,000 to 15,000 m^3/s, $\Pr[C] = 0.9$.

(a) Sketch the foregoing events.

(*b*) Show on the sketch AB, AC, and $A^c + B^c$.
(*c*) Determine the probabilities $\Pr[AB]$ and $\Pr[A^c + B^c]$.
(*d*) Determine the conditional probabilities $\Pr[A|B]$ and $\Pr[B|A]$.

2.5. Wind direction and intensity. Strong winds in a particular area come uniformly from any direction from north, $\theta = 0°$, to east, $\theta = 90°$. Wind speed V is also variable, and it can exceed 50 km/h with a probability of 0.04, and 100 km/h with a probability of 0.01.

(*a*) Sketch the sample space for wind speed and direction.
(*b*) Sketch the following events: $A \equiv \{V > 50 \text{ km/h}\}$, $B \equiv \{50 < V < 100 \text{ km/h}\}$, AB, $A + B$, $C \equiv \{30 < \theta < 60°\}$, AC, and BC.
(*c*) Find $\Pr[B]$ and $\Pr[BC]$ assuming that wind speed and direction are stochastically independent.

2.6. Irrigation water supply. A dam is designed to supply water to three separate irrigation schemes, I_1, I_2, and I_3. The demand for the first scheme I_1 is 0 or 1 or 2 m^3/s, whereas that for I_2 and I_3 is 0 or 2 or 4 m^3/s in each case.

(*a*) Sketch the sample space for I_1, I_2, and I_3 separately, and for I_1, I_2, and I_3 jointly.
(*b*) Show the following events:

- $A \equiv \{I_1 > 1 \text{ m}^3/\text{s}\}$;
- $B \equiv \{I_2 \geq 2 \text{ m}^3/\text{s}\}$;
- $C \equiv \{I_3 < 4 \text{ m}^3/\text{s}\}$;
- A^c; AB; $A + B$; $(A + B)^c$; AB^c; AC; A^cC; B^cC; B^cC^c; (where feasible).

(*c*) Assuming that the demands from the three schemes are independent of each other, and that all possible demands are equally likely to occur, find the probability that the total water demand exceeds 5 m^3/s.

2.7. Port occupancy. An experiment consists of counting the number of ships in a small harbor on a particular day and estimating the total tonnage. The maximum number of vessels permitted at a given time in the port is six, while each vessel can have a tonnage from 5,000 to 25,000. Only the total number of ships and the total tonnage is recorded.

(*a*) Sketch the sample space for this experiment.
(*b*) Indicate on the diagram the regions corresponding to the following events:

- $A \equiv$ {the number of ships is less than 5};
- $B \equiv$ {the total tonnage is less less than 35,000};
- $C \equiv$ {three ships each of maximum tonnage are present};
- $A + B$; AB; $A^c + B^c$; A^cB^c; AC; A^cC (where feasible).

2.8. Simply supported beam. A load of 200 kg is placed on a simply supported beam of length 6 m. If R_1 and R_2 denote the reactions at the left and right supports, respectively, $R_1 + R_2 = 200$ kg for any location of the load.

(*a*) Define and sketch the sample space for this experiment.
(*b*) Sketch on the diagram the following events:

- $A \equiv$ {the load is located at 1 m from support 1};
- $B \equiv$ {the load is located between 2 and 4 m from support 1};
- $C \equiv$ {the load is located between 3 and 5 m from support 1};

- $A + B$; AB; $B + C$; BC; $A^c + BC$; and $A^c B^c C^c$ (where feasible).

(*c*) If the load can vary from 100 to 400 kg, define and sketch the new sample space. Sketch on this diagram the following events:

- $D \equiv$ {a load heavier than 100 kg is located at 2 to 4 m from support 1};
- $E \equiv$ {a load heavier than 200 kg is located at 3 to 5 m from support 1};
- $D + E$; DE; and DE^c (where feasible).

2.9. Storm rainfall. Analysis of the data of Problem 1.20 indicates that the estimated probability of a storm resulting in more than 40 mm of rainfall in 1 hour is about .5. Using relative frequencies, compute the probability that in any year the same rainfall intensity is exceeded over a duration of (*a*) 20 minutes, and of (*b*) 3 hours.

If the annual 30-minute and 1-hour rainfalls refer to the same storm events, what is the conditional probability that the intensity does not decrease from 60 mm/h or more during the first 30 minutes by more than 25% during a 1 hour period?

2.10. Hydropower. Run-of-river hydroelectrical plants convert the natural potential energy of surface water in a stream into electrical energy. The plant capacities depend on natural river flow, which generally varies during the year according to season and precipitation regime. Assume that the design flow of a given power station, say, Q_D, is the natural flow, which is exceeded during 274 days in a year on average. At other times, when the river flow is lower than the design flow, the plant is nevertheless capable of producing some power if the flow is not lower than Q_0. Moreover, during floods it is not possible to convey water to the plant due to sedimentation, which occurs when the natural river flow Q exceeds Q_1.

(*a*) If $\Pr[Q < Q_0] = 0.1$ and $\Pr[Q > Q_1] = 0.05$, for how many days in a year will the plant be incapable of supplying electric energy?

(*b*) What is the probability that the plant works at full capacity?

(*c*) What is the probability that the plant fulfills its minimum target? Note that $Q_0 < Q_D < Q_1$.

2.11. Reservoir operational policy. Consider the water storage S in a reservoir as described in Example 2.1 and Fig. 2.1.1. The manager must release in a year an amount of water R that depends on the amount of the annual inflow I, the storage S at the beginning of that year, and the demand d in that year. The manager follows the following "normal operational policy" for water releases:

$$\begin{aligned} R &= d, && \text{if } d \le I + S < d + c, \\ R &= I + S, && \text{if } I + S < d, \\ R &= I + S - c, && \text{if } I + S \ge d + c, \end{aligned}$$

with c denoting the effective storage capacity of the reservoir. If $\Pr[d \le I + S \le d + c] = 0.6$, $\Pr[I + S < d] = 0.1$, and $\Pr[I + S > d + c] = 0.3$, find the probability that the demand is satisfied.

2.12. Industrial park utilities. Consider the design requirements of water supply and wastewater removal systems in a new industrial park, which consists of five independent buildings. Assume that the water demand S of each of the five industrial buldings can be 10 or 15 units, whereas the required wastewater removal capacity R can be 8, 10, or 15 units. After some interviews with potential clients, the designer has estimated that the combined requirements of the two systems are likely to occur with the following probabilities at the ith site:

	$R = 15$	$R = 10$	$R = 8$
$S = 10$	0.00	0.25	0.15
$S = 15$	0.20	0.35	0.05

Stochastic independence can also be assumed among the requirements of different buildings.

(a) What is the probability that the total water demand exceeds 60 units?

(b) What is the probability that the total wastewater removal capacity exceeds 50 units?

2.13. Construction scheduling. Consider the sequential construction scheme of Problem 2.3, and assume that both the foundation and the superstructure can be completed at three different rates, say, a, b, or c. These rates modify the probability of completion of each phase of construction as shown in the table given here. Also, monthly costs vary for the different rates.

Phase	Rate	Cost per month at rate ($)	Probability of time of completion: 4 months	5 months	6 months	7 months
Foundation	a	30,000	0.3	0.7	0	0
Foundation	b	36,000	0.5	0.5	0	0
Foundation	c	42,000	0.3	0.7	0	0
Superstructure	a	25,000	0	0.1	0.4	0.5
Superstructure	b	40,000	0	0.3	0.3	0.3
Superstructure	c	50,000	0	0.5	0.3	0.2

In addition, if the construction is not completed in 11 months, the contractor must pay a penalty of $300,000 per month.

(a) Compute the expected cost of foundation performed at rate a as the summation for all times of completion of the product between the total cost (the product of the number of required months and the cost per month) and probability.

(b) Compute all expected costs.

(c) Compute the total expected penalty for each possible strategy of completion of the whole structure.

(d) Determine the best strategy by minimizing the sum of total expected cost and penalty.

2.14. Research project ranking. A committee consisting of three independent referees (R_1, R_2, and R_3) is to rank four different research project applications (A, B, C, and D). Each referee ranks the four projects as 3 (for the best), 2, 1, and 0, and then the assigned ranks for each project are summed. Assume that the referees are unable to discriminate between projects so that the rankings are randomly assigned. What is the probability that project A will receive a total score of 4?

2.15. Probabilities of reservoir storage. Consider the water storage S in a reservoir described by a sequence of four states $\omega_1, \omega_2, \omega_3, \omega_4$, where each state describes water volumes ranging from 0 to $c/4$, from $c/4$ to $2c/4$, and so on (see Example 2.10 and Fig. 2.1.2). The reservoir manager is interested in the simple events given by $A_{i,k} \equiv \{(i-1)c/4 \leq S < ic/4\}$ for $i = 1, 2, 3, 4$ and annual time periods $k = 1, 2, 3, \ldots$.

The manager has estimated the following conditional probabilities: $\Pr[A_{j,k+1}|A_{i,k}] = 1/2$ for $j = i$, and $\Pr[A_{j,k+1}|A_{i,k}] = 1/6$ for $j \neq i$. What is the transition probability matrix p_{ij} from the ith to the jth state after one step? What is the probability that state 1 occurs in the third operational period, given that the reservoir was in state 4 in the first period?

2.16. Pumping station. Two pumps operate in parallel to provide water supply of a village located in a recreational area. Water demand is subject to considerable weekly and seasonal fluctuations. Each unit has a capacity so that it can supply the demand 80% of the time in case the other unit fails. The probability of failure of each unit is 10%, whereas the probability that both units fail is 3%. What is the probability that the village demand will be satisfied?

2.17. Analysis of reservoir lifetime. A reservoir is designed for an area with high erosional rates. The engineer is interested in determining the lifetime of the reservoir, which can come to an end either because the impounding dam can be destroyed by a flood exceeding the spillway capacity or because excessive sedimentation results in a severe loss in reservoir capacity. It is necessary to determine the probability that the structure will come to an end of its useful life in each of the years after construction. One can assume a constant probability q that in any year a flow exceeding the spillway capacity can occur, and an exponentially increasing probability p_i that reservoir sedimentation can occur in the ith year after construction, given that no significant sedimentation has occurred prior to the ith year, that is, $p_i = 1 - \exp(-\beta i)$, with $\beta > 0$.

Denote by A_n the event associated with a destructive flood occurring in the nth year after construction and by B_n that associated with excessive sedimentation.

(*a*) What is the probability that the system will survive for n years, that is

$$\Pr\left[\left(A_1^c B_1^c\right)\left(A_2^c B_2^c\right)\ldots\left(A_n^c B_n^c\right)\right]?$$

(*b*) What is the probability that the system will come to an end in the nth year, where S_n denotes survival up to the nth year,

$$\Pr[(A_n + B_n)|S_{n-1}]\Pr[S_{n-1}]?$$

(*c*) Compute the foregoing probabilities for $q = 0.01$, $\beta = 0.002$, and $n = 25$.

2.18. Highway system. To reach Grenoble, France, from Turin, Italy, one can follow either of two routes. The first directly connects Turin and Grenoble, whereas the second passes through Chambery, France. During extreme weather conditions in winter, travel between Turin and Grenoble is not always possible because some parts of the highway may not be open to traffic. Denote with A, B, and C the events that highways from Turin to Grenoble, Turin to Chambery, and Chambery to Grenoble are open, respectively. In anticipation of driving from Turin to Grenoble, a traveler listens to the next day's weather forecast. If snow is forecast for the next day over the southern Alps, one can assume (on the basis of past records) that $\Pr[A] = 0.6$, $\Pr[B] = 0.7$, $\Pr[C] = 0.4$, $\Pr[C|B] = 0.5$, and $\Pr[A|BC] = 0.4$.

(*a*) What is the probability that the traveler will be able to reach Grenoble from Turin?
(*b*) What is the probability that the traveler will be able to drive from Turin to Grenoble by way of Chambery?
(*c*) Which route should be taken in order to maximize his chance to reach Grenoble?

2.19. Wastewater treatment. The wastewater from an industrial plant requires treatment before disposal in the sea. This process consists of three sequential stages. For simplicity, define these stages as primary, secondary, and tertiary treatments, respectively. The result for each stage can be rated as unsatisfactory, incomplete, and satisfactory. Denote with A_k the event that the kth stage of the treatment process is unsatisfactory, with B_k the event that it is incomplete, and with C_k the event that it is satisfactory. The associated probabilities are given in the following table:

	$\Pr[A_k]$	$\Pr[B_k]$	$\Pr[C_k]$
$k = 1$	0.1	0.3	0.6
$k = 2$	0.2	0.3	0.5
$k = 3$	0.1	0.5	0.4

Further, assume that the three stages of the process are stochastically independent. If the satisfactory overall treatment requires that none of the three stages is unsatisfactory and at least two of these stages are satisfactory, what is the probability of this event?

2.20. Earthquake occurrence and intensity. Because of the uncertainties associated with the occurrence and intensity of earthquakes, one must consider earthquakes occurring in a given location as random phenomena. MCS intensity is a measure based on earthquake impact on the landscape, buildings, and population. In Problem 1.22 records of earthquake intensity in terms of MCS index are given for a period of about 1000 years in Rome, Italy. They are ranked from 2 to 7 for increasing intensities. In ten centuries 329 earthquakes were reported in the study area, and in only two centuries there were no occurrences. Calculate a frequency-based estimate of the probability that at least one earthquake is likely to occur in a century. What is the probability that a recorded earthquake is of intensity 7?

2.21. Air pollution control. The air pollution in Milan, Italy, is mainly caused by industrial, automobile, and heating emissions. A newly elected local government wishes to control these three sources of pollution within a period of 4 years. The chances of successfully controlling these sources are 80, 70, and 50%, respectively. The government assumes that if only one of these three sources is successfully controlled, the probability of bringing air pollution below the acceptable level would be 50% only, but this probability increases to 80% if two of them are successfully controlled. The government also assumes stochastic independence among controlling industrial, heating, and automobile exhausts. What is the probability that two of the sources of air pollution will be successfully controlled in Milan during the 4-year period?

2.22. Imperfect concrete testing. An existing reinforced concrete building must be tested for possible obsolescence. Based on professional judgement, the engineer classifies concrete quality as either 35–39.9, 40–44.9, 45–49.9, or 50–60 N/mm^2 based on a 28-day test of compressive strength of concrete cubes. The relative likelihoods assigned to these four states are 0.2, 0.3, 0.4, and 0.1, respectively. Concrete cores are to be cut and tested to help ascertain the true state, although the engineer knows that results from test scores are not conclusive. Therefore, conditional probabilities are estimated to account for the uncertainties involved in examining the cores. These probabilities describe the likelihood that the value of core strength indicated predicts a given unknown state. For example, if the true state is 35–39.9 N/mm^2, there is a 70% chance that the tested core strength also lies between 35 and 39.9 N/mm^2, but there is a 20% chance that it will lie between 40 and 44.9 N/mm^2, and a 10% chance that it lies in the range 45–49.9 N/mm^2. The conditional probabilities are tabulated next:

	State			
Core strength	x_1 35–39.9 N/mm^2	x_2 40–44.9 N/mm^2	x_3 45–49.9 N/mm^2	x_4 50–60 N/mm^2
y_1: 35–39.9 N/mm^2	0.7	0.2	0.1	0.0
y_2: 40–44.9 N/mm^2	0.2	0.6	0.2	0.1
y_3: 45–49.9 N/mm^2	0.1	0.1	0.6	0.2
y_4: 50–60 N/mm^2	0.0	0.1	0.1	0.7

If the engineer takes three subsequent cores, and the laboratory tests yield $z_{(1)} = 41$, $z_{(2)} = 49$, and $z_{(3)} = 44$ N/mm^2, respectively, what are the posterior probabilities of the four states at the end of the experiment? The required posterior probability is given by Pr[state x_i|sample$z_{(3)} = y_2$].

2.23. Highway pavement. Before any 250-m length of a pavement is accepted by the State Highway Department, the thickness of a 30 cm is monitored by an ultrasonics instrument to verify compliance to specification. Each section is rejected if the measured thickness is less than 10 cm; otherwise, the entire section is accepted. From past experience, the State Highway engineer knows that 85% of all sections constructed by the contractor comply with specifications. However, the reliability of ultrasonic thickness testing is only 75%, so that there is a 25% chance of erroneous conclusions based on the determination of thickness with ultrasonics.

(*a*) What is the probability that a poorly constructed section is accepted on the basis of the ultrasonics test?

(*b*) What is the probability that if a section is well constructed, it will be rejected on the basis of the ultrasonics test?

2.24. Remote sensing of inundated areas. Two independent satellite-borne sensors are used to determine the extension of inundated areas after a flood. Sensor *A* has a reliability of 70%, that is, the probability of detecting a pixel (picture element) whose characteristics reflect inundation is 0.7, whereas sensor *B* has a reliability of 90%. Also, the probability of both sensors detecting a pixel is 0.65.

(*a*) Find the probability that a pixel reflecting inundation is detected, that is, it is detected by at least one of the two sensors.

(*b*) Determine the probability that a pixel reflecting flooding is detected by only one sensor.

2.25. Runoff production. Characterization of the soils of a small catchment includes 40% of well-drained sand and gravel (type A hydrologic soil group), 35% of fine-textured soils (type C hydrologic soil group), and 25% of clay soils (type D hydrologic soil group). Type A and type D terrains have been contoured and are covered with small grains in poor condition, 60% of type C terrains is covered by pasture in fair condition, and the remaining type C terrain is sparsely forested land without forest litter. The engineer evaluates runoff production using the Soil Conservation Service procedure (see Soil Conservation Service (1983), "Section 4: Hydrology."). This procedure gives surface runoff R as

$$R = (P - 0.2S)^2/(P + 0.8S),$$

where P is the rainfall depth of the design storm, and S is the maximum soil potential retention, which is given by

$$S = 25.4(1000/\mathrm{CN} - 10),$$

where CN is a dimensionless parameter known as the "Curve Number." The values of CN range from 0 to 100 depending on the joint categories of "hydrologic soil group," and "land use" according to the table below. R, P, and S are measured in millimeters per unit area.

Values of CN obtained by matching hydrologic soil group with land use

	Hydrologic soil group			
Land use	A	B	C	D
Straight row crops in poor condition	72	81	88	91
Contoured row crops in poor condition	70	79	84	88
Contoured row crops in good condition	65	75	82	86
Contoured small grain in poor condition	63	74	82	85
Pasture in fair condition	49	69	79	84
Wood and forestland with thin stand, poor cover, no mulch	45	66	77	83
Woods protected from grazing with adequate brush coverage	30	55	70	77
Commercial and business areas (85% impervious)	89	92	94	95

(a) Determine the expected surface runoff caused by a heavy storm resulting in 120 mm of rainfall per unit area.

A new commercial and business area is planned (85% impervious). The site includes 40% of type A terrains but 60% of this is pastureland. The engineer has two alternatives: (1) designing a large culvert to carry runoff excess due to urbanization, or (2) improving the hydrologic conditions of the surrounding forestland (for example, by protecting woods from grazing and providing adequate brush coverage) so that the expected runoff from the catchment does not change. The design storm is 120 mm.

(b) Determine the expected excess runoff due to urbanization.

(c) Evaluate the feasibility of the second alternative under (a).

2.26. Universal soil loss equation. In the United States the prediction of upland erosion amounts is frequently made by the universal soil loss equation (USLE) developed by the U.S.D.A. Agricultural Research Service in cooperation with U.S.D.A. Soil Conservation Service and certain experimental stations (see Soil Conservation

Service (1983), "Section 3: Sedimentation."). The USLE gives the annual soil loss due to erosion in kilograms per square meter per year, say, A, as

$$A = cR \times K \times L \times S \times C \times P,$$

where c is a constant, R denotes the rainfall factor, K the soil erodibility factor, L the slope length factor, S the slope gradient factor, C the crop-management factor, and P the erosion control practice factor. The engineer must analyze the effects of crop management on the annual soil loss in a small forested catchment. From previous computations $c = 1$, $R = 185$, $K = 0.38$, $LS = 1.4$, and $P = 1$. The values of C vary from 0.0005 to 0.009 depending on the joint variation of the percentage of area covered by the canopy of trees and undergrowth C_1 and of the percentage of area covered by litter, C_2, as shown in the table below:

Values of the crop management factor, C

	$C_1 =$ 100–90·1%	90–70·1%	70–40%
$C_2 =$ 100–70.1%	0.0005	0.0008	0.0010
70–40.1%	0.0020	0.0030	0.0040
40–20%	0.0030	0.0060	0.0090

(*a*) Assuming that all the foregoing categories of crop management are equally likely, compute the probability that A exceeds 0.3 kg/m^2 per year.

(*b*) Assuming that the catchment is partitioned as in the foregoing table into nine subcatchments equal in area, each having a different crop management, compute the expected annual soil loss from the catchment.

(*c*) What is the minimum number of subcatchments where crop management must be improved in order to reduce the expected annual soil loss from the catchment to a value lower than 0.2?

Chapter 3
Random Variables and Their Properties

The objective of this chapter is to introduce fundamental theoretical concepts of random variables and probability distributions. This presentation will establish the necessary link between statistics and probability. First, random variables are formally defined. Then discrete and continuous types are treated separately, followed by a description of their properties and use. The mean, variance, skewness, and other descriptors of sample data, introduced in Chapter 1, are defined for random variables. We describe moment-generating functions and illustrate their applications with numerous examples. Methods of estimation such as moments, probability weighted moments, L moments, and maximum likelihood are considered, then concepts of entropy, jackknife, and bootstrap.

Multiple random variables are treated extensively. The derivations of joint probability distributions of discrete and continuous variables, with conditional and marginal functions and covariance, are covered. In the section on associated random variables are included properties of derived variables and contagious distributions. We also provide a brief introduction to the related subject of copulas.

Specific types of discrete and continuous models of importance in civil and environmental engineering are elaborated and classified in Chapters 4 and 7. A few of these—such as the binomial, Poisson, gamma, and normal types—are used in some of the examples in this chapter to clarify various aspects of the theory.

3.1 RANDOM VARIABLES AND PROBABILITY DISTRIBUTIONS

The concept of random variables is central to probability theory and its applications. This was introduced and discussed in Chapters 1 and 2. The presentation in this section is more formal. In particular, the specification of a random variable by a probability distribution is described.

3.1.1 Random variables

A variable, such as the strength of a concrete or any other material or physical quantity, whose value is uncertain or unpredictable or nondeterministic is called a *random variable* or a *variate* if its distribution is known. A random variable may assume some value, the magnitude of which depends on a particular occurrence or outcome (usually noted by an observation or measurement) of an experiment in which tests are made and records maintained.[1] Each outcome of a random variable, or each simple event defined with respect to the sample space, corresponds to a numerical value of the random variable; whether these values differ or not from one event to another is unimportant for this definition. A random variable can be formally viewed as a function defined on the sample space of an

[1] These range from modern Darwinism and management science to quantum mechanics and engineering.

experiment such that there is a numerical value of the random variable corresponding to each possible outcome; that is, there is a probability associated with each occurrence in the sample space.[2] In contrast to a simple event, a compound event may be associated with more than one value of the random variable. As previously noted, an uppercase letter denotes a random variable, and the corresponding lowercase letter represents the value that it assumes. For example, one may refer to the number of floods in a year or the number of vehicles passing an intersection during a given period as X; then $x = 5$, say, is a particular value that the random variable X may take.

3.1.2 Probability mass function

A random variable can be statistically specified by its distribution or probability law. That is, the probability distribution of the random variable is specified using a mathematical function. The variable can be of the discrete or continuous type. In the discrete case the variable can only assume at most a countable set of isolated values, as already stated, such as positive integers (the type of discrete variable with which one is usually concerned). Then the mathematical function is called a *probability mass function*, abbreviated pmf, which is defined as

$$p_X(x) = \Pr[X = x]. \tag{3.1.1}$$

Definition: Probability mass function, pmf. The pmf of a discrete random variable X gives the point probabilities of the values taken by X.

The axioms of probability (noted in Chapter 2) are applicable here. These are as follows:

$$0 \leq p_X(x) \leq 1, \quad \text{for all possible } x, \tag{3.1.2a}$$

$$p_X(x) = 0, \quad \text{for all unrealizable } x, \tag{3.1.2b}$$

$$\sum p_X(x) = 1, \quad \text{which is summed over all possible } x. \tag{3.1.2c}$$

If in a particular case one is certain that the outcome is c, for example,

$$p_X(c) = \Pr[X = c] = 1 \tag{3.1.2d}$$

and for mutually exclusive outcomes, $x_1, x_2, \ldots, x_n$,

$$p_X(x_1 + x_2 + \cdots + x_n) = p_X(x_1) + p_X(x_2) + \cdots + p_X(x_n). \tag{3.1.2e}$$

Example 3.1. Flood occurrences. The number of floods recorded per year at a gauging station in Italy is given in Table 1.1.1. For this data, the pmf is as follows:

$p_X(0) = .00$; $p_X(1) = 2/34 = .06$; $p_X(2) = 6/34 = .18$;
$p_X(3) = 7/34 = .20$; $p_X(4) = 9/34 = .26$; $p_X(5) = 4/34 = .12$;
$p_X(6) = 1/34 = .03$; $p_X(7) = 4/34 = .12$; $p_X(8) = 1/34 = .03$;
$p_X(x) = .00$ for $x > 8$.

Figure 3.1.1 shows a plot of the pmf.[3]

[2] More formally, a real random variable is a measurable function denoted by X or $X(\cdot)$ having its domain in the sample space Ω and its counterdomain in subsets of the real line; see also Feller (1968).

[3] It is identical to the line diagram of Fig. 1.1.1 except that the vertical axis denotes the probabilities of the corresponding values on the horizontal axis.

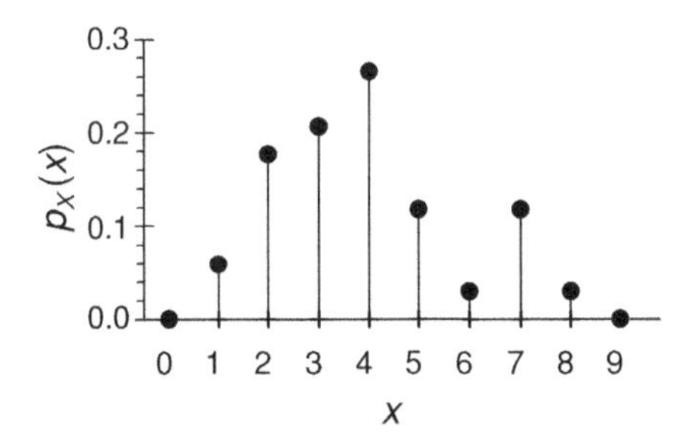

Fig. 3.1.1 Probability mass function of flood occurrences X per year at the gauging station of Calamazza on the Magra River between Pisa and Genoa, Italy, for the period 1939–1972. A flood occurrence is a discharge in excess of 300 m^3/s.

3.1.3 Cumulative distribution function of a discrete random variable

For a discrete or continuous random variable, the *cumulative distribution function*, abbreviated as cdf and denoted by $F_X(x)$, is the probability of nonexceedance of x; this is sometimes referred to as the distribution function. That is,

$$F_X(x) = \Pr[X \leq x]. \tag{3.1.3a}$$

We note that $F_x(x)$ is a monotonic function, which, by definition, increases for increasing values of X and, as previously defined,

$$0 \leq F_X(x) \leq 1, \qquad \text{for all possible } x. \tag{3.1.3b}$$

Definition and properties: Cumulative distribution function, cdf. For a discrete or continuous random variable X the cdf is the probability of nonexceedance of the value x. The cdf is a monotonic (nondecreasing) continuous function that is bounded by 0 and 1. In the discrete case it is obtained by summing over values of the pmf.

In the case of a discrete random variable, $F_X(x)$ is the sum of the probabilities of all possible values of X that are less than or equal to the argument x. That is,

$$F_X(x) = \sum_{X_k \leq x} p_X(x_k). \tag{3.1.4}$$

This is summed over all possible X_k less than or equal to x.

Example 3.2. Flood occurrences. Returning to the example of the flood occurrences per year at a gauging station in Italy given in Table 1.1.1 with pmf shown in Example 3.1, we find the cdf is as follows:

$F_X(0) = 0.00$; $F_X(1) = 0.06$; $F_X(2) = 0.24$; $F_X(3) = 0.44$;
$F_X(4) = 0.70$; $F_X(5) = 0.82$; $F_X(6) = 0.85$; $F_X(7) = 0.97$;
$F_X(8) = 1.0$; $F_X(x) = 1.00$, for $x > 8$.

Figure 3.1.2 shows the cdf of the flood occurrences.[4]

Example 3.3. Maximum potential soil absorption capacity. The absorption capacity of a portion of terrain can be described through its *curve number*. The curve number takes

[4] This is, of course, different from Fig. 1.1.6 which is for a continuous variable. For the discrete variable represented here, the graph takes the form of a step function.

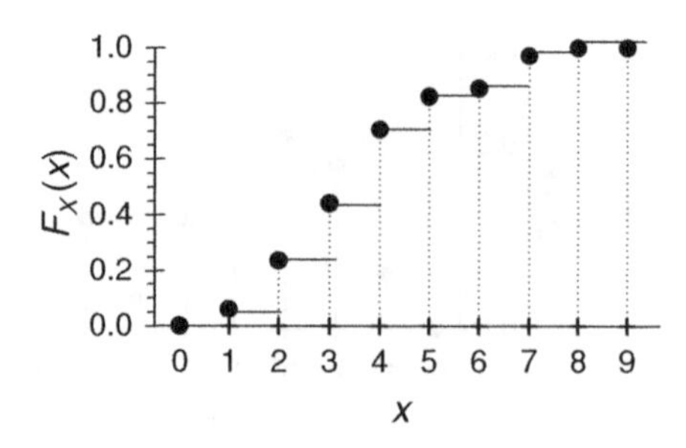

Fig. 3.1.2 Cumulative distribution function of flood occurrences X per year at the gauging station of Calamazza on the Magra River between Pisa and Genoa, Italy, for the period 1939–1972.

integer values in the range 1–100 and depends on soil properties and land use.[5] As a first approximation, values taken by the random variable CN in a region may be assumed to be equally likely (that is to say, uniformly distributed) with pmf:

$$P_{\text{CN}}(cn) = 1/100, \quad \text{for } 1 \le cn \le 100.$$

The corresponding cdf is given by

$$F_{\text{CN}}(cn) = \sum_{i=1}^{cn} \frac{1}{100} = \frac{cn}{100}, \quad \text{for } 1 \le cn \le 100;$$

$$F_{\text{CN}}(cn) = 0, \quad \text{for } cn < 1;$$

$$F_{\text{CN}}(cn) = 1, \quad \text{for } c \ge 100.$$

For example, $F_{\text{CN}}(25) = \Pr[cn \le 25] = .25$. The pmf and cdf are shown in Fig. 3.1.3*a*.

This is a step function that appears to be a curve because of numerous steps. The maximum potential soil absorption capacity S is related to the CN as follows:

$$S = 25.4[(1000/\text{CN}) - 10],$$

where S is measured in millimeters. Accordingly, S can take a value from 0 to 25,146 mm. For example, $S = 762$ mm for CN $= 25$, and $p_S(762) = 0.01$. The corresponding cdf of S is given by the sum of the probabilities of those outcomes of CN that yield a value of S less than or equal to 762 mm. This corresponds to CN ≥ 25. Hence,

$$F_S(s) = \sum_{i=cn}^{100} \frac{1}{100} = \sum_{i=\frac{25{,}400}{s+254}}^{100} \frac{1}{100} = 1 - \frac{254}{s+254},$$

for $0 \le s \le 25{,}146$ mm.

$$F_S(s) = 0, \quad \text{for } s < 0; \quad F_S(s) = 1, \quad \text{for } s > 25{,}146.$$

This cdf is shown in Fig. 3.1.3*b*.

This is also a step function. The log scale allows a clearer definition for high and low values of S.

It is often convenient to consider S as a continuous random variable that can take any real value from 0 to 25,146 mm.

3.1.4 Probability density function

A continuous variable can take any value within two limits, determined by physical or theoretical means; such a value can in theory be specified using an unlimited number of

[5] See Problem 2.25.

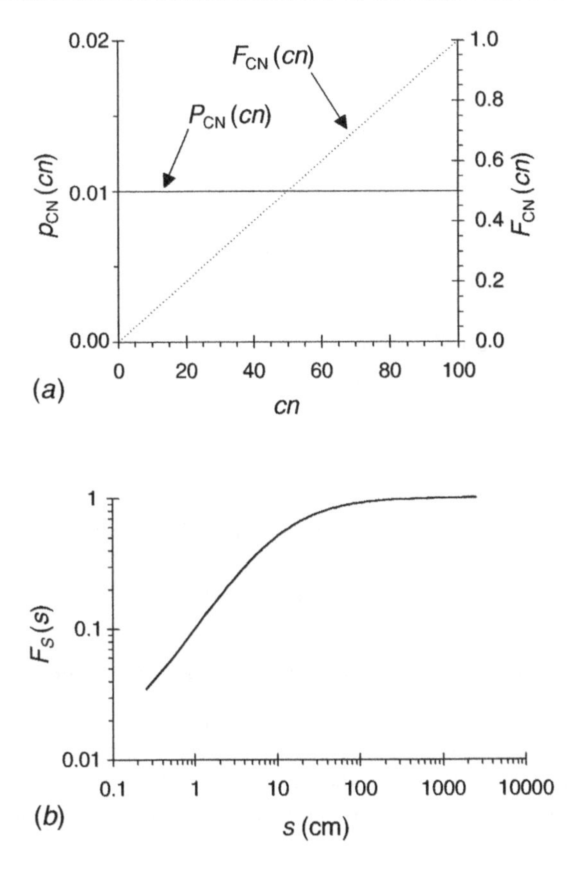

Fig. 3.1.3 (*a*) Probability density function and cumulative distribution function of equally likely values of curve number, CN. (*b*) Cumulative distribution function of maximum soil potential retention S as obtained from equally likely values of the curve number.

decimal places but is limited in practice by the accuracy of the measuring device used, such as a flow gauge or weighing device. The probability law for a continuous random variable is specified by a *probability density function*, abbreviated pdf, which represents the limiting case if the relative frequency polygon is applied to a sample of infinite size and the class widths tend to zero.[6] Thus, the pdf denoted by $f_X(x)$ is a nonnegative mathematical function that in its graphical representation usually takes the form of a continuous curve over a range of values that the random variable can possibly take. As implied by its definition, $f_X(x)$ is not dimensionless and hence by itself does not represent a probability; it merely denotes an intensity of probability or a probability rate.[7] However, the area under the curve between limits, such as x_1 and x_2, with a nonzero range gives the probability that the random variable X lies in the interval x_1 to x_2; occurrences over two or more nonoverlapping intervals are mutually exclusive events. The physical analogue to this is a unit vertical force which is continuously and (in general) nonuniformly distributed over a horizontal structural component and denoted by $f_X(x)$; here the area under the curve between the points gives the fraction of the force acting between the points. Thus,

$$f_X(x) \geq 0 \tag{3.1.5a}$$

[6] A relative frequency polygon is shown in Fig. 1.1.5.

[7] This is analogous to the statement that density = mass/volume.

and

$$\Pr[x_1 \leq X \leq x_2] = \int_{x_1}^{x_2} f_X(x)dx \leq 1. \tag{3.1.5b}$$

If x_a and x_b denote the lowest and highest values, respectively, that X can possibly take (which means of course that the pdf is zero outside these limits) then the integral in Eq. (3.1.5*b*), applied over these limits, will be equivalent to the maximum probability of 1. In other words, the area under the pdf between these limits is 1. In the case of some distributions these extreme limits are at negative and positive infinity, denoted by $-\infty$ and $+\infty$, respectively. This convention for symbolizing the complete range of a random variable is adopted in the general case. That is,

$$\int_{-\infty}^{+\infty} f_X(x)dx = 1. \tag{3.1.5c}$$

A quantity used in indexing a pdf is termed as *parameter*. For example, if $f_X(x) = \lambda \exp(-\lambda x)$, λ is a parameter. As λ takes positive values over an infinite range, the collection of pdfs is called a *parametric family* of pdfs.

3.1.5 Cumulative distribution function of a continuous random variable

For continuous random variables, Eq. (3.1.3) is also applicable for describing the cumulative distribution function or cdf, $F_X(x)$. Thus, $F_X(x)$ is the probability of nonexceedance and is the range 0–1. However, the summation in Eq. (3.1.4) for the pmf corresponds to an integral in the case of the pdf; and hence the relationship between the cdf and the pdf becomes

$$F_X(x) = \int_{-\infty}^{x} f_X(z)dz. \tag{3.1.6a}$$

It also directly follows that

$$\frac{dF_X(x)}{dx} = f_X(x). \tag{3.1.6b}$$

Example 3.4. Maximum potential soil absorption capacity. Consider the problem of determining the distribution of maximum potential soil absorption capacity S from the curve number (see Example 3.3). Because it is often assumed that S can take any real value from 0 to 25,146 mm, one can derive the pdf of S by using (3.1.6*b*) as follows:

$$f_S(s) = \frac{dF_S(s)}{ds} = \frac{d}{ds}\left(1 - \frac{254}{s+254}\right) = \frac{254}{(s+254)^2}, \quad \text{for } 0 < s \leq 25,146,$$

$$f_S(s) = 0, \quad \text{elsewhere},$$

where $f_S(s)$ is measured in units of $(\text{mm})^{-1}$.

Definition and properties: Probability density function (pdf) and cumulative distribution function (cdf) of a continuous random variable. A probability density function is defined as

$$f_X(x) \geq 0, \quad \text{for all } x, \quad \text{and} \quad \int_{-\infty}^{+\infty} f_X(x)dx = 1.$$

For a continuous random variable X, the cdf is obtained by the integral

$$F_X(x) = \int_{-\infty}^{x} f_X(z)dz. \tag{3.1.7}$$

In general, for a continuous random variable, the pdf can be discontinuous but the cdf is continuous.

Example 3.5. Timber strength. For an application of the pdf and the cdf, consider the timber strength data of Table E.1.1. To verify some of the properties given above, we may simplify the pdf as follows:

$$\begin{aligned} f_X(x) &= \frac{x}{1400}, \quad \text{for } 0 \le x \le 40; \\ &= \frac{70-x}{1050}, \quad \text{for } 40 \le x \le 70; \\ &= 0, \quad \text{for } x \le 0 \text{ and } x \ge 70. \end{aligned}$$

The units of X and $f_X(x)$ are N/mm^2 and mm^2/N, respectively. The function is shown in Fig. 3.1.4*a*. Clearly, Eq. (3.1.5*a*) applies to the function just given. One can easily verify graphically or by integration that Eq. (3.1.5*c*) is also applicable. The probability that X lies in the interval 30–60 N/mm^2 is given by the shaded area which is .631. This can also be found by substituting the foregoing function with the given limits in Eq. (3.1.5*b*). This calculation is equivalent to using Eq. (3.1.6*a*) as follows:

$$\Pr[30 \le X \le 60] = F_X(60) - F_X(30) = .952 - .321 = .631.$$

The cdf obtained by integration of the pdf is shown in Fig. 3.1.4*b*.

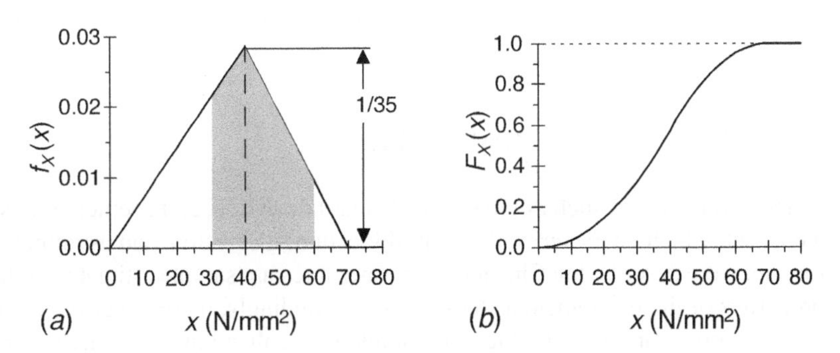

Fig. 3.1.4 Data on modulus of rupture of timber from Table E.1.1: (*a*) probability density function in which the shaded area represents the probability that X is in the interval 30–60 N/mm^2 and (*b*) cumulative distribution function.

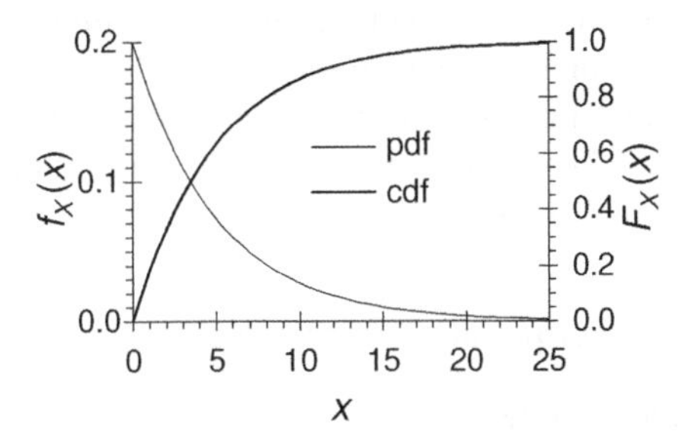

Fig. 3.1.5 Probability density function and cumulative distribution function of exponentially distributed earthquake intensities X, with $\lambda = 0.2$.

The pmf or pdf and the cdf of a random variable provide the probability model that describes a process that is subject to uncertainty. As already mentioned, the probability model of a random system is often developed in the form of a parametric function; this function and the values taken by the parameters come from the random mechanism that governs the system behavior. Practical expediency, however, demands the estimation of parameters through experimentation and observations.

Example 3.6. Earthquake intensity. For another application of the pdf and cdf, consider the occurrence of earthquakes in a region for which the cdf can be simplified to the exponential distribution

$$F_X(x) = 1 - e^{-\lambda x},$$

where λ is a parameter and the random variable X is the magnitude of an earthquake in the region in the range $0 \leq x \leq +\infty$. From Eq. (3.1.6*b*), the pdf is given by

$$f_X(x) = \lambda e^{-\lambda x}.$$

It is estimated that λ is 0.2. Equation (3.1.5*a*) is clearly applicable and it is easy to show by integration the validity of Eq. (3.1.5*c*) in this case. The probability of an earthquake exceeding 10 units, for example, is given by

$$\Pr[X \geq 10] = 1 - F_X(10) = e^{-2} = .135.$$

The pdf and cdf are shown in Fig. 3.1.5.

3.1.6 Summary of Section 3.1

In this section we have introduced the basic concept of random variables, probability mass and density functions, and cumulative distribution functions. Estimation of parameters follows in the next section.

3.2 DESCRIPTORS OF RANDOM VARIABLES

Numerical measures such as the mean, standard deviation, and coefficient of skewness of a data set, which characterize the central location, dispersion, and asymmetry of its histogram, were described in Chapter 1. The measures are sample estimates of the statistical properties of the phenomenon studied. Correspondingly, there are also similar measures of the random variables of which the sample is a realization. These are descriptors of its probability mass function, pmf, or the probability density function, pdf, in the case of a discrete or continuous variable, respectively. The descriptors summarize some important features of the behavior of random variables and are directly relevant for engineering applications. This set of features comprises *expectations* of the function, usually called *population measures*, in the form of some weighted averages of functions of the random variable. The pdf or the pmf provides the weights. Also, the constants or parameters of the cdf are shown to be related to one or more of the population measures such as the mean, standard deviation, and coefficient of skewness.

3.2.1 Expectation and other population measures

3.2.1.1 Mean or expected value

The sample arithmetic mean of a set of data is simply the average of the observed data (as noted in Chapter 1). By analogy, the centroid is the center of mass of a solid body.

Corresponding to this geometrical definition is the statistical expected value $E[X]$ of a random variable X. The expected value is the average value that is weighted according to the probability distribution. It is often called the *population mean* and is denoted by μ_X. In the discrete case, the pmf gives the weighted average as

$$\mu_X = E[X] = \sum_{\text{all } x_i} x_i p_X(x_i). \tag{3.2.1}$$

Example 3.7. Flood exceedances. Consider the simple case where the variable can only take a value of 0 or 1. This stuation can represent the occurrence of a flood at a particular site on a river, where the event 1 is the exceedance of a specified flow in the river. Let the probability of such an occurrence be p. The event 0 is the complementary event and has a probability of occurrence of $(1 - p)$. The probabilities of the two events are given by the Bernoulli distribution

$$p_X(x_j) = \Pr[X = x_j] = p^{x_j}(1-p)^{1-x_j}, \quad \text{for } x_j = 0, 1.$$

Hence

$$\mu_X = E[X] = 0p^0(1-p)^1 + 1p^1(1-p)^0 = p.$$

This gives the required expectation.

If X is a continuous random variable its pdf is used to give

$$\mu_X = E[X] = \int_{-\infty}^{+\infty} x f_X(x) dx. \tag{3.2.2}$$

In general, the difference between a sample mean and the population mean tends to be small for large sample sizes greater than, say, 30 in the case of symmetrical distributions. The mean is the most useful single number in engineering applications for theoretical, comparative, and other purposes. It is an important measure of the central tendency of the random variable and is well-suited to represent the phenomenon studied.

Example 3.8. Timber strength. For an example of the computation of the population mean, consider the triangular distribution introduced in Example 3.5 to be a first approximation to the timber strength data of Table E.1.1. From Eq. (3.2.2)

$$\mu_X = E[X] = \int_0^{40} \frac{x^2}{1400} dx + \int_{40}^{70} \frac{70x - x^2}{1050} dx = \frac{40^3}{4200} + \frac{70^3 - 40^2 \times 70}{2100} + \frac{40^3 - 70^3}{3150}$$

$$= 15.24 + 110.00 - 88.57 = 36.67 \text{ N/mm}^2.$$

The difference between this and the sample arithmetic mean of 39.09 N/mm^2 given in Table 1.2.2 is expected because the distribution we assumed is not a close approximation to the relative frequencies of the timber strength data.

It would be interesting to compare the median u, which has a probability of nonexceedance of 0.5. This calculation is carried out as follows:

$$F_X(u) = 0.5 = \int_0^u \frac{x dx}{1400} = \frac{u^2}{2800}.$$

Hence, $u = 37.42$ N/mm^2 (which can be compared with the sample median of 39.05 underlined in Table 1.1.3). For the assumed distribution, the mode is at 40 N/mm^2. The distribution is asymmetrical with a longer left tail indicating negative skewness. Thus, there is compliance with the condition

mean < median < mode,

which is applicable in the case of negative skewness.[8]

Example 3.9. Earthquake intensity. The mean strength of earthquakes in a region can be found following the procedure in Example 3.6 where the cdf was approximated by the exponential distribution

$$F_X(x) = 1 - e^{-\lambda x},$$

with the constant $\lambda = 0.2$. Then from Eq. (3.2.2)

$$\mu_X = E[X] = \int_0^{\infty} x\lambda e^{-\lambda x} dx.$$

Integrating by parts, we have

$$\mu_X = E[X] = [-xe^{-\lambda x}]_0^{\infty} + \int_0^{\infty} e^{-\lambda x} dx = 0 + \frac{1}{\lambda} = \frac{1}{\lambda},$$

where the zero term on the right-hand side is obtained through l'Hospital's rule. Thus, $\mu = 5$.

The median, u, has cdf

$$F_X(u) = 1 - e^{-\lambda u} = 0.5.$$

Hence,

$$u = \frac{-\ln(0.5)}{\lambda} = \frac{-\ln(0.5)}{0.2} = 3.47.$$

With its mode at zero, the distribution has no left tail, which indicates positive skewness. Thus, as expected, the condition

mode < median < mean

(applicable if skewness is positive) is satisfied.

3.2.1.2 Expectation operator

The mathematical expectation (that is, mean value) of a function of X, such as $g(x)$, can be obtained by substituting $g(X)$ for X on the left-hand side of Eq. (3.2.1) or (3.2.2) and $g(x)$ for x before the weighting function on the right-hand side. [Indeed, $Y = g(X)$ is a random variable.] Thus for a discrete variable,

$$E[g(X)] = \sum_{\text{all } x_i} g(x_i) p_X(x_i). \tag{3.2.3}$$

Correspondingly, for a continuous variable,

$$E[g(X)] = \int_{-\infty}^{+\infty} g(x) f_X(x) dx. \tag{3.2.4}$$

This is applicable only if the integral is absolutely convergent; that is, if $g(x)$ is replaced by its absolute value, the expectation is finite.

[8] As stipulated in Section 1.2.

Definition and properties: Expectation, $E[\cdot]$. Let X be a random variable and $g(x)$ a function of X. The expectation of the function $g(x)$ is given by

$$E[g(X)] = \sum_{\text{all } x_i} g(x_i) p_X(x_i)$$

if X is a discrete variable with mass points x_i and

$$E[g(X)] = \int_{-\infty}^{+\infty} g(x) f_X(x) dx,$$

if X is a continuous variable with pdf $f_X(x)$, provided that the series and the integral are absolutely convergent. If $h(x) = x$, the expectation is the mean of the variable. The following properties hold:

$$E[a] = a, \quad \text{for a constant } a;$$
$$E[ah(X)] = aE[h(X)], \quad \text{for a constant } a;$$
$$E[ah_1(X) + bh_2(X)] = aE[h_1(X)] + bE[h_2(X)], \quad \text{for two constants } a \text{ and } b;$$
$$E[h_1(X)] \geq E[h_2(X)] \quad \text{if } h_1(X) \geq h_2(X), \quad \text{for two functions } h_1(x) \text{ and } h_2(x).$$

Chebyshev inequality[9]

$$\Pr[h(X) \geq m] \leq m^{-1} E[h(X)], \text{ for every } m > 0 \text{ and } h(x) \geq 0.$$

Jensen inequality. $E[h(X)] \geq h(E[X])$, if $h(x)$ is a convex function; that is, a function represented by a bowl-shaped curve (which appears as if it can hold water). Alternatively, it can be said to be concave upward. The companion (reversed) inequality is applied for a concave function.[10]

Example 3.10. Jensen inequality applied to the arithmetic and geometric means. Let X be a random variable with sample realization $X_1, X_2, \ldots, X_n$ which are positive numbers. Then $h(X) = \log(X)$ is a concave function.

The geometric mean is given by[11]

$$\bar{X}_g = (X_1 \times X_2 \times \cdots \times X_n)^{1/n}$$

and hence

$$\log\left(\bar{X}_g\right) = \frac{1}{n} \sum_{i=1}^{n} \log(X_i).$$

The arithmetic mean is given by

$$\bar{X} = \frac{1}{n}(X_1 + X_2 + \cdots + X_n),$$

from which

$$\log(\bar{X}) = \log\left[\frac{1}{n}(X_1 + X_2 + \cdots + X_n)\right].$$

Because of the concavity of the log function $h(\cdot)$, we apply the companion Jensen inequality:

$$E[h(X)] \leq h[E[X]].$$

That is, [mean of the logarithm of X] $\leq$ logarithm [mean of X]. It is seen from the above that is $\log(\bar{X}_g) \leq \log(\bar{X})$. Hence, [geometric mean] < [arithmetic mean].

[9] The second e in Chebyshev is pronounced as an o and is stressed as are Gorbachev and Kruschev. This is sometimes referred to as the Bienaymé-Chebyshev inequality.

[10] Proofs of the foregoing properties are given in Appendix A, Sections A1 and A2.

[11] See Eq. (1.2.3).

3.2.1.3 Moments

As stated before, the average or expectation of a function of a random variable X can be found by weighting the function by its density or mass function. This procedure is called the method of moments. It constitutes a family of averages of the random variable, which serve as numerical descriptors of the behavior of the random variable. Accordingly, it is a standard practice to summarize a pdf or pmf by its moments. In general, μ_r^* is called the moment of order r about the point a and is defined for a *discrete variable* by

$$\mu_r^* = E[(X-a)^r] = \sum_{\text{all } x_i} (x_i - a)^r p_X(x_i), \tag{3.2.5}$$

where r denotes a positive integer. The rth-order moment about the origin, that is, for $a = 0$, is called a *absolute, crude, or raw moment*. Here it will be simply called a *moment* and denoted by μ_r'. The first moment is the mean,

$$\mu_X = \mu_1' = E[X] = \sum_{\text{all } x_i} x_i p_X(x_i). \tag{3.2.6}$$

As an analog from civil engineering, consider a rigid beam subject to a system of loads. The mean is equivalent to the distance from a given axis to the point of application of the equilibrant of the system of discrete vertical forces acting on the beam.

When a in Eq. (3.2.5) is the mean μ_X, the moment of order r is written without the superscript (asterisk) as

$$\mu_r = E[(X-\mu_X)^r] = \sum_{\text{all } x_i} (x_i - \mu_X)^r p_X(x_i). \tag{3.2.7}$$

These moments are called *central moments* [although, strictly speaking, the term central is applicable only when a is the median, that is, if $F(a) = 0.5$]. As shown shortly, μ_2 denotes the variance; also μ_1 is the mean deviation.

In the case of a discrete variable with values spaced at unit intervals, a useful concept is that of *factorial moments*. The factorial moment of order r is defined as

$$\mu_{(r)}' = \sum_{\text{all } x_i} x_i^{(r)} p_X(x_i), \tag{3.2.8}$$

where

$$x_i^{(r)} = x_i(x_i - 1)(x_i - 2)\cdots(x_i - r + 1). \tag{3.2.9}$$

and, as before, r is a positive integer.

In the case of a *continuous variable*, the moments are written as

$$\mu_r^* = E[(X-a)^r] = \int_{-\infty}^{+\infty} (x-a)^r f_X(x)dx \tag{3.2.10}$$

and

$$\mu_r = E[(X-\mu_X)^r] = \int_{-\infty}^{+\infty} (x-\mu_X)^r f_X(x)dx, \tag{3.2.11}$$

respectively. Analogously (as noted in Chapter 1), the mean μ_X is the centroid of the unit area between the pdf and the horizontal axis. By the same token, the variance given by μ_2 corresponds to the moment of inertia about the vertical axis through the centroid.

3.2.1.4 Variance, standard deviation, and coefficient of variation

As previously emphasized, the main characteristic of a random phenomenon is its variability. This may be high in some cases and low in others. It follows from Eq. (3.2.11) that the variance can be written as

$$\begin{aligned}\mathrm{Var}[X] \equiv \sigma_X^2 &= E[(X - E[X])^2] = E[X^2 - 2XE[X] + (E[X])^2] \\ &= E[X^2] - 2E[X]E[X] + (E[X])^2 = E[X^2] - (E[X])^2. \qquad (3.2.12)\end{aligned}$$

This result shows that the variance is equal to the difference between the mean of the squares and the square of the means.

Example 3.11. Reliability bounds using Chebyshev inequality. Let the squared deviation of a random variable X from its mean μ_X be represented by $h(X) = (X - \mu_X)^2$,

$$E[h(X)] = E[(X - \mu_X)^2] = \sigma_x^2.$$

Let $m = k^2\sigma_x^2$. Then by using Chebyshev inequality,

$$\Pr\left[(X - \mu_X)^2 \geq k^2\sigma_x^2\right] \leq 1/k^2, \quad \text{for } k \geq 1.$$

It follows that

$$\Pr[|X - \mu_X| < k\sigma_X] = \Pr[-k\sigma_X < X - \mu_X < k\sigma_X] \geq 1 - 1/k^2,$$

which can be written as

$$\Pr[\mu_X - k\sigma_X < X < \mu_X + k\sigma_X] \geq 1 - 1/k^2.$$

This expression states that the probability that X falls within $k\sigma_X$ units of μ_X is greater than or equal to $1 - 1/k^2$. For $k = 2$, the lower bound to the probability is 3/4, and for $k = 3$ it is 8/9. Using this rule, one can establish operational bounds without specifying the probability law of the investigated system. For example, consider a supply system subject to a random load or demand with known mean and standard deviation. An engineer who designs the capacity of this system in order to satisfy any demand ranging within two standard deviations of the mean ($k = 2$) does so with the knowledge that the reliability of this system will not be less than 75%. However, as we shall see in subsequent chapters, a higher reliability is obtained by making further assumptions—for example, that the pdf of the variable is known.

Example 3.12. Flood occurrence. For the Bernoulli probability that the flow of a river exceeds a given magnitude at a particular site, as specified in Example 3.7,

$$P_X(x_i) = \Pr[X = x_i] = p^{xj}(1 - p)^{1-xj}, \quad \text{for } x_i = 0, 1$$

can be used. If we take the second moment about the origin, then from Eq. (3.2.5)

$$E[X^2] = 0^2p^0(1 - p)^1 + 1^2p^1(1 - p)^0 = p.$$

Let $p = 0.1$. Since $\mu_X = p$, it follows from Eq. (3.2.12) that the variance is given by

$$\mathrm{Var}[X] \equiv \sigma_X^2 = p - p^2 = p(1 - p) = 0.09.$$

The standard deviation σ_X is defined as the positive square root of the variance. It is measured in the same units as the variable and is therefore more practically meaningful as a measure of dispersion than the variance, because it can be compared directly with the mean.

Example 3.13. Timber strength. For an example of the variance of a continuous variable, consider the timber strength data of Table E.1.1. The simplified triangular distribution in Examples 3.5 and 3.8 is assumed. From Eq. (3.2.10), with $a = 0$,

$$E[X^2] = \int_0^{40} \frac{x^3}{1400} dx + \int_{40}^{70} \frac{70x^2 - x^3}{1050} dx$$

$$= \frac{40^4}{5600} + \frac{70^4 - 40^3 \times 70}{3150} + \frac{40^4 - 70^4}{4200} = 1550.$$

By substituting $E[X] = 36.67$ N/mm^2 as obtained in Example 3.8, then from Eq. (3.2.12) we can write,

$$\text{Var}[X] \equiv \sigma_X^2 = E[X^2] - (E[X])^2 = 1550 - 1344.69 = 205.31 \text{ N}^2/\text{mm}^4.$$

Hence, $\sigma_X = 14.33$ N/mm^2.

The coefficient of variation $V_X = \sigma_X/\mu_X$ provides a relative measure of dispersion that is dimensionless. It is sometimes given as a percentage. Figure 3.2.1*a* and Figure 3.2.1*b* show schematically the pmfs and cdfs, respectively, of three discrete variables with different coefficients of variation.

Figure 3.2.2*a* and Figure 3.2.2*b* show the same effects on the pdfs and cdfs, respectively, of three continuous variables.

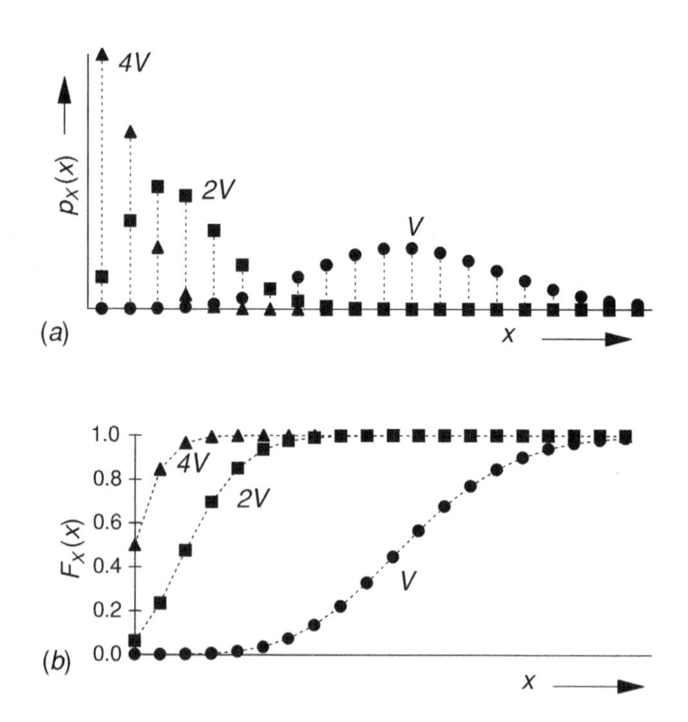

Fig. 3.2.1 Schematic diagrams of the (*a*) probability mass functions and (*b*) the cumulative distribution functions of three discrete variables X with different coefficients of variation V.

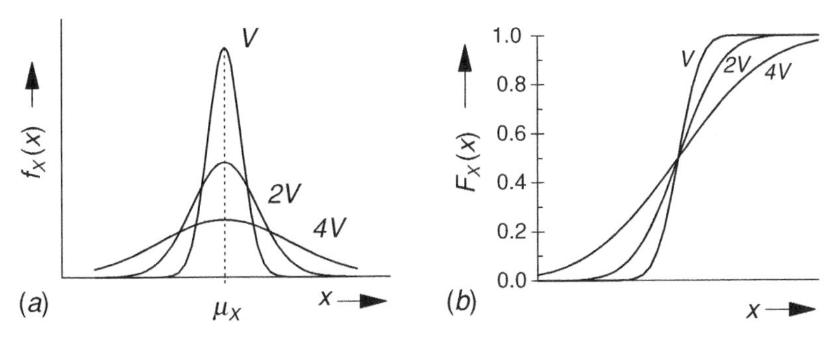

Fig. 3.2.2 Schematic diagrams of (*a*) symmetrical probability density functions and (*b*) cumulative distribution functions of three continuous variables X with different coefficients of variation V.

Example 3.14. Earthquake intensity. Another application to a continuous variate is the case of earthquakes in a region. Proceeding as in Example 3.9 for the evaluation of $E[X]$ (that is, integrating by parts and using l'Hospital's rule), we write

$$E[X^2] = \int_0^\infty \lambda x^2 e^{-\lambda x} dx = \left[x^2 e^{-\lambda x}\right]_\infty^0 + \int_0^\infty 2x e^{-\lambda x} dx = 0 + \frac{2}{\lambda^2} = \frac{2}{\lambda^2}.$$

By substituting $\mu_X = E[X] = 1/\lambda$ in Eq. (3.2.12),

$$\mathrm{Var}[X] \equiv \sigma_X^2 = E[X^2] - (E[X])^2 = \frac{2}{\lambda^2} - \frac{1}{\lambda^2} = \frac{1}{\lambda^2}.$$

3.2.1.5 Coefficient of skewness and coefficient of kurtosis

The coefficient of skewness γ_1 of a random variable is estimated by g_1 which is a measure of the asymmetry of a set of data about the mean. By using moments of order three and two about the mean, γ_1 is defined as follows:

$$\gamma_1 = \frac{\mu_3}{\sqrt{\mu_2^3}} = \frac{E[(X - E[X])^3]}{\sqrt{\{E[(X - E[X])^2]\}^3}} = \frac{E[X^3] - 3E[X^2]E[X] + 2(E[X])^3}{(E[X^2] - (E[X])^2)^{3/2}}. \tag{3.2.13}$$

The numerator is the central moment of order three and the denominator is the cube of the standard deviation. Also, the terms in the numerator on the right are obtained by expansion, taking expectations and regrouping as in Eq. (3.2.12).

For a probability distribution symmetrical about μ_X the third central moment $E[(X - \mu_X)^3] = 0$. If the values of X greater than μ_X are more widely dispersed than those that are less than μ_X, the third moment is positive, and the pmf or pdf has its dominant tail on the right. In this case the probability distribution is said to be *positively skewed* (as discussed in Section 1.2). On the other hand, for $E[(X - \mu_X)^3] < 0$ we have a *negatively skewed* probability distribution with the dominant tail on the left. Since the denominator of Eq. (3.2.13) is nonnegative, γ_1 has the same sign of $E[(X - \mu_X)^3]$, so that it is used as a dimensionless measure of the degree of skewness of a probability distribution (see Fig. 3.2.3*a* and 3.2.3*b*).

The *coefficient of kurtosis* γ_2 of a random variable as estimated by g_2 is a measure of the "peakedness" of a histogram. In the case of a symmetric distribution, the quantity $(\gamma_2 - 3)$ is sometimes referred to as the *coefficient of excess* (as a relative measure of kurtosis, considering that $\gamma_2 = 3$ for the normal distribution).

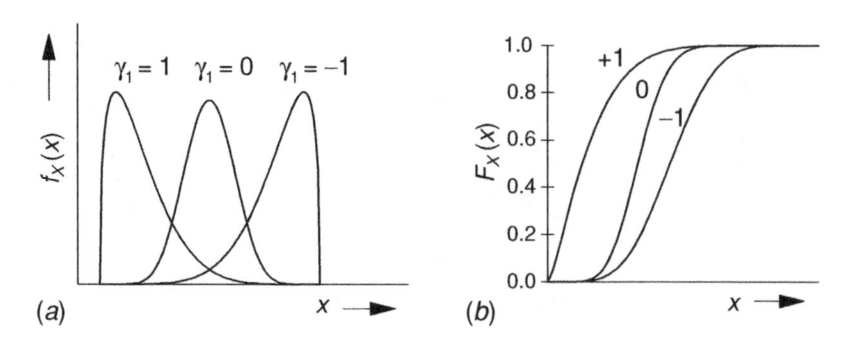

Fig. 3.2.3 Schematic diagrams of (a) the probability density functions and (b) the cumulative distribution functions of three continuous variables X with coefficients of skewness, $\gamma_1 = 1, 0,$ and -1.

By following a procedure similar to that in Eq. (3.2.13) and using moments of order four and two about the mean, γ_2 is defined by

$$\begin{aligned}\gamma_2 &= \frac{\mu_4}{\mu_2^2} = \frac{E[(X - E[X])^4]}{\{E[X - (E[X])^2\}^2} \\ &= \frac{E[X^4] - 4E[X^3]E[X] + 6E[X^2](E[X])^2 - 3(E[X])^4}{\{E[X^2] - (E[X])^2\}^2}.\end{aligned} \tag{3.2.14}$$

Example 3.15. Earthquake intensity. Consider the case of earthquakes in a region introduced in Example 3.6. Proceeding as in the evaluation of $E[X]$ and $E[X^2]$ in Examples 3.9 and 3.14 (integrating by parts and using l'Hospital's rule), one obtains for any integer r

$$E[X^r] = \int_0^\infty \lambda x^r e^{-\lambda x} dx = \frac{\Gamma(r+1)}{\lambda^r},$$

where $\Gamma(r+1) = r!$ for integer r. This result can be used in Eq. (3.2.13) to obtain

$$\begin{aligned}\gamma_1 &= \frac{E[X^3] - 3E[X^2]E[X] + 2(E[X])^3}{\{E[X^2] - (E[X])^2\}^{3/2}} = \frac{\Gamma(4)\lambda^{-3} - 3\Gamma(3)\lambda^{-3} + 2\lambda^{-3}}{\{\Gamma(3)\lambda^{-2} - \lambda^{-2}\}^{3/2}} \\ &= \frac{(6 - 6 + 2)\lambda^{-3}}{(2-1)\lambda^{-3}} = 2.\end{aligned}$$

This result means that the exponential distribution is positively skewed, as shown in Fig. 3.1.5, and $\gamma_1 = 2$ for any $\lambda > 0$. Similarly, from Eq. (3.2.14),

$$\begin{aligned}\gamma_2 &= \frac{E[X^4] - 4E[X^3]E[X] + 6E[X^2](E[X])^2 - 3(E[X])^4}{\{E[X^2] - (E[X])^2\}^2} \\ &= \frac{\Gamma(5)\lambda^{-4} - 4\Gamma(4)\lambda^{-4} + 6\Gamma(3)\lambda^{-4} - 3\lambda^{-4}}{\{\Gamma(3)\lambda^{-2} - \lambda^{-2}\}^2} = \frac{(24 - 4 \times 6 + 6 \times 2 - 3)\lambda^{-4}}{(2-1)^2\lambda^{-4}} = 9.\end{aligned}$$

3.2.1.6 Quantiles

Many engineering problems require solutions to the probability of a load exceeding a specified design level or the probability of a system designed to work within a target range. To address such problems, one needs to solve the so-called inverse problem, that is, to determine the value of a random variable that is exceeded with a given probability.

If we denote by q the required probability level, the qth quantile is the smallest number ξ satisfying the inequality

$$F_X(\xi) \geq q, \tag{3.2.15}$$

and it is denoted as ξ_q. For a continuous random variable the equality holds, so that the quantile can be defined as the smallest number ξ satisfying $F_X(\xi) = q$, and the quantile is the value of X which is exceeded with a probability $(1 - q)$. According to this definition, the median is the 0.5th quantile, that is, $\xi_{0.5}$.

Definition and properties: Quantile, ξ_q. The qth quantile of a random variable X is defined as the smallest number ξ satisfying the inequality $F_X(\xi) \geq q$, with $F_X(x)$ denoting the cdf of X. For a continuous variable the equality holds, so that ξ_q is the value of X with a probability of nonexceedance equal to q.

Example 3.16. Earthquake intensity. The mean strength of earthquakes in a region, as noted in Example 3.6, has cdf

$$F_X(x) = 1 - e^{-\lambda x}.$$

To determine the intensity of an earthquake which is exceeded with a probability of say, 0.1, we write the qth quantile, with $\lambda = 0.2$, as

$$q = 1 - \Pr[X > x] = 1 - 0.1 = 0.9.$$

Hence using Eq. 3.2.15 with the equality sign, we write

$$\xi_q = \xi_{0.9} = -(1/\lambda)\ln(1 - q) = -5\ln(0.1) = 11.5.$$

3.2.2 Generating functions

A generating function is a convenient way for compactly summarizing the information contained in a sequence, such as the moments of the sequence. Alternatively, the function can directly generate the probabilities of a discrete sequence. The function, which is usually of a quantity t, is expanded as a power series to give, for instance, the values of the moments as the coefficients. These functions have been used in theoretical statistics and probability for more than 250 years.

3.2.2.1 Moment-generating function

The moments of a pmf or pdf play an important role in theoretical and applied statistics. In fact, if all the relevant moments are known and finite, the mass or density can be determined.

It is possible in the case of some distributions to define a function, appropriately called the *moment-generating function*, mgf, existing in some neighborhood, say, $-\varepsilon < t < \varepsilon$ of the origin,

$$M_X(t) = E[e^{tX}] \tag{3.2.16}$$

that, when expanded in powers of t about zero as a Maclaurin's series, will provide the moments of the distribution as coefficients in the expansion. Thus,

$$M_X(t) = E[e^{tX}] = E\left[1 + Xt + \frac{1}{2!}(Xt)^2 + \cdots\right] = 1 + \mu_1 t + \frac{1}{2!}\mu_2 t^2 + \cdots. \tag{3.2.17}$$

The quantity t is a dummy value used merely to convey the information in the sequence.

Alternatively, a more direct approach can be taken. It follows from Eq. (3.2.16) that for discrete distributions

$$M_X(t) = \sum e^{tx_j} p_X(x_j), \quad \text{summed for all possible } x_j, \tag{3.2.18}$$

and for continuous types,

$$M_X(t) = \int_{-\infty}^{\infty} e^{tx} f_X(x)dx. \tag{3.2.19}$$

In general, if a random variable has an associated moment-generating function, then $M_X(t)$ is continuously differentiable in some neighborhood of the origin, and the moment of order m about the origin is thus generated from Eq. (3.2.16) by taking the mth derivative with respect to t and evaluating the derivative at $t = 0$. For this purpose we use Eqs. (3.2.11) and (3.2.17). That is,

$$\frac{dM_X(0)}{dt} = \left[\int_{-\infty}^{\infty} xe^{xt} f_X(x)dx\right]_{t=0} = \int_{-\infty}^{\infty} xf_X(x)dx = E[X], \tag{3.2.20a}$$

$$\frac{d^2M_X(0)}{dt^2} = \int_{-\infty}^{\infty} x^2 f_X(x)dx = E[X^2], \tag{3.2.20b}$$

and in general,

$$\frac{d^mM_X(0)}{dt^m} = \int_{-\infty}^{\infty} x^m f_X(x)dx = E[X^m]. \tag{3.2.20c}$$

Definition and properties: Moment-generating function, or mgf. The mgf of a random variable X is defined as $E[e^{tX}]$. If the mgf exists, its mth derivative at the origin ($t = 0$) is the mth-order central moment of X.

Example 3.17. Earthquake intensity. Returning to the occurrences of earthquakes (introduced in Example 3.6), we find from Eq. (3.2.19) that for the exponential distribution,

$$M_X(t) = \int_0^{\infty} e^{tx}\lambda e^{-\lambda x}dx = \frac{\lambda}{\lambda - t}$$

for $t < \lambda$. Hence, from Eq. (3.2.20a) the mean is evaluated as

$$\mu_X = E[X] = \left.\frac{\lambda}{(\lambda - t)^2}\right|_{t=0} = \frac{1}{\lambda}.$$

Also, from Eq. (3.2.20b),

$$E[X^2] = \left.\frac{2\lambda}{(\lambda - t)^3}\right|_{t=0} = \frac{2}{\lambda^2}.$$

Therefore, the variance is given by

$$\text{Var}[X] \equiv \sigma_X^2 = E[X^2] - (E[X])^2 = \frac{1}{\lambda^2}.$$

These results conform with those given in Examples 3.9 and 3.14.

If X and Y are two random variables with pdfs $f_X(x)$ and $f_Y(y)$, respectively, and their moment generating functions $M_X(t)$ and $M_Y(t)$ exist and are equal for all t in some interval $(-h < t < h)$, then the two cumulative distribution functions $F_X(x)$ and $F_Y(y)$ are equal. This property can be useful in many applications. For example, suppose that we find that a random variable X has a mgf given by $\lambda/(\lambda - t)$, we conclude that the pdf of X is $f_X(x) = \lambda e^{-\lambda x}$ for $x \geq 0$. This property is also applicable to other generating functions discussed here.

3.2.2.2 Factorial moment-generating function

In some discrete distributions, it is convenient to apply the *factorial moment-generating function*, which is defined by $E[t^X]$. The difference from an ordinary moment-generating function is that $t = 1$ is the condition of interest, as shown shortly, and not $t = 0$, which is used for the mgf.

Definition and properties: The factorial moment-generating function. The factorial mgf of a random variable X is defined as $E[t^X]$. If this function exists, its mth derivative at unity $(t = 1)$ is the mth-order factorial moment of X.

Example 3.18. Earthquake occurrence. Let X denote the number of occurrences of earthquakes in a given region in a year. A probability model that can be used for the purpose of describing such occurrences is given by the Poisson distribution:

$$p_X(x) = \frac{v^x e^{-v}}{x!}, \quad \text{where } x = 0, 1, 2, \ldots, v > 0.$$

This has the factorial mgf

$$E[t^X] = \sum_{x=0}^{\infty} \frac{t^x v^x e^{-v}}{x!} = e^{-v} \sum_{x=0}^{\infty} \frac{(vt)^x}{x!} = e^{-v} e^{vt} = e^{v(t-1)}.$$

The first and second derivatives taken at $t = 1$ are

$$\frac{d}{dt} E[t^X]_{t=1} = v$$

and

$$\frac{d^2}{dt^2} E[t^X]_{t=1} = v^2 e^{v(t-1)}|_{t=1} = v^2.$$

Also, the first and second derivatives at $t = 1$ of $E[t^X]$ are

$$\frac{d}{dt} E[t^X] = E[Xt^{X-1}]_{t=1} = E[X]$$

and

$$\frac{d^2}{dt^2} E[t^X]_{t=1} = E[X(X-1)t^{X-2}]_{t=1} = E[X^2] - E[X].$$

Hence, by equating terms,

$$E[X] = v,$$

$$E[X^2] - E[X] = v^2,$$

and

$$\text{Var}[X] \equiv \sigma_X^2 = E[X^2] - (E[X])^2 = v^2 + v - v^2 = v.$$

3.2.2.3 Cumulants

As a set of descriptive constants, the moments are useful for measuring the properties of a distribution and specifying it. In addition, there is another set of constants called *cumulants* or *semi-invariants*, which are preferable for theoretical purposes; these arise from the cumulant-generating function which is the logarithm of the mgf. Details of this set, which can be obtained by a procedure similar to that used to derive Eq. (3.2.17), are given, for example, by Stuart and Ord (1994, Chapter 3).

3.2.2.4 Characteristic function

The *characteristic function* of a random variable X is an alternative auxiliary function that is useful in many cases when the mgf does not provide estimates of moments directly. It is defined as

$$\phi_X(t) = E[e^{itx}] = M_X(it) \tag{3.2.21}$$

where $i = \sqrt{-1}$. The substitution of it for t is the modification to the moment-generating function of Eq. (3.2.16) necessary to transform it to a characteristic function. Likewise, Eqs. (3.2.17) to (3.2.19) can be converted by substituting it for t.

Example 3.19. Timber strength. Consider the gamma distribution specified by the pdf,

$$f_X(x) = \frac{\lambda^r}{\Gamma(r)} e^{-\lambda x} x^{r-1},$$

where $0 < x < +\infty; \lambda > 0;\ r > 0$. For example, this pdf can be used to model the distribution of the timber strength data of Table E.1.1 more closely than that assumed in Example 3.5. The characteristic function for this type is given by

$$\phi_X(t) = \frac{\lambda^r}{\Gamma(r)} \int_0^{\infty} e^{x(-\lambda+it)} x^{r-1} dx,$$

where, by definition, $\int_0^{\infty} e^{-z} z^{r-1} dz = \Gamma(r) = (r-1)!$ for integer r. Hence, substituting $z = x(\lambda - it)$, we find that

$$\phi_X(t) = \frac{\lambda^r}{(\lambda - it)^r \Gamma(r)} \int_0^{\infty} e^{-z} z^{r-1} dz = \left(1 - \frac{it}{\lambda}\right)^{-r}.$$

Thus, by Maclaurin's series expansion [as in Eq. (3.2.17)],

$$\phi_X(t) = 1 + r\frac{it}{\lambda} + \frac{r(r+1)}{2!}\left(\frac{it}{\lambda}\right)^2 + \frac{r(r+1)(r+2)}{3!}\left(\frac{it}{\lambda}\right)^3 + \cdots.$$

From the derivatives at $t = 0$ corresponding to Eq. (3.2.20*a*) and (3.2.20*b*)

$$\mu_X = E[X] = \frac{1}{i}\frac{d\phi_X(0)}{dt} = \frac{r}{\lambda},$$

and

$$E[X^2] = \frac{1}{i^2}\frac{d^2\phi_X(0)}{dt^2} = \frac{r(r+1)}{\lambda^2}.$$

Similarly,

$$E[X^3] = \frac{r(r+1)(r+2)}{\lambda^3},$$

and so on. Thus,

$$\text{Var}[X] \equiv \sigma_X^2 = E[X^2] - (E[X])^2 = \frac{r}{\lambda^2}.$$

The coefficients of skewness and kurtosis can be obtained using Eqs. (3.2.13) and (3.2.14).

3.2.3 Estimation of parameters

An engineer applies statistics to seek relevant information from a given sample of data. The procedure leads to conclusions regarding a population, which includes all possible observations of the process or phenomenon, and is called *statistical inference*. In this section, methods of parameter estimation called *point estimation* are introduced.[12]

One assumes for this purpose that the distribution of the population is known. However, the values of the parameters of the distribution have to be estimated from a sample of data, that is, a subset of the population. One also assumes that the sample is random.

In this context the term *bias* is related to an estimator, which is a method of obtaining the value of a parameter from a sample.[13] Different estimators are presented in this section. We begin with the method of moments, the most commonly used method; subsequently the probability weighted, L-moment, maximum likelihood, entropy, jackknife, bootstrap, and other methods are discussed.

3.2.3.1 Method of moments

The method of moments is a long-established procedure for finding point estimators. When fitting a parametric distribution to a set of data by this method, we equate the sample moments to those of the fitted distribution in order to estimate the parameters. This can be demonstrated by the following example:

Example 3.20. Timber strength. The mean, $\bar{x}$, and the standard deviation, $\hat{s}$, of the sample data of timber strength are given in Table E.1.1. It was assumed in Example 3.19 that these data are distributed according to the following gamma distribution, with pdf

$$f_X(x) = \frac{\lambda^r}{\Gamma(r)} e^{-\lambda x} x^{r-1}.$$

Also,

$$E[X] = \frac{r}{\lambda} \quad \text{and} \quad \text{Var}[X] = \frac{r}{\lambda^2}.$$

Therefore, substituting sample estimates $\bar{x}$ and $\hat{s}^2$ for the mean and variance, respectively, one can obtain the following estimates of the parameters of the gamma pdf:

$$\hat{r} = \frac{\bar{x}^2}{\hat{s}^2} \quad \text{and} \quad \hat{\lambda} = \frac{\bar{x}}{\hat{s}^2}.$$

If the timber strength data is modeled using this distribution, the statistics of Table 1.2.2 (for the full sample) give a mean of 39.09 N/mm^2 and a standard deviation of 9.92 N/mm^2, so that we obtain $\hat{r} = 15.5$, and $\hat{\lambda} = 0.40$ N/mm^2. The curve of the theoretical cdf of timber strength is compared with the cumulative relative frequency curve in Fig. 3.2.4.

In general, the numerical values of sample moments can differ greatly from those of the probability distribution from which the sample is generated. This difference occurs particularly when the sample size n is small, say, $n < 30$, and if third and higher moments are considered when, say, $n < 100$.

On the positive side, this method gives estimates that are easily obtained in most cases. They are also said to be *consistent*, where a consistent estimator is one which converges in probability as the sample size increases to the true value of the parameter (and, as already

[12] The associated methods of interval estimation, in which an unknown parameter is located (with a given assurance) within the range of two numbers, and hypothesis testing are shown in Chapter 5.

[13] More about this follows in Chapter 5.

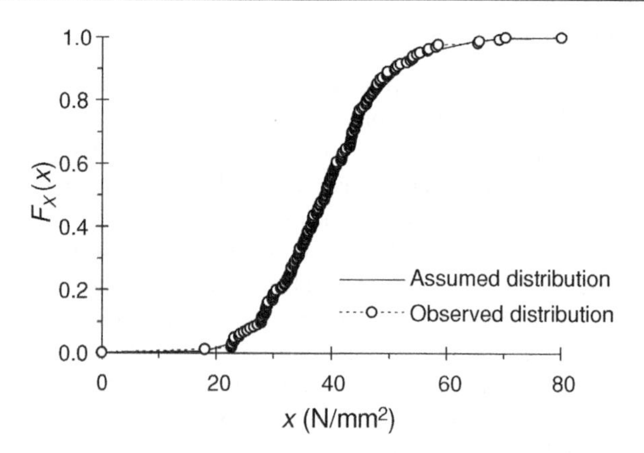

Fig. 3.2.4 Gamma cdf of modulus of rupture X of timber compared with the cumulative relative frequency curve of observed data.

stated, an estimator is a method of estimating a parameter of the parent population). However, methods based on moments are not always satisfactory. Firstly, the estimates may be *biased* in some cases; that is, if an average is taken of the estimates of the parameter from a large number of samples of the same size, say, n, the average will probably not converge to the value of the parameter. Note that, if an estimator has one of the properties of consistency and unbiasedness, the other property is not necessarily implied. Secondly, the estimator may sometimes be inefficient where an efficient estimator has the smallest variance among all possible estimators (that is, if one compares the variances of the estimates of the parameter obtained, for each estimator, from a large number of samples of size n). Furthermore, an efficient estimator is consistent and unbiased.[14] These shortcomings notwithstanding, the method of moments has survived as an effective tool and is widely used. Besides, it can be used when other methods are intractable.

3.2.3.2 Method of probability weighted and L-moments

Probability weighted moments (pwms) are expectations of functions of the quantiles and probabilities of nonexceedance of a random variable X where $E[X]$ exists. Unlike conventional moments, however, they do not have a physical connotation.[15]

Probability weighted moments characterize a distribution as in the case of conventional moments, but are expected to be less prone to adverse sampling effects. These pwms belong to the class of L estimators, which are linear functions (signified by the L) of an ordered sample.[16] This approach has had a long history in statistics. Two associated procedures are the trimmed mean and interquantile range, discussed in Chapter 1.

This method is best suited for the estimation of parameters of a distribution, the inverse form of which can be written in a closed form. The pwms are defined as

$$M_{ijk} = E[X^i\{F_x(x)\}^j\{1 - F_x(x)\}^k] = \int_0^1 \{x(F)\}^i F^j(1 - F)^k dF, \tag{3.2.22}$$

[14] A detailed description of these and other desirable properties is given in Chapter 5, with numerous examples.

[15] The subject was introduced by Greenwood et al. (1979).

[16] David and Nagaraja (2003) provide an extensive treatment on the theoretical aspects of order statistics which are statistics of ranked data. More about L moments follows in this section.

where $x(F)$ is the quantile or inverse cumulative distribution function of X. Equation (3.2.22) with $j = k = 0$ gives the moments of order i about zero in the conventional sense. In the application of pwms it is often convenient to assume $i = 1$ and either $k = 0$ or $j = 0$. Then the first m sample pwms (that is, for $j = 0, 1, \ldots, m$, or $k = 0, 1, \ldots, m$ evaluated by associating empirical probabilities with each item of data in a sample), as shown in the following example, are equated to the first m population pwms obtained from Eq. (3.2.22).

Example 3.21. Extreme storm data. Consider the following distribution which is useful in modeling extreme values such as flood discharges, earthquake intensities, wind loads, sea waves, and so on. If X is an *extreme value* variate, the cdf is given by

$$F_X(x) = \exp\left[-\exp\left(-\frac{x-b}{\alpha}\right)\right].$$

The quantile function can be expressed explicitly as follows:

$$x(F) = b - \alpha \ln[-\ln q],$$

where α and b are parameters and $q \equiv F_X(x)$. By integrating Eq. (3.2.22) for $i = 1$ and $k = 0$ after substituting for $x(F)$,

$$M_j = M_{1j0} = \frac{b}{1+j} + \frac{\alpha\,[\ln(1+j) + n_e]}{(1+j)},$$

where $n_e \approx 0.57721$ denotes Euler's number. Hence, the two parameters are estimated from

$$\alpha = \frac{2M_1 - M_0}{\ln 2}, \quad \text{and} \quad b = M_0 - n_e\alpha.$$

The probability weighted moments are estimated for $j = 0, 1$ from the sample data $x_{(i)}$, $i = 1, \ldots, n$ (arranged in increasing order of magnitude) as follows:

$$M_j = \sum_i p_i^j \frac{x_{(i)}}{n},$$

where the probabilities, p_i, are called *plotting positions*.[17]

It has been found that the following plotting position[18]:

$$p_i = \frac{i - 0.35}{n}$$

provides a close approximation for this distribution.

To analyze storm data recorded by the gauging station located on the campus of Genoa University, we apply the EV1 distribution to the annual maximum storm depth observed for a 3-hour duration, which are listed in the table under Problem 1.20. The ordered data values are given in Table 3.2.1 jointly with their plotting positions and basic statistics.

Note that M_0 is equal to the mean, so that its estimate $x = 83.7$ mm. To estimate M_1, one calculates the product between each sampling value and its plotting position; the summation of all the products is then divided by $n = 14$ to obtain $M_1 = 54.6$ mm. Thus,

$$\tilde{\alpha} = \frac{2M_1 - M_0}{\ln 2} = \frac{2 \times 54.6 - 83.7}{\ln 2} = 36.8 \text{ mm},$$
$$\tilde{b} = M_0 - n_e\alpha = 83.7 - 0.5772 \times 36.8 = 62.4 \text{ mm}.$$

The probability distribution is plotted in Fig. 3.2.5.

[17] These are used in probability plots (as introduced in Chapter 1) for assigning probabilities to ranked data. The plotting position is the probability of nonexceedance at which the x_i should be plotted after ranking in ascending order. We shall return to probability plotting in Section 5.8. and in Chapter 7.

[18] See Hosking (1990), this reference also considers L moments, discussed in the next subsection.

Table 3.2.1 Ordered data of maximum storm depth $x_{(i)}$ observed in a 3-hour duration at Genoa University[a]

Order, i	$x_{(i)}$ (mm)	p_i	$P_i\ x_{(i)}$ (mm)
1	32.3	0.046	1.500
2	41.2	0.118	4.856
3	49.9	0.189	9.445
4	51.6	0.261	13.453
5	56.0	0.332	18.600
6	56.7	0.404	22.883
7	63.8	0.475	30.305
8	66.5	0.546	36.338
9	80.6	0.618	49.799
10	90.1	0.689	62.105
11	114.2	0.761	86.874
12	122.3	0.832	101.771
13	157.3	0.904	142.132
14	188.7	0.975	183.983

[a] The corresponding plotting positions, p_i, and the product $p_i x_{(i)}$ are given in order to use the probability weighted moments method for evaluating the EV1 distribution.

3.2.3.3 L moments

L moments are summarizing functions of pwms and of the locations, scales, and other properties of a distribution and can thus be used for estimation. The first L moment estimator is the mean,

$$L_1 = E[X]. \tag{3.2.23a}$$

Let $X_{(i|n)}$ be the ith largest observation in a sample of size n. Then the second, third, and fourth L moments are defined as

$$L_2 = E[X_{(2|2)} - X_{(1|2)}]/2, \tag{3.2.23b}$$

$$L_3 = E[X_{(3|3)} - 2X_{(2|3)} + X_{(1|3)}]/3, \tag{3.2.23c}$$

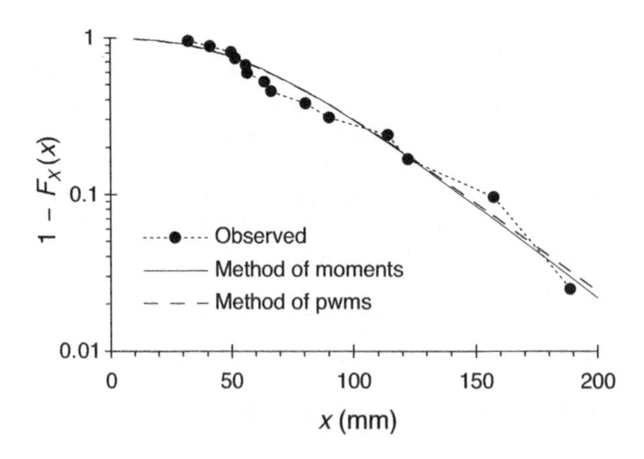

Fig. 3.2.5 Theoretical distribution of 3-hour annual maximum storm intensity in Genoa, Italy, compared with the cumulative relative frequency curve of observed data using the probability of exceedance, $\Pr[X > x] = 1 - F_X(x)$. Two methods of evaluating the probability distribution are used.

and

$$L_4 = E[X_{(4|4)} - 3X_{(3|4)} + 3X_{(2|4)} - X_{(1|4)}]/4. \quad (3.2.23d)$$

Samples estimators of L moments are *linear* combinations (hence the name *L moments*) of the ranked observations; and thus they do not involve squaring, cubing, and so on, of the observations, which one must do for the moment estimators of variance, skewness, and kurtosis. As a result, the L moment estimators of the dimensionless coefficients of variation, skewness, and kurtosis, L_2/L_1, L_3/L_2, and L_4/L_2, are much less variable than their conventional counterparts and are nearly normal in distribution.[19]

For any distribution, the L moments can be given in terms of the probability weighted moments (pwms)

$$L_1 = M_0, \quad (3.2.24a)$$

$$L_2 = 2M_1 - M_0, \quad (3.2.24b)$$

$$L_3 = 6M_2 - 6M_1 + M_0, \quad (3.2.24c)$$

and

$$L_4 = 20M_3 - 30M_2 + 12M_1 - M_0. \quad (3.2.24d)$$

3.2.3.4 Maximum likelihood procedure

The maximum likelihood, or ML, procedure is an alternative to the method of moments. As a means of finding an estimator, statisticians often give it preference. For a random variable X with a known pdf, $f_X(x)$, and observed values $x_1, x_2, \ldots, x_n$, in a random sample of size n, the likelihood function of θ, where θ represents the vector of unknown parameters, is defined as

$$L(\theta) = \prod_{i=1}^{n} f_X(x_i \mid \theta). \quad (3.2.25)$$

The objective is to maximize $L(\theta)$ for the given data set. This is easily done by taking m partial derivatives of $L(\theta)$, where m is the number of parameters, and equating them to zero. We then find the *maximum likelihood* (ML) *estimators* of the parameter set θ from the solutions of the equations. In this way the greatest probability is given to the observed set of events, provided that we know the true form of the probability distribution.

Definition and properties: Maximum likelihood estimator. The parameter set $\tilde{\theta}$ for which $L(\tilde{\theta})$ takes a maximum given a sample $x_1, x_2, \ldots, x_n$ is a maximum likelihood estimator set.

Example 3.22. Flood occurrence. Consider the following Bernoulli pmf specified in Example 3.7, where x is a discrete variable and p is a parameter, to model flood occurrences.

$$P_X(x_j) = \Pr[X = x_j] = p^{xj}(1-p)^{1-xj}, \quad \text{for } x_j = 0, 1.$$

If there are n trials or outcomes,

$$L(p) = \prod_{j=1}^{n} p^{x_j}(1-p)^{1-x_j}, \quad \text{where } x_j = 0, 1.$$

[19] L moments can also be easily interpreted in terms of order statistics. The subject is discussed further in Subsection 7.2.5.

We find that $L(p)$ is positive and that the value of p which maximizes it is the same value which maximizes $\ln L(p)$. The latter is more convenient in such cases and also when applied to a pdf with an exponential term. Thus,

$$\ln L(p) = \sum_{j=1}^{n} x_j \ln p + \left(n - \sum_{j=1}^{n} x_j\right) \ln(1-p)$$

and

$$\frac{d \ln L(p)}{dp} = \frac{\sum_{j=1}^{n} x_j}{p} - \frac{n - \sum_{j=1}^{n} x_j}{(1-p)}.$$

Hence by equating the foregoing equation to zero, the estimate of the parameter is obtained as

$$\tilde{p} = \frac{\sum_{j=1}^{n} x_j}{n}.$$

Example 3.23. Compressive strength of concrete. Assume the distribution of compressive strengths of Table 1.2.1 can be represented by using the following normal pdf:

$$f_Y(y) = \frac{1}{\sqrt{2\pi} b} \exp\left[-\frac{1}{2}\left(\frac{y-a}{b}\right)^2\right],$$

where a and b are unknown parameters, and b is positive. We seek here the ML estimators of these unknown parameters. If we denote by $y_1, y_2, \ldots, y_n$ the n observed values of compressive strength, with $n = 40$, the ML function of Eq. (3.2.25) is given by

$$L(a, b) = \left(\frac{1}{\sqrt{2\pi} b}\right)^n \exp\left[-\frac{1}{2b^2}\sum_{i=1}^{n}(y_i - a)^2\right].$$

By introducing the log-likelihood function as in the previous example,

$$\ln L(a, b) = -n \ln \sqrt{2\pi} - n \ln b - \frac{1}{2b^2}\sum_{i=1}^{n}(y_i - a)^2.$$

We set the partial derivatives equal to zero:

$$\frac{\partial \ln L}{\partial a} = \frac{1}{b^2}\sum_{i=1}^{n}(y_i - a) = 0 \quad \text{and} \quad \frac{\partial \ln L}{\partial b} = -\frac{n}{b} + \frac{1}{b^3}\sum_{i=1}^{n}(y_i - a)^2 = 0.$$

Hence,

$$\tilde{a} = \frac{1}{n}\sum_{i=1}^{n} y_i = \bar{y} \quad \text{and} \quad \tilde{b} = \sqrt{\frac{1}{n}\sum_{i=1}^{n}(y_i - \bar{y})^2} = s.$$

Thus, for the normal distribution, the ML estimators of parameters a and b are the sample mean and the sampling standard deviation, respectively. It is easy to show that the same result is obtained by the method of moments. From Table 1.2.2 $\bar{y} = 60.14$ N/mm^2 and $s = 5.02$ N/mm^2. The distribution is compared with the cumulative relative frequency curve in Fig. 3.2.6.

The maximum likelihood estimator is consistent and is the method favored by statisticians. On the other hand, large samples are necessary before the estimator becomes unbiased. Additionally, this estimator does not have a low variance in comparison with others. Furthermore, there are estimation problems because maximum likelihood solutions do not exist in some cases and may often require numerical solutions in others.

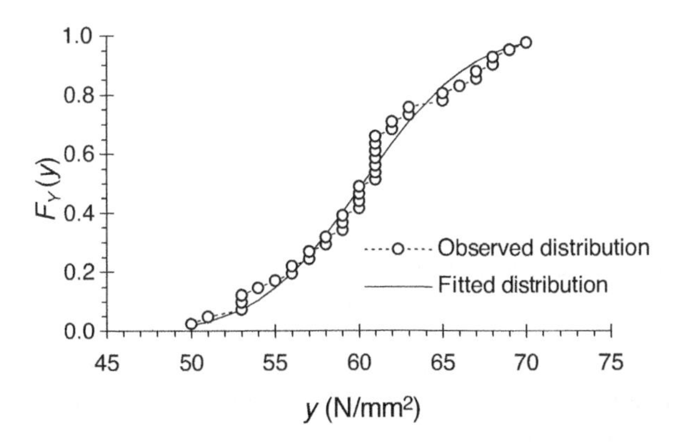

Fig. 3.2.6 Maximum likelihood fitted distribution of compressive strength of concrete Y compared with the cumulative relative frequency curve of observed data.

3.2.3.5 Bayesian estimation

The Bayesian approach is a generalization of the ML method. For this purpose, the likelihood function is multiplied by a prior pdf, which is estimated on a physical basis (that is, on the engineer's knowledge of the process) or from previous experiments. The product is then divided by a normalizing constant to satisfy the requirement for the resulting pdf. Accordingly, we can obtain the resulting distribution for a particular sample and prior density; the result is known as the posterior distribution.[20]

3.2.3.6 Least squares methods

Least squares procedures can be applied in several ways. For example, if the inverse form of a distribution can be written so that a linear relationship is obtained between a function of the variable and a function of the cdf, then a least-squares fit can give estimates of two parameters.[21]

3.2.3.7 Entropy

A physically based concept used in estimation is that of entropy which was coined around 1850 and originates from the second law of thermodynamics. The term implies a transformation and was first applied to work transformed into heat. Boltzmann (1894) derived the continuous form of entropy from kinetic theory applied to gases, an expression known as the Boltzmann H function. In general, entropy is a measure of some property of a system or process that has inherent uncertainty. With the notation already used, this takes the following summary form in the case of a continuous random variable:

$$H(x) = -\int_{-\infty}^{\infty} f_X(x) \ln f_X(x) dx. \tag{3.2.26}$$

[20] See, for example, Smith and Naylor (1987). If the prior pdf is uniform, then the Bayesian and ML methods will give the same results. Details of the method are found in Chapter 10.

[21] See Example 4.25. There are additional procedures based on quantiles and order statistics [see, for example, Stuart and Ord (1991, Chapter 19)].

For a discrete random variable

$$H(x) = -\sum_{all x_j} p_X(x_j) \ln p_X(x_j). \tag{3.2.27}$$

Definition and properties: Entropy function, *H(x)*. The entropy function gives the degree of randomness affecting a random variable X, and it is defined as the summation (or the integral) of the product between the pmf (or pdf) and its logarithm extended to all sample points of the sample space. More formally, entropy is defined as the logarithm of the number of quantum states accessible to a system.

Jaynes (1957) introduced the *principle of maximum entropy in* statistical mechanics. The term currently signifies uncertainty or randomness as in the original definition of Boltzmann; this is often equivalent to disorder, ignorance, or lack of knowledge. The concept has found numerous applications in science and engineering, particularly in information and communication theory.[22] More subjectively, it has been used in statistical analysis in order to obtain the form of the probability law and estimates of parameters by maximizing the entropy function, as shown in the following examples[23]:

Example 3.24. Density of concrete. Consider the distribution of the density of concrete given in Table E.1.2 and ordered in Table 1.2.1, denoted by X. Let the only stipulation be that $a \le X \le b$, where a and b are the lower and upper bounds, respectively. The problem is to maximize $H(x)$ in Eq. (3.2.26) subject to the constraint that is applicable to all distributions, Eq. (3.1.5*c*),

$$\int_a^b f_X(x)dx = 1.$$

The solution is found directly by the method of Lagrange multipliers. Denoting $f_X(x)$ by f and using λ as a constant multiplier, one must find f from the sum of partial derivatives:

$$-\frac{\partial}{\partial f}(f \ln f) + \lambda \frac{\partial}{\partial f}(f) = 0.$$

Thus, $-1 - \ln f + \lambda = 0$. That is, $f = e^{\lambda - 1}$, which is a constant. This represents the following distribution:

$$f_X(x) = \frac{1}{b-a}, \quad \text{where } a \le x \le b.$$

This conclusion is expected when one has insufficient prior knowledge of the distribution.

Example 3.25. Entropy and uncertainty. As a matter of particular interest in this context, the entropy for the distribution derived in Example 3.24 is found from Eq. (3.2.26) to be

$$H(x) = -\int_a^b \frac{1}{b-a} \ln \frac{1}{b-a} dx = \ln(b-a).$$

When the difference between a and b increases, the entropy, that is, the uncertainty, becomes larger. With regulated and controlled methods of workmanship (as in the case of the properties of concrete cited in Examples 3.23 and 3.24), the entropy is minimized in applications of this type.

[22] Note the pioneering work in the theory of communications by Shannon (1948).

[23] See also Sonuga (1972) and Harrop-Williams (1983).

Table 3.2.2 Jackknife estimates of the coefficients of skewness and kurtosis of the timber and concrete data of Tables 1.1.3 and 1.2.1

Data set	Coefficient of skewness[a]	Coefficient of kurtosis[a]
Timber strength (full sample)	0.13 (0.15)	4.73 (4.46)
Compressive strength of concrete	0.01 (0.03)	2.42 (2.33)
Density of concrete	0.47 (0.38)	3.79 (3.15)

[a] Values within parentheses are from Table 1.2.2.

3.2.3.8 Jackknife and bootstrap

It was shown in Section 1.2 that a simple bias correction can be applied to the variance by using the divisor $(n-1)$. In the case of other estimators, the correction can be complicated or even unknown. For the general case, Quenouille (1956) suggested an original method of adjusting for bias that John W. Tukey (whose work on exploratory data analysis was discussed in Chapter 1) termed the *jackknife*, which means a "useful tool" by implication.[24]

Definition: If $t_{n-1}^{(i)}$ is the estimator of a parameter based on a sample of size $(n-1)$ obtained by omitting the ith observation, the **jackknife estimator** is given by

$$J = nt_n - \frac{(n-1)}{n}\sum_{i=1}^{n} t_{n-1}^{(i)}, \tag{3.2.28}$$

where t_n is the estimator based on the full sample size n.

The bias in the jackknife estimator is not more than $1/n^2$, if an estimator of the same parameter based on the complete sample of size n has a bias of the order of $1/n$.

Example 3.26. Timber and concrete data. The coefficients of skewness and kurtosis of the timber and concrete data of Tables E.1.1 and E.1.2, respectively, are given with other statistics in Table 1.2.2. The skewness and kurtosis of the data were also estimated using the jackknife method of Eq. (3.2.28). These are given in Table 3.2.2.

It is interesting to compare the coefficient of skewness of the timber strengths. Whereas in Table 1.2.2, the sample data without the zero value has a skewness of 0.53 compared to that of 0.15 for the full sample, the jackknife estimate of 0.13 (using the full sample) is close to the original estimate from the full sample.

In place of the jackknife which uses n subsample of size $n-1$, Efron (1979) proposed a *bootstrap* method (originating from the expression "lifting oneself by one's bootstraps"). This method of resampling and averaging uses all possible subsamples of size m, where $1 < m < n$ and n is the original sample size. For applications, a shorter number of subsamples is suggested: of the order of $n \ln(n)$. The method is expected to give less biased estimates than the jackknife.[25]

The jackknife and bootstrap methods are closely related to the unbiased "U" statistics originated by Paul Hamos in 1946 and followed by some authors.[26]

[24] Miller (1974) provides a mathematical review.

[25] See discussion on confidence limits by Press et al. (1992, pp. 686–688) and the example by Zucchini and Adamson (1988).

[26] See, for example, Randles and Wolfe (1979). The jackknife also provides approximate confidence limits, a subject that is a part of Chapter 5.

3.2.3.9 Kernel-based methods of estimation

In recent years there has been a renewed interest in kernel-based nonparametric methods of estimation.[27] Given a random sample $X_i, i = 1, 2, \ldots, n$ with continuous univariate pdf $f(\cdot)$, the kernel estimator is defined as

$$\hat{f}(x, h) = \frac{1}{nh}\sum_{i=1}^{n} k\left(\frac{x - X_i}{h}\right),$$

where h is the window or bandwidth or smoothing parameter, and k is the kernel function. The optimal size of h is similar to that of the class width in a histogram. As in other methods of application of windows, it may cause under- or oversmoothing. With increasing n the estimator tends to converge in probability to the true pdf. The advantage of the method, as already stated, is its nonparametric nature. Also, by spreading the influence of each data point over its neighborhood and by taking each contribution in the summation, it allows the data to play a direct role in the estimation of the distribution. Accordingly, this approach can uncover special features in the data which conventional methods do not reveal.[28]

3.2.4 Summary of Section 3.2

In this section applications of the theory of random variables are presented. Expectation, variance, and other properties are shown. The relationships with moments are given. Demonstrations are made, for both continuous and discrete random variables, on the use the moment-generating function, the factorial moment-generating function, and the characteristic function. The concept of quantiles is needed to solve the inverse problem, which is of particular interest in engineering design. Classical methods of estimation—moments and maximum likelihood—are given together with introductions to alternative procedures such as probability weighted-moments, maximum entropy, the jackknife and bootstrap, and kernel-based methods of estimation. The advantages and possible limitations of these procedures are discussed.

3.3 MULTIPLE RANDOM VARIABLES

Up to this point we have considered members of a population represented by a single random variable. Univariate distributions are used to describe them. Their properties have been examined in Section 3.2. These concepts can be generalized to include populations with two or more variables that occur simultaneously. The variables are viewed jointly, and their distributions are of the multivariate type, which reduces to the bivariate case when there are only two variables. Thus treated, their probability laws are described by joint probability mass or density functions for discrete or continuous variables, respectively. An example of a continuous bivariate distribution is provided by the average rainfall over a catchment area above a flow-measuring station on a river and the volume of water passing the station during corresponding intervals of time. If one considers, in addition, other variables such as catchment saturation and groundwater flows, the distribution becomes a multivariate type with several variables.

---

[27] There are nonparametric in the sense that less rigid assumptions are made about the underlying distribution.

[28] See, for example, Silverman (1986). Kernel smoothing dates back to the work of M. Rosenblatt and E. Parzen and is often referred to as the Rosenblatt-Parzen estimator. On selecting a data-based bandwidth, see Sheather and Jones (1991); also, Granovsky and Muller (1991) review various aspects of kernel choice.

3.3.1 Joint probability distributions of discrete variables

3.3.1.1 Joint probability mass function

Let us consider the joint probability mass function, pmf, for the bivariate case. For two discrete variables X and Y, the bivariate pmf is given by the intersection probability

$$p_{X,Y}(x, y) = \Pr[(X = x) \cap (Y = y)]. \tag{3.3.1}$$

As in the univariate case it is also a condition that

$$\sum_{\text{all } x_i} \sum_{\text{all } y_j} p_{X,Y}(x_i, x_j) = 1. \tag{3.3.2}$$

A graphical representation of this function requires a three-dimensional form in which the two horizontal axes represent the two random variables and the pmf is measured vertically. The joint cumulative distribution function cdf is given by

$$F_{X,Y}(x, y) = \Pr[(X \le x) \cap (Y \le y)] = \sum_{x_i \le x} \sum_{y_j \le y} p_{X,Y}(x_i, y_j). \tag{3.3.3}$$

One can easily extend Eqs. (3.3.1) and (3.3.3) to multivariate cases by introducing additional variables W, Z, and so on. In general, if $X_1, X_2, \ldots, X_k$ are random variables defined on the same probability space, then $(X_1, X_2, \ldots, X_k)$ is defined as a *k-dimensional discrete random variable* if it can take values only at a countable number of points $(x_1, x_2, \ldots, x_k)$. One also says that the variables $(X_1, X_2, \ldots, X_k)$ are joint discrete random variables.

Definition: Joint probability mass function, pmf. The joint pmf of a k-dimensional random variable $(X_1, X_2, \ldots, X_k)$ is defined as the intersection probability of the k-tuple of events $(X_1 = x_1), (X_2 = x_2), \ldots, (X_k = x_k)$ if $(x_1, x_2, \ldots, x_k)$ are points in the k-dimensional sample space of this variable, and 0 otherwise.

Example 3.27. Wind measurements. As an example of the application of Eq. (3.3.1), consider a case in which high-intensity winds occur in a particular area. Such winds are liable to cause damage to buildings and other structures: The number of days annually that these winds occur has been observed in the recent past using a precise measuring device. These occurrences have also been recorded over a longer period of time with a less accurate instrument. As a first step to possible statistical calibration, it is necessary to estimate the joint pmf of the two instruments. The exceedance of some critical wind velocities is of interest to engineers, and this exercise is an aid to planning structural designs in the area. Table 3.3.1 gives the joint pmf of winds greater than 60 km/h with the discrete variables X and Y representing the precisely and less accurately measured number of days, respectively, with such high-intensity winds.

Table 3.3.1 Joint pmf of days of high-intensity winds measured accurately, X, and less accurately, Y, at a particular site

	$Y = 0$	$Y = 1$	$Y = 2$	$Y = 3$	$p_X(x)^a$
$X = 0$	0.2910	0.0600	0.0000	0.0000	0.3510
$X = 1$	0.0400	0.3580	0.0100	0.0000	0.4080
$X = 2$	0.0100	0.0250	0.1135	0.0300	0.1785
$X = 3$	0.0005	0.0015	0.0100	0.0505	0.0625
$p_Y(y)^a$	0.3415	0.4445	0.1335	0.0805	$\Sigma = 1.0000$

[a] The entries in the last column and bottom row are the respective marginal probabilities.

The marginal pwms of X and Y are given in the extreme right-hand and bottom rows, respectively, of Table 3.3.1. The accuracy of the second instrument, the results from which are in doubt relative to the first, which is an accurate measuring device, can be represented by the probability $P[A]$, say, that the readings are the same in all cases. This is given by

$$\Pr[A] = \sum_{\text{all } x_i} p_{X,Y}(x_i, x_i) = p_{X,Y}(0,0) + p_{X,Y}(1,1) + p_{X,Y}(2,2) + p_{X,Y}(3,3)$$
$$= 0.2910 + 0.3580 + 0.1135 + 0.0505 = 0.813.$$

3.3.1.2 Conditional probability mass function

If the value of one variable, such as Y is known (or fixed), say, for $Y = y_j$, and each of the joint probabilities $p_{X,Y}(x, y_j)$, defined for all possible X with $Y = y_j$, are divided by the sum of these joint probabilities, then one obtains a set of probabilities with the pmf of $p_{X|Y}(x \mid y)$ of X given Y, which is called a conditional probability mass function.[29] Thus, if Y is known or given (that is, it has occurred or is occurring elsewhere), then the conditional distribution of X is

$$p_{X|Y}(x|y_j) \equiv \Pr[X = x \mid Y = y_j] = \frac{\Pr[(X = x) \cap (Y = y_j)]}{\Pr[Y = y_j]}$$
$$= \frac{p_{X,Y}(x, y_j)}{\sum_{\text{all } x_i} p_{X,Y}(x_i, y_j)} = \frac{p_{X,Y}(x, y_j)}{p_Y(y_j)}, \quad \text{for all } j, \tag{3.3.4}$$

where the vertical lines on the left denote (as in Chapter 2) "given that" or "conditional to." It is conventional that the conditional pmf is zero when the denominator in Eq. (3.3.4) is zero. Also, as in the univariate case, each conditional probability is in the range 0–1, and their sum over all possible values of the variable is equal to unity:

$$0 \le p_{X|Y}(x \mid y_j) \le 1, \quad \text{for all } j, \tag{3.3.5a}$$

$$\sum_{\text{all } x_i} p_{X|Y}(x_i \mid y_j) = 1, \quad \text{for all } j. \tag{3.3.5b}$$

Example 3.28. Wind measurements. In Table 3.3.1, if $Y = 1$, for example, the joint probabilities of interest are denoted by $p_{X,Y}(x, 1)$ and are given by 0.0600, 0.3580, 0.0250, and 0.0015 for $x = 0, 1, 2$, and 3, respectively. If each of these values is divided by their sum, 0.4445, the conditional pmf, $p_{X|Y}(x \mid 1)$, is correspondingly given by 0.1350, 0.8054, 0.0562, and 0.0034, for $x = 0, 1, 2$, and 3, respectively, the sum of which is 1.0.

3.3.1.3 Marginal probability mass function

If all other variables are disregarded apart from a single variable of interest, then one can obtain the marginal pmf of that variable from the joint probabilities in the discrete case (as demonstrated in Example 3.27). The concept can easily be followed when the joint probabilities are bivariate by applying the theorem of total probability.[30] As a consequence, one obtains the marginal pmf by summing for each value of the variable of interest the

[29] This follows directly from the definition given by Eq. (2.2.10).
[30] See Eq. (2.2.15).

joint probabilities for all possible values of the other variable. For example, the marginal pmf of X (when Y is disregarded) is as follows:

$$p_X(x) \equiv \Pr[X = x] = \sum_{\text{all } y_j} \Pr[X = x \mid Y = y_j]\Pr[Y = y_j] = \sum_{\text{all } y_j} p_{X,Y}(x, y_j). \tag{3.3.6}$$

Similarly, the marginal pmf of Y (when X is disregarded) can be defined. Also, one can obtain the cdf of X from the marginal pmf of X as follows:

$$F_X(x) \equiv \Pr[X \le x] = \sum_{x_i \le x} \sum_{\text{all } y_j} p_{X,Y}(x_i, y_j). \tag{3.3.7}$$

Returning to the conditional pmf, we note that the denominator in Eq. (3.3.4) is the marginal probability of Y evaluated at a specified value. The term *marginal* is used because the marginal pmf is obtained by summing the entries (as in Table 3.3.1) horizontally or vertically, depending on whether the marginal pmf is for X or Y, respectively, and writing the sums in the margins as shown.

Example 3.29. Wind measurements. As an example, consider the case $X = 0$ in Table 3.3.1,

$$p_X(0) \equiv \Pr[X = 0] = \sum_{y=0}^{3} p_{X,Y}(0, y) = 0.2910 + 0.0600 + 0.0000 + 0.0000 = 0.3510.$$

One can easily verify that the sum of all the *other* marginal probabilities, which is equivalent $\Pr[X > 0]$, is 0.6490. The marginal pmfs of X and Y are shown in Fig. 3.3.1*a* and 3.3.1*b*, respectively.

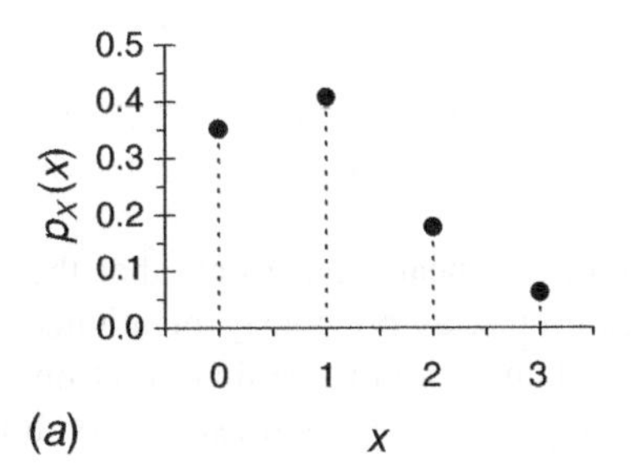

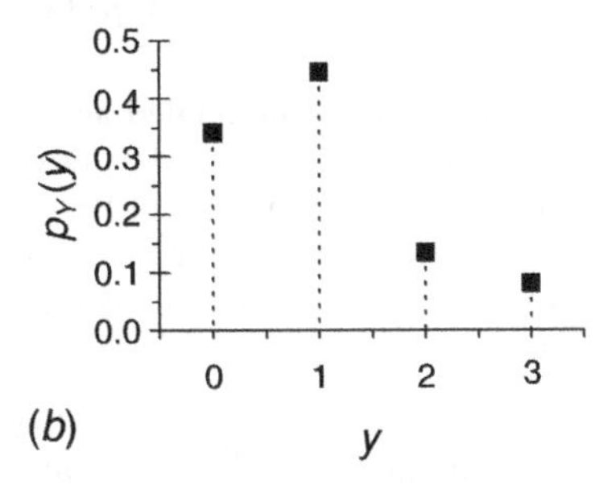

Fig. 3.3.1 Marginal probability mass function of number of days X per year of high-intensity winds recorded by (*a*) the more accurate instrument, and (*b*) the less accurate instrument, as summarized in Table 3.3.1.

3.3.1.4 Other conditional probability concepts for discrete variates

The preceding concepts can be extended to form other conditional probability distributions such as the probability distribution of X when Y is not less than y. This is given in general terms as

$$p_{X|Y\geq y} \equiv \Pr[X = x \mid Y \geq y] = \frac{\sum\limits_{y_j \geq y} p_{X,Y}(x, y_j)}{\sum\limits_{y_j \geq y} p_Y(y_j)}. \tag{3.3.8}$$

Example 3.30. Wind measurements. By summing columns 2, 3, and 4 in Table 3.3.1, $\sum_{y_j \geq 1} p_Y(y_j) = 0.6585$. Also, the terms of $\sum_{y_j \geq 1} p_{X,Y}(x, y_j)$ are 0.0600, 0.3680, 0.1685, and 0.0620 for $X = 0, 1, 2$ and 3 respectively, obtained by summing each row over columns 2, 3, and 4. Thus, the conditional probabilities $p_{X|Y}(x \mid y \geq 1)$ are obtained from Eq. (3.4.8) as 0.0911, 0.5588, 0.2559, and 0.0942 for $X = 0, 1, 2.$ and 3 respectively. [Clearly, the conditions given by Eq. (3.3.5) are satisfied.] This result shows, for instance, that there is a probability of about 0.91 that on one or more days per year the winds exceed 60 km/h at the site when the less accurate instrument records at least one occurrence for the year. The conditional pmf is shown in Fig. 3.3.2.

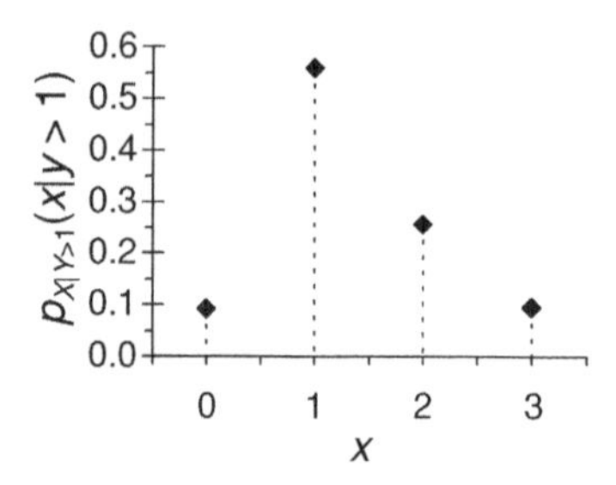

Fig. 3.3.2 Conditional probability mass function of number of days X per year of high-intensity winds when the less accurate instrument records $Y \geq 1$ such events (see Table 3.3.1).

Engineers are aware that nearly all events that they observe are conditional to the occurrences of other events, as previously mentioned. Thus, conditional probabilities may be more easily obtained in practice than joint pmfs. Therefore, it will possibly be more convenient to compute the joint pmf using a conditional pmf and the corresponding marginal pmf as follows[31]:

$$p_{X,Y}(x, y) = p_{X|Y}(x \mid y)p_Y(y) = p_{Y|X}(y \mid x)p_X(x). \tag{3.3.9}$$

Example 3.31. Remote sensing of inundated areas. A satellite-borne sensor is used to determine the area of inundation after a flood by detecting the number of pixels reflecting flooding. This sensor has a chance, in percent, of $100p$, to detect an inundated pixel. Because the outcome of the experiment is either a success or failure in detecting an inundated pixel, we can apply the two-random-variable probability model. Accordingly, the conditional pmf that M pixels are detected when N pixels are inundated is given by the binomial distribution

$$p_{M|N}(n, m) = \frac{n!}{(n-m)!m!} p^m (1-p)^{n-m} = \binom{n}{m} p^m (1-p)^{n-m}.$$

[31] This is analogous to Eq. (2.2.11).

In addition it is assumed that the distribution of the number N of inundated pixels is given by the Poisson distribution (as in Example 3.18),

$$p_N(n) = \frac{\nu^n e^{-\nu}}{n!}.$$

Therefore, from Eq. (3.3.9)

$$p_{M,N}(n,m) = \frac{\nu^n e^{-\nu}}{n!} \frac{n!}{(n-m)!m!} p^m (1-p)^{n-m} = \frac{\nu^n e^{-\nu}}{(n-m)!m!} p^m (1-p)^{n-m},$$

which is the joint probability that there are n inundated pixels and that m of these pixels are detected by the sensor. If Δa is the unit area of each pixel, the risk that an inundated area A larger that $k\Delta a$ remains undetected by the sensor is given by

$$\Pr[A > k\Delta a] = \Pr[K > k] = 1 - F_K(k) = 1 - \sum_{m=0}^{k} \sum_{n=m}^{\infty} \frac{\nu^n e^{-\nu}}{(n-m)!m!} p^m (1-p)^{n-m}.$$

3.3.1.5 Independent discrete random variables

If the events $(X = x)$ and $(Y = y)$ are statistically independent,

$$p_{X|Y}(x \mid y) = p_X(x) \quad \text{and} \quad p_{Y|X}(y \mid x) = p_Y(y).$$

From Eq. (3.3.9), therefore,[32]

$$p_{X,Y}(x, y) = p_X(x) p_Y(y). \tag{3.3.10}$$

Example 3.32. Autoroute traffic. The number X of vehicles that can travel a multilane autoroute in an hour is x_1 or $x_2 < x_1$, where x_2 depends on the probability p that one lane is closed for maintenance. Let the random variables Y_1 and Y_2 represent the numbers of vehicles per hour during the peak hours commencing at 4 and 5 p.m., respectively. Also, let A represent the failure of the system to meet the peak demands.

We write A in terms Y_1 and Y_2,

$$\Pr[A] = 1 - \Pr[Y_1 \le X, Y_2 \le X],$$

that is, the complementary probability.[33] By using the conditional probability concept and considering the possible closure of one lane between 4 and 6 p.m., this expression can be expanded as

$$\begin{aligned}\Pr[A] &= 1 - \{\Pr[Y_1 \le x_1, Y_2 \le x_1 \mid X = x_1]\Pr[X = x_1] \\ &\quad + \Pr[Y_1 \le x_2, Y_2 \le x_2 \mid X = x_2]\Pr[X = x_2]\} \\ &= 1 - [F_{Y_1,Y_2|X}(x_1, x_1, x_1) p_X(x_1) + F_{Y_1,Y_2|X}(x_2, x_2, x_2) p_X(x_2)].\end{aligned}$$

The autoroute capacity X is assumed to be independent of the Y_is, thus

$$\begin{aligned}\Pr[A] &= 1 - \{\Pr[Y_1 \le x_1, Y_2 \le x_1]\Pr[X = x_1] + \Pr[Y_1 \le x_2, Y_2 \le x_2]\Pr[X = x_2] \\ &= 1 - [F_{Y_1,Y_2}(x_1, x_1) p_X(x_1) + F_{Y_1,Y_2}(x_2, x_2) p_X(x_2)].\end{aligned}$$

If we assume that the Y_is are (1) independent, that is,

$$F_{Y_1,Y_2}(x_1, x_1) = \Pr[Y_1 \le x_1, Y_2 \le x_1] = \Pr[Y_1 \le x_1]\Pr[Y_2 \le x_1] = F_{Y_1}(x_1) F_{Y_2}(x_1),$$

and (2) identically distributed,

$$F_{Y_1,Y_2}(x_1, x_1) = [F_Y(x_1)]^2 \quad \text{and} \quad F_{Y_1,Y_2}(x_2, x_2) = [F_Y(x_2)]^2,$$

[32] The result corresponds with Eq. (2.2.14).

[33] See Eq. (2.2.5).

then

$$\Pr[A] = 1 - \{[F_Y(x_1)]^2 p_X(x_1) + [F_Y(x_2)]^2 p_X(x_2)\}.$$

Since $p_X(x_2) = p$ and $p_X(x_1) = (1 - p)$, we obtain

$$\Pr[A] = 1 - \{(1 - p)[F_Y(x_1)]^2 + p[F_Y(x_2)]^2\}.$$

The distribution of Y is assumed to be

$$F_Y(y) = \sum_{k=0}^{y} \frac{v^k e^{-v}}{k!}$$

(as in Example 3.31 for the numbers of inundated pixels). We can then write

$$\Pr[A] = 1 - (1 - p)\left(\sum_{k=0}^{x_1} \frac{v^k e^{-v}}{k!}\right)^2 - p\left(\sum_{k=0}^{x_2} \frac{v^k e^{-v}}{k!}\right)^2.$$

This approach can be used to determine the failure of a system that is subject to a succession of n random demands (for example, extreme winds, flows, sea waves, and so on). If $X_1, X_2, \ldots, X_n$ are n independent random demands with a common cdf, $F_X(x)$, the failure probability is given by

$$\Pr[A] = 1 - \sum_{\text{all } x} [F_Y(x)]^n p_X(x),$$

where $p_X(x)$ is the pdf of the capacity of the system.

3.3.2 Joint probability distributions of continuous variables

3.3.2.1 Joint pdf and cdf for continuous x and y

In the case of jointly distributed continuous random variables X and Y, the probability distribution is described by the joint probability density function $f_{X,Y}(x, y)$, which is analogous to the bivariate pmf for discrete variables. Probabilities are defined by integration of the joint pdf over the region of interest in the sample space:

$$\Pr[(x_1 \le X \le x_2) \cap (y_1 \le Y \le y_2)] = \int_{x_1}^{x_2} \int_{y_1}^{y_2} f_{X,Y}(x, y) dy dx. \tag{3.3.11}$$

Graphically, this represents the volume under the joint pdf, $f_{X,Y}(x, y)$ over the region of interest as shown schematically in Fig. 3.3.3.

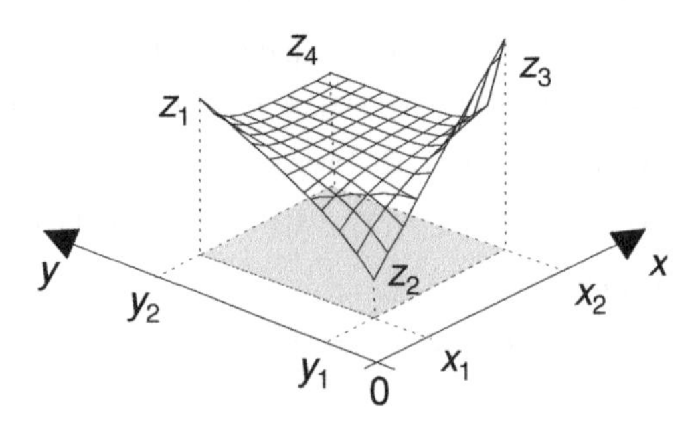

Fig. 3.3.3 Schematic diagram of bivariate pdf represented by heights of a curved surface in the given ranges of the variables.

The joint pdf has the following properties:

$$f_{X,Y}(x, y) \geq 0, \tag{3.3.12a}$$

$$\int_{-\infty}^{+\infty}\int_{-\infty}^{+\infty} f_{X,Y}(x, y)\, dx\, dy = 1. \tag{3.3.12b}$$

Also, the joint cumulative distribution function, cdf, which is analogous to Eq. (3.3.3) for the discrete case and is subject to the same conditions, is defined by

$$F_{X,Y}(x, y) \equiv \Pr[(-\infty \leq X \leq x) \cap (-\infty \leq Y \leq y)] = \int_{-\infty}^{x}\int_{-\infty}^{y} f_{X,Y}(u, v)\, du\, dv. \tag{3.3.13a}$$

This concept can be extended to any number of random variables. In general, if $X_1, X_2, \ldots, X_k$ are variables defined on the same probability space, then $(X_1, X_2, \ldots, X_k)$ is a *k-dimensional continuous random variable* if and only if there exists a function $f_{X_1,X_2,\ldots,X_k}(x_1, x_2, \ldots, x_k) \geq 0$ such that

$$F_{X_1,X_2,\ldots,X_k}(x_1, x_2, \ldots, x_k) \equiv \Pr[(X_1 \leq x_1) \cap (X_2 \leq x_2) \cap \cdots \cap (X_k \leq x_k)]$$

$$= \int_{-\infty}^{x_1}\int_{-\infty}^{x_2} \cdots \int_{-\infty}^{x_k} f_{X_1,X_2,\ldots,X_k}(u_1, u_2, \ldots, u_k) du_1 du_2, \ldots, du_k. \tag{3.3.13b}$$

where $(x_1, x_2, \ldots, x_k)$ is a k-tuple of points in the sample space. We also say that the variates $(X_1, X_2, \ldots, X_k)$ are jointly continuous random variables.

Definition: Joint cumulative distribution function, cdf. The joint cdf of a k-dimensional random variable $(X_1, X_2, \ldots, X_k)$ is defined as the intersection probability of the k-tuple of events $(X_1 \leq x_1)$, $(X_2 \leq x_2), \ldots, (X_k \leq x_k)$ where $(x_1, x_2, \ldots, x_k)$ are points in the k-dimensional sample space of this multidimensional variable.

The following partial derivative, which replaces the derivative in the univariate case for the relationship between the pdf and cdf, should exist:

$$f_{X,Y}(x, y) = \frac{\partial^2}{\partial x \partial y} F_{X,Y}(x, y). \tag{3.3.14a}$$

Note that the cdf is a differentiable function in x and y. In general, we have

$$f_{X_1,X_2,\ldots,X_k}(x_1, x_2, \ldots, x_k) = \frac{\partial^k}{\partial x_1 \partial x_2 \cdot \partial x_k} F_{X_1,X_2,\ldots,X_k}(x_1, x_2, \ldots, x_k). \tag{3.3.14b}$$

Example 3.33. Storm intensity and duration. A storm event occurring at a point in space is characterized by two variables, namely, the duration X of the storm, and its intensity Y, which is defined as the average rainfall rate. The variables X and Y are taken to be distributed as follows, with parameters a and b,

$$F_X(x) = 1 - e^{-ax}, x \geq 0, a > 0; \qquad F_Y(y) = 1 - e^{-by}, \quad y \geq 0, b > 0.$$

It is assumed that the joint cdf of X and Y is given by the exponential bivariate distribution.[34] Thus

$$F_{X,Y}(x, y) = 1 - e^{-ax} - e^{-by} + e^{-ax-by-cxy},$$

[34] See Gumbel (1960) and Bacchi et al. (1987).

with c denoting a parameter describing the joint variability of the two variates. Since $a > 0$ and $b > 0$, let us find the possible values that c can take.

To search for the lower bound for c, we note that for all bivariate distributions, $F_{X,Y}(x, y) \le F_X(x)$, because the joint probability $\Pr[X \le x, Y \le y]$ cannot exceed $\Pr[X \le x]$ independently of the value taken by Y. The inequality

$$F_{X,Y}(x, y) = 1 - e^{-ax} - e^{-by} + e^{-ax-by-cxy} \le 1 - e^{-ax} = F_X(x)$$

yields

$$-x(a + cy) \le 0.$$

Since x and y are always nonnegative, the latter inequality holds if and only if $a + cy \ge 0$.

To search for the upper bound for c, we need to determine the joint pdf. By using Eq. (3.3.14*a*), the joint pdf of X and Y can be found by differentiating the cdf once for x, and then for y. Since

$$\frac{\partial F}{\partial x} = \frac{\partial(1 - e^{-ax} - e^{-by} + e^{-ax-by-cxy})}{\partial x} = ae^{-ax} - (a + cy)e^{-ax-by-cxy},$$

$$f_{X,Y}(x, y) = \frac{\partial^2 F}{\partial x \partial y} = \frac{\partial(ae^{-ax} - (a + cy)e^{-ax-by-cxy})}{\partial y}$$
$$= [(a + cy)(b + cx) - c]e^{-ax-by-cxy}.$$

For $x = y = 0$, the joint pdf at the origin is $f_{X,Y}(0, 0) = ab - c$. Because the pdf is a nonnegative function, the inequality $(ab - c) \ge 0$ must hold; hence, the upper bound of parameter c is $c \le ab$. Therefore, the bivariate exponential distribution is defined for $0 \le c \le ab$.

Note that c is related to the degree of correlation between X and Y. This joint distribution is capable of representing the existing *negative correlation* between the duration and the associated intensity of a storm.[35]

3.3.2.2 Conditional probability density function

By assuming that the relationship for discrete variables used in Eq. (3.3.4) holds for continuous variables,[36] the conditional density function of Y given X is written as

$$f_{Y|X}(y \mid x) = \frac{f_{X,Y}(x, y)}{f_X(x)}. \tag{3.3.15}$$

Hence,

$$f_{X,Y}(x, y) = f_{Y|X}(y \mid x) f_X(x) = f_{X \mid Y}(x \mid y) f_Y(y). \tag{3.3.16}$$

Example 3.34. Storm intensity and duration. In Example 3.33, the duration X of a storm and its average intensity Y are assumed to be jointly distributed variates with bivariate exponential pdf

$$f_{X,Y}(x, y) = [(a + cy)(b + cx) - c]e^{-ax-by-cxy}.$$

where $a > 0$, $b > 0$, and $0 \le c \le 1$ are three parameters to be evaluated from rainfall data. From the data collected at the rain gauge located at the City Hall of Milan, Italy, the values of $a = 0.05\ h^{-1}$, $b = 1.2$ h/mm, and $c = 0.06\ \text{mm}^{-1}$ were estimated. For the design of a drainage system, let us find the conditional probability that a storm lasting $X = 6$ hours will exceed an average intensity of $Y = 2$ mm/h.

---

[35] See Bacchi et al. (1994).

[36] See Stuart and Ord (1994).

Since the conditional pdf of the storm intensity for a given duration is

$$f_{Y|X}(y \mid x) = \frac{f_{X,Y}(x, y)}{f_X(x)} = \frac{[(a + cy)(b + cx) - c]e^{-ax-by-cxy}}{ae^{-ax}}$$
$$= a^{-1}[(a + cy)(b + cx) - c]e^{-y(b+cx)},$$

the conditional cdf is

$$F_{Y|X}(y \mid x) = \int_0^y a^{-1}[(a + cu)(b + cx) - c]e^{-(b+cx)u}du = 1 - \frac{a + cy}{a}e^{-(b+cx)y},$$

which yields

$$\Pr[Y > 2 \mid X = 6] = 1 - F_{Y|X}(2 \mid 6) = 1 - 1 + \frac{0.05 + 0.06 \times 2}{0.05}e^{-(1.2+0.06\times 6)2} = 0.15.$$

3.3.2.3 Independent continuous random variables

If the events $X = x$ and $Y = y$ are stochastically independent, then $f_{X|Y}(x \mid y) = f_X(x)$ and $f_{Y|X}(y \mid x) = f_Y(y)$. From Eq. (3.3.15), therefore,[37]

$$f_{X,Y}(x, y) = f_X(x)f_Y(y). \tag{3.3.17}$$

The assumption of independence and its justification is important because one needs to simplify the application of probability to engineering problems when several variables are involved, a situation which is encountered frequently.

Example 3.35. Storm intensity and duration. Returning to Example 3.33, let us take both the duration X of a storm and its average intensity Y as exponentially distributed variates with parameter a and b, respectively. If one neglects the joint variability of X and Y, the joint pdf is given by

$$f_{X,Y}(x, y) = f_X(x)f_Y(y) = ae^{-ax}be^{-by} = ab\, e^{-ax-by}.$$

This pdf is that of the bivariate exponential distribution used in Example 3.33 with $c = 0$.

3.3.2.4 Marginal probability density function

Extension of the total probability theorem gives[38]

$$f_X(x) = \int_{-\infty}^{\infty} f_{X|Y}(x|y)f_Y(y)dy = \int_{-\infty}^{\infty} f_{X,Y}(x, y)dy \tag{3.3.18a}$$

and

$$f_Y(y) = \int_{-\infty}^{\infty} f_{Y|X}(y|x)f_X(x)dx = \int_{-\infty}^{\infty} f_{X,Y}(x, y)dx. \tag{3.3.18b}$$

Example 3.36. Storm intensity and duration. In Example 3.33, the duration X of a storm and its average intensity Y are assumed to have a joint bivariate exponential distribution with pdf

$$f_{X,Y}(x, y) = [(a + cy)(b + cx) - c]e^{-ax-by-cxy}.$$

---

[37] See Eqs. (2.2.13a,b) and (2.2.14).

[38] See Eq. (2.2.15).

From Eq. (3.3.18*a*) therefore

$$f_X(x) = \int_0^\infty f_{X,Y}(x, y)dy = \int_0^\infty [(a + cy)(b + cx) - c]e^{-ax-by-cxy}dy = ae^{-ax},$$

and from Eq. (3.3.18*b*)

$$f_Y(y) = \int_0^\infty f_{X,Y}(x, y)dx = \int_0^\infty [(a + cy)(b + cx) - c]e^{-ax-by-cxy}dx = be^{-by}.$$

This expression confirms the initial assumptions made in Example 3.33 regarding the marginal distributions.

Example 3.37. Density and compressive strength of concrete. As a further application of the joint pdf for continuous variables, of interest to a civil engineer, consider the densities and compressive strengths of concrete given in Table E.1.2 and ranked in Table 1.2.1. For these data, the bivariate histogram is shown in Fig. 3.3.4.

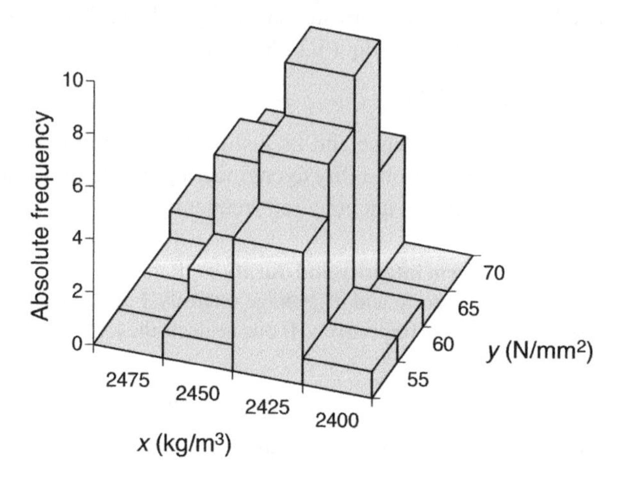

Fig. 3.3.4 Bivariate histogram of compressive strength Y and density X of concrete from Table 1.2.1.

Let us assume that the density X has a uniform marginal distribution and the compressive strength Y has a triangular marginal distribution. The bivariate distribution is written as

$$\begin{aligned} f_{X,Y}(x, y) &= \frac{1}{2000}\frac{y - 40}{20}, && \text{for } 2400 \le x \le 2500 \quad \text{and} \quad 40 \le y \le 60, \\ &= \frac{1}{2000}\left(1 - \frac{y - 60}{20}\right), && \text{for } 2400 \le x \le 2500 \quad \text{and} \quad 60 \le y \le 80, \\ &= 0, && \text{elsewhere,} \end{aligned}$$

and is shown in Fig. 3.3.5.

From Eq. (3.3.18*b*), the marginal distribution of Y is

$$\begin{aligned} f_Y(y) &= \int_{2400}^{2500} \frac{1}{2000}\frac{y - 40}{20}dx = 0.05\frac{y - 40}{20}, && \text{for } 40 \le y \le 60; \\ &= \int_{2400}^{2500} \frac{1}{2000}\left(1 - \frac{y - 60}{20}\right)dx = 0.05\left(1 - \frac{y - 60}{20}\right), && \text{for } 60 \le y \le 80; \\ &= 0, && \text{elsewhere.} \end{aligned}$$

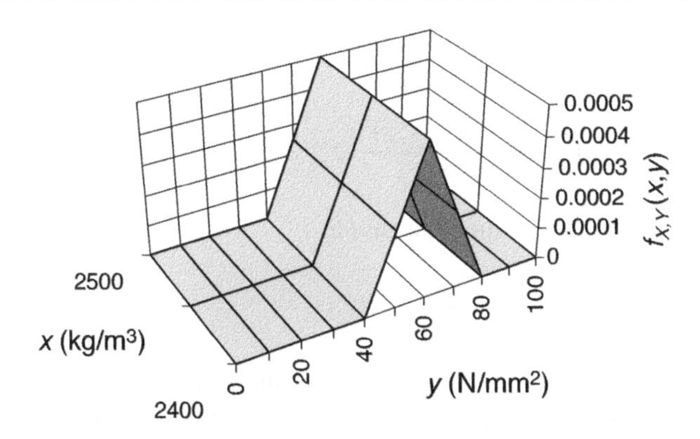

Fig. 3.3.5 Assumed bivariate pdf of compressive strength Y and density X of concrete from Table 1.2.1.

The marginal distribution of X is more easily obtained as follows from Eq. (3.3.18*a*):

$$\begin{aligned} f_X(x) &= \frac{1}{100}, && \text{for } 2400 \le x \le 2500; \\ &= 0, && \text{elsewhere.} \end{aligned}$$

The marginal distributions of X and Y are shown in Fig. 3.3.6*a* and 3.3.6*b*, respectively.
The conditional distribution of Y given X is

$$\begin{aligned} f_{Y|X}(y \mid x) &= \frac{f_{Y,X}(y,x)}{f_X(x)} = \frac{100}{2000}\frac{y-40}{20} = 0.05\frac{y-40}{20}, && \text{for } 40 \le y \le 60; \\ &= \frac{100}{2000}\left(1 - \frac{y-60}{20}\right) = 0.05\left(1 - \frac{y-60}{20}\right), && \text{for } 60 \le y \le 80. \\ &= 0, && \text{elsewhere.} \end{aligned}$$

This result is the same as the marginal distribution of Y, because X has a uniform distribution and thus the joint distribution is not a function of X. We see that the information from the simplified probability densities has not altered an engineer's knowledge of the distribution of the compressive strengths.

3.3.2.5 Joint distributions involving more than two variables

The foregoing concepts can easily be extended to three or more variables. When there are three variables involved, for example, Eq. (3.3.14*b*) is replaced by

$$f_{X,Y,Z}(x,y,z) = \frac{\partial^3}{\partial x \partial y \partial z} F_{X,Y,Z}(x,y,z) \tag{3.3.19}$$

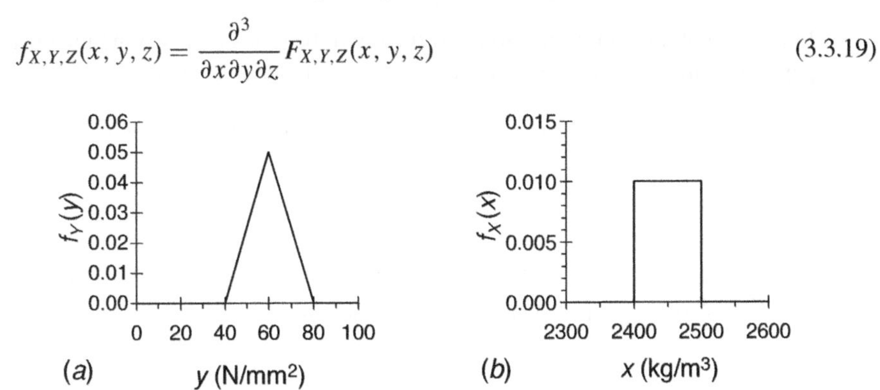

Fig. 3.3.6 Marginal pdfs of (*a*) the compressive strength of concrete Y and of (*b*) the density of concrete X approximated from Table 1.2.1.

to give the partial derivatives representing the joint pdfs. Also, conditional pdfs [Eq. (3.3.15)] are written as

$$f_{X|Y,Z}(x \mid y, z) = \frac{f_{X,Y,Z}(x, y, z)}{f_{Y,Z}(y, z)} \tag{3.3.20}$$

in the simple case, and in the joint case as

$$f_{X,Y|Z}(x, y|z) = \frac{f_{X,Y,Z}(x, y, z)}{f_Z(z)}. \tag{3.3.21}$$

If events $X = x$, $Y = y$, and $Z = z$, for example, are independent, the right-hand side Eq. (3.2.20) will be simply $f_X(x)$. Furthermore, it easily follows that conditional pdfs become marginal pdfs in the simple case [Eq. (3.2.20)], and in the joint case Eq. (3.3.21) becomes a corresponding joint pdf.

3.3.3 Properties of multiple variables

In Section 3.2 the fundamental properties and measures of a random variable were introduced. These properties are also applicable to each compenent of a multidimensional variable. Moreover, some additional properties and measures can be introduced to describe the joint variability of two or more components of a multiple variable. These properties are determined by using the concept of expectation.

3.3.3.1 Covariance and correlation

The expectation operator introduced in Subsection 3.2.1 for a single variable can be extended to two or more variables. For example, the expectation of a linear combination of two variables is given by

$$E[aX_1 + bX_2] = aE[X_1] + bE[X_2], \tag{3.3.22}$$

for any constants a and b. This result follows from the linearity property of the expectations operation and can be applied to more than two variables.

The variance of the sum of two random variables is given by the sum of terms representing the variance of each of the variables and a third term representing their covariance:

$$\text{Var}[aX_1 + bX_2] = a^2\text{Var}[X_1] + b^2\text{Var}[X_2] + 2ab\text{Cov}[X_1, X_2]. \tag{3.3.23}$$

In the case of three or more variables, there will be terms representing the covariances of pairs of all the variables plus the variances of each variable. Also,

$$\text{Cov}[X_1, X_2] = E[(X_1 - E[X_1])(X_2 - E[X_2])] = E[X_1X_2] - E[X_1]E[X_2]. \tag{3.3.24}$$

and

$$E[X_1X_2] = \int_{-\infty}^{+\infty}\int_{-\infty}^{+\infty} x_1x_2 f_{X_1,X_2}(x_1, x_2)\, dx_1\, dx_2. \tag{3.3.25}$$

Definition: Covariance, Cov[X_1, X_2]. The covariance of two random variables is defined as the expectation of the product between the respective deviations from their mean.

If X_1 and X_2 are independent,

$$E[X_1X_2] = \int_{-\infty}^{+\infty}\int_{-\infty}^{+\infty} x_1 f_{X_1}(x_1)x_2 f_{X_2}(x_2)\,dx_1\,dx_2$$

$$= \left[\int_{-\infty}^{+\infty} x_1 f_{X_1}(x_1)\,dx_1\right]\left[\int_{-\infty}^{+\infty} x_2 f_{X_2}(x_2)\,dx_2\right] = E[X_1]E[X_2]. \quad (3.3.26)$$

Hence if X_1 and X_2 are independent,

$$\text{Var}[aX_1 + bX_2] = a^2\text{Var}[X_1] + b^2\text{Var}[X_2], \quad \text{because} \quad \text{Cov}[X_1, X_2] = 0.$$

It is important to note that if the condition $\text{Cov}[X_1, X_2] = 0$ is shown to be true, it does not necessarily follow that X_1 and X_2 are independent. It is only when either Eq. (3.3.10) or Eq. (3.3.17) holds are X_1 and X_2 independent.

Example 3.38. Storm intensity and duration. In Example 3.35 the joint density function of independently distributed storm duration X and intensity Y was found to be

$$f_{X,Y}(x, y) = f_X(x)f_Y(y) = ae^{-ax}be^{-by} = ab\,e^{-ax-by},$$

where a^{-1} is the mean of X, and b^{-1} is that of Y. Using Eq. (3.3.25) one gets

$$E[XY] = \int_0^\infty\int_0^\infty xyf_{X,Y}(x, y)dxdy = \int_0^\infty\int_0^\infty xyab\,e^{-ax-by}dxdy = \frac{1}{ab},$$

which yields

$$\text{Cov}[X, Y] = E[XY] - E[X]E[Y] = \frac{1}{ab} - \frac{1}{a}\frac{1}{b} = 0.$$

This confirms that there is no covariance between the two variables X and Y which are assumed to be independent.

It can be seen from Eq. (3.3.24) that $\text{Cov}[X_1, X_2]$ is large and positive when X_1 and X_2 tend to be both large, or both small, with respect to their means. On the other hand, if when one variable is large and the other tends to be small, the covariance is large and negative. Furthermore, if there is no relationship between the two variables, the covariance does not exist. Thus, $\text{Cov}[X_1, X_2]$ is a measure of the linear interrelationship between X_1 and X_2.

The coefficient of linear correlation ρ is the normalized covariance between two variables, say, X_1 and X_2:

$$\rho = \frac{\text{Cov}[X_1, X_2]}{\sigma_{X_1}\sigma_{X_2}}. \quad (3.3.27)$$

It can be shown that[39]

$$-1 \leq \rho \leq +1.$$

Example 3.39. Water supply. Two neighboring communities need extra water, for industrial and other purposes, beyond that normally delivered from surface sources. These supplementary supplies are made available by pumping from wells having a maximum total output

[39] See, for example, Popoulis (2001). For purposes of estimation from given data, the corresponding sample correlation coefficient is given by Eq. (1.4.3).

of two units per day; each unit is equivalent to 10^6 L. On account of unforeseen demands, which fluctuate from day to day, the pumped water to the two communities, X and Y, say, are treated as random variables with equal marginal pdfs. The bivariate pdf is given by

$$f_{X,Y}(x, y) = \frac{3}{4}(2 - x - y), \quad \text{for} \quad 0 \leq x, y \leq 2$$

as estimated from a joint histogram of observed data. After this function is verified initially for its validity as a pdf, it is of interest to determine the corrrelation ρ between X and Y.

The bivariate pdf is defined only within the triangular space bounded by the x and y axes (because supplies are positive by definition) and the straight line $x + y = 2 \times 10^6$ as stipulated. This is shown in Fig. 3.3.7.

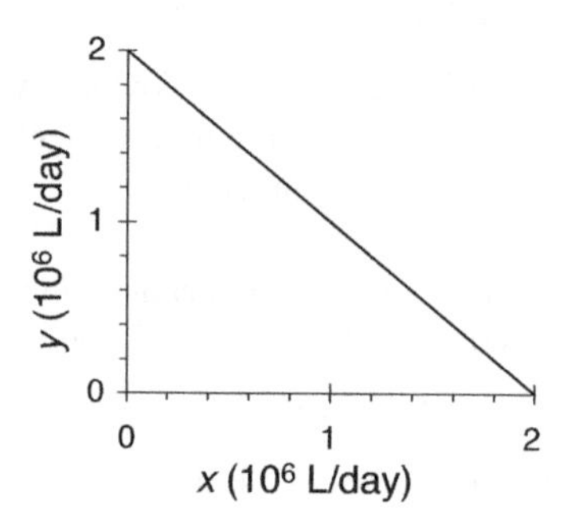

Fig. 3.3.7 Region of definition of bivariate pdf for supplementary water supplies X and Y to two regions.

A double integral is applied to the bivariate pdf with the first variable X bounded by 0 and 2. On account of the given constraint, the second variable Y should be bounded by 0 and $(2 - x)$. Hence,

$$\int_{x=0}^{x=2}\int_{y=0}^{y=2-x} \frac{3}{4}(2 - x - y)dydx = \frac{3}{8}\int_{0}^{2}(2 - x)^2 dx = \frac{3}{8}\left[4x - 2x^2 + \frac{x^3}{3}\right]_0^2 = 1.$$

This proves that the volume represented by the pdf with form and limits as specified is unity [as in Eq. (3.3.12*b*)]. The bivariate pdf is shown in Fig. 3.3.8.

It is a linear function bounded by four planes. These are the horizontal plane on which the two axes lie, the two vertical planes passing through the axes (considering only the positive quadrant) and an inclined plane with the highest elevation at 3/2 units above the origin (corresponding to the maximum pdf of this magnitude) and sloping linearly to zero elevation

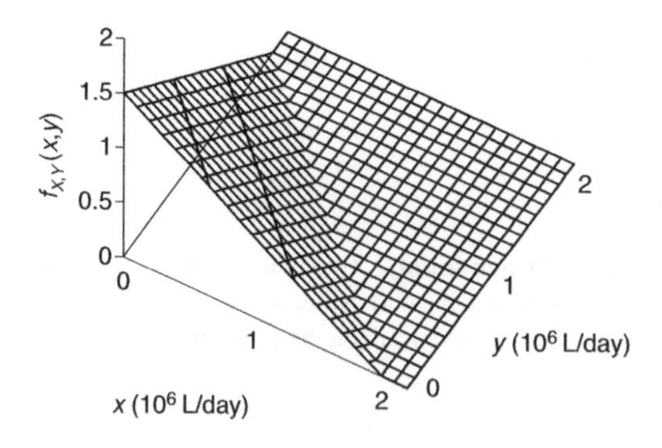

Fig. 3.3.8 Bivariate pdf for supplementary water supplies X and Y to two regions.

along the straight line $x + y = 2$ on the horizontal plane as shown in Figs. 3.3.7 and 3.3.8. (It is easy to verify geometrically that the enclosed volume is unity.)

The marginal pdf of X is given by

$$f_X(x) = \frac{3}{4}\int_0^{2-x}(2 - x - y)dy = \frac{3}{8}(2 - x)^2, \quad \text{for } 0 \le x < 2,$$
$$= 0, \quad \text{elsewhere.}$$

Likewise, the marginal pdf of Y is

$$f_Y(y) = \frac{3}{8}(2 - y)^2, \quad \text{for } 0 \le y < 2,$$
$$= 0, \quad \text{elsewhere.}$$

From Eq. (3.2.2)

$$E[X] = \int_{-\infty}^{\infty} x f_X(x)dx = \int_0^2 \frac{3}{8}x(2 - x)^2 dx = \frac{3}{8}\left[2x^2 - \frac{4x^3}{3} + \frac{x^4}{4}\right]_0^2 = \frac{1}{2} = E[Y]$$

and

$$E[X^2] = \int_{-\infty}^{\infty} x^2 f_X(x)dx = \int_0^2 \frac{3}{8}x^2(2 - x)^2 dx$$
$$= \frac{3}{8}\left[\frac{4x^3}{3} - x^4 + \frac{x^5}{5}\right]_0^2 = \frac{2}{5} = E[Y^2].$$

From Eq. (3.3.25)

$$E[XY] = \int_{x=0}^{x=2} x \int_{y=0}^{y=2-x} \frac{3}{4}y(2 - x - y)dydx = \int_0^2 \frac{1}{8}x(2 - x)^3 dx$$
$$= \frac{1}{8}\left[4x^2 - 4x^3 + \frac{3x^4}{2} - \frac{x^5}{5}\right]_0^2 = \frac{1}{5}.$$

From Eq. (3.3.24)

$$\text{Cov}[X, Y] = \left(\frac{1}{5}\right) - \left(\frac{1}{2}\right)\left(\frac{1}{2}\right) = -\frac{1}{20}.$$

From Eq. (3.2.12)

$$\text{Var}[X] = \text{Var}[Y] = \left(\frac{2}{5}\right) - \left(\frac{1}{2}\right)^2 = \frac{3}{20},$$

and from Eq. (3.3.27) the coefficient of linear correlation is

$$\rho = \frac{(-1/20)}{(\sqrt{3/20}\sqrt{3/20})} = -\frac{1}{3}.$$

On the basis of the assumed model, this shows that there is negative correlation between the excess water supplies to the two communities. This is a consequence of the constraint of the total maximum output from the wells supplying the communities.

3.3.3.2 Joint moment-generating function

Similar to the way in which one specifies the moment-generating function of a random variable [see Eq. (3.2.16)], one can define the joint moment-generating function of a multiple variable $X_1, X_2, \ldots, X_n$ as

$$M_{X_1,\ldots,X_k}(t_1, \ldots, t_k) = E\left[\exp\left(\sum_{i=1}^{k} t_i X_i\right)\right]. \qquad (3.3.28)$$

Thus the rth moment of X_i can be determined by differentiating the joint moment-generating function r times with respect to t_i and then evaluating the derivative with $t_i = 0$. Also, the mixed moments of X_i and X_j, say, $E[X_i^r X_j^s]$, can be generated from the joint mgf by differentiating r times with respect to t_i and s times with respect to t_j, and then letting $t_i = t_j = 0$.

If X and Y are statistically *independent*,

$$M_{X,Y}(t_1, t_2) = E[e^{t_1 X + t_2 Y}] = E[e^{t_1 X}]E[e^{t_2 Y}] = M_X(t_1)M_Y(t_2), \qquad (3.3.29)$$

and it can be shown that if the joint mgf of two variables equals the product of their individual mgfs, the two variables are statistically independent.

Example 3.40. Storm intensity and duration. From Example 3.38 the joint density function of independently distributed duration X of a storm and average intensity Y is given by

$$f_{X,Y}(x, y) = abe^{-ax-by}.$$

This is shown in Fig. 3.3.9 for particular values of parameters.

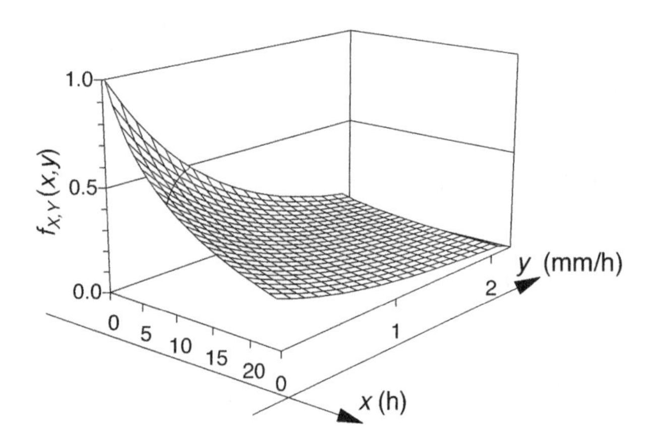

Fig. 3.3.9 Joint pdf of storm intensity Y and storm duration X. The values taken by the parameters are those given in Example 3.34 ($a = 0.05$ h^{-1}, $b = 1.2$ h/mm, $c = 0.06$ mm^{-1}).

The joint mgf of X and Y is given by

$$M_{X,Y}(t_1, t_2) = E[e^{t_1 X + t_2 X}] = \frac{ab}{(a - t_1)(b - t_2)},$$

which equals the product of the mgfs of the two variables with means $1/a$ and $1/b$, respectively (see Example 3.17 for the derivation of the univariate mgf).

3.3.3.3 Conditional expectation

Similar to the way in which one specifies the expectation or mean of a random variable [see Eqs. (3.2.1) and (3.2.2)], one can define the conditional expectation following the concepts of conditional distributions already introduced in this chapter.

If X and Y are discrete random variables with a bivariate pmf $p_{X,Y}(x, y)$, the conditional mean of X for a given value of Y and vice versa are defined, respectively, as follows:

$$E[X \mid Y = y_j] = \sum_{\text{all } i} x_i p_{X|Y}(x_i \mid y_j) \tag{3.3.30}$$

and

$$E[Y \mid X = x_i] = \sum_{\text{all } j} y_j p_{Y|X}(y_j \mid x_i). \tag{3.3.31}$$

Example 3.41. Wind measurements. Returning to the data in Table 3.3.1 regarding joint probabilities of the occurrences of specified high winds per year recorded by two measuring devices, consider the case when the less accurate measuring device with output represented by the random variable Y records a value of zero. The joint probabilities are 0.2910, 0.0400, 0.0100, and 0.0005 when the random variable X, which corresponds to the accurate measuring device, takes values of 0, 1, 2, and 3, respectively. Dividing these by the marginal distribution $p_Y(0) = 0.3415$, one obtains the conditional distributions, $p_{X|Y}(x \mid 0)$, 0.8521, 0.1171, 0.0293, and 0.0015 for $X = 0$, 1, 2, and 3, respectively. Hence, from Eq. (3.3.30)

$$E[X \mid Y = 0] = 0 \times 0.8521 + 1 \times 0.1171 + 2 \times 0.0293 + 3 \times 0.0015 = 0.1802.$$

For the case $Y = 3$, the conditional distributions $p_{X|Y}(x|3)$ are 0.0000, 0.0000, 0.3727, and 0.6273, respectively. Then

$$E[X \mid Y = 3] = 0 \times 0.0000 + 1 \times 0.0000 + 2 \times 0.3727 + 3 \times 0.6273 = 2.6273.$$

These two numbers are the mean (or expected) values of the true number of specified winds when the less accurate instrument records 0 and 3 occurrences, respectively, per year at the measuring station. The conditional distributions and expectations of X for all values of Y are given in Table 3.3.2.

The last row gives marginal probabilities of Y given X, obtained from Table 3.3.1 and used in the preceding calculations.

If X and Y are statistically *independent*, it follows from Eq. (3.3.10) that the two conditional expectations of Eqs. (3.3.30) and (3.3.31) reduce to $E[X]$ and $E[Y]$, respectively.

Table 3.3.2 Conditional distributions and expectations, evaluated from Table 3.3.1, of the number of occurrences of specified winds per year X given numbers Y recorded by a less accurate instrument

	$Y = 0$	$Y = 1$	$Y = 2$	$Y = 3$
$X = 0$	0.8521	0.1350	0.0000	0.0000
$X = 1$	0.1171	0.8054	0.0749	0.0000
$X = 2$	0.0293	0.0562	0.8502	0.3727
$X = 3$	0.0015	0.0034	0.0749	0.6273
Σ	1.0000	1.0000	1.0000	1.0000
$E[X \mid Y = y]$	0.1802	0.9280	2.0000	2.6273
$p_Y(y)$[a]	0.3415	0.4445	0.1335	0.0805

[a] The last row gives marginal probabilities of Y given X, obtained from Table 3.3.1.

This result is also obtained in determining the *expectation of the conditional expectation*, for example, of the random variable $E[X \mid Y]$:

$$E[E[X \mid Y]] = \sum_{\text{all } y_j} E[X \mid Y = y_j] p_Y(y_j) \tag{3.3.32a}$$

[where, unlike in Eq. (3.3.4), Y does not take a *single fixed* value]. From Eq. (3.3.30),

$$E[E[X \mid Y]] = \sum_{\text{all } y_j} \sum_{\text{all } x_i} x_i p_{X|Y}(x_i \mid y_j) p_Y(y_j), \tag{3.3.32b}$$

and from Eqs. (3.3.4), (3.4.6), and (3.2.1)

$$E[E[X \mid Y]] = \sum_{\text{all } x_i} x_i p_X(x_i) = E[X]. \tag{3.3.32c}$$

Thus, from Eq. (3.3.32*a*) and (3.3.32*c*) one can also state that

$$E[X] = \sum_{\text{all } y_j} E[X \mid Y = y_j] p_Y(y_j). \tag{3.3.33}$$

(This expression is simply a further application of the total probability theorem of Chapter 2.)

Example 3.42. Wind measurements. From Table 3.3.2,

$$\begin{aligned} E[X] &= 0.1802 \times 0.3415 + 0.9280 \times 0.4445 + 2 \times 0.1335 \\ &\quad + 2.6273 \times 0.0805 = 0.9525. \end{aligned}$$

This is the expected (that is mean) number of specified winds per year at the recording station.

Turning now to the continuous case, the conditional expectation of X given that $Y = y$ is

$$E[X \mid Y = y] = \int_{-\infty}^{+\infty} x f_{X|Y}(x \mid y) dx. \tag{3.3.34}$$

Also, it follows from Eq. (3.3.33) that

$$E[X] = \int_{-\infty}^{+\infty} E[X \mid Y = y] f_Y(y) dy. \tag{3.3.35}$$

Example 3.43. Density and compressive strength of concrete. Consider again the densities and compressive strengths of concrete given in Table 1.2.1 and the assumed model of Example 3.37. The expected value of Y for a value of X in the range 2400–2500 kg/m^3 is obtained using Eq. (3.3.24) by substituting the conditional distribution of Y given X of Example 3.37 as follows:

$$\begin{aligned} E[Y \mid X = x] &= \int_{40}^{60} 0.05y \frac{y - 40}{20} dy + \int_{60}^{80} 0.05y \frac{80 - y}{20} dy \\ &= 0.0025 \left[\frac{y^3}{3} - 20y^2 \right]_{40}^{60} + 0.0025 \left[40y^2 - \frac{y^3}{3} - \right]_{60}^{80} \\ &= 26.67 + 33.33 = 60 \text{ N/mm}^2. \end{aligned}$$

Also, the (unconditional) mean value of Y is obtained as follows using the marginal pdf of X, which is uniform in the range 2400–2500 kg/m^3 and equal to 0.01, and zero elsewhere as shown in Fig. 3.3.6*b*:

$$E[Y] = \int_{2400}^{2500} \frac{60}{100} dx = 60 \text{ N/mm}^2.$$

This result of course is expected if one assumes that the marginal distribution for Y is triangular in the range 40–80 with peak at 60 N/mm^2 and is zero elsewhere. Although a simplified distribution was adopted here, the result is only marginally different from the sample mean, which is 60.14 N/mm^2, as given in Table 1.2.2.

Example 3.44. Storm intensity and duration. Returning to Example 3.34, the conditional expectation of storm intensity Y for a given duration X of the storm is obtained [from Eq. (3.3.34)] as

$$E[Y \mid X = x] = \int_0^\infty \frac{y}{a}[(a + cy)(b + cx) - c]e^{-y(b+cx)} dy = \frac{b + cx + c/a}{(b + cx)^2}.$$

(We may write $d = b + cx$ and solve the integral as in Examples 3.9 and 3.14.)

We note that the conditional expectation of storm intensity decreases with increasing duration of the storm.

Similarly, we obtain the conditional moment of second order of Y as a function of x:

$$E[Y^2 \mid X = x] = \int_0^\infty \frac{y^2}{a}[(a + cy)(b + cx) - c]e^{-y(b+cx)} dy = \frac{2(b + cx + 2c/a)}{(b + cx)^3},$$

which gives the conditional variance

$$\text{Var}[Y \mid X = x] = \frac{1}{(b + cx)^2} + \frac{c(2ab + 2acx - c)}{a^2(b + cx)^4}.$$

The corresponding conditional standard deviation is

$$\sigma_{Y|X=x} = \sqrt{\text{Var}[Y \mid X = x]} = \frac{1}{(b + cx)}\sqrt{1 + \frac{c(2ab + 2acx - c)}{a^2(b + cx)^2}},$$

which decreases less rapidly than the conditional expectation for increasing duration. Therefore, the variability of the conditional storm intensity increases with increasing storm durations. The conditional mean and standard deviation of the storm intensity Y are shown as a function of storm duration X in Fig. 3.3.10 with the values of parameters of Example 3.34.

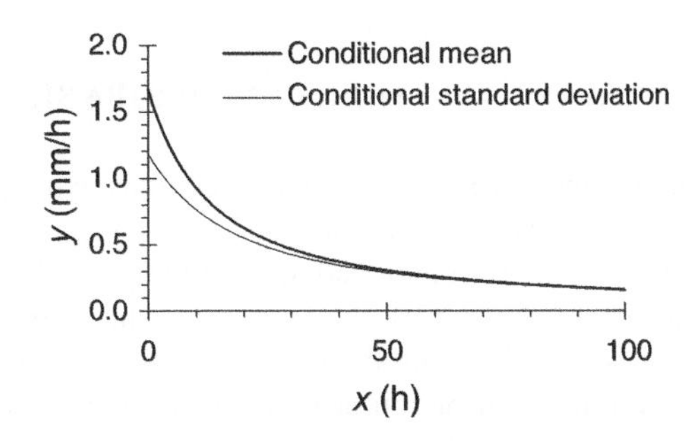

Fig. 3.3.10 Conditional mean and standard deviation of storm intensity Y as related to the duration of the storm X. The values taken by the parameters are those given in Example 3.34 ($a = 0.05$ h^{-1}, $b = 1.2$ h/mm, $c = 0.06$ mm^{-1}).

By using the properties of expectation, some further properties of the conditional expectation can be obtained. For instance, Eq. (3.3.32*c*) states that the mean of a variable X is the expectation or mean of the conditional mean of X, that is,

$$E[X] = E[E[X \mid Y]]. \tag{3.3.36}$$

More generally,

$$E[g(X)] = E[E[g(X) \mid Y]]. \tag{3.3.37}$$

where $g(\cdot)$ is a function with argument X, and the conditional expectation $E[g(X) \mid Y]$ is generally a function of Y. The conditional variance of X given $Y = y$ is defined as

$$\text{Var}[X \mid Y] = E[X^2 \mid Y] - E^2[X \mid Y] \tag{3.3.38}$$

and application of Eq. (3.3.37) to $E[\text{Var}[X \mid Y]]$ gives

$$\begin{aligned} E[\text{Var}[X \mid Y]] &= E[E[X^2 \mid Y]] - E[E^2[X \mid Y]] \\ &= E[X^2] - E^2[X] - E[E^2[X \mid Y]] + E^2[X] \\ &= \text{Var}[X] - E[E^2[X \mid Y]] + E^2[E[X \mid Y]] \\ &= \text{Var}[X] - \text{Var}[E[X \mid Y]]. \end{aligned}$$

Hence,

$$\text{Var}[X] = E[\text{Var}[X \mid Y]] + \text{Var}[E[X \mid Y]]. \tag{3.3.39}$$

Accordingly, the variance of X is the expectation or mean of the conditional variance of X, augmented by the variance of the conditional mean of X. If two variables X and Y are independent, and the conditional mean $E[X \mid Y] = \mu_X$ does not depend on Y, then $E[\text{Var}[X \mid Y]] = \text{Var}[X]$.

3.3.4 Summary of Section 3.3

Following the discussion of random variables and the univariate applications in Sections 3.1 and 3.2, the basic concepts of probability mass and density functions for multiple random variables are introduced here. Joint cumulative distribution functions and some important properties of multiple variables are also discussed.

3.4 ASSOCIATED RANDOM VARIABLES AND PROBABILITIES

Engineers are often concerned with functional relationships that exist between variables so that if the value of one variable is observed then the value of another can be predicted. Such relationships are widely sought and applied in hydraulics, soil mechanics, strength of materials, and other branches of civil and environmental engineering. Consider the case of two variables X and Y. Initially one establishes the form of association between X and Y; then, given the distribution or probability law that governs X, one determines the distribution of Y by a procedure given in this section. Furthermore, if another variable Z has a relationship with X and Y, it is possible to derive the distribution of Z, unless the calculus becomes intractable. These concepts can be extended to the case where several variables are involved.

3.4.1 Functions of a random variable

One can obtain the pmf or pdf for a function of a random variable X by applying a transformation technique. The transformation is usually obtained readily in the case of discrete variables. For example, let the random variable X with pmf $p_X(x)$ be transformed such that $Y = g(X)$. It is necessary to find the inverse function of Y which may be written as

$$X = g^{-1}(Y) = h(Y), \tag{3.4.1}$$

corresponding to

$$Y = g(X). \tag{3.4.2}$$

Then the pmf of Y, $p_Y(y)$, is given by

$$p_Y(y) = p_X[h(y)], \tag{3.4.3}$$

where $h(y)$ denotes the inverse function such as $x = h(y)$ if $y = g(x)$. In order to apply the transformation, there should be one-to-one correspondence between X and Y, for example, when $y = 1 + 9x$ but not when $y = \sin x$, in which case more than one value of X will give the same value of Y. In other words [if we leave out the often difficult and involved cases of multiple roots of $y = g(x)$], it is required that $h(Y)$ is a monotonic increasing or decreasing function of the variable X.

Example 3.45. Flood occurrence. An example of this is the application of the geometric distribution:

$$p_X(x) = p(1-p)^{x-1}, \quad \text{for } x = 1, 2, 3, \ldots; 0 \le p \le 1,$$

to the occurrences of a design flood at the site of a cofferdam which is to be constructed to protect the work on a large dam across a river. The design flood causes failure of the cofferdam and has a probability of occurrence of p in any year. The foregoing distribution gives the probability that failure will occur in year x and not before. Suppose that because of concern regarding possible failure, the cofferdam originally constructed has been strengthened, so that it may not fail for a much longer period of years. The improvements are such that it is now appropriate to use the function $Y = 3X$ for the purpose of predicting failure.

The inverse function is $X = Y/3$ and the pdf for Y is given by

$$p_Y(y) = p(1-p)^{[(y/3)-1]}, \quad \text{for } y = 3, 6, 9, \ldots; 0 \le p \le 1;$$

$$p_Y(y) = 0, \quad \text{elsewhere.}$$

Comparing $p_X(x) = p(1-p)^{x-1}$ and $p_Y(y) = p(1-p)^{[(y/3)-1]}$, we see that the probabilities of failure under the old scheme X after 1, 2, 3, and 4 years, for example, is equivalent to the probabilities of failure after 3, 6, 9, and 12 years, respectively, under the new scheme Y.

For continuous variables, the transformation from the pdf of X to that of Y involves firstly the substitution of the inverse function of Y solved for X [Eq. (3.4.1)] in the pdf of X, a procedure similar to that adopted in the discrete case [as in Eq. (3.4.3)]. Secondly, the pdf of X so defined should be multiplied by the absolute value of the first derivative of the inverse function, say, $h(y)$. This first derivative is called the *Jacobian* of the transformation and it is denoted by J. Thus,

$$f_Y(y) = \left|\frac{dx}{dy}\right| f_X[h(y)] = \left|\frac{dh(y)}{dy}\right| f_X[h(y)] = |J| f_X[h(y)]. \tag{3.4.4}$$

Multiplication by the absolute J term ensures that $f_Y(y)$ integrates to unity, a requirement for any pdf, as already noted; this is evident by rearranging the left and middle terms of Eq. (3.4.4) and integrating.

Property: The probability density function (pdf) of a derived variable by one-to-one transformation. Let X be a random variable with continuous pdf, $f_X(x)$. The pdf of the random variable Y defined by the one-to-one transformation $Y = g(X)$ is given by

$$f_Y(y) = \left|\frac{dh(y)}{dy}\right| f_X[h(y)] = |J| f_X[h(y)],$$

where $h(y)$ denotes the inverse function such as $x = h(y)$ if $y = g(x)$. This property requires that the Jacobian of the transformation is nonzero.

For example, if $Y = mX + c$, the inverse function is $X = (Y - c)/m$, from which $dx/dy = 1/m$. Hence,

$$f_Y(y) = \left|\frac{1}{m}\right| f_X\left(\frac{y-c}{m}\right). \tag{3.4.5}$$

Example 3.46. Bacterial growth. Consider the population of bacteria, C, in a small lake. Under ideal conditions, the population increases exponentially with time T, commencing with an initial population c_0 as

$$C = c_0 e^{\lambda T},$$

where $\lambda > 0$ is the growth rate. However, because of the uncertain effects of various extraneous factors, the time T allowed for the increase in the bacterial population is a random variable with distribution function $F_T(t)$, where $t \geq 0$. Thus the bacteria immediately prior to flushing has a population with distribution function $F_C(c)$ given by

$$F_C(c) = \Pr[c_0 e^{\lambda T} \leq c] = \Pr\left[\lambda T \leq \ln\frac{c}{c_0}\right] = F_T\left(\frac{1}{\lambda}\ln\frac{c}{c_0}\right),$$

where $c \geq c_0$. Thus,

$$t = \frac{1}{\lambda}\ln\frac{c}{c_0},$$

and the first derivative of the inverse function is

$$\frac{dt}{dc} = \frac{1}{\lambda c}.$$

Therefore, from Eq. (3.4.4),

$$f_C(c) = \frac{1}{\lambda c} f_T\left(\frac{1}{\lambda}\ln\frac{c}{c_0}\right),$$

where $c \geq c_0$. Now if $f_T(t)$ is known, then $f_C(c)$ can be found from the foregoing equation.
For example, if T is exponentially distributed with pdf

$$f_T(t) = ae^{-at},$$

with $a > 0$, one gets

$$f_C(c) = \frac{1}{\lambda c} ae^{-[(a/\lambda)(\ln(c/c_0))]} = \frac{a}{\lambda c}\left(\frac{c_0}{c}\right)^{a/\lambda} = \frac{\theta c_0^{\theta}}{c^{\theta+1}},$$

where $\theta = a/\lambda > 0$, and $c \geq c_0$. The cdf of C is given by[40]

$$F_C(c) = 1 - \left(\frac{c_0}{c}\right)^{\theta},$$

with $\theta > 0$ and $c \geq c_0$.

3.4.2 Functions of two or more variables

3.4.2.1 Sum and difference of two or more variables

Scientists and engineers often encounter more than one random variable in relation to a particular problem, and therefore they seek a function of two or more variables. In the case of two variables X and Y, consider the case $Z = X + Y$.

$$F_Z(z) = \Pr[Z \leq z] = \Pr[X + Y \leq z]$$

$$= \Pr[Y \leq Z - X] = \int_{-\infty}^{\infty} \int_{-\infty}^{z-x} f_{XY}(x, y)\, dy dx \tag{3.4.6}$$

$$\Rightarrow f_Z(z) = \frac{\partial}{\partial z} F_Z(z) = \int_{-\infty}^{+\infty} f_{XY}(x, z - x)\, dx = \int_{-\infty}^{+\infty} f_{XY}(z - y, y)\, dy \tag{3.4.7}$$

If X and Y are independent, it follows from Eq. (3.3.17) that,

$$f_Z(z) = \int_{-\infty}^{\infty} f_X(z - y) f_Y(y) dy. \tag{3.4.8}$$

We can see from Eqs. (3.4.7) and (3.4.8) that the variables X and Y are interchangeable. Equation (3.4.8) in its general form is sometimes known as the *convolution integral.*

Example 3.47. Ferry and train transportation. Consider the case where pedestrians reaching a ferry station have to await the arrival of a ferry. After traveling in the ferry, they must remain at a railway station until a train arrives. Circumstances are such that the times in hours, X and Y, spent by passengers awaiting ferries and trains, respectively, are random variables. Let us assume that for the particular short segments traversed, the travel times in the two modes of transport are constant. The pdfs are given respectively by

$$f_X(x) = 0.7e^{-0.7x} \quad \text{and} \quad f_Y(y) = 0.5e^{-0.5y}.$$

If we also assume that the arrivals of the ferries and trains are independent, the pdf of the total time spent by a passenger in awaiting transport, $Z = X + Y$, is obtained from Eq. (3.4.8) as follows:

$$f_Z(z) = \int_{0}^{\infty} f_X(z - y) 0.5e^{-0.5y} dy,$$

zero being the lower limit of integration because negative times are not possible. Also, the argument $(z - y)$ of f_X cannot be negative; and thus because the variate is Y, the upper limit

[40] This is the Pareto distribution.

of integration becomes z. Hence,

$$f_Z(z) = \int_0^z 0.7e^{-0.7(z-y)}0.5e^{-0.5y}dy = 0.35e^{-0.7z}\int_0^z e^{-0.2y}dy$$

$$= \frac{0.35}{0.7-0.5}(e^{-0.5z} - e^{-0.7z}) = 1.75(e^{-0.5z} - e^{-0.7z}),$$

which gives the pdf of the total time Z in hours spent by a foot passenger at the ferry and train stations. Figure 3.4.1 gives the pdfs of X, Y, and Z.

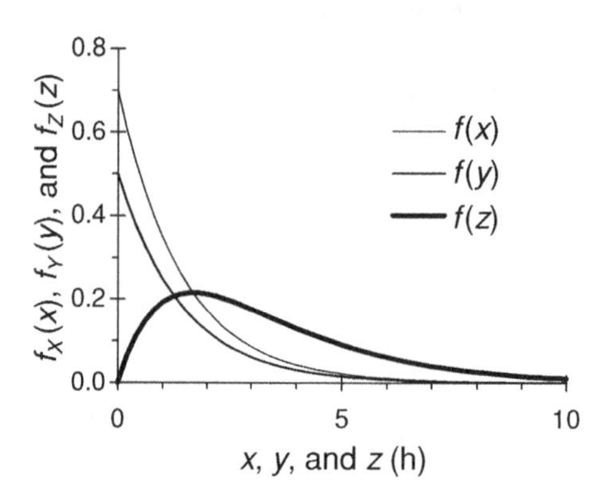

Fig. 3.4.1 Probability density functions of waiting time in hours for ferries, X; trains, Y; and cumulative for ferry plus train, $Z = X + Y$.

If one considers the case of the difference between two variables, say, $W = X - Y$, the marginal pdf of W is found using the same arguments, thus obtaining

$$f_W(w) = \int_{-\infty}^{\infty} f_{W,Y}(w, y)dy = \int_{-\infty}^{\infty} f_{X,Y}(w + y, y)dy. \tag{3.4.9}$$

This yields, for independent X and Y,

$$f_W(w) = \int_{-\infty}^{\infty} f_X(w + y)f_Y(y)dy. \tag{3.4.10}$$

3.4.2.2 Maximum and minimum of two or more variables

In many applications the engineer is concerned with the maximum or the minimum of a certain number of variables. In the case of n variables, say, $X_1, X_2, \ldots, X_n$, define $Y = \max[X_1, X_2, \ldots, X_n]$. The cdf of Y will be given by the joint probability that each variable X_i is less or equal to y, that is,

$$\Pr[Y \le y] = \Pr[X_1 \le y, X_2 \le y, \ldots, X_n \le y] = F_{X_1,X_2,\ldots,X_n}(y, y, \ldots, y)$$

If $X_1, X_2, \ldots, X_n$ are independent of each other, this probability is equal to the product of the individual probabilities, so that

$$F_Y(y) = \prod_{k=1}^{n} \Pr[X_k \le y] = \prod_{k=1}^{n} F_{X_k}(y), \tag{3.4.11}$$

which becomes

$$F_Y(y) = \prod_{k=1}^{n} F_{X_k}(y) = [F_X(y)]^n \tag{3.4.12}$$

for $F_{X_1}(\cdot) = \cdots\ldots = F_{X_n}(\cdot) = F_X(\cdot)$; that is, the n variables have a common probability distribution. The corresponding pdf is found by differentiating Eq. (3.4.12), thus obtaining

$$f_Y(y) = \frac{d}{dy} F_Y(y) = n[F_X(y)]^{n-1} f_X(y). \tag{3.4.13}$$

The same procedure can be used to derive the probability distribution of the minimum, say, $Z = \min[X_1, X_2, \ldots, X_n]$. If the n variables $X_1, X_2, \ldots, X_n$ are independent of each other,

$$F_Z(z) = 1 - \prod_{k=1}^{n} \Pr[X_k > z] = 1 - \prod_{k=1}^{n} [1 - F_{X_k}(z)], \tag{3.4.14}$$

which becomes

$$F_Z(z) = 1 - \prod_{k=1}^{n} [1 - F_{X_k}(z)] = 1 - [1 - F_X(z)]^n, \tag{3.4.15}$$

for $F_{X_1}(\cdot) = \cdots = F_{X_n}(\cdot) = F_X(\cdot)$; that is, the n variables have a common probability distribution. The corresponding pdf is given by

$$f_Z(z) = \frac{d}{dz} F_Z(z) = n[1 - F_X(z)]^{n-1} f_X(z). \tag{3.4.16}$$

Example 3.48. Earthquake intensity. In Example 3.6, and subsequently, the pdf of earthquake intensity in a region was assumed to be exponentially distributed as follows:

$$f_X(x) = \lambda e^{-\lambda t},$$

with $\lambda = 0.2$. If five earthquakes are observed in a century, let us find the probability distribution of earthquakes of the maximum intensity.

If one assumes that the intensities of the earthquakes are independent of each other, Eq. (3.4.12) yields

$$F_Y(y) = [F_X(y)]^n = (1 - e^{-0.2y})^5,$$

and from Eq. (3.4.13)

$$f_Y(y) = n[F_X(y)]^{n-1} f_X(y) = 5(1 - e^{-0.2y})^4 e^{-0.2y}.$$

Also, from Eq. (3.4.15) the minimum earthquake intensity Z has the cdf

$$F_Z(z) = 1 - [1 - F_X(z)]^n = 1 - (e^{-0.2y})^5 = 1 - e^{-y}.$$

This expression has the same type of distribution as the intensity X with parameter $n\lambda = 5 \times 0.2 = 1$.

3.4.2.3 Product and quotient of two random variables

Some engineering problems require the product or the quotient of random variables be evaluated. For example, the safety factor Z of a system with random capacity, subject to a random load Y, is defined as the ratio between capacity and load, that is, $Z = X/Y$. If benefit and costs are random variables, the benefit-to-cost ratio is also a random variable with its distribution depending on the joint distribution of cost and benefit. If X is the demand of the system and Y is the cost per unit demand, the project engineer needs to evaluate the total cost as $Z = XY$. The same occurs if X is a variable measured in the laboratory or in the field and Y is the multiplicative error associated with such a measurement.

In the case of two variables X and Y, the probability distribution of $Z = XY$ is determined by observing that $\Pr[Z \leq z]$ can be obtained by integrating the two dimensional pdf of X and Y for those values of the product xy that are less than or equal to z. This means that $f_{X,Y}(x, y)$ must be integrated over the shaded region shown in Fig. 3.4.2 between the curves and the two axes. Thus,

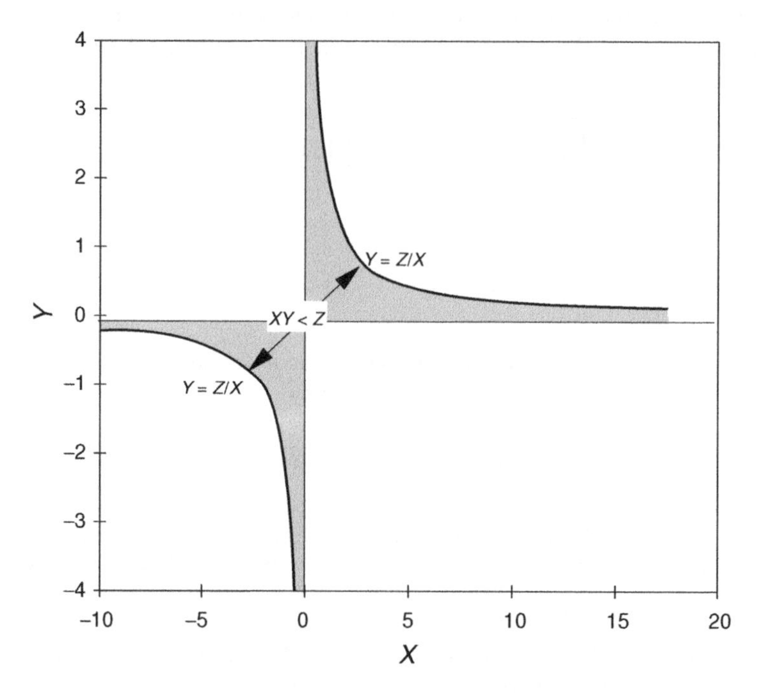

Fig. 3.4.2 Region for integration of the joint pdf of X and Y to obtain the cdf of $Z = XY$.

$$F_Z(z) = \Pr[Z \leq z] = \iint\limits_{xy \leq z} f_{X,Y}(x, y)\, dx dy$$

$$= \int_{-\infty}^{0} \left[\int_{z/x}^{+\infty} f_{X,Y}(x, y) dy \right] dx + \int_{0}^{+\infty} \left[\int_{-\infty}^{z/x} f_{X,Y}(x, y) dy \right] dx$$

and if $t = xy$,

$$F_Z(z) = \int_{-\infty}^{0} \left[\int_{z}^{+\infty} f_{X,Y}\left(x, \frac{t}{x}\right) \frac{dt}{x} \right] dx + \int_{0}^{+\infty} \left[\int_{-\infty}^{z} f_{X,Y}\left(x, \frac{t}{x}\right) \frac{dt}{x} \right] dx$$

$$= \int_{-\infty}^{z} \left[\int_{-\infty}^{+\infty} \frac{1}{|x|} f_{X,Y}\left(x, \frac{t}{x}\right) dx \right] dt,$$

which can be differentiated with respect to z to get

$$f_Z(z) = \int_{-\infty}^{+\infty} \frac{1}{|x|} f_{X,Y}\left(x, \frac{z}{x}\right) dx = \int_{-\infty}^{+\infty} \frac{1}{|y|} f_{X,Y}\left(\frac{z}{y}, y\right) dy. \tag{3.4.17}$$

If X and Y are independent, the product of their marginals can be substituted for the bivariate pdf, thus obtaining

$$f_Z(z) = \int_{-\infty}^{+\infty} \frac{1}{|x|} f_X(x) f_Y\left(\frac{z}{x}\right) dx = \int_{-\infty}^{+\infty} \frac{1}{|y|} f_X\left(\frac{z}{y}\right) f_Y(y) dy. \tag{3.4.18}$$

Example 3.49. Storm intensity and duration. In Example 3.40 it was assumed that the duration X of a storm and its average intensity Y are independent variates with joint pdf

$$f_{X,Y}(x, y) = f_X(x) f_Y(y) = ae^{-ax} be^{-by} = abe^{-ax-by}.$$

Let us determine the probability distribution of the total amount of rainfall delivered in a storm. Define rainfall depth as $Z = XY$.

Using Eq. (3.4.18) one gets

$$F_Z(z) = \int_{-\infty}^{z} \int_{-\infty}^{+\infty} \frac{1}{|y|} f_{X,Y}\left(\frac{\zeta}{y}, y\right) dy d\zeta$$

$$= \int_{0}^{z} \int_{0}^{+\infty} \frac{ab}{y} e^{-[(a\zeta/y)+by]} dy d\zeta = 1 - 2\sqrt{abz} K_1(2\sqrt{abz}),$$

where $K_1(\xi)$ is a modified Bessel function with argument ξ of the first order.[41]

Using the same arguments, one can derive the pdf of the quotient $W = X/Y$ of two random variables X and Y. This is given by

$$f_W(w) = \int_{-\infty}^{+\infty} |y| f_{X,Y}(wy, y) dy. \tag{3.4.19}$$

which yields, for independent X and Y,

$$f_W(w) = \int_{-\infty}^{\infty} |y| f_X(wy) f_Y(y) dy. \tag{3.4.20}$$

Figure 3.4.3 shows, for example, the pdf of the product and quotient of two independent random variables, X and Y, uniformly distributed on 0–1.

3.4.2.4 Transformation of two variables

The procedures for the transformation of single variables given by Eqs. (3.4.1) to (3.4.5) can be applied to two variables (and similarly extended to three or more variables with joint pdfs). Let $f_{X_1,X_2}(x_1, x_2)$ represent the bivariate pdf of X_1 and X_2.

Given a random variable Y, the distribution of which one is interested in, where $Y = g_1(X_1, X_2)$, a new random variable $Z = g_2(X_1, X_2)$ is introduced in order to solve the

[41] See Eagleson (1978) and Abramowitz and Stegun (1964).

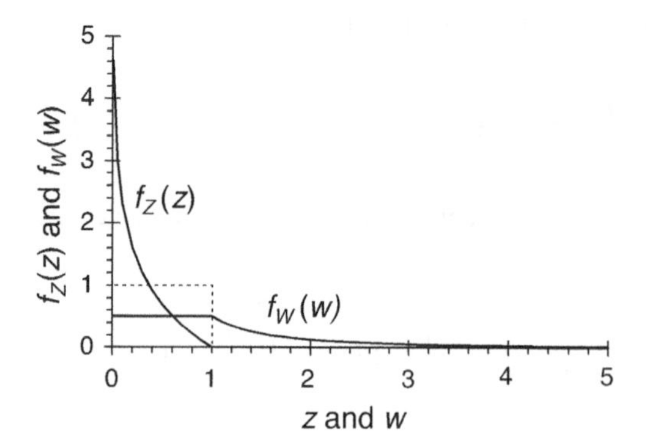

Fig. 3.4.3 The pdf of the product Z and quotient W of two uniform (0, 1) variates. The dotted line shows the common pdf of X and Y.

problem. This step is taken so that both Y and Z define a one-to-one transformation between (x_1, x_2) and (y, z), thus making it possible to define the inverse functions of x_1 and x_2 without ambiguity. We then proceed as follows:

(1) Convert the limits for X_1 and X_2 to corresponding limits for the variables Y and Z.

(2) Solve the inverse functions $x_1 = h_1(y, z)$, and $x_2 = h_2(y, z)$.

(3a) If the variables are discrete, the bivariate distribution of the transformed variables is

$$p_{Y,Z}(y, z) = p_{X_1,X_2}(h_1(y, z), h_2(y, z)). \tag{3.4.21}$$

The pmf of Y is the marginal pmf of Y from the foregoing bivariate pmf and is obtained using Eq. (3.3.6).

(3b) If the variables are continuous, the following partial derivatives should exist in a continuous form:

$$\frac{\partial x_1}{\partial y}, \frac{\partial x_1}{\partial z}, \frac{\partial x_2}{\partial y}, \text{ and } \frac{\partial x_2}{\partial z}.$$

The bivariate pdf of Y and Z denoted by $f_{Y,Z}(y, z)$ is given by

$$f_{Y,Z}(y, z) = |J| f_{X_1,X_2}[h_1(y, z), h_2(y, z)] \tag{3.4.22}$$

where $J \neq 0$ is the Jacobian corresponding to Eq. (3.4.4). Whereas in Eq. (3.4.4) only two variables were involved and the J term is the derivative, in Eq. (3.4.22) it becomes the determinant of partial derivatives:

$$J = \begin{vmatrix} \dfrac{\partial x_1}{\partial y} & \dfrac{\partial x_1}{\partial z} \\ \dfrac{\partial x_2}{\partial y} & \dfrac{\partial x_2}{\partial z} \end{vmatrix}. \tag{3.4.23}$$

Note that the variables in the numerators in the first and second rows (and likewise for more than two variables) of J correspond, respectively, to the original variables in $f_{X1,X2}(x_1, x_2)$. Likewise the variables in the denominators of the first and second columns correspond, respectively, to the new variables Y and Z. As in Eq. (3.4.4) the absolute value of the Jacobian term is used in Eq. (3.4.22).

To obtain the distribution of Y, one obtains its marginal distribution by integrating out Z in Eq. (3.4.22) following the procedure of Eq. (3.3.18). This procedure can be extended,

Table 3.4.1 Bivariate pmf of the output per day in numbers completed, X_1, and X_2, by two types of pile drivers

	$X_2 = 0$	$X_2 = 1$	$X_2 = 2$	$p_{X_1}(x)$
$X_1 = 0$	0.05	0.10	0.15	0.30
$X_1 = 1$	0.10	0.15	0.25	0.50
$X_1 = 2$	0.01	0.08	0.11	0.20
$p_{X_2}(x)$	0.16	0.33	0.51	$\Sigma = 1.00$

as follows, to $n \times k$-dimensional problems involving an n-dimensional random variable which is function of k variates. The case $k = n$ is sufficient for discussion.[42]

Property: General one-to-one transformation. Let $X_1, \ldots, X_k$ be a multiple random variable with continuous pdf $f_{X_1,\ldots X_k}(x_1, \ldots, x_k)$. The pdf of the multiple random variable $Y_1, \ldots, Y_n$ defined by a set of one-to-one transformations $Y_i = g_i(X_1, \ldots, X_k)$ for $i = 1, \ldots, k$ is given by

$$f_{Y_1,\ldots,Y_n}(y_1, \ldots, y_n) = |J| f_{X_1,\ldots,X_k}[h_1(y_1, \ldots, y_n), \ldots, h_{ki}(y_1, \ldots, y_n)],$$

where $x_1 = h_{1i}(y_1, \ldots, y_n), \ldots, x_k = h_{ki}(y_1, \ldots, y_n)$ denote the inverse transformations, and

$$J = \begin{bmatrix} \dfrac{\partial x_1}{\partial y_1} & \dfrac{\partial x_1}{\partial y_2} & \cdots & \dfrac{\partial x_1}{\partial y_n} \\ \dfrac{\partial x_2}{\partial y_1} & \dfrac{\partial x_2}{\partial y_2} & \cdots & \dfrac{\partial x_2}{\partial y_n} \\ \vdots & \vdots & \ddots & \vdots \\ \dfrac{\partial x_k}{\partial y_1} & \dfrac{\partial x_k}{\partial y_2} & \cdots & \dfrac{\partial x_k}{\partial y_n} \end{bmatrix}$$

the Jacobian of the transformation, provided that the partial derivatives in J are continuous, and the determinant $|J|$ is nonzero.

In a discrete case where the joint probabilities are evaluated numerically (from observed data rather than by theoretical means), the marginal pmf can be obtained directly as in the following example:

Example 3.50. Pile drivers. An engineer uses two types of pile drivers, the first of which produces X_1 piles a day. The second, an older version that is still indispensable, can complete X_2 units during the same period of time. Mechanical, site, soil, and climatic conditions are such that X_1 and X_2 are random variables. The bivariate distribution of X_1 and X_2, each of which are in the range 0–2, is estimated as given in Table 3.4.1.

Three units of the first type are obtained to be used with the available unit of the second type. Consequently, the engineer is interested in the distribution of the random variable $Y = 3X_1 + X_2$. The marginal pmf of Y can be evaluated from Table 3.4.1 within the sample space of Y, that is (0, 1, 2, 3, 4, 5, 6, 7, 8) obtained for all possible combinations of X_1 and X_2. Table 3.4.2 gives the required marginal distribution.

[42] Papoulis (2001).

Table 3.4.2 The marginal pmf of $Y = 3X_1 + X_2$ from the output per day X_1 and X_2 by two types of pile drivers

Y	X_1	X_2	$p_Y(y)$
0	0	0	0.05
1	0	1	0.10
2	0	2	0.15
3	1	0	0.10
4	1	1	0.15
5	1	2	0.25
6	2	0	0.01
7	2	1	0.08
8	2	2	0.11
			$\Sigma = 1.00$

In the continuous case, Eqs. (3.4.22) and (3.4.23) are applied as shown in the following example:

Example 3.51. Project costs and benefits. The benefits X_1 and costs X_2 of a particular scheme are treated as random variables on account of numerous factors considered to be unpredictable. From some trial calculations based on past projects, it is possible to approximate the joint pdf as follows, that is, by the bivariate negative exponential distribution:

$$f_{X_1,X_2}(x_1, x_2) = e^{-(x_1+x_2)}.$$

The engineer is interested in the benefit-to-cost ratio X_1/X_2. Introducing two random variables Y and Z, let $y = x_1/x_2$ and $z = x_1 + x_2$; the supplementary equations are chosen to simplify the inverse functions while maintaining one-to-one correspondence between the two sets of variables, as already discussed. Hence the inverse relationships, as functions of y and z, are $x_1 = zy(1 + y)^{-1}$ and $x_2 = z(1 + y)^{-1}$. Thus, from the foregoing bivariate pdf and from Eq. (3.4.22),

$$f_{Y,Z}(y, z) = |J| f_{X_1,X_2}(zy(1 + y)^{-1}, z(1 + y)^{-1}) = |J| e^{-z}.$$

The four partial derivatives required for the Jacobian of Eq. (3.4.23) are

$$\frac{\partial x_1}{\partial y} = z(1 + y)^{-2}; \quad \frac{\partial x_1}{\partial z} = y(1 + y)^{-1}; \quad \frac{\partial x_2}{\partial y} = -z(1 + y)^{-2}; \quad \frac{\partial x_2}{\partial z} = (1 + y)^{-1}.$$

Substituting these results in Eq. (3.4.23), the determinant of the 2×2 matrix simplifies to

$$|J| = z(1 + y)^{-2}.$$

Hence, the bivariate pdf of Y and Z is given by

$$f_{Y,Z}(y, z) = e^{-z} z(1 + y)^{-2}.$$

The marginal pdf of Y, which is the benefit-to-cost ratio, is given by

$$f_Y(y) = \int_0^{\infty} ze^{-z}(1 + y)^{-2} dz = \int_0^{\infty} f_{YZ}(y, z) dz = (1 + y)^{-2}, \quad \text{for } y \geq 0,$$
$$= 0, \quad \text{elsewhere.}$$

The lower limit of integration is zero because the variables cannot take negative values. The foregoing result is obtained after integrating by parts and applying l'Hospital's rule. (We can also solve this without Jacobians.)

3.4.3 Properties of derived variables

3.4.3.1 Expectation and moments of derived variables

A random variable that is a function of other random variables and its probability distribution are sometimes defined as a *derived variable* and a *derived distribution*, respectively. Although the distribution can be derived theoretically from the probability distribution of the basic variates, such derivations are often difficult, especially if the function is nonlinear. However, the moments of the derived variate can provide useful information under these circumstances. For the purpose, we can use the properties of expectations introduced in Section 3.2.

If $Z = g(X_1, X_2, \ldots, X_k)$ is a derived variable from a k-dimensional random variable $X_1, X_2, \ldots, X_k$, the expected value of Z is written as

$$E[Z] = E[g(X_1, X_2, \ldots, X_k)] = \int_{-\infty}^{+\infty} \cdots \int_{-\infty}^{+\infty} g(x_1, x_2, \ldots, x_k) f_{X_1,X_2,\ldots,X_k}(x_1, x_2, \ldots, x_k) dx_1 dx_2 \ldots dx_k. \tag{3.4.24}$$

More generally, the rth-order moment of Z is given by

$$E[Z^r] = E[[g(X_1, X_2, \ldots, X_k)]^r] = \int_{-\infty}^{+\infty} \cdots \int_{-\infty}^{+\infty} [g(x_1, x_2, \ldots, x_k)]^r f_{X_1,X_2,\ldots,X_k}(x_1, x_2, \ldots, x_k) dx_1 dx_2 \ldots dx_k. \tag{3.4.25}$$

For example, in the case of a linear function $Z = aX + b$, with constants a and b, the properties of the expectation operator yield $E[Z] = aE[X] + b$, that is, $\mu_Z = a\mu_X + b$. The variance is obtained as follows:

$$\mathrm{Var}[Z] = \mathrm{Var}[aX + b] = E[(aX + b - aE[X] - b)^2] = a^2 \int_{-\infty}^{+\infty} (x - \mu_X)^2 f_X(x) dx = a^2 \mathrm{Var}[X] = a^2 \sigma_X^2. \tag{3.4.26}$$

In the case of a linear function of two variables, say, $Z = aX + bY$, we obtain

$$E[Z] = E[aX + bY] = aE[X] + bE[Y] = a\mu_X + b\mu_Y \tag{3.4.27}$$

and

$$\begin{aligned}\mathrm{Var}[Z] &= \mathrm{Var}[aX + bY] = E[(aX + bY - aE[X] - bE[Y])^2] \\ &= E[a^2(X - \mu_X)^2 + 2ab(X - \mu_X)(Y - \mu_Y) + b^2(Y - \mu_Y)^2] \\ &= a^2\mathrm{Var}[X] + 2ab\mathrm{Cov}[X, Y] + b^2\mathrm{Var}[Y] = a^2\sigma_X^2 + 2ab\rho_{X,Y}\sigma_X\sigma_Y + b^2\sigma_Y^2,\end{aligned} \tag{3.4.28}$$

where $\rho_{X,Y}$ denotes the correlation coefficient of X and Y. More generally,

$$E[Z] = E\left[\sum_{i=1}^{k} a_i X_i + b_i\right] = \sum_{i=1}^{k} a_i E[X_i] + b_i, \tag{3.4.29}$$

$$\mathrm{Var}[Z] = \mathrm{Var}\left[\sum_{i=1}^{k} a_i X_i + b_i\right] = \sum_{i=1}^{k} a_i^2 \mathrm{Var}[X_i] + \sum_{i=1}^{k} \sum_{j=1; j \neq i}^{k} a_i a_j \mathrm{Cov}[X_i, X_j]. \tag{3.4.30}$$

Example 3.52. Toll station. Assume that the interarrival time X of a vehicle approaching a toll station of a bridge has an exponential pdf with parameter λ. Accordingly, the mean and the variance of X are λ^{-1} and λ^{-2}, respectively (see Examples 3.9 and 3.14). If k toll lines are available, let us determine the mean arrival time of k vehicles and the coefficient of variation of this arrival time. Assume that the arrivals are independent of each other.

If one defines the total arrival time as

$$Z = \sum_{i=1}^{k} X_i,$$

with X_i denoting the arrival time of the ith vehicle, then

$$E[Z] = \sum_{i=1}^{k} E[X_i] = \sum_{i=1}^{k} \lambda^{-1} = k\lambda^{-1}$$

and $$\mathrm{Var}[Z] = \sum_{i=1}^{k} \mathrm{Var}[X_i] = \sum_{i=1}^{k} \lambda^{-2} = k\lambda^{-2}.$$

Hence the coefficient of variation,

$$V_Z = \sqrt{\frac{\mathrm{Var}[Z]}{E^2[Z]}} = \sqrt{\frac{k\lambda^{-2}}{k^2\lambda^{-2}}} = \sqrt{\frac{1}{k}},$$

decreases quite rapidly with increasing k.

In the case of the product of two variates, X and Y, the expected value is closely related to the covariance of the variables. If $Z = XY$,

$$E[Z] = E[XY] = \mathrm{Cov}[X, Y] - E[X]E[Y] = \rho_{X,Y}\sigma_X\sigma_Y + \mu_X\mu_Y. \qquad (3.4.31)$$

Since

$$XY = \mu_X\mu_Y + \mu_Y(X - \mu_X) + \mu_X(Y - \mu_Y) + (X - \mu_X)(Y - \mu_Y),$$

one can compute $E[XY]$ and $E[(XY)^2]$ to obtain the variance of $Z = XY$ as

$$\begin{aligned}\mathrm{Var}[Z] = \mathrm{Var}[XY] &= \sigma_X^2\mu_Y^2 + \sigma_Y^2\mu_X^2 \\ &+ 2\rho_{X,Y}\sigma_X\sigma_Y\mu_X\mu_Y - \rho_{X,Y}^2\sigma_X^2\sigma_Y^2 + E[(X - \mu_X)^2(Y - \mu_Y)^2] \\ &+ 2\mu_X E\lfloor(X - \mu_X)(Y - \mu_Y)^2\rfloor + 2\mu_Y E\lfloor(X - \mu_X)^2(Y - \mu_Y)\rfloor, \end{aligned} \qquad (3.4.32)$$

where $\rho_{X,Y}\sigma_X\sigma_Y = \mathrm{Cov}[X, Y]$.

If $X_1, X_2, \ldots, X_k$ are k mutually independent random variables, the paired correlation coefficient of X_i and X_j is null, so that the mean of their product, $Z = X_1, X_2, \ldots, X_k$, equals the product of their individual means. Thus,

$$E[Z] = E[X_1X_2 \ldots X_k] = E[X_1]E[X_2] \ldots E[X_k] = \mu_{X_1}\mu_{X_2} \cdots \mu_{X_k}. \qquad (3.4.33)$$

Since

$$E[Z^2] = E[(X_1X_2 \ldots X_k)^2] = E\left[X_1^2\right]E\left[X_2^2\right] \ldots E\left[X_k^2\right],$$

the variance of the product of k independent variables is given by

$$\begin{aligned}\mathrm{Var}[Z] &= E[Z^2] - E^2[Z] = E\left[X_1^2\right]E\left[X_2^2\right] \ldots E\left[X_k^2\right] - (E[X_1]E[X_2] \ldots E[X_k])^2 \\ &= \left(\sigma_{X_1}^2 + \mu_{X_1}^2\right)\left(\sigma_{X_2}^2 + \mu_{X_2}^2\right) \ldots \left(\sigma_{X_k}^2 + \mu_{X_k}^2\right) - \mu_{X_1}^2\mu_{X_2}^2 \cdots \mu_{X_k}^2. \end{aligned} \qquad (3.4.34)$$

For two independent variables X and Y, Eq. (3.4.34) yields

$$\begin{aligned}\mathrm{Var}[Z] = \mathrm{Var}[XY] &= (\sigma_X^2 + \mu_X^2)(\sigma_Y^2 + \mu_Y^2) - \mu_X^2\mu_Y^2 \\ &= \sigma_X^2\sigma_Y^2 + \sigma_X^2\mu_Y^2 + \sigma_Y^2\mu_X^2.\end{aligned} \tag{3.4.35}$$

Example 3.53. Storm rainfall total. The duration X and the average rainfall rate Y of a storm at a given location are assumed to be independent, exponentially distributed variates with parameter a and b, respectively, as specified in Example 3.33. Accordingly, their means are a^{-1} and b^{-1}, respectively, and their variances are a^{-2} and b^{-2}, respectively. In Example 3.49 the total amount of water delivered by a storm was defined as $Z = XY$, and its cdf was derived in the form of a Bessel-type function. Although this cdf is somewhat cumbersome, the mean and variance of Z are easily found as

$$E[Z] = \mu_X\mu_Y = (ab)^{-1}$$

and

$$\mathrm{Var}[Z] = \sigma_X^2\sigma_Y^2 + \sigma_X^2\mu_Y^2 + \sigma_Y^2\mu_X^2 = (ab)^{-2} + (ab)^{-2} + (ab)^{-2} = 3(ab)^{-2}.$$

It is noted that the coefficient of variation of Z is

$$V_Z = \sqrt{\frac{\mathrm{Var}[Z]}{E^2[Z]}} = \sqrt{\frac{3(ab)^{-2}}{(ab)^{-2}}} = \sqrt{3},$$

and it does not depend on the statistics of the underlying distribution of storm duration and intensity.

In the general case of a derived variable Z which is given as a function $g(\cdot)$ of some other basic variables, say $X_1, X_2, \ldots, X_k$, the mean and variance can be derived using Eq. (3.4.25) for $r = 1, 2$. However, the capability of deriving the second-order statistics of Z strongly depends on the combination between the functional relationship between the variables, $g(X_1, X_2, \ldots, X_k)$, and the joint pmf or pdf of the basic variables.

Example 3.54. Safety factor. Consider a system subject to a random load, with pdf given by the gamma distribution

$$f_Y(y) = \frac{b^\eta}{\Gamma(\eta)} y^{\eta-1} e^{-by},$$

where η and b are known parameters. The shape parameter η and the scale parameter b of the pdf are related to the mean and variance of Y through

$$\mu_Y = \frac{\eta}{b},$$

$$\sigma_Y^2 = \frac{\eta}{b^2},$$

respectively (see Example 3.19). The capacity of the system is also uncertain; we assume that it is represented by a similarly distributed variate X with known parameters γ and a. Accordingly, the mean and variance of X are

$$\mu_X = \frac{\gamma}{a}$$

and

$$\sigma_X^2 = \frac{\gamma}{a^2},$$

respectively.

Define the safety factor of the system as the ratio between capacity and load, say, $Z = X/Y$. The engineer must evaluate the expected safety factor and its variance. If he assumes that capacity and load are independent, Eq. (3.4.24) yields

$$E[Z] = E\left[\frac{X}{Y}\right] = \int_0^{+\infty}\int_0^{+\infty} \frac{x}{y} \frac{a^\gamma x^{\gamma-1} e^{-ax}}{\Gamma(\gamma)} \frac{b^\eta y^{\eta-1} e^{-by}}{\Gamma(\eta)} dxdy = \frac{\gamma}{\eta-1}\frac{b}{a},$$

which exists only for $\eta > 1$. The second-order moment of Z is found by using Eq. (3.4.25), which yields

$$E[Z^2] = E\left[\left(\frac{X}{Y}\right)^2\right] = \int_0^{+\infty}\int_0^{+\infty} \left(\frac{x}{y}\right)^2 \frac{a^\gamma x^{\gamma-1} e^{-ax}}{\Gamma(\gamma)} \frac{b^\eta y^{\eta-1} e^{-bx}}{\Gamma(\eta)} dxdy$$

$$= \frac{\gamma+\gamma^2}{(\eta-1)(\eta-2)}\left(\frac{b}{a}\right)^2,$$

which exists only for $\eta > 2$. The variance of Z is then found as

$$\text{Var}[Z] = E[Z^2] - E^2[Z] = \frac{\gamma(\gamma+\eta-1)}{(\eta-1)^2(\eta-2)}\left(\frac{b}{a}\right)^2,$$

which also exists only for $\eta > 2$. The coefficient of variation of Z, which also exists only for $\eta > 2$, is given by

$$V_Z = \sqrt{\frac{\text{Var}[Z]}{E^2[Z]}} = \sqrt{\frac{\gamma(\gamma+\eta-1)}{(\eta-1)^2(\eta-2)}\left(\frac{\eta-1}{\gamma}\right)^2\frac{(b/a)^2}{(b/a)^2}} = \sqrt{\frac{(\gamma+\eta-1)}{\gamma(\eta-2)}}.$$

We note that this result depends only on the two shape parameters γ and η of the underlying gamma distributions of system capacity and load, respectively.

In some cases it is difficult to derive the distribution or the moments of a dependent variate by analytical methods. However, the required moments of $Z = g(X_1, \ldots, X_n)$ can be determined by using Taylor's series expansion about the means of variables $X_1, \ldots, X_n$. For example, the mean and variance of a random variable $Z = g(X, Y)$ that is a function of two variables X and Y with means μ_X and μ_Y, respectively, can be approximated by

$$E[Z] \approx g(\mu_X, \mu_Y) + \frac{1}{2}\frac{\partial^2 g}{\partial x^2}\text{Var}[X] + \frac{1}{2}\frac{\partial^2 g}{\partial y^2}\text{Var}[Y] + \frac{\partial^2 g}{\partial x \partial y}\text{Cov}[X, Y], \quad (3.4.36)$$

$$\text{Var}[Z] \approx \left(\frac{\partial g}{\partial x}\right)^2 \text{Var}[X] + \left(\frac{\partial g}{\partial y}\right)^2 \text{Var}[Y] + 2\left(\frac{\partial g}{\partial x}\frac{\partial g}{\partial y}\right)^2 \text{Cov}[X, Y], \quad (3.4.37)$$

where the derivatives of $g(x, y)$ are computed for $x = \mu_X$, and $y = \mu_Y$, and all terms of order higher than 2 in the expansion are excluded. Eqs. (3.4.36) and (3.4.37) are simplified for independent variates X and Y, because $\text{Cov}[X, Y] = 0$.

Example 3.55. Safety factor. Returning to Example 3.54, consider a system subject to the combined effect of random capacity and load. Although the probability distributions of X and Y are undetermined, the engineer has estimated the mean and standard deviation of these variables, say, μ_X, μ_Y, σ_X, and σ_Y. Based on this information, the engineer must evaluate the mean and standard deviation of the *safety factor* $Z = X/Y$ of the system.

Since $g(x, y) = x/y$, the derivatives in the approximated moment equations are

$$\frac{\partial g}{\partial x} = \frac{1}{y}; \quad \frac{\partial g}{\partial y} = -\frac{x}{y^2}; \quad \frac{\partial^2 g}{\partial x^2} = 0; \quad \frac{\partial^2 g}{\partial y^2} = \frac{2x}{y^3}; \quad \frac{\partial^2 g}{\partial x \partial y} = -\frac{1}{y^2}.$$

These are used in Eqs. (3.4.36) and (3.4.37) to obtain

$$E[Z] \approx \frac{\mu_X}{\mu_Y} + \frac{\mu_X}{\mu_Y^3}\mathrm{Var}[Y] - \frac{1}{\mu_Y^2}\mathrm{Cov}[X, Y]$$

and

$$\mathrm{Var}[Z] \approx \left(\frac{\mu_X}{\mu_Y}\right)^2 \left(\frac{\mathrm{Var}[X]}{\mu_X^2} + \frac{\mathrm{Var}[Y]}{\mu_Y^2} - \frac{2\mathrm{Cov}[X, Y]}{\mu_X \mu_Y}\right).$$

For example, assume that X and Y are independent of each other, and have similar distributions as shown in Example 3.54. Since

$$\mu_Y = \eta/b, \quad \sigma_Y^2 = \eta/b^2, \quad \mu_X = \gamma/a, \quad \sigma_X^2 = \gamma/a^2,$$

one gets

$$E[Z] \approx \frac{\mu_X}{\mu_Y} + \frac{\mu_X \sigma_Y^2}{\mu_Y^3} = \frac{\gamma b}{\eta a} + \frac{\gamma \eta b^3}{\eta^3 a b^2} = \frac{\gamma(\eta+1)}{\eta^2}\frac{b}{a}$$

and

$$\mathrm{Var}[Z] \approx \left(\frac{\mu_X}{\mu_Y}\right)^2 \left(\frac{\sigma_X^2}{\mu_X^2} + \frac{\sigma_Y^2}{\mu_Y^2}\right) = \left(\frac{\gamma b}{\eta a}\right)^2 \left(\frac{\gamma a^2}{\gamma^2 a^2} + \frac{\eta b^2}{\eta^2 b^2}\right) = \frac{\gamma(\gamma+\eta)}{\eta^3}\left(\frac{b}{a}\right)^2.$$

The coefficient of variation of Z is then given by

$$V_Z = \sqrt{\frac{\sigma_Z^2}{\mu_Z^2}} \approx \sqrt{\frac{\eta(\gamma+\eta)}{\gamma(\gamma+1)^2}} = \frac{\sqrt{\eta + \eta^2/\gamma}}{\eta+1}.$$

By comparing these results with those obtained in the following Example 3.56, we can evaluate the accuracy of the approximate estimates of the required second-order statistics. We note that the accuracy of these approximations depends on the values taken by the parameters of the underlying distributions, namely, by the shape parameters of the two pdfs describing the distribution of capacity and load. Therefore, the accuracy of this method depends on the values taken by the moments of the basic variates.

3.4.3.2 Moment-generating function of derived variables

Although the determination of the probability distribution of a derived variable from those of the basic variables is not an easy task, the evaluation of its moments can provide some useful information on the variable in question. However, the computation of Eq. (3.4.25) can also be cumbersome in certain circumstances, and the approximation by a Taylor's series expansion [see Eqs. (3.4.36) and (3.4.37)] should be carefully verified for accuracy, because it does not provide accurate estimates of second-order statistics in all cases. An alternative, but powerful technique is then provided by the use of the moment-generating function.

In Section 3.2 the moment-generating function of a random variable Z was defined as $M_Z(t) = E[e^{tZ}]$, and this definition was extended to multidimensional variables in Section 3.3. If Z is a derived random variable from k basic variables $X_1, X_2, \ldots, X_k$, that is $Z = g(X_1, X_2, \ldots, X_k)$, the mgf of Z can be determined as

$$\begin{aligned} M_Z(t) &= E[e^{tZ}] = E[e^{tg(X_1,X_2,\ldots,X_k)}] \\ &= \int_{-\infty}^{+\infty} \cdots \int_{-\infty}^{+\infty} e^{tg(x_1,x_2,\ldots,x_k)} f_{X_1,X_2,\ldots,X_k}(x_1, x_2, \ldots, x_k) dx_1 dx_2 \ldots dx_k. \end{aligned} \quad (3.4.38)$$

After the integration of Eq. (3.4.38) is performed, the moments of any order of the derived variable Z can be determined by computing the derivatives of $M_Z(t)$ for $t = 0$, as indicated by Eq. (3.2.20).

Example 3.56. Wave pressure on coastal structures. The impact pressure of sea waves on coastal structures may be evaluated as $Z = cX^2$, where X is the horizontal velocity of the advancing wave and c is a constant. Because of the uncertainty involved in the evaluation of X, we consider this to be a random variable; Z is thus a derived variable from X.

Assume X has mean μ_X and standard deviation σ_X and has the normal pdf

$$f_X(x) = \frac{1}{\sigma_X\sqrt{2\pi}}e^{-0.5[(x-\mu_X)/\sigma_X]^2}.$$

An engineer needs to evaluate the second-order statistics of the impact pressure on the coastal structure.

Let $Y = (X - \mu_X)/\sigma_X$, which has zero mean, unit variance, and pdf

$$f_Y(y) = \frac{1}{\sqrt{2\pi}}e^{-0.5y^2}.$$

Thus,

$$Z = cX^2 = c(\mu_X + \sigma_X Y)^2 = c\mu_X^2 + 2c\mu_X\sigma_X Y + c\sigma_X^2 Y^2$$

and one must find the mean and variance of $W = Y^2$.

By substituting $f_Y(y)$ in Eq. (3.4.38),

$$M_W(t) = \int_{-\infty}^{+\infty} e^{ty^2} f_Y(y)dy = \int_{-\infty}^{+\infty} e^{ty^2}\frac{1}{\sqrt{2\pi}}e^{-0.5y^2}dy = \int_{-\infty}^{+\infty}\frac{1}{\sqrt{2\pi}}e^{-0.5y^2(1-2t)}dy$$

$$= \int_{-\infty}^{+\infty}\frac{1}{\sqrt{2\pi}}e^{-0.5y^2(1-2t)}dy = \frac{1}{\sqrt{1-2t}}.$$

(using the transformation $z = y\sqrt{1-2t}$ and because the area under the $f_Y(y)$ curve is unity.)

The mean of W is given by the first derivative of the mgf at the origin, that is,

$$\mu_W = \frac{\partial M_W(t)}{\partial t}|_{t=0} = [(1-2t)^{-3/2}]_{t=0} = 1.$$

The second-order moment of W is given by the second derivative of the mgf at the origin, that is,

$$E[W^2] = \frac{\partial^2 M_W(t)}{\partial t^2}|_{t=0} = [3(1-2t)^{-5/2}]_{t=0} = 3.$$

Hence, the variance of W is

$$\sigma_W^2 = E[W^2] - E^2[W] = 3 - 1 = 2.$$

The mean of the required impact pressure Z follows immediately from the linear property of expectation as

$$\mu_Z = c\mu_X^2 + 2c\mu_X\sigma_X\mu_Y + c\sigma_X^2\mu_W = c\left(\mu_X^2 + \sigma_X^2\right),$$

since $\mu_Y = 0$ and $\mu_W = 1$. The variance can be then found using Eq. (3.4.30), so

$$\sigma_Z^2 = (2c\mu_X\sigma_X)^2\text{Var}[Y] + \left(c\sigma_X^2\right)^2\text{Var}[W] = 2c^2\sigma_X^2\left(2\mu_X^2 + \sigma_X^2\right).$$

If $V_X = \sigma_X/\mu_X$ denotes the coefficient of variation of horizontal velocity of the advancing wave X, the mean of Z can be written as

$$\mu_Z = c\mu_X^2\left(1 + V_X^2\right),$$

which equals $g(E[X])$ augmented by a factor of $(1 + V_X^2)$. The variance of Z is

$$\sigma_Z^2 = 2\left(1 + \frac{2}{V_X^2}\right)c^2\sigma_X^4,$$

which equals $g(\text{Var}[X])$ augmented by a factor of $2(1 + 2/V_X^2)$.

Compared with the derivation of moments, the additional advantage of the mgf is that it is capable of indicating the probability distribution of the derived variable if the mgf of a derived variable can be recognized as the mgf of some known distribution.[43] This property descends from the property of the mgf reported in Section 3.2, which states that the mgf of a random variable, when it exists, is unique and uniquely determines its probability distribution.

Example 3.57. Surveying errors. The ground elevation measured by using remote-sensing data is affected by two independent sources of random errors that are described by two normal variates X and Y with means μ_X and μ_Y, and standard deviations σ_X and σ_Y, respectively, and with normal pdf for X as in Example 3.56. Let us denote by $Z = X + Y$ the overall error of the estimated elevation and find the distribution of Z by using the moment-generating function technique.

The moment-generating function of Z is given by

$$M_Z(t) = E[e^{tZ}] = E\left[e^{t(X+Y)}\right] = E[e^{tX}]E[e^{tY}] = M_X(t)M_Y(t).$$

Since the moment-generating function of X or Y is

$$M(t) = \exp\left(\mu t + \frac{1}{2}\sigma^2 t^2\right),$$

(see Appendix A.5), the moment-generating function of Z becomes

$$M_Z(t) = e^{\mu_X t + \frac{1}{2}\sigma_X^2 t^2} e^{\mu_{Y_t} + \frac{1}{2}\sigma_Y^2 t^2} = e^{(\mu_X+\mu_Y)t + \frac{1}{2}(\sigma_X^2+\sigma_Y^2)t^2}.$$

The first derivative of the mgf at the origin gives the mean of Z, that is,

$$\mu_Z = \left.\frac{\partial M_Z(t)}{\partial t}\right|_{t=0} = \left[\left[(\mu_X + \mu_Y) + t\left(\sigma_X^2 + \sigma_Y^2\right)\right] e^{(\mu_X+\mu_Y)t + \frac{1}{2}(\sigma_X^2+\sigma_Y^2)t^2}\right]_{t=0}$$
$$= \mu_X + \mu_Y.$$

The second derivative of the mgf at the origin gives the second-order moment of Z, that is,

$$E[Z^2] = \frac{\partial^2 M_Z(t)}{\partial t^2}|_{t=0} = \sigma_X^2 + \sigma_Y^2 + (\mu_X + \mu_Y)^2.$$

Accordingly, the variance of Z is given by

$$\sigma_Z^2 = E[Z^2] - E^2[Z] = \sigma_X^2 + \sigma_Y^2 + (\mu_X + \mu_Y)^2 - (\mu_X + \mu_Y)^2 = \sigma_X^2 + \sigma_Y^2.$$

Therefore, the mean of the overall error Z equals the sum of the means of the individual errors, and its variance equals the sum of the variances of the individual errors. If both sources of error do not involve the presence of systematic errors, one can assume that $\mu_X = \mu_Y = 0$, so that the overall error in ground elevation is a zero-mean variable with standard deviation.[44]

$$\sigma_Z = \sqrt{\sigma_X^2 + \sigma_Y^2}.$$

[43] In Example 3.56, the mgf of W can be recognized as the moment-generating function of a gamma-distributed variate (see Example 3.54) with shape parameter 1/2 and scale parameter equal to 1/2. Therefore, one can conclude that the transformation $W = Y^2$ of a standard normal variate Y results in a gamma-distributed variate W, which in this case is a $\chi^2(1)$ variate.

[44] Considering the form of the mgf of Z, one notes that $M_Z(t)$ can be the mgf of a normal variate with mean μ_Z, and standard deviation σ_Z (see Appendix A.5). Therefore, we conclude that the overall error found as the sum of normal individual errors is also a normal error. In addition, we note that the moment-generating function of the sum of two or more independent normal variates equals the product of the individual moment-generating functions.

The preceding example shows another important property of the mgf technique. If a random variable Z is the sum of k independent basic variates $X_1, X_2, \ldots, X_k$, its mgf can be determined as

$$M_Z(t) = E[e^{tZ}] = E\left[\exp\left(t\sum_{i=1}^{k} X_i\right)\right] = E\left[\prod_{i=1}^{k} e^{tX_i}\right] = \prod_{i=1}^{k} M_{X_i}(t). \quad (3.4.39)$$

This property states that the mgf of the sum of any number k of independent random variables equals the product of the individual moment-generating functions. This result is independent of the underlying distributions of the random variables.

Example 3.58. Toll bridge. In Example 3.52 the arrival time X of a vehicle approaching the toll station of a bridge was assumed to be exponentially distributed with parameter λ. Since k toll lines are available, let us determine the pdf of the arrival time of k vehicles, assuming that the arrivals are independent of each other.

One must find the pdf of

$$Z = \sum_{i=1}^{k} X_i,$$

which is the sum of k independent random variables. Since

$$M_{X_i}(t) = \frac{\lambda}{\lambda - t}$$

(see Example 3.17) the mgf of the Z variable is given by[45]

$$M_z(t) = \prod_{i=1}^{k} M_{X_i}(t) = \left(\frac{\lambda}{\lambda - t}\right)^k.$$

Therefore,

$$E[Z] = k\lambda^{-1} \quad \text{and} \quad \mathrm{Var}[Z] = k\lambda^{-2}.$$

These results agree with the derived mean and variance of Z of Example 3.52.

Note that Eq. (3.3.39) also holds for discrete variables, as shown in the following example:

Example 3.59. Flood occurrence. In Example 3.7 the occurrence of a flood exceeding a given design level was considered as a two-valued Bernoulli variate X with probability of occurrence p. Accordingly, X can take the value of 1 with probability p and the value of 0 with probability $1 - p$. Suppose that $X_1, X_2, \ldots, X_k$ is a sequence of k independent variables with common probability p that describes the occurrence of k floods at a river site, with the assumption that the occurrences are independent of each other. Therefore, the occurrence of N floods exceeding the design level is given by

$$N = \sum_{i=1}^{k} X_i,$$

which is the sum of k independent random variables. Since

$$M_{X_i}(t) = 1 + \mathrm{p}(e^t - 1),$$

---

[45] This mgf can be recognized as that of a gamma-distributed variate with shape parameter k and scale parameter λ, as shown in Chapter 4.

the mgf of the N is given by[46]

$$M_N(t) = \prod_{i=1}^{k} M_{X_i}(t) = [1 + p(e^t - 1)]^k.$$

The properties of the mgf of the sum of independent random variables can also be extended to the difference between variables.

Example 3.60. Safety margin. Suppose the capacity X of a water supply system is distributed as specified for the X variable in Example 3.56, with mean $\mu_X = 5$ and $\sigma_X = 0.75$ units per year; one unit is 10^6 L. The city's estimated annual water demand Y is normally distributed with mean $\mu_Y = 4$ and $\sigma_Y = 1$ units. Find the probability distribution of the safety margin of system operation which is defined as $Z = X - Y$, if X and Y are independent of each other.

To solve the problem, one can make use of the auxiliary variable $W = g(Y) = -Y$, which has mean $\mu_W = -\mu_Y$ and standard deviation $\sigma_W = \sigma_Y$ because of the properties of expectation operator. From Eq. (3.4.4) the pdf of W is

$$f_W(w) = \left|\frac{dy}{dw}\right| f_Y[h(w) = |-1|] f_Y(-w)$$

$$= \frac{1}{\sigma_Y\sqrt{2\pi}} e^{[-0.5(-w-\mu_Y)^2/\sigma_Y^2]} = \frac{1}{\sigma_W\sqrt{2\pi}} e^{[-0.5(w-\mu_W)^2/\sigma_W^2]},$$

that is, W is a normal variate with mean $\mu_W = -\mu_Y$ and standard deviation $\sigma_W = \sigma_Y$. Therefore, the moment-generating function of $Z = X + W$ is given by

$$M_Z(t) = M_X(t)M_W(t) = e^{\mu_X t + 0.5\sigma_X^2 t^2} e^{\mu_W t + 0.5\sigma_W^2 t^2} = e^{(\mu_X - \mu_Y)t + 0.5(\sigma_X^2 + \sigma_Y^2)t^2},$$

which indicates that Z is a normal variate with mean[47]

$$\mu_Z = \mu_X - \mu_Y = 1 \text{ unit},$$

and standard deviation

$$\sigma_Z = \sqrt{\sigma_X^2 + \sigma_Y^2} = \sqrt{0.75^2 + 1^2} = 1.25 \text{ units}.$$

One is usually interested in the probability that the safety margin is positive, that is, $\Pr[Z > 0] = 1 - F_Z(0)$.

3.4.4 Compound variables

3.4.4.1 Contagious distributions

Let X be a random variable with a density function $f_X(x|\theta)$. If θ is not a constant, but it takes values randomly in a given interval or set, say, Ω_Θ, the probability distribution of X is of the *contagious* or *compound* or *mixture* type. This distribution can be found by

[46] This can be recognized as the mgf of a binomial variate with parameters k and p, namely, the number of trials in the sequence and the probability of success of each individual trial.

[47] See Appendix A.5.

integration or summation (depending on the type of variable) for all possible values Ω_Θ of θ. In the continuous case

$$f_X(x) = \int_{\Omega_\Theta} f_X(x, \theta) f_\Theta(\theta) d\theta,$$

where $f_\Theta(\theta)$ denotes the density function of θ. If θ is a discrete random variable,

$$f_X(x) = \sum_{\Omega_\Theta} f_X(x, \theta) p_\Theta(\theta), \tag{3.4.40}$$

where $p_\Theta(\theta)$ denotes the mass function of θ. The cdf of X will be found as

$$F_X(x) = \int_{\Omega_\Theta} F_X(x, \theta) f_\Theta(\theta) d\theta, \tag{3.4.41}$$

or

$$F_X(x) = \sum_{\Omega_\Theta} F_X(x|\theta) p_\Theta(\theta), \tag{3.4.42}$$

where $F_X(x|\theta)$ denotes the family of cdfs of X parameterized by θ.

Example 3.61. Earthquake intensity. In Example 3.48 the cdf of the maximum earthquake intensity observed in a century was found to be

$$F_Y(y) = [F_X(y)]^n = (1 - e^{-\lambda y})^n, \qquad y > 0,$$

where λ is the parameter of the distribution of the intensity of earthquakes and n is the number of earthquakes occurring in a century. Because of the randomness of earthquake occurrences, the number of occurrences should be viewed as a random variable, say, N. Therefore, the foregoing distribution can be viewed as a contagious distribution of variable Y with a parameter n, that is,

$$F_Y(y|n) = (1 - e^{-\lambda y})^n.$$

It seems reasonable to assume that N is Poisson distributed as in Example 3.18, that is,

$$p_N(n) = \frac{\nu^n e^{-\nu}}{n!}, \qquad \text{for } n = 0, 1, 2, \ldots,$$

where ν is the expected number of earthquakes occurring in a century. The probability distribution of the maximum earthquake intensity becomes

$$F_Y(y) = \sum_{n=0}^{\infty} F_Y(y|n) p_N(n) = \sum_{n=0}^{\infty} \frac{(1 - e^{-\lambda y})^n \nu^n e^{-\nu}}{n!}$$

$$= e^{-\nu} \sum_{n=0}^{\infty} \frac{[\nu(1 - e^{-\lambda y})]^n}{n!} = \exp(-\nu) \exp[\nu(1 - e^{-\lambda y})] = \exp(-\nu e^{-\lambda y}).$$

By substituting α for $1/\lambda$, and b for $\lambda^{-1} \ln \nu$, one gets[48]

$$F_Y(y) = \exp[-e^{-(y-b)/\alpha}].$$

Physical or technical considerations can suggest to the engineer that a mixture model is required. This assumption may be made, for example, to account for the inherent

[48] This is the probability law introduced in Example 3.21, namely, the extreme value type I (EV1) distribution; this distribution will be discussed in Chapter 7.

uncertainties in the parameterization adopted to model a particular random variable, or to account for the additional uncertainties involved in the estimation of the parameters from the outcomes of the experiment. A simple case of a mixture is given by compounding two or more distributions $f_{X_i}(x)$ weighted by the probability p_i that the outcome x of the investigated random variable X comes from the ith population. In this case, the pdf of X is given by

$$f_X(x) = \sum_{all\ i} p_i f_{X_i}(x). \tag{3.4.43}$$

This probability model can be used, for example, to describe the annual flood flows at a point of observation in a river by using the seasonal floods, which are described by different distributions.

3.4.4.2 Properties of contagious variables

The following example is an illustration of some of the basic properties of contagious variables. More follows in Chapter 7.

Example 3.62. Seasonal rainfall. In previous illustrations (see Examples 3.49 and 3.53), a rainfall event occurring at a point in space was described by two variables, namely, the duration X of the storm, and its intensity Y, which is defined as the average rain rate. Assume that X and Y are independent variables with means of

$$\mu_X = 10 \text{ mm/h}, \quad \mu_Y = 3 \text{ hours},$$

and standard deviations of

$$\sigma_X = 5 \text{ mm/h}, \quad \sigma_Y = 2 \text{ hours},$$

respectively. The total amount of rainfall delivered in a storm is $Z = XY$. Using Eq. (3.4.31), we find the mean of this variable is

$$\mu_Z = \mu_X \mu_Y = 10 \times 3 = 30 \text{ mm};$$

using Eq. (3.4.35) we compute its standard deviation as

$$\sigma_Z = \sqrt{\mu_X^2 \sigma_Y^2 + \mu_Y^2 \sigma_X^2 + \sigma_X^2 \sigma_Y^2} = \sqrt{10^2 \times 2^2 + 3^2 \times 5^2 + 5^2 \times 2^2} = \sqrt{725} \approx 27 \text{ mm}.$$

If n storm events occur in a season, the rainfall total T in that season will be given by the sum of the rainfall amounts delivered by each of the events. If the storms are independent of each other,

$$T = \sum_{i=1}^{n} Z,$$

so that its mean is $n\mu_Z$, and its variance equals

$$\text{Var}[T] = \text{Var}\left[\sum_{i=1}^{n} Z\right] = n\text{Var}[Z] = n\sigma_Z^2.$$

Because of the randomness of storm occurrences, the number of such occurrences should be considered as a random variable, say, N. Therefore, this can be viewed as a contagious or mixture variable T with a parameter N. Thus, one must seek the mean and variance of T conditional to a given outcome n of occurrences, and then account for all possible values taken by N. Since

$$E[T|n = N] = E\left[\sum_{i=1}^{n} Z\right] = n\mu_Z,$$

the conditional mean is given by $E[T \mid N] = N\mu_Z$. Therefore,

$$E[T] = E[E[T \mid N]] = E[N\mu_Z] = E[N]\mu_Z = \mu_N \mu_Z,$$

where μ_N denotes the mean number of storms in a season. Since

$$E[T^2 \mid N = n] = \text{Var}[T \mid N = n] + E^2[T \mid N = n] = \text{Var}\left[\sum_{i=1}^{n} Z\right] + (n\mu_Z)^2$$

$$= n\sigma_Z^2 + n^2\mu_Z^2,$$

the second-order (raw) moment conditional to N is given by

$$E[T^2 \mid N] = N\sigma_Z^2 + N^2\mu_Z^2.$$

So that

$$E[T^2] = E[E[T^2 \mid N]] = E\left[N\sigma_Z^2 + N^2\mu_Z^2\right] = E[N]\sigma_Z^2 + E[N^2]\mu_Z^2,$$

$$= \mu_N\sigma_Z^2 + \left(\mu_N^2 + \sigma_N^2\right)\mu_Z^2.$$

The required variance of T is thus given by

$$\sigma_T^2 = E[T^2] - E^2[T] = \mu_N\sigma_Z^2 + \left(\mu_N^2 + \sigma_N^2\right)\mu_Z^2 - \mu_N^2\mu_Z^2 = \mu_N\sigma_Z^2 + \sigma_N^2\mu_Z^2.$$

3.4.5 Summary of Section 3.4

In this section we have provided more advanced concepts of associated random variables. These include functions of two or more variables, products, transformations, and compound variables. The mgfs of derived variables and contagious distributions are also discussed.

3.5 COPULAS

In Section 3.3 we provided details of multiple random variables and their joint probability distributions. This was followed by the functional relationships between random variables and transformation techniques of Section 3.4, in which we gave details of methods of obtaining the distributions of associated and transformed jointly distributed random variables. In this section we introduce briefly the related subject of copulas. Our aim is to obtain joint distributions of associated series of random variables. The presentation is limited to bivariate probability distributions, and hence 2-copulas, but the method can be extended to multivariate distributions.

The term derives from the latin verb *copulare* which essentially translates as "to join together" (Sklar, 1959). In its simplest form, a copula function is a bivariate distribution function with uniform marginal distributions. At the next level of modeling one may transform the two continuous random variables, that we consider, in such a way that their marginal distributions are uniform over a certain interval of length; this involves standardization of the parameters of the bivariate distribution under scale invariance (see Subsection 7.3.1).

For the statistician, copulas provide a means of studying scale-free measures of dependence between random variables and as a prelude to the construction of families of bivariate distributions, with possible use in simulations. The origins of the subject can be traced back to three seminal papers by Wassily Hoeffding, written in German in the early 1940s (see translations by Fisher and Sen, 1994). In his first paper, Hoeffding stated that all the properties of a mutivariate distribution that pertain to the topic of correlation can be divided into two classes, depending on whether or not they are invariant to arbitrary changes of scale (see one-to-one correspondence introduced in Subsection 3.4.1 and the probability integral transform of Subsection 8.1.2). Let us now provide some specifics.

The determination of a probability model for dependent bivariate observations $(X_1, Y_1), \ldots, (X_n, Y_n)$ from a population with a nonnormal distribution function $F_{X,Y}(x, y)$ can be simplified by expressing $F_{X,Y}(x, y)$ in terms of its marginals, $F_X(x)$ and $F_Y(y)$ and an associated dependence function $\mathbf{C}$, called a 2-copula, implicitly defined through the functional identity $F_{X,Y}(x, y) = \mathbf{C}(F_X(x), F_Y(y))$. A natural way of studying bivariate data thus consists of separately estimating the dependence function and the marginals. This two-step approach to stochastic modeling is often convenient, since many tractable models are readily available for the marginal distributions. It is clearly appropriate when the marginals are known, and it is invaluable as a general strategy for data analysis in that it enables the dependence structure to be investigated independently of marginal effects.

Let $\mathbf{I} = [0, 1]$. A 2-copula function is a bivariate function $\mathbf{C} : \mathbf{I} \times \mathbf{I} \to \mathbf{I}$ such that

(1) for all $u, z \in \mathbf{I}$,

$$\mathbf{C}(u, 0) = 0, \mathbf{C}(u, 1) = u, \mathbf{C}(0, z) = 0, \text{ and } \mathbf{C}(1, z) = z;$$

(2) for all $u_1, u_2, z_1, z_2 \in \mathbf{I}$ such that $u_1 \leq u_2$ and $z_1 \leq z_2$,

$$\mathbf{C}(u_2, z_2) - \mathbf{C}(u_2, z_1) - \mathbf{C}(u_1, z_2) + \mathbf{C}(u_1, z_1) \geq 0.$$

[For mathematical details that are omitted in this Section see Joe (1997), Nelsen (1999), and Salvadori et al. (2007).]

The link between 2-copulas and bivariate distributions is provided by Sklar's theorem: Let X and Y be two continuous random variables, and let $F_{X,Y}(x, y)$ be their bivariate distribution function with marginals $F_X(x)$ and $F_Y(y)$. Then there exists a unique 2-copula $\mathbf{C}$ such that

$$F_{XY}(x, y) = \mathbf{C}(F_X(x), F_Y(y)), \quad \text{for all } x, y. \tag{3.5.1}$$

Conversely, if $\mathbf{C}$ is a 2-copula and $F_X(x)$ and $F_Y(y)$ are distribution functions, then $F_{X,Y}(x, y)$ is a bivariate distribution function with marginals $F_X(x)$ and $F_Y(y)$.

The interesting point is that the properties of $F_{X,Y}(x, y)$ can be discussed in terms of the structure of $\mathbf{C}$. It is precisely the 2-copula that captures many of the features of a bivariate distribution. Also, it enables one to investigate measures of association and dependence properties between random variables. Furthermore, a 2-copula describes exactly and models the dependence structure between two random variables, independently of the marginal laws of the variables involved. Clearly, this provides freedom in choosing the univariate marginal distributions once the desired dependence framework has been selected, and it usually makes it easier to formulate bivariate (and hence multivariate) models. Incidentally, we observe that all the bivariate models seen in the literature can easily be described in terms of appropriate 2-copula.

For example, the Gumbel family of 2-copula has the following analytical expression:

$$\mathbf{C}_\delta(u, z) = \exp\{-[(-\ln u)^\delta + (-\ln z)^\delta]^{1/\delta}\}, \tag{3.5.2}$$

where $u, z \in \mathbf{I}$ and $\delta \in [1, \infty)$. Here δ represents the dependence parameter. The (limit) case $\delta = 1$ corresponds to independent variables, with $\mathbf{C}_1(u, z) = uz$; the (limit) case

$\delta \to \infty$ corresponds to complete dependence between the variables. Note that this family of 2-copula models positively dependent variables.

The Frank family of 2-copulas has the following expression:

$$C_\delta(u,z) = -\frac{1}{\delta}\ln\left(1 + \frac{(e^{-\delta u}-1)(e^{-\delta z}-1)}{e^{-\delta}-1}\right), \tag{3.5.3}$$

where $u, z \in \mathbf{I}$ and $\delta \in (-\infty, +\infty)$. The case $\delta < 0$ corresponds to a negative dependence, the case $\delta > 0$ corresponds to a positive dependence, and the (limit) case $\delta = 0$ corresponds to independent variables, with $C_0(u,z) = uz$; thus, members of the Frank family of 2-copulas can model both negatively and positively dependent variables.

Let us assume that the two continuous random variables of interest, X and Y, have exponential and Pareto marginals, respectively:

$$F_X(x) = 1 - e^{-\lambda x}, \quad x \ge 0 \tag{3.5.4}$$

and

$$F_Y(y) = 1 - \left(\frac{y_0}{y}\right)^\theta, \quad y \ge y_0, \tag{3.5.5}$$

where $\lambda > 0$ is the parameter of the exponential distribution, and $\theta > 1$ and y_0 are the shape parameter and lower bound of the Pareto distribution, respectively.

If the dependence between X and Y is described by a 2-copula from Gumbel's family (including the GEV distribution of Subsection 7.2.5), by using Sklar's theorem it is easy to calculate the bivariate distribution $F_{X,Y}(x,y)$:

$$\begin{aligned} F_{X,Y}(x,y) &= \mathbf{C}_\delta(F_X(x), F_Y(y)) = \exp\{-[(-\ln F_X(x))^\delta + (-\ln F_Y(y))^\delta]^{1/\delta}\} \\ &= \exp\left\{-\left[(-\ln(1-e^{-\lambda x}))^\delta + \left(-\ln\left(1-\left(\frac{y_0}{y}\right)^\theta\right)\right)^\delta\right]^{1/\delta}\right\}. \end{aligned} \tag{3.5.6}$$

The dependence parameter δ of the 2-copula is generally expressed in terms of some measure of association such as Kendall's tau, τ, or Spearman's rho, ρ. These measures were introduced to generalize the linear coefficient of correlation, also called Pearson's product-moment correlation coefficient [see Eq. (1.4.5)]. This coefficient has been used extensively as a measure of dependence between random variables, even though it is not the best measure of dependence for nonnormal random variables.

Kendall's τ rank correlation coefficient can be expressed as a one-to-one function of δ as

$$\tau(\delta) = 4\int_0^1\int_0^1 C_\delta(F_X(x), F_Y(y))dC(F_X(x), F_Y(y)) - 1. \tag{3.5.7}$$

For Gumbel's family, this becomes

$$\tau(\delta) = \frac{(\delta - 1)}{\delta}. \tag{3.5.8}$$

For Frank's family, the following approximate relationship holds:

$$\tau(\delta) = \frac{1}{9}\delta - \frac{1}{900}\delta^3 + \frac{1}{52,920}\delta^5 - \frac{1}{2,721,600}\delta^7 + \cdots. \tag{3.5.9}$$

This provides a good approximation to τ for $|\delta| < 5$.

Once an estimate of τ is obtained (see, for example, Salvadori et al., 2007, p. 229), it is then possible to calculate an estimate of δ from Eqs. (3.5.7) to (3.5.9). Alternatively, an estimate of δ can be obtained using the maximum likelihood method.

In summary, the 2-copula method provides a convenient means of obtaining the bivariate distribution of two continuous associated random variables with known marginals. It can be extended to multivariate distributions. For applications of copulas to geophysical problems see Salvadori et al. (2007).

3.6 SUMMARY FOR CHAPTER 3

The concept of a random variable which was initially discussed in Chapters 1 and Chapter 2 is formerly presented in this chapter, and its relevance to diverse practical applications is shown. Discrete variables associated with counting processes and continuous variables used to model various observations of engineers are considered throughout. The mean, variance, and moments of higher order are given as descriptors of random variables. The moment-generating and characteristic functions are presented. Methods of estimation introduced here include the classical moments and maximum likelihood procedures and alternative methods such as probability weighted moments. Distributions of functions of random variables are determined. Jointly distributed random variables are presented. Joint, conditional, and marginal distributions are examined; among the topics discussed are correlation and independence. Details of methods of obtaining the distributions of associated and transformed jointly distributed random variables are also given. We then introduce the related subject of copulas, an alternative method of modeling joint distributions. Thus the fundamental tools required by a civil or environmental engineer who applies statistical and probabilistic methods are shown in this chapter, with detailed methods of application; this chapter is an essential sequel to the basic concepts outlined in Chapter 2.

REFERENCES

General. The following references are given for further reading as necessary:

Ang, A. M. S., and W. H. Tang (1975). *Probability Concepts in Engineering Planning and Design, Vol. 1: Principles*, John Wiley and Sons, New York. Theory and practice well blended with engineering appeal and many worked examples for civil engineers.

Benjamin, J. R., and C. A. Cornell (1970). *Probability, Statistics and Decision for Civil Engineers*, McGraw-Hill, New York. Still the classic book for civil engineers, with numerous examples, case studies, and problems to solve.

Blake, I. F. (1979). *An Introduction to Applied Probability*, John Wiley and Sons, New York. Excellent introduction.

Blank, L. T. (1980). *Statistical Procedures for Engineering, Management and Science,* McGraw-Hill, New York. A good mix of well-explained theory and practical examples; ample coverage of random variables.

Evans, D. H. (1992). *Probability and its Application for Engineers*, Marcel Dekker, New York. Chapters 4 to 6 on random variables make this a useful introduction.

Feller, W. (1968). *An Introduction to Probability Theory and Its Applications*, Vol. 1, 3rd. ed., John Wiley and Sons, New York. A classic in probability theory suitable for the discerning student.

Hahn, G. J., and S. S. Shapiro (1967). *Statistical Methods in Engineering*, John Wiley and Sons, New York. Reprinted in 1994 as one of the Wiley classics in the subject area. Recommended for understanding the basics as applied to engineering.

Hald, A. (1952). *Statistical Theory with Engineering Applications*, John Wiley and Sons, New York. Reference book for engineers with an introduction to joint densities.

Harr, M. E. (1987). *Reliability-Based Design in Civil Engineering*, McGraw-Hill New York. Modern text oriented to design and reliability theory; particularly useful for civil engineers.

Helstrom, C. W. (1991). *Probability and Stochastic Processes for Engineers*, Macmillan, New York. Theory well explained, though somewhat advanced; a wide coverage of multiple random variables.

Hines, W. H., and D. C. Montgomery (1990). *Probability and Statistics in Engineering and Management Science,* 3rd ed., John Wiley and Sons, New York. Comprehensive book of 700 pages, 19 chapters (3 on random variables,) 300 examples and 626 problems.

John, P. W. M. (1990). *Statistical Methods in Engineering and Quality Assurance*, John Wiley and Sons, New York. Many well-selected examples; Chapters 3 and 4 are on discrete and continuous *variables.*

Kelley, D. G. (1994). *Introduction to Probability,* Macmillan, New York. For the beginner, a suggested introduction to be read before proceeding to advanced books.

Kennedy, J. B. and A. M. Neville (1986). *Basic Statistical Methods for Engineers and Scientists*, 3rd ed., Harper and Row, New York. A commendable basic introduction.

Larsen, R. J., and M. L. Maw (1986). *An Introduction to Mathematical Statistics and its Applications*, 2nd. ed., Prentice Hall, Englewood Cliffs, NJ. Excellent chapter 3 on random variables; explanations and applications of moment-generating function.

Metcalfe, A. V. (1994). *Statistics in Engineering—A Practical Approach*, Chapman and Hall, London. An intermediate-level reference book with many civil engineering applications.

Mood, A. M., F. A. Graybill, and D. C. Boes (1974). *Introduction to the Theory of Statistics*, 3rd ed., McGraw-Hill, New York. Chapters 2 and 4 on random variables recommended for above-average students.

Neville, A. M., and J. B. Kennedy (1964). *Basic Statistical Methods for Engineers and Scientists*, Intext Publishing Co., New York. A beginner's guide.

Parzen, E. (1960). *Modern Probability Theory and Its Applications*, John Wiley and Sons, New York. Best suited for students requiring further knowledge in probability theory.

Smith, G. N. (1986). *Probability and Statistics in Civil Engineering, An Introduction*, William Collins & Sons, Ltd, London. Basic examples with emphasis later on reliability methods, soil mechanics, and structures.

Vardeman, S. B. (1994). *Statistics for Engineering Problem Solving*, PWS Publishing Company, Boston. Very good introduction to random variables (Chapter 5) with basic treatment of joint random variables.

Wackerly, D. D., W. Mendenhall, and R. L. Scheaffer (2002). *Mathematical Statistics with Applications*, 6th ed., Duxbury, Pacific Grove, CA. Explanations of moment-generating functions and bivariate distributions.

Zuzek, W. H. (ed.) (1990). *Complexity, Entropy, and the Physics of Information*, Addison-Wesley, Redwood City, CA. A wide range of applications of entropy.

Additional references quoted in text

Abramowitz, M., and I. A. Stegun (eds.) (1964). *Handbook of Mathematical Functions*, National Bureau of Standards (United States), Applied Mathematics Section, Publication No. 55, Dover, New York. Much used handbook.

Bacchi, B., G. Becciu, and N. T. Kottegoda (1994). "Bivariate exponential model applied to intensities and durations of extreme rainfall," *J. Hydrol.*, Vol. 155, pp. 225–236. Storm modeling.

Bacchi, B., R. Rosso, and P. La Barbera (1987). "Storm characterization by Poisson models of temporal rainfall," *Proc. XXII Congr. Int. Assoc. Hydraul. Res.*, Lausanne, August 31–September 4, Vol. 4, pp. 35–40. Storm modeling.

Boltzmann, L. (1894). "Zur Integration der Diffusionsgleichung bei variablen Diffusionskoeffizienten," *Ann.der Phys. Lpz.*, Vol. 53, pp. 959–965. Pioneering concepts which led to entropy theory, translated to English.

David, H. A., and H. N. Nagaraja (2003). *Order Statistics*, 3nd ed., John Wiley and Sons, New York. Standard (advanced level) text for order statistics.

Eagleson, P. S. (1978). "Climate, soil and vegetation: 2. The distribution of annual precipitation derived from observed storm sequence," *Water Resour. Res.*, Vol. 14 No. 5, pp. 713–771. An in-depth study.

Efron (1979). "Bootstrap methods: another look at the jackknife," *Ann. Stat.*, Vol. 7, pp. 1–26. Original reference given for the bootstrap.

Feller, W. (1968). *An Introduction to Probability Theory and its Applications*, Vol. 1, 3rd ed., John Wiley and Sons, New York. A classic in probability theory suitable for the discerning student.

Fisher, N. I., and P. K. Sen (1994). *The Collected Works of Wassily Hoeffding*, Springer, New York. Translations of Hoeffding's original work related to copulas.

Granovsky, B., and H. G. Muller (1991). "Optimizing kernel methods: A unifying variational principle," *Int. Statistic. Rev.*, Vol. 59, pp. 373–388. A review of kernel choice in nonparametric estimation of a pdf.

Greenwood, J. A., J. M. Landwehr, N. C. Matalas, and J. R. Wallis (1979). "Probability weighted moments compared with some traditional techniques in estimating Gumbel parameters and quantiles," *Water Resour. Res.*, Vol. 15 No. 5, pp. 1049–1054. Original reference on probability weighted moments.

Gumbel, E. J. (1960). "Bivariate exponential distributions," *J. Am. Stat. Assoc.*, Vol. 55, pp. 698–707. Original work on this family of distributions.

Harrop-Williams, K. (1983). "Some geotechnical applications of entropy," in *Proceedings of the Applications of Statistics and Probability in Soil and Structural Engineering*, 4th Int. Conf. 13–17 June 1983, edited by G. Augusti, A. Borri, and G. Vannucchi, pp. 1643–1654, Civil Eng. Dept., Univ. of Firenze, Italy, Pitagora Editrice, Bologna. Practical examples with comparison of conventional variance and entropy analysis of soil properties.

Hosking, J. R. M. (1990). "L-moments: Analysis and estimation of distribution using linear combinations of order statistics," *J. R. Stat. Soc.*, B, Vol. 52, pp. 105–124. Reference given for the theory of L moments.

Jaynes, E. T. (1957). "Information theory and statistical mechanics, II," *Phys. Rev.*, Vol. 108, pp. 620–630. Useful review on the principles and application of the entropy concept.

Joe, H. (1997). *Multivariate Models and Dependence Concepts*, Chapman and Hall, London. Introduction to copulas.

Miller, R. G. (1974). "The jackknife—a review," *Biometrika*, Vol. 61, pp. 1–15. A mathematical review of the jackknife method.

Nelsen, R. B (1999). *An Introduction to Copulas*, Springer, New York. On copulas.

Papoulis, A. (2001). *Probability, Random Variables and Stochastic Processes*, 4th. ed., McGraw-Hill, New York. A theoretically advanced book suitable for those proceeding to higher studies.

Press, W. H., S. A. Teukolsky, W. T. Vetterling, and B. P. Flannery (1992). *Numerical Recipes in Fortran: The Art of Scientific Computing*, 2nd. ed., Cambridge University Press, Cambridge. Bootstrap confidence limits with example (pp. 686–688).

Quenouille, M. H. (1956). "Notes on bias in estimation," *Biometrika*, Vol. 43, pp. 353–360. Original work on which the jackknife method is based.

Randles, R. H., and D. A. Wolfe (1979) *Introduction to the Theory of Nonparametric Statistics*, John Wiley and Sons, New York. Reference for "U" Statistics.

Salvadori G., C. De Michele, N. T. Kottegoda, and R. Rosso (2007). *Extremes in Nature: An Approach Using Copulas*, Springer, Dordrecht. Applications in geophysics.

Shannon, C. E. (1948). "The mathematical theory of communications, I and II," *Bell Syst. Tech. J.*, Vol. 27, pp. 379–423. On entropy.

Sheather, S. J., and M. C. Jones (1991). "A reliable data-based bandwidth selection method for kernel density estimation," *J. R. Stat. Soc.*, B, Vol. 53, pp. 683–690. Reference given for kernel density estimation.

Silverman, B. W. (1986). *Density Estimation for Statistics and Data Analysis*, Chapman and Hall, London. A standard reference for density estimation.

Sklar, A. (1959). "Fonctions de répartition à *n* dimensions et leurs marges," *Publ. Inst. Statist. Univ. Press*, Vol. 8, pp. 229–231.

Smith, R. L., and J. C. Naylor (1987). "A comparison of maximum likelihood and Bayesian estimation for the three-parameter Weibull distribution," *Appl. Stat.*, Vol. 36, pp. 358–369. Reference for comparative methods of estimating the parameters of the Weibull distribution.

Sonuga, J. O. (1972). "Principle of maximum entropy in hydrologic frequency analysis," *J. Hydrol.*, Vol. 17, pp. 177–191. Reference cited in the application of the principle of maximum entropy.

Stuart, A., and J. K. Ord (1994). *Kendall's Advanced Theory of Statistics*, Vol. 1, 6th ed., Edward Arnold Ltd, London. See Chapter 3 for cumulants.

Zucchini, W., and P. T. Adamson (1988). "Confidence limits using bootstrap," *Hydrol. Sci. J.*, Vol. 34 No. 1, pp. 41–48. Hydrological application of bootstrap in assessing standard errors.

PROBLEMS

3.1. Sea waves. The pmf of the observed number of days per month of high-amplitude waves acting on a sea pier is given below.

X =	0	1	2	3	4	5	6	≥ 7
$p_X(x)$ =	0.38	0.22	0.18	0.13	0.09	0.06	0.03	0.01

Determine the expected value and variance of X.

3.2. Tensile strength. The tensile strength in a structural material is found to be highly variable, although tests showed that there is an increasing number of specimens of high strengths with a possible limit of 20 N/mm^2 in strength. Based on observations and as a first approximation, the pdf of tensile strength X is represented by the function $f_X(x) = ax^2, 0 \leq x \leq 20$ N/mm^2.
(a) Determine the constant a in the function.
(b) What is the probability of $X > 10$ N/mm^2?

3.3. Wind load. A tower is subject to a horizontal force caused by high winds. An important factor which should be taken into account when strengthening the tower is the duration of the winds. The duration T of winds in the area is a random variable with a maximum of 18 hours. From observations of wind data, the pdf of T can be approximated to the form $f_T(t) = ct^{1.5}$, with a maximum ordinate of k.
(a) Evaluate c and k.
(b) Find the mean and coefficient of variation of T.
(c) What is the probability of a wind lasting more than 9 hours?

3.4. Flood exceedance. A flow of magnitude 40 m^3/s is exceeded at a particular site on a river once in 3 months on average. What is the probability of having at least one such flood in a year? State assumptions made.

3.5. Compressive strength of concrete. The expected value of the compressive strength of a particular concrete is 60 N/mm^2 and the coefficient of variation is 10%. Assuming that the theoretical probability distribution is symmetrical but is unknown, calculate the probability that the compressive strength will be greater than 50 N/mm^2.

3.6. Highway accidents. Highway accidents along a busy highway leading away from a city have the following pmf (see Example 3.18 for this Poisson pmf):

$$p_X(x) = v^x \frac{e^{-v}}{x!}, \quad \text{for } x = 0, 1, 2, \ldots.$$

Originally v has been estimated as 0.9. Subsequently, the exit road was widened and the parameter was estimated as 0.5. Plot the pmf in each case and determine the probabilities of $\Pr[X > 0]$.

3.7. Earthquake occurrence. During a period of 125 years, 16 major earthquakes have occurred in the San Francisco area. Assuming these are Poisson events (see Problem 3.6 and Example 3.18), determine
(*a*) the probability of more than one such earthquake during a 5-year period, and
(*b*) the mean time between such earthquakes.

3.8. Computer system failure. The times to failure in months of several identical computer systems are observed as follows: 21, 53, 43, 56, 18, 17, 40, 14, 13. Assuming these are distributed as $F_T(t) = 1 - e^{-\lambda t}$, estimate the parameter λ by the method of maximum likelihood. Repeat the procedure using the method of moments.

3.9. Maximum flows. In some applications the exponential distribution of Problem 3.8 is written with a lower bound ε and this makes $F_T(t) = 1 - \exp[-\lambda(t - \varepsilon)]$. Show how the parameters may be estimated using the probability weighted moments procedure.

3.10. Occurrence of volcanic eruptions. There are frequent volcanic eruptions at a particular site. The times of the occurrences are unpredictable. From past observations, the pmf of occurrences X over periods of 10 years is as follows:

X	=	0	1	2	3
$p_X(x)$	=	0.1	0.3	0.4	0.2

What entropy does this distribution represent? What is the maximum possible entropy for the four values of probability?

3.11. Pipe settlement. Three subcontractors laid water pipes running through a flat part of a city. Excavations made at 3-meter intervals along the pipelines after a period of 5 years showed that settlements had taken place from the original lends. The following table gives the settlements in millimeters at each excavation:

Subcontractor 1	181	190	71	55	105
Subcontractor 2	99	78	25	50	198
Subcontractor 3	23	23	197	75	189

If in a particular case, the settlements had been the same at each point of observation along the pipeline, no problem will arise with regard to the system. On the basis of entropy, determine the relative settlement of the pipes laid by each subcontractor. Which system has the least relative settlement? What is the entropy of a particular system with no relative settlement?

3.12. Project scheduling. In a building project, the construction of the foundations takes time T_1 and the construction of the superstructure takes time T_2. On account of inclement weather, labor problems, and other factors, T_1 and T_2 behave like random variables with empirical pmfs as follows:

Time in weeks	1	2	3	4	5	6	7
$p_{T1}(t_1) =$	0.1	0.3	0.4	0.2	0.0	0.0	0.0
$p_{T2}(t_2) =$	0.0	0.0	0.0	0.1	0.5	0.4	0.0

(*a*) Calculate the mean times taken for the foundations and the superstructure.
(*b*) Evaluate the pmf of the total time spent on the foundations and superstructure.
(*c*) What is the probability that the total work will be completed in less than 7 weeks?

3.13. Sea pier construction. With reference to the data given in Problem 3.1, a contractor is assigned to work on an extension to the sea pier. The contractor finds that the profits Y of the job are directly decreased by the number of days per month X of high-amplitude waves acting on the sea front. It is estimated that $Y = 10{,}000(10 - X)$. Determine the pmf of Y and the mean and variance of Y.

3.14. Head loss in a pipe. The head loss H in a pipe is related to the mean velocity of flow V as $H = kV^2$, where k is a constant depending on pipe length, diameter, and roughness. In a particular case, V varies randomly between limits v_1 and v_2. Assuming a symmetrical triangular pdf for V, derive the pdf of H.

3.15. Joint wind measurements. For the joint pdf of the number of days of occurrences of high winds recorded by two instruments and given in Table 3.3.1, evaluate the probability that the differences between the observations by the two instruments are not greater than 1.

3.16. Contract analysis. A contractor's financial outlay X and labor force Y are random variables with bivariate pdf given by:

$$f_{X,Y}(x, y) = kxy, \quad \text{for } 10{,}000 < x < 100{,}000 \quad \text{and} \quad 10 < y < 20,$$

$$\text{and} \quad = 0, \quad \text{elsewhere.}$$

(*a*) Evaluate constant k.
(*b*) Determine the marginal pdf of X and Y.

3.17. Welding legs. The joint pdf of the lengths of horizontal and vertical legs, X and Y, of welding joints (similar to the ones referred to in Example 1.4) is given by

$$f_{X,Y}(x, y) = \frac{1}{16}xy, \quad \text{for } 4.0 < x, y < 8.0$$

$$\text{and} \quad = 0, \quad \text{elsewhere.}$$

Determine the probability $\Pr[5.5 < X < 6.5; 5.5 < Y < 6.5]$.

3.18. Density and compressive strength of concrete. Estimate the correlation in the case of the simplified joint distribution of concrete density and compressive strength given in Example 3.37 and shown in Fig. 3.3.5 from the data of Table E.1.2.

3.19. Contractor's profits, financial outlay, and labor force. For the pdf given in Problem 3.16, the contractor's profits P may be assumed to be related to his financial outlay X and labor force Y as follows:

(1) $P = 1.3\,X + 15{,}000$

(2) $P = 1.2\,X + 1000\,Y + 10{,}000$

Determine the pdf of P in each case.

3.20. Rivet production. Two machines produce rivets for a factory job. The numbers of substandard rivets per hour by the two machines are random variables denoted by X_1 and X_2. The bivariate pmf of X_1 and X_2 is given by the following table:

	$X_2 = 0$	$X_2 = 1$	$X_2 = 2$	$X_2 = 3$	$p_{X_1}(x_1)$
$X_1 = 0$	0.07	0.05	0.02	0.01	0.15
$X_1 = 1$	0.05	0.10	0.12	0.02	0.35
$X_1 = 2$	0.02	0.12	0.17	0.05	0.36
$X_1 = 3$	0.01	0.01	0.05	0.07	0.14
$p_{X_2}(x_2)$	0.15	0.34	0.36	0.15	$\sum = 1.00$

(*a*) Determine the probability that the number of substandard rivets produced do not differ by more than 1 between one machine and the other.

(*b*) Determine the conditional distribution of $P_{X_2|X_1}(x_2 \mid x_1)$.

(*c*) The factory manager estimated that an older machine, which was replaced, produced $X_1 + X_2$ substandard rivets per hour. Estimate its marginal pmf.

3.21. Earthquake hazard. Two adjoining regions are subject to earthquakes at irregular intervals. The first region experiences X_1 earthquakes over a period of time, and X_2 earthquakes occur in the second region over the same period, where X_1 and X_2 are random variables. It is estimated that the joint distribution of earthquakes over the two regions is as follows:

$$p_{X_1,X_2}(x_1, x_2) = \frac{x_1 + x_2}{21}, \quad \text{for } x_1 = 0, 1, 2 \quad \text{and} \quad x_2 = 2, 3,$$

$$\text{and} \qquad = 0, \qquad \text{elsewhere.}$$

Determine the probabilities $p_{X_1|X_2}(x_1 \mid x_2)$ and the expected values $E[X_1 \mid X_2]$.

3.22. Water treatment plant. A water treatment plant has two units which are designed to perform with identical characteristics. The consequenses of both units failing simultaneously are severe on the community. The times to failure in days are denoted by X_1 and X_2 and their bivariate pdf is given by

$$f_{X_1,X_2}(x_1, x_2) = ae^{-b(x_1+x_2)} \quad \text{or} \quad x_1, x_2 \geq 0$$

(*a*) What is the relationship between the constants a and b?

(*b*) How may they be estimated in practice?

(*c*) What is the chance that both units will fail within a year?

3.23. Water treatment plant. In Problem 3.22 a change in design is made so that only one of the units needs to operate at a time. The second will be brought into operation only on failure of the first, whenever that happens. What is the probability that the plant will be inoperative within a year?

3.24. Sewerage pollution discharge. Two sewage plants serving different communities discharge a pollutant into a stream. The concentrations of the respective discharges are measured as X and Y parts per million. Suppose the bivariate distribution is given by

$$f_{X,Y}(x, y) = 2 - x - y,$$

for $0 \leq X, Y \leq 1$, and 0 elsewhere.
(a) Determine the joint probability $\Pr[X < 0.5, Y < 0.6]$.
(b) If $X \leq 0.5$, determine the distribution of Y.
(c) Determine the correlation between X and Y.

3.25. Dam construction. The times spent in months by a contractor, engaged in the construction of small earthen dams, on the substructure and conduit on the one hand and the dam itself on the other are random variables (on account of frequent interruptions by weather and other unpredictable factors) denoted by X_1 and X_2, respectively. These times have common expectations, and past experience suggests that the bivariate pdf can be approximated by

$$f_{X_1,X_2}(x_1, x_2) = ax_1x_2e^{-b(x_1+x_2)}, \quad \text{for } x_1, x_2 > 0.$$

Determine the probability that the time spent on the earthwork is greater than or equal to 1.5 times that on the earthwork.

3.26. Maximum annual flood. Flood flows at a given river site are assumed to be independent identically distributed variables. The peak flow X for each flood exceeding a level of a is assumed to have a distribution with cdf

$$F_X(x) = 1 - \left(\frac{a}{x}\right)^\theta$$

with $x > a$ (Pareto with parameters a and θ). Since flood events occur randomly, the number N of flood flows exceeding a in a year is assumed to be distributed as

$$p_N(n) = \frac{\nu^n e^{-\nu}}{n!}$$

for $n = 0, 1, 2, \ldots$ (Poisson with parameter ν). Show that the probability distribution of the annual maximum peak flow, Y is

$$F_Y(y) = \exp\left[-\left(\frac{x_0}{x}\right)^\beta\right],$$

[EV2 (Fréchet) distribution with parameters x_0 and β]. Find the relationships linking parameters x_0 and β with ν, θ, and a.

Chapter 4
Probability Distributions

The physical problems that confront an engineer are associated with random factors that can greatly influence outcomes. Hence, the application of probability models in seeking engineering solutions is often a necessity. The assumptions that one needs to make in practical situations depend on the phenomenon studied. The parameters of the model will of course vary from one case to another, but the distributions are often identifiable. They have been given common names such as normal, binomial, and Poisson. There are instances, however, when the answers may not be straightforward because of uncertainties regarding the underlying physical mechanisms.

The purpose of this chapter is to identify, describe, derive, and show the use of several types of probability models. The models are formulated through considerations of real-world phenomena. Together they constitute a set, or form of tool kit, which is sufficient for empirical usage once the model has been justified on physical grounds, not merely on the basis of the closeness of fit (as discussed in Chapter 5). Thus, the decision is made firstly through prior reasoning and secondly in a confirmatory manner from the shape of the histogram or the visible properties of other investigative diagrams described in Chapter 1. Accordingly, the presentation in this chapter is geared to facilitate an understanding of the laws governing each distribution. This can then be matched with the physical situation that the engineer encounters.

In this chapter we deal with statistical distributions which one needs to model populations. As in Chapter 3, the models are classified separately according to whether the variables are continuous or discrete. Discrete variables can only assume isolated numbers such as integers. Thus, the derivation of their distributions is necessarily different from that of continuous variables, which can have any value between two limits.

4.1 DISCRETE DISTRIBUTIONS

A discrete distribution is used to model a random variable X that has, at most, a countable sample space over a range of values. Our interest is in integer-valued outcomes of X although the theory is not confined to such.

There are times when interest is focused on an experiment consisting of a single trial, the outcome of which is deemed to belong to one of two categories. For example, such a trial might be the determination of whether a concrete sample will fail in compression when subjected to a specified load, or whether the flood stage of a river is exceeded, or whether a rivet made on a structure satisfies given specifications, or whether a soil boring encounters rock. In such cases the probability model that describes the event is associated with a simple discrete trial. Arising from the so-called Bernoulli process associated with such discrete trials, the binomial, geometric, and the negative binomial distributions all belong to the same category and are among those initially considered. Next, the Poisson process is extensively covered with numerous illustrations, because of its importance in probability and statistics. The log series and hypergeometric distributions are also discussed in this section as well as discrete variables that can be in one of several categories, such as the multinomial case.

4.1.1 Bernoulli distribution

The simplest type of experiment is one in which there is only a single trial. Consider, for example, one toss of a coin or simply one test which gives only a "yes" or "no" answer. In general, when referring to persons or objects, those with a particular attribute are distinguished from those who do not have it. Such a trial has only two possible outcomes, which are *mutually exclusive and collectively exhaustive.*

These concern the occurrence of an event, which is a "yes" result, usually called a *success*, and its nonoccurrence, or *failure*. The results from such a trial constitute a two-sided Bernoulli random variable (as described by James Bernoulli of Switzerland around the year 1700, when probability theory was first applied to games of chance). The distribution of the probabilities of the two outcomes is called a Bernoulli distribution. Furthermore, a series of these trials are said to constitute a Bernoulli process, *where a process is characterized by the behavior of the underlying system over time or space*. The governing criteria are these:

(1) There are only two possible outcomes, called a *success* or *failure*;
(2) The probability of occurrence of a success (or a failure) is constant;
(3) The trials are independent (that is, the outcome of a trial does not depend on the outcome of another trial).

The Bernoulli pmf is defined as

$$p_X(x) \equiv \Pr[X = x|p] = p^x(1-p)^{1-x}, \quad \text{for } x = 0, 1 \text{ and } 0 \le p \le 1$$
$$= 0, \quad \text{otherwise.} \tag{4.1.1}$$

where p denotes the probability of a success, that is, a "yes" result. An example of a Bernoulli distribution shown in Fig. 4.1.1, for $p = 0.7$.

By using expectation measures (defined in Subsection 3.2.1), the mean, variance, and moment-generating function of a Bernoulli-distributed variable are obtained as follows:

$$E[X] = (1)p + (0)(1-p) = p, \tag{4.1.2a}$$
$$\text{Var}[X] = (1-p)^2 p + (0-p)^2(1-p) = (1-p)p \tag{4.1.2b}$$

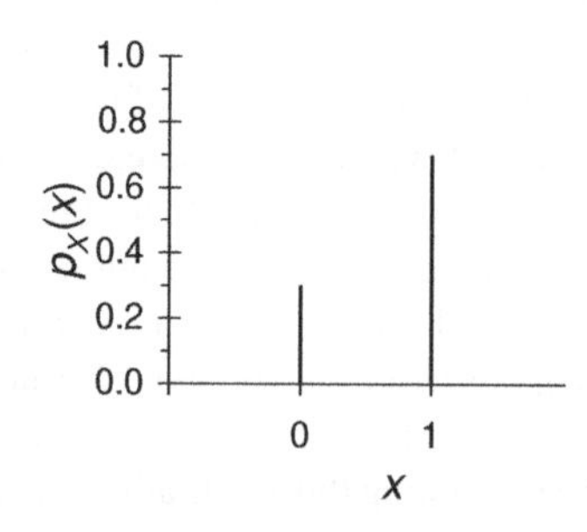

Fig. 4.1.1 Bernoulli pmf, $p_X(x)$, for $p = 0.7$.

and

$$M_X(t) = \sum_{x=0}^{1} e^{tx} p_X(x) = (1 - p) + pe^t, \tag{4.1.2c}$$

after substituting from Eq. (4.1.1)[1] [t is a dummy value, introduced in Eq. (3.2.17)].

The result for the mean [Eq. (4.1.2*a*)] confirms that the average number or proportion of times a success is achieved is the probability of a success. Less obviously from Eq. (4.1.2*b*) when this probability is 0.5 the variance of the Bernoulli variable is at a maximum of 0.25, and the variance decreases nonlinearly to zero as the probability tends to 0 or 1.

Definition and notation: A Bernoulli trial has only two possible outcomes: a *success* or a *failure*, with constant probabilities p and $(1 - p)$, respectively. The outcomes of a series of such trials are independent. The Bernoulli random variable X has pmf $p^x(1 - p)^{1-x}$ for $x = 0, 1$, and $0 \leq p \leq 1$, mean p, variance $(1 - p)p$, and mgf $(1 - p) + pe^t$.

The abbreviation $X \sim$ Bernoulli (p) will be used, as in this case, to differentiate between different types of distribution where the symbol $\sim$ means "distributed as."

4.1.2 Binomial distribution

Let us continue the preceding experiment (performing it under the same conditions) of determining whether an outcome of a trial is a success or failure. Also, as stipulated before, let the outcomes be independent of one another with a constant probability p of success. Then the sequence of such experiments is called a set of *Bernoulli trials*. For example, one may take n water samples to determine whether a particular pollutant is detectable or not and observe that there are m successes in all the trials. (It would seem that the term "success" is a misnomer because it is not a desirable happening if the pollutant is present; its use is merely conventional.)

Example 4.1. Water pollution. When monitoring a water pollutant consider these cases: (*a*) $n = 4$; $m = 1$; and (*b*) $n = 4$; $m = 2$, where n denotes the number of trials and m the number of successes.

(*a*) If the first trial is a success, the probability of occurrence of the compound event is given as follows on the assumption of independence of the trials[2]:

$$p(1 - p)(1 - p)(1 - p).$$

(*b*) Likewise, if the first two trials are successes, the probability of occurrence of the compound event is

$$p^2(1 - p)(1 - p).$$

In case *a* of Example 4.1, there can be four sequences or arrangements, each corresponding to a different compound event (depending on the order in which the two types of outcomes occur) but having the same joint probability of occurrence. Hence, the

---

[1] As a matter of interest, the Bernoulli distribution is applicable to other pairs of values which the random variable X can take apart from 1 to 0 (with different yet related means, variances, and mgfs).

[2] See Eq. (2.2.14).

probability of one success and three failures in four trials is obtained from the addition rule as follows[3]:

$$p_X(1) \equiv \Pr[X = 1|4, p] = \binom{4}{1} p(1-p)^3 = \frac{4!}{1!3!} p(1-p)^3 = 4p(1-p)^3.$$

In general, the result can be written as the pmf of a binomial random variate:

$$p_X(x) \equiv \Pr[X = x|n, p] = \binom{n}{x} p^x (1-p)^{n-x}, \quad \text{for } x = 0, 1, \ldots, n;\ 0 \le p \le 1$$
$$= 0, \quad \text{otherwise}, \qquad (4.1.3)$$

where

$$\binom{n}{x} = \frac{n!}{x!(n-x)!}$$

is the total number of possible combinations when selecting x objects from n objects.[4] Thus the random variable X, which represents the total number of successes in n trials, has a binomial distribution with parameters p (the probability of a success) and n.

The binomial cdf is given by

$$F_X(x) = \sum_{k=0}^{x} \binom{n}{k} p^k (1-p)^{n-k}. \qquad (4.1.4)$$

The maximum value of the cdf is at $X = n$, for which the right-hand side of Eq. (4.1.4) is the binomial expansion $[p + (1-p)]^n$, with $n + 1$ terms representing the probabilities from $k = 0$ to $k = n$. Because the maximum is equal to 1.0 and the cdf is positive and nondecreasing, Eq. (4.1.4) satisfies the requirements for a cdf.

In summary, for a random variable to have a binomial distribution, the following conditions are necessary:

(1) A series of Bernoulli trials is made, each of which has only one of two possible outcomes: a *success* or a *failure*.
(2) The trials are conducted under the same conditions and the probability p of a success is constant.
(3) The number of trials n is fixed.
(4) The outcomes of the trials are independent.
(5) The random variable X is the total number of successes in n trials and the order in which the events in the trials occur is immaterial.

Because each trial can only have one of two outcomes (a success or failure), the distribution is termed *binomial.*[5] For $n = 1$, the binomial random variable, is a simple Bernoulli random variable.

Example 4.2. Flooding of a road. Suppose a road is flooded with probability $p = 0.1$ during a year and not more than one flood occurs during a year. What is the probability that it will be flooded at least once during a 5-year period?

One needs to determine the probability of having no floods and subtracting this from unity, which is the sum of the probabilities of having 0, 1, 2, 3, 4, or 5 floods during the 5-year period. This procedure is followed when it is easier to compute the probability of the

[3] See Eq. (2.2.6).

[4] For long factorials, Stirling's formula can be applied: $n! \approx n^{n+1/2}\sqrt{2\pi}/\exp(n)$; see Feller (1968, pp. 52–53).

[5] When more than two outcomes are possible, the distribution is termed *multinomial* as shown in Subsection 4.1.6; the binomial is a special type of the multinomial distribution.

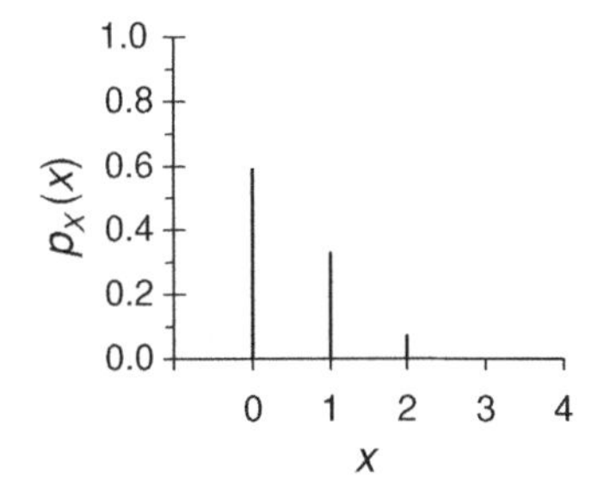

Fig. 4.1.2 Binomial pmf, $p_X(x)$, for $n = 5$ and $p = 0.1$.

complementary event than the probability of the stated event. Thus, the probability that the road will be flooded at least once is

$$1 - 0.1^0(1 - 0.1)^5 \approx 1 - 0.59 = 0.41.$$

The probabilities of having 0, 1, 2, 3, 4, and 5 floods of probability 0.1 during the 5-year period are $0.9^5 = 0.59049$; $5(0.1)(0.9)^4 = 0.32805$; $10(0.1)^2(0.9)^3 = 0.07290$; $10(0.1)^3(0.9)^2 = 0.00810$; $5(0.1)^4(0.9) = 0.00045$; and $0.1^5 = 0.00001$, respectively; the sum of these probabilities is unity.

Figure 4.1.2 shows the binomial probability distribution for all six possible events. As a matter of interest, this is also given for $p = 0.5$ in Fig. 4.1.3.

From Figs. 4.1.2 and 4.1.3 it is seen that the shape of the probability distribution depends on the values of the parameters p and N. If $p = 0.9$, the probabilities of Fig 4.1.2 are reversed thus changing the positively skewed distribution to one of negative skew.[6]

4.1.2.1 Mean and variance of a binomial vatiate

The mean of the binomial variate can be obtained from Eq. (3.2.6) as follows:

$$\begin{aligned}
E[X] &= \sum_{\text{all } x_j} x_j p_X(x_j) = \sum_{x=0}^{n} x \binom{n}{x} p^x (1-p)^{n-x} \\
&= \sum_{x=1}^{n} \frac{n!}{(x-1)!(n-x)!} p^x (1-p)^{n-x} \\
&= np \sum_{x=1}^{n} \frac{(n-1)!}{(x-1)!(n-x)!} p^{x-1} (1-p)^{n-x}.
\end{aligned}$$

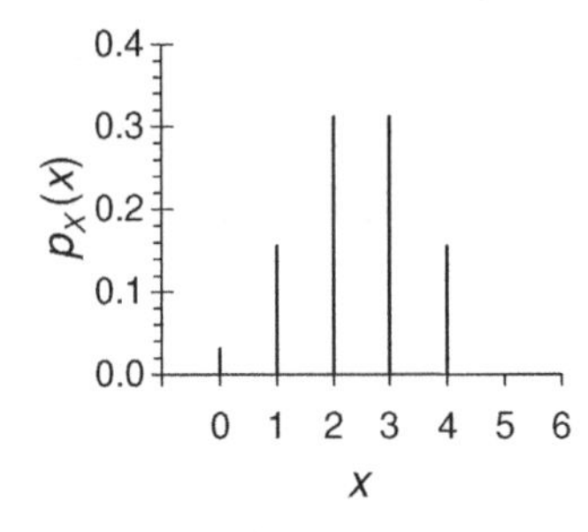

Fig. 4.1.3 Binomial pmf, $p_X(x)$, for $n = 5$ and $p = 0.5$.

[6] When the distribution is symmetrical, as in Fig. 4.1.3, its shape can be approximated by the continuous normal distribution; this is shown in Subsection 4.2.6.

Changing the variable on the right-hand side by writing $y = (x - 1)$, we have

$$E[X] = np \sum_{y=0}^{n-1} \frac{(n-1)!}{y!(n-1-y)!} p^y (1-p)^{n-1-y}.$$

Because the summation over all n items in Eq. (4.1.4) is unity, the summation over all $(n-1)$ items in the foregoing equation is also unity. Hence,

$$E[X] = np. \tag{4.1.5a}$$

The same result can be obtained by applying the factorial moment generating function, $E[t^x]$ of Subsection 3.2.2.2. Thus,

$$E[t^X] = \sum_{x=0}^{n} t^x \frac{n!}{x!(n-x)!} p^x (1-p)^{n-x}$$

$$\frac{d}{dt} E[t^X]|_{t=1} = E[Xt^{X-1}]|_{t=1} = E[X]$$

$$= \frac{d}{dt} \sum_{x=0}^{n} t^x \frac{n!}{x!(n-x)!} p^x (1-p)^{n-x} |_{t=1}$$

$$= np \sum_{x=1}^{n} \frac{(n-1)!}{(x-1)!(n-x)!} p^{x-1} (1-p)^{n-x} = np$$

[as in the derivation of Eq. (4.1.5a)].

To determine the variance through the factorial moment-generating function, one takes the second derivative of $E[t^X]$ at $t = 1$. Thus,

$$\frac{d}{dt^2} E[t^X]|_{t=1} = E[X(X-1)t^{X-2}]|_{t=1} = E[X^2] - E[X].$$

Also,

$$\frac{d}{dt^2} \sum_{x=0}^{n} t^x \frac{(n)!}{(x)!(n-x)!} p^x (1-p)^{n-x} |_{t=1}$$

$$= n(n-1)p^2 \sum_{x=2}^{n} \frac{(n-2)!}{(x-2)!(n-x)!} p^{x-2} (1-p)^{n-x}$$

$$= n(n-1)p^2,$$

because the foregoing summation is equal to unity, after substituting $y = x - 2$, for the reasons given before.

$$\begin{aligned} \text{Var}[X] &= E[X^2] - (E[X])^2 = (E[X^2] - E[X]) + E[X] - (E[X])^2 \\ &= (n^2p^2 - np^2) + np - n^2p^2 = np(1-p). \end{aligned} \tag{4.1.5b}$$

The moment-generating function is obtained from Eqs. (3.2.18) and (4.1.3) as

$$\begin{aligned} M_X(t) &= \sum_{x=0}^{n} e^{tx} \binom{n}{x} p^x (1-p)^{n-x} = \sum_{x=0}^{n} \binom{n}{x} (pe^t)^x (1-p)^{n-x} \\ &= [pe^t + (1-p)]^n, \end{aligned} \tag{4.1.5c}$$

which derives from the binomial theorem.

As already stated, a binomial variate X, which represents the number of successes in n trials, arises from a sequence of Bernoulli trials, each with a constant probability of success p. It follows, for example, that if a variate X_1 is binomially distributed with parameters

n_1 and p and another variate X_2 is binomially distributed with parameters n_2 and p, then $Y = X_1 + X_2$ is binomially distributed with parameters $n = n_1 + n_2$ and p. Similarly, this statement can be extended to more than two binomial variates.[7]

In this way the results of Eq. (4.1.5) can be obtained directly from those of Eq. (4.1.2) by considering that the binomial variate X is the sum of n (independent) Bernoulli variates each of which has a mean of p and a variance $p(1-p)$. The expectation operators are then applied as in Eqs. (3.3.22) and (3.3.23).

Definition and properties: In a series of independent Bernoulli trials, the outcome of each of which is either a *success* or *failure*, the random variable $X \sim$ binomial (n, p) is the number of successful trials out of a total of n trials. The number of trials n and the probability p of a success are constant. The pmf of X is

$$p_X(x) = \binom{n}{x} p^x (1-p)^{n-x}, \qquad \text{for } x = 0, 1, 2, \ldots, n;\ 0 \le p \le 1.$$

The variate X has mean np, variance $np(1-p)$, and mgf $[pe^t + (1-p)]^n$. A random variable Y which is the sum of two binomial variates X_1 with parameters n_1 and p and X_2 with parameters n_2 and p, respectively, is a binomial variate with parameters $n = n_1 + n_2$ and p.

Example 4.3. Bacterial count. A count of a particular type of bacterium is taken over a series of 10 tests. The numbers of positive results are as follows:

$$17, 21, 25, 23, 17, 26, 24, 19, 21, 17.$$

The mean and variance of the positive results are 21 and 10.6, respectively.

Assume that the tests are so conducted that conditions $1-5$ (stipulated earlier) for a binomial distribution apply. Then estimated values of n (the maximum number of organisms that a test sample can possibly have) and p (the probability of finding an organism at each trial applied to a portion of a test sample) can be obtained from the mean and variance using Eq. (4.1.5*a*) and (4.1.5*b*). Thus, $(1-p) = 10.6/21.0 = 0.505$. Hence, $p = 0.495$ and because the mean $= 21 = np$, $n = 43$.

Note that the conditions assumed in the foregoing examples may not always be found in practice. For example, the probability of a specific flood in a given year at a particular site often changes over time on account of climatic and geomorphological factors. In other applications, bacteria may be found in clusters, and outputs from machines will diminish in quality with wear. The approximations, however, are deemed to be close enough in many instances for the application of the theory of Bernoulli trials.

4.1.3 Poisson distribution

If p is small and n is large, the binomial can be approximated by the Poisson distribution (named after Poisson, a prominent French mathematician of the nineteenth century). This distribution is also based on the assumptions of independence and identical distribution. Let $\nu = np$ be the mean or expected number of successes in a series of n Bernoulli trials with probability p. Equation (4.1.3) can then be written as

$$\begin{aligned} p_X(x) &= \binom{n}{x}\left(\frac{\nu}{n}\right)^x \left(1-\frac{\nu}{n}\right)^{n-x} \\ &= \frac{n(n-1)(n-2)\cdots(n-x+1)}{n^x x!}\nu^x \left(1-\frac{\nu}{n}\right)^n \left(1-\frac{\nu}{n}\right)^{-x}. \end{aligned} \tag{4.1.6a}$$

[7] As stated in Subsection 3.1.1, we shall use the term *variate* when the distribution of the random variable is specified.

If x and v are fixed and finite,

$$\lim_{n\to\infty} \frac{n(n-1)(n-2)\cdots(n-x+1)}{n^x} = 1 \quad \text{and} \quad \lim_{n\to\infty} \left(1-\frac{v}{n}\right)^{-x} = 1.$$

If, for example, z is some parameter, where $0 < z < 1$, consider the division of 1 by $(1-z)$ which results in the series

$$\frac{1}{1-z} = 1 + z + z^2 + z^3 \cdots.$$

Integrating both sides of the equation, one obtains the series

$$\ln\left(\frac{1}{1-z}\right) = -\ln(1-z) = z + \frac{z^2}{2} + \frac{z^3}{3} + \frac{z^4}{4} + \cdots, \tag{4.1.6b}$$

which can also be obtained by considering $f(z) = \ln(1-z)$ as a Taylor series.[8] Hence, if $d = \left(1-\frac{v}{n}\right)^n$,

$$\ln(d) = n\ln\left(1-\frac{v}{n}\right) = -v - \frac{v^2}{2n} - \frac{v^3}{3n^2} - \cdots.$$

If n tends to infinity in the preceding series, only the first term is nonzero. Therefore,

$$\lim_{n\to\infty}\left(1-\frac{v}{n}\right)^n = e^{-v}.$$

Thus, when p is small and n tends to infinity such a way that $np = v$ in [Eq. (4.1.6a)], the random variable X is Poisson-distributed with parameter v and pmf:

$$\begin{aligned} p_X(x) \equiv \Pr[X = x|v] &= \frac{v^x e^{-v}}{x!}, \quad \text{for } x = 0, 1, 2, \ldots, \quad \text{and } v > 0 \\ &= 0, \qquad \text{otherwise.}^9 \end{aligned} \tag{4.1.7}$$

Examples of Poisson pmfs are given for $v = 0.2, 2.5$, and 5.0 in Figs. 4.1.4, 4.1.5, and 4.1.6, respectively. Note that the Poisson pmf, which for small values of v is positively skewed, tends to become symmetrical as v increases.

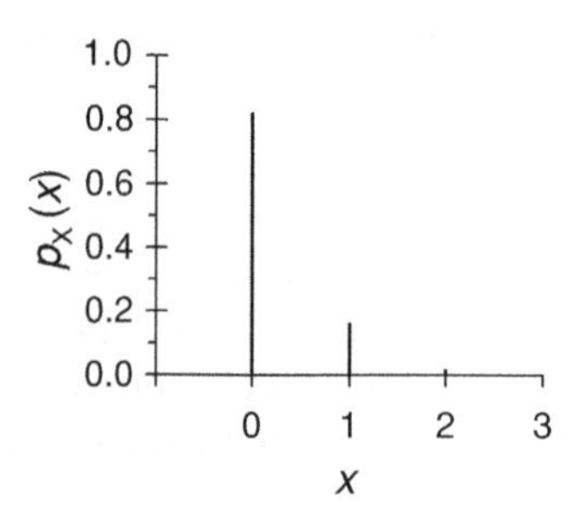

Fig. 4.1.4 Poisson pmf, $p_X(x)$, for $v = 0.2$.

[8] That is, $f(z) = f(a) + (z-a)\frac{df(z)}{dz}\Big|_{z=a} + \frac{(z-a)^2}{2}\frac{d^2f(z)}{dz^2}\Big|_{z=a} + \cdots +$ (remainder). The infinite series converges in this case. We finally substitute $a = 0$.

[9] Hald (1990, pp. 213–217) gives the historical background.

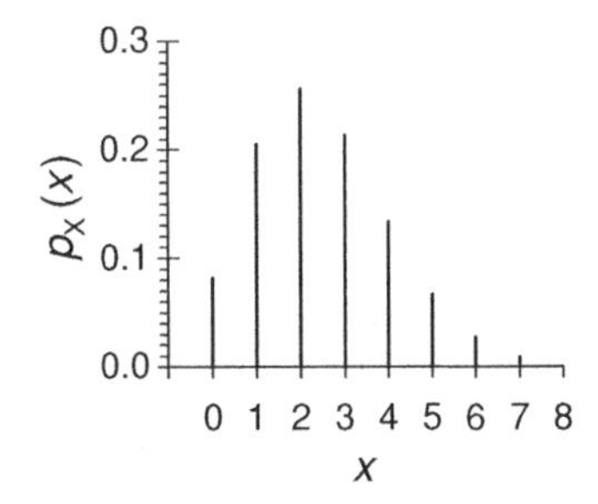

Fig. 4.1.5 Poisson pmf, $p_X(x)$, for $v = 2.5$.

The Poisson cdf is

$$F_X(k) = \sum_{k=0}^{k} \frac{v^x e^{-v}}{x!}, \quad \text{for } k = 0, 1, 2, \ldots. \tag{4.1.8}$$

Then,

$$F_X(0) = e^{-v} \quad \text{and} \quad F_X(\infty) = e^{-v} \sum_{x=0}^{\infty} \frac{v^x}{x!}.$$

From the Taylor series for e^v,

$$F_X(\infty) = e^{-v} e^v = 1.$$

The terms on the right-hand side Eq. (4.1.8) are positive increments, and thus it satisfies the requirements for a cdf. As shown in Example 3.18, the mean and variance of the Poisson variate X are as follows:

$$E[X] = v \tag{4.1.9a}$$

and

$$\text{Var}[X] = v. \tag{4.1.9b}$$

The fact that $E[X] = v$ is anticipated because the Poisson distribution derives from a limiting process that keeps the mean equal to v (as discussed shortly).

The two results from Eq. (4.1.9) suggest one method of verifying whether a variable is Poisson-distributed, that is, by testing a sample of data with respect to the mean and variance. However, it should not preclude other tests. More importantly, and as discussed shortly, prior physical justification is a requirement.

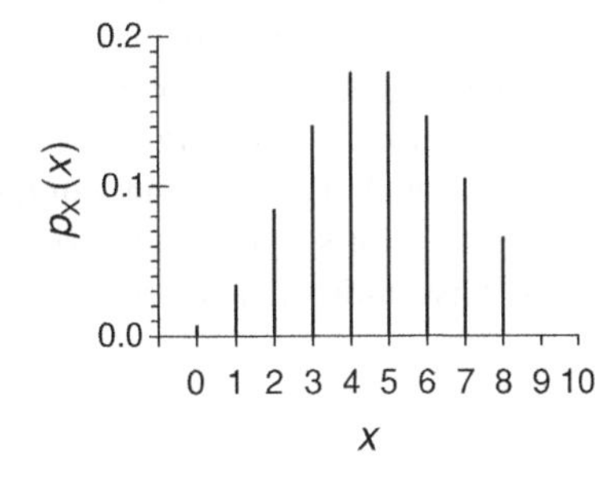

Fig. 4.1.6 Poisson pmf, $p_X(x)$, for $v = 5$.

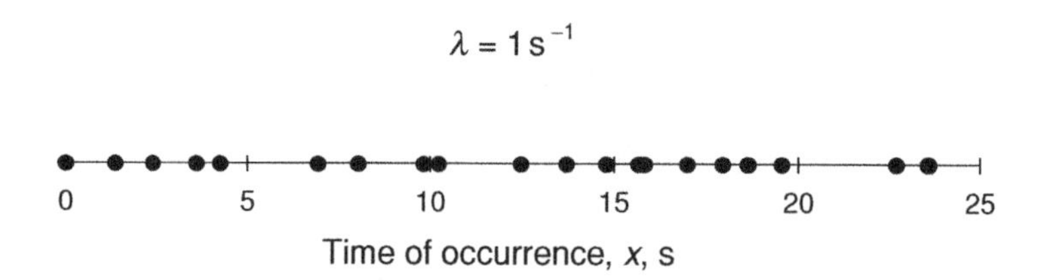

Fig. 4.1.7 Occurrences of a Poisson process in time.

The stipulations and developments just given lead to the derivation of the Poisson distribution [Eq. (4.1.7)] as a limiting form of the binomial. This is also referred to as the Poisson approximation to the binomial, because of the assumptions made.

The Poisson parameter ν is considered to be the mean number of arrivals or happenings in a time interval, say, of length t; it is also applicable to an interval in space, as discussed shortly. Let λ equal the *mean rate, or hazard rate, or intensity*, of occurrence of a Poisson happening or arrival. If the generating process is *homogeneous*, the rate λ and mean ν are constant for all time intervals; we discuss the nonhomogeneous case shortly. One can then divide the time interval t into small subintervals each of length Δt, such that $n\Delta t = t$. Accordingly, the probability of exactly one arrival is approximately $\lambda \Delta t$. When Δt tends to zero, this probability becomes exact. Then such a subinterval can have at most a single success. Furthermore, when the Poisson assumptions hold there is no relationship between the outcomes in one subinterval and another disjoint subinterval; and the occurrence over Δt is that of a Bernoulli trial. These conditions can be summarized by the following postulates:

(1) If one partitions a time interval t into sufficiently small subintervals of equal length Δt, the probability of exactly one arrival over any such subinterval tends to $\lambda \Delta t$.
(2) The probability of more than one arrival in the subinterval then becomes zero.
(3) The occurrence of an arrival within a subinterval is independent of occurrences in other disjoint subintervals.

From the assumptions made in deriving Eq. (4.1.7), and from the foregoing stipulations, the relationship between the mean count ν and the mean rate λ, over an interval t is as follows:

$$\nu = np = \frac{t}{\Delta t}\lambda \Delta t = \lambda t. \tag{4.1.10}$$

A randomly constructed experiment based on the preceding assumptions gives a *homogeneous or stationary Poisson process* $X(t)$ with rate λ and occurrences shown, for example, in Fig. 4.1.7.

In the application of a Poisson process such as the one illustrated, the random variable is the count of the number of arrivals or happenings or incidents over an interval of time. The assumptions made in the theoretical development are justifiable when one considers numerous types of empirical evidence. The Poisson process is a particular form of a *stochastic or random process* which denotes a random function in time or space.[10]

The sum of two independent Poisson random variables X_1 and X_2 with parameters ν_1 and ν_2, respectively, is Poisson-distributed with parameter $\nu = \nu_1 + \nu_2$. This can be

[10] We shall discuss this further under *renewal and point processes* at the end of Subsection 4.2.2. The Poisson distribution, obtained through the binomial approximation in Eq. (4.1.7), is derived independently in Appendix A.3 by considering the number of arrivals or counts in intervals (t, $t + \Delta t$, and so on).

shown as follows:

$$\Pr[X_1 + X_2 = n] = \sum_{k=0}^{n} \Pr[X_1 = k]\Pr[X_2 = n-k] = \sum_{k=0}^{n} \frac{v_1^k e^{-v_1}}{k!}\frac{v_2^{n-k} e^{-v_2}}{(n-k)!}.$$

Hence, from the binomial expansion of $(v_1 + v_2)^n$,

$$\Pr[X_1 + X_2 = n] = \frac{e^{-(v_1+v_2)}}{n!}\sum_{k=0}^{n}\frac{n!}{k!(n-k)!}v_1^k v_2^{n-k} = \frac{e^{-(v_1+v_2)}(v_1+v_2)^n}{n!}.$$

Apart from using time as a variable, the Poisson theory is applicable when other intervals such as those in length, area, volume, or space are used. Several illustrations are provided later in this section.

Note that the Poisson distribution is applied in situations where a large number of objects are distributed over a large area. For example, the number of defects per unit of some material used in engineering, the number of organisms per unit volume of some fluid, algal counts from a similar sample for monitoring the quality of lake water, bacterial counts on petri plates, and the arrivals of hurricane events can be assumed to be Poisson-distributed. However, the theory is not applicable if there is clustering and the chance of finding an object in a particular location (or time) is *not the same* as finding it elsewhere. One such example is the spatial distribution of pollutants near a seashore stemming from discharges by small communities located along the coast. Likewise, the Poisson distribution does not apply when there is interdependence between events. Therefore, events of interest to the engineer that occur in groups or clusters—such as interrelated rainfalls in certain regions and imperfections in materials which are found over particular locations but not over other areas—should be excluded.

Definition: The homogeneous **Poisson** model is defined as follows:

(1) The random variable $X(t) \sim$ Poisson $(v = \lambda t)$ is the number of arrivals that occur in an interval t (such as time, length, area, or space) of a given sequence.
(2) The parameter v is the expected number of arrivals in the interval t and $\lambda = v/t$ is the constant rate of occurrence of the events.
(3) If one partitions the interval t into subintervals of equal length Δt that are sufficiently small, the probability of exactly one arrival over any such subinterval tends to $\lambda\Delta t$.
(4) The probability of more than one arrival in the subinterval then becomes zero.
(5) The occurrence of an arrival within a subinterval is independent of occurrences in other disjoint subintervals.

The Poisson pmf is given by

$$p_X(x) = \frac{v^x e^{-v}}{x!}, \quad \text{for } x = 0, 1, 2, \ldots, \quad \text{and} \quad v > 0.$$

The mean and variance of X are equal to v.

Example 4.4. Atmospheric pollution. Atmospheric dust particles at a particular location cause an environmental problem. The number of particles within a unit volume is observed by focusing a powerful microscope on the particles and making counts. The results of tests on 100 such volumes are shown in Table 4.1.1.

By using Eq. (3.2.1), the estimated mean of the number of dust particles within each volume is calculated as follows:

$$\bar{x} = \frac{13}{100}\times 0 + \frac{24}{100}\times 1 + \frac{30}{100}\times 2 + \frac{18}{100}\times 3 + \frac{7}{100}\times 4 + \frac{8}{100}\times 6 = 2.14.$$

The theoretical Poisson frequencies of occurrence shown in Table 4.1.1 are obtained from

$$p_X(X = x|2.14) = 2.14^x e^{-2.14}/x! \quad \text{for } x = 0, 1, 2, 3, 4, 6.$$

Table 4.1.1 Poisson distribution of dust particles in the atmosphere

	Particles in unit volume					
	0	1	2	3	4	>4
Observed frequency	13	24	30	18	7	8
Poisson frequency	12	26	27	19	10	6

Example 4.5. Reliability of machinery at a treatment plant. All the pumps at a water treatment plant have been made to the same specifications by a single manufacturer. From tests made over 4-week period, it has been ascertained that there are on average two breakdowns during each period. A new plant manager assumes that the problem is not serious if there are no more than four breakdowns over a period of 4 weeks. What is the probability p of such an occurrence?

It is assumed that the failures occur randomly in time, the occurrences are independent, and the rate of failure is constant. Thus, the Poisson model is applicable with parameter $\nu = 2$ and

$$F_X(4) = \sum_{x=0}^{4} \frac{2^x e^{-2}}{x!} = 0.135 + 0.271 + 0.271 + 0.180 + 0.090 = 0.947.$$

Hence, $p = 0.947$.

Example 4.6. Closure of causeway caused by high flows. High flows result in the closure of a causeway. From past records, the road is closed for this reason on 10 days during a 20-year period. At an adjoining village, there is concern about the closure of the causeway because it provides the only access. The villagers assume that the probability of a closure of the road for more than one day during a year is less than 0.10. Is this correct?

The conditions seem to be satisfied for the application of the Poisson distribution. The mean number of closures per year is 0.5. Hence, $\nu = 0.5$, and

$$F_X(1) = \sum_{x=0}^{1} \frac{0.5^x e^{-0.5}}{x!} = 0.607 + 0.303 = 0.910.$$

Thus,

$$F_X(X \geq 2) = 1 - 0.91 = 0.09.$$

The villagers are therefore apparently justified in their assumption.

Example 4.7. Strength of timber. From past data, an engineer has estimated a probability of $p = 0.01$ that timber delivered at a construction site from a particular source is below specification. If 150 joists of timber are necessary for a particular construction job, determine the minimum number which should be ordered so that the chance of not having the required number of suitable joists is less than 10%.

Selecting suitable timber for the construction work constitutes a series of Bernoulli trials. Let $150 + x$ joists be ordered to meet the required stipulation where x is the number of defective joists. Using the Poisson approximation with $n = 150 + x$ and $p = 0.01$, the parameter $\nu = (150 + x)p \approx 150p = 1.5$. The Poisson cdf [Eq. (4.1.8)] is used here to obtain the probability of finding no more than k defective joists. Thus,

$$F_X(k) = \sum_{x=0}^{k} \frac{1.5^x e^{-1.5}}{x!} \geq 0.9.$$

For values of k equal to 2 and 3, the summation is equal to 0.81 and 0.93, respectively. Therefore, 153 joists should be ordered.

Example 4.8. Floorboards. Floorboards supplied by a contractor have some imperfections. A builder decides that two imperfections per 40 m^2 is acceptable. Is there at least a 95% chance of meeting such requirements, if from previous experience with the same material, an average of one imperfection per 65 m^2 has been found?

It is assumed that imperfections per unit area have a Poisson distribution and the same material as before is supplied. For this purpose, Eq. (4.1.8) is used in conjunction with Eq. (4.1.10) in which time t is replaced by area m^2. Thus one makes the approximation that

$$\nu = \frac{40}{65} = 0.615.$$

Hence,

$$F_X(2) = \sum_{k=0}^{2} \frac{e^{-0.615} 0.615^k}{k!} = 0.540 + 0.333 + 0.102 = 0.975.$$

Therefore, the answer is yes.

Example 4.9. Coliform bacteria in wastewater. A public health engineer finds the most probable number (MPN) of the concentration of coliform bacteria per 100 mL after testing the wastewater and applying the Poisson distribution. Five samples of wastewater are taken, and three different dilutions are made of each sample and tested in a total of fifteen 100-mL tubes. [For a general description of such procedures, see Greenberg et al. (2005).] Samples are incubated in MacConkey broth, an essential part of the experiment, for 24 hours at a temperature of 40°C. At the end of this period, the presence of coliform organisms in a particular test tube is indicated by gas bubbles and a yellowish tint in the liquid. Conversely, the result is treated as negative if there is no gas and the color remains purple.

The maximum likelihood, or MPN value, that gives the observed distribution of positive and negative results from the 15 tubes is an estimate of the coliform concentration in the raw water. The MPN represents the unknown Poisson rate or intensity parameter λ. Let c be a constant used in scaling the probabilities so that their sum is unity, and let λ represent the mean bacterial concentration per mL that one wishes to estimate from the test results. Let ν_1, ν_2, and ν_3 be the dilutions in mL of raw water per 100 mL total in the test tubes, and let n_1, n_2, and n_3 be the numbers of positive results from the five test tubes for each of the three dilutions, respectively.

It is assumed that the bacteria have a Poisson distribution in the wastewater. Thus Eq. (4.1.7) is used in conjunction with Eq. (4.1.10). The probability of a negative result for dilution i is $e^{-\nu_i \lambda}$ and, correspondingly, if a test tube indicates the presence of bacteria, the probability is taken as $1 - e^{-\nu_i \lambda}$. In considering that the tests are independent events, the joint probability of the 15 outcomes is

$$p = \frac{1}{c} \prod_{i=1}^{k} (e^{-\nu_i \lambda})^{m - n_i} (1 - e^{-\nu_i \lambda})^{n_i},$$

Table 4.1.2 Data input for coliform tests

i	1	2	3
ν_i	100.0	10.0	1.0
n_i	5	4	3

ν_i = dilution volume in mL/100 mL; n_i = number of positive results.

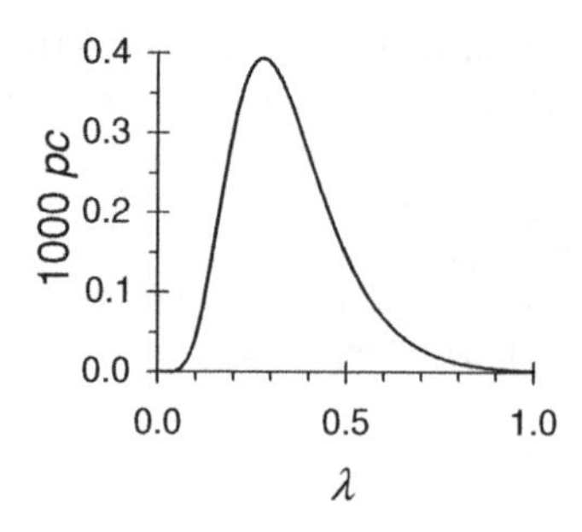

Fig. 4.1.8 Variation of pc with Poisson rate or intensity parameter λ; p = probability of given joint outcomes of wastewater tests conditional to given value of Poisson parameter, c = scaling constant for probabilities.

where $i = 1, 2, \ldots, k$ represents the dilutions and m = the number of samples for each dilution.

In this application $k = 3$ and $m = 5$. For the purpose at hand it is not essential to evaluate c. A maximum value of p in the above equation is found empirically by varying λ over an appropriate range. This gives the MPN value of the concentration, which equals 100λ, where λ corresponds to the maximum value of pc.

A sample of wastewater is taken and three dilutions of 100, 10, and 1 mL/100 mL are made. The numbers of positive results are shown in Table 4.1.2. Values of 1000 pc are computed for a range of λ from 0 to 1.0 (where 1000 is an additional scaling factor). The results are shown graphically in Fig. 4.1.8. Also, seven values around the maximum are given in Table 4.1.3. The maximum value of pc is 0.000394 and the corresponding value of λ is 0.28 coliform organisms per milliliter. Thus the MPN = 28 organisms per 100 mL.

4.1.3.1 Truncated Poisson process

There are applications when the case $X = 0$ is not an acceptable happening or is not of interest, but the other conditions for a Poisson process hold. Some possible examples are in the study of groups of certain objects or happenings such as the numbers of accidents or hurricanes. A group of zero is not realizable (or is not relevant) and therefore $X = 0$ is eliminated from the frequency distribution. In such cases, a truncated Poisson distributed is used as follows:

$$\begin{aligned}
\Pr[X = x | X \geq 1] = \frac{\Pr[X = x | X \geq 0]}{\Pr(X \geq 1)} &= \frac{v^x e^{-v}}{x!} \frac{1}{(1 - e^{-v})} \\
&= \frac{v^x}{x!(e^v - 1)}, \quad \text{for } x = 1, 2, 3, \ldots \\
&= 0, \quad \text{otherwise.} \qquad (4.1.11)
\end{aligned}$$

Table 4.1.3 Computations of pc for coliform tests for different values of λ

λ	0.25	0.26	0.27	0.28	0.29	0.30	0.31
1000 pc	0.383	0.389	0.393	0.394	0.392	0.388	0.382

λ = average number of coliform bacteria per mL; pc = probability multiplied by normalizing constant.

The kth factorial moment is given by

$$E[X(X-1)\cdots(X-k+1)] = \sum_{x=1}^{\infty} x(x-1)\cdots(x-k+1)\frac{v^x}{x!(e^v-1)}$$
$$= \sum_{x=k}^{\infty} \frac{v^k v^{x-k}}{(x-k)!(e^v-1)}$$
$$= \frac{v^k}{e^v-1}\sum_{j=0}^{\infty}\frac{v^j}{j!} = \frac{v^k e^v}{e^v-1}.$$

Hence, from the first and second factorial moments

$$E[X] = \frac{ve^v}{e^v-1} \tag{4.1.12a}$$

and

$$\mathrm{Var}[X] = \frac{ve^{2v} - v(v+1)e^v}{(e^v-1)^2}. \tag{4.1.12b}$$

Definition and properties: *A truncated Poisson variate* X that takes positive integer values greater than zero, has pmf

$$\Pr[X = x|X \geq 1] = \frac{v^x}{x!(e^v-1)}, \quad \text{for } x = 1, 2, 3, \ldots, \text{and } v > 0,$$

with mean $E[X] = \dfrac{ve^v}{e^v-1}$ and variance $\mathrm{Var}[X] = \dfrac{ve^{2v} - v(v+1)e^v}{(e^v-1)^2}$.

Example 4.10. Groups of thunderstorms. Thunderstorms that occur over metropolitan areas raise pond levels and cause high sewer flows. During summer, thunderstorms tend to occur in groups. An engineer seeks to determine the Poisson frequency distribution of the areal extent of the groups as an aid to design. Some observations are given in Table 4.1.4.

We estimate the parameter v numerically. One simple procedure is as follows. The sample mean is estimated by dividing the sum of the product of columns 1 and 2 by the sum of column 2. This is substituted in Eq. (4.1.12*a*) which is rearranged so that the Poisson parameter is on the left and the sample mean and the exponential terms are on the right. Commencing with an initial value such as 1.0 of the unknown parameter and substituting on the right, we obtain a new approximate value. Proceeding in this way, we solve the equation by successive approximations until the change in the estimate from two successive steps is negligible. Using the estimated value of the parameter, the theoretical frequencies are computed and are given in column 3. It is seen that these are close to the observed frequencies.

Table 4.1.4 Truncated Poisson distribution of groups of thunderstorms

Size of group	Observed frequency	Theoretical frequency
1	17	15.8
2	15	15.7
3	10	10.4
4	4	5.1
5	2	2.0
6	2	0.7

Sample average = 2.300; truncated Poisson parameter = 1.984.

4.1.3.2 Nonhomogeneous poisson process

We have heretofore considered homogeneous or stationary Poisson processes. However, there are many instances when these conditions are not applicable. In such a nonhomogeneous or nonstationary Poisson process $X(t)$, the rate λ varies with time and is denoted by $\lambda(t)$. Then an increment $X(t_1) - X(t_2)$ of the process which gives the number of events in the interval (t_1, t_2) has a Poisson distribution with parameter

$$\nu = \int_{t_1}^{t_2} \lambda(t)dt.$$

These increments constitute a sequence of independent random variables.

Example 4.11. Distributions of rainfalls with variable Poisson rates. Unrelated occurrences of rainfall in a particular locality constitute a nonhomogeneous Poisson process with variable mean rates $\lambda(t)$, $t = 1, 2, \ldots, 13$, during an annual cycle, where t represents a period of 4 weeks. Periods 3–7 experience more rain on average than the other periods. The rates for the year are as follows:

$$\begin{aligned} \lambda(t) &= \frac{2t}{3}, && \text{for } 0 \le t < 3, \\ &= 2, && \text{for } 3 \le t < 8, \\ &= \frac{13-t}{3}, && \text{for } 8 \le t \le 13. \end{aligned}$$

(1) What is the probability of having three or more rainfalls during the first five periods of the year?

(2) What is the probability of having no more than one rainfall during the periods 8, 9, and 10 and no more than one rainfall during the last three periods of the year?

For part 1, the mean count for the first five periods is

$$\nu = \int_0^3 \frac{2t}{3}dt + \int_3^5 2dt = 7.$$

Hence,

$$\Pr[X(5) \ge 3] = 1 - \Pr[X(5) \le 2] = 1 - \left(e^{-7} + 7e^{-7} + \frac{7^2 e^{-7}}{2}\right) = 0.97.$$

For part 2, the independence of $[X(10) - X(7)]$ and $[X(13) - X(10)]$ is used. The mean count for the periods 8, 9, and 10 is

$$\nu = \frac{1}{3}\int_7^{10} (13 - t)dt = 4.5.$$

The mean count for the last three periods is

$$\nu = \frac{1}{3}\int_{10}^{13} (13 - t)dt = 1.5.$$

Hence,

$$\begin{aligned} &\Pr[X(10) - X(7) \le 1, X(13) - X(10) \le 1] \\ &= (e^{-4.5} + 4.5e^{-4.5})(e^{-1.5} + 1.5e^{-1.5}) \\ &= 5.5 \times 2.5e^{-6} = 0.034. \end{aligned}$$

In summary, the Poisson process is one of the important random or stochastic processes that are essential for the development and applications of probability theory. The role played by the Poisson distribution is crucial in many aspects of probability and statistics. As noted before, the binomial is also a fundamental probability distribution. These are two of the three principal distributions viewed as fundamental.[11]

4.1.4 Geometric and negative binomial distributions

The geometric and negative binomial distributions have a special role in statistics. Let us first consider a series of Bernoulli trials that are continued until exactly r successes occur and let X trials be required for the purpose. That is, after $(X - 1)$ trials there are exactly $(r - 1)$ successes. As before, let p denote the probability of a success. Then, by using the property of independence, the probability of occurrence of the given sequence is $p^{r-1}(1-p)^{x-r}$.

The total number of combinations possible when selecting $(r - 1)$ objects from $(x - 1)$ objects is

$$\binom{x-1}{r-1} \equiv \frac{(x-1)!}{(r-1)!(x-r)!}.$$

Hence, the probability of the rth success occurring at the xth trial is given by the *negative binomial distribution* with pmf

$$\begin{aligned} p_X(X = x|r, p) &= \binom{x-1}{r-1} p^r(1-p)^{x-r}, \quad && \text{for } x = r, r+1, r+2, \ldots, \\ &= 0, && \text{otherwise.} \end{aligned} \qquad (4.1.13)$$

Example 4.12. Delivery of treatment plants for a water supply system. A company has bid to supply standardized treatment plants for a rural water supply system in a region, having quoted a low price for the job. However, the supervising engineer has estimated from previous experience that 10% of plant delivered by this company are defective in some way. If five items are required, determine the minimum number of plants to be ordered to be 95% sure that a sufficient number of nondefective plants are delivered. It is assumed that the delivery of a plant is an independent trial and any fault that may occur in one plant is not related to possible faults in other plants. The probability of a success, $p = 1 - 0.1 = 0.9$.

The problem is solved after one or more trials. From Eq. (4.1.11) the cdf for $X = 7$ is given by

$$\sum_{x=5}^{7} p_X(X = x|5, 0.9) = \sum_{x=5}^{7} \binom{x-1}{4} 0.9^5 0.1^{x-5} = 0.9^5(1.0 + 0.5 + 0.15) = 0.97.$$

Hence the required number is 7. The pmf for the negative binomial distribution is given in Fig. 4.1.9.

4.1.4.1 The alternative form of the negative binomial distribution

The negative binomial can be defined in another form in which the random variable is the number of failures before the rth success, denoted by $Y = X - r$. The pmf is obtained by

[11] The third is the normal distribution, which follows in Section 4.2.

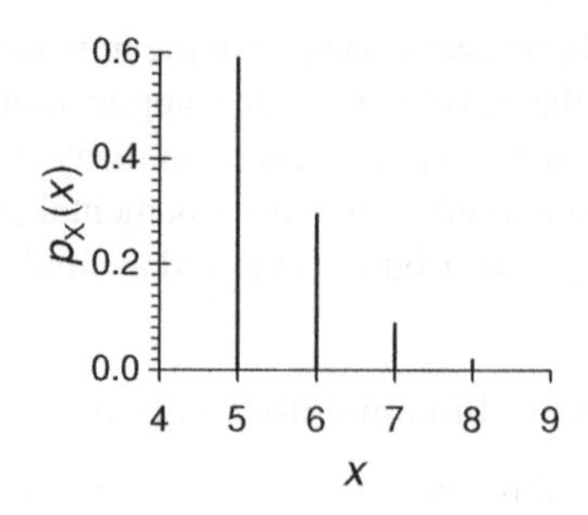

Fig. 4.1.9 Negative binomial pmf, $p_X(x)$, for $r = 5$ and $p = 0.9$.

substituting Y in Eq. (4.1.13):

$$\begin{aligned} P_Y(Y = y) &= \binom{r+y-1}{y} P^y(1-p)^y \\ &= (-1)^y \binom{-r}{y} p^y(1-p)^y, \quad \text{for } y = 0, 1, 2, \ldots \\ &= 0, \quad \text{otherwise.} \end{aligned} \tag{4.1.14}$$

The resemblance between this and the binomial given by Eq. (4.1.3) is the reason for the name *negative binomial.*[12]

If $r = 1$, that is, the number of successes required is one, then the negative binomial of Eq. (4.1.13) becomes the *geometric distribution* with pmf

$$\begin{aligned} p_X(X = x|p) &= p(1-p)^{x-1}, \quad \text{for } x = 1, 2, 3, \ldots \\ &= 0, \quad \text{otherwise,} \end{aligned} \tag{4.1.15}$$

where the parameter p satisfies the condition $0 < p \leq 1$.

The name geometric is given to this distribution because the values taken by the pmf form a geometric series. Figure 4.1.10 shows the geometric distribution for $p = 0.7$.

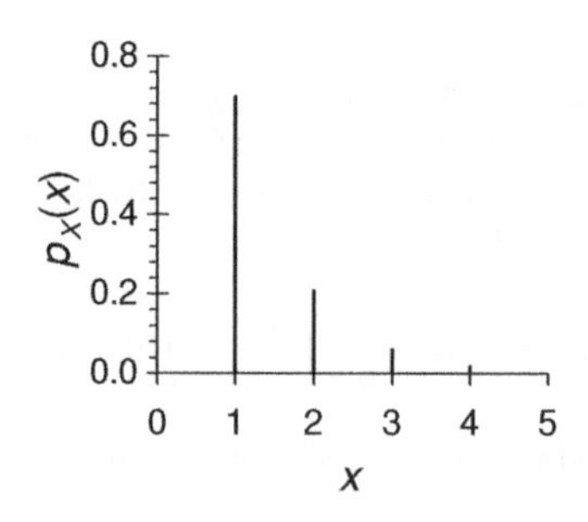

Fig. 4.1.10 Geometric pdf, $p_X(x)$, for $p = 0.7$.

[12] Note that

$$\begin{aligned} (-1)^y \binom{-r}{y} &= (-1)^y \frac{(-r)(-r-1)\cdots(-r-y+1)}{y!} \\ &= (-1)^y(-1)^y \frac{r(r+1)\cdots(r+y-1)}{y!} = \binom{r+y-1}{y}. \end{aligned}$$

Likewise one can obtain the pmf for values of p in the interval $0 < p < 1$; however, the mode is at unity. A random variable that is geometrically distributed is said to represent a discrete waiting time (that is, the length of time one needs to wait) before a success occurs.

4.1.4.2 Mean and variance of a geometric variate

The sum of a converging geometric series with first term and multiplier equal to $(1 - p)$, where $0 < p < 1$, is

$$(1-p)+(1-p)^2+(1-p)^3+\cdots=\sum_{x=1}^{\infty}(1-p)^x=\frac{1-p}{p}=\frac{1}{p}-1. \quad (4.1.16)$$

Let the first derivatives with respect to p be equated and both sides multiplied by $(-p)$. Hence by treating X as the random variable, the mean of a geometric variate is

$$E[X]=\sum_{x=1}^{\infty}xp(1-p)^{x-1}=\frac{1}{p}. \quad (4.1.17a)$$

To obtain the variance, one uses the second factorial moment. That is,

$$E[X(X-1)]=E[X^2]-E[X]=\sum_{x=1}^{\infty}x(x-1)p(1-p)^{x-1}. \quad (4.1.17b)$$

The second derivatives with respect to p of the middle and right terms of Eq. (4.1.16) are equated and both sides are multiplied by $(1 - p)p$. Hence, by using the Eq. (4.1.17b) it follows that

$$E[X^2]-E[X]=(1-p)p\frac{2}{p^3}=\frac{2}{p^2}-\frac{2}{p}.$$

Using this result and Eq. (4.1.17a), the variance of a geometric variate is

$$\mathrm{Var}[X]=E[X^2]-[E[X]]^2=\frac{2}{p^2}-\frac{2}{p}+\frac{1}{p}-\frac{1}{p^2}=\frac{1-p}{p^2}. \quad (4.1.17c)$$

4.1.4.3 Return period

The concept of a return period is important particularly when one is dealing with extreme events such as floods, droughts, and wind speeds. Let the random variables Y_i represent in one instance, the maximum flood in year i, $i = 1, 2, 3, \ldots$, respectively. Then if the Y_i are independent, the probability that the time interval $\tilde{T}$ between exceedances of a flood of magnitude y equals n is

$$\begin{aligned}\Pr[\tilde{T}=n]&=\Pr[Y_1<y]\Pr[Y_2<y]\cdots\Pr[Y_{n-1}<y]\Pr[Y_n>y]\\&=\{\Pr[Y<y]\}^{n-1}\Pr[Y>y], \qquad \text{for } n=1,2,\ldots,\end{aligned}$$

assuming that the Y_i are identically distributed as in Eq. (4.1.15), which is the pmf of the geometric distribution.

Hence, when considering high events as in floods, we define the *return period* as the mean time interval between exceedances of a specified value:

$$E[\tilde{T}]=\frac{1}{\Pr[Y>y]}. \quad (4.1.17d)$$

That is, *the return period is the reciprocal of the probability of exceedance.*

Conversely, when considering low events as in droughts, we define the return period as the mean time interval between nonexceedances of a specified value. Then *the return period is the reciprocal of the probability of nonexceedance.* That is,

$$E[\tilde{T}] = \frac{1}{\Pr[Y \leq y]} \tag{4.1.17e}$$

4.1.4.4 Mean and variance of a negative binomial variate

As in the relationship between the parameters of Bernoulli and binomial variates, the mean and variance of a negative binomial variate are related to those of a geometric variate as follows:

$$E[X] = \frac{r}{p} \tag{4.1.18a}$$

and

$$\mathrm{Var}[X] = \frac{r(1-p)}{p^2}. \tag{4.1.18b}$$

In the alternative form of the negative binomial, in which the variable $Y = X - r$ is the number of failures before the rth success,

$$E[Y] = E[X] - r = \frac{r}{p} - r = \frac{r(1-p)}{p} \tag{4.1.18c}$$

and

$$\mathrm{Var}[Y] = \mathrm{Var}[X] = \frac{r(1-p)}{p^2}. \tag{4.1.18d}$$

The variance of the binomial variate is less than its mean. In the case of a Poisson variate, the mean and variance are equal. This equality of mean and variance applies also to a geometric or negative binomial X variate only if $p = 0.5$, but if $p < 0.5$ the variance is greater than the mean for these two types and if $p > 0.5$ the variance is less than the mean.

Definition: In a series of Bernoulli trials, the random variable X which denotes the trial at which the rth success occurs, where r is a fixed positive integer, has a *negative binomial pmf*:

$$p_X(X = x|r, p) = \binom{x-1}{r-1} p^r (1-p)^{x-r}, \quad \text{for } x = r, r+1, r+2, \ldots, \quad \text{and } 0 \leq p \leq 1,$$

$$\text{mean } E[X] = \frac{r}{p} \quad \text{and} \quad \text{variance} \quad \mathrm{Var}[X] = \frac{r(1-p)}{p^2}.$$

If $r = 1$, $X \sim$ geometric (p).

Example 4.13. Design return period. The level of the road in Example 4.2 is raised so that the chance that the road is flooded (at least once) during a 5-year period is 20%. What should the design return period be, that is, the number of years on average between exceedances of a critical river level that causes the flooding of the road?

The probability that the road is not flooded during a 5-year period consequent to a raise in the level is 0.8. From Eq. (4.1.3), $(1-p)^5 = 0.8$. Hence, $p = 0.04$. Thus, the design return period, which is the reciprocal of p, is 25 years.

Example 4.14. Mean and variance of road floods. On the basis of data in Example 4.13, what are the mean and variance of the number of 25-year floods over a 5-year period?

$$E[X] = Np = 5 \times 0.04 = 0.2.$$

$$\mathrm{Var}[X] = Np(1-p) = 5 \times 0.04 \times 0.96 = 0.192.$$

Example 4.15. Waiting time for rare floods. A cofferdam is constructed to enable work on the main dam, downstream on the same river, to be completed up to a safe level. If the cofferdam is designed to withstand a 10-year flood, what is the probability that at least two or more years will elapse before the occurrence of the design flood?

The 10-year flood has a 1 in 10 year chance of occurrence, that is, $p = 0.1$.

$$p_X(X = 1|p) = p \quad \text{and} \quad p_X(X = 2|p) = p(1-p).$$

Hence, the required probability is

$$1 - p - p(1-p) = 1 - 2p - p^2 = 0.79.$$

The engineer would be well-advised to strengthen the cofferdam, which means that a substantially larger design period should be adopted.

4.1.4.5 Memoryless property of the geometric distribution

A special property of the geometric distribution is that if, say, k successive failures have occurred after k or more trials, the distribution of the total number of trials, say, $k + l$, required before the first success occurs does not change. This is shown as follows by using conditional probabilities and the properties of an infinite geometric series as in Eq. (4.1.16):

$$\begin{aligned}\Pr[X \geq k + l | X \geq k] &= \frac{\Pr[X \geq k + l]}{\Pr[X \geq k]} = \frac{\sum_{n=k+l}^{\infty} p(1-p)^n}{\sum_{n=k}^{\infty} p(1-p)^n} \\ &= \frac{(1-p)^{k+l}}{(1-p)^k} = (1-p)^l = \Pr(X \geq l).\end{aligned} \tag{4.1.19}$$

In contrast, but in common with other discrete distributions, the log-series distribution (which follows) does not have this property.

4.1.5 Log-series distribution

Consider, for example, the series given by Eq. (4.1.6*b*). Substituting $(1 - p)$ for z and dividing both sides by $-\ln(p)$, one obtains the infinite log series

$$-\frac{(1-p)}{\ln(p)} - \frac{(1-p)^2}{2\ln(p)} - \frac{(1-p)^3}{3\ln(p)} \cdots.$$

By the given operation (that is, dividing a series by its sum), the series must sum to one and also the other necessary conditions for a probability distribution must be satisfied. This leads to the pmf of the log-series distribution:

$$p_X(X = x|p) = -\frac{(1-p)^x}{x\ln(p)}, \quad \text{for } x = 1, 2, \ldots; \; 0 < p < 1, \tag{4.1.20}$$

which represents the probability of a success after x failures. In the geometric case, the probability of a success after x failures decreases proportionately by $(1 - p)$ as the run length increases from $(1 - x)$ to x, as seen from Eq. (4.1.19). For the log-series distribution, however, the corresponding decrease is $(1 - p)(1 - x)/x$ which depends on x; and for low values of x the factor is low.[13] An example of a log-series distribution is given in Fig. 4.1.11.

[13] This property is found in many natural phenomena as shown, for example, by Johnson et al. (1992). The log-series distribution was developed in the 1940s by R. A. Fisher in relation to the distribution of certain species. It can be shown that the log series is a limiting form of the negative binomial distribution without the zero class.

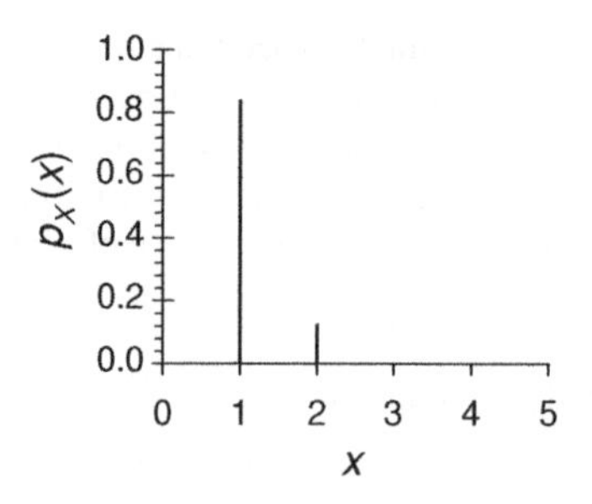

Fig. 4.1.11 Log-series pmf, $p_X(x)$, for $p = 0.7$.

4.1.5.1 Mean and variance of the log-series distribution

It follows from Eq. (4.1.20) that

$$-\ln(p)E[X] = \sum_{x=1}^{\infty} \frac{x(1-p)^x}{x} = \sum_{x=1}^{\infty} (1-p)^x$$
$$= \sum_{x=1}^{\infty} (1-p) + (1-p)^2 + (1-p)^3 + \cdots. \qquad (4.1.21a)$$

Hence, from Eq. (4.1.16)

$$E[X] = -\frac{1}{\ln(p)} \frac{1-p}{p}. \qquad (4.1.21b)$$

Likewise,

$$-\ln(p)E[X^2] = \sum_{x=1}^{\infty} \frac{x^2(1-p)^x}{x}$$
$$= (1-p) + 2(1-p)^2 + 3(1-p)^3 + \cdots. \qquad (4.1.21c)$$

If one divides the series on the right-hand side of Eq. (4.1.21c) by $(1-p)$ and subtracts the series of Eq. (4.1.21a) from the result, it follows from Eq. (4.1.21a), (4.1.21b), and (4.1.21c) that,

$$\text{Var}[X] = E[X^2] - (E[X])^2 = \frac{E[X]}{p} - (E[X])^2. \qquad (4.1.21d)$$

Definition and properties: The variate X with a **log-series distribution** has a pmf

$$p_X(X = x|p) = \frac{(1-p)^x}{-x\ln(p)}, \quad \text{for } x = 1, 2, \ldots; \ 0 < p < 1.$$

Its mean and variance are

$$E[X] = -\frac{1}{\ln(p)} \frac{1-p}{p} \quad \text{and} \quad \text{Var}[X] = \frac{E[X]}{p} - (E[X])^2,$$

respectively.

Example 4.16. Wet and dry spells of rainfall. A period of consecutive rainy days is called a *wet run* if the day immediately before the period and the day immediately after the period are dry. Similarly, a period of days on which no rainfall is experienced is called a *dry run* if wet days precede and succeed it.

Table 4.1.5 Distributions of wet runs[a]

Length of wet run in days, (1)	Observed distribution, (2)	Product of (1) × (2), (3)	Geometric distribution, (4)	Log-series distribution, (5)
1	194	194	179.6	224.3
2	101	202	109.2	90.6
3	66	198	66.4	48.8
4	30	120	40.3	29.6
5	26	130	24.5	19.1
6	11	66	14.9	12.9
7	13	91	9.1	8.9
8	7	56	5.5	6.3
9	5	45	3.3	4.5
10	2	20	2.2	3.3
11 + (15.3)	3	46		
Totals	458	1168		

Observed from January 1958 to May 1965 at Kew, London, England.
[a] *Note*: 15.3 represents mean length of 3 runs longer than 10.
Wet run: a period of consecutive rainy days for which the day immediately before the period and the day immediately after the period are dry.
Source: Data from Chatfied (1966); obtained with the permission of the publishers.

The distribution of wet runs observed from January 1958 to May 1965 at Kew in London, England, is shown in Table 4.1.5 (from Chatfield, 1966).[14]

The probability distributions of wet and dry spells are important to engineers for planning and design purposes. One approach is to use a memoryless geometric distribution to model wet runs. On the other hand, observations from many parts of the world suggest that, as the length of a dry spell increases, the probability increases that the day following a dry run will also be dry. Thus the log-series distribution can be advantageous for modeling dry runs.

Consider the observed distribution of wet runs in Table 4.1.5. The data can be used to compare the fit of the geometric and log-series distributions, thus making a preliminary verification of the above hypothesis. By dividing the total of column 3 by the total of column 2, the mean run length of 2.55 days is obtained for the wet spells. The reciprocal of this gives $\hat{p} = 0.392$, using Eq. (4.1.17*a*), as the estimate of the parameter of the geometric distribution applied to these data. Hence, from Eq. (4.1.15) the geometric distribution is obtained and results are given in column 4. The parameter of the log-series distribution is estimated numerically from Eq. (4.1.21*b*) as $\hat{p} = 0.192$ using the calculated mean run length of 2.55 days. Hence by using Eq. (4.1.20), the log-series distribution is applied and the results are presented in column 5.

The procedure is repeated for the dry spells for which the data and results shown in Table 4.1.6. The mean run length of the dry spells is 3.50 days.

Estimates of parameter p for the geometric and log-series distributions are 0.286 and 0.118, respectively.

From Fig. 4.1.12 we see that the geometric distribution provides a generally closer fit to the observed distribution in the case of the wet runs. However, as expected, the log-series distribution is a better candidate for the dry runs as seen in Fig. 4.1.13.

4.1.6 Multinomial distribution

The binomial distribution can be extended to the general case in which the sample space of an experiment is partitioned into r mutually exclusive events with probabilities of

[14] This is from the Journal *Weather* and is reproduced with permission from the Royal Meteorological Society, 104, Oxford Rd, Reading, Berkshire, RG1 7LJ, England.

Table 4.1.6 Distributions of dry runs

Length of dry run in days, (1)	Observed distribution, (2)	Product of (1) × (2), (3)	Geometric distribution, (4)	Log-series distribution, (5)
1	176	176	123.4	178.2
2	81	162	88.2	78.6
3	44	132	63.4	46.2
4	28	112	45.0	30.6
5	21	105	32.1	21.6
6	17	102	22.9	15.9
7	12	84	16.4	12.0
8	15	120	11.7	9.3
9	9	81	8.4	7.3
10	3	30	6.0	5.8
11	3	33		
12	4	48		
13	5	65		
14–20 (17)	11	187		
21 + (25)	3	75		
Totals	432	1512		

Observed from January 1958 to May 1965 at Kew, London, England.
Note: 17 represents mean length of 11 runs of length 14–20; 25 represents mean length of 3 runs longer than 20.
Dry run: a period of consecutive dry days for which the day immediately before the period and the day immediately after the period experience rain.
Source: Data from Chatfied (1966); obtained with the permission of the publishers.

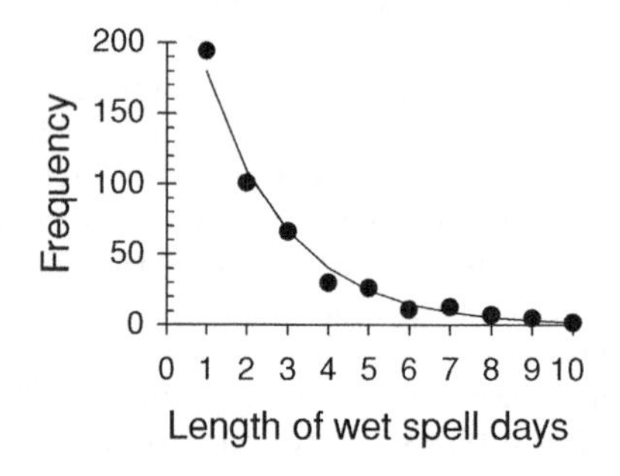

Fig. 4.1.12 Observed distribution of wet spells at Kew, London, from January 1958 to May 1965 fitted with geometric distribution.

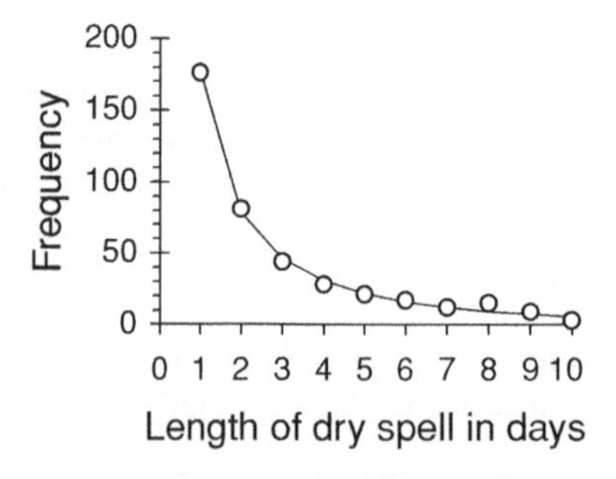

Fig. 4.1.13 Observed distribution of dry spells at Kew, London, from January 1958 to May 1965 fitted with log-series distribution.

occurrence $p_1, p_2, \ldots, p_r$ and the corresponding frequencies of occurrence are equal to $x_1, x_2, \ldots, x_r$ after n independent trials. The probabilities are represented by the *multinomial distribution* as follows:

$$p(X_1 = x_1, X_2 = x_2, \ldots, X_r = x_r | n, p_1, p_2, \ldots, p_r) = \left(\frac{n!}{x_1!x_2!\cdots x_r!}\right) p_1^{x_1} p_2^{x_2} \cdots p_r^{x_r}. \tag{4.1.22}$$

In the case of the binomial distribution based on Bernoulli trials, $r = 2$.

The ith class of event, say, has a marginal binomial probability distribution with mean

$$E[X_i] = np_i \tag{4.1.23a}$$

and variance

$$\mathrm{Var}[X_i] = np_i(1 - p_i). \tag{4.1.23b}$$

Definition and properties: A random experiment consists of n independent trials with outcomes classified into r classes, where the probability of an outcome in class i is p_i. This generates random variables $X_1, X_2, \ldots X_r$ which denote the possible numbers of trials with outcomes in the respective classes and which have a **multinomial distribution** with joint pmf

$$p(X_1 = x_1, X_2 = x_2, \ldots, X_r = x_r | n, p_1, p_2, \ldots p_r) = \left(\frac{n!}{x_1!x_2!\cdots x_r!}\right) p_1^{x_1} p_2^{x_2} \cdots p_r^{x_r},$$

for $x_1 + x_2 + \cdots + x_r = n$ and $p_1 + p_2 + \cdots + p_r = 1$.

The marginal pmf of X_i is binomial with mean $E[X_i] = np_i$ and variance

$$\mathrm{Var}[X_i] = np_i(1 - p_i).$$

Example 4.17. Bids for contracts. A city engineer invites separate bids for widening four roads. Three contractors submit their quotations. The first contractor is usually successful in getting 60% of similar work in the area, whereas the other two have equal chances of 15%. What is the probability that the first contractor will be given at least three of the jobs on the basis of past performances?

From Eq. (4.1.22) the required probability is given by

$$\frac{4!}{3!1!0!}(0.60)^3(0.15)^1(0.15)^0 + \frac{4!}{3!0!1!}(0.60)^3(0.15)^0(0.15)^1 + \frac{4!}{4!0!0!}(0.60)^4(0.15)^0(0.15)^0 = 0.389.$$

4.1.7 Hypergeometric distribution

The multinomial case previously described follows a procedure in which sampling is made with replacement. If samples are taken without replacement, on the other hand, the outcomes have the *hypergeometric distribution.* For example, consider a sample of 50 floorboards of which 5 are known to be defective. If 1 out of the 50 is selected at random, the probability that it will be satisfactory is 45/50. The probability that a second floorboard drawn randomly is also satisfactory becomes 44/49 if the first board is not replaced before the draw, but the probability remains the same if the first board is replaced.

This concept can be extended to the general case, with more than two types of outcomes. Suppose a sample of size n is drawn without replacement, as part of a random experiment, from a population which contains m elements. Within the sample space of m elements, let there be a total of m_1 elements of one type, m_2 elements of another type, and, similarly, elements of other types to include finally m_r elements of type r, thus comprising a mutually exclusive and exhaustive set of elements.

The probability of having in the randomly drawn sample of size n, x_1 elements of the first type, x_2 elements of the second type, and, similarly, other types including finally x_r elements of type r is given by the hypergeometric distribution. By extending the basic combinatorial concepts used for the binomial and negative binomial distributions as already described, the pmf is written as:

$$p(X_1 = x_1, X_2 = x_2, \ldots, X_r = x_r | m_1, m_2, \ldots, m_r, n, m) = \frac{\binom{m_1}{x_1}\binom{m_2}{x_2}\cdots\binom{m_r}{x_r}}{\binom{m}{n}}, \tag{4.1.24}$$

where $x_1 + x_2 + \cdots + x_r = n$ and $m_1 + m_2 + \cdots + m_r = m$; also, $x_1 \le m_1, x_2 \le m_2, \ldots, x_r \le m_r$ and $n \le m$.

In many applications, only the case $r = 2$ is considered, as in the basic example of floorboards of two classes just discussed. The solution to this problem can be obtained directly from Eq. (4.1.24). However, we shall retain the binomial classification for demonstrating that under certain conditions the two-variable hypergeometric distribution can be closely approximated by the binomial distribution. Accordingly, in a sample space (population) of m, let there be mp successes and mq failures where $p + q = 1$ and both mp and mq are integers. Let the random variable X represent the number of successes in n trials. Its pmf is

$$p_X(X = x) = \frac{\binom{mp}{x}\binom{mq}{n-x}}{\binom{m}{n}} \tag{4.1.25}$$

$$= \frac{(mp)!(mq)!(n)!(m-n)!}{(x)!(mp-x)!(n-x)!(mq-n+x)!(m)!}$$

$$= \binom{n}{x}\left[\left(\frac{mp}{m}\right)\left(\frac{mp-1}{m-1}\right)\cdots\left(\frac{mp-x+1}{m-x+1}\right)\right]$$

$$\times\left[\left(\frac{mq}{m-x}\right)\left(\frac{mq-1}{m-x-1}\right)\cdots\left(\frac{mq-n+x+1}{m-n+1}\right)\right].$$

After dividing the numerator and denominator of each of the terms in parenthesis by m, one can show that, as m tends to infinity,

$$p(X = x) = \binom{n}{x} p^x (1-p)^{n-x}.$$

This limiting form is the binomial pmf, as given by Eq. (4.1.3), with parameter p, where the random variable X is the number of successes in n trials; it is applicable in the case of sampling with replacement. When sampling is made without replacement, the binomial distribution that is valid for sampling with replacement may be used as an approximation to represent the two-parameter hypergeometric distribution when m is large, say, $m \geq 200$, and n is relatively very small, say, $m \geq 20n$. Likewise, the pmf given by Eq. (4.1.24), for the general case in which the population of size m contains several classes of elements, approaches that of the multinomial distribution.

4.1.7.1 Mean and variance of the hypergeometric distribution

Consider again the case $r = 2$ in Eq. (4.1.24), that is, two types of events corresponding with the Bernoulli case. Thus, the total number of X successes in n trials can be interpreted as the sum

$$X = Y_1 + Y_2 + \cdots + Y_n, \tag{4.1.26a}$$

where Y_k, in general, indicates that the kth trial is a success or failure by taking a value of 1 or 0, respectively. As in the binomial case, the Y_k, $k = 1, 2, \ldots, n$, are independent and identically distributed Bernoulli (0, 1) random variables with $E[Y] = p$ as in Eq. (4.1.2a). Hence,

$$E[X] = E[Y_1] + E[Y_2] + \cdots + E[Y_n] = np, \tag{4.1.26b}$$

which, of course, is the same as for the binomial distribution, given by Eq. (4.1.5a).

In considering the multivariate case represented by Eq. (4.1.24) and the marginal distributions of the component variables, the mean of the ith random variable X_i is

$$E[X_i] = n\frac{m_i}{m}. \tag{4.1.26c}$$

To determine the variance, one proceeds by computing the square of the random variable X [see Eq. (4.1.26a)], taking expectations, and grouping terms:

$$E[X^2] = \sum_{k=1}^{n} E\left[Y_k^2\right] + 2\sum_{k<m} E\left[Y_k Y_m\right]. \tag{4.1.26d}$$

The first term on the right-hand side represents a sum of the squares of n Bernoulli variates, which clearly sums to the same as that on the right-hand side of Eq. (4.1.26b). Each product in the second term on the right-hand side becomes a zero if both or either of the two terms are zero (failure), but is unity (success) otherwise. In the summation of the second term we choose 2 out of n events. The probability that the first event is a success is p. Because we sample without replacement, the probability of the second variable is $(mp - 1)/(m - 1)$. Thus,

$$E[X^2] = np + 2\binom{n}{2}\frac{mp(mp-1)}{m(m-1)} = np + 2\binom{n}{2}\frac{p(mp-1)}{(m-1)}.$$

Hence,

$$\text{Var}[X] = E[X^2] - (E[X])^2 = np(1-p)\frac{m-n}{m-1}, \tag{4.1.26e}$$

which is obtained after simplifying and rearranging terms. For the multivariate case represented by Eq. (4.1.24),

$$\text{Var}[X_i] = n\frac{m_i}{m}\left(1 - \frac{m_i}{m}\right)\left(\frac{m-n}{m-1}\right). \tag{4.1.26f}$$

From Eq. (4.1.26e) one can see that the variance of the hypergeometric distribution can be obtained by multiplying the variance of the binomial variate, as given by Eq. (4.1.5b), by

$(m-n)/(m-1)$. This step effects the reduction of the variance that results from sampling without replacement.

Definition and properties: A random variable X which denotes the number of successes in a sample of n items selected at random *without replacement* from a population of m items, in which there are mp items classified as successes and $mq = m(1-p)$ items classified as failures, has a **hypergeometric** pmf

$$p_X(X=x) = \frac{\binom{mp}{x}\binom{m(1-p)}{n-x}}{\binom{m}{n}}, \quad \text{for} = 0, 1, 2, \ldots, \min(mp, n),$$

where $0 \le p \le 1$. Also, $E[X] = np$ and $\text{Var}[X] = np(1-p)(m-n)/(m-1)$.

Example 4.18. Components for water pumps. Standard components for pumps ordered for a water supply scheme have the same specifications but it is found that p percent are defective. A consignment of 100 items has been received. For this consignment to be accepted, no more than one item in a lot of 10 items selected at random can be defective. Compare the probabilities by the binomial and hypergeometric distributions, assuming $p = 0.02$.

Binomial:

$$\Pr[X \le 1] = \sum_{k=0}^{1}\binom{10}{k}p^k(1-p)^{10-k} = (0.98)^{10} + 10(0.02)(0.98)^9 = 0.984.$$

Hypergeometric:

$$\Pr[X \le 1] = \frac{\binom{100p}{0}\binom{100(1-p)}{10} + \binom{100p}{1}\binom{100(1-p)}{9}}{\binom{100}{10}}$$

$$= \frac{\binom{2}{0}\binom{98}{10} + \binom{2}{1}\binom{98}{9}}{\binom{100}{10}} = \frac{981}{990} = 0.991$$

(after many common terms are eliminated from the numerator and denominator).

As m increases and n decreases the two results become closer. In practice the engineer does not know the value of p. This can, however, be estimated by repeated sampling, for example, by means of data from supplies made elsewhere, and using Eq. (4.1.24*b*) or (4.1.24*c*) for the mean.

4.1.8 Summary of Section 4.1

The distributions given in this section play a useful role in engineering applications. It is important to determine whether the assumptions are met to a reasonable degree so that an appropriate choice can be made. A summary of the distributions is given in Table 4.1.7. For a wider range of discrete distributions, see Johnson and Kotz (1992) or Evans et al. (2000). In recent years there has been a movement to many generalizations (sometimes of the multimodal types) of the classical discrete distributions; Kemp (1997) gives 87 references.

Table 4.1.7 Summary of discrete distributions

Distribution	pmf, $p_X(X = x)^a$	Mean, E[X]	Variance, Var[X]	mgf
Bernoulli	$p^x(1-p)^{1-x}$, for $x = 0, 1; 0 \le p \le 1$	p	$p(1-p)$	$(1-p) + pe^t$
Binomial	$\binom{n}{x} p^x(1-p)^{n-x}$, for $x = 0, 1, 2, \ldots, n; 0 \le p \le 1$	np	$np(1-p)$	$[pe^t(1-p)]^n$
Poisson	$\frac{v^x e^{-v}}{x!}$, for $x = 0, 1, 2, \ldots, n; v \ge 0$	v	v	$\exp[v(e^t - 1)]$
Geometric	$p(1-p)^{x-1}$, for $x = 1, 2, 3, \ldots, n; 0 \le p \le 1$	$\frac{1}{p}$	$\frac{(1-p)}{p^2}$	$\frac{pe^t}{1-(1-p)e^t}$
Negative binomial	$\binom{x-1}{r-1} p^r(1-p)^{x-r}$, for $r = 0, 1, 2, \ldots, n$; $x = r, r+1, r+2, \ldots; 0 \le p \le 1$	$\frac{r}{p}$	$\frac{r(1-p)}{p^2}$	$\frac{pe^t}{1-(1-p)e^t}$
Alternative negative binomial	$\binom{r+y-1}{y} p^r(1-p)^y$, for $y = x - r$; for $y = 0, 1, 2, \ldots$	$\frac{r(1-p)}{p}$	$\frac{r(1-p)}{p^2}$	$\left[\frac{p}{1-(1-p)e^t}\right]^r$
Log-series	$\frac{(1-p)^x}{-x \ln(p)}$, for $x = 1, 2, 3, \ldots, n; 0 \le p \le 1$	$-\frac{1}{\ln(p)} \frac{1-p}{p}$	$\frac{E[X]}{p} - \{E[X]\}^2$	$\frac{\ln(1-pt)}{\ln(1-p)}$
Multinomial	$\left(\frac{n!}{x_1!x_2!\ldots x_r!}\right) p_1^{x_1} p_2^{x_2} \cdots p_r^{x_r}$	See binomial for marginal distributions	See binomial for marginal distributions	Not useful
Hypergeometric	$\frac{\binom{mp}{x}\binom{m(1-p)}{n-x}}{\binom{m}{n}}$, for $x = 0, 1, 2, \min(mp, n)$	np	$np(1-p)\left(\frac{m-n}{m-1}\right)$	Not useful; see Johnson et al. (1992)

[a] Change to Y variable for alternative negative binomial; some authors reverse the definitions for negative binomial and its alternative form.

4.2 CONTINUOUS DISTRIBUTIONS

Several continuous distributions play useful roles in engineering as in numerous other disciplines. The more important ones are the uniform, exponential, gamma, beta, Weibull, normal, and lognormal distributions. These are introduced in this section.

As noted before, continuous distributions are applicable when the random variable can take any value within some range. Examples of this type of random variable are the strength of a concrete or the flow in a river. A continuous variable is in contrast to a discrete variable, which is confined in occurrence to integer or other specific values.

4.2.1 Uniform distribution

The simplest type of continuous distribution is the *uniform*. As implied by the name, the pdf is constant over a given interval—for example, from a to b, where $a < b$—and takes the form

$$\begin{aligned} f_X(x) &= \frac{1}{b-a}, \quad \text{for } a \le x \le b, \\ &= 0, \quad \text{otherwise.} \end{aligned} \tag{4.2.1}$$

It is also called a rectangular distribution because of the shape of the density function. The parameters a and b are real-valued constants. All values of the variate between the lower limit a and the upper limit b are equally frequent or equally likely to occur. An example of a uniform pdf is given in Fig. 4.2.1 for $a = 0.5$ and $b = 2.5$.

Thus, we can say that a variable is uniformly distributed between limits a and b if the probability that it lies within any interval (c, d) between a and b is proportional to that interval. That is,

$$\Pr[c < X < d] = \frac{d-c}{b-a}.$$

The integral of the pdf between the specified limits is unity, which together with its other properties satisfies the stated requirements [Eq. (3.1.5a), (3.1.5b), and (3.1.5c)]. The cdf corresponding to Fig. 4.2.1 is given in Fig. 4.2.2.

By making $a = 0$ and $b = 1$, one obtains the standard or unit uniform distribution. This is used in generating random variates for all types of probability distributions for purposes of simulation.[15] Also of importance is the fact that the uniform pdf, as given by Eq. (4.2.1),

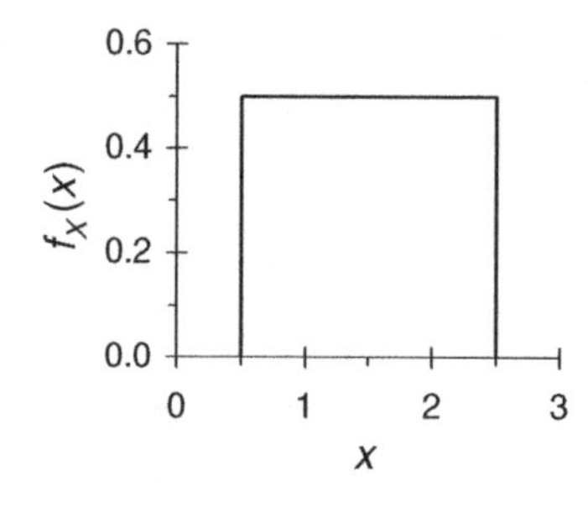

Fig. 4.2.1 The pdf, $f_X(x)$, for a uniform(0.5, 2.5) variate.

[15] See Chapter 8.

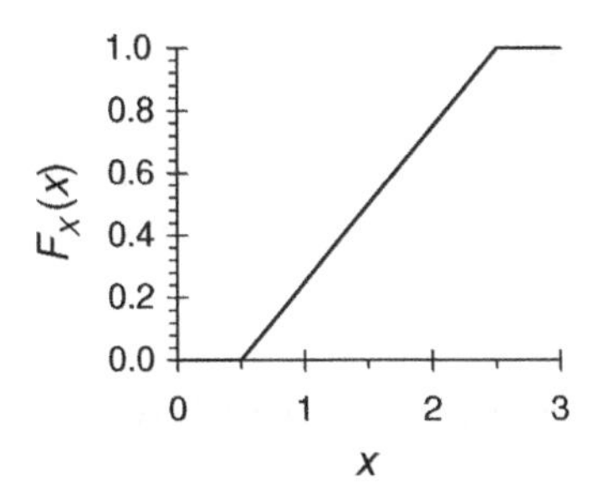

Fig. 4.2.2 The cdf, $F_X(x)$, for a uniform(0.5, 2.5) variate.

is widely used to give equal likelihoods to the values of a random variable when prior reasoning or available information does not indicate otherwise.

4.2.1.1 Mean, variance, and moment-generating function of a uniform variate

$$E[X] = \int_a^b \frac{x dx}{b-a} = \frac{b+a}{2}, \tag{4.2.2a}$$

$$\text{Var}[X] = \int_a^b \frac{x^2 dx}{b-a} - \{E[X]\}^2 = \frac{(b-a)^2}{12}, \tag{4.2.2b}$$

and

$$M_X(t) = E[e^{tX}] = \int_a^b e^{tX} \frac{1}{b-a} dx = \frac{e^{tb} - e^{ta}}{t(b-a)}. \tag{4.2.2c}$$

Definition: A uniformly distributed random variate can have any value in an interval *a* to *b* with equal likelihood. The pdf, mean, and variance are

$$f_X(x) = \frac{1}{b-a}, \quad a \le x \le b, \quad E[X] = \frac{b+a}{2}, \quad \text{and} \quad \text{Var}[X] = \frac{(b-a)^2}{12}.$$

Example 4.19. Density of concrete. Consider the data of Table 1.2.1. This shows the densities and compressive strengths of concrete samples. It may be assumed (as in Chapter 3) that the marginal pdf of the densities of concrete can be approximated by a uniform distribution. The pdf is shown in Fig. 3.3.6*b* assuming $a = 2400$ kg/m^3 and $b = 2500$ kg/m^3. From the data, the lowest of the 40 values is 2411 and the highest value is 2488 kg/m^3; also, the estimated mean and standard deviation are 2445 kg/m^3 and 15.99 kg/m^3 (from Table 1.2.2), respectively. If one uses the method of moments to estimate the two parameters, then by solving from the two relationships for the mean and variance just given [Eq. (4.2.2*a*) and (4.2.2*b*)],

$$\hat{a} = \bar{x} - \sqrt{3}\hat{s} = 2417 \text{ kg/m}^3 \quad \text{and} \quad \hat{b} = \bar{x} + \sqrt{3}\hat{s} = 2473 \text{ kg/m}^3.$$

Because some of the data are outside the range given by the values of the parameters, these estimates by the method of moments are unacceptable. See Example 3.24 based on the method of entropy.

4.2.2 Exponential distribution

From the Poisson process discussed in Subsection 4.1.3, the probability of no occurrences of the random variable X, which denotes a success, arrival, count, happening, or any other type of specified event, during a time interval t is

$$p_X(X = 0) = e^{-\lambda t}.$$

Using this result and considering the time T between occurrences as the random variable, we find the cdf of variable T is

$$F_T(T \leq t) = 1 - e^{-\lambda t}.$$

This means that the *waiting time* between successive events of a Poisson process has an *exponential distribution.* We note that, as in the Poisson, the distribution is applicable for other variables such as length and space in addition to time. Also, it follows that the exponential in continuous time corresponds to the geometric distribution in discrete time.

The pdf of the exponential distribution is written as follows by differentiating the expression just given with respect to t and by replacing T by a general variable X:

$$\begin{aligned} f_X(x) &= \lambda e^{-\lambda x}, && \text{for } x \geq 0, \lambda > 0, \\ &= 0, && \text{otherwise.} \end{aligned} \tag{4.2.3a}$$

For the same conditions, the cdf is

$$F_X(x) = 1 - e^{-\lambda x}. \tag{4.2.3b}$$

This is also referred to as the *negative exponential* because of the negative term in the exponent.

4.2.2.1 Mean, variance, and moment-generating function

The mean of the exponential distribution is obtained as follows, integrating by parts, and using l'Hospital's rule:

$$E[X] = \int_0^\infty xe^{-\lambda x}dx = [-xe^{-\lambda x}]_0^\infty + \int_0^\infty e^{-\lambda x}dx = \frac{1}{\lambda}. \tag{4.2.4a}$$

For the Poisson process, λ is the rate at which events occur, whereas $1/\lambda$, as just shown, is the average time between events. In relation to reliability analysis, it is often referred to as the *mean life time* or *time to failure.* Proceeding further,

$$E[X^2] = [-x^2e^{-\lambda x}]_0^\infty + \frac{2}{\lambda}\int_0^\infty x\lambda e^{-\lambda x}dx = \frac{2}{\lambda^2}.$$

Hence, the variance is

$$\text{Var}[X] = E[X^2] - \{E[X]\}^2 = \frac{1}{\lambda^2}. \tag{4.2.4b}$$

It is interesting to note that the coefficient of variation is

$$V_X = \frac{\sqrt{\text{Var}[X]}}{E[X]} = 1. \tag{4.2.4c}$$

The moment-generating function (as seen in Example 3.17) is

$$M_X(t) = E[e^{tx}] = \int_0^\infty e^{tx}\lambda e^{-\lambda x}dx = \frac{\lambda}{\lambda - t}, \qquad \text{for } t < \lambda. \tag{4.2.4d}$$

From the foregoing one can also show that the coefficients of skewness and kurtosis are 2 and 9, respectively (see Example 3.15).

Definition: The **exponential** distribution models the time (or length or area) between Poisson events. It has pdf $f_X(x) = \lambda e^{-\lambda x}$ for $x \geq 0$ and $\lambda > 0$, mean $E[X] = 1/\lambda$, variance $\text{Var}[X] = 1/\lambda^2$ and mgf $M_X(t) = \lambda/(\lambda - t)$, for $t < \lambda$.

Example 4.20. Floods affecting construction. An engineer constructing a bridge across a river is concerned of the possible occurrence of a flood exceeding 100 m^3/s which can seriously affect his work. If a flow of such magnitude is exceeded once in 5 years on average, on the basis of recorded data, what is the chance that the work which is scheduled to last 14 months can proceed without interruption or detrimental effects?

Assume floods exceeding the given magnitude are independent and identically distributed events. The estimated mean time interval between flood events, $\bar{x} = 5.0$ years. Hence $\hat{\lambda} = 1/5$ and from the exponential distribution,

$$\Pr\left[X \geq \frac{14}{12}\right] = e^{-1/5 \times 14/12} = .79.$$

The risk, $1 - 0.79 = 0.21$ seems to be rather high. A shorter time schedule should be adopted for the work. Alternatively, one should use a different method of construction, one that allows the engineer to cope with the imposed flood threat.

Example 4.21. Intervals of time between vehicles passing an observation point. In traffic engineering one is concerned with the length of time between vehicles passing a given point. If the intervals are too short there will be halts and interruptions, which the engineer attempts to minimize. The numbers of vehicles that pass a point during a time interval tend to be Poisson-distributed in sections where there are no obstructions, congestions, or stoppages. Under these conditions, the time intervals between vehicles passing a point of observation are exponentially distributed.

The following set of data gives the ordinates of a histogram of the intervals in seconds between the passing of vehicular traffic. There are 30 equal class intervals of 6 seconds each, from 0 to 3 minutes, as observed in Dorset, England, and reported by Leeming (1963):

54	23	16	10	16	16	12	8	8	7	4	5	4	5	1
2	0	3	1	2	2	2	0	0	1	0	0	1	1	0

The data are given in Table 4.2.1 in which, for convenience, class intervals of 0.2 minutes (12 seconds) are used.

The mean time interval is 0.551 minute, which is estimated by multiplying the midpoints of the class intervals by the relative frequencies given in column 3 and summing. Hence from Eq. (4.2.4*a*), $\hat{\lambda} = 1/0.551 = 1.81$. After substituting for λ in Eq. (4.2.3*a*), the expected relative frequencies are obtained by multiplying $f_X(x)$ by the class width of 0.2 minute for each x, which represents the midpoint of a class interval. The observed and expected relative frequencies are shown in Fig. 4.2.3.

The theoretical function shown in Fig. 4.2.3 has maximum ordinate $f(x) = \lambda$ at $x = 0$. As λ becomes smaller, the curvature decreases and at the limit, $\lambda = 0$, the $f(x)$ becomes a uniform pdf defined over $(0, +\infty)$.

The observed frequencies indicate that the distribution is close to the exponential type. On this basis and prior to considerations of future road improvements, the engineer may need to

Table 4.2.1 Fitting exponential distribution to intervals of time between vehicles passing a point of observation

Upper class limit in minutes, (1)	Observed number in class, (2)	Observed number divided by total, (3)	Cumulative distributions	
			Observed, (4)	Exponential, (5)
0.2	77	0.377	0.377	0.166
0 4	26	0.127	0.505	0.420
0.6	32	0.157	0.662	0.596
0.8	20	0.098	0.760	0.719
1.0	15	0.074	0.833	0.805
1.2	9	0.044	0.877	0.864
1.4	9	0.044	0.922	0.906
1.6	3	0.015	0.936	0.934
1.8	3	0.015	0.951	0.954
2.0	3	0.015	0.966	0.968
2.2	4	0.020	0.985	0.978
2.4	0	0.000	0.985	0.985
2.6	1	0.005	0.990	0.989
2.8	1	0.005	0.995	0.993
3.0	1	0.005	1.000	0.995
Total	204			

Source: Data from Leeming (1963).

make various computations. One may estimate, for example, the percentage of vehicles that have interarrival times less than 12 seconds. This is given by

$$100F_X(X \leq 0.2) = 100[1 - \exp(-1.84)(0.2)] = 30.8\%.$$

4.2.2.2 Memoryless property of the exponential distribution

As in the case of the geometric analogue of the discrete category, the exponential distribution models a behavior, arising from the parental Poisson process, that is independent of present or past occurrences. This can be shown as follows:

$$\Pr[X > x_1 + x_2 | X > x_1] = \frac{\Pr[X > x_1 + x_2]}{\Pr[X > x_1]} = \frac{e^{-\lambda(x_1+x_2)}}{e^{-\lambda x_1}} \tag{4.2.5}$$
$$= e^{-\lambda x_2} = \Pr[X > x_2].$$

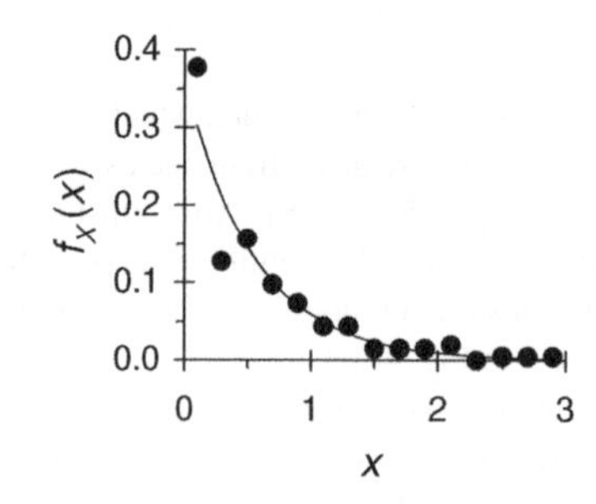

Fig. 4.2.3 Exponential distribution ($\lambda = 1.81$) fitted to time intervals between vehicles passing a point of observation.

4.2.2.3 General exponential distribution

A more general form of the exponential is obtained when the lowest value the random variable X takes is, say, ε, which is different from zero. In this way, the distribution is sometimes called a *shifted exponential* with pdf

$$\begin{aligned} f_X(x) &= \lambda e^{-\lambda(x-\alpha)}, \qquad \text{for } x \geq \varepsilon, \lambda > 0, \\ &= 0, \qquad \text{otherwise.} \end{aligned} \tag{4.2.6}$$

4.2.2.4 Other applications

The exponential distribution has wide applications in all areas of science and engineering. In Chapter 3 the times between earthquakes were estimated in this way. Similarly, one can apply the distribution to model time to failure and probabilities of survival of various design components of a mechanical or electrical nature, for example, in a water treatment plant.

Even when the assumptions of a Poisson process are not fully met, the exponential model may be adopted as a reasonable approximation. For example, arrival times of vehicles may not be independent because minimum distances are maintained between vehicles; this fact may account for the less-than-satisfactory fit to the exponential at the left extremity in Example 4.21 (Fig. 4.2.3). Causes for the departure from a Poisson process can also be attributed to speed limits, slow-moving vehicles, and other restrictions or impediments. Likewise in applying risk and reliability to design components, slow deterioration of machine parts will violate the conditions for a constant rate of risk over time. Nevertheless, the exponential is adoptable in a wide variety of situations, even if it is sometimes done as a first approximation. The exponential is equivalent to the Pearson Type X distribution.

4.2.2.5 Renewal and point processes

Consider the following operative strategy concerning, for example, light bulbs or electronic devices or components of a water treatment plant or other units that have an unpredictable lifetime. Suppose all the units of interest are identical and that the first unit u_1 is placed in operation at time zero. We let this unit remain in operation up to a random instant T_1 (corresponding to a random lifetime X_1) until it fails and then immediately replace it with another unit u_2. This second unit is also made to survive up to failure at a random instant T_2, and we continue likewise indefinitely.

The (positive) lifetimes of successive units $X_i, i = 1, 2, \ldots$ represent a sequence of independent identically distributed random variables. Hence the name *renewal process* is given to these occurrences. The random variables $T_1, T_2, \ldots$ represent the times of failure of the units $u_1, u_2, \ldots$, and the link between the series $\{X_i\}$ and $\{T_i\}$ is given by $T_n = \sum_{i=1}^{n} X_i$ where $T_0 = 0$. If we also consider that the random variables $N(t), t > 0$ count the (random) number of renewals in the interval $(0, t]$, then $\{N(t), t > 0\}$ is called a counting renewal process. Statisticians do not often make a distinction between a renewal process and the associated counting process.

For example, the Poisson process with rate of occurrence λ is a renewal counting process with interarrival times that are exponentially distributed with common parameter λ. Some practical situations that can be modeled are the successive failures of electronic or mechanical components and the successive times or distances between cars passing a given point of observation on an uninterrupted single-lane road.

The discrete series $T_0, T_1, \ldots$ constitute a *point process*.

4.2.3 Erlang and gamma distribution

In the case of the number of discrete independent trials required for r successes, we noted that the negative binomial distribution is an extension of the geometric distribution, which concerns a single success. Similarly, when dealing with continuous variables, one may be interested in the distribution of time to the rth arrival of a Poisson process. This results in the *Erlang distribution*, the pdf (that is the probability per unit time) of which can be obtained by multiplying the arrival rate by the probability that the $(r-1)$th arrival occurs around time t. (A. K. Erlang was a Danish engineer well known for his work on congestion in telephone lines around 100 years ago.) Hence,

$$f_X(x) = \frac{\lambda(\lambda x)^{r-1}e^{-\lambda x}}{(r-1)!}, \quad \text{for } 0 < x \text{ and } r = 1, 2, 3\ldots,$$
$$= 0, \quad \text{otherwise.} \tag{4.2.7}$$

More generally, we can consider that the times between arrivals, say, $T_i, i = 1, 2, 3, \ldots, r$ are independent and exponentially distributed with parameter λ. The distribution of $X = T_1, T_2, \ldots, T_r$ can be shown, using Eq. (3.4.8) to take the form given by Eq. (4.2.7).

The denominator, $(r-1)!$, is the product of the first $(r-1)$ natural numbers. It can be written in the form $\Gamma(r)$. This is applicable also to noninteger values of r, and is called the *complete (standard)gamma function* when defined as

$$\Gamma(r) = \int_0^{\infty} t^{r-1}e^{-t}dt, \quad \text{for } r > 0,$$
$$= 0, \quad \text{otherwise.} \tag{4.2.8}$$

Integrating by parts, it can be shown that $\Gamma(r+1) = r\Gamma(r)$ for any $r > 0$. Also, $\Gamma(1) = 1$, which follows directly by substituting $r = 1$ in Eq. (4.2.8). Also,[16] $\Gamma(1/2) = \sqrt{\pi}$.

The *standard gamma* pdf is written as

$$f_T(t) = \frac{t^{r-1}e^{-t}}{\Gamma(r)}, \quad \text{for } 0 \le t \text{ and } r > 0,$$
$$= 0, \quad \text{otherwise.} \tag{4.2.9}$$

The parameter r is known as the *shape parameter* because it is related to the shape of the pdf. This characteristic can be seen from the graphs of the standardized gamma pdfs for $r = 1, 2,$ and 5 as shown in Fig. 4.2.4*a*; the corresponding cdfs are given in Fig. 4.2.4*b*.

When $r = 1$, the distribution is exponential. As r increases, the skewness decreases; that is, the distribution tends to become more symmetrical.

The *incomplete gamma function ratio* takes the form:

$$F_T(t) = \int_0^{t} \frac{x^{r-1}e^{-x}}{\Gamma(r)}dx, \quad \text{for } r > 0, 0 \le t,$$
$$= 0, \quad \text{otherwise.} \tag{4.2.10}$$

[16] See, for example, Olkin et al., (1980, p. 537).

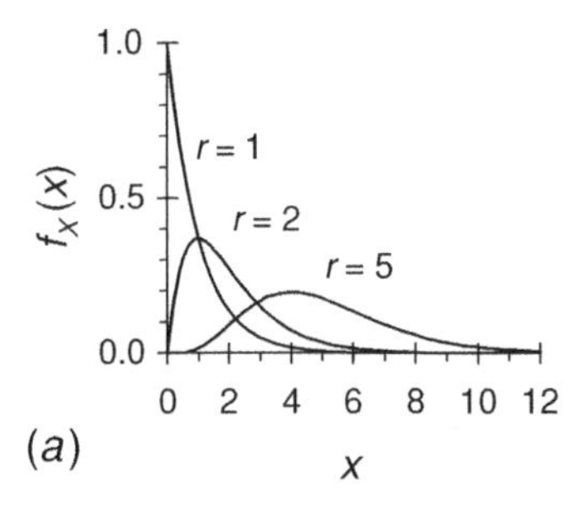

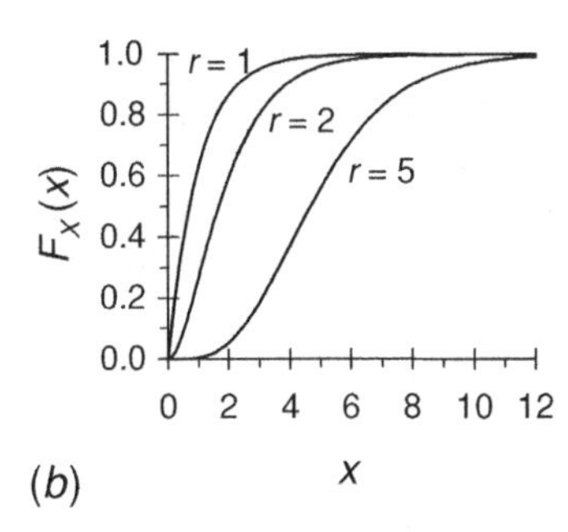

Fig. 4.2.4 Standard gamma distribution (*a*) pdfs and (*b*) cdfs for $r = 1, 2$, and 5.

Extensive tables of the foregoing integral and its inverse have been published in the past, for example, by Harter (1964). Since then, values are obtained from available computer software programs or through a suitable computer algorithm.[17]

The *gamma distribution* is usually defined with two parameters arising directly from Eq. (4.2.7) except that parameter r also takes noninteger values. Hence it is written as

$$f_X(x|\lambda, r) = \frac{\lambda^r x^{r-1} e^{-\lambda x}}{\Gamma(r)}, \quad \text{for } 0 \leq x \quad \text{with } \lambda > 0 \quad \text{and } r > 0, \tag{4.2.11}$$
$$= 0, \qquad \text{otherwise.}$$

The term λ is the *scale parameter*, which scales the variable and makes it dimensionless.

4.2.3.1 Mean and variance of the two-parameter gamma distribution

As shown in Example 3.19, the mean and variance are as follows:

$$E[X] = \frac{r}{\lambda} \tag{4.2.12a}$$

and

$$\text{Var}[X] = \frac{r}{\lambda^2}. \tag{4.2.12b}$$

These equations provide estimates of the two parameters by the method of moments. With regard to the moment-generating function, if one treats the gamma variates X as the sum of r independent exponentially distributed variates where r is an integer, then the mgf

[17] See, for example, Shea (1988).

follows from the mgf of the exponential [see Eq. (4.2.4*d*)]. Alternatively, one obtains the mgf directly as follows:

$$M_X(t) = E[e^{Xt}]$$
$$= \int_0^\infty e^{xt} \frac{\lambda^r}{\Gamma(r)} x^{r-1} e^{-\lambda x} dx = \left(\frac{\lambda}{\lambda - t}\right)^r \int_0^\infty \frac{(\lambda - t)^r}{\Gamma(r)} x^{r-1} e^{-(\lambda - t)x} dx.$$

The integral is the area under a gamma $(r, \lambda - t)$ function and is equal to 1. Hence,

$$M_X(t) = \left(\frac{\lambda}{\lambda - t}\right)^r. \tag{4.2.12c}$$

From the physical viewpoint, engineers and scientists have found that the empirical distributions of many natural and structural processes closely resemble the gamma. The gamma distribution is especially important in statistics because the *chi-squared distribution* is a particular form of the gamma with parameter $r = v/2$, where v denotes the degrees of freedom, and $\lambda = 1/2$. The cdf of this distribution is

$$F(\chi^2) = \frac{1}{2} \int_0^{\chi^2} \frac{(t/2)^{(v/2)-1} e^{-t/2}}{\Gamma(v/2)} dt, \tag{4.2.12d}$$

where t is a dummy variable. The cdf of the chi-squared distribution is given in Table C.3 in the Appendix C.[18]

It is also possible to introduce [as in Eq. (4.2.6) for the exponential distribution] a location parameter in Eq. (4.2.11), say, $\varepsilon \geq 0$, by changing the variable x twice to $(x - \varepsilon)$, which means that $\varepsilon \leq x$. Written as such with three parameters (or sometimes in the simpler two-parameter form), the distribution is also referred to as a Pearson Type III.[19]

Definition: The gamma distribution models the waiting time between the nth and $(n + r)$th Poisson events. In the two-parameter form, it has pdf

$$f_X(x|\lambda, r) = \frac{\lambda^r x^{r-1} e^{-\lambda x}}{\Gamma(r)}, \qquad \text{for } 0 \geq x \text{ with } \lambda > 0 \text{ and } r > 0.$$

The mean and variance are

$$E[X] = \frac{r}{\lambda} \quad \text{and} \quad \text{Var}[X] = \frac{r}{\lambda^2}.$$

The mgf is

$$M_X(t) = \left(\frac{\lambda}{\lambda - t}\right)^r.$$

Example 4.22. Pumps for water supply. At a remote pumping station two pumps are operated so that, when there is a breakdown of one pump, the other pump (which serves as a standby) is switched on automatically. The pumps are identical and have a mean time between breakdowns of 300 days. Determine the probability density function of the time, in days, during which the system operates until a complete breakdown.

[18] For large v, $\chi_p^2 \approx 1/2(z_p + \sqrt{2v - 1})^2$ where χ_p^2 and z_p are the pth quantiles of the chi-squared and normal distributions, respectively. If $v \geq 30$, for example, the error is less than 1%.

[19] See Chapter 7.

It is assumed that the occurrences of breakdowns are Poisson-distributed, and the operating time span of a pump has a pdf

$$f_X(x) = \frac{1}{300} e^{-x/300}, \quad \text{for } x \geq 0,$$
$$= 0, \quad \text{otherwise.}$$

Also, the performance of one pump is assumed to be independent of the other, so that the life of the system, Z, which is the sum of the lives of the units X which comprise it, has a gamma pdf

$$f_Z(z) = \frac{1}{300}\left(\frac{z}{300}\right) e^{-z/300}, \quad \text{for } z \geq 0,$$
$$= 0, \quad \text{otherwise.}$$

Example 4.23. Design with timber. For the data pertaining to the modulus of rupture of timber in Table E.1.1, determine the value which is not exceeded more than 5% of the time on the basis of a two-parameter gamma model. Also, determine the probability of exceedance of 20 N/mm^2.

From Table 1.2.2, the estimated mean and standard deviation of the timber strength data are $\bar{x} = 39.09$ and $\hat{s} = 9.92$ N/mm^2, respectively. From Eq. (4.2.12*a*) and (4.2.12*b*), the estimates of the two parameters are

$$\hat{\lambda} = \frac{\bar{x}}{\hat{s}^2} = \frac{39.09}{9.92^2} = 0.397\ (\text{N/mm}^2)^{-1} \quad \text{and} \quad \hat{r} = \hat{\lambda}\bar{x} = 15.5.$$

We obtain approximate solutions by using the inverse of the chi-squared distribution [Eq. (4.2.12*d*)] given in Table C.3 in the Appendix. For the chi-squared distribution, we note the following:

(*a*) The shape parameter r and the degrees of freedom v have the relationship $v = 2r$.
(*b*) The scale parameter $\lambda = 1/2$.

Therefore, we refer to $v = 31$, because of (*a*), and $F = 0.05$ in Table C.3. The entry of 19.3 is then multiplied by 1/2, because of (*b*), to give a value of 9.65 as the standard gamma variate t of Eq. (4.2.10). To apply the two-parameter gamma distribution of Eq. (4.2.11), one must divide the t value by $\hat{\lambda} = 0.397$ to give 24.3 N/mm^2 as the answer to the first part of the question. That is, we are 95% confident that the timber strength exceeds 24.3 N/mm^2.

For the second part of the question,

$$x = 20.00\ \text{N/mm}^2,$$

and thus the corresponding gamma standard variate

$$t = x\hat{\lambda} = 7.94.$$

If one divides this by the chi-squared scale parameter of 1/2, one gets the required entry of 15.88 in Table C.3 for $v = 31$. We see that $F_X(15.88)$ is approximately 0.01; that is the probability of exceedance is 0.99. Thus, if the design is based on a timber strength of 20 N/mm^2, we can reasonably expect this value to be exceeded in 99 cases out of 100. An engineer may decide to lower the design strength or select a different timber, if more stringent conditions are to be imposed.

4.2.4 Beta distribution

The beta distribution models a random variable that takes values in the interval given by 0–1. The distribution plays a special role in decision methods.

Definition: The beta pdf is given by

$$f_X(x|\alpha, \beta) = \frac{1}{B(\alpha, \beta)} x^{\alpha-1}(1-x)^{\beta-1}, \quad \text{for } 0 < x < 1, \alpha > 0, \beta > 0, \qquad (4.2.13a)$$
$$= 0, \quad \text{otherwise.}$$

The expression

$$B(\alpha, \beta) = \int_0^1 x^{\alpha-1}(1-x)^{\beta-1}dx = \frac{\Gamma(\alpha)\Gamma(\beta)}{\Gamma(\alpha+\beta)} \qquad (4.2.13b)$$

denotes the beta function with parameters α and β.

It follows from the two preceding equations that the nth moment is

$$E[X^n] = \frac{B(\alpha+n, \beta)}{B(\alpha, \beta)}. \qquad (4.2.14a)$$

Hence, by using the relationship with the gamma function in Eq. (4.2.13b) and using the previously mentioned relationship $\Gamma(r+1) = r\Gamma(r)$,

$$E[X] = \frac{\alpha}{\alpha+\beta}. \qquad (4.2.14b)$$

Also, because $\Gamma(r+2) = (r+1)\Gamma(r+1) = r(r+1)\Gamma(r)$ and by simplifying,

$$\text{Var}[X] = \frac{\alpha\beta}{(\alpha+\beta)^2(\alpha+\beta+1)}. \qquad (4.2.14c)$$

Graphs of the beta distributions are given in Fig. 4.2.5 for beta(1, 4), (4, 4), and (2, 6).

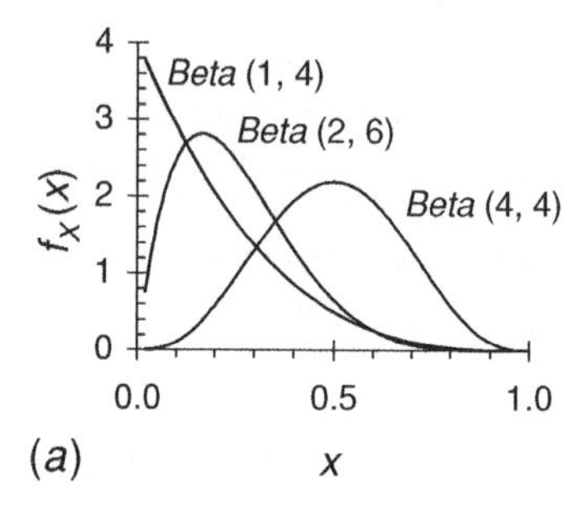

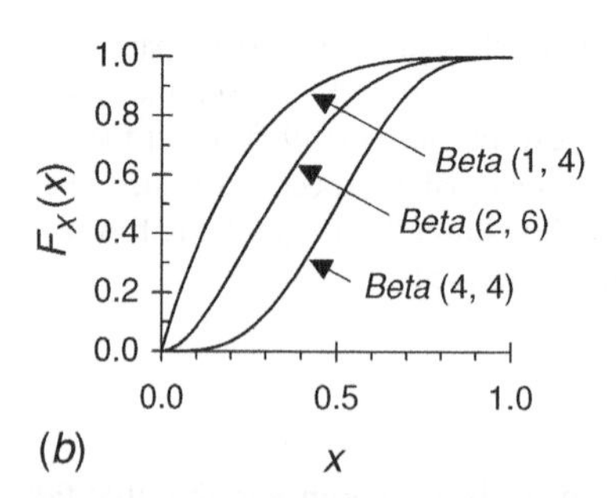

Fig. 4.2.5 Standard beta distribution (a) pdfs and (b) cdfs for three sets of shape parameters.

Note that for equal values of the parameters, the beta pdf is symmetrical. From Eq. (4.2.13) it follows that beta(1, 1) $\sim$ uniform(0, 1). Furthermore, the beta is related to the normal and F distributions (details of which are given in Chapter 5). Thus, the beta is a very versatile distribution and can be applied in diverse situations. The beta, in the form given here, is equivalent to the Pearson Type I distribution.

Example 4.24. Maintenance of major roads. In a certain country, there are 10 major roads in State A and a similar number and length of roads in State B. The proportion of roads that require substantial maintenance works during an annual period can be approximated by beta(4, 3) and beta(1, 4) distributions, respectively, in the two states.

(1) Which State should spend more on annual maintenance?
(2) What is the probability that not more than two roads will require substantial maintenance work in State B during an annual period?

For the first question, the mean number of roads that require extensive maintenance work is $10\mu = 10\alpha/(\alpha + \beta)$. By substituting the values of the parameters given, it is seen that state A should need to spend on approximately six roads. In State B this number is reduced to two. Therefore, State A needs to spend more.

To answer the second question, we use Eq. (4.2.13b) for the beta(1, 4) distribution. The proportion of roads is 0.2. Hence from the beta cdf,

$$F(0.2) = \int_0^{0.2} \frac{x^{\alpha-1}(1-x)^{\beta-1}}{B(\alpha, \beta)} dx = \frac{\Gamma(1+4)}{\Gamma(1)\Gamma(4)} \int_0^{0.2} (1-x)^3 dx$$

$$= 4\left[x - \frac{3x^2}{2} + \frac{3x^3}{3} - \frac{x^4}{4}\right]_0^{0.2} = 0.59.$$

Because this probability is not very high, an engineer may provide for more than two roads to be maintained in a given year.

4.2.5 Weibull distribution

The Weibull distribution (named after the Swedish physicist W. Weibull, who applied it when studying material strengths such as yield strengths of Bofors steel and fiber strength of Indian cotton, and other phenomena; see Weibull, 1951) provides a close approximation to the probability laws of many natural phenomena. It has been used to model, for example, the time to failure of electrical and mechanical systems. In reliability engineering, attention is often focused on a threshold level below which a system or component or basic material has an unacceptable probability of failure. The exponential is sometimes used as a basic model. However, the Weibull distribution has greater flexibility and closer fit to failure strengths and times of failure, and it is also one of the asymptotic distributions of general extreme value theory [see Gumbel (1954), and discussions in Chapter 7]. Numerous publications have been made on the theory and application of the Weibull distribution. As shown in this section, it is closely linked to the exponential distribution which can be viewed as a (simplified) special case of the Weibull.

Definition: The two-parameter Weibull pdf with parameters β and $\lambda > 0$ is given by

$$f_X(x) = \frac{\beta}{\lambda}\left(\frac{x}{\lambda}\right)^{\beta-1} \exp\left[-\left(\frac{x}{\lambda}\right)^{\beta}\right], \quad \text{for } x > 0, \qquad (4.2.15)$$

$$= 0, \quad \text{otherwise.}$$

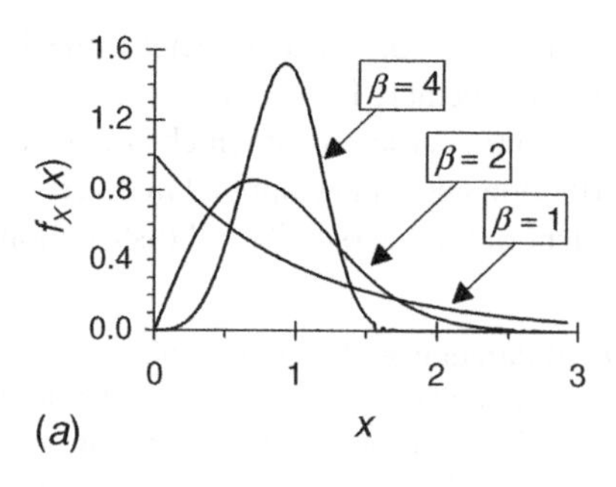

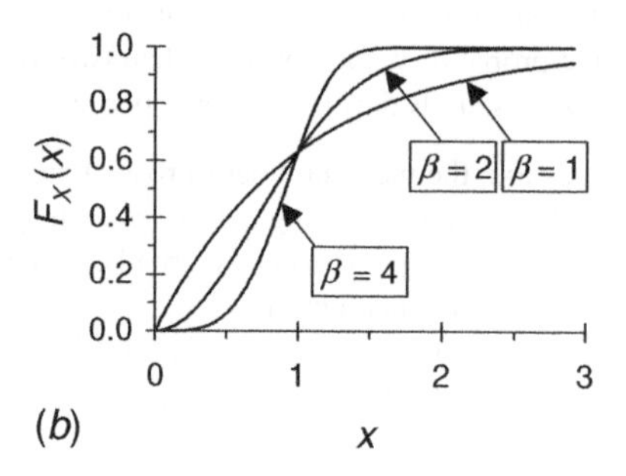

Fig. 4.2.6 Weibull distribution (*a*) pdf and (*b*) cdf for $\beta = 1, 2$, and 4.

The cdf takes the form

$$F_X(x) = 1 - \exp\left[-\left(\frac{x}{\lambda}\right)^{\beta}\right], \quad \text{for } x > 0, \tag{4.2.16}$$
$$= 0, \quad \text{otherwise.}$$

From the preceding definition we see that if a random variable $X \sim$ Weibull (β, λ), then $Y = (X/\lambda)^{\beta} \sim$ exponential (1), that is, $f_Y(y) = \exp(-y)$.

Graphs of the Weibull pdf and cdf are given in Fig. 4.2.6 for values of the shape parameter β equal to 1, 2, and 4, with the scale parameter $\lambda = 1$. For higher values of β, the pdf tends to become symmetrical, as seen for $\beta = 4$.

By using the relationship between the exponential and the Weibull distribution already described, one can show that the mean and variance of the Weibull distribution are given by

$$E[X] = \lambda\Gamma\left(1 + \frac{1}{\beta}\right), \tag{4.2.17a}$$

and

$$\text{Var}[X] = \lambda^2\left[\Gamma\left(1 + \frac{2}{\beta}\right) - \left(\Gamma\left(1 + \frac{1}{\beta}\right)\right)^2\right]. \tag{4.2.17b}$$

Also, the coefficient of variation V_X has the relationship:

$$V_X^2 = \frac{\Gamma[1 + (2/\beta)]}{[\Gamma[1 + (1/\beta)]]^2} - 1 \tag{4.2.17c}$$

(see Tables C.5 and C.6 in Appendix C and Problem 4.21).

It is clear from Eq. (4.2.16) and Fig. 4.2.6 that for $\beta = 1$, the distribution is exponential. Confidence limits can be constructed (as shown in the next example) on the value of β because

$$X^2 = \frac{2n\beta}{\hat{\beta}} \tag{4.2.18}$$

has an approximately chi-squared distribution with $\nu = 2n$ degrees of freedom, where n is the sample size.[20]

Example 4.25. Estimation of low flows. Ten years of annual minimum 10-day minimum flow data—using the average of 10 consecutive low flows in the driest sequence of each year—from the River Pang at Pangbourne in hydrometric area 39 in England are ranked and given below in cubic meters per second:

13.4 25.7 32.2 35.9 40.0 40.0 40.4 50.7 58.2 71.4.

The data are from the Institute of Hydrology, Wallingford. The Weibull distribution is associated with extreme value theory applied to minima and, therefore, using it to model the distribution of low flows is justifiable. The present sample size does not warrant the application of a distribution with more than two parameters. Using the method of moments, one can estimate the mean and variance of the data and solve Eq. (4.2.17) iteratively for the shape parameter β and then for the scale parameter λ. This requires an algorithm for the complete gamma function, or tables of this function and particular ratios of the function. The maximum likelihood method, as discussed in Subsection 3.2.3, is also iterative.[21] Alternatively, one can use a least-squares procedure, as given here. It follows from Eq. (4.2.16) that

$$\ln(x_i) - \ln(\lambda) = \frac{\{\ln[-\ln(1 - F_X(x_i))]\}}{\beta}.$$

We apply the plotting position $F_X(x_i) = (i - 0.35)/n$ (as in Example 3.21) and let[22]

$$z_i = \ln(x_i)$$

and

$$y_i = \ln\{-\ln[1 - F_X(x_i)]\}.$$

Hence, we obtain the following linear relationship:

$$z_i = \frac{y_i}{\beta} + \ln(\lambda) + \varepsilon$$

where i is the rank of the data in ascending order, n is the number of items of data, and ε is an error term in the regression (that is a deviation from the linear model).[23] We have from a least-squares fit to the equation just given, for $n = 10$,

$$\hat{\beta} = \frac{\sum_i^{10} (y_i - \bar{y})^2}{\sum_i^{10} (y_i - \bar{y})(z_i - \bar{z})} = 2.59,$$

and

$$\hat{\lambda} = \exp\left(\bar{z} - \bar{y}/\hat{\beta}\right) = 44.85 \text{ m}^3/\text{s},$$

[20] See Crow (1974).
[21] See Van der Auwera et al. (1980) and Smith and Naylor (1987).
[22] As discussed further in Chapter 5, the plotting position refers to the probability at which $x_{(i)}$ should be plotted.
[23] Details of regression are given in Chapter. 6.

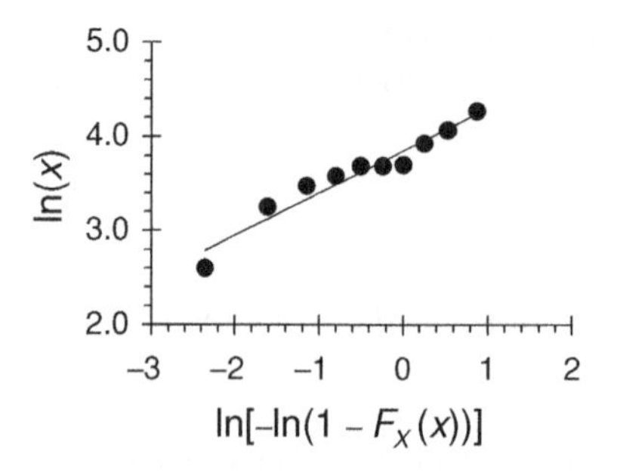

Fig. 4.2.7 Ten-day low flows at Pangbourne, on the Pang River, England. Weibull probability plot using plotting position $(i - 0.35)/n$; i is rank of data in ascending order; and number n of items of data is 10.

where

$$\bar{z} = \frac{1}{10}\sum_{i=1}^{10} z_i,$$

and

$$\bar{y} = \frac{1}{10}\sum_{i=1}^{10} y_i.$$

The preceding linear relationship between z and y (without the error term) representing the least squares fit of the data to the two-parameter Weibull distribution is shown in Fig. 4.2.7.

From Eq. (4.2.18) the approximate lower 99% confidence limit for β is

$$\frac{\hat{\beta}}{2n}x^2_{2n,0.99} = \frac{2.59}{20} \times 8.26 = 1.07.$$

Because this is greater than 1, the distribution is significantly different from the exponential.

Quantile estimates can be obtained using the estimates of the Weibull parameters and the preceding equations. For example, the annual minimum 10-day low flow with a return period of 10 years can be estimated as follows:

$$y_{10} = \ln\left\{-\ln\left[1 - \frac{1}{10}\right]\right\} = -2.25,$$

and

$$x_{10} = \exp\left[\left(\frac{y_{10}}{\hat{\beta}}\right) + \ln(\hat{\lambda})\right] = \exp\left[\left(\frac{-2.25}{2.59}\right) + \ln(44.85)\right] = 18.8 \text{ m}^3/\text{s}.$$

Example 4.26. Wind speeds. Engineers must deal with problems and opportunities related to wind speed. For example, tall structures must be designed to resist the force of high-speed winds (as discussed in Chapter 7). Wind power is an inexpensive energy resource.

The Weibull distribution is frequently used to model wind speeds.[24] Thus if U denotes the wind speed at a point in space, the pdf is

$$f_U(u) = \frac{\beta}{\lambda}\left(\frac{u}{\lambda}\right)^{\beta-1}\exp\left[-\left(\frac{u}{\lambda}\right)^{\beta}\right],$$

where λ and β denote the scale and shape parameters, respectively.

[24] See Van der Auwera et al. (1980).

If $\beta = 2$, the pdf is said to have the *Rayleigh distribution*. In fact the values of β are close to 2 for the fitted Weibull distributions in Northern Europe.[25] This property is noted from the estimates of the Weibull parameters at 175 meteorological stations using twelve 30° sectors, five heights, and four classes of surface roughness. However, problems have been encountered in fitting data pertaining to very low or very high wind speeds.

If U is Weibull distributed with λ and β as stated, then U^m is also Weibull distributed with corresponding parameters λ^m and β/m.[26] The available power wind density (that is, expected wind power per unit area) can be determined using this property. Neglecting the effects of turbulent fluctuations, we find power density is related to the wind speed as follows:

$$P = \rho_a \frac{U^3}{2},$$

where P denotes power density in watts per square meter and ρ_a is the density of air (approximately 1.2 kg/m^3 at a temperature of 15°C). From the properties of the Weibull distribution, the mean power density is then

$$E[P] = 0.5\rho_a \lambda^3 \Gamma\left(1 + \frac{3}{\beta}\right).$$

Furthermore, from the differential of the Weibull pdf [Eq. (4.2.15)], one can show that the modal value of wind speed is given by

$$\lambda\left(\frac{\beta - 1}{\beta}\right)^{1/\beta}.$$

Thus for the Rayleigh distribution, the modal wind speed is $\lambda/\sqrt{2}$.

Further developments of the Weibull distribution are given in Chapter 7 in relation to extreme values.

4.2.6 Normal distribution

The normal distribution arose originally in the study of experimental *errors*. Such errors pertain to unavoidable differences between observations when an experiment is repeated under similar conditions. An alternative term is *noise*, which is used in telecommunication engineering and elsewhere when referring to the difference between the true state of nature and the signal received. The uncertainties which are manifest in the errors may arise from different causes that are not easily identifiable. One must exclude from this definition of errors any mistakes in measurement or recording, which should be rectified beforehand. As noted in Chapter 1, when a sequence of observations is made, they generally tend to cluster around a central value, with smaller deviations occurring more frequently than larger errors.

The normal distribution is an ideal candidate to represent such errors when they are of an additive nature. The normal curve originated in the eighteenth century and was developed later by the German mathematician Gauss and others. By *normal* one used to imply that any data set that does not comply is exceptional. On the other hand, the contributory causes of the aforementioned errors may have a multiplicative effect or their

[25] See Troen and Peterson (1989).

[26] See Subsection 3.4.1 on transformations.

squares may be additive, in which cases the lognormal distribution (see Subsection 4.2.7) and the gamma distribution (see Subsection 4.2.3), respectively, are appropriate.

The pdf of the normal distribution is given by

$$\phi(x) = \frac{1}{\sigma\sqrt{2\pi}} \exp\left[-\frac{1}{2}\left(\frac{x-\mu}{\sigma}\right)^2\right], \quad \text{for } -\infty < x < \infty \tag{4.2.19}$$

and specified by two parameters. As shown in Example 3.23, these are the mean μ, the location parameter, and the standard deviation σ, the scale parameter, of the population.[27] The notation $N(\mu, \sigma^2)$ is used to indicate such a distribution. The normal curve is shown in Fig. 4.2.8 for two sets of parameter values—these should be in the units of the variable, such as N/mm^2 in the case of the strength of a material used in civil engineering.

We see that the mean μ locates the mode, whereas the variance σ^2 governs the spread. The cdf of the normal distribution can only be evaluated by numerical methods. In practice one uses the standardized curve for the purpose of evaluation with the transformation of the variate X to Z as follows:

$$Z = \frac{X-\mu}{\sigma}. \tag{4.2.20}$$

Thus Z is an $N(0, 1)$ variate with pdf

$$\phi(z) = \frac{1}{\sqrt{2\pi}} \exp\left(-\frac{z^2}{2}\right). \tag{4.2.21}$$

The pdf of the standard normal variate is shown in Fig. 4.2.9.

The cdf of X can then be written as

$$\begin{aligned} F_X(x) &= \Pr[X \le x] = \Pr\left[Z \le \frac{x-\mu}{\sigma}\right] = \Pr[Z \le z] \\ &= \Phi(z) = \int_{-\infty}^{z} \frac{1}{\sqrt{2\pi}} \exp\left(-\frac{t^2}{2}\right) dt, \quad \text{for } -\infty < z < +\infty. \end{aligned} \tag{4.2.22}$$

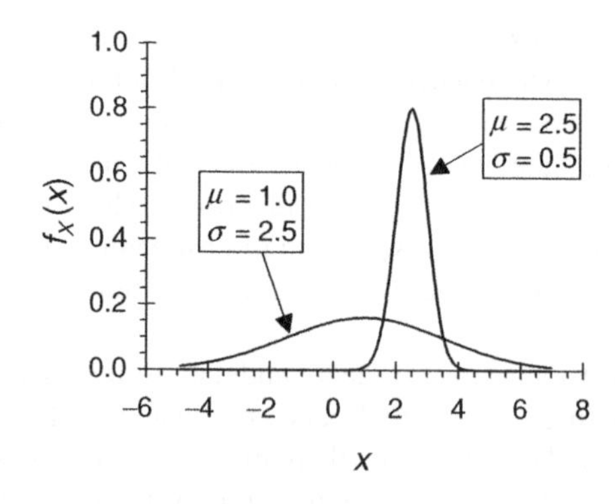

Fig. 4.2.8 Normal probability density functions.

[27] The normal distribution is derived in Appendix A.4.

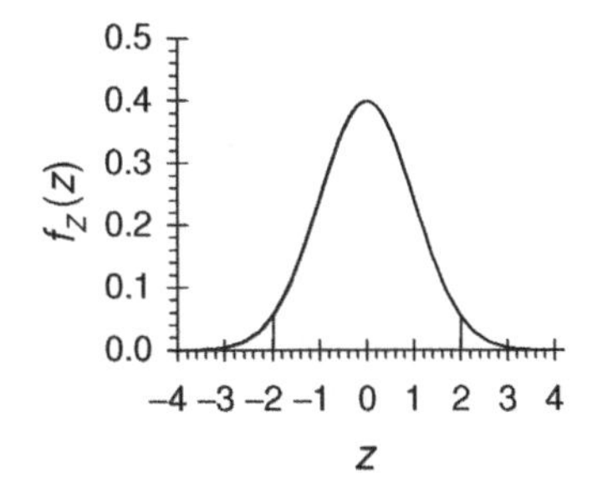

Fig. 4.2.9 Standard normal density function.

Table C.1 in Appendix C gives numerical solutions of this integral for values of the argument from 0 to 4.5. Because of the symmetry of the pdf as shown by Eq. (4.2.21),

$$\Phi(-z) = 1 - \Phi(z), \tag{4.2.23}$$

which allows one to find solutions for negative arguments.

Example 4.27. Compressive strength of concrete. In Table 1.2.2 the mean and standard deviation of the compressive strengths of 40 concrete cubes from a particular mix are given as 60.14 and 5.02 N/mm^2. As seen from the stem-and-leaf plot of Fig. 1.3.1, the box plot of Fig. 1.3.2, and the estimated values of skewness and kurtosis in Table 1.2.2, it is reasonable to assume that the compressive strengths have a normal distribution. In fact this is confirmed from the results of numerous studies on this phenomenon.[28] The answers to the following questions will, therefore, be based on this assumption:

(*a*) What value of compressive strength is exceeded in 19 tests out of 20?

We need to find the value of x which satisfies the condition

$$\Pr[X > x] = 1 - F_X(x) = 1 - \Phi(z) = \frac{19}{20} = .95.$$

Thus,

$$\Phi(z) = 1 - 0.95 = 0.05.$$

From Table C.1 and Eq. (4.2.23), $z = -1.645$ satisfies the condition. Hence using Eq. (4.2.20) with sample estimates and values of random variables, we find

$$x = \bar{x} + z \times \hat{s} = 60.15 - 1.645 \times 5.02 = 51.9 \text{ N/mm}^2.$$

(*b*) What is the probability that a test cube will fail when subjected to a compressive strength of 45 N/mm^2 or less?

For the assumed $N(60.14, 5.02)$ population, one obtains from Eq. (4.2.22),

$$\Pr[X < 45] = F_X(45) = \Phi\left(\frac{45 - 60.14}{5.02}\right) = \Phi(-3.02) = 1 - 0.9987 = 0.0013,$$

by using Eq. (4.2.23) and Table C.1.

Although this value is close to zero, an engineer may decide to adopt a lower design value, particularly if there is doubt about quality control on-site. It is, however, important to consider the size and nature of the work, the risks involved, and the consequences of failure in decisions of this nature.

[28] See, for example, Neville (1995, pp. 637–641).

(*c*) What is the probability that the compressive strengths are in the range 50.11–70.19 N/mm^2, that is, two standard deviations from the mean?

$$\begin{aligned}\Pr[50.11 \le X \le 70.19] &= \Phi\left(\frac{70.19 - 60.14}{5.02}\right) - \Phi\left(\frac{50.11 - 60.14}{5.02}\right)\\ &= \Phi(2) - \Phi(-2) = \Phi(2) - [1 - \Phi(2)]\\ &= 2\Phi(2) - 1 = 2 \times .97725 - 1 = .9545.\end{aligned}$$

The probability is the area under the curve between the two vertical lines in Fig. 4.2.9.

Example 4.28. Settlement of bridge foundations A proposed bridge across a stream is supported at the two ends and on a center pier. Although the design allows for relative settlements of the foundations, the engineer needs to keep these within limits. Settlements caused by the sum of numerous dead loads and impacts of moving vehicles may be assumed to be normally distributed, with the variability arising from the effects on the soil by the foundations. By correlating with results of tests on similar structures and soil conditions, the means and standard deviations of the settlements are estimated at the left end, center pier, and right end as 3.0, 5.0, and 3.0 cm and 1.0, 1.5, and 1.0 cm, respectively, with the higher values pertaining to the foundation of the center pier. It is also assumed as an initial approximation that the settlements are independent.

(*a*) What is the probability that the maximum settlement is in excess of 7.5 cm?

Because of the assumed independence between the three settlements,

$$\begin{aligned}\Pr[X_{\max} > 7.5] &= 1 - \Pr[(X_1 \le 7.5) \cap (X_2 \le 7.5) \cap (X_3 \le 7.5)]\\ &= 1 - \Phi\left(\frac{7.5 - 3}{1.0}\right)\Phi\left(\frac{7.5 - 5}{1.5}\right)\Phi\left(\frac{7.5 - 3}{1.0}\right)\\ &= 1 - \Phi(4.5)\Phi(1.67)\Phi(4.5)\\ &= 1 - 0.952 = .048.\end{aligned}$$

(*b*) Specify the maximum relative settlement of the center pier for which the engineer should design on the basis that this will be exceeded with a probability of 0.0001. Assume that the two adjacent foundations on the banks of the stream are stabilized so that their settlements are negligible.

$$\Pr[X_2 > x] = 0.0001.$$

$$\Pr[X_2 \le x] = \Phi(z) = \Phi\left(\frac{x - 5}{1.5}\right) = 0.9999.$$

Hence, $z = 3.72$ and $x = 1.5 \times 3.72 + 5 = 10.58$ cm.

4.2.6.1 Some properties of the normal distribution

(1) A linear transformation $Y = a + bX$ of an $N(\mu, \sigma^2)$ random variable X makes Y an $N(a + b\mu, b^2\sigma^2)$ random variable.

(2) If $X_i, i = 1, 2, \ldots, n$ are independent and identically distributed random variables from a population with mean μ and standard deviation σ, then the random variable

$$\bar{X}_n = \sum_{i=1}^{n} \frac{X_i}{n},$$

that is the sampling mean, from a random sample of size n from the same population, tends to have an $N(\mu, \sigma^2/n)$ distribution as n approaches infinity.

This important result is called the *Central Limit Theorem.*[29] If X_i is from a $N(\mu, \sigma^2)$ population, then the result holds regardless of the sample size n. The theorem states that even if the distribution of the random variable X_i is not normal, the so-called sampling distribution of its mean will tend to normality asymptotically. It also means that a large sample size n is required if the X_i are moderately nonnormal to obtain this result approximately, say, $n > 30$, but n should be much larger for greater departures from normality.

Example 4.29. Concrete densities. The mean and standard deviation of the densities of concrete from a particular mix given in Table 1.2.2 are (very close to) 2445 and 16 N/mm^2, respectively. We do not know the values pertaining to the population. However, let us assume that the standard deviation of 16 N/mm^2 is the true value. (We shall consider the case of an unknown standard deviation, Chapter 5.) From the values of skewness and kurtosis in Table 1.2.2, it seems reasonable to assume that the density of concrete represented here is from a normal population; a uniform distribution is an approximation used, for instance, in Example 3.37. Then the distribution of $\bar{X}_n = \sum_{i=1}^{n} X_i/n$ will be approximately $N(\mu, 16^2/n)$ even for small values of n. We note from Table C.1 that, for example, $\Phi(2.575) = 0.995$. Thus we can say, using Eq. (4.2.20),

$$\Pr[-2.575 \times 16/\sqrt{n} \leq (\bar{X}_n - \mu_X) \leq 2.575 \times 16/\sqrt{n}] = .99.$$

This means that if we have a sample size, say, $n = 16$,

$$\Pr[-10.3 \leq (\bar{X}_n - \mu) \leq 10.3] = 0.99.$$

The implication is that even with a small sample of 16, we are 99% confident that the mean can be estimated within 10 N/mm^2 of the true value. Of course, if the variance is larger (or if the sample size is smaller), the difference will be greater. There will be more about these aspects in Chapter 5.

4.2.6.2 Binomial and Poisson approximations to normality

In Section 4.1 we noted that if X is a binomial (N, p) variate, we may treat X as the sum of n Bernoulli (p) variates. Hence from the Central Limit Theorem, X/n tends to have a normal distribution when n becomes large. Recalling that the mean and variance of the binomial X are np and $np(1-p)$, the distribution of the random variable

$$Z = \frac{X - np}{\sqrt{np(1-p)}} \tag{4.2.24a}$$

will tend to that of an $N(0, 1)$ variate as n increases to infinity. If p is close to 0 or 1 (which means large departures from normality), greater values of n are required for the result to hold approximately than when p is close to 0.5.

Likewise, we noted that the Poisson distribution has mean and variance ν. If $X \sim$ Poisson (ν), then for increasing values of the mean count ν, the standardized variable

$$Z = \frac{X - \nu}{\sqrt{\nu}} \tag{4.2.24b}$$

tends to be $N(0, 1)$ distributed; $\nu > 5$ is a reasonable, close approximation.

---

[29] See Appendix A.6. Furthermore, the *Weak Law of Large Numbers* states that $|\bar{X}_n - \mu|$ is ultimately small as n increases (which can be shown using the Chebyshev inequality when the variance of X exists). It might be large infrequently for some n. The *Strong Law of Large Numbers* states that the probability of such an event is extremely small. See Feller (1968, Chapter 10). Problems 4.24 and 4.25 concern the application of the Central Limit Theorem.

In general, if one approximates a discrete random variable by a continuous variable, the probability mass (discrete or point value) when X takes the value x is spread over the interval $(x - 1/2)$ to $(x + 1/2)$ in the continuous case; in so doing we replace a line by an area under a curve. Thus, an appropriate correction should improve results and the rectification for the spread is termed the *continuity correction*.

Example 4.30. Computer breakdowns. An engineer's consulting office has a large number of computers. However, in recent months there have been an excessive number of breakdowns, averaging 5 per week. If the number out of use exceeds 7, not all the staff can work, and the engineer needs to reinvest in new computers. What is the probability of such an occurrence?

We shall assume that the breakdowns are independent (on the hypothesis that if the defects were related, diagnosis would have been possible earlier) and identically distributed. Hence considering the weekly breakdowns B as occurrences of a Poisson random variable with mean $v = 5$,

$$\Pr[B > 7] = 1 - \sum_{0}^{7} \frac{e^{-5}5^b}{b!} = 1 - 0.866 = 0.134.$$

We now apply the normal approximation with the continuity correction. Noting that the variable has a mean and variance of 5 and using Eq. (4.2.24*b*) and Table C.1 for the distribution of the standard normal variable, we find

$$\Pr[B > 7] = 1 - \Phi\left(\frac{7 + 0.5 - 5}{\sqrt{5}}\right) = 1 - 0.868 = 0.132.$$

We see that virtually the same result as from the Poisson distribution is obtained in a small fraction of the time by using the normal approximation.

Definition: Normal distribution. The pdf of a normal variate, X, with mean μ and standard deviation σ, that is, an $N(\mu, \sigma^2)$ variate, is

$$\phi(x) = \frac{1}{\sigma\sqrt{2\pi}} \exp\left[-\frac{1}{2}\left(\frac{x-\mu}{\sigma}\right)^2\right], \quad \text{for } -\infty < x < +\infty.$$

The cdf of the (transformed) standard normal variate $Z = (X - \mu)/\sigma$, that is, a $N(0, 1)$ variate, is

$$\Phi(z) = \int_{-\infty}^{z} \frac{1}{\sqrt{2\pi}} \exp\left(-\frac{t^2}{2}\right) dt, \quad \text{for } -\infty < z < +\infty,$$

which can be read from Table C.1. Some important properties include the following:

(1) A linear transformation $Y = a + bX$ of an $N(\mu, \sigma^2)$ variate X makes Y an $N(a + b\mu, b^2\sigma^2)$ variate.

(2) The mgf is given by $M_X(t) = \exp(\mu t + \sigma^2 t^2/2)$ (see Appendix A.5).

(3) *Central Limit Theorem*: If a random variable X_i is from a population with mean μ and standard deviation σ, then the random variable $\bar{X}_n = \sum_{i=1}^{n} X_i/n$ from a random sample of size n tends to have an $N(\mu, \sigma^2/n)$ distribution as n approaches infinity. If X_i is normally distributed, the result holds regardless of sample size (see Appendix A.6).

(4) The mean $\bar{X}$ and variance S^2 of n independent normal variates, $X_1, X_2, \ldots, X_n$, are independent.[30]

[30] See, for example, Hoel (1984, pp. 394–396).

4.2.6.3 Truncated normal distribution

In Section 4.1, the truncated Poisson process was discussed for appropriate discrete variates. Similarly for the normal and other continuous distributions, a population may be truncated because, for example, a stipulation is applied in the manufacturing process. If the upper limit is x_μ, then the resulting distribution of the variate X with original pdf $f_X(x)$ and cdf $F_X(x)$ is represented by the conditional pdf:

$$\begin{aligned} f_{X|X\leq x_a}(x) &= af_X(x), \quad \text{for } x \leq x_u, \\ &= 0, \quad \text{otherwise,} \end{aligned} \tag{4.2.25a}$$

where the constant a is given by

$$a = 1/F_X(x_\mu). \tag{4.2.25b}$$

If a lower bound x_l is introduced, the inequality in Eq. (4.2.25*a*) is reversed and Eq. (4.2.25*b*) modified accordingly. It is also possible to introduce lower and upper bounds at the same time, thus truncating the pdf in both directions. Then,

$$a = 1/[F_X(x_\mu) - F_X(x_l)].$$

4.2.7 Lognormal distribution

We noted in Section 4.2.6 that the addition of a large number of small random effects will tend to make the distribution of the aggregate approach normality. Likewise, if a phenomenon arises from the multiplicative effect of a large number of uncorrelated factors, the distribution tends to be *lognormal* (or *logarithmic normal*), that is, the logarithm of the variable becomes normally distributed. There are numerous examples in nature, such as the distribution of small particle sizes in sediment transport, the crushing of aggregates, the strength or yield stress of some materials used in construction, and the magnitudes and interarrival times of earthquakes (see Aitchison and Brown, 1957). This type of reasoning can also be extended to the occurrences of floods and droughts.

Let X be a positive random variable and define

$$Y = \ln(X). \tag{4.2.26}$$

Also let Y have an $N(\mu_Y, \sigma_Y^2)$ distribution; then we can say that X has a lognormal distribution, that is, $LN[\mu_{\ln(X)}, \sigma^2_{\ln(X)}]$. Using the one-to-one transformation given by Eq. (3.4.4), the pdf of the lognormal distribution is obtained from Eqs. (4.2.19) and (4.2.26) as

$$f_X(x) = \frac{1}{x\sigma_{\ln(X)}\sqrt{2\pi}} \exp\left\{-\frac{1}{2}\left[\frac{\ln(x) - \mu_{\ln(X)}}{\sigma_{\ln(X)}}\right]^2\right\}, \quad \text{for } 0 \leq x < +\infty. \tag{4.2.27}$$

Plots of the lognormal pdf are shown in Fig. 4.2.10 for two different pairs of values of the parameters.

Because $Y = \ln(X)$ has a normal distribution, we can use the tables of the normal distribution (Appendix C, Table C.1) to determine a probability or solve an inverse problem as shown shortly.

By using the moment-generating function with Eqs. (4.2.26) and (4.2.27), the expectation operator can be written as

$$E[X^r] = \int_0^\infty x^r f_X(x)dx = \frac{1}{\sigma_{\ln(X)}\sqrt{2\pi}} \int_{-\infty}^{\infty} e^{ry} \exp\left\{-\frac{1}{2}\left[\frac{y - \mu_{\ln(X)}}{\sigma_{\ln(X)}}\right]^2\right\} dy.$$

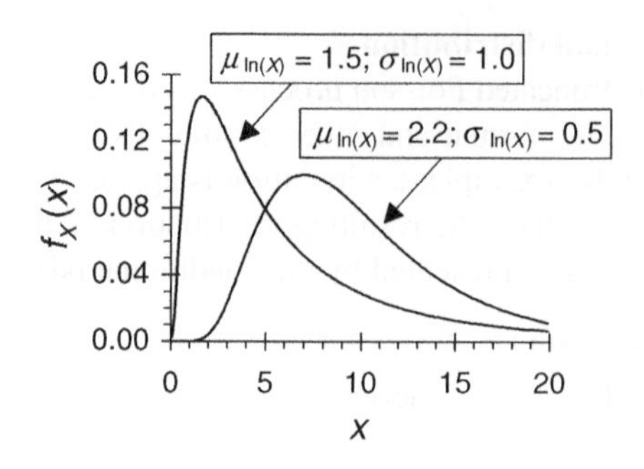

Fig. 4.2.10 Lognormal density functions.

Thus,

$$E[X^r] = E[e^{Yr}]; Y \sim N(\mu_{\ln(X)}, \sigma^2_{\ln(X)}).$$

Therefore,

$$E[X^r] = M_Y(r) = \exp\left[r\mu_{\ln(X)} + \tfrac{1}{2}r^2\sigma^2_{\ln(X)}\right].$$

Hence the mean is

$$\mu_X = E[X] = \exp\left[\mu_{\ln(X)} + \tfrac{1}{2}\sigma^2_{\ln(X)}\right]. \quad (4.2.28a)$$

Similarly, the variance takes the form

$$\sigma^2_X \equiv E[X^2] - \mu^2_X = \exp\left\{2\left[\mu_{\ln(X)} + \sigma^2_{\ln(X)}\right]\right\} - \exp\left\{2\left[\mu_{\ln(X)} + \tfrac{1}{2}\sigma^2_{\ln(X)}\right]\right\}.$$

Using Eq. (4.2.28*a*), we can write this as

$$\sigma^2_X = \mu^2_x\left\{\exp\left[\sigma^2_{\ln(X)}\right] - 1\right\}. \quad (4.2.28b)$$

Equation (4.2.28*b*) gives the coefficient of variation:

$$V_X = \frac{\sigma_X}{\mu_X} = \left\{\exp\left[\sigma^2_{\ln(X)}\right] - 1\right\}^{1/2}. \quad (4.2.28c)$$

By denoting the median by m, it follows directly from the properties of the normal distribution [as shown by Eq. (4.2.22) and Table C.1] that

$$F_{\ln(X)}\left[\frac{m_{\ln(X)} - \mu_{\ln(X)}}{\sigma_{\ln(X)}}\right] = 0.5,$$

that is, $m_{\ln(X)} = \mu_{\ln(X)}$. Also, we note that the median of the log-transformed X population equals the logarithm of the median of the original X population. (This result occurs because the median is the middle term in a ranked series and the transformation does not change the ranks—but such a property is not applicable to the mean or any other parameter.) Hence, using the previous result, we write

$$m_{\ln(X)} = \ln(m_X) = \mu_{\ln(X)}. \quad (4.2.28d)$$

From Eq. (4.2.28*a*), we have

$$\ln(\mu_X) = \mu_{\ln(X)} + \tfrac{1}{2}\sigma^2_{\ln(X)}.$$

Also, from Eq. (4.2.28*c*),

$$\tfrac{1}{2}\sigma^2_{\ln(X)} = \ln\left[\left(V^2_X + 1\right)^{1/2}\right].$$

It follows from the two foregoing equations that

$$\mu_{\ln(X)} = \ln(m_X) = \ln\left[\frac{\mu_X}{\left(V_X^2+1\right)^{1/2}}\right]. \tag{4.2.28e}$$

The lognormal is equivalent to the Johnson S_L distribution.

Example 4.31. Lognormal distribution of timber strengths. Without the zero value, the summary data representing the modulus of rupture in Table 1.2.2 has a coefficient of skewness of 0.53. Although this is not highly significant, we may, in the first instance, fit a lognormal distribution to the data. Also, the mean and coefficient of variation of the data are 39.33 N/mm^2 and 0.24, respectively.

From Eq. (4.2.28*c*), we have

$$\sigma_{\ln(X)} = \sqrt{\ln\left(V_X^2+1\right)} = \sqrt{\ln(0.24^2+1)} = 0.237.$$

From Eq. (4.2.28*e*)

$$\mu_{\ln(X)} = \ln\left[\frac{\mu_X}{\left(V_X^2+1\right)^{1/2}}\right] = \ln\left[\frac{39.33}{(0.24^2+1)^{1/2}}\right] = 3.644.$$

Also from Eq. (4.2.28*d*), the median is

$$m_X = \frac{39.33}{(0.24^2+1)^{1/2}} = 38.24 \text{ N/mm}^2.$$

For example, to determine the modulus of rupture that is exceeded 95% of the time, we solve the inverse of $\Phi(x) = 0.05$ using Table C.1 of the normal distribution. Thus, $z = -1.645$ for $\Phi(z) = 0.05$ and the corresponding y is given by

$$\frac{y - 3.644}{0.237} = -1.645.$$

Hence, $x = \exp(y) = \exp(3.644 - 0.237 \times 1.645) = 25.9$ N/mm^2.

We may also be interested to know, for example, the probability that the modulus of rupture of a randomly selected timber is not less than 20 N/mm^2. This is given by

$$\Pr[X \geq 20] = 1 - F_X(20) = 1 - F_Z\left[\frac{\ln(20) - 3.644}{0.237}\right] = 1 - F_z(-2.735) = .997$$

4.2.8 Summary of Section 4.2

In this section we have introduced some of the more commonly used continuous distributions. These are summarized in Table 4.2.2. For other types and more advanced concepts, refer to books cited at the end of the chapter. In the next section we discuss some types of multivariate distributions.

4.3 MULTIVARIATE DISTRIBUTIONS

Jointly distributed random variables were introduced in Section 3.3. In this section we examine the multivariate type of distribution, which can model the common occurrences of different types of events. Indeed many types of multivariate models can be formed on the basis of the univariate distributions described in Sections 4.1 and 4.2. See, for instance, Example 3.31 in which the number of pixels detected in the remote sensing of flooded areas has a binomial distribution and the total number of inundated pixels is Poisson-distributed, thus yielding a Poisson-binomial distribution for the probability of not detecting a part

Table 4.2.2 Summary of continuous distributions

Distribution	Pdf	Mean, E[X]	Variance, Var[X]	Mgf
Uniform	$\frac{1}{b-a}$, for $a \le x \le b; -\infty < a < b < \infty$	$\frac{a+b}{2}$	$\frac{(b-a)^2}{12}$	$\frac{e^{bt}-e^{at}}{t(b-a)}$
Exponential	$\lambda e^{-\lambda x}$, for $x \ge 0; \lambda > 0$	$\frac{1}{\lambda}$	$\frac{1}{\lambda^2}$	$\frac{\lambda}{\lambda-t}$, for $t < \lambda$
Gamma	$\frac{\lambda^r x^{r-1} e^{-\lambda x}}{\Gamma(r)}$, for $x \ge 0; \lambda, r > 0$	$\frac{r}{\lambda}$	$\frac{r}{\lambda^2}$	$\left(\frac{\lambda}{\lambda-t}\right)^r$, for $t < \lambda$
Beta	$\frac{1}{B(\alpha,\beta)} x^{\alpha-1}(1-x)^{\beta-1}$, for $0 < x < 1; \alpha, \beta > 0$	$\frac{\alpha}{\alpha+\beta}$	$\frac{\alpha\beta}{(\alpha+\beta)^2(\alpha+\beta+1)}$	Not in a useful form
Weibull	$\frac{\beta}{\lambda}\left(\frac{x}{\lambda}\right)^{\beta-1} \exp\left[-\left(\frac{x}{\lambda}\right)^{\beta}\right]$, for $x, \beta, \lambda > 0$	$\lambda\Gamma\left(1+\frac{1}{\beta}\right)$	$\lambda^2\left\{\Gamma\left(1+\frac{2}{\beta}\right) - \left[\Gamma\left(1+\frac{1}{\beta}\right)\right]^2\right\}$	$\lambda^t\left(1+\frac{t}{\beta}\right)$
Normal	$\frac{1}{\sigma\sqrt{2\pi}} \exp\left[-\frac{1}{2}\left(\frac{x-\mu}{\sigma}\right)^2\right]$, for $-\infty < x, \mu < \infty; \sigma > 0$	μ	σ^2	$\exp\left(\mu t + \frac{1}{2}\sigma^2 t^2\right)$
Lognormal[a]	$\frac{1}{x\sigma\sqrt{2\pi}} \exp\{-[\ln(x)-\mu]^2/2\sigma^2\}$, for $x, \sigma > 0; -\infty < \mu < \infty$	$\exp(\mu + \sigma^2/2)$	$\exp(2\mu + 2\sigma^2) - \exp(2\mu + \sigma^2)$	Does not exist

[a] *Note*: μ and σ are the mean and standard deviations, respectively, of ln(x).

of an inundated area. Also, the exponentially distributed storm intensity and duration of Example 3.33 lead to a bivariate exponential distribution of intensity and duration, taking into account the correlation between these two variables. Furthermore, the density and compressive strength of concrete is assumed to have a bivariate distribution (Example 3.43). In addition, the multinomial distribution which is used in counting multiple events was derived in Section 4.1 as a generalization of the binomial distribution, replacing the two types of Bernoulli events with several types. Thus there can be numerous possible multivariate distributions.[31] Our coverage is limited to the bivariate normal distribution, but we discuss other types briefly.

4.3.1 Bivariate normal distribution

The joint distribution of two random variates, say, X and Y, each normally distributed is termed the bivariate normal distribution. The pdf in standardized form is given as follows applied to corresponding variates, Z_1 and Z_2:

$$f_{Z_1,Z_2}(z_1, z_2) = [2\pi(1-\rho^2)^{1/2}]^{-1} \exp\left[\frac{-\left(z_1^2 - 2\rho z_1 z_2 + z_2^2\right)}{(2-2\rho^2)}\right], \tag{4.3.1}$$

where $-\infty < z_1, z_2 < +\infty$ and ρ, constrained by $-1 \le \rho \le 1$, is the coefficient of linear correlation between the two variates. Also, $Z_1 = (X - \mu_X)/\sigma_X$ and $Z_2 = (X - \mu_Y)/\sigma_Y$, and the normalizing constants are constrained by $-\infty < \mu_X, \mu_Y < +\infty$, and $\sigma_X, \sigma_Y > 0$. In fact, these sets of parameters can be shown to be the means and standard deviations of X and Y, respectively.

The bivariate normal distribution of the X and Y variates takes the form

$$f_{X,Y}(x, y) = \frac{1}{2\pi\sigma_X\sigma_Y\sqrt{1-\rho^2}} \exp\left\{-\frac{1}{2(1-\rho^2)}\left[\left(\frac{x-\mu_X}{\sigma_X}\right)^2 - 2\rho\frac{x-\mu_X}{\sigma_X}\frac{y-\mu_Y}{\sigma_Y} + \left(\frac{y-\mu_Y}{\sigma_Y}\right)^2\right]\right\}, \tag{4.3.2}$$

where $-\infty < x, y < +\infty$. The volume enclosed by the surface represented by Eq. (4.3.2) is unity, which is also applicable to Eq. (4.3.1). The surface of a typical bivariate normal distribution is shown in Fig. 4.3.1.

It is easy to show that the marginal distributions are the univariate distributions for X and Y, and the conditional distributions can be obtained as in Eq. (3.3.15).

Example 4.32. Bivariate normal distribution of compressive strengths and densities of concrete. In Example 3.37 we considered marginal triangular and uniform distributions for the compressive strengths Y and densities X, respectively, of concrete listed in Table E.1.2. Because of the assumption that the densities are uniformly distributed, the conditional and marginal distributions of compressive strengths are equivalent; that is, the information from the densities does not alter the engineer's knowledge of the distribution of compressive strengths. However, a closer representation can be made using marginal normal distributions for both variables. Thus, the bivariate distribution is given by Eq. (4.3.2) and the marginal distribution of the densities is

$$\phi(x) = \frac{1}{\sigma_X\sqrt{2\pi}} \exp\left[-\frac{1}{2}\left(\frac{x-\mu_x}{\sigma_x}\right)^2\right], \quad \text{for } -\infty < x < +\infty.$$

[31] For details of many types of multivariate distributions, see Johnson and Kotz (1975).

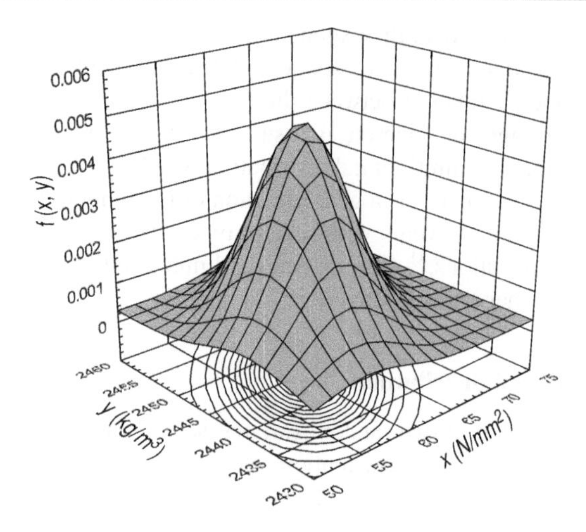

Fig. 4.3.1 Bivariate normal pdf.

From Eq. (3.3.15), the conditional pdf of the compressive strengths is

$$f_{Y|X}(y|x) = \frac{f_{Y,X}(y,x)}{f_X(x)} = \frac{\sigma_X\sqrt{2\pi}}{\exp[-(1/2)((x-\mu_X)/\sigma_X)^2]} \frac{1}{2\pi\sigma_X\sigma_Y\sqrt{(1-\rho^2)}}$$

$$\times \exp\left\{-\frac{1}{2(1-\rho^2)}\left[\left(\frac{x-\mu_X}{\sigma_X}\right)^2 - 2\rho\frac{x-\mu_X}{\sigma_X}\frac{y-\mu_Y}{\sigma_Y} + \left(\frac{y-\mu_Y}{\sigma_Y}\right)^2\right]\right\}$$

$$= \frac{1}{\sigma_Y^*\sqrt{2\pi}} \exp\left[-\frac{1}{2}\left(\frac{y-\mu_Y^*}{\sigma_Y^*}\right)^2\right].$$

If, as in this application, the correlation is nonnegative, the variance of the conditional distribution of compressive strengths reduces from that of the marginal to

$$\text{Var}[Y|x] = \sigma_Y^{*2} = \sigma_Y^2(1-\rho^2).$$

Correspondingly, the mean is given by

$$E[Y|x] = \mu_Y^* = \mu_Y + \rho\frac{\sigma_Y}{\sigma_X}(x-\mu_X).$$

These relationships enable us to obtain the conditional pdf of the compressive strengths Y, given the marginal pdf, the mean of the densities X, the correlation between the two variables, and a measurement of X. If this measurement yields a value greater than the mean and correlation is positive, then the conditional mean of compressive strengths is greater than that of the marginal mean, as seen from the second relationship. However, the conditional variance will be less than the marginal variance for all nonzero values of correlation; more about this follows.

For the data referred to (from Table 1.2.2), $\bar{y} = 60.14$ N/mm^2, $s_Y = 5.02$ N/mm^2, $\bar{x} =$ 2445 kg/m^3, and (as given in Example 1.29) $\hat{\rho} = 0.44$. As assumed, the marginal pdf of $Y \sim N(60.14, 5.02^2)$. For an observed X value of 2550 kg/m^3, the conditional pdf of $Y \sim N(74.64, 4.51^2)$, which follows from the two relationships above. For example, prior to this observation,

$$\Pr[Y < 55] = 1 - \Phi\left(\frac{60.14-55}{5.02}\right) = 1 - \Phi(1.023) = 0.153$$

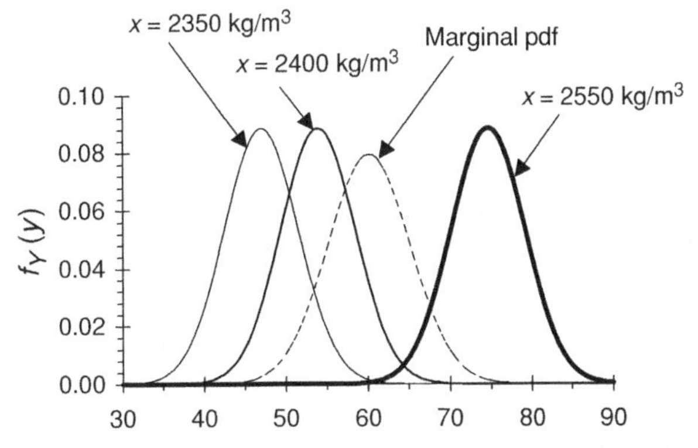

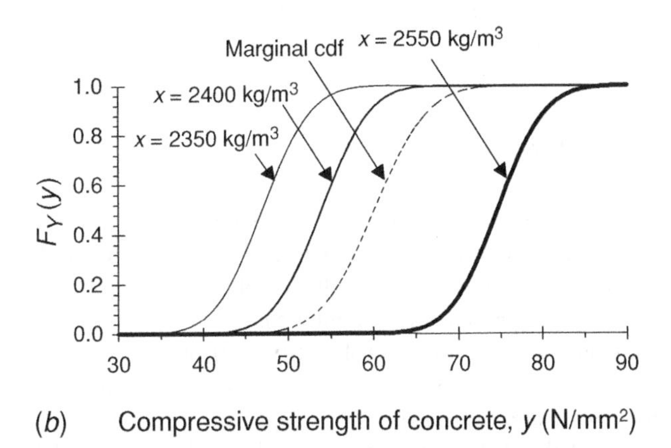

Fig. 4.3.2 Marginal and conditional (*a*) pdfs and (*b*) cdfs of compressive strength of concrete y for three values of concrete density x.

and after the observation,

$$\Pr[Y < 55] = 1 - \Phi\left(\frac{74.64 - 55}{4.51}\right) = 1 - \Phi(4.354) = 0.00001.$$

In Fig. 4.3.2, the marginal pdf of Y is compared to three conditional pdfs for X values of 2350, 2400, and 2550 kg/m^3. Also shown are the corresponding cdfs. Results and diagrams such as these can be useful for design purposes.

Example 4.32 and Fig. 4.3.2 show how observations of one variable X can be used to predict the performance of another, Y. We note that the correlation between the two variables has reduced the spread of each of the conditional pdfs $f_{Y|X}(y|x)$ compared to that of the marginal pdf $f_Y(y)$. This observation reflects a reduction in the variance and the mean square error of prediction. The reduction increases with the correlation regardless of its sign; on the other hand, with little or no correlation between the variables this exercise is ineffectual. If the correlation is positive and the observation of X is greater (lower) than its expected value, then the prediction of Y is revised upward (downward). With negative correlation these conditions are reversed. This procedure is applicable in many spheres of civil and environmental engineering.

4.3.2 Other bivariate distributions

Several bivariate distributions such as the lognormal distribution are useful in science and engineering. The cdf of the bivariate exponential distribution is given by

$$F_{X,Y}(x, y) = 1 - e^{-ax} - e^{-by} - e^{-ax-bx-cxy}, \quad \text{for } x, y \geq 0; 0 < a, b, c.$$

We demonstrated its use in Chapter 3.[32]

Another example is the *bivariate logistic distribution*, the basic form of which is

$$F_{X,Y}(x, y) = (1 + e^{-x} + e^{-y})^{-1}.$$

For other types and for the estimation of parameters, refer to one of the texts cited at the end of this chapter.[33]

4.4 SUMMARY FOR CHAPTER 4

As shown here, it is desirable to represent a set of observations by a specific type of distribution rather than a histogram (Chapter 1), which can then be treated as a preliminary graph. These distributions will enhance objectivity through computer simulation or hypothesis testing on parameters (Chapter 5).

In this chapter, several useful probability distributions are described and developed. Important properties have been listed. The physical relevance of the various types and the assumptions made are discussed.

Summaries of the distributions are given in Tables 4.1.7 and 4.2.2, which pertain to discrete and continuous distributions, respectively. The lists are not comprehensive. For example, details of extreme value distributions are included in Chapter 7. Furthermore, an example of the Rayleigh distribution is given in this chapter (Example 4.26), the Pareto distribution is illustrated in Chapter 3 (Example 3.46) and elsewhere, and the Cauchy pdf is given by Eq. (5.9.2). Further details of the distributions analyzed here and several additional types are found in the cited references (see, for example, Evans et al., 2000).

We have shown in this chapter and elsewhere that there are theoretical links between different discrete and continuous distributions. For example, the exponential distribution is associated with the gamma and Weibull, whereas the normal is central to many types such as the Poisson, binomial, gamma, lognormal, and beta distributions.[34]

Many of the frequent distributions can be represented by so-called families of distributions. These include useful types such as the Johnson system of distributions.[35]

---

[32] See Examples 3.33, 3.34, 3.35, and 3.36, and 3.44.

[33] See, for example, Johnson and Kotz (1975). The bivariate lognormal distribution is applied by Kottegoda and Natale (1994) in irrigation.

[34] Leemis (1986) presents a chart showing most of these relationships.

[35] See, for example, Hahn and Shapiro (1967), Johnson et al. (1994, 1995), Kottegoda (1980, 1987).

REFERENCES

General. The following references are given for further reading:

Balakrishnan, N. (ed.) (1992). *Handbook of the Logistic Distribution*, Marcel Dekker, New York. Valuable reference for the logistic distribution; all in 62 pages.

Casella, G., and R. L. Berger (2002). *Statistical Inference*, 2nd ed., Pacific Grove, Duxbury. An advanced text recommended for point estimation.

Evans, M., N. Hastings, and B. Peacock (2000). *Statistical Distributions*, 3rd ed., John Wiley and Sons, New York. Excellent summaries of distributions of random variables.

Johnson, N. L., and S. Kotz (1975). *Multivariate Distributions*, John Wiley and Sons, New York. Advanced book on multivariate distributions.

Johnson, N. L., S. Kotz, and A. W. Kemp (1992). *Univariate Discrete Distributions*, 2nd ed., Wiley International, New York. Advanced authoritative reference book on discrete distributions. This is a substantially revised edition.

Johnson, N. L., S. Kotz, and N. Balakrishran (1994, 1995). *Continuous Univariate Distributions*, Vols. 1 and 2, 2nd ed., John Wiley and Sons, New York. Comprehensive reference book for those needing further information on continuous distributions.

Johnson, R. A. (1994). *Miller and Freund's Probability and Statistics for Engineers*, 5th ed., Prentice Hall International, Englewood Cliffs, NJ. Elementary aspects of probability distributions are included.

Kocherlakota, S., and K. Kocherlakota (1992). *Bivariate Discrete Distributions*, Marcel Dekker, New York. Advanced reference book.

Montgomery, D. C., and G. C. Runger (1994). *Applied Statistics and Probability for Engineers*, John Wiley and Sons, New York. Probability distributions, point estimation, and other aspects treated at a basic level.

Olkin, I., L. J. Gleser, and C. Derman (1980). *Probability Models and Applications*, Macmillan, New York. Discrete, continuous, and multivariate distributions can be easily understood.

Ord, J. K. (1972). *Families of Frequency Distributions*, Griffin, London. Useful summaries of distributions.

Additional references quoted in text

Aitchison, J., and J. A. C. Brown (1957). *The Lognormal Distribution*, Cambridge University Press, Cambridge, England. Reference for lognormal distribution.

Chatfield, C. (1966). "Wet and dry spells," *Weather*, Vol. 21, pp. 308–310. Data used in illustration of geometric and log series distributions.

Crow, L. H. (1974). "Reliability analysis for complex repairable systems," in *Reliability and Biometry*, edited by F. Proschan and R. J. Serfling, pp. 379–410, Society for Industrial and Applied Mechanics, Philadelphia, PA. Test for Weibull distribution in relation to the exponential.

Feller, W. (1968). *An Introduction to Probability Theory and Its Applications*, Vol. 1, 3rd ed., John Wiley and Sons, New York. Strong and weak laws of large numbers, Chapter 10.

Gumbel, E. J. (1954). "Statistical theory of droughts," *Proc. Am. Soc. Civ. Engrs.*, Vol. 80, Issue 439, Sept. Reference for Weibull distribution.

Greenberg, A. E., L. S. Clesceri, and A. D. Eaton (eds.) (2005). *Standard Methods for the Examination of Water and Wastewater*, 21st ed., American Public Health Association, Washington, DC.

Haan, C. T. (1977). *Statistical Methods in Hydrology*, Iowa University Press, Ames, IA. Reference for data in Problem 4.19.

Hahn, G. J., and S. S. Shapiro (1967). *Statistical Methods for Engineers*, John Wiley and Sons Inc., New York. Reference for Johnson and other families of distributions, Chapter 6.

Hald A. (1990). *A History of Probability and Statistics and Their Applications before 1750*, John Wiley and Sons, New York. History of Poisson distribution etc.

Harter, H. L. (1964). *New Tables of the Incomplete Gamma-Function Ratio and of Percentage Points of the Chi-Squared and Beta Distributions*, Aerospace Research Laboratories, U.S. Air Force, Superintendent of Documents, U.S. Government Printing Press, Cambridge, MA. References for tables of the gamma distribution.

Hoel, P. G. (1984). *Introduction to Mathematical Statistics*, 5th ed., John Wiley and Sons, New York. Proof of independence of $\bar{X}$ and S^2 of normal variables pp. 394–396.

Kemp, A. W. (1997). "Generalizations of Classical Discrete Distributions," in *Encyclopedia of Statistical Sciences*, Vol. 1, edited by S. Kotz, C. B. Read, and D. L. Banks, pp. 93–110, John Wiley and Sons, New York.

Kottegoda, N. T. (1980). *Stochastic Water Resources Technology*, Macmillan, London. Reference for families of distributions, Chapter 3.

Kottegoda, N. T. (1987). "Fitting Johnson S_B curve by the method of maximum likelihood to annual maximum daily rainfalls," *Water Resour. Res.*, Vol. 23, pp. 728–732. Reference on Johnson distributions.

Kottegoda, N. T., and L. Natale (1994). "Two-component log-normal distribution of irrigation-affected low flows," *J. Hydrol.*, Vol. 158, pp. 187–199. Reference for data in Problem 4.29.

Leeming, J. J. (1963). *Statistical Methods for Engineers*, Blackie, London. Reference for data used in Example 4.21.

Leemis, L. M. (1986). "Relationship among common univariate distributions," *Am. Stat.*, Vol. 40, pp. 143–146. Chart showing relationships among distributions.

Markovic, R. D. (1965). "Probability functions of best fit to distributions of annual precipitation and runoff," *Hydrology Papers*, no. 8, Colorado State University, Fort Collins, CO. Reference for data in Problem 4.28.

Neville, A. M. (1995). *Properties of Concrete*, 4th ed., Langman Group Ltd., Harlow, England. Normal distribution of concrete strengths.

Shea, B. L. (1988). "Chi-squared and incomplete gamma integral," *Appl. Stat.*, Vol. 37, pp. 466–473. Numerical method of integrating the gamma function.

Smith, R. L., and J. C. Naylor (1987). "A comparison of maximum likelihood and Bayesian estimators for the three-parameter Weibull Distribution," *Appl. Stat.*, Vol. 36, pp. 358–369. Reference for comparative methods of estimating the Weibull distribution.

Troen, I., and E. L. Peterson (1989). *European Wind Atlas*, Commission of the European Communities Directorate-General for Science, Research and Development Brussels, Belgium; Riso National Laboratory, Roskilde, Denmark. Extensive analysis of wind data using the Weibull distribution.

Van der Auwera, L., F. De Meyer, and L. M. Malet (1980). "The use of the Weibull three-parameter model for estimating mean wind power densities," *J. Am. Meteorol. Soc.*, Vol. 19, pp. 819–825. Reference given for the use of the Weibull distribution. See also earlier references given here for wind power generation using this distribution.

Weibull, W. (1951). "A statistical distribution function of wide applicability," *ASME J. Appl. Mech.*, Vol. 18, pp. 293–297. Includes 7 case studies.

PROBLEMS

4.1. Protective sea embankment. To counteract the effects of erosion and damage caused by sea waves, an embankment wall is built alongside a railway line. From recorded data, the annual maximum wave height exceeds that of the embankment, on average, once in 8 years. What is the probability that the embankment will be overtopped at least once during the next 10 years? Assume that the events are independent and identically distributed.

4.2. Dam design. Determine the return period that should be used in a design for a small dam so that the design flood is exceeded with a probability of not more than 0.05

during a 50-year economic time horizon. Assume that the events are independent and identically distributed.

4.3. Bridge design. A bridge is to be constructed over a river. The design criterion is that a flood should rise above the high-level marks on the piers in not more than once in 25 years with a probability not exceeding 0.1. What return period should be used in the flood design? Assume that the events are independent and identically distributed.

4.4. Revision of dam design. In a situation similar to that of Problem 4.3, supposing the engineer adopts a 100-year design period, determine (a) the probability that the design flood level is not exceeded during a 100-year period (b) the probability that the design flood is exceeded just after the tenth year but not during the first 10 years.

4.5. Frequent flooding. Calculate the probability of having two 10-year flows in a 5-year period assuming that the events are independent and identically distributed.

4.6. Storm sewer design. For a storm sewer design, an engineer uses the annual maximum 1-hour rainfall with a 5-year return period as a design criterion. As shown on the city plan, sewer A drains one area of the city and sewer B drains the remaining area. However, there is no correlation in the intensive rainfalls which occur in the two parts of the city, although the storm characteristics are the same. What is the probability that there will not be more than two design events in the city during a 5- year period?

4.7. Vehicle count. The following count is made on the number of vehicles that pass an observation point every 10 minutes for 1 hour. What counts are expected theoretically if the distribution is Poisson?

Count i	0	1	2	3	4	5	6
Frequency f	220	94	23	11	4	2	1

4.8. Machine failure. The probability that a certain make of piling machine breaks down is 0.00002 per 100 m of piles made. What is the probability of having one breakdown after 1000 m and before 1010 m of piles?

4.9. First-time failure. Taking the probability given in Problem 4.8 as the probability of failure during a week's work and an average weekly production of 1000 m of piles, determine the probability of failure for the first time after 3 months. How does the first-time probability of failure vary with time?

4.10. Transportation. An operator runs a small bus which conveys people from a town center to a large shopping complex. The bus leaves as soon as 12 people have arrived. If we assume that the passenger arrivals are independent and are at a mean rate of 9 per hour, what is the probability that the time between two consecutive departures is more than 60 minutes? Assume that there are no delays caused by the nonarrival of the bus because standby buses are available.

4.11. Traffic: number of cars waiting to turn. For the control of vehicles at a traffic light, one needs to determine the length of the left-turn lane (right-turn lane in countries where vehicles are driven on the left). The occurrences of left (right) turns are assumed to have a Poisson distribution in time. The mean uninterrupted rate of left (right) turns is 160 per hour and the red light is on for 50 seconds. What is the expected number of vehicles awaiting a left (right) turn?

4.12. Traffic: length of lane. In Problem 4.11 the design criterion for the length of the left (right) lane is that it should be sufficient for 95% of the time. What should be the minimum length of the lane as a multiple of the average length of a vehicle?

4.13. Wet spells. The following distribution of wet spells was observed at the Dharamjaigarh rainfall station in central India during the monsoon season:

i, length of wet spell in days	1	2	3	4	5	6	7	8
O_i, observed number of spells	161	52	32	17	8	6	4	1

What is the minimum length of wet spell which is exceeded with probability less than 0.05 assuming a geometric distribution?

4.14. Stream pollution. Traces of toxic wastes from an unknown source are found in a stream. From tests made on the water the mean concentration is found to be 1 mg/L. What is the probability that the concentration of the pollutant will be in the range 0.5–2 mg/L assuming the distribution is (*a*) exponential (*b*) normal?

4.15. Failure of pumps installed in parallel. A pumped storage power supply scheme has five pumps of identical specification installed in parallel. The mean life span of a pump is estimated as 10 years from previous experience. What is the minimum number of pumps required in parallel so that the probability of not having a failure of the system during a 3-year period is more than 0.95?

4.16. Failure of pumps in a compound system. Suppose that in the scheme described in Problem 4.15, two pumps are placed in parallel, one of which must work. This subsystem is combined in series with another identical pump. Determine the probability of not having a failure of the system in any year.

4.17. Traffic accidents. From experience it is found that there are about three accidents per year at an intersection. If the occurrences are Poisson-distributed, what is the pdf of the time till the fourth accident?

4.18. Defective valves. A manufacturer supplies nine valves for a pumping scheme. Two faulty valves were included in the consignment. However, the scheme had been completed using three of the nine valves. What is the probability that no faulty valves were used?

4.19. Gamma-distributed annual runoff. The annual runoffs in the Cave Creek, near Fort Spring, Kentucky, U.S.A., are given as follows in millimeters over an 18-year period:

337 84 385 394 361 538 196 448 582 480 326 294 385 264 458 413 299 455.

Assuming independence and a gamma distribution for the annual runoff, determine the probability that the runoff will be greater than 100 mm in a given year. Data from C. T. Haan (1977), *Statistical Methods in Hydrology*, Iowa University Press, Ames, IA; used with permission, copyright 1977, The Iowa State University Press.

4.20. Low river flows in a tropical region. The following annual minimum mean daily flows, given in m^3/s, were recorded at the proposed Bango diversion site in the Hasdo subcatchment of the Mahanadi basin in India over a 22-year period:

2.78	2.47	1.64	3.91	1.95	1.61	2.72	3.48	0.85	2.29	1.72
2.41	1.84	2.52	4.45	1.93	5.32	2.55	1.36	1.47	1.02	1.73

Assuming a two-parameter Weibull distribution, determine the probability that the annual minimum low flow will be less than 2 m^3/s over a 2-year period.

4.21. Low river flows in a temperate region. Ten years of annual minimum daily mean low-flow data from the River Pang at Pangbourne in hydrometric area 39 in England are ranked and given here in cubic meters per second:

11.5 23.6 29.1 32.7 34.5 37.0 39.8 49.0 54.6 53.5.

Fit a Weibull distribution to the data, estimating the parameters using Eqs. (4.2.17) and Tables C.5 and C.6 in Appendix C, noting that $\Gamma(r+1) = \Gamma(r)$. If it is not permissible to pump water from the river when the daily mean low flow is less than 20 m^3/s, estimate the return period of such an event. Data are used by permission from Institute of Hydrology (1980), "Low flow studies report," Institute of Hydrology, Wallingford.

4.22. Ferry transport schedule. A ferry boat is designed to carry 35 passengers across a lagoon from station A during the busy hours of the day. If the passengers arrive at an average rate of two per 5 minutes and ferries leave every 70 minutes, what is the probability that there will be more than the stipulated number of passengers waiting to take the boat? How often should a ferry be scheduled to leave station A if the chance of an excess is to be less than 5%? Assume that the arrivals of the passengers constitute a Poisson process.

4.23. Distribution of concrete strengths. The compressive strengths of concrete in Table 1.2.2 have an estimated mean of 60.14 N/mm^2 and a standard deviation of 5.02 N/mm^2 and are assumed to be normally distributed. What is the probability that in ten random tests the compressive strength will be in the range 45–75 N/mm^2?

4.24. Ferry transport: weight restriction. Suppose there is a weight restriction of 2900 kg for a ferryboat. Random tests carried out on a large number of incoming passengers establish a mean weight of 75 kg per person and a standard deviation of 25 kg. What is the probability that the total weight of an incoming batch of 35 passengers will exceed the limit?

4.25. Monthly rainfalls. Monthly rainfalls in a locality are independent and identically distributed normal variates with mean 20 cm and variance 12 cm^2. Determine the probability that 220 cm of rainfall occurs over a period of 6 months. What is the probability of having less than 18 cm rainfall each month for a period of 6 months?

4.26. Relationship between rainfall and runoff. Annual rainfall is usually normally distributed over many river basins around the world. In a particular catchment, annual rainfall X has a mean of 1000 mm and a standard deviation of 200 mm. The annual runoff Y is related to the rainfall as follows:

$$Y = 100 + 0.4X.$$

Specify the complete distribution of Y. What is the probability that Y will be less than 350 mm in a year?

4.27. River diversion. A river with annual flows $X \sim N(300, 50)$ is joined by a major tributary with annual flows $Y \sim N(150, 75)$ at point P. At point Q on the river below P there is a diversion with annual flows $Z \sim N(100, 25)$. The units are in 1000 m^3. Below Q, suppose the annual flows are denoted by R. If X, Y, and Z are independent, determine the following:
(a) the distribution of R
(b) $\Pr(R > 300)$
Recalculate (a) and (b) if there are miscellaneous withdrawals and net losses affecting X and Y which total 15% in each case.

4.28. Lognormal distribution of annual river flows. The annual flows, in cubic meters per second, at the Weldon River at Mill Grove, Missouri, for the period 1930–1960 are averaged as follows:

3.06	1.52	16.60	2.78	1.15	13.39	2.74	6.16	1.21	5.90	
4.06	2.66	11.29	8.46	7.04	12.51	10.91	16.09	3.46	4.28	
6.92	11.35	6.95	3.23	18.70	3.75	1.25	2.06	3.83	18.02	14.41.

Fit the lognormal distribution to this data. What is the probability that the annual river flow is in the range 2–15 m^3/s? These data are from R. D. Markovic (1965), "Probability functions of best fit to distributions of annual precipitation and runoff," *Hydrology Papers*, no. 8, Colorado State University, Fort Collins, CO, and are used with permission of Colorado State University.

4.29. Lognormal distribution of low flows in the Po River, Italy. Low flows in the Po River basin in northern Italy are affected by irrigation releases and return flows. The following are the annual minimum low flows in cubic meters per second occurring at Pontelagoscuro during the period 1 October to 14 April, a period that is outside the irrigation season. There are 18 occurrences during the period 1920–1991:

735	429	742	828	554	855	787	668	655
830	732	577	1030	650	620	561	588	635

The low flows in the lower reaches of the Po River have a two-component log-normal distribution on account of the intervention caused by irrigation (from N. T. Kottegoda and L. Natale (1994), "Two-component log-normal distribution of irrigation-affected low flows," *J. Hydrol.*, Vol. 158, pp. 187–199). For the data given, which represents one component, determine the probability that an annual minimum of 400 m^3/s can be maintained in the Po at Pontelagoscuro over a 3-year period?

4.30. Ratios of densities of concrete. Densities of concrete (such as those given in Table 1.2.1) can be approximated by a uniform distribution. Taking data from two similar mixes of concrete, determine the distribution of the ratios of the densities, after transformation to $U(0, 1)$.

4.31. Relationship between strengths of construction materials. The strength of a construction material X, in newtons per square millimeter, is found to be normally distributed. It is claimed that a new material Y can be produced that is proportional in strength to the square of the strength of X. Derive the distribution of Y assuming that X is standardized to zero mean and unit variance.

Chapter 5
Model Estimation and Testing

In Chapters 1 and 3, we stressed the importance of statistical inference. This allows one to make conclusions regarding a population, which represents the aggregate of all conceivable measurements, on the basis of information contained in a sample of data from the population. For the engineer, inference and decision-making are the ultimate aims of statistical analysis. As a crucial part of the inference, methods of estimating one or more parameters of a probability model characterizing the population were discussed in Subsection 3.2.3. We encountered several point estimators by which a value is found, using a given sample of observations, to represent a parameter.

The second phase of the inferential procedure, which we deal with in this chapter, involves verification of the model. Thus, we can ascertain whether our initial hypothesis is correct. We also test the significance of a batch of data.

With the intention of emphasizing the importance of the basic concepts, we begin this chapter with a review of the definitions of terms related to random sampling and clarification of the properties of point estimators introduced in Subsection 3.2.3. We then proceed to interval estimation, a method of obtaining, at a given level of confidence (or probability), two statistics which include within their range an unknown but fixed parameter. The significance of estimated parameters using sample data and the differences between estimated parameters from two or more samples are then discussed together with the appropriate sampling distributions. On the contrary, nonparametric methods described next are not based on sampling distributions.

An important area of statistical inference is the analysis of variance of random variables. As shown, it deals with associations and causal factors. The analysis of variance is related to the design of experiments, the aim of which is to view the state of nature, or systems devised by humans, that is partially obscured by random effects.

We also discuss model fitting and goodness-of-fit tests, supplemented by various graphical methods, and followed by methods of detecting and coping with outliers.

5.1 A REVIEW OF TERMS RELATED TO RANDOM SAMPLING

A *population* consists of all conceivable observations of a process or attribute of a component (such as the density of a batch of concrete sampled in Table E.1.2). A population may consist of elements that do not exist (in a physical sense); it is then said to be conceptual. A *sample*, such as the values listed in Table E.1.1, is a subset of a population. A *random* sample is one that is representative of the population.[1] A *random variable* is a real-valued function defined on a sample space. Whether a random variable is continuous or discrete depends on how the sample space is defined.

[1] More formally, a random sample is a collection $X_1, X_2, \ldots, X_n$ of random variables taken from a population with density $f(\cdot)$ if the joint density $f_{X1,X2,\ldots,Xn}(x_1, x_2, \ldots, x_n) = f_1(x_1)f_2(x_2)\cdots f_n(x_n)$.

If the population is known or assumed to have a distribution such as the normal distribution discussed in Chapter 4 but the value of a parameter θ is unknown, then we need a random sample of observations, say, $X_1, X_2, \ldots, X_n$ of size n, to *estimate* θ. The joint distribution of $X_1, X_2, \ldots, X_n$ is known as the *sampling distribution* of $X_1, X_2, \ldots, X_n$. Any function of the observations that is quantifiable and does not contain any unknown parameter is called a *statistic*. A statistic is a random variable that gives us a means of *estimation*. We can determine a single number to represent θ or we can determine two numbers, which include θ within their range, at a given level of probability. These procedures are discussed in the next two sections, respectively. It is also important to distinguish between an *estimator* and an *estimate*. The first is the rule or method of estimation; for example, the sample mean $\bar{X}$ is a point estimator of μ, the population mean; the second is the value which the estimator yields in a particular application.

5.2 PROPERTIES OF ESTIMATORS

An important field of statistical inference is the estimation of parameters. Alternative types of estimators, which have properties that are more or less desirable than others, can be used for such a purpose, as discussed initially in Subsection 3.2.3. In this section we summarize and exemplify these properties.

5.2.1 Unbiasedness

Given a sample of observations, our objective here is to estimate the value of a parameter θ. The observations are random variables, say, $X_1, \ldots, X_n$; hence an estimate of the parameter obtained from them, which is a statistic and a function of the observations, is also a random variable. In most cases, such a statistic can differ considerably from the true value of the parameter regardless of the method of estimation. However, we seek to find an estimator that will, on average (that is, after repeated sampling), give satisfactory results. That is, the estimator will produce statistics that are distributed according to a certain law. This law is the sampling distribution to which we referred earlier. Some types of these sampling distributions will be considered in this chapter. The law must have some desirable attributes if the estimator is to be acceptable for our purpose. For instance, if the mean value of this distribution is θ, then the estimator has the property of *unbiasedness*.

Definition and properties: A point estimator $\hat{\theta}$ is an **unbiased estimator** of the population parameter θ if $E[\hat{\theta}] = \theta$. If the estimator is biased, the **bias** $= E[\hat{\theta}] - \theta$.

Example 5.1. Mean and variance of the sample mean. Let us show that the sample mean $\bar{X}$ and the sample variance

$$\hat{S}^2 = \frac{1}{n-1}\sum_{i=1}^{n}(X_i - \bar{X})^2$$

are unbiased estimators of μ and σ^2.

The first result follows immediately by taking expectations of a random sample of size n,

$$\bar{X} = \frac{1}{n}(X_1 + X_2 + \cdots + X_i \ldots + X_n)$$

which yields

$$E[\bar{X}] = \frac{1}{n}(nE[X_i]) = \frac{1}{n}(n\mu) = \mu.$$

For the variance [as in Eq. (1.2.7)]

$$E[\hat{S}^2] = \frac{1}{n-1} E\left[\sum_{i=1}^{n}(X_i - \bar{X})^2\right] = \frac{1}{n-1}\left(\sum_{i=1}^{n} E\left[X_i^2\right] - nE[\bar{X}^2]\right)$$

$$= \frac{1}{n-1}\left\{\sum_{i=1}^{n} E[(X_i - \mu)^2] - nE[(\bar{X} - \mu)^2]\right\} = \frac{1}{n-1}\left\{\sum_{i=1}^{n} \sigma^2 - n\mathrm{Var}[\bar{X}]\right\}$$

$$= \frac{1}{n-1}\left(n\sigma^2 - n\frac{\sigma^2}{n}\right) = \sigma^2.$$

Unfortunately, many estimators are biased but have other desirable properties. Methods of correcting or reducing the bias such as the jackknife and bootstrap were discussed in Subsection 3.2.3.

There are also three other properties that our ideal estimator should have. These are consistency, efficiency, and sufficiency, concepts introduced by the English statistician Fisher.

5.2.2 Consistency

A *consistent estimator* of a parameter θ produces statistics that converge to θ, in terms of probability. Thus we can define consistency as follows:

Definition and properties: An estimator $\hat{\theta}_n$, based on a sample size n, is a consistent estimator of a parameter θ, if for any positive number ε,

$$\lim_{n\to\infty} \Pr[|\hat{\theta}_n - \theta| \leq \varepsilon] = 1. \tag{5.2.1}$$

One finds, however, that sometimes an unbiased estimator may not be consistent. This case is illustrated as follows:

Example 5.2. Unbiasedness and consistency. A simple example of a consistent estimator that does not necessarily have the property of unbiasedness is found in Subsection 1.2.2 in which we considered two methods of estimating the variance σ^2. The estimators are written as random variables, firstly by

$$S^2 = \frac{1}{n}\sum_{i=1}^{n}(X_i - \bar{X})^2,$$

and secondly by

$$\hat{S}^2 = \frac{1}{n-1}\sum_{i=1}^{n}(X_i - \bar{X})^2.$$

As noted in Example 5.1, the second equation gives the unbiased estimator of the variance. However, it can be shown (by considering the entire population as implied by the original Fisher definition of consistency) that the first equation gives a consistent estimator. Because inconsistency in this case is considered to be of less importance, we prefer to use the second equation.

5.2.3 Minimum variance

In practice we seldom have more more than one sample, but if we had a number of samples with high variability, we may find that a single statistic that gives an estimate of a population parameter θ is quite different from the true value even if the estimator is unbiased. So we must seek an estimator that is also comparatively low in variance. Among unbiased estimators, the one with the smallest variance is called the *minimum variance unbiased estimator.*

Furthermore, it has been found that some types of estimators have a bound that is exceeded by the variance. This type is known as a minimum variance bound (mvb) estimator. The lower bound is found by what is known as the Cramer-Rao inequality.[2] Hence we obtain the relationship:

$$\frac{\partial \ln L}{\partial \theta} = g(\theta)\{\hat{\theta} - f(\theta)\}, \tag{5.2.2}$$

where $\ln L$ is the log-likelihood function discussed in Section 3.2, $g(\theta)$ and $f(\theta)$ are functions independent of the sample of observations, and $f(\theta)$ is in a simple form such as θ or θ^2, which is relevant to the sampling distribution. Thus if an equation of the form of Eq. (5.2.2) can be obtained, $\hat{\theta}$ is a minimum variance bound estimator, mvb, of $f(\theta)$. It can also be shown from Eq. (5.2.2) that

$$\text{Var}[\hat{\theta}] = f'(\theta)/g(\theta), \tag{5.2.3}$$

so that if $f(\theta) \equiv \theta$, $\text{Var}[\hat{\theta}] = 1/g(\theta)$.

Definition and properties: A minimum variance unbiased estimator is the estimator with the smallest variance out of all unbiased estimators.

If the derivative of the log-likelihood function $\ln L$ can be put in the form

$$\frac{\partial \ln L}{\partial \theta} = g(\theta)\{\hat{\theta} - f(\theta)\},$$

then $\hat{\theta}$ is a *minimum variance bound estimator*, mvb, of $f(\theta)$, with variance

$$\text{Var}[\hat{\theta}] = f'(\theta)/g(\theta).$$

Example 5.3. Minimum variance bound of the location parameter and the square of the scale parameter of the normal distribution. The pdf of the normal distribution is given by

$$\phi(x) = \frac{1}{b\sqrt{2\pi}} \exp\left[-\frac{1}{2}\left(\frac{x-a}{b}\right)^2\right], \quad \text{for } -\infty < x < +\infty.$$

The estimator of the square of the scale parameter b is the variance as seen from Example 3.23. Also,

$$\frac{\partial \ln L}{\partial a} = \frac{n}{b^2}(\bar{X} - a).$$

Thus from Eq. (5.2.2), $f(\theta) \equiv f(a) = a$; and $\bar{X}$ is an mvb estimator of a with variance b^2/n, from Eq. (5.2.3).

To estimate the variance statistic of b^2, we use the estimator for a just given. From Example 3.23 and Eq. (5.2.2), with $f(\theta) \equiv f(b) = b^2$,

$$\frac{\partial \ln L}{\partial b} = -\frac{n}{b} + \frac{\sum_{i=1}^{n}(X_i - \bar{X})^2}{b^3} = \frac{n}{b^3}\left[\frac{1}{n}\sum_{i=1}^{n}(X_i - \bar{X})^2 - b^2\right].$$

[2] See, for example, Stuart and Ord (1991, pp. 614–618).

Then from Eq. (5.2.2) we find that

$$S^2 = \frac{1}{n}\sum_{i=1}^{n}(X_i - \bar{X})^2$$

is an mvb estimator of b^2. Its variance is $2b^4/n$ from Eq. (5.2.3).

5.2.4 Efficiency

The term *efficiency* is used as a relative measure of the variance of the sampling distribution, with the efficiency increasing as the variance decreases. One may search unbiased estimators to find the one with the smallest variance and call it the most efficient. It seems, however, desirable to combine the properties of unbiasedness and minimum variance because an estimator can have minimum variance but it may be biased, albeit to a small degree. This combination can be accomplished by means of the *mean square error* (mse) criterion. Thus if A is an estimator of θ, the mse is

$$\begin{aligned} E[(A-\theta)^2] &= E[\{(A - E[A]) - (\theta - E[A])\}^2] \\ &= E[(A - E[A])^2] + (\theta - E[A])^2 \\ &= \text{Var}[A] + (\text{bias})^2. \end{aligned}$$

(In the first equation, we see that the terms of the cross-product, $2E[(A - E[A])(\theta - E[A])]$, sum to zero). Thus the estimator becomes more efficient as the mse decreases.

Definition and properties: An estimator that has minimum mean square error among all possible unbiased estimators is called an efficient estimator. The mean square error of an estimator, which is equivalent to the sum of its variance and the square of its bias, can be used as a relative measure of efficiency when comparing two or more estimators.

Example 5.4. Relative efficiencies of the estimators of the mean of concrete densities. From Tables 1.2.1 and 1.2.2, the mean of the densities of 40 concrete test cubes is 2445 kg/m^3. However, if we had only the first five test cubes, the estimated mean would be 2431 kg/m^3. Both estimators are unbiased as seen in Example 5.1. Hence the relative efficiency, as given by the ratio of the mse values, bears inversely with the ratio of variances:

$$\frac{\sigma^2/40}{\sigma^2/5} = \frac{1}{8}.$$

This merely confirms what we already know, that is, the large-sample estimator for the mean is more efficient than that based on a smaller sample. The efficiency is seen to be proportional to the sample size n.

The outcome of the Example 5.4 notwithstanding, the minimization of variance will generally give different results from the minimization of mse.

5.2.5 Sufficiency

Properties such as unbiasedness, consistency, and minimum mean square error guide us to select the most suitable estimators. To complete the discussion, we now discuss the important concept of sufficiency. A *sufficient estimator* gives as much information as possible about a sample of observations so that no additional information can be conveyed by any other estimator. This can also be defined more formally as follows:

Definition and properties: Let a sample $X_1, X_2, \ldots, X_n$ be drawn randomly from a population having a probability distribution with unknown parameter θ. Then the statistic

$T = f(X_1, X_2, \ldots, X_n)$ is said to be sufficient for estimating θ if the distribution of $X_1, X_2, \ldots, X_n$ conditional to the statistic T is independent of θ.

We can see, for example, that the median, taken as a measure of mean density or central tendency as discussed on Subsection 1.2.1, does not contain all the information in a sample. The median is the middle value of the sample; if any other value is changed the mean changes but the median is unaltered. It is therefore not a sufficient statistic for the purpose, unlike the mean which is discussed in the following example:

Example 5.5. Normal and uniform variates. From Table 1.2.2, the mean and standard deviation of the compressive strengths of concrete are 60.14 and 5.02 N/mm^2. We also made the hypothesis (in Example 4.27) that these strengths are normally distributed (which is subject to verification later in this chapter but is confirmed by numerous other studies). If the sample variance is the true value of the variance σ^2, then the sample mean $\bar{X} = (1/n)\sum_{i=1}^{n} X_i$ is a sufficient statistic for the location parameter of the normal distribution, which is the population mean μ. On the other hand, if the sample mean is the true value of μ, the sample variance $S^2 = (1/n)\sum_{i=1}^{n}(X_i - \bar{X})^2$ is a sufficient statistic for σ^2, the square of the scale parameter of the normal distribution

In practice, both parameters are unknown. However, if $\bar{X}$ and S^2 are considered jointly, these two statistics are jointly sufficient for μ and σ^2. This is because no other estimators can provide any more information for the population mean and variance.

Also consider a uniform$(0, \theta)$ distribution. Let us draw a random sample $X_1, X_2, \ldots, X_n$ from this distribution. Then for estimating θ, $X_{\max} = \max[X_1, X_2, \ldots, X_n]$ is sufficient.[3]

Example 5.6. Poisson variates. Suppose $X_1, X_2, \ldots, X_n$ is a random sample of Poisson (v) variates. Then it can be shown as follows that $T = \sum_{i=1}^{n} X_i$ is a sufficient statistic for v.

The joint sampling pdf of the variates is

$$f(x_1, x_2, \ldots, x_n \mid v) = \prod_{i=1}^{n}\left(\frac{v^{x_i} e^{-v}}{x_i!}\right) = \frac{v^{S_n} e^{-nv}}{M_n}$$

where

$$S_n = \sum_{i=1}^{n} x_i$$

and

$$M_n = \prod_{i=1}^{n} x_i!.$$

The sum of n Poisson (v) variates is Poisson (nv) distributed (as shown in Subsection 4.1.3). Thus the pdf of T is $p(t \mid v) = (nv)^t e^{-nv}/t!$ for $t = 0, 1, 2, \ldots$

If $S_n = t$, the joint conditional sampling pdf of the variates is

$$h(x_1, x_2, \ldots, x_n \mid t, v) = \frac{v^{S_n} e^{-nv}/M_n}{(nv)^t e^{-nv}/t!} = \frac{t!}{n^t M_n}.$$

Because the result does not depend on v, T is a sufficient statistic for v.

5.2.6 Summary of Section 5.2

This formal summary of the desirable properties of point estimators is intended to provide insight to the various methods of estimation. These were discussed initially in Section 3.2

[3] See, for example, Casella and Berger (2002, pp. 281–282).

of Chapter 3 and used in Chapter 4. They will be applied in one form or other throughout the book.

5.3 ESTIMATION OF CONFIDENCE INTERVALS

In Chapters 3 and 4, we discussed and applied methods of estimating the values of one or more parameters of a population; in Section 5.2 we examined more closely the properties of the resulting estimators. We have seen that point estimates can be erroneous; in reality the probability that an estimate is equal to an unknown parameter is zero. The resulting uncertainty can be quantified by the relative variances or mean square errors of the estimators. Because of this uncertainty, the next step of inference is interval estimation. Here we determine two numbers, say, a and b, that are expected to include within their range an unknown parameter θ in a specified percentage of cases after repeated experimentation under identical conditions. That is, in place of one statistic that estimates θ, we find a range specified by two statistics, which includes it at a given level of probability. The end points a and b of this range are known as *confidence limits*, and the interval (a, b) is known as the *confidence interval*. We do not have the precision as for a point estimator but we have confidence (without absolute certainty) that the assertion is right.

5.3.1 Confidence interval estimation of the mean when the standard deviation is known

Let C_l and C_u be the lower and upper confidence limits that include an unknown but invariable parameter θ. Although there is some uncertainty associated with this statement, we will be right in a proportion, say, $1 - \alpha$, of the samples taken from the same population, on average, when we make the assertion that the given interval includes θ. Thus we can say, by adopting the long-run frequency interpretation of probability,

$$\Pr[C_l \leq \theta \leq C_u] = 1 - \alpha, \quad \text{for } 0 < \alpha < 1. \tag{5.3.1a}$$

As noted, θ is a constant, but the estimator $\hat{\theta}$ and the confidence limits C_l and C_u are random variables (the values of which depend fully or partly on observations); examples of $\hat{\theta}$ such as $\bar{X}$ and $\hat{S}^2$ have been encountered in Examples 5.1 to 5.3. The probability $(1 - \alpha)$ is known as the *confidence level* or *confidence coefficient*. It often takes values such as 0.99, 0.95, and 0.90. The confidence limits C_l and C_u depend on the sampling distribution of $\hat{\theta}$. The standard deviation of the statistic $\hat{\theta}$ is called its standard error.

Equation (5.3.1a) is used to establish *two-sided confidence limits*. In some situations, we may require one-sided confidence limits. The *lower and upper one-sided confidence limits* are specified, respectively, by

$$\Pr[C_l \leq \theta] = 1 - \alpha, \quad \text{for } 0 < \alpha < 1 \tag{5.3.1b}$$

and

$$\Pr[\theta \leq C_u] = 1 - \alpha, \quad \text{for } 0 < \alpha < 1. \tag{5.3.1c}$$

In the first case, the upper limit is considered to be at the upper limit of the sampling distribution; in the second case, the lower limit is at the lower limit of such a distribution.

5.3.1.1 Sampling distribution of the mean

From the Central Limit Theorem of Section 4.2, we noted that, regardless of the distribution of a variable X, the standard error of the sample mean, $\bar{X}$, tends to

$$\sigma_{\bar{X}} = \sigma/\sqrt{n} \tag{5.3.2}$$

as the sample size n tends to ∞. The sampling distribution of $\bar{X}$ is close to normal if $n > 30$, even when X is not normally distributed. However, if X is normal, the distribution of $\bar{X}$ is exactly normal regardless of sample size. Hence by using Eq. (5.3.2), we can apply Eq. (5.3.1) to find confidence limits for the population mean μ. Knowing that if $X \sim N(\mu, \sigma^2)$, we may write

$$Z = \frac{\bar{X} - \mu}{\sigma/\sqrt{n}} \sim N(0, 1). \tag{5.3.3}$$

For example, using Eqs. (5.3.1*a*) and (5.3.3) and Table C.1 of Appendix C for the standard normal cdf, we have for $(1 - \alpha) = 0.95$,

$$\Pr[-1.96 \le Z \le 1.96] = \int_{-1.96}^{1.96} F_Z(z)dz = 0.95. \tag{5.3.4}$$

This result stems from the symmetry of the normal distribution; that is, for $\alpha = 0.05$, $\Phi(1.96) - \Phi(-1.96) = 0.975 - 0.025 = 0.95$. Note that the entries required in Table C.1, and in the following tables, are $(1 - \alpha/2)$ for two-sided confidence limits or $(1 - \alpha)$ for a one-sided confidence limit.

This means that the probability that the interval -1.96 to 1.96 includes the random variable Z as defined by Eq. (5.3.3) is 0.95. Similarly, intervals can be provided with other values of α. We can also construct an endless number of *noncentral* intervals that contain Z with probability 0.95, but the interval given by Eq. (5.3.4) is *central* and is the shortest of all possible intervals because of the symmetry of the distribution. Confidence intervals are summarized by the following definitions, and the statistical interpretations are given in the examples that follow.

Definition and properties: Let $\bar{X}$ be the mean of a random sample of size n drawn from a population with known standard deviation σ. The $100(1 - \alpha)$ percent central two-sided confidence interval for the population mean μ is

$$(\bar{X} - z_{\alpha/2}\sigma/\sqrt{n},\ \bar{X} + z_{\alpha/2}\sigma/\sqrt{n})$$

where $z_{\alpha/2}$ is a standard normal variate that is exceeded with probability $\alpha/2$. The one-sided upper and lower $100(1 - \alpha)$ percent confidence limits for the population mean μ are, respectively, as follows:

$$\bar{X} + z_{\alpha}\sigma/\sqrt{n} \quad \text{and} \quad \bar{X} - z_{\alpha}\sigma/\sqrt{n} \le \mu.$$

Figure 5.3.1 is an illustration of the pdf of the standard normal variate Z as defined in Eq. (5.3.3) with the locations of $z_{\alpha/2}$ and $-z_{\alpha/2}$ required for a $100(1 - \alpha)$ percent central two-sided confidence interval on the mean when the standard deviation is known.

Example 5.7. Confidence limits for the mean of concrete strengths. From Table 1.2.2, the mean and standard deviation of the compressive strengths of 40 test cubes of concrete are 60.14 and 5.02 N/mm^2, respectively. We also assume that the compressive strengths are

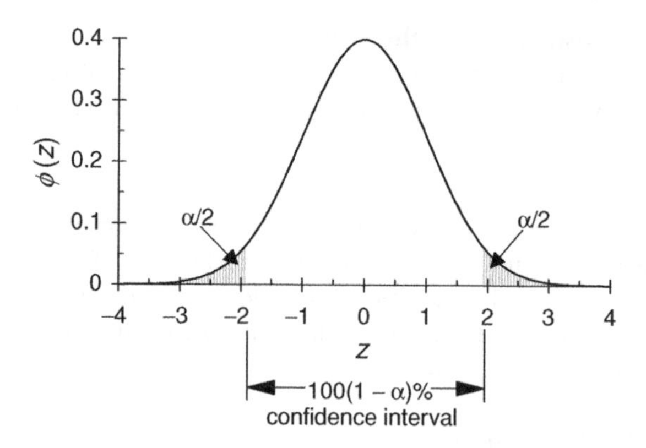

Fig. 5.3.1 Standard normal pdf showing two-sided confidence interval.

normally distributed.[4] To facilitate the application, let us assume that the estimated standard deviation of 5.02 N/mm^2 is the true value. Then from Eqs. (5.3.3) and (5.3.4),

$$\Pr\left[-1.96 \leq \frac{\bar{X} - \mu}{5.02/\sqrt{40}} \leq 1.96\right] = 0.95,$$

and hence

$$\Pr[-1.56 \leq \bar{X} - \mu \leq 1.56] = 0.95.$$

Thus we can say with 95% confidence, before estimating the mean value of the concrete strengths by experimentation, that the sample mean will be greater or less than 1.56 N/mm^2 of the true value of the mean. Let us now substitute the sample value $\bar{x} = 60.14$ for $\bar{X}$. The value 60.14 is then subtracted throughout. Finally we reverse the negative signs and, because this calls for a reversal of the inequalities,

$$\Pr[58.58 \leq \mu \leq 61.70] = 0.95.$$

We know that the true mean μ, which is fixed but unknown, cannot vary from one experiment to another on the same concrete, as already mentioned. The correct interpretation of the above stipulation, on a frequency basis, is that if one were to repeat the same experiment and the standard deviation be constant, 95% of the intervals constructed in this way will, on average, include the true mean μ. With the single experiment at hand, we are 95% confident that the interval (58.58, 61.70) includes μ.

Example 5.8. One-sided confidence limit for the mean time interval between vehicles passing an observation point. In Example 4.21 we found it reasonable to assume that the time interval between vehicles at a given location is exponentially distributed. In that case study the average time interval was calculated as 0.551 minute. Let us suppose we are concerned that an excessively low mean time interval between vehicles will cause traffic congestion on the road. As an initial step in the investigation, we can provide a one-sided 99% confidence limit on the mean rate. The mean and variance of the exponential distribution of the time interval between vehicles with cdf

$$F_X(x) = 1 - e^{-\lambda x}$$

are $1/\lambda$ and $1/\lambda^2$, respectively, as noted in Subsection 4.2.2. For a large sample size, we can assume that the sampling distribution of the estimator of the mean is approximately normally distributed with mean $1/\lambda$ and variance $1/(n\lambda^2)$, where n is the number ofobservations.

[4] As we said in Chapter 4, there is much evidence to support this assumption. See, for example, Problem 5.34.

We make use of the symmetry of the normal distribution and calculate the one-sided 99% confidence limit of the mean time interval as follows:

$$\Pr\left[\frac{1/\hat{\lambda} - 1/\lambda}{1/\lambda\sqrt{n}} \geq 2.326\right] = 0.01.$$

which can be written as

$$\Pr\left[\frac{\lambda}{\hat{\lambda}} \geq \left(1 + \frac{2.326}{\sqrt{n}}\right)\right] = 0.01.$$

Substituting $\hat{\lambda} = 1/0.551$ and $n = 204$ from Example 4.21, we write

$$\Pr\left[\lambda \geq \frac{1.162851}{0.551}\right] = 0.01.$$

This is equivalent to

$$\Pr\left[\frac{1}{\lambda} \leq 0.474\right] = 0.01.$$

Hence, the 99% lower confidence limit of the mean time interval is 0.474 minute.

5.3.2 Confidence interval estimation of the mean when the standard deviation is unknown

Quite often in practice the mean and standard deviation are both unknown. Under such conditions we must modify our approach. We assume normality of the variate X as before, but the consequences of a nonnormal distribution are minor for small to moderate departures from normality, if the sample size is large, say, $n > 30$. In this situation we apply the Student's t distribution.[5] The T variable represents the mean $\bar{X}$ standardized as in Eq. (5.3.3) except that the sample standard deviation $\hat{S}$ is used in place of σ. Thus for a sample size n, the variable

$$\frac{\bar{X} - \mu}{\hat{S}/\sqrt{n}} \sim t_{n-1}; \tag{5.3.5}$$

that is, it has the Student's t distribution with $\nu = (n-1)$ degrees of freedom (as defined in Subsection 1.2.2). The pdf of Student's t distribution is given by

$$f_T(t) = \frac{\Gamma[(\nu+1)/2]}{\sqrt{\pi\nu}\,\Gamma(\nu/2)} \frac{1}{[(t^2/\nu)+1]^{(\nu+1)/2}}, \qquad \text{for } -\infty < t < \infty. \tag{5.3.6}$$

It is equivalent to the Pearson Type VII distribution. The derivation is shown in Appendix A.7.

Graphs of the pdf of Student's t distribution are given in Fig. 5.3.2 for degrees of freedom $\nu = n - 1$ equal to 2, 4, and 10. Also, comparison is made with the standard normal pdf, which has a smaller spread.

These graphs show that the Student's t distribution approaches the standard normal distribution as the sample size increases. The approximation is found to be quite close for $n > 30$.

[5] Originated by William S. Gosset, who worked for the Guinness brewery in Dublin around 1900 and wrote under the pseudonym *Student*. His correspondence with R. A. Fisher in subsequent years is a fascinating part of the history of mathematical statistics (see Box, 1981).

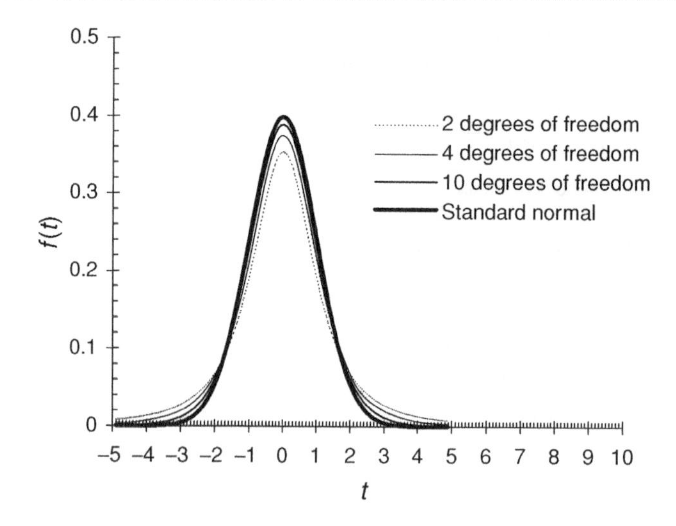

Fig. 5.3.2 Graphs of pdfs of Student's t distribution with 2, 4, and 10 degrees of freedom compared with standard normal pdf.

The cdf of Student's t distribution is tabulated in Appendix C.2. Because the distribution is a function of $v = n - 1$, the tables are different from those of the normal distribution. For selected values of $F(t)$, t values are given for a range of values of v.

Definition and properties: Let $\bar{X}$ and $\hat{S}$ be the mean and standard deviation of a random sample of size n drawn from a normal distribution with unknown standard deviation σ. The $100(1 - \alpha)$ percent central two-sided confidence interval for the population mean μ is as follows:

$$(\bar{X} - t_{n-1,\alpha/2}\,\hat{S}/\sqrt{n},\ \bar{X} + t_{n-1,\alpha/2}\,\hat{S}/\sqrt{n})$$

where $t_{n-1,\alpha/2}$ is a Student's t variate with $n - 1$ degrees of freedom and probability $\alpha/2$ of exceedance.

Example 5.9. The 95% confidence limits for the mean of concrete strengths with unknown standard deviation. We return to Example 5.7 based on the data of Table 1.2.2 in which the estimated mean and standard deviation of the compressive strengths of 40 test cubes of concrete are 60.14 and 5.02 N/mm^2, respectively. We assume that the compressive strengths are normally distributed as before. The same exercise as in Example 5.7 is repeated to find the 95% confidence limits but we use Student's t distribution because in reality the standard deviation is unknown, and therefore an estimated value is used. Referring to Table C.2 of Appendix C, for $v = n - 1 = 39$ degrees of freedom and $F(t) = 1 - \alpha/2 = 0.975$ (because $\alpha = 1 - 0.95 = 0.05$ and the sampling distribution is symmetrical like the normal), we obtain the t variate as 2.023 approximately by interpolation. Thus from Eqs. (5.3.1) and (5.3.5),

$$\Pr\left[-2.023 \leq \frac{\bar{X} - \mu}{5.02/\sqrt{40}} \leq 2.023\right] = 0.95.$$

Hence,

$$\Pr[58.53 \leq \mu \leq 61.75] = 0.95.$$

We note that the confidence limits so obtained are very close to those in Example 5.7 where we assumed that the standard deviation is known; for a smaller sample size the difference is greater.

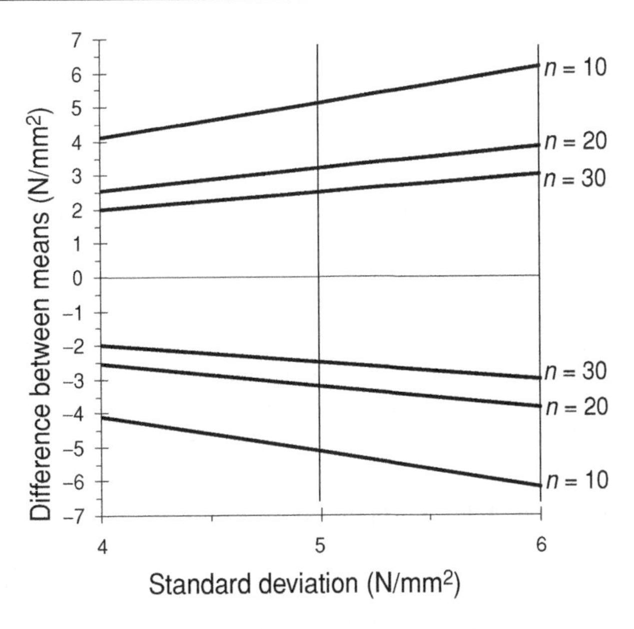

Fig. 5.3.3 The 99% confidence limits of difference between population and sample means of compressive strengths of concrete using sample standard deviation.

Let us, however, restate the problem, firstly by applying 99% confidence limits to the difference between the true mean and sample mean for n equal to 10, 20, and 30 and then by considering a range of values from 4 to 6 N/mm^2 for the unknown standard deviation (around its estimated value of 5.02 N/mm^2). To find the confidence limits, we commence with the general condition from Eq. (5.3.5):

$$\Pr\left[-t_{n-1,0.005}\frac{\hat{S}}{\sqrt{n}} \leq \bar{X} - \mu \leq t_{n-1,0.005}\frac{\hat{S}}{\sqrt{n}}\right] = 0.99.$$

From Table C.2, values of $t_{n-1,0.005}$ for n equal to 10, 20, and 30 are 3.250, 2.861, and 2.756, respectively. The 99% confidence limits are presented in Fig. 5.3.3 for the three specified values of n as graphs of the difference $\bar{X} - \mu$ against the standard deviation $\hat{S}$.

For example, if $n = 20$ and the estimated standard deviation $\hat{S} = 6$ N/mm^2, the 99% confidence limits for the difference in means are 3.84 and −3.84 N/mm^2. These differences become smaller for increasing values of n and smaller values of the sample standard deviation.

Example 5.10. Sample sizes required for one-sided confidence interval on the mean concrete strength. In Example 5.9 we considered different sample sizes and different two-sided confidence intervals for $\bar{X} - \mu$ over a range of values of the standard deviation. Let us now consider a one-sided lower 99% confidence interval on the mean. In this example we find the minimum sample size required for tests, when the confidence interval and the coefficient of variation are specified functions of the mean.

From the results given in Table 1.2.2, the coefficient of variation $V = 8\%$ approximately. In the first instance, let us use this value in our application. We also stipulate that the lower confidence limit does not exceed 16% of the mean. That is,

$$\frac{\hat{S}}{\sqrt{n}}t_{n-1,0.01} \leq 0.16\bar{X}.$$

Substituting $V = \hat{S}/\bar{X} = 0.08$, we get the following condition:

$$\frac{t_{n-1,0.01}}{\sqrt{n}} \leq 2.$$

From Table C.2, the minimum sample size to meet this condition is $n = 5$.

In our second example, let us consider that the concrete strengths are more variable and assume that $V = 16\%$. We also specify that the lower confidence limit does not exceed 8% of the mean. This leads to the condition

$$\frac{t_{n-1,0.01}}{\sqrt{n}} \leq 0.5.$$

From Table C.2 the minimum sample size required to meet this condition is $n = 25$.

One practical approach is to substitute the coefficient of variation that is appropriate for the particular type of concrete tested and put limits on the confidence interval as a function of the mean.

5.3.3 Confidence interval for a proportion

Consider a two-sided Bernoulli variate as in Subsection 4.1.1 that occurs with probability p; that is, a given trial is a success with probability p whereas the failure probability is $(1 - p)$. The mean and variance of such a variate are p and $p(1 - p)$, respectively. If an experiment involving a Bernoulli variate is repeated n times, the standard error of the estimated proportion of a success is

$$\sigma_{\hat{p}} = \sqrt{\frac{p(1-p)}{n}}. \qquad (5.3.7)$$

For a large sample size, say, $n > 30$, and for $np > 5$, and $n(1 - p) > 5$, the sampling distribution is very nearly normal. In practice we substitute the sample values for p in the right-hand side of Eq. (5.3.7).

Example 5.11. Leaks from water pipes. A survey of leaks from water pipes in a city water distribution system, conducted over a representative area, shows that substantial loss occurs in 7 out of 37 pipes tested. We can find 95% confidence limits for the proportion of leaking pipes in the city first by substituting $7/37$ for p in Eq. (5.3.7) and then by using the normal approximation and Eq. (5.3.1):

$$\frac{7}{37} \pm 1.96\sqrt{\frac{7}{37} \times \frac{30}{37} \times \frac{1}{37}} = 0.19 \pm 0.13.$$

That is, we can say with 95% confidence that the interval (0.06, 0.32) includes the true proportion of pipes in the city from which there is a substantial waste.

5.3.4 Sampling distribution of differences and sums of statistics

Let S_1 and S_2 be independent statistics from two populations (based on sample sizes n_1 and n_2). Also, let the respective means and standard errors of the sampling distributions of the two statistics be μ_{S_1}, μ_{S_2} and $\sigma_{S_1}, \sigma_{S_2}$, respectively.

The differences between the statistics (after repeated sampling) from the two populations have a sampling distribution with mean

$$\mu_{S_1-S_2} = \mu_{S_1} - \mu_{S_2} \qquad (5.3.8a)$$

and standard error

$$\sigma_{S_1-S_2} = \sqrt{\sigma_{S_1}^2 + \sigma_{S_2}^2}. \qquad (5.3.8b)$$

The sampling distribution of the sums of the statistics $S_1 + S_2$ has a mean

$$\mu_{S_1+S_2} = \mu_{S_1} + \mu_{S_2} \tag{5.3.8c}$$

and a standard error

$$\sigma_{S_1+S_2} = \sqrt{\sigma_{S_1}^2 + \sigma_{S_2}^2}, \tag{5.3.8d}$$

which is the same as for the differences of the two statistics.

Example 5.12. Confidence limits for the differences between two means of the annual rainfalls at two stations. Measurements of rainfall have been taken at a particular location, say, station 1, over a period of $n_1 = 50$ years, and the annual mean and standard deviation are estimated as $\bar{x}_1 = 900$ mm and $\hat{s}_1 = 80$ mm, respectively. At another location, say, station 2, measurements cover a period of $n_2 = 40$ years, from which estimates of the annual mean and standard deviation are $\bar{x}_2 = 825$ mm and $\hat{s}_2 = 90$ mm, respectively. Because annual rainfalls are the result of a large number of small causes, such an additive effect makes it plausible to assume that the distribution is normal (see Central Limit Theorem of Appendix A.6). There is sufficient empirical evidence to support this claim; and although in some cases nonnormal distributions seem to be appropriate, the approximation is a reasonable one. Thus, the sampling distribution of the difference between the two means is also normal. The $(1-\alpha)$ percent confidence limits for the difference between the two means are found using Eqs. (5.3.1), (5.3.2), (5.3.8a), and (5.3.8b) as follows:

$$(\bar{x}_1 - \bar{x}_2) \pm z_{\alpha/2}\sqrt{\frac{\sigma_1^2}{n_1} + \frac{\sigma_2^2}{n_2}},$$

where $z_{\alpha/2}$ is the standard normal variate which is exceeded with probability $\alpha/2$. Hence, substituting $\hat{s}_1$ for σ_1 and $\hat{s}_2$ for σ_2, the 95% confidence limits are approximately

$$75 \pm 1.96\sqrt{\frac{80^2}{50} + \frac{90^2}{40}} = 75 \pm 36 \text{ mm} \Rightarrow (39 \text{ mm}, 111 \text{ mm}).$$

In applications such as the foregoing, the results are approximate, if the population is nonnormal. The error is of course less for smaller departures from normality or for larger sample sizes. Furthermore, when the standard deviations are unknown, Student's t distribution is appropriate for the purpose, but as seen in Example 5.9 in comparison with Examples 5.7 and, the differences are very small for $n > 30$.

5.3.5 Interval estimation for the variance: chi-squared distribution

Consider a set of normally and independently distributed random variates $X_i, i = 1, 2, \ldots, \nu$, with means μ_i and variances σ_i^2. Suppose we standardize the X_i by subtracting the mean and dividing by the standard deviation, respectively, to a set $Z_i, i = 1, 2, \ldots, \nu$, with zero mean and unit variance. Consider the random variable formed by the sum of squares of the Z variates, say,

$$\chi^2 = Z_1^2 + Z_2^2 + \cdots + Z_\nu^2. \tag{5.3.9}$$

The mgf of χ^2 is given by

$$M_{\chi^2}(t) = E[\exp(t\chi^2)] = \prod_{i=1}^{\nu} E\left[\exp\left(tZ_i^2\right)\right].$$

Because

$$E\left[\exp\left(tZ_i^2\right)\right] = \int_{-\infty}^{\infty} e^{tz^2} \frac{e^{-z^2/2}}{\sqrt{2\pi}} dz = \frac{1}{\sqrt{1-2t}} \int_{-\infty}^{\infty} \frac{\sqrt{1-2t}}{\sqrt{2\pi}} e^{-(1-2t)z^2/2} dz,$$
$$\text{for } i = 1, 2, \ldots, v, \text{ and } t < 1/2,$$

and the integral on the right-hand side is unity, being a representation of the area under a normal curve with a mean of zero and variance $1/(1-2t)$,

$$M\chi^2(t) = \left[\frac{1/2}{(1/2) - t}\right]^{v/2}.$$

If we compare this mgf with that of Eq. (4.2.12*c*) for the gamma distribution, we can say that the sum of squares of independent standard normal variates has the chi-squared distribution with v degrees of freedom and cdf

$$F(\chi^2) = \frac{1}{2} \int_0^{\chi^2} \frac{(u/2)^{(v/2)-1} e^{-u/2}}{\Gamma(v/2)} du, \tag{5.3.10}$$

[as given by Eq. (4.2.12*d*)]. Figure 5.3.4 shows pdfs of the chi-squared distribution for $v = 2$, 6, and 12.

Consider a random sample $X_1, X_2, \ldots, X_n$ of independent normally distributed variables with common mean μ, variance σ^2, and sample mean $\bar{X}$. As shown in Example 5.1,

$$\hat{S}^2 = \frac{1}{n-1} \sum_{i=1}^{n} (X_i - \bar{X})^2 \tag{5.3.11}$$

is an unbiased estimator of the variance σ^2. Let us now consider the quantity

$$\sum_{i=1}^{n} (X_i - \mu)^2 = \sum_{i=1}^{n} [(X_i - \bar{X}) + (\bar{X} - \mu)]^2. \tag{5.3.12}$$

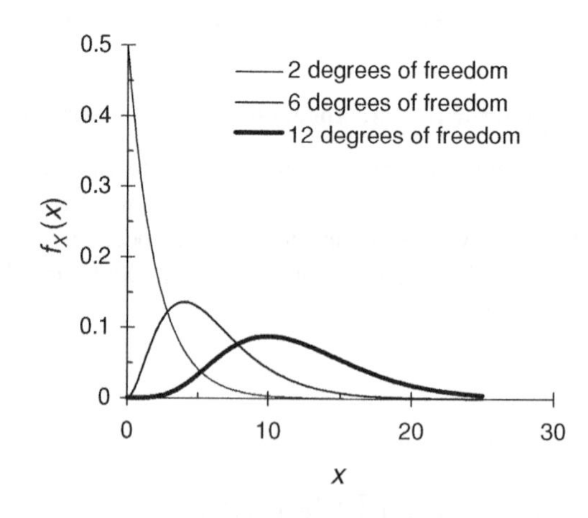

Fig. 5.3.4 Chi-squared distributions for degrees of freedom equal to 2, 6, and 12.

[Because $\sum_{i=1}^{n}(X_i - \bar{X}) = 0$, the cross-product term of Eq. (5.3.12) is zero.] Hence.

$$\sum_{i=1}^{n}(X_i - \mu)^2 = \sum_{i=1}^{n}(X_i - \bar{X})^2 + n(\bar{X} - \mu)^2.$$

Dividing throughout by σ^2 and substituting $(n-1)\hat{S}^2$ from Eq. (5.3.11) for the middle term, we find

$$\frac{\sum_{i=1}^{n}(X_i - \mu)^2}{\sigma^2} = \frac{(n-1)\hat{S}^2}{\sigma^2} + \frac{(\bar{X} - \mu)^2}{\sigma^2/n}. \tag{5.3.13}$$

Following Eq. (5.3.9), the first term of Eq. (5.3.13) is distributed as χ_n^2; likewise, and with reference to the Central Limit Theorem (see Appendix A.6), the last term is distributed as χ_1^2. Therefore, taking account of the additive property of the gamma and chi-squared variates, we can say

$$(n-1)\frac{\hat{S}^2}{\sigma^2}$$

has a χ_{n-1}^2 distribution, that is, with $\nu = n - 1$ degrees of freedom.

This important result is used in finding confidence limits for the variance σ^2 of a normal population. The chi-squared cdf is given in Table C.3 of Appendix C, for various values of the degrees of freedom ν. For example, the $(1-\alpha)$ percent two-sided confidence interval for σ^2 is found from

$$\Pr\left[\chi_{n-1,1-\alpha/2}^2 \leq \frac{(n-1)\hat{S}^2}{\sigma^2} \leq \chi_{n-1,\alpha/2}^2\right] = 1 - \alpha. \tag{5.3.14a}$$

This probability is represented by the area between the two vertical lines of Fig. 5.3.5*a*, which shows the pdf.

Equation (5.3.14*a*) corresponds with Eq. (5.3.4) which is the basis for the confidence interval for the mean μ of a normal population. The chi-squared values in Eq. (5.3.14*a*) are, from left to right, the values that a χ_{n-1}^2 variate exceeds with probabilities $(1-\alpha/2)$ and $\alpha/2$, respectively. Let us take reciprocals of the three terms on the left-hand side of Eq. (5.3.14*a*), and therefore reverse the directions of the inequalities. Hence, after multiplying the three terms by $(n-1)\hat{S}^2$, we obtain the following probability which corresponds with Eq. (5.3.1*a*):

$$\Pr\left[\frac{(n-1)\hat{S}^2}{\chi_{n-1,\alpha/2}^2} \leq \sigma^2 \leq \frac{(n-1)\hat{S}^2}{\chi_{n-1,1-\alpha/2}^2}\right] = 1 - \alpha. \tag{5.3.14b}$$

This is used to set the *equal-tails* confidence interval for the variance σ^2. See Fig. 5.3.5*a* and 5.3.5*b*. We can also obtain a shorter interval with unequal tails. However, this needs a numerical solution and is therefore not practicable.

Similarly, we can set lower and upper confidence limits for the variance σ^2 as in Eq. (5.3.1*b*) and (5.3.1*c*), respectively.

Definition and properties: Let $\hat{S}^2$ be the variance of a random sample of size n drawn from a normal distribution with unknown variance. The $100(1-\alpha)$ percent equi-tailed two-sided confidence interval for the population variance σ^2 is as follows:

$$\left(\frac{(n-1)\hat{S}^2}{\chi_{n-1,\alpha/2}^2}, \frac{(n-1)\hat{S}^2}{\chi_{n-1,1-\alpha/2}^2}\right),$$

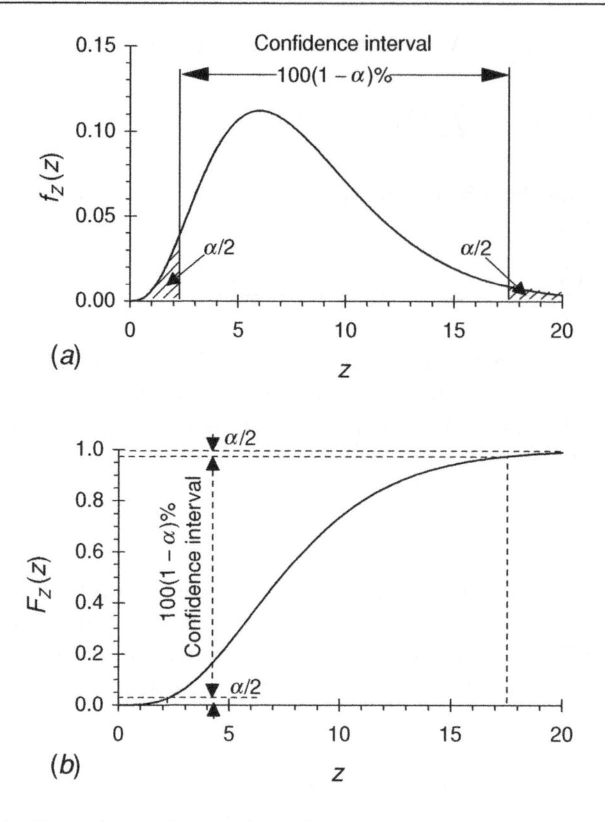

Fig. 5.3.5 Equal-tails confidence interval for variance: (*a*) pdf and (*b*) cdf of chi-squared distribution.

where $\chi^2_{n-1,\alpha/2}$ and $\chi^2_{n-1,1-\alpha/2}$ are the values that a χ^2_{n-1} variate exceeds with probabilities $\alpha/2$ and $(1-\alpha/2)$, respectively.

The corresponding one-sided upper confidence limit for σ^2 is defined as

$$\frac{(n-1)\hat{S}^2}{\chi^2_{n-1,1-\alpha}}.$$

Example 5.13. Upper 99% confidence limit for the standard deviation of compressive strengths of concrete test cubes. In the data of Table 1.2.2, we noted that the compressive strengths of 40 test cubes have an estimated standard deviation of 5.02 N/mm^2. This comes from an assumed normal population as justified in Example 5.37. As in Eq. (5.3.14*b*) but corresponding to Eq. (5.3.1*c*), we can obtain a one-sided upper 99% confidence limit for the population variance as follows:

$$\Pr\left[\sigma^2 \leq \frac{(n-1)\hat{S}^2}{\chi^2_{n-1,1-\alpha}}\right] = 1-\alpha.$$

Hence from an approximate interpolation from Table C.3 for $\alpha = 0.01$, the upper confidence limit for σ^2 is found from

$$\Pr\left[\sigma^2 \leq \frac{39 \times 5.02^2}{21.5}\right] = 0.99.$$

As confidence limits for the standard deviation, we can take the square roots of the corresponding limits of the variance because of the monotonic relationship between the variance and the standard deviation. Hence the 99% upper confidence limit for σ is 6.76 N/mm^2.

5.3.6 Summary of Section 5.3

Confidence intervals are obtained in this section for the estimation of parameters of different types under several conditions. In setting a random interval that at a preassigned level of probability includes a fixed but unknown population parameter, the long-run frequency approach to probability is adopted. Some situations require two-sided confidence intervals whereas in other cases one-sided confidence limits can be more appropriate. Three other statistical intervals are discussed elsewhere in the book. In Sections 6.1 and 6.2, we determine a prediction interval that will contain at a given level of confidence a future observation of a population. The notion of tolerance limits (as applied to a proportion p of a population) is illustrated in Subsection 8.1.3. In Subsection 9.3.1, we apply prediction intervals in reliability assessments.

The use of three important sampling distributions, the normal, Student's t, and chi-squared are shown in this section. The F distribution is introduced in the next section.

5.4 HYPOTHESIS TESTING

A second major area of statistical inference is the testing of hypotheses. This is closely related to the determination of confidence limits discussed in Section 5.3. Either subject could have been considered initially. What we are examining concerns the parameters or form of the probability distribution that yields the observations. This involves making a declaration or statement called a *hypothesis* about a population. It should be noted that the statement is not about the available sample. The discussion is focused on certain assumptions (regarding, for instance, the mean of the population) without initially making any commitment on assuming them. Acceptance of the hypothesis is on the basis of a statistical test. The consequent action and decision-making are called hypothesis testing.

From a philosophical viewpoint, our conception of the actual state of nature may not be correct. Besides, the testing procedure will have its shortcomings. However, we have a sample of data that represents the physical reality, and although our initial hypothesis may be somewhat tentative, we can revise the concept each time new information becomes available and thus come closer to the true state of nature.

The approach we follow is akin to the verification of some of the scientific hypotheses of scientists and engineers. We cannot, however, test statistically whether, for instance, a highway will be safe under the influence of an earthquake that exceeds a particular magnitude, except in a rare case in which there are sufficient data to test the hypothesis. On the contrary, it is usually feasible to ascertain whether a new method of road surfacing increases the lifetime of a highway, or whether a procedure for treating wastewater makes a change in the quality of the effluent, as measured, for example, by the biochemical oxygen demand discussed in Example 1.30. As another example, there may be reason to believe that the distribution of high flows in a river has changed on account of climatic effects or because the flow regime has been altered. All of these concern random variables, observable in sufficient numbers, as opposed to a single event.

In each of these cases just mentioned, and similar ones, a random sample is taken from the population and statistical hypotheses, called null and alternative, are declared. Then a statistical test is made. If the observations do not support the model or theory postulated, the null hypothesis is rejected in favor of the alternative one, which may be considered to be true. However, if the observations are in agreement, then the null hypothesis is not rejected. This does not necessarily mean that it is accepted. It suggests that there is insufficient evidence against the null hypothesis in favor of the alternative one.

Associated with the decision is a *level of significance*, α. This is complementary to the probability introduced in Eq. (5.3.1a) for setting confidence limits. An engineer may question whether there has been a significant change, for instance, in a mean rate of traffic flow on a highway or an average strength of a material. On the contrary, such a variation may sometimes be due to sampling differences without any change in the mean or other aspect of the population. It is then said to be *not significant.*

It is important to note that the initial assumption of a significance level removes any subjectivity in our decision-making so that two or more investigators will reach the same conclusion. Hypothesis testing thus concerns procedures for rejecting a statement or not rejecting it and the chances of making incorrect decisions of either kind. It also involves the use of a particular function of the sample measurements. If we assume that the observations come from a normal or other specified population, then the test is called *parametric. Nonparametric tests*, which are discussed in the next section, are not based on such assumptions. Furthermore, in this section we deal only with hypotheses concerning parameters of a distribution. Those hypotheses used in testing whether a random variable has a particular distribution will be examined in Section 5.5.

5.4.1 Procedure for testing

As just outlined, hypothesis testing concerns one or more parameters and also the related probability distribution. The basic steps, which are common to innumerable tests of this type, are as listed here. They involve an engineer's assumed model and whether the available observations provide any contradictory evidence. In testing, two different hypotheses are compared.

(1) The first step is to declare a *null hypothesis*, H_0. This is the hypothesis to be tested. It assumes that the observed results are entirely due to chance. For example, we may wish to verify whether the initial mean μ_1 of the annual maximum flows in a river at a point of observation is indeed the same as a projected new mean μ_2 consequent to changes in the flow regime that may have proved sufficient for a change in the population mean. If our null hypothesis is true, any observed difference in means is merely the consequence of sampling variation from the same population. That is, the difference is attributed in full to differences in random sampling. The hypothesis is expressed as

$$H_0: \mu_1 - \mu_2 = 0.$$

(2) In the next step an *alternative hypothesis* H_1 is declared. The term may be confusing, because this is what we really wish to test. In our case it is

$$H_1: \mu_1 - \mu_2 \neq 0.$$

(3) The third step is to determine or specify a *test statistic*. In the example cited, it is based on the difference between the respective observed means $\bar{X}_1$ and $\bar{X}_2$.

(4) Then we need to know the distribution of the test statistic, through sampling theory as introduced in Section 5.3 and discussed further shortly. The sampling distribution depends also, as noted in Section 5.3, on the distribution to which the observations belong. Therefore, we may need to make an assumption regarding the underlying distribution.

(5) As a fifth step, we must define a *rejection region*, also called a *critical region*, for the test statistic. For this definition it is necessary to preassign a level of significance α as defined shortly.

(6) Finally, we use the observed data to verify whether the computed value of the test statistic is within or outside the rejection region. If this is within the rejection region, we say that the difference (or whatever is tested) is *significant at the* 100α *percent level.*

Note that in the cited example (of annual maximum flows in a river subject to changes in the flow regime) with the foregoing steps, we are testing whether one mean is equal to another, and so our critical region will cover both tails of the sampling distribution. Accordingly, we call the procedure *a two-tailed test.* On the other hand if we were testing whether one mean is greater than the other, then a *one-tailed test* is required because the critical region will cover only one of the tails.

If the test statistic and rejection region are defined as T and R, respectively, the probability of rejecting the null hypothesis H_0 is given by

$$\Pr[T \in R \mid H_0] = \alpha. \tag{5.4.1}$$

The foregoing notation merely implies that H_0 is true. The probability α provides the link between the confidence intervals of Section 5.3 and hypothesis testing.

Rejection is the same as stating that the test statistic is statistically significant. That is, if the engineer notes that the observed difference can occur by chance less than, say, once in a hundred such tests (with $\alpha = 0.01$), then the results are treated as statistically significant. In other words, this should be a convincing demonstration, beyond a reasonable doubt that the null hypothesis is false.

Quite often, failure to reject leads to further verification and not to immediate acceptance, as already discussed; so the opposite test result may thus simply mean that the hypothesis is not rejected. Because some engineers do not like this kind of ambiguity, we shall continue to use the term "acceptance" to signify that a hypothesis has not been rejected. We now discuss two types of possible errors arising out of these tests.

If a hypothesis is rejected when it should be accepted, because the null hypothesis H_0 *is true, the error we make is of Type I.* The probability of doing so is equal to the level of significance α as defined by Eq. (5.4.1). Thus,

$$\Pr[\text{Type I error}] = \Pr[T \in R \mid H_0] = \alpha. \tag{5.4.2}$$

Alternatively, we may fail to reject the null hypothesis H_0 *when it is not true and thus make what is called a Type II error.* If the acceptance region of the test statistic is denoted A, which does not of course have anything in common with the rejection region R, and A and R taken together comprise the parameter space Ω for T,

$$\Pr[\text{Type II error}] = \Pr[T \in A \mid H_1] = \beta. \tag{5.4.3}$$

In either of these cases we are wrong in our judgment. Now consider the two cases in which we are right. *In the first case, when we correctly fail to reject the null hypothesis, the probability is complementary to that of making a Type I error.* This is the same probability associated with the confidence interval in Eq. (5.3.1*a*). That is,

$$\Pr[T \in A \mid H_0] = 1 - \alpha. \tag{5.4.4}$$

The probability of correctly rejecting the null hypothesis H_0 *when it is not true is the complement of that given by Eq. (5.4.3),*

$$\Pr[T \in R \mid H_1] = 1 - \beta. \tag{5.4.5}$$

The complement of β is also called the *power* of the test of the null hypothesis H_0 versus the alternative H_1. This important criterion is used, for example, in determining a minimum sample size to restrict the aforementioned types of errors and in comparing two tests.

Because a decrease in one type of error increases the other error, we must design our decision rules so that errors are minimized, usually by reaching a compromise between the two types. The probability of risking a Type I error is the level of significance α of the test, as already stated. We can limit this by choosing α suitably. Quite often in practice, $\alpha = 0.05$ seems reasonable, and this corresponds to an incorrect rejection once in 20 times on average. If the consequences of a Type I error are more serious, we choose a smaller value such as $\alpha = 0.01$ or even $\alpha = 0.001$ or less. For minimizing β, the Type II error (that is, a wrong acceptance) the procedure is not straightforward, because β is conditioned on H_1 and depends on α, and the sample size n, *in addition to the true value of the parameter tested.* It is customary to draw graphs showing β or the power of the test $(1 - \beta)$, which is the chance of avoiding this type of error. These are called *operating characteristic or power function curves*, respectively, and are illustrated shortly.

Example 5.14. Tests for proportions using the binomial approximation to the normal distribution. Two types of plant are used to treat the sewage effluent from two similar areas of a city. Of 90 tests made on the output from plant X, 33 tests show that the pollution has been reduced significantly, whereas 44 tests out of 100 on the output from plant Y show that the pollution has been reduced to the same or lower levels. Are the effects of the plants in reducing pollution different?

It is generally reasonable to assume that the number of tests that show significant results is binomially distributed, with parameters represented by the numbers of tests and the population proportions of successful results. These are given by $(n_X = 90, p_X)$ and $(n_Y = 100, p_Y)$ for plants X and Y, respectively.

The null hypothesis H_0: $p_X = p_Y = p$.
The alternative hypothesis H_1: $p_X \neq p_Y$.
Level of significance: $\alpha = 0.05$.

Calculations: Because we do not specify which plant is more effective, if there is any significant difference between the two, a two-tailed test is called for. On the basis of the null hypothesis, we can obtain an estimate of p using data from both plants. This gives

$$\hat{p} = \frac{33 + 44}{90 + 100} = \frac{77}{190} = 0.405.$$

We then take the difference between the observed proportions $\hat{p}_X$ and $\hat{p}_Y$ as the test statistic, in order to examine the possible difference in effectiveness of the two plants:

$$\hat{p}_X - \hat{p}_Y = \frac{33}{90} - \frac{44}{100} = -0.073.$$

The variance of the difference between the observed proportions is

$$\text{Var}[\hat{p}_X - \hat{p}_Y] = \text{Var}[\hat{p}_X] + \text{Var}[\hat{p}_Y] = \frac{p_X(1 - p_X)}{n_X} + \frac{p_Y(1 - p_Y)}{n_Y},$$

using Eqs. (5.3.7) and (5.3.8). On the basis of the null hypothesis, we substitute the estimate of p, that is, 0.405, obtained above for p_X and p_Y. Hence,

$$\text{Var}[\hat{p}_X - \hat{p}_Y] = 0.405 \times 0.595 \left(\frac{1}{90} + \frac{1}{100}\right) = 0.005087.$$

On the basis of the null hypothesis and with reference to Subsection 4.2.6, $(\hat{p}_X - \hat{p}_Y)$ has an approximate N(0, 0.005087) distribution. Thus, the estimated standard normal score (that is the resulting variate) is

$$z = \frac{-0.073 - 0}{\sqrt{0.005087}} = -1.03.$$

At the 5% level of significance, this is within the acceptance region of −1.96 to +1.96, from Table C.1.

Decision: The null hypothesis is thus not rejected and the evidence is insufficient to show that there is a difference in effectiveness of the treatment plants.

Example 5.15. Change in the mean annual maximum flow with known standard deviation and prior mean. Annual maximum flows in the Pond Creek catchment area in the eastern United States are listed below in cubic meters per second for two periods of 12 years from 1945 to 1968. It is thought that changes in the flow regime during the middle of this period have modified the distribution of flows resulting in higher maximum flows. This is equivalent to stating that the mean flow has increased from the first period to the second period.

First period:	2000	1740	1460	2060	1530	1590	1690	1420	1330	607	1380	1660
Second period:	2290	2590	3260	2490	3080	2520	3360	8020	4310	4380	3220	4320

Null hypothesis H_0: $\mu_2 = \mu_1$, that is, the new mean is equal to the past mean, where the mean flow for the first and second periods are μ_1 and μ_2, respectively.

Alternative hypothesis H_1: $\mu_2 > \mu_1$, that is, the new mean is greater than the past mean.
Level of significance: $\alpha = 0.01$.

Calculations: During the first period, that is, from 1945 to 1956 the mean and standard deviation are estimated as 1539 and 372 m^3/s, respectively. We assume initially that these represent the population mean μ and standard deviation σ prior to 1957. We also assume as a first approximation that the standard deviation is the same before and after 1956. We use a one-tailed test (because we are testing whether the flow has increased).

Let $\bar{X}_2$ be the estimated new mean, *after 1956*, based on a sample size $n = 12$. If we assume that the annual maximum flows are normally distributed, also as a first approximation, the standard normal score is

$$Z = \frac{(\bar{X}_2 - \mu)}{(\sigma/\sqrt{n})}.$$

This becomes our test statistic. For $\alpha = 0.01$, our decision rule is

(1) Reject H_0 if the z score is greater than 2.326 [see Table C.1 for $\Phi(z) = 1 - \alpha = 0.99$].
(2) Otherwise do not reject H_0.

Substituting for μ, σ, and n, $z = (\bar{x}_2 - 1539)/(372/\sqrt{12} = (\bar{x}_2 - 1539)/107.4$. From the foregoing data, the mean for the second period of 12 years is $\bar{x}_2 = 3653$ m^3/s. The z score of 19.7 is far greater than the critical value of 2.326.

Decision: The null hypothesis H_0 is therefore rejected.

Example 5.16. Change in the mean annual maximum flow with known prior mean and unknown but constant standard deviation. In practice we do not know the standard deviation in Example 5.15. So our test will be based on the sample value and thus it is appropriate to apply the t distribution.

Null hypothesis H_0: $\mu_2 = \mu_1$, that is, the new mean is equal to the past mean.
Alternative hypothesis H_1: $\mu_2 > \mu_1$, that is, the new mean is greater than the past mean.
Level of significance: $\alpha = 0.01$.

Calculations: As before we assume (1) that the standard deviation is constant and (2) that the annual maximum flows are normally distributed. Thus the test statistic is

$$T = \frac{(\bar{X}_2 - \mu)}{(\sigma/\sqrt{n})}.$$

For $\alpha = 0.01$, our decision rule is

(1) Reject H_0 if the t score is greater than $t_{11,0.01} = 2.718$ [see Table C.2 for $F(t) = 1 - \alpha = 0.99$ and $v = n - 1 = 11$].
(2) Otherwise do not reject H_0.

The sample t score is 19.7, the same as the z score in Example 5.15, which is also far greater than the critical value $t_{11,0.01} = 2.718$.

Decision: Here too the null hypothesis H_0 is rejected.

5.4.1.1 Testing the difference between two means using known variances

If $\bar{X}_1$ and $\bar{X}_2$, say, are two estimated means, of μ_1 and μ_2 obtained from two small samples of sizes n_1 and n_2, respectively, and if the two populations are normal with known variances σ_1^2 and σ_2^2, then the random variable

$$Z = \frac{(\bar{X}_1 - \bar{X}_2) - (\mu_1 - \mu_2)}{\sigma_{\bar{X}_1 - \bar{X}_2}} \qquad (5.4.6)$$

has an $N(0, 1)$ distribution. The denominator of the variable is the standard error of the difference between two statistics as defined by Eq. (5.3.8*b*). In this case it is given by

$$\sigma_{\bar{X}_1 - \bar{X}_2} = \sqrt{\frac{\sigma_1^2}{n_1} + \frac{\sigma_2^2}{n_2}}. \qquad (5.4.7)$$

5.4.1.2 Testing the difference between two means when the variances are unknown but equal

Let the sample variances be $\hat{S}_1^2$ and $\hat{S}_2^2$. Because both are estimates of the constant variance σ^2, a pooled estimate can be formed by

$$\hat{S}_p^2 = \frac{(n_1 - 1)\hat{S}_1^2 + (n_2 - 1)\hat{S}_2^2}{n_1 + n_2 - 2}. \qquad (5.4.8)$$

Then the random variable

$$T = \frac{(\bar{X}_1 - \bar{X}_2) - (\mu_1 - \mu_2)}{\hat{S}_p\sqrt{1/n_1 + 1/n_2}}$$

$$= \frac{(\bar{X}_1 - \bar{X}_2) - (\mu_1 - \mu_2)}{\sqrt{(n_1 - 1)\hat{S}_1^2 + (n_2 - 1)\hat{S}_2^2}} \sqrt{\frac{(n_1 + n_2 - 2)n_1 n_2}{n_1 + n_2}} \qquad (5.4.9)$$

has the t distribution with $n_1 + n_2 - 2$ degrees of freedom. In the case of small samples, we may make the assumption that the population variances are equal when there are insufficient grounds to assume the contrary.

Example 5.17. Differences in the mean compressive strengths of concrete from two batches with assumed equal but unknown variances. Results of tests on concrete strengths made on two batches of concrete are summarized as:

Batch 1: $\bar{x}_1 = 60.34$ $\hat{s}_1 = 5.72$ N/mm^2 $n_1 = 12$.
Batch 2: $\bar{x}_2 = 54.23$ $\hat{s}_2 = 7.01$ N/mm^2 $n_2 = 10$.

Suppose we wish to test whether the strength of the first batch is significantly higher than that of the second.

Null hypothesis H_0: $\mu_1 = \mu_2$.
Alternative hypothesis H_1: $\mu_1 > \mu_2$.
Level of significance: $\alpha = 0.05$.

Calculations: From Eq. (5.4.9) on the basis of the null hypothesis,

$$t = \frac{60.34 - 54.23}{\sqrt{11 \times 5.72^2 + 9 \times 7.01^2}} \sqrt{\frac{20 \times 12 \times 10}{22}} = 2.25.$$

The degrees of freedom are $v = 12 + 10 - 2 = 20$. From Table C.2, $t_{20,0.05} = 1.725$ and $t_{20,0.01} = 2.528$.

Decision: Hence the mean strength of the first batch of concrete is higher at the 5% level of significance but not at the 1% level. Further tests are recommended.

When, as in the data sited in Example 5.17, the variances seem to be totally different, the assumptions related to Eq. (5.4.9) are not justified. However, some modification can be made to the test procedure. Tests on variances will follow in the next subsection.

5.4.1.3 Testing the difference between two means when the variances are unknown and unequal

There are many instances when it is not correct to assume that the variances corresponding to the means tested are equal. However when the variances are unequal, the statistic does not have a t distribution. This is called the *Behrens-Fisher Problem.* The sampling distribution is rather complicated (see Stuart and Ord, 1991, pp. 785–786). It is possible to find an approximation as indicated, for example, by Casella and Berger (2002, pp. 409–410) and Brownlee (1965, pp. 299–302). In the case of observations taken from normal populations with unknown and unequal variances, the statistic

$$T' = \frac{(\bar{X}_1 - \bar{X}_2) - (\mu_1 - \mu_2)}{\sqrt{\hat{S}_1^2/n_1 + \hat{S}_2^2/n_2}} \tag{5.4.10}$$

has an approximate t distribution with

$$v = \frac{\left[\hat{S}_1^2/n_1 + \hat{S}_2^2/n_2\right]^2}{\left(\hat{S}_1^2/n_1\right)^2/(n_1 - 1) + \left(\hat{S}_2^2/n_2\right)^2/(n_2 - 1)} \tag{5.4.11}$$

degrees of freedom.

Example 5.18. Change in the mean annual maximum flow with unknown and unequal variances. We return again to the data of Example 5.15. The two main parameters are μ_1 and μ_2.

Null hypothesis H_0: $\mu_1 = \mu_2$.
Alternative hypothesis H_1: $\mu_1 > \mu_2$.
Level of significance: $\alpha = 0.01$.

Calculations: We use a one-tailed test. From the data of Example 5.15,

$\bar{x}_1 = 1539$, $\hat{s}_1 = 372$, $\bar{x}_2 = 3653$, $\hat{s}_2 = 1563\ \text{m}^3/\text{s}$, $n_1 = n_2 = 12$.

Hence from Eqs. (5.4.10) and (5.4.11),

$$v = \frac{[(372^2/12) + (1563^2/12)]^2}{[(372^2/12)^2/11] + [(1563^2/12)^2/11]} \approx 12.$$

$$T' = \frac{(3653 - 1539)}{\sqrt{(372^2/12) + (1563^2/12)}} = 4.56.$$

From Table C.2, $t_{12,0.01} = 2.681$.

Decision: Hence the null hypothesis is rejected (as in Example 5.15).

5.4.2 Probabilities of Type I and Type II errors and the power function

As already noted, the designer chooses the Type I error probability α, but the Type II error probability depends on α, the sample size, and the true value of the parameter tested. Consider the two-tailed test used for testing the mean of a $N(\mu, \sigma^2)$ population:

Null hypothesis H_0: $\mu = \mu_0$.
Alternative hypothesis H_1: $\mu \neq \mu_0$.

Suppose that it is correct to reject the null hypothesis because the true value of μ is $\mu_0 + c$, where c is a constant. It is also correct (apparently) to accept the alternative hypothesis with standardized test statistic

$$Z \sim (c\sqrt{n}/\sigma, 1)$$

where n is the size of a test sample; that is, the mean has been displaced by $c/(\sigma/\sqrt{n})$ standardized units. For example, Fig. 5.4.1 is a graph of the pdfs of Z for the null hypothesis with zero mean and the alternative hypothesis with the mean displaced by three standardized units; the displacement is positive here but it can also be negative.

A Type II error [defined initially by Eq. (5.4.3)] is made, that is, when H_1 is true, only if $-z_{\alpha/2} \leq Z \leq z_{\alpha/2}$, where $-z_{\alpha/2}$ and $z_{\alpha/2}$ are the values which a standard normal deviate exceeds with probabilities of $(1 - \alpha/2)$ and $\alpha/2$, respectively. The probability β of a Type II error is determined after adjusting the specified limits by the mean of the Z variate. Accordingly, it is given by

$$\beta = \Phi\left(z_{\alpha/2} - \frac{c\sqrt{n}}{\sigma}\right) - \Phi\left(-z_{\alpha/2} - \frac{c\sqrt{n}}{\sigma}\right)$$

where $\Phi(\cdot)$ is the cdf of the standard normal distribution. See Fig. 5.4.1.

Thus we see that the probability β of a Type II error is dependent on α, n, and c/σ. Curves representing β are called *characteristic curves*. These are shown in Fig. 5.4.2*a* for a two-tailed normal test with level of significance $\alpha = 0.05$, sample sizes n from 1 to 100 and $c/\sigma = 0.25, 0.50, 0.75$, and 1.0.

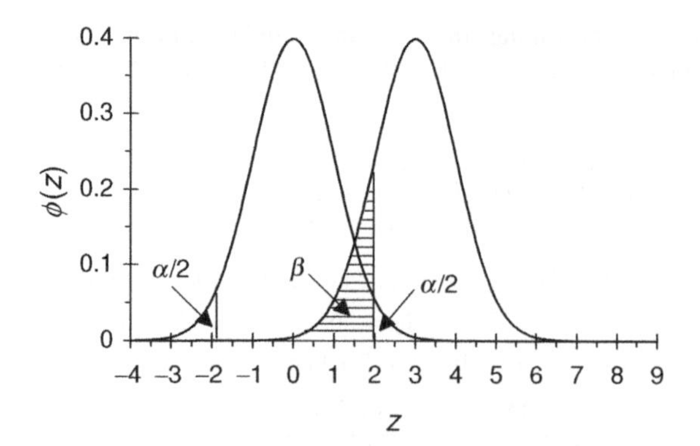

Fig. 5.4.1 Significance test on the sample mean. The normal (0, 1) pdf on the left is the assumed model in standardized form. Suppose the normal (3, 1) pdf on the right represents the true model. Then the Type I error is α and the Type II error is β. The test is applicable regardless of the sign of the shift in the mean. The magnitude of β changes with the absolute magnitude of the shift; α is invariant here.

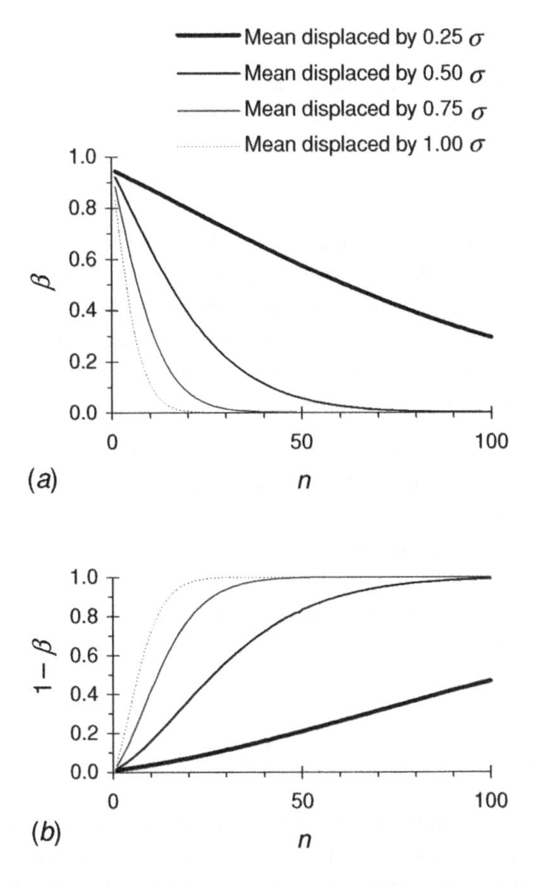

Fig. 5.4.2 (*a*) Operating characteristic curves with sample sizes n from 1 to 100 and absolute displacements of the mean from 0.25 to 1.00 σ for a two-sided normal test with a level of significance $\alpha = 0.05$. (*b*) Power curves with the sample sizes and absolute displacements but with a level of significance $\alpha = 0.01$.

From these curves we see that, for the same sample size n, β increases as c/σ decreases. That is, small differences in the mean are more difficult to detect and lead to a higher probability of incorrect acceptance. Also, as expected, there is an increase in β as the sample size n decreases.

The complement of β is the *power function.* As already stated it is the *probability of rejecting the null hypothesis when it is not true* (which is of course the right thing to do) in favor of the alternative hypothesis. It is equivalent to

$$\text{Power} = 1 - \beta = 1 - \left[\Phi\left(z_{\alpha/2} - \frac{c\sqrt{n}}{\sigma}\right) - \Phi\left(-z_{\alpha/2} - \frac{c\sqrt{n}}{\sigma}\right)\right].$$

Power curves are shown in Fig. 5.4.2*b* for a two-tailed normal test with level of significance $\alpha = 0.01$, sample sizes n from 1 to 100 and $c/\sigma = 0.25, 0.50, 0.75$, and 1.0. Clearly, the power increases with c/σ and sample size n.

The characteristic and power curves are complementary. These can be easily changed for different values of α. Similar procedures can be applied for a one-sided test in which the alternative hypothesis is H_1: $\mu > \mu_0$ or H_1: $\mu < \mu_0$.

5.4.3 Neyman-Pearson lemma

(1) As mentioned earlier, statistical hypotheses are more restrictive than those in scientific applications such as in particle physics. Here we are concerned with the behavior of observable random variables. For example, if a random sample is taken from a distribution with specific parameters, then a *simple hypothesis* is one that uniquely defines the distribution from which the sample is taken against an alternative hypothesis. The other type of hypothesis is called a *composite hypothesis.*
(2) A critical (rejection) region with power not less than that of any other region of the same size used in testing the null hypothesis H_0 against the alternative hypothesis H_1 is called a *best critical region* (bcr). Here H_0 and H_1 are simple hypotheses if we are choosing between two specified distributions. A test involving a bcr is called a *most powerful* (mp) test.

If our test statistics are devised such that the probability of a Type I error does not exceed a constant α, called the size of the test, then in effect we are applying the classical Neyman-Pearson theory, which is discussed briefly here. It is named after Jerzy Neyman and Egon Pearson. Accordingly, we keep the probability of a Type I error fixed and search for the test statistic that maximizes $(1 - \beta)$ where β is the probability of a Type II error. Suppose that our hypotheses concern a distribution with only one parameter θ. Let H_0 and H_1 denote $\theta = \theta_0$ and $\theta = \theta_1$ respectively. Then to construct a bcr we consider the corresponding likelihoods

$$L_0 = \prod_{i=1}^{n} f(x_i \mid \theta_0) \quad \text{and} \quad L_1 = \prod_{i=1}^{n} f(x_i \mid \theta_1).$$

The ratio L_0/L_1 should be low for points within the critical region R, thus minimizing the Type I error α and maximizing the power $(1 - \beta)$. It should be high, on the other hand, for points in the (complementary) acceptance region A. Thus the probability of a correct decision is high under H_0. In this approach one sees again the fundamental trade-off in hypothesis testing. We now outline a procedure for application.

Definition: Neyman-Pearson lemma. Let R be a critical (rejection) region of size α, when testing the null hypothesis H_0: $\theta = \theta_0$ against the alternative hypothesis H_1: $\theta = \theta_1$. We define a constant k_α such that in the region R, $(L_0/L_1) \le k_\alpha$, and in the acceptance region A, $L_0/L_1 > k_\alpha$, where $L_0 = \prod_{i=1}^{n} f(x_i \mid \theta_0)$ and $L_1 = \prod_{i=1}^{n} f(x_i \mid \theta_1)$. Then R is the best critical region of size α. Such a test involving a bcr is called a *most powerful (mp) test.*

Example 5.19. A simple likelihood test on a normal population with unit variance. Let $X_1, X_2, \ldots, X_n$ be a random sample from an $N(\mu, 1)$ population. Suppose our objective is to test the null hypothesis H_0: $\mu = \mu_0$ against the alternative hypothesis H_1: $\mu = \mu_1 > \mu_0$. The ratio of the likelihoods can be used to find the bcr of size α.

The ratio of the likelihoods is given by

$$\frac{L_0}{L_1} = \frac{(1/\sqrt{2\pi})^n \exp[-(1/2)\sum_{i=1}^{n}(X_i - \mu_0)^2]}{(1/\sqrt{2\pi})^n \exp[-(1/2)\sum_{i=1}^{n}(X_i - \mu_1)^2]}$$

$$= \exp\left[\left(\sum_{i=1}^{n} X_i\right)(\mu_0 - \mu_1) + \frac{n}{2}\left(\mu_1^2 - \mu_0^2\right)\right].$$

(The likelihoods are as in Example 3.23 but with unit variance.)

We define a constant k_α, which depends on α, as a limiting value of the likelihood ratio, and a critical region R of size α within which

$$\exp\left[\left(\sum_{i=1}^{n} X_i\right)(\mu_0 - \mu_1) + \frac{n}{2}\left(\mu_1^2 - \mu_0^2\right)\right] \le k_\alpha.$$

Outside R,

$$\exp\left[\left(\sum_{i=1}^{n} X_i\right)(\mu_0 - \mu_1) + \frac{n}{2}\left(\mu_1^2 - \mu_0^2\right)\right] \ge k_\alpha.$$

Taking logarithms and simplifying, we find for H_1: $\mu = \mu_1 > \mu_0$, the bcr is defined for

$$\bar{X} \ge \frac{1}{2}(\mu_0 + \mu_1) - \frac{\ln(k_\alpha)}{[n(\mu_1 - \mu_0)]} = c_\alpha.$$

In applications we determine the constant c_α using α.

If the alternative hypothesis is stipulated as H_1: $\mu = \mu_1 < \mu_0$, the bcr is then defined for

$$\bar{X} \le \frac{1}{2}(\mu_0 + \mu_1) + \frac{\ln(k_\alpha)}{[n(\mu_0 - \mu_1)]}.$$

The foregoing procedure is applicable to simple hypotheses. A more general method can be applied to composite hypotheses in which more than one parameter is considered and where nuisance parameters with unknown values, which are not crucial for purposes of inference, are present. These tests are termed likelihood ratio tests but are not necessarily the most powerful under the circumstances.[6]

5.4.4 Tests of hypotheses involving the variance

In Eq. (5.3.14*a*) we noted that a quantity such as $\sum_{i=1}^{n}(X_i - \bar{X})^2/\sigma^2$ has a chi-squared distribution with $(n - 1)$ degrees of freedom. However, if the true mean μ is known, this replaces $\bar{X}$ in the preceding sum of squares and the degrees of freedom are n. Thus with the knowledge of the sampling distribution, a hypothesis test can be made on the variance.

Example 5.20. A significance test on the change in variance. From a long series of annual river flows, the variance is found to be 49 units. This can be treated as the population variance. However, a new sample of 25 years gives a value $\hat{s}^2 = 81$ units.

Null hypothesis H_0: $\sigma^2 = 49$.
Alternative hypothesis H_1: $\sigma^2 > 49$.
Level of significance: $\alpha = 0.05$.

Calculations: The quantity $n\hat{s}^2/\sigma^2$ has a chi-squared distribution with n degrees of freedom. From Table C.3, $\chi^2_{25,0.05} = 37.7$ and $n\hat{s}^2/\sigma^2 = 25 \times 81/49 = 41.3$.

Decision: Thus the null hypothesis H_0 is rejected.

[6] Reference may be made to one of the books cited at the end of this chapter such as Stuart and Ord (1991), which also have proofs of the Neyman-Pearson and other lemmas. The classical or frequentist approach is presented here. We shall return to this subject in Chapter 10 in relation to the alternative Bayesian method, in which the prior probabilities of the occurrences of H_0 and H_1 are assumed to be known.

5.4.5 The F distribution and its use

Consider the random variable formed by the ratio of two independent chi-squared random variables divided by the corresponding degrees of freedom m and n,

$$F = \frac{\chi_1^2/m}{\chi_2^2/n}. \tag{5.4.12}$$

This is sometimes known as the *variance ratio*. It has the Snedecor's F distribution (named after R. A. Fisher) with pdf,

$$h_F(f) = \frac{\Gamma[(m+n)/2]}{\Gamma(m/2)\Gamma(n/2)}\left(\frac{m}{n}\right)^{m/2}\frac{f^{(m/2)-1}}{[1+(m/n)f]^{(m+n)/2}}, \quad \text{for } 0 < f < \infty, \tag{5.4.13}$$

with m and n—numerator and denominator, respectively—degrees of freedom. Its derivation, which is similar to that of Student's t, is shown in Appendix A.8.

Recall from Eq. (5.3.13) that $(v/\sigma^2)\hat{S}^2$ has a χ_v^2 distribution (that is, with v degrees of freedom) where $\hat{S}^2$ is the estimated variance of a sample of size n, using $v = n - 1$ as the divisor [as in Eq. (5.3.11)]. If two populations are normal and independent samples of sizes $(m+1)$ and $(n+1)$ are drawn from them, then

$$F = \frac{\hat{S}_1^2/\sigma_1^2}{\hat{S}_2^2/\sigma_2^2} \tag{5.4.14}$$

also has an F distribution with m and n—numerator and denominator, respectively—degrees of freedom.

This distribution has practical use in significance testing of variances. In Table C.4, F values are given for a range of numerator and denominator degrees of freedom, m and n, respectively, and corresponding to values of cdf from 0.9 to 0.995. The tabulation gives

$$G(F) = \Pr[F \le F_{m,n;\alpha}] = 1 - \alpha,$$

for α ranging from 0.005 to 0.1. To find a corresponding value with complementary probability, that is, for α ranging from 0.9 to 0.995 we use the property which follows from Eq. (5.4.12) that if $U \sim F_{m,n}$, then $V = 1/U \sim F_{n,m}$. If we let ξ_p and ξ'_p be the pth quantiles of U and V, respectively,

$$p = \Pr[U \le \xi_p] = \Pr\left[\frac{1}{U} \ge \frac{1}{\xi_p}\right] = \Pr\left[V \ge \frac{1}{\xi_p}\right] = 1 - \Pr\left[V \le \frac{1}{\xi_p}\right].$$

However, we also know that

$$1 - p = \Pr[V \le \xi'_{1-p}].$$

Thus,

$$\xi'_{1-p} = \frac{1}{\xi_p}.$$

So to find a quantile with probability p in the range 0.005–0.1, we refer to Table C.4 and find the one with probability $(1-p)$ *with reversed degrees of freedom and then take its reciprocal.*

Graphs of two F pdfs are shown in Fig. 5.4.3.

Example 5.21. Comparing variances of test results on an effluent. A constituent in an effluent is analyzed seven and nine times through procedures X and Y, respectively. Test results have standard deviations of 1.9 and 0.8 mg/L, respectively, by the two procedures. It

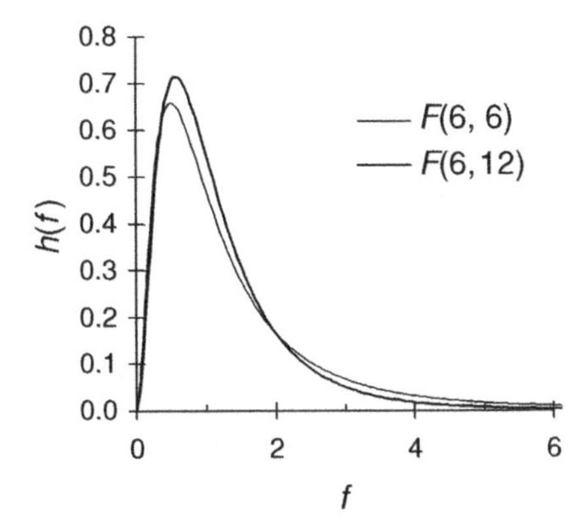

Fig. 5.4.3 Two examples of the F density function. The numerator and denominator degrees of freedom are given within parentheses.

is important to know whether the second method is more precise than the first (that is, with less variance in the outcome).

Null hypothesis $H_0: \sigma_1^2 = \sigma_2^2$.
Alternative hypothesis $H_1: \sigma_1^2 > \sigma_2^2$.
Level of significance: $\alpha = 0.05$.

Calculations: We use a one-tailed test. The data are as follows:

$$n_1 = 7, \quad n_2 = 9, \quad s_1^2 = 1.9^2, \quad s_2^2 = 0.8^2.$$

We apply Eq. (5.4.14) under the null hypothesis to give $F = 1.9^2/0.8^2 = 5.64$. From Table C.4, after some extrapolation, $F_{6,8,0.05} = 3.58$; note also that $F_{6,8,0.01} = 6.37$.

Decision: We reject the null hypothesis. Thus we may conclude that the second method is more precise at the 5% level of significance. We note that there is no difference between the methods at the 1% level of significance. Investigations of the differences should therefore continue.

5.4.6 Summary of Section 5.4

In Section 3.2 we discussed methods of estimating parameters. Point estimates were considered there, whereas in Section 5.3 interval estimates were investigated. In this section another important aspect of the inferential procedure which concerns statistical hypotheses with respect to the parameters of a distribution has been examined. A summary of significance tests is given in Table 5.4.1.

The testing of hypotheses hinges on assumptions about the probabilities associated with the null hypothesis. The method is obviously not infallible, so we must proceed with care. When judiciously applied, however, it is one of the best ways of differentiating chance effects from real differences. Type I and Type II errors, which are the chances of incorrectly rejecting a hypothesis or incorrectly accepting it, respectively, and the power of a test, which is the complement of the Type II error, are emphasized in this section. Considerations of errors and the power of a test give validity to the inferential procedure. However, we have made distributional assumptions in hypothesis testing that are not always tenable. Nonparametric methods, discussed in the next section, do not have this limitation.

Table 5.4.1 Summary[a] of significance tests of the mean and variance of normal variates, X_i

Null hypothesis, H_0	Other parameters	Estimate of test statistic	Distribution of test statistic
$\mu = \mu_0$	Known σ^2	$\frac{\bar{X}-\mu_0}{\sigma/\sqrt{n}}$	$N(0, 1)$
$\mu = \mu_0$	Unknown σ^2	$\frac{\bar{X}-\mu_0}{\hat{S}/\sqrt{n}}$	t_{n-1}
$\sigma^2 = \sigma_0^2$	Known μ	$\sum_{i=1}^{n}(X_i - \mu)^2/\sigma_0^2$	χ_n^2
$\sigma^2 = \sigma_0^2$	Unknown μ	$\sum_{i=1}^{n}(X_i - \bar{X})^2/\sigma_0^2$	χ_{n-1}^2
$\mu_1 = \mu_2$	Known $\sigma_1^2 = \sigma_2^2 = \sigma^2$	$\frac{(\bar{X}_1-\bar{X}_2)}{\sigma\sqrt{1/n_1+1/n_2}}$	$N(0, 1)$
$\mu_1 = \mu_2$	Known σ_1^2, σ_2^2	$\frac{(\bar{X}_1-\bar{X}_2)}{\sqrt{\sigma_1^2/n_1+\sigma_2^2/n_2}}$	$N(0, 1)$
$\mu_1 = \mu_2$	Unknown $\sigma_1^2 = \sigma_2^2 = \sigma^2$	$\frac{(\bar{X}_1-\bar{X}_2)}{\hat{S}_p\sqrt{1/n_1+1/n_2}}$ Eq. (5.4.8) gives $\hat{S}_p$	$t_{n_1+n_2-2}$
$\mu_1 = \mu_2$	Unknown $\sigma_1^2 \neq \sigma_2^2$	Approximation by Eq. (5.4.10)	See Eq. (5.4.11)
$\frac{\sigma_1^2}{\sigma_2^2} = 1$	Known μ_1, μ_2	$\frac{\sum_{i=1}^{m}(X_{i,1} - \mu_1)^2/(m-1)}{\sum_{i=1}^{n}(X_{i,2} - \mu_2)^2/(n-1)}$	$F_{m,n}$
$\frac{\sigma_1^2}{\sigma_2^2} = 1$	Unknown μ_1, μ_2	$\frac{\sum_{i=1}^{m}(X_{i,1} - \bar{X}_1)^2/(m-1)}{\sum_{i=1}^{n}(X_{i,2} - \bar{X}_2)^2/(n-1)}$	$F_{m-1,n-1}$

[a] Sample mean $\bar{X}$ and variance $\hat{S}^2$ estimated from n values.

5.5 NONPARAMETRIC METHODS

As noted in Chapter 4, the normal distribution plays a central role in the theory of statistics and probability and its use is supported by the Central Limit Theorem. However, there are many instances when one is uncertain of the true form of the probability distribution of the variable involved and the available sample is small, say, less than 30. It is therefore desirable to have methods that can be applied regardless of distribution. Such procedures are *called distribution-free methods*. For instance, one may set confidence limits (on a parameter) which do not depend on the form of the underlying distribution. Furthermore, there are experiments in which the measurements cannot be quantified in numerical terms and are expressed, for example, as a success or failure, a form of goodness, a color, and so on.

The term *nonparametric methods* is generally used to describe the wide-ranging procedures used in analyzing all the aforementioned types of data. In these methods we are not concerned with the parameters of populations. In contrast, parametric methods pertain to those techniques where the distribution is known but the values of the parameters are not specified. Furthermore, nonparametric methods do not require much computational effort and are thus easier to apply. Also, small samples can be used. Although the methods are less powerful in detecting Type I and Type II errors, the loss of power is sometimes minimal and is outweighed because the assumptions are less restrictive. Initially, hypotheses pertaining to parameters, such as the means, in one or more samples will be examined. Tests of randomness and association are then considered.

5.5.1 Sign test applied to the median

The nonparametric sign test is an alternative to the tests on the mean using the normal or the t distribution as sampling distributions, as discussed in Section 4.4. *The one sample sign test is applicable when we sample any continuous distribution that has a median, M.* We test the hypothesis that the median is equal to a specific value M_0.

We commence the application of the sign test by subtracting M_0 from each of the observations and noting the signs of the differences. Sometimes an observation may be equal to the specified median, in which case we discard that observation. Let n be the number of nonzero differences. We then count the number of positive differences, say, k out of a total of, say, n nonzero differences. If $M = M_0$, the expected number of positive differences is equal to the expected number of negative differences. The null hypothesis to be tested is

$$H_0: M = M_0.$$

The alternative hypothesis is *one* of the following:

(1) $H_1: M > M_0$.
(2) $H_1: M < M_0$.
(3) $H_1: M \neq M_0$.

Under the null hypothesis, and on account of sampling variations, the observed plus and minus signs are binomially $(n, 1/2)$ distributed. We use the normal approximation to the binomial except when sample sizes are very small, say, less than 10. The mean and variance of a binomial (n, p) variable are np and $np(1-p)$ as noted from Eq. (4.1.5a) and (4.1.5b), respectively; these are, respectively, $n/2$ and $n/4$ for $p = 1/2$. Also, because of the discrete approximation, a continuity correction is necessary.

Having chosen a level of significance, α, we find z_α from Table C.1 for the normal distribution, noting that for the alternative hypotheses 1 and 2 a one-tailed test is required, whereas a two-tailed test is used for 3. Thus the critical region is defined.

Definition and properties: The one-sample sign test applied to the median. Let n be the number of nonzero differences between the sample values and the assumed median, which is true under the null hypothesis, and let k be the number of positive differences. Then by the normal approximation to the binomial, the following variate has a standard normal distribution under the null hypothesis:

$$z = \frac{(k + 1/2) - n/2}{(\sqrt{n})/2}, \quad \text{if } k < \frac{n}{2}, \tag{5.5.1a}$$

$$z = \frac{(k - 1/2) - n/2}{(\sqrt{n})/2}, \quad \text{if } k > \frac{n}{2}. \tag{5.5.1b}$$

(The addition and subtraction of $1/2$ from k are continuity corrections, introduced in Chapter 4.) The z_α value from Table C.1 is compared with the z score calculated from one of the preceding equations to decide whether the null hypothesis should be rejected at a level of significance α.

Example 5.22. Sign test on the median of the compressive strengths of concrete. We apply the test to the 40 compressive strengths listed in Table E.1.2.

Null hypothesis H_0: $M = 61$ N/mm^2.
Alternate hypothesis H_1: $M < 61$ N/mm^2.
Level of significance: $\alpha = 0.05$.

Calculations: Replacing each value exceeding 61 N/mm^2 with a plus sign and each value less than 61 N/mm^2 by a minus sign, we get

$$- - - - - + + - - - + + + + - + + + + - - - - - - - - - - + - - + - - - + + - - -$$

The number of positive differences $k = 14$.
The number of nonzero differences $n = 40$.

Because $k < n/2$,

$$z = \frac{(14 + 1/2) - 40/2}{(\sqrt{40})/2} = -1.74.$$

Using the left tail only, we find $z_\alpha = -1.64$.

Decision: The null hypothesis is rejected at the 5% level of significance.

Note that if the assumption that the concrete strengths are normally distributed is correct (and this seems justifiable, as noted previously, from empirical evidence in numerous case studies), then the more powerful parametric tests on the mean and other parameters are preferable.

The sign test can also be applied to *paired samples.* A plus or minus sign takes the place of each pair of sample values if the difference between the paired observations is positive or negative, respectively, in such a test. This method is analogous in a sense to the parametric two-sample t test on the mean, but here the null hypothesis is that the two populations which we sample are continuous and symmetrical and have equal means. The null hypothesis is therefore less binding in the assumptions.[7]

5.5.2 Wilcoxon signed-rank test for association of paired observations

In the sign test we make use of only the signs of the differences between the observations and the median or of the differences between the paired observations in the two-sample test. The Wilcoxon signed-rank test utilizes the magnitudes of the differences and is thus intuitively a more powerful test compared with the much simpler sign test. Both procedures can be used in testing differences in means.

In this test we note the magnitudes and the signs of the paired differences from two samples of observations and then rank the absolute values of the differences. Thus the smallest difference is assigned rank 1 and the largest difference gets the highest rank, which is equal to the number of pairs n with nonzero differences (with the signs disregarded in each case). If the absolute values of two or more of such differences are equal, we give each a rank equal to the mean of the ranks they would otherwise receive. The ranks attributed to positive and negative differences are then summed in separate columns. These sums are denoted T^+ and T^-, respectively. The null hypothesis is that the means of the two populations are equal. The alternate hypothesis can take one of three forms as already specified for the sign test.

For small sample sizes, the test statistic is $T = \min(T^+, T^-)$. Its quantiles are provided in special tables for different levels of significance.[8] If $T > T_0$, where T_0 is the critical quantile, the null hypothesis is rejected.

---

[7] See Problem 5.15.

[8] See, for example, Siegel and Castellan, Jr. (1988, pp. 332–334) or Conover (1998) or Gibbons and Chakraborti (2003).

Definition and properties: Wilcoxon signed-rank test. Under the null hypothesis we can use either T^+ or T^-, the sum of the ranks applied to the absolute values of the positive and negative differences, respectively, between the paired observations for the test statistic T. For large sample sizes n, T can be approximated by the standard normal variate

$$Z = \frac{T - \mu_T}{\sigma_T},$$

where

$$\mu_T = \frac{n(n+1)}{4} \quad (5.5.2a)$$

$$\sigma_T^2 = \frac{n(n+1)(2n+1)}{24}. \quad (5.5.2b)$$

(See Appendix A.9 for a proof.) The approximation is sufficiently accurate for $n \geq 15$. The test procedure is similar to that for the sign test.

Example 5.23. Comparing strengths of materials made by two different methods using the Wilcoxon signed-rank test. The following are the compressive strengths of a material in N/mm^2, manufactured by two different methods A and B.

A: 60.3 50.2 56.5 60.6 59.3 49.7 50.8 59.8 52.5 57.4 55.8 54.5 53.6 56.7 57.1
B: 56.0 56.2 55.1 59.2 62.3 54.5 56.5 57.1 56.2 56.1 58.5 63.5 58.2 48.9 53.0

We apply the signed-rank test to investigate whether method B is superior to method A as claimed, assuming that the values represent independent random variables.

Null hypothesis H_0: $\mu_A = \mu_B$.
Alternate hypothesis H_1: $\mu_B > \mu_A$.
Level of significance: $\alpha = 0.05$.

Calculations: Differences between corresponding pairs, B − A, are as follows:

−4.3 6.0 −1.4 −1.4 3.0 4.8 5.7 −2.7 3.7 −1.3 2.7 9.0 4.6 −7.8 −4.1.

Ranking the absolute differences in increasing order, we get the ranks of the positive differences:

13 6 11 12 7 4.5 15 10.

Hence $T^+ = 78.5$.

$$\mu_T = \frac{15 \times 16}{4} = 60 \quad \text{and} \quad \sigma_T^2 = \frac{15 \times 16 \times 31}{24} = 310.$$

From which,

$$z = \frac{78.5 - 60}{\sqrt{310}} = 1.05.$$

Decision: Because z is less than $z_\alpha = 1.645$, the null hypothesis is not rejected at the $\alpha = 5\%$ level of significance.

[Note that the ranks of the negative differences are:

9 2.5 2.5 4.5 1 14 8

Hence $T^- = 41.5$; because $T^+ + T^- = n(n+1)/2$ this gives $z = -1.05$ and the result corresponds with that for T^+.]

A nonparametric alternative to the two-sample t test, which can be applied to two samples *that are unequal in size* is given by the *Wilcoxon rank-sum* test or, equivalently, the *Mann-Whitney U* test.[9] We deal next with a generalized form of this test called the *Kruskal-Wallis* test, applicable to three or more samples.

5.5.3 Kruskal-Wallis test for paired observations in k samples

The test is applied to, say, k independent random samples of sizes n_i, $i = 1, 2, \ldots, k$, with a total of n observations. It is assumed that all samples are random samples from their individual populations and that there is independence within the samples and between them. The null hypothesis is that the samples come from the same continuous population. The alternate hypothesis is that at least one of the populations tends to produce comparatively larger values than the others.

Following a procedure similar to that in the Wilcoxon signed-rank and the Mann-Whitney tests, the pooled data are ranked from the lowest to the highest as if they belong to one sample. However, one notes against each rank the original sample from which the item of data comes (as in the Wilcoxon signed-rank test).

Definition and properties: Kruskal-Wallis test for paired observations. If R_i is the sum of the ranks of the data in the ith sample of size n_i and n is the total sum of the k samples, the normalized test statistic is

$$H = \frac{12}{n(n+1)} \sum_{i=1}^{k} \frac{R_i^2}{n_i} - 3(n+1). \qquad (5.5.3a)$$

Under the null hypothesis that the samples come from the same population, H has an approximate chi-squared distribution with $(k-1)$ degrees of freedom.

The sum of the ranks of the pooled data is $n(n+1)/2$ (the sum of an arithmetic series) which gives the mean rank of $(n+1)/2$ for all the data, and R_i/n_i is the mean rank of the values of the ith sample.[10] Consequently, it is not difficult to show that the normalized test statistic of Eq. (5.5.3a) is equivalent to

$$H = \frac{12}{n(n+1)} \sum_{i=1}^{k} n_i \left[\frac{R_i}{n_i} - \frac{n+1}{2} \right]^2. \qquad (5.5.3b)$$

That is, H is the weighted sum of the squared differences between the mean rank from the ith sample and the mean rank for all the data. Thus large values of H lead to rejection of the null hypothesis that all the samples are realizations of one population. Therefore, from Eq. (5.3.9) and if $n_i \geq 5$ for $i = 1, 2, \ldots, k$, the test statistic can be approximated by a χ^2_{k-1} distribution.[11] We provide a general example followed by a more detailed application.

Example 5.24. Kruskal-Wallis test on the association of storm patterns with depths and durations of rainfalls. Kottegoda and Kassim (1991) presented an original method of classifying storms into three or more types. The suggested storm-structure classifications are shown in Fig. 5.5.1.

These are based on the form of the variation of the percentage cumulative depth with the percentage storm duration. In each case, a storm type with constant intensity, that is one

---

[9] See, for example, Wackerly et al. (2002, pp. 812–813) or Gibbons and Chakraborti (2003) or Conover (1998).

[10] See, for example, Kruskal and Wallis (1952).

[11] For the statistic represented by Eq. (5.5.3b) without the normalizing constant $12/(n(n+1))$, tables including values of $n_i < 5$ are provided by Gibbons and Chakraborti (2003), Lehmann (1975), and some of the other references.

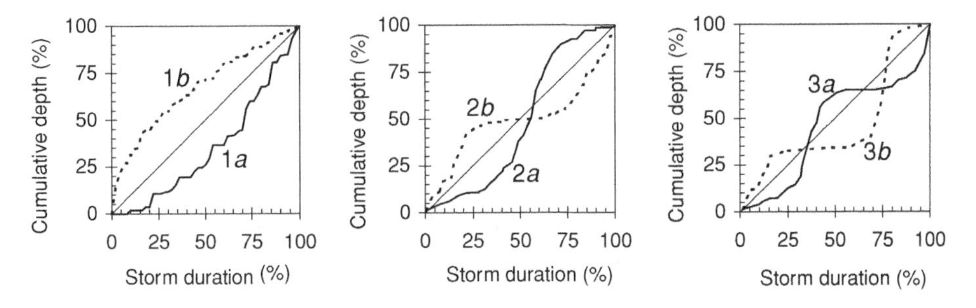

Fig. 5.5.1 Classification of storm structure types.

with a uniform rate of rainfall, is identified by a diagonal straight line. The curves represent storm profiles or structures found in practice. If, as shown by the first diagram, the profile does not cross the line of constant intensity, the storm structure is classified as type 1. The next two diagrams show storm profiles that cross the mean line once and twice, respectively, within the storm duration. These are classified correspondingly as types 2 and 3. The profiles can deviate in many ways from one storm to another; in particular, they can be reversed in form with respect to the lines of constant intensity as seen, for example, by types 1*a* and 1*b*.

A particular aspect of the study was to determine whether storm types are related to the amounts and durations of measured rainfalls. The Kruskal-Wallis test was used for the purpose. Application was made to observations recorded from 1985 to 1988 at 13 rain gauges around Birmingham, England, in the Severn-Trent catchment area.

The null hypothesis H_0: The mean depths under the three types of storms are equal at each station. This hypothesis is also applied to the storm durations.

The alternate hypothesis H_1: The three means are not equal.

Level of significance: $\alpha = 0.05$.

Calculations: Observations made during 49 storms at the 13 rain gauge stations over a 2-year period are used in the study. Table 5.5.1 shows the computations and the H values at each of the rain gauge stations. The number of storms recorded at a station is denoted by n. Note that all storms were not recorded at every station.

Initially, the storms are ranked in ascending order at each station first with respect to the depths and second with respect to the durations. The different types of storms are then identified at each station. Next, the numbers and ranks falling into each type are then grouped and the mean ranks are calculated. Last, Eq. (5.5.3*b*) is used to estimate the H statistics for depths and durations separately.

Because the storms are of three types, $k = 3$ and hence the critical region is $H \geq \chi^2_{2,0.05} = 5.99$ from Table C.3.

Decision: In Table 5.5.1 we see that 3 out of 13 of the H values fall into the critical region for the rainfall depths and also for the durations. We note that the third type of storm is inadequately represented because, as stipulated, the number of items should not be less than five for each type for the chi-squared approximation. However, reference to tables for small samples already cited confirms the findings.

Therefore, one cannot come to a definite conclusion regarding the null hypothesis and further studies are required after collection of additional data. It is also possible that such storm patterns are not associated with depths and durations except in some areas that experience special climatic or seasonal effects.

Example 5.25. Kruskal-Wallis test applied to determine whether three samples of compressive strengths of concrete come from the same population. Consider the concrete cube test records of Table E.1.2. As noted these tests were made during the period 8 July

Table 5.5.1 Kruskall-Wallis test applied to rainfall data

Station	1	2	3	4	5	6	7	8	9	10	11	12	13
N	41	45	35	34	27	38	49	37	30	38	44	35	38
n_1^a	22	25	17	19	14	15	21	19	17	19	15	19	22
$\sum r_{ia}/n_1$	18.0	19.8	15.9	15.7	12.2	11.1	19.5	14.4	14.2	21.4	21.4	17.2	17.0
$\sum r_{ib}/n_1$	23.5	20.6	14.4	18.1	14.7	17.2	19.5	16.9	16.1	21.4	21.6	17.8	18.0
n_2	17	18	16	13	12	21	24	14	9	15	26	15	14
$\sum r_{ia}/n_2$	22.9	26.2	18.8	18.8	15.5	24.7	28.4	23.8	15.7	17.9	21.3	18.3	21.0
$\sum r_{ib}/n_2$	16.4	25.6	20.1	15.8	13.6	20.8	28.5	22.9	15.8	16.2	21.7	17.2	21.0
n_3	2	2	2	2	1	2	4	4	4	4	3	1	2
$\sum r_{ia}/n_3$	37.0	34.0	29.7	26.2	20.0	28.2	33.8	24.1	20.6	16.2	38.3	28.5	25.0
$\sum r_{ib}/n_3$	33.1	29.3	32.3	23.4	9.7	24.5	33.6	19.5	12.2	23.9	3.7	34.5	20.4
H_a	5.4	4.0	3.4	2.4	1.7	14.4	6.0	7.1	1.7	1.2	4.9	1.2	1.9
H_b	5.5	2.0	6.7	1.1	0.4	1.3	6.0	2.5	0.7	2.5	6.6	2.7	0.7

[a] Suffixes 1, 2, and 3 refer to storm types; *a* and *b* refer to depths and durations of rainfall, respectively.

1991 to 21 September 1992. Let us suppose the composition of the aggregates of concrete were changed on 18 September 1991, and again on 4 December 1991. If so, one should consider the possibility that the resulting three samples do not come from the same population. Our interest here is in the compressive strengths of concrete. We use the Kruskal-Wallis test to examine any differences in the mean strengths.

The null hypothesis H_0: The mean compressive strengths of the samples taken before September 18, 1991, those taken during the period September 18 to December 4, 1991, and those taken after December 4, 1991 are equal.

Alternate hypothesis H_1: The three means are not equal.

Level of significance: $\alpha = 0.05$.

Calculations: Sample 1 has 12 compressive strengths, taken prior to 18 September 1991. The ranks in chronological order are obtained using Tables 1.2.1 and E.1.2 as follows:

14 6 15 40 32.5 9 7 3 30 16 2 27

Note that in the case of ties, mean ranks are given. The total of the ranks is 201.5. The mean rank $= 201.5/12 = 16.79$.

Sample 2 has 14 compressive strengths, taken from 18 September to 4 December 1991. The ranks are as follows in chronological order:

4 8 21 13 10 18.5 23 17 20 28 34 38 37 23.

The total of the ranks is 294.5. The mean rank $= 294.5/14 = 21.04$.

Sample 3 has 14 compressive strengths, taken after 4 December 1991. The ranks are as follows in chronological order:

31 32.5 35 29 25.5 12 1 39 36 11 5 18.5 25.5 23.

The total of the ranks is 324. The mean rank is $324/14 = 23.14$.

[We verify that the total sum of the ranks $= 201.5 + 294.5 + 324 = 820.0$, which tallies with the theoretical total of $n(n + 1)/2 = 40 \times 41/2 = 820$.]

The critical region is $H \geq \chi^2_{0.05,2} = 5.99$ from Table C.3.
From Eq. (5.5.3*a*)

$$H = \frac{12}{n(n+1)} \sum_{i=1}^{3} \frac{R_i^2}{n_i} - 3(n+1)$$
$$= \frac{12}{40 \times 41} \left[\frac{201.5^2}{12} + \frac{294.5^2}{14} + \frac{324^2}{14} \right] - 3 \times 41 = 1.95.$$

[The same result is obtained from Eq. (5.5.3*b*).]
Decision: Because $H \leq \chi^2_{0.05,2} = 5.99$, the null hypothesis is not rejected.

5.5.4 Tests on randomness: runs test

In the applications of statistical methods, it is often assumed that a sample is taken at random and that the values of a sample represent variables that are themselves random. Some tests are devised specifically to detect trends; others are used to investigate periodic behavior in a series. A technique based on the number of runs shown by a series is applicable as a general test on randomness. By definition, a run is a sequence of variables of a particular kind that is preceded and followed by a sequence of variables of a different kind or by no variables at all (in the first and last positions). For example, magnitudes greater than the median can be of one kind, in which case magnitudes below the median will constitute the other kind.

As an illustration, consider the following series which represents 35 pollutant levels in a stream measured at regular intervals of time denoted by either A, acceptable, or U, unacceptable:

$$U\,U\,A\,U\,A\,U\,U\,U\,U\,U\,A\,A\,U\,U\,U\,U\,A\,U\,A\,A\,A\,A\,U\,U\,A\,A\,A\,A\,A\,U\,A\,A\,U\,U\,A.$$

By the given definition, this comprises 16 runs where, for example, each run of one or more Us is preceded and followed by a run of one or more As or no measurements at all. In such cases, the total number of runs relative to the sample size provides an indicator of randomness. Too few runs suggest a grouping, clustering, trend, or periodic behavior, whereas too many runs lead to the suspicion that there are high-frequency oscillations. A random series should have neither too few nor too many runs.

Definition and properties: Runs test for randomness. A run is a sequence of variables of a particular kind that is preceded and followed by a sequence of variables of a different kind or no variables at all. In a sequence containing n variables of one kind and m variables of another kind, the sampling distribution of the total number of runs, R, can be closely approximated by the normal distribution with[12]

$$\mu_R = 1 + \frac{2nm}{(n+m)} \tag{5.5.4a}$$

and

$$\text{Var}[R] = \frac{2nm(2nm - n - m)}{(n+m)^2(n+m-1)}. \tag{5.5.4b}$$

The approximation is quite close for $m, n > 9$.

[12] Tables are provided by Wackerly et al. (2002, pp. 814–815) and Gibbons and Chakraborti (2003) for small samples. See also Wald and Wolfowitz (1940).

Because a discrete distribution is approximated by a continuous distribution, a continuity correction, as in Eq. (5.5.1a) and (5.5.1b), should be applied to the observed number of runs, r, as shown in Example 5.22.

Example 5.26. Runs test on annual river flows. Annual flows of the Derwent at Yorkshire Bridge, England, for the period 1938–1967 are tabulated below in millimetres of equivalent runoff over the catchment area above the site.

946	1074	867	1058	838	837	1133	815	1138	869
910	868	927	1193	969	742	1386	737	1113	955
1143	665	1187	947	955	891	763	1288	1302	1029.

Our objective is to ascertain whether the population is random using the runs test.

Null hypothesis H_0: The population is random.
Alternate hypothesis H_1: The population is not random.
Level of significance: $\alpha = 0.05$.

Calculations: The median is 951 mm. Runs above the median are underlined as follows:

946	$\underline{1074}$	867	$\underline{1058}$	838	837	$\underline{1133}$	815	$\underline{1138}$	869
910	868	927	$\underline{1193}$	$\underline{969}$	742	$\underline{1386}$	737	$\underline{1113}$	$\underline{955}$
$\underline{1143}$	665	$\underline{1187}$	947	$\underline{955}$	891	763	$\underline{1288}$	$\underline{1302}$	$\underline{1029.}$

The number of values n above the median = the number of values m below the median = 15. The total number of observed runs $r = 20$.

The test statistic is

$$Z = \frac{(R \pm (1/2)) - \mu_R}{\sqrt{\mathrm{Var}[R]}}.$$

By using a two-tailed test with $\alpha = 0.05$, the critical region is Z (positive with $R + 1/2$) > 1.96 and Z (negative with $R - 1/2$) < −1.96, from Table C.1.

Sample estimates of the mean and variance of the total number of runs are obtained using Eq. (5.5.4a) and (5.5.4b):

$$\hat{\mu}_R = 1 + \frac{2 \times 15 \times 15}{30} = 16 \quad \text{and} \quad \mathrm{Var}[R] = \frac{2 \times 15 \times 15(450 - 30)}{30^2 \times 29} = 7.24.$$

Hence,

$$z = \frac{(20 + 1/2) - 16}{\sqrt{7.24}} = 1.49.$$

Decision: Thus the null hypothesis that the population is random is not rejected.

5.5.5 Spearman's rank correlation coefficient

In Chapter 3 we studied the correlation between two series [see Eq. (3.3.27)], but (as discussed further in Chapter 6) the associated significance test is based on assumptions that may be restrictive at times. The nonparametric alternative is called *Spearman's rank correlation coefficient*; it is often referred to simply as the *rank correlation coefficient*. For this purpose we use ranks as in some of the tests of location already described. That is, the data are converted to ranks. In this way the correlation coefficient does not depend on the actual values and, furthermore, the ranks do not vary if one makes a monotonic transformation (which is unambiguous and does not change the order) of the variables. Unlike Pearson's product-moment correlation coefficient, this does not require that the relationship between the variables is linear.

Definition and properties: Spearman's rank correlation coefficient is estimated as follows for a set of paired data, (x_i, y_i), $i = 1, 2, \ldots, n$, that are ranked separately so that, for each data set, the highest value has rank 1 and rank n is that of the lowest value:

$$r_s = 1 - \frac{6\sum_{i=1}^{n} d_i^2}{n(n^2-1)}, \tag{5.5.5}$$

where d_i is the difference between the ranks given to x_i and y_i. Alternatively, the statistic is denoted by the Greek letter ρ.

Under the null hypothesis of no correlation between the X and Y series, the distribution of r_s can be closely approximated by the normal distribution with

$$\mu_{r_s} = 0 \quad \text{and} \quad \text{Var}[r_s] = 1/(n-1).$$

If we return to the product-moment correlation coefficient estimated by Eq. (1.4.3) and substitute the ranks for the actual paired observations (x_i, y_i), the formula reduces algebraically to Eq. (5.5.5). (See Appendix A.10 for proof.)[13]

Example 5.27. Rank correlations of concrete densities and strengths. Table E.1.2 gives densities and strengths of concrete from 40 samples taken over a period of 15 months. The correlation is tested using Spearman's rank correlation method.

Null hypothesis H_0: $\mu_{r_s} = 0$, the densities and strengths are uncorrelated.
Alternate hypothesis H_1: $\mu_{r_s} \neq 0$, the densities and strengths are correlated.
Level of significance: $\alpha = 0.01$.

Calculations: The data are presented in Table 5.5.2. This table also includes the ranked data, commencing with the highest (rank 1) in each data set, as well as the proper ranks allocated with adjustments for ties. Finally, the differences in ranks are shown. The sum of squared differences is 6189. Hence, from Eq. (5.5.5)

$$r_s = 1 - \frac{(6 \times 6189)}{\{40(40^2 - 1)\}} = 0.42.$$

The critical region: $z \geq z_{\alpha/2} = 2.575$ or $z \leq z_{\alpha/2} = -2.575$.

From the given properties,

$$z = \frac{r_S - \mu_{rS}}{\text{Var}[r_s]} = \frac{0.42 - 0}{\sqrt{1/(40-1)}} = 2.62.$$

Decision: The null hypothesis is rejected and there is a positive correlation. We note also that the value of 0.42 compares closely with the coefficient of 0.44 obtained from the product-moment correlation coefficient using Eq. (1.4.3).

5.5.6 Summary of Section 5.5

In this section we examined several nonparametric tests. They can be advantageous because the tests are not based on restrictive assumptions. The procedures ranged from the sign tests on the media to comparisons of the distributions of two or more samples, tests on randomness and correlation tests. Some of the methods given elsewhere, such as the Kolmogorov-Smirnov two-sample test in Subsection 5.6.3 and the jackknife and kernel-based methods of Subsection 3.2.3 are also nonparametric. A few additional tests are provided in the references cited.

[13] Tables for small sample are provided by Wackerly et al. (2002, p. 816) and Gibbons and Chakraborti (2003).

Table 5.5.2 Spearman's rank correlation test of concrete densities and compressive strengths[a]

Item	Density x	Strength y	Density, ranked	Strength, ranked	Rank of x	Rank of y	Difference in ranks d	d^2
1	2437.0	60.5	2411.0	49.9	25.5	18.0	7.5	56.3
2	2437.0	60.9	2415.0	50.7	25.5	15.5	10.0	100.0
3	2425.0	59.8	2425.0	52.5	38.0	22.5	15.5	240.3
4	2427.0	53.4	2427.0	53.2	36.5	36.0	0.5	0.3
5	2428.0	56.9	2427.0	53.4	35.0	30.0	5.0	25.0
6	2448.0	67.3	2428.0	54.4	15.5	5.0	10.5	110.3
7	2456.0	68.9	2429.0	54.6	8.5	2.0	6.5	42.3
8	2436.0	49.9	2433.0	55.8	28.5	40.0	−11.5	132.3
9	2435.0	57.8	2435.0	56.3	31.5	29.0	2.5	6.3
10	2446.0	60.9	2435.0	56.7	19.0	15.5	3.5	12.3
11	2441.0	61.9	2436.0	56.9	23.5	12.0	11.5	132.3
12	2456.0	67.2	2436.0	57.8	8.5	6.0	2.5	6.3
13	2444.0	64.9	2436.0	57.9	22.0	8.5	13.5	182.3
14	2447.0	63.4	2436.0	58.8	17.5	10.0	7.5	56.3
15	2433.0	60.5	2437.0	58.9	33.0	18.0	15.0	225.0
16	2429.0	68.1	2437.0	59.0	34.0	4.0	30.0	900.0
17	2435.0	68.3	2441.0	59.6	31.5	3.0	28.5	812.3
18	2471.0	65.7	2441.0	59.8	4.0	7.0	−3.0	9.0
19	2472.0	61.5	2444.0	59.8	3.0	13.0	−10.0	100.0
20	2445.0	60.0	2445.0	60.0	20.5	21.0	−0.5	0.3
21	2436.0	59.6	2445.0	60.2	28.5	24.0	4.5	20.3
22	2450.0	60.5	2446.0	60.5	13.0	18.0	−5.0	25.0
23	2454.0	59.8	2447.0	60.5	11.5	22.5	−11.0	121.0
24	2449.0	56.7	2447.0	60.5	14.0	31.0	−17.0	289.0
25	2441.0	57.9	2448.0	60.9	23.5	28.0	−4.5	20.3
26	2457.0	60.2	2448.0	60.9	7.0	20.0	−13.0	169.0
27	2447.0	55.8	2449.0	61.1	17.5	33.0	−15.5	240.3
28	2436.0	53.2	2450.0	61.5	28.5	37.0	−8.5	72.3
29	2458.0	61.1	2454.0	61.9	6.0	14.0	−8.0	64.0
30	2415.0	50.7	2454.0	63.3	39.0	39.0	0.0	0.0
31	2448.0	59.0	2455.0	63.4	15.5	25.0	−9.5	90.3
32	2445.0	63.3	2456.0	64.9	20.5	11.0	9.5	90.3
33	2436.0	52.5	2456.0	64.9	28.5	38.0	−9.5	90.3
34	2469.0	54.6	2457.0	65.7	5.0	34.0	−29.0	841.0
35	2455.0	56.3	2458.0	67.2	10.0	32.0	−22.0	484.0
36	2473.0	64.9	2469.0	67.3	2.0	8.5	−6.5	42.3
37	2488.0	69.5	2471.0	68.1	1.0	1.0	0.0	0.0
38	2454.0	58.9	2472.0	68.3	11.5	26.0	−14.5	210.3
39	2427.0	54.4	2473.0	68.9	36.5	35.0	1.5	2.3
40	2411.0	58.8	2488.0	69.5	40.0	27.0	13.0	169.0
							Sum	6189.0

[a] Density in kg/m^3; strength in N/mm^2.

5.6 GOODNESS-OF-FIT TESTS

In Section 5.4 hypothesis-testing procedures related to the parameters of a distribution were discussed. Another important type of hypothesis concerns the form of a probability distribution. It may, for example, be necessary to test whether a discrete variable has a

Poisson distribution or whether a continuous variable is normally distributed. For these purposes, we make an overall comparison of observed and hypothetical frequencies that fall into specified classes as discussed in Section 1.1. Alternatively, we may compare observed and theoretical cumulative frequencies. These constitute two main types of goodness-of-fit tests. The Anderson-Darling goodness-of-fit test is a special type devised to give heavier weighting to the tails of a distribution; we discuss applications when the model parameters are estimated from the same sample that is used for the test. Other methods for testing the goodness-of-fit to a normal distribution are also given. In Section 5.8, we describe some graphical techniques that are highly useful in fitting probability distributions.

5.6.1 Chi-squared goodness-of-fit test

The chi-squared test is a test of significance based on the chi-squared statistic with cdf given by Eq. (4.2.12*d*) and used in this chapter in various applications. As shown by Eq. (5.3.9), the statistic is derived by the sum of squares of independent standard normal variates. The main steps are the ranking of a sample of data, division into a number of classes depending on the magnitudes and the range, and the fitting of a probability distribution. The statistic comes from the weighted sum of squared differences between the observed and theoretical frequencies.

Consider, for example, the modulus of rupture data of a certain type of timber presented in Tables E.1.1 and 1.1.3. These are ranked and divided into classes with intervals of 5 N/mm^2. In Table 1.1.4 are shown the observed relative frequencies within each class and the corresponding histogram is given by Fig. 1.1.4. Let a normal distribution, as shown for example by the pdfs in Fig. 4.2.8, be fitted and the expected relative frequencies obtained. The observed frequencies O_i and expected frequencies E_i are found by multiplying the relative frequencies, for each class i from a total of l classes, by the sample size n. To test whether the differences between the observed and expected frequencies are significant, we use the statistic

$$X^2 = \sum_{i=1}^{l} \frac{(O_i - E_i)^2}{E_i}. \tag{5.6.1}$$

A large value of this statistic indicates a poor fit; so we need to know what values are acceptable. The sampling distribution of X^2 tends, as n approaches infinity, to a χ^2_ν distribution, where $\nu = l - k - 1$ represents the degrees of freedom and k is the number of parameters estimated from the same data used for the test. (The test is not exact for finite samples because the statistic is an approximation obtained by taking only the first term in a logarithmic series expansion consequent to maximizing the likelihood of the joint occurrences of events in each interval. For practical use, however, it is sufficient that the distribution is approximately of the χ^2 form.)[14]

A possible shortcoming, which may reduce the effectiveness of the test, is that X^2 is positive and that no consideration is given to the signs and locations of the differences in Eq. (5.6.1). Nevertheless, the test gives satisfactory results when there is no significant dependence between the variables, if $n \geq 50$ and for each class i, $n_i \geq 5$. It is versatile and does not require one to know the values of the parameters before the test, as in the classical form of the Kolmogorov-Smirnov goodness-of-fit test, which is discussed next.

[14] See Johnson and Leone (1977, pp. 274–277).

Example 5.28. Goodness-of-fit of wet runs of rainfall. In Example 4.16 we considered the distributions of wet and dry runs of daily rainfall at Kew in London, England, from January 1958 to May 1965. It was seen in Table 4.1.5 and Fig. 4.1.12 that the geometric distribution provides a close fit to the observed distribution of wet runs. For a formal assessment of these fits, we can use the chi-squared goodness-of-fit criterion, as approximated by Eq. (5.6.1), using the given data.

Null hypothesis H_0: The random variable (wet runs) has a geometric distribution with $p = 0.392$.

Alternate hypothesis H_1: The random variable does not have the specified distribution.

Level of significance: $\alpha = 0.05$.

Critical region: $X^2 \geq \chi^2_{8,0.05} = 15.5$ (from Table C.3), noting that in Table 4.1.5 there are $l = 10$ classes and $k = 1$ for the application of Eq. (5.6.1), which gives $v = 10 - 1 - 1 = 8$.

Calculations: From Table 4.1.5,

$$
\begin{aligned}
X^2 &= \frac{(194 - 179.6)^2}{179.6} + \frac{(101 - 109.2)^2}{109.2} + \frac{(66 - 66.4)^2}{66.4} \\
&\quad + \frac{(30 - 40.3)^2}{40.3} + \frac{(26 - 24.5)^2}{24.5} + \frac{(11 - 14.9)^2}{14.9} \\
&\quad + \frac{(13 - 9.1)^2}{9.1} + \frac{(7 - 5.5)^2}{5.5} + \frac{(5 - 3.3)^2}{3.3} + \frac{(2 - 2.2)^2}{2.2} \\
&= 8.49.
\end{aligned}
$$

Decision: Because $X^2 < \chi^2_{8,0.05} = 15.5$, the null hypothesis is not rejected. It can be concluded that the wet runs have a geometric distribution as specified.

In a practical sense we should be aware, however, that the true distribution of wet runs cannot be exactly the same as the hypothesized distribution. What we have found is a good approximation to the true distribution.

Regarding the dry runs, the logarithmic series distribution seems to provide a better fit, as seen from Table 4.1.6 and Fig. 4.1.13. The chi-squared test can also be applied to this hypothesis.

For the application of the chi-squared test to a continuous variable, as represented, for example, by the timber strength data of Table 1.1.3, the expected frequencies E_i are the products of the total sample size n and the areas under the pdf, as specified by the null hypothesis, between the bounds of each class i. Calculations of areas under the normal and gamma pdfs are demonstrated in Examples 4.27 and 4.23, respectively.

The choice of classes will affect the power of the test. Furthermore, it is not the best approach to have equal class intervals for the purpose. An equitable allocation of the frequencies is obtained if we divide the total area under the pdf into equal areas and hence find the class boundaries. This is the equal-probabilities method of constructing classes as proposed by Mann and Wald (1942) and clarified by Williams (1950) who also suggest for a level of significance $\alpha = 0.05$, values of classes $l = 39, 35, 30, 23, 15, 12$, and 9 for total sample sizes $n = 2000, 1500, 1000, 500, 200, 100$, and 50, respectively. For other values of n and α, we use the formula

$$l = 2\left[\frac{2(n-1)^2}{z_a^2}\right]^{0.2}, \tag{5.6.2}$$

where z_α is the value which a standard normal variate exceeds with probability α (see Table C.1). Also $n/l \geq 5$; although this requirement is relaxed somewhat by recent authors, too many classes tend to reduce the power of the test.[15]

[15] An alternative form of Eq. (5.6.2) and other information are given by Stuart and Ord 1991, pp. 1172–1182.

Definition and properties: Chi-squared goodness-of-fit test. The test statistic is

$$X^2 = \sum_{i=1}^{l} \frac{(O_i - E_i)^2}{E_i},$$

where O_i and E_i are the observed and expected frequencies, respectively, for each class i out of a total of l classes into which an ordered sample of n observations is placed. A hypothesized theoretical distribution gives the expected frequencies. The sampling distribution of X^2 tends, as n approaches infinity, to a X^2 to a χ^2_ν distribution, where $\nu = l - k - 1$ represents the degrees of freedom and k is the number of parameters estimated from the same sample as used in the test. The test is applicable to discrete and continuous variables, with a minimum of 5 values in each class.

5.6.2 Kolmogorov-Smirnov goodness-of-fit test

The Kolmogorov-Smirnov goodness-of-fit test is a nonparametric test that relates to the cdf rather than the pdf of a continuous variables. It is not applicable to discrete variables.[16] The test statistic, in a two-sided test, is the maximum absolute difference (that is, usually the vertical distance) between the empirical and hypothetical cdfs.

For a continuous variate X let $x_{(1)}, x_{(2)}, \ldots x_{(n)}$ represent the order statistics of a sample of size n, that is, the values arranged in increasing order. The empirical or sample distribution function $F_n(x)$ is a step function. This gives the proportion of values not exceeding x and is defined as

$$\begin{aligned} F_n(x) &= 0, \quad &&\text{for } x < x_{(1)}, \\ &= k/n, \quad &&\text{for } x_{(k)} \le x \le x_{(k+1)}; k = 1, 2, \ldots, n-2, \\ &= 1, \quad &&\text{for } x \ge x_{(n)}. \end{aligned} \tag{5.6.3}$$

Let $F_0(x)$ denote a completely specified theoretical continuous cdf. The null hypothesis H_0 is that the true cdf of X is the same as $F_0(x)$. That is, under the null hypothesis

$$\lim_{n \to \infty} \Pr[F_n(x) = F_0(x)] = 1.$$

The test criterion is the maximum absolute difference between $F_n(x)$ and $F_0(x)$, formally defined as

$$D_n = \sup_x |F_n(x) - F_0(x)|. \tag{5.6.4a}$$

The foregoing measure of deviation is for a two-sided test which is commonly used. If for some reason a one-sided test is required to test whether, for instance, $F_n(x) > F_0(x)$, then the statistic is modified as

$$D_n^+ = \sup_x [F_n(x) - F_0(x)]. \tag{5.6.4b}$$

Likewise one can define the statistic D_n^-. One of the advantages of the test is that the test statistic is distribution-free, unlike those applied in hypothesis testing in Section 5.4.

For large values of n, Smirnov (1948) gives the limiting distribution of $\sqrt{n} D_n$, as

$$\lim_{n \to \infty} \left[\Pr\left[\sqrt{n} D_n \le z\right]\right] = \left(\frac{\sqrt{2\pi}}{z}\right) \sum_{k=1}^{\infty} \exp\left[-(2k-1)^2 \frac{\pi^2}{(8z^2)}\right]. \tag{5.6.5}$$

[16] See, however, the modifications suggested for such a test by Conover (1998).

Thus, one can compute that the critical values $D_{n,\alpha}$ for large samples, say, $n > 35$, are $1.3581/\sqrt{n}$ and $1.6276/\sqrt{n}$ for $\alpha = 0.05$ and 0.01, that is, for probabilities of 0.95 and 0.99 in Eq. (5.6.5), respectively. [From simulations we note that these results hold closely for the summation of even the first five terms in Eq. (5.6.5).] For smaller sample sizes, critical values $D_{n,\alpha}$ are given in Table C.7, of Appendix C.

The test is applied on the assumption that $F_0(x)$ denotes a completely specified theoretical continuous cdf, that is with *known parameters*.[17] In the next example *we do not estimate* the parameters of the distribution from the same sample that is used for the test.

Example 5.29. Testing goodness-of-fit of timber strengths. In Table E.1.1, modulus of rupture data from 50 mm × 150 mm Swedish redwood and whitewood timber are given in newtons per square millimeter. Let us suppose that the first 100 items were delivered by one supplier and a second lot of 64 items came from another batch. (Let us ignore the zero item at the end.) We assume that the distributions of the two lots are identical; but we will examine this in the next example. Because of the positive skewness shown in Table 1.2.3, the distribution can be one of several types such as the gamma described in Chapter 4. The Weibull seems to be an ideal candidate, however, because it was originally devised to model material strengths and similar effects and has been used for such purposes for over 75 years. We apply, in the first instance, the two-parameter Weibull distribution with cdf, as presented before through Eq. (4.2.16),

$$F_X(x) = 1 - \exp\left[-\left(\frac{x}{\lambda}\right)^{\beta}\right].$$

Following the least squares estimation procedure of Example 4.25, we estimate the following parameters from the first 100 items of data:

$$\hat{\beta} = 5.39 \quad \text{and} \quad \hat{\lambda} = 42.55.$$

We use these estimates to model the empirical distribution of the next 64 items and hence apply the Kolmogorov-Smirnov goodness-of-fit test.

Null hypothesis H_0: The random variable (representing the modulus of rupture of 50 mm × 150 mm Swedish redwood and whitewood timber) has a Weibull distribution as specified earlier.

Alternate hypothesis H_1: The random variable has a different distribution.

Level of significance: $\alpha = 0.05$.

Calculations: Critical region: D_n, as defined by Eq. (5.6.4*a*), is

$$> D_{n,\alpha} = \frac{1.3581}{\sqrt{n}} = \frac{1.3581}{\sqrt{64}} = 0.17.$$

By setting this value above and below the sample cdf, confidence limits can be drawn as shown in Fig. 5.6.1.

It is also shown that the observed value of the maximum absolute difference between the theoretical and step functions $d_n = 0.1008$, which is less than the critical value.

Decision: The null hypothesis is not rejected.

5.6.3 Kolmogorov-Smirnov two-sample test

The preceding test can be adapted to ascertain whether two samples come from the same distribution.

---

[17] When parameters are estimated from the same sample as used in the test, see Table C.8, in Appendix C. Also, reference may be made to Lilliefors (1967), Gibbons and Chakraborti (2003), and the discussion by Stuart and Ord (1991, pp. 1191–1192). A corrected table, for use with the normal distribution, is given by Dallal and Wilkinson (1986). (See Problem 5.28.)

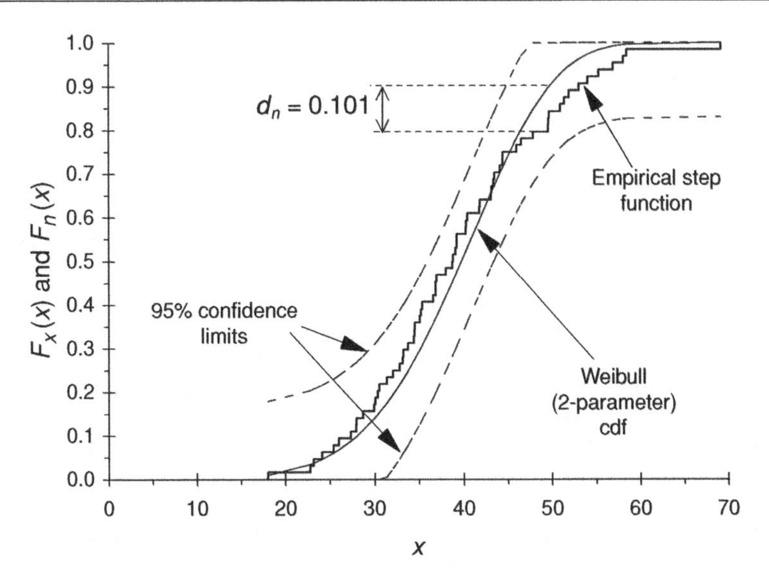

Fig. 5.6.1 Kolgomorov-Smirnov one-sample goodness-of-fit test.

Let us define the maximum absolute difference between two empirical distribution functions as $D_{m,n}$. Let these functions be represented by the step functions $F_m(x)$ and $G_n(x)$ based on two samples of sizes m and n, respectively. That is,

$$D_{m,n} = \sup_x |F_m(x) - G_n(x)|. \tag{5.6.6}$$

This is applicable to a two-sided test and corresponds to Eq. (5.6.4a). Also, with the test statistic given by Eq. (5.6.4b) we can adopt a one-sided test.[18] For large values of m and n,

$$\lim_{m,n\to\infty}\left[\Pr\left(\sqrt{\frac{mn}{m+n}}D_{m,n} \le z\right)\right] = \left(\frac{\sqrt{2\pi}}{z}\right)\sum_{k=1}^{\infty}\exp\left[-(2k-1)^2\frac{\pi^2}{(8z^2)}\right], \tag{5.6.7}$$

which is the same as the distribution given by Eq. (5.6.5). That is, for large values of m and n, $\sqrt{mn/(m+n)}$ can be substituted for $\sqrt{n}$ in Eq. (5.6.5), and the test procedure applied.

Example 5.30. Testing whether two samples come from the same population. We refer again to the modulus of rupture data from 50 mm × 150 mm Swedish redwood and whitewood timber presented in Table E.1.1. In Example 5.29 we supposed that the first 100 items were delivered by one supplier and a second lot of 64 items also came from the same supplier but from another batch. Let us verify whether the two samples are from the same population. We apply the Kolmogorov-Smirnov two-sample test for the purpose.

Null hypothesis H_0: The random variables sampled by the first 100 values (of modulus of rupture of 50 mm × 150 mm Swedish redwood and whitewood timber presented in Table E.1.1) and the random variables sampled by the next 64 values have the same distribution.
Alternate hypothesis H_1: The random variables have different distributions.
Level of significance: $\alpha = 0.05$.

Calculations: In Table 5.6.1 the data from each sample are ranked separately with values of the step functions $F_m(x)$ and $G_n(x)$.

[18] Tables are provided by Gibbons and Chakraborti (2003) for the application of the two-sample test to small samples. For equal sample sizes less than 40, tables are given by Birnbaum (1952).

Table 5.6.1 Data for Kolmogorov-Smirnov two-sample test applied to timber strengths

Rank	$F_m(x)$	x^a	$F_n(x)$	x	Rank	$F_m(x)$	x	$F_n(x)$	x
1	0.01	22.67	0.01563	17.98	51	0.51	39.34	0.79688	47.83
2	0.02	22.75	0.03125	22.74	52	0.52	39.60	0.81250	49.57
3	0.03	23.16	0.04688	23.14	53	0.53	39.62	0.82813	49.59
4	0.04	23.19	0.06250	24.09	54	0.54	39.77	0.84375	49.65
5	0.05	24.25	0.07813	25.39	55	0.55	39.93	0.85938	50.91
6	0.06	24.84	0.09375	25.98	56	0.56	39.97	0.87500	51.39
7	0.07	26.63	0.10938	27.31	57	0.57	40.53	0.89063	51.90
8	0.08	28.00	0.12500	27.90	58	0.58	40.71	0.90625	53.00
9	0.09	28.13	0.14063	27.93	59	0.59	40.85	0.92188	53.99
10	0.10	28.46	0.15625	28.71	60	0.60	40.85	0.93750	55.23
11	0.11	28.69	0.17188	29.90	61	0.61	41.64	0.95313	56.80
12	0.12	28.76	0.18750	30.02	62	0.62	41.75	0.96875	57.99
13	0.13	28.83	0.20313	30.33	63	0.63	41.85	0.98438	58.34
14	0.14	28.97	0.21875	30.53	64	0.64	42.31	1.00000	69.07
15	0.15	28.98	0.23438	31.33	65	0.65	42.47		
16	0.16	29.11	0.25000	32.03	66	0.66	43.26		
17	0.17	29.93	0.26563	32.76	67	0.67	43.33		
18	0.18	30.05	0.28125	33.14	68	0.68	43.48		
19	0.19	31.60	0.29688	33.18	69	0.69	43.48		
20	0.20	32.02	0.31250	33.71	70	0.70	43.64		
21	0.21	32.40	0.32813	34.40	71	0.71	43.99		
22	0.22	32.48	0.34375	34.44	72	0.72	44.00		
23	0.23	32.68	0.35938	34.49	73	0.73	44.30		
24	0.24	33.06	0.37500	35.03	74	0.74	44.51		
25	0.25	33.19	0.39063	35.17	75	0.75	44.54		
26	0.26	33.47	0.40625	35.30	76	0.76	44.59		
27	0.27	33.61	0.42188	36.53	77	0.77	44.78		
28	0.28	33.92	0.43750	36.84	78	0.78	44.78		
29	0.29	34.12	0.45313	36.85	79	0.79	45.19		
30	0.30	34.56	0.46875	36.92	80	0.80	45.54		
31	0.31	34.63	0.48438	38.00	81	0.81	45.92		
32	0.32	35.43	0.50000	38.71	82	0.82	46.01		
33	0.33	35.58	0.51563	38.81	83	0.83	46.33		
34	0.34	35.67	0.53125	39.05	84	0.84	46.86		
35	0.35	35.88	0.54688	39.20	85	0.85	46.99		
36	0.36	35.89	0.56250	39.21	86	0.86	47.25		
37	0.37	36.00	0.57813	40.20	87	0.87	47.42		
38	0.38	36.38	0.59375	40.27	88	0.88	47.61		
39	0.39	36.47	0.60938	40.39	89	0.89	47.74		
40	0.40	36.81	0.62500	41.72	90	0.90	48.37		
41	0.41	36.88	0.64063	41.78	91	0.91	48.39		
42	0.42	37.51	0.65625	43.07	92	0.92	48.78		
43	0.43	37.65	0.67188	43.12	93	0.93	50.98		
44	0.44	37.69	0.68750	43.33	94	0.94	53.63		
45	0.45	37.78	0.70313	43.41	95	0.95	54.04		
46	0.46	38.05	0.71875	44.07	96	0.96	54.71		
47	0.47	38.16	0.73438	44.36	97	0.97	56.60		
48	0.48	38.64	0.75000	44.36	98	0.98	65.35		
49	0.49	39.15	0.76563	45.97	99	0.99	65.61		
50	0.50	39.33	0.78125	46.50	100	1.00	70.22		

Note: In practice $F_n(x)$ can be reduced to, say, 3 significant figures.
[a] Modulus of rupture, in N/mm^2.

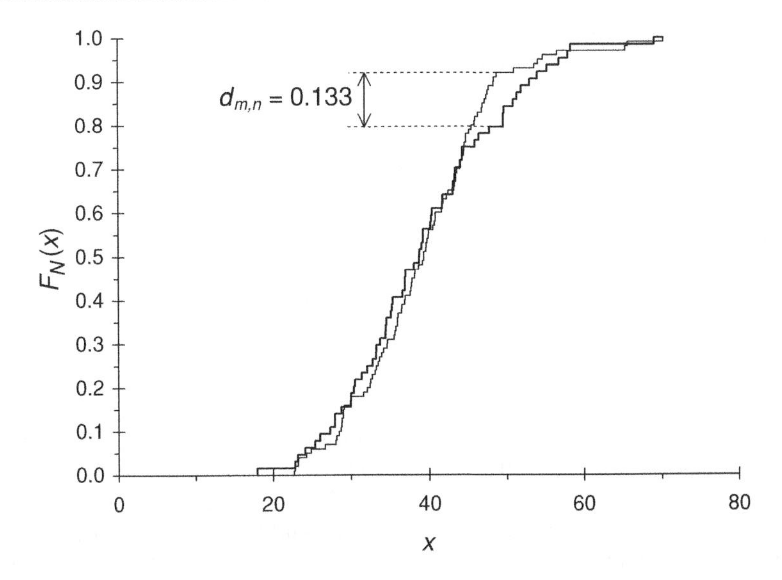

Fig. 5.6.2 Kolgomorov-Smirnov two-sample goodness-of-fit test.

The vertical positions of the items indicate the ranks of the combined data, from top to bottom. Also, the step functions are plotted in Fig. 5.6.2.

We find that the observed maximum absolute difference between the two empirical distribution functions, $d_{m,n}= 0.133$. The test statistic

$$\sqrt{\frac{mn}{m+n}}D_{m,n} = \sqrt{\frac{100 \times 64}{100 + 64}} \times 0.133 = 0.831.$$

This is less than the numerator of the corresponding critical value of $1.36/\sqrt{n}$ (for a one-sample test) of Table C.7.

Decision: The null hypothesis is not rejected.

Definition and properties: Kolmogorov-Smirnov tests. The test statistic for a two-sided goodness-of-fit is D_n, the maximum absolute difference between the empirical and hypothetical distribution functions. For large sample sizes,

$$\Pr\left[\sqrt{n}D_n \leq 1.3581\right] = 0.95 \quad \text{and} \quad \Pr\left[\sqrt{n}D_n \leq 1.6276\right] = 0.99$$

These results are quite close for sample sizes greater than 35. Other critical values can be obtained from the limiting sampling distribution,

$$\lim_{n\to\infty}\left[\Pr\left(\sqrt{n}D_n \leq z\right)\right] = \left(\frac{\sqrt{2\pi}}{z}\right)\sum_{k=1}^{\infty}\exp\left[-(2k-1)^2\frac{\pi^2}{(8z^2)}\right].$$

Values for smaller sample sizes are as tabulated.

The test statistic for a two-sided two-sample test is $D_{m,n}$, the maximum absolute difference between two empirical distribution functions. For large values of m and n, $\sqrt{mn/(m+n)}$ can be substituted for $\sqrt{n}$ in the preceding equations.

5.6.4 Anderson-Darling goodness-of-fit test

The Anderson-Darling test is devised to give heavier weighting to the tails of a distribution where unexpectedly high or low values, called outliers as discussed in Chapter 1, are sometimes located. This is made possible if one divides the difference between the empirical cdf $F_n(x)$ and theoretical cdf $F_0(x)$ to be tested [that is, $(F_n(x) - F_0(x))$—which approaches zero in each tail] by $\sqrt{F_0(x)[1 - F_0(x)]}$; see Anderson and Darling (1954). After squaring the test statistic becomes

$$A^2 = \int_{-\infty}^{\infty} [F_n(x) - F_0(x)]^2 \frac{1}{F_0(x)[1 - F_0(x)]} F_0(x)dx, \tag{5.6.8}$$

where $f_0(x)$ is the hypothetical pdf. It is shown that this is equivalent to

$$A^2 = -n - \sum_{i=1}^{n} \frac{(2i - 1)(\ln F_0[x_{(i)}] + \ln\{1 - F_0[x_{(n-i+1)}]\})}{n}, \tag{5.6.9}$$

where $x_{(1)}, x_{(2)}, \ldots, x_{(n)}$ are the observations ordered in ascending order. Because the cdfs are in the range 0–1, their logarithms are negative and hence the summation on the right-hand side of Eq. (5.6.9) is negative. The absolute value of the summation is also greater than n, thus resulting in a positive value of A^2.

For large values of the test statistic A^2, the null hypothesis that $F_n(x)$ and $F_0(x)$ have the same distribution is rejected. Let us assume that the distribution is completely specified. Accordingly, critical values A_α^2 at five α levels of significance, at the upper tail, are $A_{0.10}^2 = 1.933$, $A_{0.05}^2 = 2.492 =$, $A_{0.025}^2 = 3.070$, and $A_{0.01}^2 = 3.857$, respectively. These asymptotic values hold approximately for $n > 10$. They apply to tests on any type of distribution. However, if the parameters are estimated from the sample used in the test, the test statistic is modified to $A^* = A^2(1.0 + 0.75/n + 2.25/n^2)$ and $A^* = A^2(1.0 + 0.3/n)$ in the case of the normal and exponential distributions, respectively. In this situation, the corresponding upper tail critical values are 0.631, 0.752, 0.873, and 1.035 for the normal and 1.062, 1.321, 1.591, and 1.959 for the exponential distributions, respectively.

Tests for other distributions, such as extreme value Weibull, logistic, and gamma, are discussed by Kotz and Johnson (1982) and detailed in the references cited; see, in particular, the work of M. A. Stephens. See also Marsaglia and Marsaglia (2004).

Example 5.31. Distribution of concrete densities. Let us reconsider the distribution of the concrete densities given in Table 1.2.1. In Chapter 3 we assumed a marginal uniform pdf as a first approximation. However, a normal distribution seems to be a better choice. We now apply the Anderson-Darling test to verify the hypothesis, using the estimated mean and standard deviation of 2445 and 15.99 kg/m^3 from Table 1.2.2.

Null hypothesis H_0: The random variables sampled by the 40 concrete densities in Table 1.2.1 have a normal distribution with the given parameters.

Alternate hypothesis H_1: The random variables do not have the specified distribution.

Level of significance: $\alpha = 0.05$.

Calculations: The calculations are shown in Table 5.6.2*a*.

The summation of the numerator of Eq. (5.6.9) is given here. The estimated A^2 is

$$-40 - \left(-\frac{1615.74}{40}\right) = 0.3935$$

After the previously specified modification, $A^* = A^2(1.0 + 0.75/n + 2.25/n^2)$, for the normal distribution, $A^* = 0.4014$. This is less than the critical value of 0.752.

Decision: The null hypothesis is not rejected.

As a matter of interest we repeat the exercise using the uniform distribution. The data range from 2411 to 2488 kg/m^3. With reference to the uniform pdf given by Eq. (4.2.1), we

Table 5.6.2*a* Anderson-Darling test applied to concrete densities: normal hypothesis

Rank	Concrete density (kg/m^3)	First F term	Second $(1-F)$ term	Summation term
1	2411	0.01675	0.00359	−9.72
2	2415	0.03032	0.03997	−29.87
3	2425	0.10550	0.04565	−56.54
4	2427	0.13014	0.05197	−91.52
5	2427	0.13014	0.06668	−134.24
6	2428	0.14385	0.20811	−172.84
7	2429	0.15850	0.22649	−216.19
8	2433	0.22649	0.24576	−259.42
9	2435	0.26587	0.24576	−305.80
10	2435	0.26587	0.26587	−356.14
11	2436	0.28678	0.28678	−408.60
12	2436	0.28678	0.28678	−466.05
13	2436	0.28678	0.37726	−521.65
14	2436	0.28678	0.40123	−580.03
15	2437	0.30844	0.42558	−638.92
16	2437	0.30844	0.42558	−701.86
17	2441	0.40123	0.45022	−758.33
18	2441	0.40123	0.45022	−818.23
19	2444	0.47506	0.47506	−873.31
20	2445	0.50000	0.50000	−927.37
21	2445	0.50000	0.50000	−984.21
22	2446	0.52494	0.52494	−1039.63
23	2447	0.54978	0.59877	−1089.63
24	2447	0.54978	0.59877	−1141.86
25	2448	0.57442	0.69156	−1187.09
26	2448	0.57442	0.69156	−1234.18
27	2449	0.59877	0.71322	−1279.27
28	2450	0.62274	0.71322	−1323.91
29	2454	0.71322	0.71322	−1362.44
30	2454	0.71322	0.71322	−1402.32
31	2455	0.73413	0.73413	−1440.02
32	2456	0.75424	0.73413	−1477.26
33	2456	0.75424	0.77351	−1512.29
34	2457	0.77351	0.8415	−1541.06
35	2458	0.79189	0.85615	−1567.87
36	2469	0.93332	0.86986	−1582.67
37	2471	0.94803	0.86986	−1596.74
38	2472	0.95435	0.8945	−1608.61
39	2473	0.96003	0.96968	−1614.12
40	2488	0.99641	0.98325	−1615.74

assume that $a = 2410$ and $b = 2490$ kg/m^3. The results are shown in Table 5.6.2*b*. The estimated A^2 is

$$-40 - \left(-\frac{1740.27}{40}\right) = 3.5068$$

which is greater than the critical value of 2.492 for a 5% level of significance but less than the 3.857 for a 1% level of significance, as given before, and assuming that the parameters are not estimated from the same sample used in the test. Thus one might use the uniform distribution as a first approximation.

Table 5.6.2*b* Anderson-Darling test applied to concrete densities: uniform hypothesis

Rank	Concrete density (kg/m^3)	First F term	Second $(1-F)$ term	Summation term
1	2411	0.01	0.03	−8.07
2	2415	0.06	0.21	−21.04
3	2425	0.19	0.23	−36.86
4	2427	0.21	0.24	−57.77
5	2427	0.21	0.26	−83.74
6	2428	0.23	0.40	−110.23
7	2429	0.24	0.41	−140.43
8	2433	0.29	0.43	−171.97
9	2435	0.31	0.43	−206.29
10	2435	0.31	0.44	−244.09
11	2436	0.33	0.45	−284.46
12	2436	0.33	0.45	−328.68
13	2436	0.33	0.50	−374.11
14	2436	0.33	0.51	−422.50
15	2437	0.34	0.53	−472.69
16	2437	0.34	0.53	−526.33
17	2441	0.39	0.54	−578.11
18	2441	0.39	0.54	−633.02
19	2444	0.43	0.55	−686.80
20	2445	0.44	0.56	−741.48
21	2445	0.44	0.56	−798.96
22	2446	0.45	0.58	−857.09
23	2447	0.46	0.61	−913.85
24	2447	0.46	0.61	−973.13
25	2448	0.48	0.66	−1029.79
26	2448	0.48	0.66	−1088.75
27	2449	0.49	0.68	−1147.66
28	2450	0.50	0.68	−1207.40
29	2454	0.55	0.68	−1263.88
30	2454	0.55	0.68	−1322.34
31	2455	0.56	0.69	−1380.30
32	2456	0.58	0.69	−1438.76
33	2456	0.58	0.71	−1496.77
34	2457	0.59	0.76	−1550.57
35	2458	0.60	0.78	−1603.41
36	2469	0.74	0.79	−1641.99
37	2471	0.76	0.79	−1679.22
38	2472	0.78	0.81	−1713.91
39	2473	0.79	0.94	−1737.27
40	2488	0.98	0.99	−1740.27

Definition and properties: Anderson-Darling goodness-of-fit test statistic

$$A^2 = -n - \sum_{i=1}^{n} \frac{(2i-1)(\ln F_0[x_{(i)}] + \ln\{1 - F_0[x_{(n-i+1)}]\})}{n},$$

in which $x_{(1)}, x_{(2)}, \ldots, x_{(n)}$ are the observations ordered in increasing order, has asymptotic values of 1.933, 2.492, 3.070, and 3.857 for $\alpha = 0.10$, 0.05, 0.025, and 0.01 levels of significance, respectively. These values are good approximations for $n > 10$ for a completely specified distribution of any type. However, if the parameters are estimated from the sample used for the test, the test statistic is modified to $A^* = A^2(1.0 + 0.75/n + 2.25/n^2)$ and $A^* = A^2(1.0 + 0.3/n)$ in the case of the normal and exponential distributions, respectively. In this situation, the corresponding critical values are 0.631, 0.752, 0.873, and 1.035 for the normal and 1.062, 1.321, 1.591, and 1.959 for the exponential distributions, respectively.

5.6.5 Other methods for testing the goodness-of-fit to a normal distribution

The following is a summary of additional procedures for verification of the fit of a normal distribution to an observed distribution.

5.6.5.1 Sample skewness g_1 and kurtosis g_2

As previously noted, for the null hypotheses $E[G_1] = 0$ and $E[G_2] = 3$. Also, as shown by R. A. Fisher, the asymptotic variance of skewness is $\text{Var}[G_1] = \{6n(n-1)/[(n-2)(n+1)(n+3)]\}$, which for large samples tends to $\text{Var}[G_1] = 6/n$. Under the null hypothesis, the distribution of the sample skewness G_1 is normal whereas that of the sample kurtosis G_2 is nonnormal. The asymptotic standard error of kurtosis is $3 + (24/n)^{1/2}$. The probability distribution of G_2 can be obtained through computer simulation. For sample sizes of 20, 30, 40, 50, 75, 100, 150, and 200 items, the upper and lower 95% confidence limits are 4.68, 1.73; 4.57, 1.84; 4.46, 1.99; 4.36, 2.06; 4.17, 2.19; 4.03, 2.27; 3.86, 2.37; 3.74, 2.44, respectively.[19] Thus the tests based on skewness and kurtosis can be applied without much computation. However, there are more powerful tests as discussed in this chapter.

Example 5.32. Testing strengths and density of materials for normality by using the skewness and kurtosis statistics. We refer to the data summaries in Table 1.2.2.

Null hypothesis H_0: The random variables sampled by each of the strengths and the density in Tables 1.1.3 and 1.2.1 have a normal distribution.

Alternate hypothesis H_1: The random variables do not have the specified distribution.

Level of significance: $\alpha = 0.05$.

Calculations:

(a) Timber strengths

$$\text{Var}[G_1] = \left[\frac{6 \times 165 \times 164}{163 \times 166 \times 168}\right] = 0.189^2$$

for the skewness of the timber strengths using the full sample $n = 165$; the variance is equal to 0.190^2 for $n = 164$. Thus the critical values of skewness are $\pm 1.96 \times 0.189 = \pm 0.37$. We see that in both cases the critical values are exceeded with respect to either skewness or kurtosis (see values for kurtosis given before for $n = 150$ and $n = 200$).

(b) Concrete strengths and densities

$$\text{Var}[G_1] = \left[\frac{6 \times 40 \times 39}{38 \times 41 \times 43}\right] = 0.374^2.$$

The large sample critical value of skewness is $\pm 1.96 \times 0.374 = \pm 0.73$. We see that in both cases, the critical values are not exceeded either with respect to skewness or kurtosis (see values for kurtosis given before for $n = 50$).

Decision: The null hypothesis is rejected in the case of timber strengths but not in the cases of concrete strength and densities.

[19] See, for example, D'Agostino and Pearson (1973). For other values of n and α, reference may be made to tables provided by Barnett and Lewis (1994) or Rosner (1975) or Ferguson (1961).

5.6.5.2 Shapiro and Wilk's W-statistic

This is the ratio of the sum of the squares of a specified linear combination of all the ordered sample values to the sum of the squared deviations of each of the values from the sample mean $\bar{x}_n$. It is given by

$$W = \frac{\left\{\sum_{i=1}^{[m]} a_{n,n+1-i}[x_{(n+1-i)} - x_{(i)}]\right\}^2}{\sum_{i=1}^{n} [x_i - \bar{x}_n]^2}.$$

The highest integer less than $n/2$ becomes the upper limit of summation $[m]$ in the numerator.[20] Theoretically, the test statistic is the same as

$$W = \int_{-\infty}^{+\infty} [F_n(x) - F_0(x)]^2 dF_0(x),$$

using the terminology of the Kolmogorov-Smirnov and Anderson-Darling tests. The test should be used in conjunction with a normal Q-Q plot.

5.6.5.3 Filliben's correlation coefficient

This is a correlation coefficient of the ordered observations and the order statistic medians from a standard normal population and is discussed in Section 5.8.

5.6.6 Summary of Section 5.6

In this section we have examined several goodness-of-fit tests. The chi-squared test has the least power, that is, the probability of incorrectly accepting a hypothesis is the highest in this case; simply stated, it is somewhat insensitive in detecting differences between models. Also, the grouping can affect the results of the test. However, it is easier to apply than the Kolmogorov-Smirnov tests and can be used when the estimates of the distribution and the test statistic are found from the same data.[21] However, when testing for normality the Kolmogorov-Smirnov test can be modified as shown in Table C.8 of Appendix C, when parameters are estimated from the same sample used in the test. The Anderson-Darling test is more powerful, particularly when detecting differences in the tails, and in situations in which suspected outliers are present; cases in which parameters are estimated from the same sample used in the test are considered separately, together with nonnormal distributions. We have also discussed additional methods of testing for normality such as the test based on skewness and kurtosis and also the Shapiro-Wilk Wtest; more, including tests for exponentiality, will be found in Section 5.9 in which we deal with outliers. The complementary practical methods of probability plotting, which are very useful in graphical verification of goodness-of-fit, will be highlighted in Section 5.8.

---

[20] The weights $a_{n,n+1-i}$ are tabulated by Shapiro and Wilk (1965), who give selected percentiles of W for $3 \leq n \leq 50$. Royston (1982) provides a Fortran algorithm based on an approximation to the null distribution of W for $n \leq 2000$. See also Barnett and Lewis (1994). It has been found to be a powerful test for normality by Pearson et al. (1977) but not by Tiku (1975). Besides, Shapiro et al. (1968) give the powers of various tests. Furthermore, Shapiro and Wilk (1972) provide a test for exponentiality.

[21] See comparisons by Stuart and Ord, 1991, pp. 1189–1190.

5.7 ANALYSIS OF VARIANCE

In this section the methods introduced in Sections 5.4 and 5.5 are generalized. However, contrary to what is implied by the title, which is usually abbreviated to ANOVA, we are concerned only with the variation in the means through the analysis of sums of squares. The methods discussed here vary from a single-factor ANOVA to the design and analysis of results of experiments involving two or more factors with interactions between them. Emphasis will of course be made on aspects of statistical inference. The analysis will be based on the fundamental statistical principles of randomization, replication, and unbiased measurement.

On the subject of hypothesis testing in Section 5.4, the question arose whether any known differences between sample means can be attributed to chance or whether the means of the sampled populations are different. For instance, it may be important to know whether three methods of aggregation (or, alternatively, three ways of curing) result in significantly different compressive strengths of concrete. In another application, one may wish to determine whether there are real changes in the lasting qualities of a road to which four different methods of surfacing are applied over various sections. A common feature in such experiments is that there is only one factor involved but it can take several levels. These levels are called *treatments*, and each treatment usually has numerous observations called *replicates*.[22]

We should seriously consider the possibility that any differences observed in the illustrations just cited arise from causes that are not accounted for in our postulations. In the road experiment, for example, the substructures or bearing capacities may not be the same over the sections tested; these factors can easily affect the durability of the road. We shall return to this subject in Example 5.34. Proper *experimental design* is necessary so that such extraneous causes do not exert an undue effect on the variable studied. For this purpose the general procedure is to perform an experiment in which we compare treatments, obtain variability through repeated observations, and use a measurement process that is kept under statistical control.

In another instance, suppose we are interested in the long-lasting qualities of water pipes made out of different materials, buried at various depths in soils of several types of chemical compositions and interacting with groundwater in which the acidity is not constant. Then one approach is to *randomize* the multiple effects and investigate the performance of each pipe under all possible exposures. *To perform the randomization procedure without bias, one must allocate experimental units to factors or treatments ideally by a random process using equal probabilities.* If we can include all the variations from extraneous factors under the subtotal for chance variation, then we have a *completely randomized design*.

There are practical and computational limitations however in randomizing out, on an equal probability basis, all causal factors that are not of direct interest even if we could identify every one of them. For instance, when working with past observations that cannot be repeated, data limitations are such that these conditions are not always realized. Nevertheless, any resulting differences can be minimized so that the variable studied is not *confounded* with other influences that are not under control. We shall discuss such aspects further in the examples that follow.

[22] The generic term *treatment* comes from the original methods of statistical experimental design, most notably by R. A. Fisher who collaborated with F. Yates, in agricultural and in genetic research; it represents fertilizers, soils, acidity levels, and so on.

The alternative to randomization is to conduct a *controlled experiment* in which all the factors except the one of interest are fixed. In the pipe experiment, for instance, if depth is the only variable with which we are concerned, we can conduct the experiment repeatedly in the same type of soil in localities where groundwater has similar interactive characteristics.

On the other hand, the extraneous effects can also be included as different types of factors if they are thought to be relevant to the design. Thus the analysis of variance also extends to the design of experiments with more than one factor. These aspects will be discussed under two-way analysis of variance in Subsection 5.7.2.

If we choose the values of a factor in predetermined ways then we have a *fixed-effects* model. The alternative method, which leads to a *random-effects* model, signifies that the choice is made in a random manner from a large population.

Here is another important point. Hitherto, single hypothesis has been tested as in Sections 5.4 and 5.5. The ANOVA procedure can accommodate the simultaneous testing of multiple hypotheses.

5.7.1 One-way analysis of variance

A simple type of problem with which one deals concerns k independent random samples of sizes $n_i, i = 1, 2, \ldots, k$, from k treatments. We denote the jth value from the ith treatment as X_{ij}. Thus the model for the one-way ANOVA is of the linear form

$$X_{ij} = \mu = + \theta_i + \varepsilon_{ij}, \quad i = 1, 2, \ldots, k, \quad j = 1, 2, \ldots, n_i, \tag{5.7.1}$$

where μ is the overall mean and θ_i is the ith treatment effect. Also, we assume that the ε_{ij} are independent and normally distributed random errors with zero mean and finite variance σ^2, which is the same for all treatments. The model given by Eq. (5.7.1) is also called the *one-way classification* because we are concerned with only one factor.

The null hypothesis in our relatively simple model as just defined is that the treatment effects are equal. The opposing alternative hypothesis is that this is not true, at least in one case. That is,

$$\begin{aligned} H_0 &: \theta_i = 0 \quad && \text{for } i = 0, 1, 2, \ldots, k, \\ H_1 &: \theta_i \neq 0, && \text{for at least one } i. \end{aligned} \tag{5.7.2}$$

For this type of test we analyze the total variability of the X_{ij}, by which we mean m var(X_{ij}) where $m = \sum_{i=1}^{k} n_i$ To simplify further, let us assume that the n_i in Eq. (5.7.1) are equal so that $m = nk$. Thus the variability, which is the sum of the squared differences from the overall mean, is estimated by

$$\sum_{i=1}^{k}\sum_{j=1}^{n}(x_{ij} - \bar{x}_{.})^2 \quad \text{where} \quad \bar{x}_{.} = \frac{1}{nk}\sum_{i=1}^{k}\sum_{j=1}^{n} x_{ij}.$$

If H_0 is true, this variability is totally attributed to chance effects. Under H_1, on the contrary, a part of the variability can be attributed to differences between the treatment effects, which we estimate by

$$\bar{x}_i = \frac{1}{n}\sum_{j=1}^{n} x_{ij}. \tag{5.7.3}$$

The test procedure applicable to Eq. (5.7.2) is called *analysis of variance* because we partition the total variability in the data (as given by sums of squared differences from the means) into different parts. The division is as follows:

$$\begin{aligned}\sum_{i=1}^{k}\sum_{j=1}^{n}(x_{ij}-\bar{x}_{.})^2 &= \sum_{i=1}^{k}\sum_{j=1}^{n}[(\bar{x}_i-\bar{x}_{.})+(x_{ij}-\bar{x}_i)]^2\\ &= \sum_{i=1}^{k}\sum_{j=1}^{n}(\bar{x}_i-\bar{x}_{.})^2+2\sum_{i=1}^{k}\sum_{j=1}^{n}(\bar{x}-\bar{x}_{.})(x_{ij}-\bar{x}_i)\\ &\quad+\sum_{i=1}^{k}\sum_{j=1}^{n}(x_{ij}-\bar{x}_i)^2\\ &= n\sum_{i=1}^{k}(\bar{x}_i-\bar{x}_{.})^2+\sum_{i=1}^{k}\sum_{j=1}^{n}(x_{ij}-\bar{x}_i)^2. \end{aligned} \tag{5.7.4a}$$

We note that the cross-product term is zero; this arises from the fact that for each i the summation of the j terms $\sum_{j=1}^{n}(x_{ij}-\bar{x}_i)$ is zero, which follows from Eq. (5.7.3).

We can write Eq. (5.7.4a) as

$$SS_T = SS_{Tr} + SS_E. \tag{5.7.4b}$$

The first term in Eq. (5.7.5b) is the total sum of squares, or SS_T. Then on the right-hand side we have, respectively, the sum of squares attributed to the differences in the treatments, or SS_{Tr} (also called the *between-samples sum of squares*), and the sum of squares arising from chance or experimental effects called *errors*, or SS_E.

Suppose we denote by $T_.$ the grand total of all kn values. Then noting that $\bar{x}_. = T_./kn$ we write

$$SS_T = \sum_{i=1}^{k}\sum_{j=1}^{n}(x_{ij}-\bar{x}_.)^2 = \sum_{i=1}^{k}\sum_{j=1}^{n}x_{ij}^2 - 2\bar{x}_.T_. + nk\bar{x}_.^2 = \sum_{i=1}^{k}\sum_{j=1}^{n}x_{ij}^2 - \frac{1}{kn}T_.^2 \tag{5.7.5a}$$

Also, if we denote by T_i the sum of the values for the ith treatment, and because $\bar{x}_i = T_i/n$,

$$SS_{Tr} = n\sum_{i=1}^{k}(\bar{x}_i-\bar{x}_.)^2 = n\sum_{i=1}^{k}\bar{x}_i^2 - 2\bar{x}_.n\sum_{i=1}^{k}\bar{x}_i + nk\bar{x}_.^2 = \frac{1}{n}\sum_{i=1}^{k}T_i^2 - \frac{1}{kn}T_.^2. \tag{5.7.5b}$$

Thus Eq. (5.7.5a) and (5.7.5b) provides rapid methods, alternative to those of Eq. (5.7.4a) and (5.7.4b), for calculating the total sum of squares and the sum of squares of the treatments.

It follows from Eq. (5.3.13) that $\sum_{j=1}^{n}(x_{ij}-\bar{x}_i)^2/\sigma^2$ is a χ^2_{n-1} variate, where σ^2 was defined for Eq. (5.7.1). This is applicable to each treatment effect $i = 1, 2, \ldots, k$. Hence by the rule governing the addition of chi-squared variables as given by Eq. (5.3.9),

$$\frac{\sum_{i=1}^{k}\sum_{j=1}^{n}(x_{ij}-\bar{x}_i)^2}{\sigma^2}$$

is a $\chi^2_{k(n-1)}$ variate. It also follows that if H_0 is true, the $\bar{x}_i$, $i = 1, 2, \ldots, k$ are independent and identically distributed $N(\mu, \sigma^2/n)$ variates; hence $\sum_{i=1}^{k}(\bar{x}_i-\bar{x}_.)^2/(\sigma^2/n)$ is a χ^2_{k-1}

Table 5.7.1 Analysis of variance for a one-way classification

Source of variation	Degrees of freedom	Sum of squares	Mean square	F value
Treatment	$k-1$	SS_{Tr}	$MS_{Tr} = \frac{SS_{Tr}}{(k-1)}$	$\frac{MS_{Tr}}{MS_E}$
Error	$k(n-1)$	SS_E	$MS_E = \frac{SS_E}{k(n-1)}$	
Total	$kn-1$	SS_T		

variate. Then from the properties of chi-squared variates, and using Eq. (5.7.4a) and (5.7.4b), we know that $SS_E/k(n-1)$ and $SS_{Tr}/(k-1)$ are each estimates of σ^2 under the null hypothesis. On the contrary, if the null hypothesis is false, a part of $SS_{Tr}/(k-1)$ arises from differences in the treatment effects as defined by Eq. (5.7.2). In such a situation this term will have a higher value than that of $SS_E/k(n-1)$. Taking the ratio of two chi-squared variables divided by their respective degrees of freedom, we can use the F distribution as defined by Eq. (5.4.14) to test the hypothesis; we assume that the variates are independent which is true under the null hypothesis.[23] That is,

$$F = \frac{SS_{Tr}/[(k-1)\sigma^2]}{SS_E/[(k(n-1)\sigma^2]} = \frac{k(n-1)SS_{Tr}}{(k-1)SS_E}. \tag{5.7.6}$$

A summary of the procedure adopted in dividing and analyzing the total sum of squares is given in Table 5.7.1.

Table 5.7.2 Table of concrete densities d, dates t, and strengths s from Table E.1.2

Treatment: density d, date t, strength s		Observations					Total	Mean
1	$d < 2430$ kg/m^3	2411	2415	2429	2428	2425		
	t	8/7	13/9	3/12	31/3[a]	26/6[a]		
	s	58.8	50.7	68.1	56.9	59.8	294.3	58.86
2	$2430 \leq d < 2440$ kg/m^3	2436	2436	2433	2436	2437		
	t	6/9	9/10	4/12	7/2[a]	21/9[a]		
	s	52.5	59.6	60.5	49.9	60.5	283.0	56.60
3	$2440 \leq d < 2450$ kg/m^3	2445	2447	2445	2446	2448		
	t	9/9	23/9	14/10	18/12	19/3[a]		
	s	63.3	55.8	60.0	60.9	67.3	307.3	61.46
4	$2450 \leq d < 2460$ kg/m^3	2454	2455	2454	2456	2456		
	t	12/7	2/9	3/10	6/12	9/3[a]		
	s	58.9	56.3	59.8	67.2	68.9	311.1	62.22
5	$d \geq 2460$ kg/m^3	2488	2473	2469	2472	2471		
	t	23/8	29/8	3/9	18/10	22/10		
	s	69.5	64.9	54.6	61.5	65.7	316.2	63.24
Total T and mean of s							1511.9	60.5

[a]Denotes that the year of testing is 1992; the other data (without superscript a) pertain to 1991 tests; the date $t = 31/3$, for example, denotes March 31.

[23] See, in particular, H. Scheffé, (1959, Chapter 7).

Example 5.33. Compressive strength and density of concrete. We refer to the densities and compressive strengths of concrete listed in Table E.1.2. Our objective is to test whether compressive strength is a function of density. Suppose we treat density d as a single factor affecting the compressive strength and divide the densities into five levels or treatments as follows:

(1) $d < 2430\ \text{kg/m}^3$,
(2) $2430 \leq d < 2440\ \text{kg/m}^3$,
(3) $2440 \leq d < 2450\ \text{kg/m}^3$,
(4) $2450 \leq d < 2460\ \text{kg/m}^3$,
(5) $d \geq 2460\ \text{kg/m}^3$.

Five "replicates" or observations are chosen for each treatment. Because experiments performed during a particular week (or longer period) may have some undesirable influences (such as operator bias) on the results, we shall try to spread the dates of observation as far as possible in the choice of observations from the set of 40 items. As seen, however, from Table 5.7.2 this effort is only partially successful in the case of the fifth treatment. Of course, this problem will not arise under a controlled experiment. We can in such a case randomize the effects of the experimenter, machinery, temperature, and other causal factors such as the moisture content of the concrete mix.

Null hypothesis H_0: The mean compressive strengths of concrete are the same for each of the five levels of density.
Alternate hypothesis H_1: There are differences in the means.
We note that the two hypotheses are formally stipulated by Eqs. (5.7.1) and (5.7.2).
Level of significance: $\alpha = 0.05$.

Calculations: The calculations are based on Tables 5.7.1 and 5.7.2 and Eq. (5.7.5*a*) and (5.7.5*b*). Thus,

$$SS_T = \sum_{i=1}^{5}\sum_{j=1}^{5} x_{ij}^2 - \frac{1}{25}T_{\cdot}^2 = 92{,}156.89 - \frac{1512.9^2}{25} = 723.23.$$

Also,

$$SS_{Tr} = \frac{1}{5}\sum_{i=1}^{5} T_i^2 - \frac{1}{25}T_{\cdot}^2 = \frac{457{,}900.43}{5} - \frac{1511.9^2}{25} = 146.42.$$

Hence,

$$SS_E = 723.23 - 146.42 = 576.81.$$

The results are summarized in Table 5.7.3.

Decision: The F value is 1.27 which is less than $F_{4,20,0.05} = 2.87$ [where the degrees of freedom are $m = 5 - 1 = 4$ and $n = 5(5 - 1) = 20$]. Therefore, the null hypothesis is not rejected.

Table 5.7.3 Analysis of variance for compressive strengths of concrete as a function of density

Source of variation	Degrees of freedom	Sum of squares	Mean square	F value
Density of concrete	4	146.42	$\frac{146.42}{4} = 36.61$	$\frac{36.61}{28.84} = 1.27$
Error	20	576.81	$\frac{576.81}{20} = 28.84$	
Total	24	723.23		

5.7.2 Two-way analysis of variance

As mentioned in the introduction to this section, many situations arise in which it is of practical interest to vary a particular extraneous factor (which we think has a significant influence on the variable studied) as much as possible so that its variation can be isolated. In this way the error sum of squares is reduced. The experiment is called a *two-factor experiment.* We then set out to do a two-way analysis of variance. In the one-way analysis of variance we partitioned the total sum of squares into two components, one of which was allocated to the experimental error. Now we have the treatment component as before and a new *block* component. A block, in this context, is a level at which an extraneous factor is held constant so that its effect on the total sum of squares can be estimated. A *complete block design* is one in which there are observations for each treatment within each block. If we randomize the treatments throughout each block, we have a *randomized block design.*

In a multifactor application one conducts a *factorial experiment* wherein all possible combinations of the levels of the factors are examined. Furthermore, for each combination of factors it is beneficial to have several trials or replicates as shown in the next example. In practice, however, tests of hypotheses are often limited to the most relevant factors in order to minimize the dimensions of the problems; this is termed an *incomplete block or factorial design.*

Alternatively, we can call the treatments and blocks *row and column variables* or factors I and II.

In addition, two factors may interact in the sense that if the factors are materials and forms, for example, one material can have a marked effect on a particular form whereas another material will be associated with another form. In such situations therefore, which may occur frequently, one needs to model the *interaction* between a pair of variables and test its significance.

The linear model for the two-way ANOVA is written as

$$X_{ijk} = \mu + \alpha_i + \beta_j + (\alpha\beta)_{ij} + \varepsilon_{ijk}, \text{ for } i = 1, 2, \ldots, l; j = 1, 2, \ldots, m; \\ k = 1, 2, \ldots, n. \tag{5.7.7}$$

Here μ is the overall mean, α_i is the treatment effect at level i, β_j is the block effect at level j, $(\alpha\beta)_{ij}$ is the interaction between the treatments and blocks at the stated levels and we assume that ε_{ijk} are independent $N(0, \sigma^2)$ variates. As stipulated by Eq. (5.7.7), there are l levels of treatments and m levels of blocks. Also, we have n replicates, one for each combination of treatment and block. Thus, we have a total of lmn observations. If we conduct the experiment so that these are obtained in a random order, we will then have a *completely randomized block design.*

Without the interaction term, Eq. (5.7.7) represents a simple linear *additive* model. The three sets of hypotheses are as follows:

(1) $H_0\colon \alpha_i = 0,$ for $i = 1, 2, \ldots, l.$

$H_1 : \alpha_i \neq 0,$ for at least one value of i. (5.7.8*a*)

(2) $H_0 : \beta_j = 0,$ for $j = 1, 2, \ldots, m.$

$H_1 : \beta_j \neq 0,$ for at least one value of j. (5.7.8*b*)

(3) $H_0 : (\alpha\beta)_{ij} = 0,$ for $i = 1, 2, \ldots, l; j = 1, 2, \ldots, m.$

$H_1 : (\alpha\beta)_{ij} \neq 0,$ for at least one pair of values of i and j. (5.7.8*c*)

The test is based on an analysis of the total variability, that is, the total sum of squared differences from the means, which is divided into different parts as follows:

$$\sum_{i=1}^{l}\sum_{j=1}^{m}\sum_{k=1}^{n}(x_{ijk} - \bar{x}_{.})^2$$
$$= \sum_{i=1}^{l}\sum_{j=1}^{m}\sum_{k=1}^{n}[(\bar{x}_i - \bar{x}_{.}) + (\bar{x}_j - \bar{x}_{.}) + (\bar{x}_{ij} - \bar{x}_i - \bar{x}_j + \bar{x}_{.}) + (x_{ijk} - \bar{x}_{ij})]^2$$
$$= mn\sum_{i=1}^{l}(\bar{x}_i - \bar{x}_{.})^2 + ln\sum_{j=1}^{m}(\bar{x}_j - \bar{x}_{.})^2$$
$$+ n\sum_{i=1}^{l}\sum_{j=1}^{m}(\bar{x}_{ij} - \bar{x}_i - \bar{x}_j + \bar{x}_{.})^2 + \sum_{i=1}^{l}\sum_{j=1}^{m}\sum_{k=1}^{n}(x_{ijk} - \bar{x}_{ij})^2. \tag{5.7.9a}$$

This result comes from the square of four terms so there are $3 \times 2 = 6$ cross-product terms. Each of which will sum to zero for reasons similar to those applicable to Eq. (5.7.4*a*). As before, we can symbolize the above equation by writing

$$SS_T = SS_{Tr} + SS_B + SS_{TrB} + SS_E \tag{5.7.9b}$$

with the terms corresponding to those of Eq. (5.7.9*a*).

To find solutions more expediently, let us denote by T_i the sum of the values for the ith treatment, by T_j the sum of the values for the jth block, by T_{ij} the sum of the values common to the ith treatment and the jth block, and by $T_{.}$ the grand total of all lmn values. Then, as in the case of Eq. (5.7.5*a*) and (5.7.5*b*), we can write the sum of squares in the alternative forms:

$$SS_T = \sum_{i=1}^{l}\sum_{j=1}^{m}\sum_{k=1}^{n}x_{ijk}^2 - \frac{T_{.}^2}{lmn}, \tag{5.7.10a}$$

$$SS_{Tr} = \sum_{i=1}^{l}\frac{T_i^2}{mn} - \frac{T_{.}^2}{lmn}, \tag{5.7.10b}$$

$$SS_B = \sum_{j=1}^{m}\frac{T_j^2}{ln} - \frac{T_{.}^2}{lmn}, \tag{5.7.10c}$$

$$SS_{TrB} = \sum_{i=1}^{l}\sum_{j=1}^{m}\frac{T_{ij}^2}{n} - \frac{T_{.}^2}{lmn} - SS_{Tr} - SS_B, \tag{5.7.10d}$$

$$SS_E = SS_T - SS_{Tr} - SS_B - SS_{TrB}. \tag{5.7.10e}$$

The testing procedure, which involves the use of the F distribution, is an extension of the methods applicable to the one-way analysis of variance as specified in the text leading up to Eq. (5.7.6). This is summarized in Table 5.7.4.

Example 5.34. Road rutting—a two-way classification experiment. Road-wearing experiments, with alternate base designs, were conducted at eight sites on busy highways in central England to investigate the behavior of the expected deterioration under heavy traffic. In particular, measurements were made at approximately 6-month intervals on the rutting or lowering of road surfaces. This example is based on site 6, which is 16 km northeast of Birmingham. Beneath the surface courses, which are about 130 mm (5 in.) in total depth, the thickness of the base layer was made equal to approximately 152 mm (6 in.) or 229 mm (9 in.) or 305 mm (12 in.) over the experimental sections. The systematic design also included

Table 5.7.4 Analysis of variance for a two-way classification with interaction

Source of variation	Degrees of freedom	Sum of squares	Mean square	F value
Treatments	$l-1$	SS_{Tr}	$MS_{Tr}=\frac{SS_{Tr}}{(l-1)}$	$\frac{MS_{Tr}}{MS_E}$
Blocks	$m-1$	SS_B	$MS_B=\frac{SS_B}{(m-1)}$	$\frac{MS_B}{MS_E}$
Interaction	$(l-1)(m-1)$	SS_{TrB}	$MS_{TrB}=\frac{SS_{TrB}}{(l-1)(m-1)}$	$\frac{MS_{TrB}}{MS_E}$
Error	$lm(n-1)$	SS_E	$MS_E=\frac{SS_E}{lm(n-1)}$	
Total	$lmn-1$	SS_T		

two types of base material. These were dense bituminous macadam (DBM) and hot-rolled asphalt (HRA). (Thus there were six sections, each with identical specifications at site 6. For all the experiments there was a total of 74 sections.) Below this layer there was a subbase of compacted stones; this was at least 229 mm (9 in.) in depth and did not have any type of added material such as bitumen or asphalt; it is not considered further. Measurements of the rutting of the road surface were made at 30 locations at site 6. At each location there were three types of base thickness and two types of base material, taken over six sections. At each section, measurements were taken at five locations called *replicates*, but not in the usual sense of the term which implies repeated measurements at the same spot. The experimental layout at the site is shown in Fig. 5.7.1.

Rutting data for the six combinations of base thickness and base material are presented in Tables E.5.1 to E.5.6 with dates of observations. There seem to be some shortcomings in the data. First, the time series of rutting are not monotonically increasing as they should be. Great care had been exercised in making measurements, taken at approximately 6-month intervals, to return to the same spots on the road, as located by numerous markings. However, the surface layers are known to shift from time to time with the impact of heavy vehicles and under the influence of extreme types of weather. These account for the oscillations in the data. Also, there are a few missing observations; these were simply infilled from neighboring values.[24] Alternatively, we considered leaving out the missing values but this will make the replicates different in length. The five time series of observations from Section 2, which is typical, with base thickness of 229 mm (9 in.) and DBM material is shown in Fig. 5.7.2.

Most of the data series show overall increases in time with no visible nonlinear behavior. Therefore, simple linear relationships were used for the rutting depth and time of measurement. (Attempts to fit nonlinear models did not result in any significant differences.) The linear model is written in the form

$$X_i = \beta_0 + \beta_1 u_i + \varepsilon_i$$

and the gradient β_1 of the fitted straight line (discussed further in Chapter 6) is estimated by the least squares method as

$$\hat{\beta} = \frac{\sum_{i=1}^{n}(u_i - \bar{u})(x_i - \bar{x})}{\sum_{i=1}^{n}(u_i - \bar{u})^2},$$

where x_i denotes a measurement at time i; $\bar{x}$ is the mean of the first n measurements, u_i is equivalent to integer i, and $\bar{u}$ is the mean of the first n integers. For the purpose of this exercise the assumption is made that the data points are equispaced in time, although this is not strictly correct. Linear models were applied and gradients estimated from each of the

[24] More elaborate infilling procedures are provided by Kottegoda and Elgy (1977).

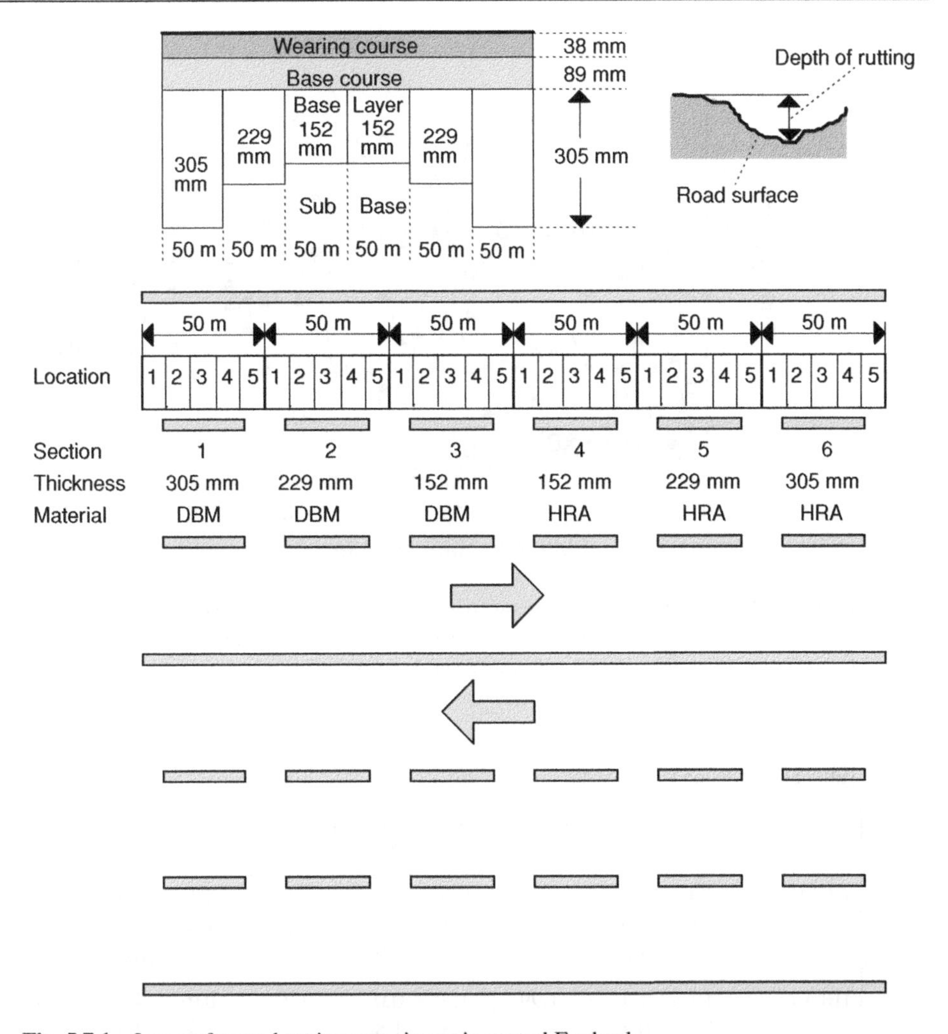

Fig. 5.7.1 Layout for road rutting experiment in central England.

30 series such as those shown in Fig. 5.7.2. For each of the 30 locations, the total increase in rutting over the entire measurement period is estimated by

$$(n-1)\hat{\beta}_1,$$

where n is the number of measurements. There are two negative values from the six series of measurements at site 6, and these are set to zero. The data are presented in Table 5.7.5.

For each combination of base thickness and material, called a cell, there are five replicates (called locations in Fig. 5.7.1). Table 5.7.5 also shows the cell means.

Hypotheses are as stipulated by Eq. (5.7.8).
Level of significance: $\alpha = 0.05$.

Calculations: Results of the two-way classification ANOVA with interaction using Eqs. (5.7.9) and (5.7.10) are shown in Table 5.7.6.

Critical region: The estimated F values are given in the last column. From Table C.4 the corresponding critical values for $\alpha = 0.05$ are $F_{1,24,0.05} = 4.26$ applicable to the treatments and $F_{2,24,0.05} = 3.40$ applicable to the blocks and interactions. We see that the estimated F values are not significant.

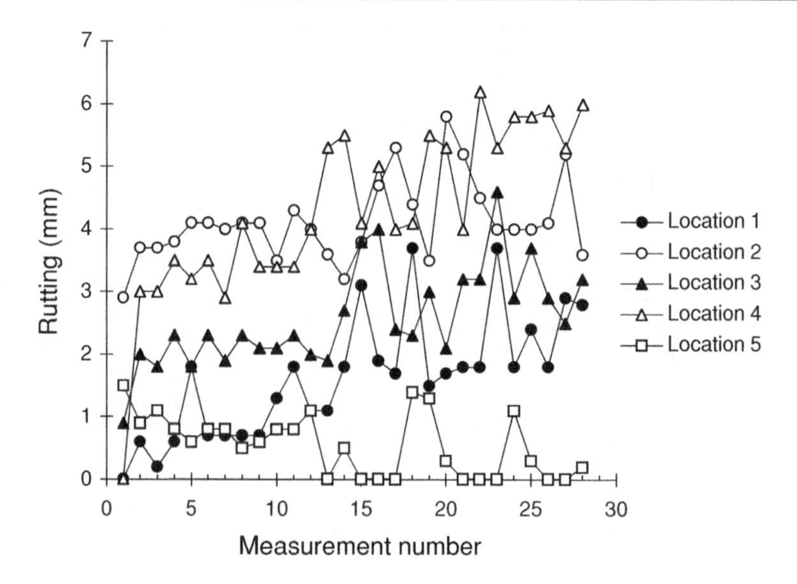

Fig. 5.7.2 Rutting of road surface at site 6, on highway 16 km northeast of Birmingham, England; 305-mm base thickness; base material: DBM (dense bituminous macadam). Measurement time series shown at five adjacent locations.

Table 5.7.5 Road rutting data

	6-inch base $j = 1$		cell mean value	9-inch base $j = 2$		Cell mean value	12-inch base $j = 3$		Cell mean value
DBM $i = 1$	4.4	3.5	2.94	0.6	2.9	2.09	2.4	0.9	1.75
	1.8	2.0		3.1	2.6		1.8	3.7	
		3.0			1.2			0.0	
HRA $i = 2$	4.1	1.4	1.99	1.3	0.0	1.35	0.9	4.9	3.06
	1.3	0.4		1.3	1.8		3.3	2.5	
		2.7			2.4			3.8	

Note: These data have been extracted from Tables E.5.1 to E.5.6 and Fig. 5.7.2.

Table 5.7.6 Analysis of variance for a two-way classification of road rutting data of Table 5.7.5 for treatments $i = 1, 2$ and blocks $j = 1, 2, 3$

Source of variation	Degrees of freedom	Sum of squares	Mean square	F value
Treatments	1	0.12	0.12	0.08
Blocks	2	3.45	1.73	1.08
Interaction	2	7.80	3.90	2.44
Error	24	38.36	1.60	
Total	29	49.73		

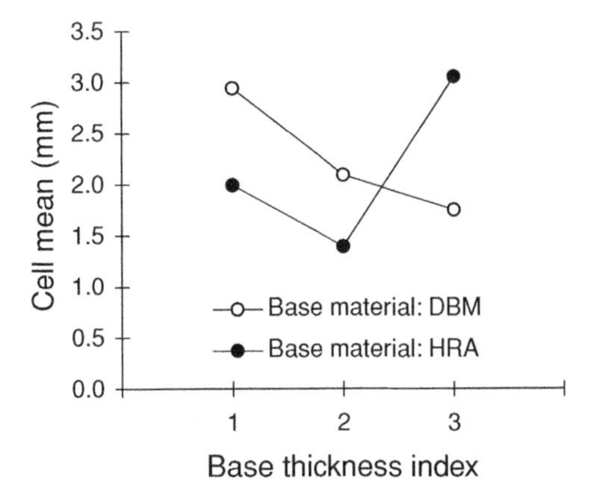

Fig. 5.7.3 Variation of mean rutting depth with base thickness indices (1–152 mm; 2–229 mm; 3–3.05 mm) for two types of base material.

In Fig. 5.7.3 are shown the variation of the cell means of the three types of base thickness for each of the two types of base material. For the base material DBM, there is expectedly a decrease in the rutting as base thickness increases. However, for the highest base thickness with the HRA material there is a large increase. The phenomenon of crossing in this graph is an indication of interaction that if supported by the ANOVA should be further investigated.[25]

We also analyzed the residuals for any unusual behavior which would violate the assumptions on which the ANOVA is made. The residuals are denoted as follows when there are two factors in the experiment:

$$\varepsilon_{ijk} = x_{ijk} - \bar{x}_{ij}.$$

These differences can be obtained directly from Table 5.7.5. The residuals are given in Table 5.7.7 and are seen to range from −2.2 to 2.1 mm.

A histogram of the residuals is shown in Fig. 5.7.4. The histogram does not show any marked nonnormal behavior. We also studied the residuals separately for each type of base thickness. From the results shown in Fig. 5.7.5, these are symmetrically distributed about zero; also, there are no large differences in the variances although the intermediate base thickness has residuals with less variance.

Table 5.7.7 Residuals of road surface data

Type of base material	6-inch base $j = 1$	9-inch base $j = 2$	12-inch base $j = 3$
DBM $i = 1$	1.5 1.6 −1.2 −1.0 0.1	−1.5 0.8 1.0 0.5 −0.9	0.6 −0.9 0.0 2.0 −1.7
HRA $i = 2$	2.1 −0.6 −0.7 −1.5 0.7	−0.1 −1.3 0.0 0.4 1.0	−2.2 1.8 0.2 −0.6 0.8

[25] See for example, Duncan (1955), who compares the range of any set of means with an approximate least significant range obtained from tables that are also given with an example by Johnson (1994, pp. 416–417). See also Montgomery and Runger (1994, pp. 650–652), Keuls (1952), Walpole and Myers (1993), and more advanced work by Scheffé (1959, Section 3.7)

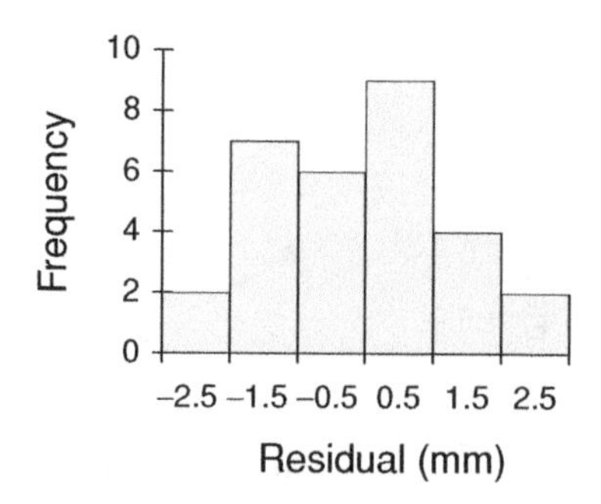

Fig. 5.7.4 Histogram of residuals from road rutting data.

The study was repeated for the two different base materials. The results are shown in Fig. 5.7.6. Here too there are no significant differences in the means and variances.

Decision: As stated, the estimated F values are not significant and the residuals do not show any unusual behavior in distributional properties. Hence the means of the variables used in the experiment, which are the base thickness and base material, are not significantly different within each type. Thus the rutting does not seem to be influenced by the thickness of the base or its material.

At other sites the outcomes may be quite different. Given the nature of the experiment this seems to be likely. The analysis was therefore repeated with data collected at site 8, where the layout is similar to site 6. The materials of the base are DBM and HRA with depths of 140 mm (5.5 in.), 216 mm (8.5 in.), and 292 mm (11.5 in.). Computations showed that the variation in the blocks, corresponding to the depths of the base layer, is significant for $\alpha = 0.05$ but not when $\alpha = 0.01$; the treatments (base material) and interactions are not significant. However, the residual analysis showed that the distribution has a very high kurtosis, and the variances for each base material and for each thickness are different by several magnitudes. Because this violates the assumptions of the model, one cannot conclude from the results at this site that the thickness of the base material affects the rutting of the road. Further investigations are necessary.

5.7.3 Summary of Section 5.7

In this section we have introduced some basic methods of analysis of variance and outlined concepts in experimental design. The approaches discussed are important to the civil and environmental and engineer.

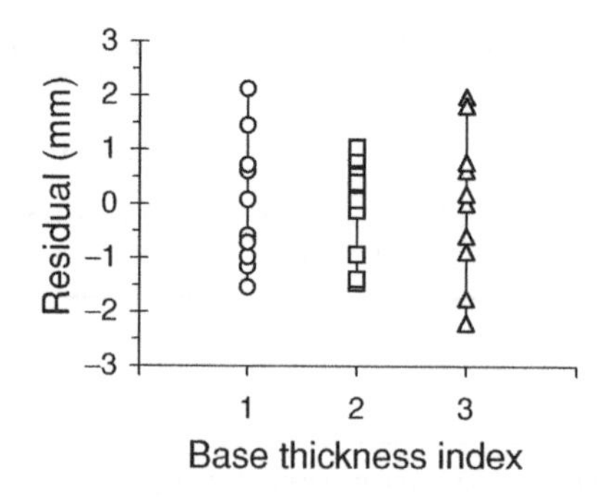

Fig. 5.7.5 Residuals from road rutting data for three values of base thickness: 1–152 mm; 2–229 mm; 3–305 mm.

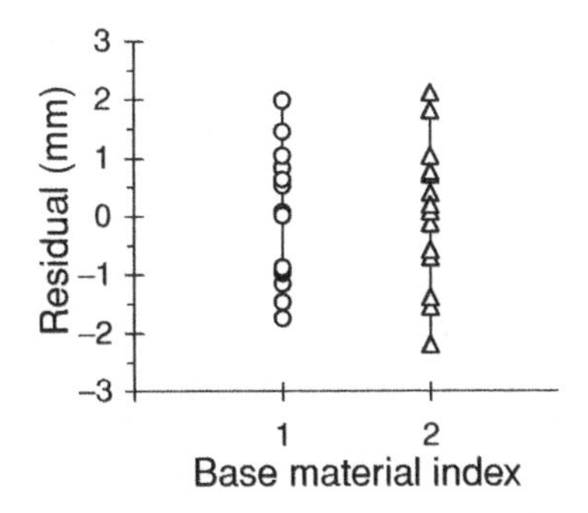

Fig. 5.7.6 Residuals from road rutting data for each type of base material: 1–dense bituminous macadam (DBM); 2–hot-rolled asphalt (HRA).

We have not discussed here the use of Latin squares, for example, discussed originally by Euler 200 years ago and adopted in experimental design by Fisher (1966) (and also included in numerous texts cited at the end of this chapter). This approach can be used when there is an important third factor, the levels of which are then symbolized by letters that appear once in each row and once in each column of a square; the rows and columns represent the first two factors and the objective is to remove from the experimental error the variation from these two factors.

As discussed, ANOVA is often applied in the presence of multiple factors. Also, there are situations in which an analysis of covariance (ANCOVA) is called for. This includes complex interactions between various factors and known or unknown influences between factors or sources; the purpose is to make "fair" comparisons between treatments or blocks and also to reduce the total residual variance for an overall assessment.[26]

In all applications, however, as in the case of other statistical techniques, the data and how they are obtained should be carefully considered. Furthermore, if the observations are not the result of a designed experiment [as in Fisher (1966)], there is a higher risk of reaching incorrect decisions.

5.8 PROBABILITY PLOTTING METHODS AND VISUAL AIDS

Graphical procedures form a very useful visual method of verifying whether a theoretical distribution fits an empirical distribution. The graphs are known as *probability plots* and are complementary to the goodness-of-fit tests described in Section 5.6. As stated in Subsection 1.4.3, a probability plot is a form of Q-Q plot in which one axis represents an empirical distribution and the other corresponds to a hypothesized theoretical distribution. The main advantage is that the plot shows us quite easily how well a theoretical distribution fits an empirical distribution. Such a graph is therefore widely accepted by engineers as a form of presentation of data, usually for a confirmation of an analysis; readily adaptable computer software or widely available computer facilities are very helpful. Some pioneering work in manual probability plotting was made by R. W. Powell, a civil engineer (see Powell, 1943); see also Cox (1978) who, in his history of graphical

[26] Details of these methods are found in Stuart and Ord (1991, pp. 1150–1152) and other references cited at the end of this chapter.

methods in statistics, states that probability plotting was suggested by Francis Galton in 1899.

For probability plotting by hand, one needs special types of graph paper called *probability paper.* The grid on one axis (the horizontal is generally preferred) of the paper is modified to suit the cdf of a particular distribution. Thus when the distribution function is plotted against the variate, which is scaled on the other axis, a linear relationship is obtained if the observations are from the hypothetical distribution. A common practice is to draw a best-fitting line by eye, which is sometimes called an *eye-ball* fit. Commercialized graph papers standardized accordingly are available for the normal, lognormal, exponential, extreme value, and Weibull continuous distributions and also for the binomial and Poisson discrete distributions.

In this section we shall examine in some detail some of the more useful types. Although the method is subjective, it can be supplemented (as shown here) by an associated goodness-of-fit criterion to support a relationship or the lack of one. A probability plot can indicate with respect to the pdf where the fit is not sufficiently good (for example, in the left tail, right tail, or the mode) and whether another type of distribution is more suitable. More importantly, it can show unusual features such as a change in a distribution at a point in time, or the presence of outliers. The plot also carries information on location spread, shape, and percentiles without the grouping problems one associates with a histogram. Furthermore, the parameters can be estimated from the intercept and gradient of the line, as shown in the examples that follow.

5.8.1 Probability plotting for uniform distribution

Consider a sample of n observations from a uniform(0, 1) distribution. The random variables $X_{(1)}, X_{(2)}, \ldots, X_{(n)}$ arranged in increasing order are termed *order statistics* and will be discussed further in Chapter 7. The expectation of a typical order statistic is

$$E(X_{(i)}) = \frac{i}{n+1} \tag{5.8.1}$$

(see Example 7.5).

If we plot the values of the ordered observations $x_{(1)}, x_{(2)}, \ldots, x_{(n)}$ from a uniform(0, 1) distribution against their expected values $[i/(n+1)]$, $i = 1, 2, \ldots, n$, we should find a linear relationship.

Example 5.35. Probability plot of concrete densities assuming a uniform distribution. Let us examine the distribution of the 40 concrete densities listed in Table E.1.2 using a probability plot. Initially, we shall compare the ordered observations and the expected values from a uniform(0, 1) distribution as defined by Eq. (5.8.1). The distribution we hypothesize is uniform(2411, 2488 kg/m^3). It follows that if the hypothesis is true, the ordered observations will form a linear relationship with the order statistics (expected values) of a uniform(0, 1) distribution. A plot is shown in Fig. 5.8.1.

The diagonal line shown here represents the hypothesized distribution. The shape of the curve is an S type, and this indicates a lack of fit in the tails. Thus the uniform distribution can only be viewed as a coarse approximation.

Example 5.36. Plotting of random uniform variates. Using a random number generator two samples of size 40 and two samples of size 200 were obtained. (Details follow in Subsection 8.2.1.) These sets are plotted in Fig 5.8.2. For the smaller data sets sampling fluctuations make the plots deviate in a random manner. They are different from the probability plot of the concrete data shown in Fig. 5.8.1 and do not seem to diverge from the diagonal so much.

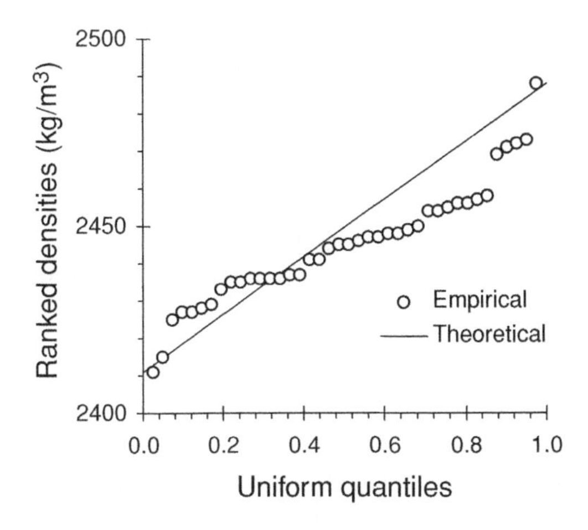

Fig. 5.8.1 Uniform probability plot of densities of concrete.

Note that as the sample size increases the deviations in the plots become less. For a theoretical sample of infinite size, one should expect a perfect linear plot to coincide with the diagonal.

5.8.2 Probability plotting for normal distribution

For the normal and other distributions of practical importance, the expected values (or other measures of location) of the order statistics do not exist in a closed form as defined by Eq. (5.8.1) for the uniform(0, 1) distribution. Therefore one often uses an approximation such as

$$E[X_{(i)}] = F^{-1}\left(\frac{i-c}{n-2c+1}\right), \quad \text{for } i = 1, 2, \ldots, n, \tag{5.8.2}$$

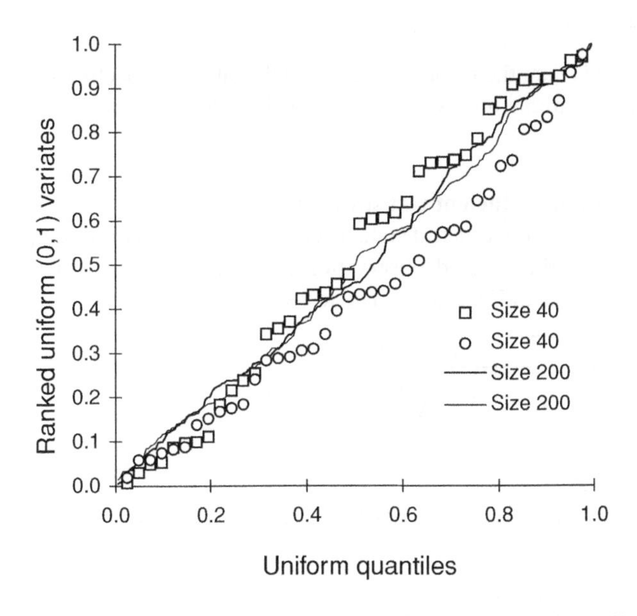

Fig. 5.8.2 Probability plots of uniform (0, 1) random variates: samples of size 40 and 200.

where c is a constant depending on the distribution $F(x)$. This expectation is of course the $[(i - c)/(n - 2c + 1)]$th quantile of the distribution. It is sometimes generalized to the $[(i + a)/(n + b)]$th quantile, where a and b are constants. The choice is known as a plotting position. This refers to the probability at which the $x_{(i)}$ should be plotted on a graph.

Much effort has been made in the past to obtain theoretically acceptable plotting positions, particularly with respect to bias. Filliben (1975), for instance, took the approximation $c = -0.3175$ and $b = 1 + 2c$ for the normal distribution; thus the ith plotting position becomes

$$p_i = \frac{i - 0.3175}{n + 0.365}, \tag{5.8.3}$$

except that $p_n = 0.5^{1/n}$ and $p_{(1)} = 1 - p_n$.

Our experience has been that the effects on the plots using different plotting positions for the same hypothetical distribution are not practically different except for small samples, say, less than 30 in size.[27] Therefore, let us consider the following plotting position:

$$p_i = \frac{i - 0.5}{n}. \tag{5.8.4}$$

This is named after Hazen, a civil engineer. It has been widely used by engineers and in recent textbooks.[28] However, to conform to results obtained elsewhere, we shall substitute the constant 0.35 for 0.5 in the above formula for the generalized Pareto, generalized extreme value, and related distributions such as the EV1 (Gumbel) and Weibull.[29]

In a normal probability plot, one axis, which is usually the horizontal one, has the standard normal quantiles z on a linear scale. These quantities are also called the *z scores* and defined as $E[Z(i)] = \Phi^{-1}(p_i)$. In commercialized graph paper, z values are replaced by $\Phi(z)$ on a nonlinear scale. The values of the ordered observations x are plotted against probabilities p_i; this has the effect of stretching the scale of ordinates toward the high and low probabilities. In either case the x versus z relationship that we plot is linear; it represents

$$x = \mu + \sigma z,$$

where μ and σ are the mean and standard deviation, respectively, of the X variable.

Many algorithms are available for the inverse function $z = \Phi^{-1}(p_i)$ of the normal distribution.[30]

5.8.2.1 Correlation coefficient test statistic

The following correlation test statistic can be used to determine empirically the goodness-of-fit. We use the ordered observations, $x_{(1)}, x_{(2)}, \ldots, x_{(n)}$ and their plotting positions $p_1, p_2, \ldots, p_n$ as in Eq. (5.8.4). The test statistic is given by

$$r = \frac{\sum_{i=1}^{n} (x_{(i)} - \bar{x})(p_i - \bar{p})}{\sqrt{\sum_{i=1}^{n} (x_{(i)} - \bar{x})^2}\sqrt{\sum_{i=1}^{n} (p_i - \bar{p})^2}}, \tag{5.8.5}$$

---

[27] See also, Wilk and Gnanadesikan (1968); they refer to a sample size less than 16 that affects the choice of plotting position.

[28] See Chambers et al. (1983).

[29] See, for example, Hosking (1990).

[30] See, for example, Abramowitz and Stegun [1964, p. 193, Eq. (26.2.23)].

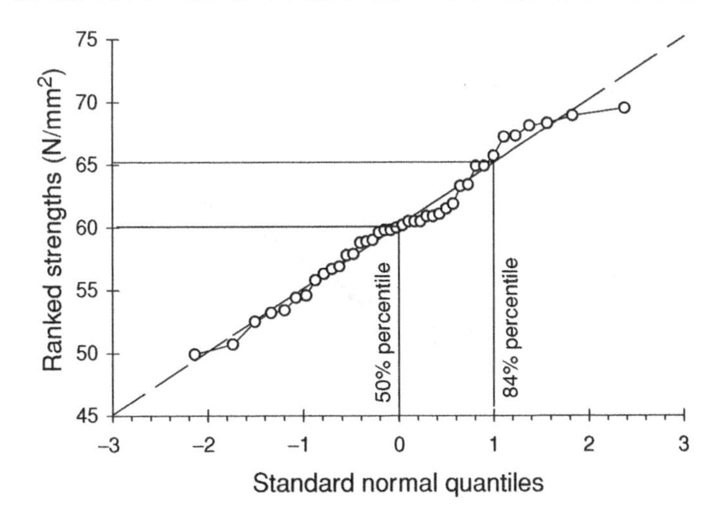

Fig. 5.8.3 Normal probability plot of concrete strengths.

where $\bar{x}$ and $\bar{p}$ are the means of the x and p series, respectively. This has been used, for example, by Filliben (1975).

Example 5.37. Normal probability plot of concrete strengths. A normal probability plot of the compressive strengths of concrete, which are ordered in Table 1.2.1, is shown in Fig. 5.8.3. The plotting position given by Eq. (5.8.3) is used. A straight line is fitted by eye, which is the usual practice. We see that the normal distribution provides a good fit to the data. For the mean, we read the ordinate from the straight line corresponding to $z = 0$, which is 60 N/mm^2 approximately. The standard deviation can be estimated from the difference in ordinates for $z = 0$ and $z = 1$ which is 6 N/mm^2. These are nearly equal to the results in Table 1.2.2.

The test correlation coefficient statistic from Eq. (5.8.5) is $r = 0.990$. From the table based on computer simulations provided by Filliben (1975) for a sample of size $n = 40$, these values have a probability level of about 0.65. This has the obvious interpretation that even if we had set our level of significance as high as $\alpha = 0.35$, the null hypothesis of normality cannot be rejected.

Example 5.38. Plot of normal and uniform variates on normal probability paper. Figure 5.8.4 shows a plot of a set of computer-generated standard normal variates plotted on normal paper. Also shown are uniform variates standardized by subtracting the mean and dividing by the standard deviation. Both are from samples of size $n = 40$ as in Figure 5.8.3. These plots are compared with the straight line that represents a theoretical normal distribution. The plot of the normal variates seems to be similar to that of the concrete strengths in Figure 5.8.3. However, the standardized uniform variates show discrepancies in both tails of the distribution, highlighting differences in the two distributions. This illustrates, in general, an important practical use of probability plots, that is, to uncover unexpected types of behavior that do not appear to be caused by sampling differences. This aspect is emphasized more clearly in the following example:

Example 5.39. Normal probability plot of observations with outliers. Probability plots are also very useful in showing possible outliers. We have seen in Chapter 1 that their presence can also be revealed (albeit at a much simpler level) by a box and whiskers plot (see Fig. 1.3.2). The U.S. Geological Survey, in common with other organizations, has records of numerous data sets of annual maximum river flows with suspected outliers. The

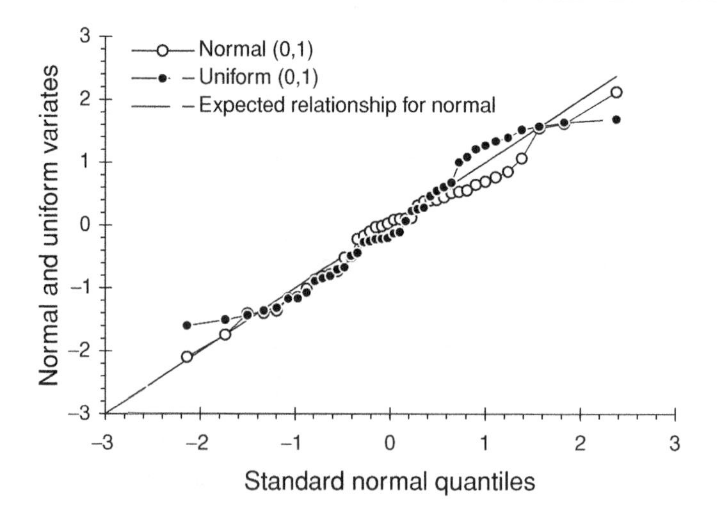

Fig. 5.8.4 Normal and uniform random variates on a normal plot.

Little River at Buffumville, MA, Station 01124500 (see data set 6 of Table E.5.7) is one of them. A probability plot of a 38-year sample from this observation station is shown in Fig. 5.8.5.

We see that the data conform to a normal distribution if one does not take into account the two highest observations. In the next section we shall test such data sets systematically for possible outliers.

Discrepancies are usually evident in the tails. Prominent curvatures indicate violation of the normal assumption. For example, if the distribution is uniform an S curve is seen as in Fig. 5.8.4; if the distribution is exponential, a J-shaped curve appears.

5.8.3 Probability plotting for Gumbel or EV1 distribution

In Chapter 3 we briefly discussed the Gumbel distribution (see Example 3.21) which is applied to extreme values, such as maximum or minimum events. This type of distribution

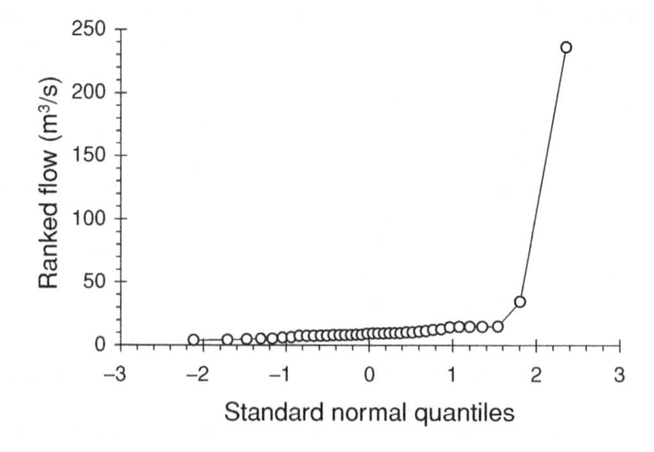

Fig. 5.8.5 Normal probability plot of annual maximum flows in the Little River, Buffumville, MA, with one or two suspected outliers.

is also known as the EV1 distribution and is examined in detail in Chapter 7. The cdf has the following form:

$$F_X(x) = \exp\left[-e^{-(x-b)/\alpha}\right],$$
$$\text{for } -\infty < x < \infty, -\infty < b < \infty, -\infty < \alpha < \infty, \quad (5.8.6)$$

where α and b are two parameters. Hence the reduced variate

$$y = -\ln[-\ln\ F(x)] = \frac{x-b}{\alpha} \quad (5.8.7)$$

plotted against the observations x will show a linear relationship if this distribution fits the data. Thus we have on a Gumbel or EV1 probability plot the ranked observations scaled on, say, the vertical axis with the reduced variates y on the horizontal axis. In commercial graph paper, this grid usually shows in addition the corresponding probabilities as given by the relationships of Eqs. (5.8.6) and (5.8.7).

Example 5.40. Gumbel extreme value probability plot of annual maximum flows. A plot of annual maximum flows in the Tevere (Tiber) at Ripetta, Roma, (see Table E.5.8) is shown on Gumbel extreme value paper in Fig. 5.8.6. We see that the Gumbel extreme value distribution provides a good approximation to the distribution of annual maximum flows.

If a straight line is fitted by eye, as shown, it follows from Eq. (5.8.7) that the parameters b and α can be estimated by the intercept at $y = 0$ and slope of the straight line, respectively. Thus $\hat{b} = 900$ m^3/s and $\hat{\alpha} = 380$ m^3/s.

5.8.4 Probability plotting of other distributions

We consider briefly other distributions which are suitable for probability plotting.

5.8.4.1 Exponential distribution

The exponential cdf takes the form

$$F_X(x) = 1 - \exp(-\lambda x)$$

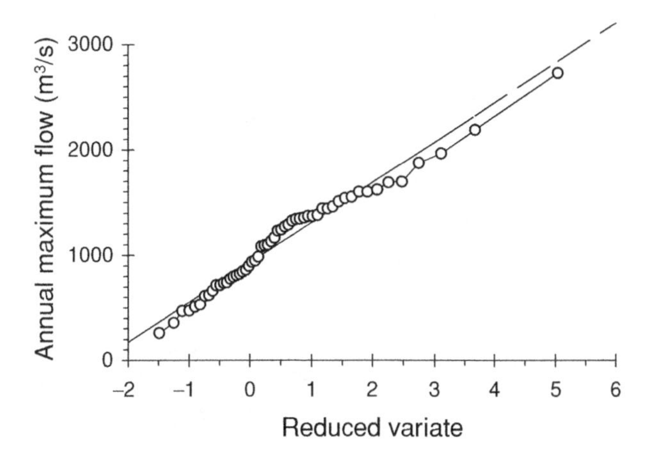

Fig. 5.8.6 Gumbel probability plot of annual maximum flows in the Tevere River at Ripetta, Rome, Italy.

[see Eq. (4.2.3*b*)]. Because

$$\ln[1 - F(x)] = -\lambda x,$$

a plot of x versus $\ln[1 - F(x)]$ will show a linear relationship for observations which are exponentially distributed (with negative slope). The parameter λ can be estimated from the reciprocal of the slope of the fitted straight line. A necessary condition is that this line must pass through the origin.

Example 5.41. Time intervals between vehicles. Consider the traffic data used in Example 4.21 (and summarized in Table 4.2.1). The ordinates of a histogram of the intervals between the passing of vehicular traffic with 30 equal class intervals of 6 seconds from 0 to 3 minutes as observed in Dorset, England, and reported by Leeming (1963), are given as follows:

54	23	16	10	16	16	12	8	8	7	4	5	4	5	1
2	0	3	1	2	2	2	0	0	1	0	0	1	1	0.

We can draw a probability plot of the data using the plotting position given by Eq. (5.8.4) and the central positions in the 30 classes or cells of the histogram. Thus,

$$F_X(x_{(i)}) = \frac{i - 0.5}{n}.$$

Accordingly, as shown above, we plot x versus $\ln[1 - F(x_{(i)})]$. The plot is shown in Fig. 5.8.7 for the aforementioned 30 values of i from the sample of $n = 204$.

It is seen that a straight line can be fitted to the data. However, for the longer gaps between vehicles for which data are few, with many zero items as just shown, there are some deviations. A straight line fitted by eye passes through the second point from the left (–4.95, 170). Hence,

$$\lambda = \frac{4.95}{170} \times 60 = 1.75 \text{ min}^{-1}.$$

As expected from the subjective nature of this approach, the result is somewhat different from the more accurate estimate of 1.81 min^{-1} obtained in Example 4.21 by the method of moments.

As already shown, graphical methods make it possible to take account of any peculiarities in the data such as values called outliers. This is the subject of the next section.

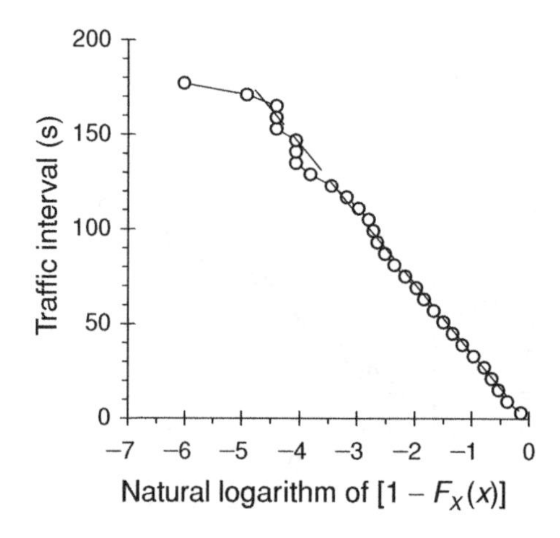

Fig. 5.8.7 Exponential probability plot of traffic interval data (fitted line excludes the two highest points).

In general, Cox (1978) compares ten graphical methods for assessing consistency with the exponential distribution and finds that most are transformations of one another. He states that the choice is partly "a matter of taste."

5.8.4.2 Lognormal distribution

Special types of normal probability paper are available with a logarithmic scale on one axis so that the observations are effectively transformed to their logarithms when plotted. We can alternatively transform the observations to logarithms and use normal probability paper. In a similar manner we can test whether the observations have a Johnson type of distribution (see Chapter 4) and in general whether any transformation of the random variable to normality is possible.

5.8.4.3 Weibull distribution

From Eq. (4.2.15), the Weibull cdf is given by

$$F_X(x) = 1 - \exp\left[-\left(\frac{x}{\lambda}\right)^{\beta}\right], \qquad \text{for } x > 0.$$

By writing $z = \ln(x)$, rearranging terms, and taking natural logarithms twice,

$$z = (1/\beta)\ln\{-\ln[1 - F_X(x)]\} + \ln(\lambda).$$

Therefore, if the x values are Weibull-distributed then $\ln(x)$ will plot as a straight line against $\ln\{-\ln[1 - F_X(x)]\}$. This result was demonstrated in Example 4.25 and Fig. 4.2.7. In commercial graph paper, a logarithmic scale is shown on one axis for plotting values of observations. The other axis is scaled linearly with respect to $\ln\{-\ln[1 - F_X(x)]\}$ but corresponding probabilities (for which a plotting position is used as in Example 4.25) are given.

5.8.4.4 Poisson distribution

In Poisson graph paper there are curves for each value of the random variable x from, say, 0 to 15. Each curve shows the relationship of the Poisson probability, on the vertical axis, with the value of the parameter ν, on the horizontal axis [as given by Eq. (4.1.7)]. One plots a dot on each curve corresponding to the empirical probability obtained from the data for that value of x (as in Example 4.4 and Table 4.1.1, for example). If these dots are generally close to a vertical straight line, the Poisson is a good approximation.[31] The position of the best-fitting vertical line with respect to the horizontal scale gives the expected value of the parameter ν.

5.8.5 Visual fitting methods based on the histogram

Some additional methods can provide further insight to the fit or lack of fit of a distribution. These are based on the histogram introduced in Chapter 1.

5.8.5.1 Hanging histogram

This histogram is a visual display of the differences between the numbers of observed and fitted values. Because this is an input to the chi-squared goodness-of-fit test, the

[31] See, for example, Volk (1969) and Problem 5.36.

resulting diagrams should be used together with the chi-squared statistic.[32] Our notation corresponds with that of the frequency histogram of Section 1.1, which gives the observed values O_i within each class i. Assuming that a particular theoretical distribution fits the data, we can also find the expected values E_i for class i. For instance, if x_i is the right boundary (maximum value) of the class and x_{i-1} is the right boundary of the previous (smaller) class,

$$E_i = n[F(x_i) - F(x_{i-1})],$$

where n is the number of observations and $F(x)$ denotes the value of the fitted cdf at x; methods of calculating $F(x)$ are discussed in Subsection 5.6.2. The hanging histogram is a plot of the difference in frequencies $(O_i - E_i)$ against the class i.

5.8.5.2 Variance-adjusted hanging histogram

Useful as it may be for diagnostic purposes, the hanging histogram is affected by the unequal variances in the classes or cells, arising particularly from the variability of the O_i. The expected frequencies E_i are also variable but to a much smaller magnitude. Therefore, and by using the properties of the binomial distribution [see Eq. (4.1.5*b*)],

$$\text{Var}[O_i - E_i] \approx \text{Var}[O_i] = np_i(1 - p_i),$$

where p_i is the probability of occurrence of an event in class i. Considering also that p_i^2 is small in comparison with p_i, we may write

$$\text{Var}[O_i - E_i] \approx np_i = E_{i.}$$

If we rescale the difference $(O_i - E_i)$ by using

$$\frac{O_i - E_i}{\sqrt{E_i}},$$

the resulting plot may be called the *variance-adjusted hanging histogram*. We also see that the foregoing variables can be squared and summed to give the chi-squared statistic [Eq. (5.6.1)]; this, as noted in Subsection 5.6.1, is also an approximation. Thus a plot of the adjusted differences can be viewed as a goodness-of-fit measure in unison with the chi-squared test.

Example 5.42. Hanging histogram of timber strength data. Consider the histogram of Fig. 1.1.4 which is based on Table 1.1.4. To demonstrate an application of the hanging histogram, we shall fit first the Weibull and second the normal distributions. However, to examine the details more closely, we shall use smaller intervals of 2.5 N/mm^2. The difference in frequencies $(O_i - E_i)$ is plotted first in Fig. 5.8.8*a* for the Weibull and second in Fig. 5.8.8*b* for the normal distribution. We notice that in general the differences are somewhat smaller for the Weibull confirming our intuition based on Table 1.2.2 that this is a more appropriate distribution.

The corresponding variance adjusted differences

$$\frac{O_i - E_i}{\sqrt{E_i}}$$

are plotted in Fig. 5.8.9*a* and 5.8.9*b*. A feature of the discrepancy between the two diagrams is the extended right tail of the normal distribution. This shows some lack of fit and in a positively skewed distribution (as seen from Table 1.2.2) the right tail is what we look at initially.

[32] These methods were originated by J. W. Tukey; see Velleman and Hoaglin (1981) and Rice (1995).

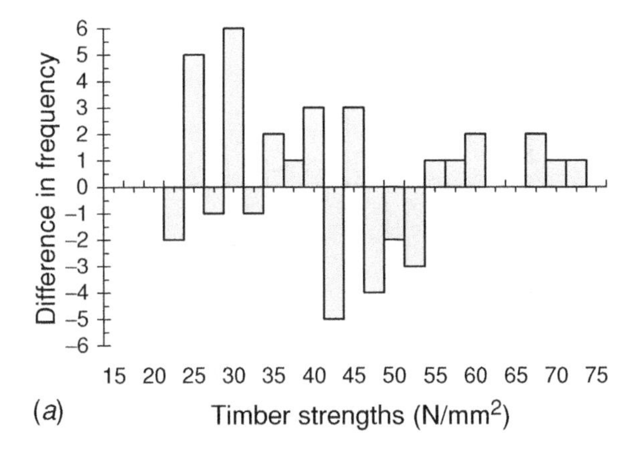

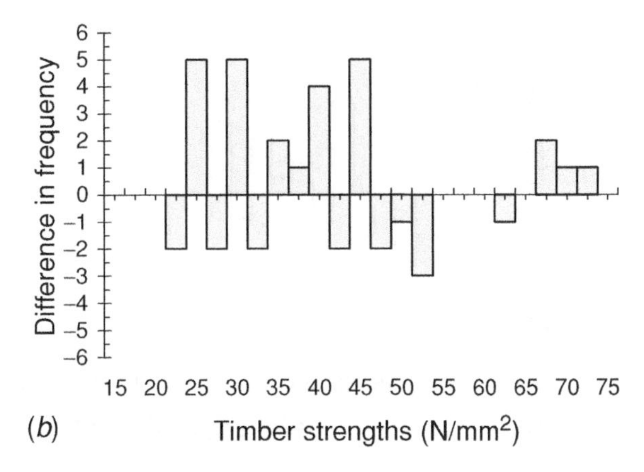

Fig. 5.8.8 Hanging histogram for distributions of timber strength data: (*a*) Weibull and (*b*) normal.

5.8.6 Summary of Section 5.8

In this section we have discussed and examined some of the methods of probability plotting. Their use by engineers is becoming increasingly common. Some methods based on the histogram were also discussed. However, they should be treated as an aid in the determination of the distribution of a random variable and be used in conjunction with a goodness-of-fit criterion. In practice, we can determine only whether a particular distribution provides a close approximation to the empirical distribution; there will always be some uncertainty. The probability plot and associated methods highlight features such as the fit in the tails of the distribution and the presence of discordant values or outliers.

5.9 IDENTIFICATION AND ACCOMMODATION OF OUTLIERS

In any sample of observations, there is the possibility of having one or more unexpectedly high or low values. These values are so far distant in magnitude from the other observations that they do not seem to be representative of the sample; that is, they do not apparently have the same distribution. Such unexpectedly high or low values are called *outliers* or

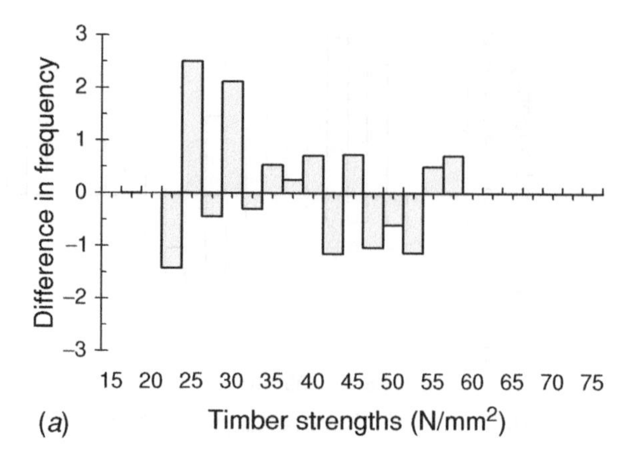

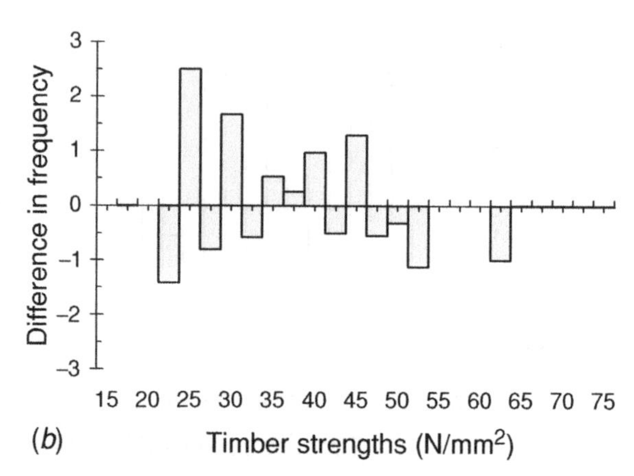

Fig. 5.8.9 Variance-adjusted hanging histogram for distributions of timber strength data (*a*) Weibull and (*b*) normal.

discordant observations. Sometimes the terms *rogue* or *spurious* are used to describe outlying observations, but we prefer to use these names for other types of suspected observations.

5.9.1 Hypothesis tests

The prior detection of outliers is often based on probability plots or box plots, and depends on the type of data and how they are presented. After initial detection, precise identification is possible through tests of homogeneity. These are formulated in order to determine whether under the null hypothesis any suspected observations, say, k in number, cannot be rejected as part of a homogeneous set of size n (in which all members are identically distributed) produced by a specified model G. The alternative hypothesis is that the extreme observations are the outcome of a different generating mechanism L. That is,

$$H_0 : x_i \in G, \qquad i = 1, 2, \ldots, n$$

and

$$\begin{aligned} H_1 : x_i \in G, \quad & i = 1, 2, \ldots, n-k; \\ x_i \in L \neq G, \quad & i = n-k+1, n-k+2, \ldots, n. \end{aligned} \tag{5.9.1}$$

Here G can be the normal distribution or one which is normal by transformation. Although by invoking normality we have various tests at our disposal, our hypotheses need not be stereotyped in this way.

H_1 in Eq. (5.9.1) is referred to as the *mixture alternative*, in which L is a different type of (contaminant) distribution or may consist of more than one type. The number k is not usually known in advance and is usually guessed; it may be regarded as a binomial variable. In the *slippage alternative*, the distribution L may only have a difference in mean from G, or variance, or perhaps both. When $k = n$, we call H_1 *the distributional alternative.* In this case L has a different distribution that governs all the observations. The distributional alternative includes the thick-tailed types, such as the Cauchy distribution used, for example, in describing some types of economic data, and having the pdf,

$$f_X(x) = \frac{1}{\pi(1 + x^2)}. \tag{5.9.2}$$

In the following discussion we shall not consider the possibility of the distributional alternative with $k = n$, for an appropriate goodness-of-fit procedure can be applied in such a situation. Also, we shall confine ourselves to high outliers. These usually arouse greater practical interest, as seen in flood studies. It does not mean that possible outliers at the lower end of a sample of observations are not important; on the contrary they can significantly change sample estimates of parameters, particularly if one adopts a log-transformation. In both cases, similar tests are applicable.

5.9.2 Test statistics for detection of outliers

Under the normal null hypothesis there are several tests which can be used. Methods such as the Shapiro and Wilk's W-statistic and the skewness and kurtosis statistics have been described in Subsection 5.6.5. Also included was the Anderson-Darling test, which gives heavy weighting to the tails and should have superiority in detecting outliers. However, this test is applicable to any distribution and is therefore less preferable, in terms of power, than those which are devised to probe departures from the normal or any other specified distribution.

It should also be emphasized that tests described here are related to different aspects of nonnormality. Thus, it is quite possible not to have the same outcome after applying different tests to one set of observations (as will be seen in Example 5.44).

In the identification of outliers under the null hypothesis of normality, a procedure which is found to be sensitive is the Studentized deviate

$$B_j = \frac{|x_{(n-j+1)} - \bar{x}_{n-j+1}|}{s_{n-j+1}}, \tag{5.9.3}$$

in which $\bar{x}_{n-j+1}$ and s_{n-j+1} are the mean and standard deviation, respectively, of the (ordered) sample, $x_{(1)} < x_{(2)} < \cdots < x_{(n-j+1)}$.. The application is shown shortly.[33]

In general, the approach we adopt initially is to suspect the presence of k(high) outliers in the ordered set $x_{(1)} < x_{(2)} < \cdots < x_{(n-j+1)} < \cdots < x_{(n-1)} < x_{(n)}$. The number k is preassigned by examining any extraordinarily large gaps in magnitude between adjacent values at the upper end in relation to similar gaps elsewhere in the ordered set. The choice of k should be sufficiently generous to eliminate the possibility of underestimating the actual number of outliers present.

[33] See Kottegoda (1984); tables are provided by Jain (1981).

In returning to Eq. (5.9.3), the notation adopted here is such that $j = 1$ signifies that the full sample is considered and that the highest value is tested as an outlier; however, with $j = 2$ we are testing for two outliers, and to calculate the mean and standard deviation in this case we delete the highest value in the original sample. Critical values are given in Table C.9 for levels of significance $\alpha = 0.05$ and 0.01 and sample sizes ranging from 20 to 100. Similarly, with $j = k$ we commence by deleting the highest $k - 1$ values. For the statistical testing, a stepwise backward elimination procedure is adopted, commencing with the stated k outliers that are initially suspected. Starting with $j = k$, in each case a suspected outlier $x_{(n-j+1)}$ with corresponding test statistic B_j is tested against the $(n - j)$ observations that are less extreme. If B_j is not significant, then j is reduced by 1, and so on. The procedure is stopped when a test statistic B_j shows significance, for $j = l$, say, at a chosen level α (when compared with the critical values given in Table C.9). The observations $x_{(n-l+1)}, x_{(n-l+2)}, \ldots, x_{(n)}$ are then treated as outliers at the given level of significance α. That is, we declare l outliers, as shown in the next example; it is, of course, possible to finish with $l = 0$. In summary, in testing k initially suspected outliers, the test points out l outliers, where $0 \leq l \leq k$.

Example 5.43. Testing high flood outliers with Studentized statistic. In Table E.5.7 are listed series of observations of annual maximum flows in North American rivers. The unique feature of these sets of data are that there are, in each case, one or more suspected outliers. (These data were provided by the kind courtesy of the U.S. Geological Survey.) Consider, for example, series 10 for Quinebaug River at Quinebaug, CT, with $n = 45$. As shown in the probability plot of Fig. 5.9.1, this seems to be lognormally distributed (that is, the normal distribution provides a good approximation to the logarithms of the observations) except that there are four suspected outliers. For our use of the Studentized test, we chose $k = 4$.

Null hypothesis H_0: The distribution of the observation $x_{(n-3)}$ is the same as the lognormal distribution of the observations $x_{(1)}, x_{(2)}, \ldots, x_{(n-4)}$.
Alternate hypothesis H_1: The distribution of $x_{(n-3)}$ is different from the lognormal distribution of the observations $x_{(1)}, x_{(2)}, \ldots, x_{(n-4)}$.
Level of significance: $\alpha = 0.05$.

Calculations: Using the test statistic of Eq. (5.9.3) with the stepwise backward testing procedure, we find

$$B_1 = 4.25; \quad B_2 = 3.94; \quad B_3 = 3.70; \quad B_4 = 3.10; \quad B_5 = 1.71.$$

If we had abided by the expected generosity in our choice of k, we should have commenced with $k = 5$. Then we would have compared $B_5 = 1.71$ with the last entries 2.47 and 2.52 for $n = 40$ and 50, respectively, under $k = 5$. So, in this case, we cannot reject the null hypothesis (different from our null hypothesis for $k = 4$) that the fifth-largest observation $x_{(n-4)}$ has the same distribution as that of the smaller observations (as we had suspected from Fig. 5.9.1 and decided to use $k = 4$). In other words, we cannot declare that the fifth-largest observation $x_{(n-4)}$ is an outlier.

Returning to the hypotheses as given earlier, we compare $B_4 = 3.10$ with the lowest entries 2.55 and 2.59 in Table C.9 for $n = 40$ and 50, respectively, under $k = 4$. Clearly, $B_4 = 3.10$ is significant and, as seen from the table, so are B_3, B_2, and B_1.

Decision: We reject the null hypothesis. Thus the fourth-largest observation $x_{(n-3)}$ is an outlier and the three largest observations are also outliers. Accordingly, we declare that there are four outliers.

Example 5.44. Comparison of tests for outliers. We applied the skewness, kurtosis, and Studentized tests to detect outliers in the annual maximum flows in 11 North American rivers, as listed in Table E.5.7. By initially excluding the suspected outliers, the normal distribution provided a sufficiently close approximation to seven series, and for the remaining four series

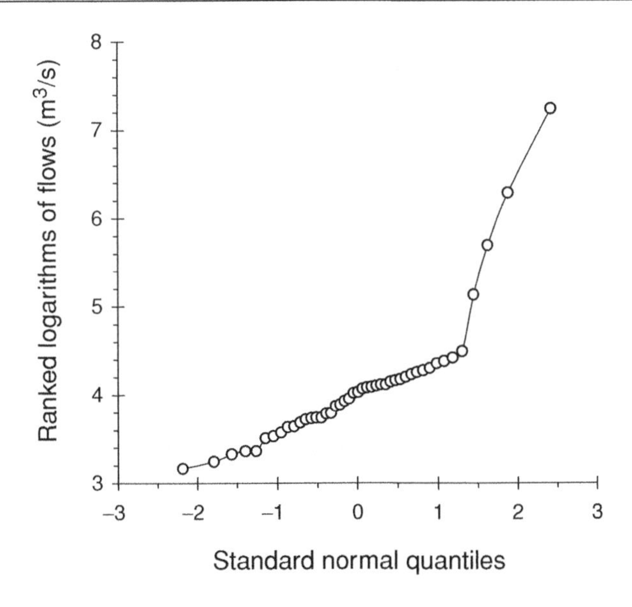

Fig. 5.9.1 Normal probability plot of log-transformed annual maximum flows in the Quinebaug River, Connecticutt, with four suspected outliers.

a simple log-transformation was similarly effective. The results of the test statistics, for the first nine series, are given in Table 5.9.1 for B_j, $j = 1, 2, 3$, for the Studentized deviates with reference to Eq. (5.9.3) and likewise for skewness and kurtosis. Because the last two series have more than two suspected outliers, the test statistics were extended using a maximum value $j = 6$.

Critical values for skewness and kurtosis statistics are obtained as shown in Section 5.6 and those for the Studentized deviates from Table C.9. The test statistics that exceed the critical values are underlined and the numbers of outliers declared, l, are given in the last column. Results are the same for the three tests except in the last two series: In these two cases, four outliers are declared on the basis of the more sensitive Studentized test, but only three are found by the other tests.

5.9.3 Dealing with nonnormal data

Subsection 5.9.2 was confined to normally distributed variables or to those variables that can be simply transformed to normality through logarithms. Similarly, if a random variable X has a two-parameter gamma distribution [Eq. (4.2.11)], then by the transformation of Wilson and Hilferty (1931), $X^{1/3}$ is normal in distribution. We can also extend these methods to cope with log-gamma-distributed variates. In all of these it is seen that the procedures for outlier detection does not depend on prior estimation of any (distributional) parameters.

However, if the lognormal or gamma distributions, for instance, need a location or shift parameter for its specification—such as $(X - \varepsilon)$ as in Eq. (4.2.6)—then the procedure becomes exploratory. This statement can also be said if the original observations are seen to be Gumbel-distributed (that is, EV1-distributed) as in Example 3.21, which is appropriate for flood flows. In this case the log-transformed variable $\exp[-(x - b)/\alpha]$ is exponentially distributed. It is necessary to estimate the parameters α and b prior to transformation. Methods of detecting outliers in exponential samples differ from those used for normal data. In the first instance, the coefficients of skewness and kurtosis may be compared to the theoretical values of 2 and 9, respectively (see Example 3.15). The

Table 5.9.1 Upper test statistics, $j = 1, 2, \ldots, 6$, from 11 series of annual maximum flows in North American rivers, with suspected outliers

			Skewness			Kurtosis			Studentizeddeviates			
Station	Distribution	n	$j = 1, 4$	$j = 2, 5$	$j = 3, 6$	$j = 1, 4$	$j = 2, 5$	$j = 3, 6$	$j = 1, 4$	$j = 2, 5$	$j = 3, 6$	l
1	1	25	$\underline{4.65}$	0.39	0.23	$\underline{22.75}$	2.67	2.59	$\underline{4.88}$	2.23	2.19	1
2	2	42	$\underline{2.43}$	$\underline{2.32}$	0.28	$\underline{9.89}$	$\underline{11.52}$	2.13	$\underline{3.94}$	$\underline{4.61}$	2.03	2
3	1	60	$\underline{6.99}$	−0.56	−0.65	$\underline{52.62}$	3.07	3.08	$\underline{7.50}$	1.96	1.69	1
4	1	17	$\underline{3.61}$	−0.28	−0.32	$\underline{14.43}$	2.04	2.02	$\underline{3.96}$	1.54	1.31	1
5	2	65	$\underline{1.74}$	$\underline{1.28}$	−0.16	$\underline{8.87}$	$\underline{8.46}$	3.26	$\underline{4.12}$	$\underline{4.59}$	2.44	2
6	1	38	$\underline{5.75}$	$\underline{3.07}$	0.40	$\underline{34.69}$	$\underline{15.58}$	2.50	$\underline{6.03}$	$\underline{4.88}$	1.88	2
7	1	17	$\underline{3.59}$	0.22	−0.66	$\underline{14.32}$	3.26	2.26	$\underline{3.95}$	2.39	1.22	1
8	2	63	$\underline{1.90}$	$\underline{1.36}$	0.10	$\underline{9.00}$	$\underline{7.84}$	2.87	$\underline{4.26}$	$\underline{4.48}$	2.29	2
9	1	60	$\underline{5.96}$	0.36	0.16	$\underline{42.68}$	2.71	2.29	$\underline{7.11}$	2.75	2.07	1
10	2	45	$\underline{2.36}$	$\underline{1.81}$	$\underline{1.15}$	$\underline{9.74}$	$\underline{7.79}$	$\underline{6.03}$	$\underline{4.25}$	$\underline{3.94}$	$\underline{3.70}$	
10			0.28	−0.38	−0.42	3.65	2.24	2.22	$\underline{3.10}$	1.71	1.58	4
11	2	43	$\underline{1.42}$	$\underline{1.19}$	$\underline{1.00}$	$\underline{5.08}$	$\underline{4.61}$	$\underline{4.47}$	$\underline{3.28}$	$\underline{3.05}$	$\underline{3.28}$	
11			0.51	0.17	−0.07	3.18	2.45	2.01	$\underline{2.83}$	2.48	1.74	4

Note: Distribution 1: normal, 2: log normal, n = number of data, l = number of outliers detected; $\alpha = .05$. Significant values are underlined.

Lewis and Fieller (1979) statistics may be used for the purpose. Note that upper outliers from the Gumbel distribution become lower outliers after the transformation. The test statistics are

$$T_j = \frac{x_{(j)}}{\sum_{i=j}^{n} x_{(i)}} \tag{5.9.4}$$

for the detection of $j = 1, 2, 3, \ldots$ outliers. Barnett and Lewis (1994) give some tables of significance levels. In addition, the Shapiro and Wilk (1972) goodness-of-fit test can be used. The W-exponential statistic, which pertains to this test, is given by the following ratio of squared differences from the means:

$$WE_j = \frac{(n-j+1)[\bar{x}_{n-j+1} - x_{(1)}]^2}{\left[(n-j)\sum_{i=j}^{n}[x_{(i)} - \bar{x}_{n-j+1}]^2\right]}. \tag{5.9.5}$$

We must reiterate that the application of Eqs. (5.9.4) and (5.9.5) requires the estimation of two parameters before the log-transformation to exponentiality. The procedure is thus data-based and is of an exploratory nature.

5.9.4 Estimation of probabilities of extreme events when outliers are present

Assigning probabilities to outliers is a difficult problem. First, one does not know the value of k in the hypotheses of Eq. (5.9.1). Tables such as C.9 for the Studentized deviates are not exact with respect to the level of significance α. Type I errors (incorrect rejection) can therefore be higher than indicated. Further, there is a possibility that some values closer to the central observations also belong to the same population as the outliers. It is evident that we are dealing with at least two different populations, the first having only one or a few visible observations and some items possibly mixed with those of the second. If, for example, pn values come from distribution L with pdf $f_1(x)$ where $0 < p \leq 1/2$ and the remainder comes from distribution G with pdf $f_2(x)$, the mixed population has pdf

$$f(x) = pf_1(x) + (1-p)f_2(x). \tag{5.9.6}$$

Difficulties arise in estimating p, pdf $f_1(x)$, and its parameters. The methods of cluster analysis of Chapter 6 may be helpful here. Alternatively, there is the Bayesian approach to consider, its inherent problems of estimation notwithstanding.[34]

The ancient French custom of Winsorization (Hampel, 1974) which can be used under the hypotheses of Eq. (5.9.1) requires the shifting of all outliers to a prefixed position closer to the central observations, prior to the estimation of parameters, thus moderating their effect. There are various robust forms of estimation.[35] The trimmed mean discussed in Subsection 1.2.1 is an example and likewise one can have a trimmed standard deviation; similarly, the estimation of location can be based on a few chosen order statistics. In addition, there are the jackknife methods of Subsection 3.2.3.

Outliers have a high influence in flood risk analysis for engineering design. The problem has been solved empirically in the vast majority of cases. For example, in the method adopted by the American Water Resources Council, the database is increased whenever possible by taking account of historical evidence.[36] This includes information provided by senior citizens, marks on bridge piers, newspaper accounts, and ancient chronicles.

---

[34] See, for example, Box and Tiao (1968) and Hawkins (1980), for example.

[35] See, for example, Huber (1972) and Staudte and Sheather (1990).

[36] See Water Resources Council (1981) and Problem 5.39.

5.9.5 Summary of Section 5.9

This section has highlighted the importance of outliers in the estimation of probabilities of extreme values. When the variable is approximately normally distributed or can be transformed to normality, several methods are available, of which the Studentized deviate seems particularly sensitive. Under other distributional assumptions, however, the procedure is of an exploratory nature because of the prior necessity to estimate parameters. For purposes of risk analysis, the presence of outliers presents problems that have hitherto been resolved empirically.

5.10 SUMMARY OF CHAPTER 5

This chapter should be considered as an essential sequel to Chapters 3 and 4, in which statistical properties of distributions and probability models were discussed. We opened with a discussion and summary of the properties of estimators encountered in the two previous chapters. Our main task was to deal initially with the uncertainties in model parameters and, at a subsequent stage, in the models themselves; the analysis was based on observations of the phenomena studied. If we sacrifice some precision in our estimates (which will almost always be imprecise), we can set up confidence limits for the parameters. Our assertion is that these confidence limits represent interval estimates or bands that include an unknown parameter; we expect this statement to be true a specified number of times out of 100, that is, if we were to conduct an experiment repeatedly. As the available data increases, our confidence interval decreases in width. Alternatively, we may be concerned with only one such boundary, that is, either the upper or lower confidence limit.

We proceed to various tests of hypotheses. These are methods of extracting significance from available samples. We define a null hypothesis in terms of a test statistic and a rejection region. If this is rejected in favor of an alternative hypothesis, because the test statistic falls in the rejection region, it means that there is a change or a difference in the mean or other statistic that is greater than that which is likely to be caused by random fluctuations. The difference is then said to be statistically significant; in other words it is beyond the limit that is exceeded only 100α percent of the time through expected behavior under the null hypothesis. Accordingly, the test is said to have a level of significance α. There is the probability α that we may incorrectly reject the null hypothesis or, on the contrary, incorrectly accept it with probability β; these are called the Type I and Type II errors, respectively, and $(1 - \beta)$ is called the power of the test. The procedure implies that the sampling distribution of the test statistic is known and appropriate types of sampling distributions are discussed in this chapter. That is, we need to make assumptions about the distributions of the basic variables, and therefore hypothesis testing may seem to be restrictive. On the other hand, nonparametric tests have no such limitations although they may be lower in power. We discuss a wide range of nonparametric tests.

Goodness-of-fit criteria are used to verify, again at a level of significance α, the form of a probability distribution and include the chi-squared, Kolmogorov-Smirnov, and Anderson-Darling tests. Some criteria that are specific to the normal distribution are then discussed.

In the analysis of variance, we consider changes in the means of variables through the sums of their squares; we also outlined concepts in experimental design. These methods make it possible, by means of significant tests, to attribute causal factors to phenomena investigated by an engineer or conclude that there is a lack of influence.

The probability plotting methods and visual aids of the penultimate section should be highly appealing to engineers. They supplement the hypothesis-testing methods and have the advantages that a fit or a lack of fit can be directly observed; for example, one can see where any discrepancy occurs, such as near the mode or in the tails of a distribution.

We then deal with unexpectedly high or low values called outliers. These are the problem-causing observations that do not seem to have the distribution of the other items. Methods of detection, identification, and coping with outliers are discussed.

REFERENCES

General. The following references are given for further reading as required:

Barnett, V., and T. Lewis (1994). *Outliers in Statistical Data*, 3rd ed., John Wiley and Sons, New York. Standard reference on outliers; contains tables for skewness and kurtosis tests for normality; table for Shapiro-Wilks and other tests.

Box, G. E. P., W. G. Hunter, and J. S. Hunter (1978). *Statistics for Experimenters*, John Wiley and Sons, New York. Valuable reference for engineers written at an introductory level.

Cochran, W. G., and G. M. Cox (1957). *Experimental Designs*, 2nd ed., John Wiley and Sons, New York. An authoritative book.

Conover, W. J. (1998). *Nonparametric Statistics*, 3rd ed., John Wiley and Sons, New York. Recommended introductory text. Tables for Mann-Whitney test, Kolmogorov test for discrete variables, and Wilcoxon signed-rank test for small samples.

Fisher, R. A. (1966). *The Design of Experiments*, 8th ed., Oliver and Boyd, Edinburgh. A classic, not to be missed by readers at all levels.

Freund, J. E. (1992). *Mathematical Statistics*, 5th ed., Prentice Hall, Englewood Cliffs, NJ. An intermediate-level text.

Gibbons, J. D., and S. Chakraborti (2003). *Nonparametric Statistical Inference*, 4th ed., Marcel Dekker, New York. An advanced-level book; includes runs test and tables for two-sample Kolmogorov-Smirnov test on small samples and numerous tables listed in this chapter.

Hahn, G. J., and W. Q. Meeker (1991). *Statistical Intervals—A Guide for Practitioners*, John Wiley and Sons, New York. Recommended for discussions of confidence limits and other intervals.

Hahn, G. J., and S. S. Shapiro (1967). *Statistical Models for Engineering*, John Wiley and Sons, New York. Reprinted in 1994 as a Wiley Classic in applied statistics for engineers. Includes an extensive chapter on probability plotting.

Hines, W. H., and D. C. Montgomery (1990). *Probability and Statistics in Engineering and Management Science*, 3rd. ed., John Wiley and Sons, New York. See in particular the chapters on hypothesis testing and analysis of variance.

Hinkelmann, K., and O. Kempthorne (1994). *Design and Analysis of Experiments, Vol. I: Introduction to Experimental Design*, John Wiley and Sons, New York. An advanced-level book.

Hoel, P. G. (1984). *Introduction to Mathematical Statistics*, 5th ed., John Wiley and Sons, New York. Engineers will find the book appealing.

Hollander, M., and D. A. Wolf (1999). *Nonparametric Statistical Methods*, 2nd. ed., John Wiley and Sons, New York. Suggested reading.

Johnson, R. A. (1994). *Miller and Freund's Probability and Statistics for Engineers*, 5th ed., Prentice Hall, Englewood Cliffs, NJ. Written at a comprehensible level; example and table for multicomparison test for ANOVA.

Kendall, M., A. Stuart, and J. K. Ord (1983). *The Advanced Theory of Statistics, Vol. 3: Design and Analysis of Time Series*, 4th ed., Charles Griffin, London. Advanced-level supplementary reading, especially Chapter 37 (ANOVA) and Chapter 38 (design of experiments).

Lehmann, E. (1975). *Nonparametrics: Statistical Methods Based on Ranks*, Holdin-Day, Oakland, CA. For further reading on rank tests; tables for Kruskal-Wallis test using small samples.

Montgomery, D. C., and G. C. Runger (1994). *Applied Statistics and Probability for Engineers*, John Wiley and Sons, New York. Chapters 11 and 12 on the analysis of variance, written at a basic level; multicomparison tests in ANOVA.

Rice, J. A. (1995). *Mathematical Statistics and Data Analysis*, 2nd ed., Duxbury Press, Belmont. A clearly written book; practical applications including bootstrap and hanging histogram.

Snedecor, G. W., and W. G. Cochran (1989). *Statistical Methods*, 8th ed., Iowa University Press, Ames, IA. Chapters 11–16 on design of experiments. Advanced-level book.

Walpole, R. E., and R. H. Myers (1993). *Probability and Statistics for Engineers and Scientists*, 5th ed., Macmillan, New York. Commendable introductory reference. Control and cusum (cumulative sum) charts; design of experiments. Tukey's multiple range test, Duncan's test, Duncan's multiple range test, Plackett-Burman, and Taguchi's robust parameter test.

Additional references quoted in text

Abramowitz, M., and I. A. Stegun (eds.) (1964). *Handbook of Mathematical Functions*, National Bureau of Standards (U.S.), Appl. Math. Sect., Publ. No. 55, Dover, New York. Reference for approximations to the normal integral among others.

Anderson, T. W., and D. A. Darling (1954). "A test of goodness of fit," *J. Am. Stat. Assoc.*, Vol. 49, pp. 765–769. Anderson-Darling goodness-of-fit test.

Birnbaum, Z. W. (1952). "Numerical tabulation of the distribution of Kolmogorov's statistic for finite sample size," *J. Am. Stat. Assoc.*, Vol. 47, pp. 425–441. Tables for the Kolmogorov-Smirnov two-sample test applied to equal samples less than 40.

Box, G. E. P., and G. C. Tiao (1968). "A Bayesian approach to some outlier problems," *Biometrika*, Vol. 55, pp. 119–129. Use of Bayesian methods to investigate outliers.

Box, J. F. (1981). "Gosset, Fisher, and the *t* distribution," *Amer. Stat.*, Vol. 35, pp. 61–66. A fascinating discussion of Gosset's correspondence with R. A. Fisher.

Brownlee, K. A. (1965). *Statistical Theory and Methodology in Science and Engineering*, John Wiley and Sons, New York. Behrens-Fisher problem, pp. 299–302.

Casella, G., and R. L. Berger (2002). *Statistical Inference*, 2nd ed., Wadsworth & Brooks/Cole, Pacific Grove, CA. Reference on a sufficient estimator for the uniform distribution, pp. 277–278; the approximate solution to the Fisher-Behrens problem, pp. 409–410.

Chambers, J. M., W. S. Cleveland, B. Kleiner, and P. A. Tukey (1983). *Graphical Methods for Data Analysis*, Wadsworth, Belmont, CA. Use of Hazen plotting position, Eq. (5.8.4).

Cox, D. R. (1978). "Some remarks on the role in statistics of graphical methods," *Appl. Stat.*, Vol. 27, No. 1, pp. 4–9. Gives the history of graphical methods in statistics and compares methods for assessing consistency with the exponential distribution.

D'Agostino, R., and E. S. Pearson (1973). "Tests for departure from normality. Empirical results for the distribution of b_2 and $\sqrt{b_1}$," *Biometrika*, Vol. 60, pp. 613–622. Goodness-of-fit test for normality.

Dallal, G. E., and L. Wilkinson (1986). "An analytic approximation to the distribution of Lilliefors' test statistic for normality," *Am. Stat.*, Vol. 40, pp. 294–296. Adaptation of Kolmogorov-Smirnov test.

Duncan, D. B. (1955). "Multiple range and multiple F tests," *Biometrics*, Vol. 11, pp. 1–42. Compares the range of any set of means with an appropriate least significant range obtained from tables, which are given with an example.

Ferguson, T. S. (1961). "On the rejection of outliers," in *Proc. 4th Berkeley Symposium on Mathematical Statistics and Probability*, Vol. 1, edited by J. Neyman, University of California Press, Berkeley, CA, pp. 253–287. Tables for the use of skewness and kurtosis as outlier detection criteria.

Filliben, J. J. (1975). "The probability plot correlation coefficient test for normality," *Technometrics*, Vol. 17, pp. 111–117. Goodness-of-fit test for normality.

Hampel, F. R. (1974). "The influence curve and its role in robust estimation," *J. Am. Stat. Assoc.*, Vol. 69, pp. 383–393. On robust statistics.

Hawkins, D. M. (1980). *Identification of Outliers*, Chapman and Hall, London. Bayesian approach to outlier detection.

Hosking, J. R. M. (1990). "L-moments: Analysis and estimation of distribution using linear combinations of order statistics," *J. R. Stat. Soc., B*, Vol. 52, pp. 105–124. Reference on plotting positions.

Huber, P. J. (1972). "Robust statistics: A review (the 1972 Wald Lecture)," *Ann. Math. Stat.*, Vol. 43, pp. 1041–1067. On robust statistics.

Jain, R. B. (1981). "Percentage points of many-outlier detection procedures," *Technometrics*, Vol. 23, pp. 71–75. Tables of Studentized deviates for use in detection of outliers associated with a normal distribution.

Johnson, N. L., and F. C. Leone (1977). *Statistics and Experimental Design in Engineering and the Physical Sciences*, Vol. 1, 2nd ed., John Wiley and Sons, New York. The chi-squared approximation, pp. 274–277.

Keuls, M. (1952). "The use of the Studentized range in connection with an analysis of variance," *Euphytica*, Vol. 1, p. 112. Of importance in the interpretation of ANOVA results.

Kottegoda, N. T. (1984). "Investigation of outliers in annual maximum flow series," *J. Hydrol.*, Vol. 72, pp. 105–137. Methods of identifying and treating outliers.

Kottegoda, N. T., and J. Elgy (1977). "Infilling flow data," in *Proceedings of the 3rd International Hydrology Symposium*, edited by H. J. Morel-Seytoux, Fort Collins, Colorado, Water Resources Publications, Highlands Ranch, CO. Several procedures for infilling data.

Kottegoda, N. T., and A. H. M. Kassim (1991). "Classification of storm profiles using crossing properties," *J. Hydrol.*, Vol. 127, pp. 37–53. Storm profile study using Kruskal-Wallis test.

Kotz, S., and N. L. Johnson (eds.) (1982). *The Encyclopedia of Statistical Sciences*, Vol. 7, John Wiley and Sons, New York. Anderson-Darling goodness-of-fit test.

Kruskal, W. H., and W. A. Wallis (1952). "Use of ranks in one criterion variance analysis," *J. Am. Stat. Assoc.*, Vol. 47, pp. 583–621. Nonparametric test.

Leeming, J. J. (1963). *Statistical Methods for Engineers*, Blackie, London. Road traffic data used in Example 4.21.

Lewis, T., and N. R. J. Fieller (1979). "A recurrence algorithm for null distribution for outliers, 1. Gamma samples," *Technometrics*, Vol. 21, pp. 371–375. Outliers associated with a gamma distribution.

Lilliefors, H. W. (1967). "On the Kolgomorov-Smirnov test for normality with mean and variance unknown," *J. Am. Stat. Assoc.*, Vol. 62, pp. 399–402. Kolmogorov-Smirnov test for normality with parameters estimated from the same sample.

Mann, H. B., and A. Wald (1942). "On the choice of the number of class intervals in the application of the chi square test," *Ann. Math. Stat.*, Vol. 13, pp. 306–317. An original reference on the chi-squared test.

Marsaglia, G., and J. C. W. Marsaglia (2004). "Evaluating the Anderson-Darling Distribution," *J. Stat. Softw.*, Vol. 9, No. 2. Further work on the Anderson-Darling test.

Pearson, E. S., R. B. D'Agostino, and K. O. Bowman (1977). "Tests for departure from normality: Comparison of powers," *Biometrika*, Vol. 64, pp. 231–246. Provides justification for the Shapiro-Wilk test.

Powell, R. W. (1943). "A simple method of estimating flood frequencies," *Civ. Eng.*, Vol. 13, pp. 105–106. Pioneering work on probability plotting.

Rosner, B. (1975). "On the detection of many outliers," *Technometrics*, Vol. 17, pp. 221–227. Skewness and kurtosis tests for normality and other methods.

Royston, J. P. (1982). "An extension of Shapiro and Wilk's *W* test for normality to large samples," *Appl. Stat.*, Vol. 31, pp. 115–124. An algorithm for the Shapiro-Wilk test.

Scheffé, H. (1959). *The Analysis of Variance*, John Wiley and Sons, New York. Section 3.7, Chapter 7, reference on multicomparison tests, use of the F distribution in ANOVA.

Shapiro, S. S., and M. B. Wilk (1965). "An analysis of variance test for normality (complete samples)," *Biometrika*, Vol. 52, pp. 591–611. The Shapiro-Wilk test for normality.

Shapiro, S. S., and M. B. Wilk (1972). "An analysis of variance test for the exponential distribution (complete samples)," *Technometrics*, Vol. 14, pp. 355–370. The Shapiro-Wilk test for exponentiality.

Shapiro, S. S., M. B. Wilk, and M. J. Chen (1968). "A comparative study of various tests for normality," *J. Am. Stat. Assoc.*, Vol. 63, pp. 1343–1372. *Power of various tests.*

Siegel, S., and N. J. Castellan, Jr. (1988). *Nonparametric Statistics for Behavioral Sciences*, 2nd ed., McGraw-Hill, New York. Tables for Wilcoxon signed-rank test, pp. 332–334.

Smirnov, N. (1948). "Table for estimating the goodness of fit of empirical distributions," *Ann. Math. Stat.*, Vol. 19, pp. 279–281. Tables for two-sample test.

Staudte, R. G., and Sheather, S. J. (1990). *Robust Estimation and Testing*, John Wiley and Sons, New York. Robustness in depth with bootstrap; less mathematical than others in the field.

Stuart, A., and J. K. Ord (1991). *Kendall's Advanced Theory of Statistics*, Vol. 2, 5th ed., Edward Arnold Ltd., London. Advanced reference; Cramer-Rao inequality, pp. 614–616; Behrens-Fisher problem, pp. 786–788; choice of class intervals for the chi-squared test, pp. 1172–1182; likelihood ratio tests; ANCOVA pp. 1150–1152; Kolmogorov- Smirnov test, pp. 1191–1192.

Tiku, M. L. (1975). "A new statistic for testing suspected outliers," *Commun. Stat.*, Vol. 4, pp. 737–752. Comments on Shapiro-Wilk test.

Velleman, P. F., and D. C. Hoaglin (1981). *Applications, Basics, and Computing of Exploratory Data Analysis*, Duxbury Press, Boston, MA. Rootograms (hanging histograms): Chapter 9.

Volk, W. (1969). *Applied Statistics for Engineers*, McGraw-Hill, New York. Reference on Poisson probability paper; see Example 24.5*a* and 24.5*b*.

Wackerly, D. D., W. Mendenhall, and R. L. Scheaffer (2002). *Mathematical Statistics with Applications*, 6th ed., Duxbury, Pacific Grove, CA. Tables for Wilcoxon, runs and rank correlation tests.

Wald, A., and J. Wolfowitz (1940). "On a test whether two samples are from the same population," *Ann. Math. Stat.*, Vol. 2, pp. 147–162. Runs test.

Water Resources Council (1981). *Guidelines for Determining Flood Flow Frequency*, Bulletin 17B, Resources Council, Washington, DC, p. 222. Deals with estimation of probabilities of outliers with historical information.

Weibull, W. (1951). "A statistical distribution function of wide applicability," *J. Appl. Mech.*, Vol. 18, pp. 293–297. Origin of Weibull distribution.

Wilk, M. B., and R. Gnanadesikan (1968). "Probability plotting methods for the analysis of data," *Biometrika*, Vol. 55, pp. 1–17. Useful reading on probability-plotting.

Williams, C. A., Jr. (1950). "On the choice of the number and width of classes for the chi-square test of goodness-of-fit," *J. Am. Stat. Assoc.*, Vol. 45, pp. 47–86. Important reference on chi-squared test.

Wilson, E. B., and M. M. Hilferty (1931). "The distribution of chi-square," *Proc. Natl. Acad. Sci. U S A*, Vol. 17, pp. 684–688. Approximation to the chi-squared distribution.

PROBLEMS

5.1. Piling failures. A contractor involved in driving piles for foundations in a region has a good record of success. Nevertheless, some piles have been unsuccessful. The following failures have been recorded from 50 driven piles in each set:

Set number, i	Number unsuccessful
1	2
2	3
3	1
4	2
5	4
6	0
7	1
8	3
9	0
10	2

Assume the probability of failure is a constant and the trials are independent.

(*a*) What type of statistical process generates the numbers given in the second column?

(*b*) What is the distribution of the average failure rate, for various i, when based on large sizes of sets?

(*c*) What is the estimated fraction of failures p from all the sets?

(*d*) Provide 95% confidence limits on the true value of p, stating the assumptions made.

(*e*) Draw a line diagram of the observed and theoretical distributions based on the above table and state whether the data are compatible with it.

5.2. Confidence limits for concrete densities. Suppose that only the top 20 of the concrete densities listed in Table E.1.2 are available.

(*a*) Assuming a normal population, provide 95% confidence limits for the mean density of concrete.

(*b*) Revise the confidence limits for the mean density if the population standard deviation is 16 kg/m^3.

5.3. Minimum sample size for estimating mean dissolved oxygen (DO) concentration. Monitoring of pollution levels of similar streams in a region indicates that the standard deviation of DO is 1.95 mg/L over a long period of time.

(*a*) What is the minimum number of observations required to estimate the mean DO within ± 0.5 mg/L with 95% confidence?

(*b*) If only 30 observations are taken, what should be the percentage level in the confidence limits for the same difference in means?

5.4. Yield strength of steel rods. Tests done on a new make of steel rods indicated that, on average, loads up to 1990 kg can be withstood before exceeding the yield strength. This value is based on estimates from 50 specimens chosen at random. The standard deviation of the load is 183 kg. If a more stringent design is based on a 99.9 lower confidence limit, determine the mean yield strength to meet this specification.

5.5. Confidence intervals on the variance of concrete densities. For the data of Problem 5.2*a*, provide 95% confidence limits on the population variance.

5.6. Confidence limits on proportions of wet days. A building contractor who works in a relatively dry area is planning to acquire additional work in a newly developing area but is somewhat doubtful of progress because of the adverse effects of rainfall in many months of the year. However, the contractor knows that March is a month of low rainfall with independently distributed daily rainfalls and no apparent relationship between the weather on successive days. Therefore, the thought is that this may be a suitable month to work on the foundations. The proportion of wet days in March is 0.10 from data of the past 3 years. Suppose it is possible to put off the decision for some time in order to make further observations of daily rainfalls in March. Determine the total number of years of data necessary before one can be 95% confident of estimating the true proportion of wet days to within 0.05.

5.7. Significance of change in temperature. A water supply engineer is concerned that possible climatic change with respect to temperature may have an effect on forecasts for future demands for water to a city. The long-period mean and standard deviation of the annual average temperature measured at midday are 33 and 0.75°C. The alarm is caused by the mean temperature of 34.3°C observed for the previous year. Does this suggest that there is an increase in the mean annual temperature at a 5% level of significance.

5.8. Time intervals between passing vehicles. In Example 4.21 the parameter of the fitted exponential distribution was estimated as 1.81 min^{-1} for the time gaps between vehicles in traffic from 204 observations. By probability-plotting methods, this is estimated in Example 5.41 as 1.75 min^{-1}. If these were field estimates over different time periods, do they constitute a significant difference in the mean time intervals, using $\alpha = 0.05$?

5.9. Comparing outputs of waste water plants. Two treatment plants are built in an area to treat wastewater from a city. Their relative performances are compared from the results of BOD tests made on the outputs. Eight preliminary results are listed below as differences in BOD between plant 1 and 2.

Test	1	2	3	4	5	6	7	8
Difference in BOD (mg/L)	+1.2	+0.2	−1.6	+0.7	+1.3	−0.9	−0.1	−1.9

Test the difference in the outputs at the 5% level of significance.

5.10. Change in the mean and variance of flood flows. Annual maximum flows of the Tevere (Tiber) River recorded at Ripetta in Rome are given in Table E.5.8 for the period 1921–1974. The observation of numerous low maximum flows during the last 20 years led to a suspicion that the flow regime or climatic conditions had changed. Divide the record into two halves. Determine if the mean annual maximum flow in the second half is lower than those in the first half at a level of significance $\alpha = 0.01$ under the following conditions:

(*a*) If the standard deviation is 450 m^3/s.
(*b*) If the standard deviation is estimated from the data but is assumed to be constant.
(*c*) If the standard deviations are estimated separately for the two halves and are assumed to be different.
(*d*) Using the estimated variances in part *c*, above, determine whether the change in the variance is significant for $\alpha = 0.01$?

5.11. Control chart for quality control of concrete. Control charts were introduced in 1924 by Walter A. Shewhart [see W. A. Shewhart (1931), *Economic Control of Quality of Manufactured Products*, D. Van Nostrand, New York] in order to detect and control any unwanted deviations in a process so that quality can be maintained.

Suppose tests based on compressive strengths have been made on concrete cubes to determine the ultimate loads that can be carried by concrete being used at a construction site. From past data the mean and standard deviation are estimated as 61.1 and 4.9 N/mm^2, respectively, and each day five test cubes are tested at random and the mean is computed. The following results are obtained from the tests of 12 working days:

Batch number, i	Mean compressive strength (N/mm^2)
1	58.1
2	60.9
3	62.5
4	59.9
5	56.1
6	58.7
7	61.5
8	61.9
9	63.5
10	58.1
11	67.1
12	60.1

Draw control charts using bands that are two standard errors from the mean. (Three standard errors are commonly used.)

(a) Do any of the above results suggest that corrective action is necessary?

(b) What is the probability that a Type I error, is made, that is, action as in (a) is taken without any need for it?

(c) What is the probability of making one or more of such errors during a 6-day working week?

(d) What is the probability of making a Type II error, if the use of aggregates of lower quality has reduced the mean strength to 57.5 N/mm^2?

(e) How does one reduce the foregoing errors?

5.12. Power curve for concrete strengths. In Example 5.9, 95% confidence limits of 58.53 and 61.76 N/mm^2 were provided for the 40 concrete strengths with mean and standard deviation 60.14 and 5.02 N/mm^2 listed in Table 1.2.2. Determine the Type II errors made if the population values are equal to each of the following values, all in newtons per square millimeter:

60.5 61.5 62.0 62.5 63.5.

Draw the power curve for the corresponding points.

5.13. Irrigation and rain. Irrigation usually commences on 15 April in the Po River basin, Italy. An engineer is interested in the probability of rain during the 7 days from April 15 to 21. From rainfall data of the past 100 years in a particular area, the following distribution of rainy days is obtained for the period:

Rainy days	0	1	2	3	4	5, 6, 7	Total
Frequency	57	30	9	3	1	0	100

The binomial model $B(M = m \mid 7, 0.1)$ is postulated. Can this be justified at the 5% level of significance on the basis of a chi-squared test?

5.14. One-sample sign test on flows. The following is a sample from the recorded annual flows in the St Lawrence River which runs out from the Great Lakes of North America. The data are in standardized units obtained by dividing the original observations by the annual mean. Test the null hypothesis that the median is 1.006 against the alternative hypothesis that it is greater or less than this value, at the 5% level of significance.

0.942 0.947 1.005 0.988 1.001 1.013 1.013 1.088 1.000 0.959.

5.15. Sign test applied to paired observations. We reexamine the concrete densities listed in Table E.1.2. Divide the record into two samples of equal length and apply the sign test to correspondingly paired observations from the two halves. Test the hypothesis that the mean density is unchanged at the 5% level of significance.

5.16. Wilcoxon signed-rank test on flows. Use the Wilcoxon signed-rank test to ascertain whether the mean of the annual maximum flows of the Tevere River has changed from the first half to the second half of the period given in Table E.5.8. Test the null hypothesis that the means are the same against the alternative hypothesis that the mean flow is less in the second half at the 1% level of significance.

5.17. Runs test on the Wolfer sunspot numbers. Wolfer sunspot numbers are an index of activity on the solar surface. They have been investigated for their impact on terrestrial climate and for the resulting environmental effects. Twenty annual observations are listed below for the period 1770–1789:

101	82	66	35	31	7	20	92	154	125
85	68	38	23	10	24	83	132	131	118

Apply a runs test for randomness. Do these represent a random series at the 5% level of significance?

5.18. Spearman rank correlation test on the DO-BOD relationship. With reference to the data in Table E.1.3, determine the rank correlation coefficient for the relationship between DO and BOD. Compare with the result in Example 1.30.

5.19. Poisson distribution of numbers of days of high waves. High waves in a coastal area where further development is planned cause property damage and erosional problems but measurements of wave heights are scanty. A researcher has obtained the following information of the number of days of high waves in a year from local chronicles and residents:

Number of days of high waves	0	1	2	3	4	5
Frequency	26	13	6	3	2	0

Sketch a histogram of the number of days of high waves recorded in the area during a 50-year period.

Test the hypothesis that the occurrence of high waves is Poisson distributed at the 5% level of significance using the chi-squared test. What is the probability that the mean rate will be more than 1 day per year?

5.20. All-red phase of traffic lights. At 12 four-way junctions in London, England, brief "all red" phases were introduced. The numbers of accidents causing injuries were recorded for 2 years before and after the installation as given here:

Site	1	2	3	4	5	6	7	8	9	10	11	12
Before	27	4	18	20	17	12	18	24	18	19	3	8
After	20	9	14	14	16	3	13	4	9	11	3	6

With the kind courtesy of the Transport and Road Research Laboratory, England.

Test the reduction in the number of accidents at the 1% level of significance. It is thought that sites with high rates of accidents are highly weighted. At a given site the variance is expected to be proportional to the mean over consecutive time periods. Taking the square roots of the numbers will reduce the differences in variances. Repeat the test at the 5 and 1% levels of significance for the variance-adjusted data.

5.21. Speed limit: USA. Speeds of cars were estimated on rural interstate roads in the United States during 1973 and 1975. The numbers of cars within certain categories of speeds are listed here.

	Less than 45 mph	45–55 mph	55–70 mph	70–85 mph	Total
Upper limits (kph)	72.5	88.5	112.6	137	
1973	0	7	63	30	100
1975	1	28	69	2	100

Determine whether there is a significant decrease, at the 1% level, of the proportion of cars exceeding the speed limit of 55 mph (88.5 kph) between the two years.

Data are from D. B. Kamerud, "The 55 mph speed limit: Costs benefits, and implied trade-offs," *Transp. Res.*, Vol. 17, pp. 51–64, Copyright (1983) with the kind permission from Elsevier Science Ltd, The Boulevard, Langford Lane, Kidlington, OX5 1GB, England.

5.22. Speed limit: England. To test the effect of a 65 km/h (40 miles/h) speed limit on the A 4123 road in England, speeds of vehicles were calculated from observations taken at sites during one day before and one day after the introduction of the limit. The following results were obtained:

		Mean speed in kph of private cars	
Day and site		Before	After
Monday	Northbound	68.3 (42.4)	63.4 (39.4)
1	Southbound	61.4 (38.1)	58.9 (36.6)
Tuesday	Northbound	72.8 (45.2)	64.1 (39.8)
2	Southbound	69.9 (43.4)	64.6 (40.1)
Wednesday	Northbound	61.4 (38.1)	56.8 (35.3)
3	Southbound	59.1 (36.7)	55.1 (34.2)

Note: Values in parentheses are in mph as originally calculated.
With the kind courtesy of the Transport and Road Research Laboratory, England.

In considering that there may be other factors, such as weather, that could have caused the differences, observations were also made on the same days over a similar part of the road where no speed limit was imposed. The following changes in mean speeds [before-after] were recorded in the same units:

Monday	−2.42	(−1.5)
Tuesday	−0.16	(−0.1)
Wednesday	−1.29	(−0.8)

Test the change at the 5% level of significance.

5.23. Machine failures. The following are intervals in hours between failures of the air conditioning system of a Boeing 720 jet airplane:

23, 261, 87, 7, 120, 14, 62, 47, 225, 71, 246, 21, 42, 20, 5, 12, 120, 11, 3, 14, 71, 11, 14, 11, 16, 90, 1, 16, 52, 95.

Test whether the data are exponentially distributed at the 5% level of significance. Draw a probability plot.

Data with the kind courtesy of the publishers from F. Proshan (1963), "Theoretical explanation of observed decreasing failure rate," *Technometrics*, Vol. 5, pp. 375–383.

5.24. Bacterials counts. The following are counts of the number of fields of bacteria reported by C. Bliss and R. A. Fisher (1953), "The negative binomial distribution to biological data," *Biometrics*, Vol. 9, pp. 176–196.

Bacteria for field	0	1	2	3	4	5	6	7	8	9	10	11	12 and more
Number of fields	11	17	31	24	29	18	19	16	13	17	6	8	31

As a preliminary step fit the geometric distribution to these data. Apply a chi-squared goodness-of-fit test at a level of significance $\alpha = 0.05$, combining the counts for fields 0 and 1.

Data used with the kind courtesy of the International Biometric Society, 808 17th Street NW, Suite 200, Washington, DC.

5.25. Pump failures. Two manufacturers, A and B, supply pumps to the same specification of 500 hours on average to failure. Twenty pumps of each manufacturer have been installed and the times to failure for each pump are as follows:

A	510	450	478	512	506	485	501	481	452	494
	514	507	487	467	502	508	503	492	502	499
B	510	513	497	506	493	501	547	514	487	490
	495	497	508	493	522	502	527	486	531	497

(*a*) Test whether the mean time of failure for A is less than that for B.

(*b*) Test whether the proportion of pumps not reaching specification is less for B than for A.

Use $\alpha = 0.01$ and an appropriate test in each case.

5.26. Groundwater quality. The following are measurements of concentrations of chloride in milligrams per liter in a shallow unconfined aquifer taken at intervals of 3 months [from J. Harris, J. C. Loftis, and R. H. Montgomery (1987), "Statistical models for characterizing ground-water quality," *Groundwater*, Vol. 25, pp. 185–193]:

38	40	35	37	32	37	37	32	45	38
33.8	14	39	46	48	41	35	49	64	73
	67	67	59	73	92.5	45.5	40.4	33.9	28.1

Compute the coefficients of skewness and kurtosis and make an approximate test for normality, using $\alpha = 0.05$. (Data used with the kind courtesy of the publishers.)

5.27. Kolmogorov-Smirnov two-sample test on flows. Annual rainfall from 1918 to 1978 in the Po River basin of northern Italy are given in the penultimate column of Table E.7.2. Divide the record into two parts of 30 and 31 years. Determine whether the rainfall regime has changed by testing whether the two parts belong to the same population at the 5% level of significance using the Kolmogorov-Smirnov two-sample test.

5.28. Lilliefor's test. The following are the ranked annual inflows in 10^6, for the period 1950–1974, to the Warragamba reservoir, which supplies water to the city of Sydney, Australia:

724	1,505	3,310	6,551	6,915	7,114	7,811	8,962	9,219	9,664
9,840	10,134	10,299	10,924	11,953	12,566	13,969	14,941	15,449	16,800
17,601	18,250	18,483	19,081	20,242					

(By kind courtesy of the University of New South Wales, Sydney.)

Test whether the distribution is normal using Table C.8 of Appendix C, which is Lilliefors' test for normality corrected by G. E. Dallal and L. Wilkinson (1986), "An analytic approximation to the distribution of Lilliefors' test statistic for normality," *Am. Stat.*, Vol. 40, pp. 294–296, for the purpose.

5.29. Chi-squared test. We transformed the Warragamba annual flows (introduced in Problem 5.28 but extended over a 103-year period) to natural logarithms. The data were ranked and sorted into ten classes with equal class intervals as follows:

Class boundary	11.61	12.07	12.52	12.98	13.44	13.89	14.35	14.81	15.26	∞
Number of class	2	6	8	14	19	16	17	11	8	2

The total is $n = 103$, the estimated mean is 13.50, and the variance is 0.874.

Using the chi-squared goodness-of-fit procedure, test the hypothesis that the log-transformed flows are normally distributed with $\alpha = 0.05$.

5.30. Anderson-Darling test. Reconsider the data of Problem 5.28. Test the hypothesis of normality using the Anderson-Darling test at the 5% level of significance.

5.31. Harmonic coefficients. Monthly inflows into Warragamba reservoir in New South Wales, Australia, were computed over the period 1881–1983 in units of 1000 m^3.

It is proposed to fit the following harmonic model to the periodicity in the means:

$$\mu_\tau = \mu + \sum_{i=1}^{6} \alpha_i \sin\left(\frac{2\pi i \tau}{6}\right) + \sum_{i=1}^{6} \beta_i \cos\left(\frac{2\pi i \tau}{6}\right),$$

where μ_τ is the harmonic mean in month τ, $\tau = 1, 2, \ldots, 12$; 1 denotes January and so on; μ is the annual mean; $\alpha_i, \beta_i, i = 1, 2, \ldots, 6$ are harmonic coefficients. The following harmonic coefficients have been computed in m^3:

	Harmonic i					
	1	2	3	4	5	6
α_i	6,066	8,568	−12,629	−2,135	8,954	0
β_i	−39,393	7,062	−16,877	6,586	−5,204	875
SS	9,817	762	2,746	296	663	5

Note: The last row gives the sum of squares associated with each harmonic in units $10^8(\text{m}^3)^2$.

Determine by an analysis of variance how many of the harmonics are significant using $\alpha = 0.05$. There are 1236 items of data and the total sum of squares is 457, 175 $\times 10^8$ $(\text{m}^3)^2$.

Calculate the fitted means using the significant harmonics.

5.32. Analysis of variance of road data. Using the data from Table 5.7.5 for the road-rutting experiment of Example 5.34, test whether the base thickness has a significant effect on the depth of the rutting. For this test combine the results from the two types of base material.

5.33. Analysis of variance of dynamic effect of vehicles. We return again to the road rutting-experiment of Example 5.34. From Fig. 5.7.2 and other data sets in Tables E.5.1 to E.5.6, it is suggested that the movement of heavy vehicles can have a dynamic effect (that is, time-dependent) on the road surface. This can change with road material and thickness. Analyze the variance of this effect by using the estimated error sum of squares $(x_i - \hat{\beta}_0 - \hat{\beta}_1 u_i)^2$ to represent it. Hence determine whether it is significant with $\alpha = 0.05$. Assume (*a*) that the intercepts and gradients for the 30 series studied in Example 5.34 are variable; (*b*) assume a constant intercept and gradient.

5.34. Normal plots. In Table E.1.2 and Problems 1.6 and 1.14 of Chapter 1, there are lists of compressive strengths of concrete in newtons per square millimeter obtained by testing three lots of test cubes. Make comparative normal probability plots of these three sets of test results. Comment on the results.

5.35. Lognormal probability plotting of sunspot data. Make normal and lognormal probability plots the Wolfer sunspot data of Problem 5.17. Comment on the results.

5.36. Number of vehicles passing using a Poisson probability plot. Draw curves to represent for each Poisson occurrence, say, from 0 to 6, the relationship between the probability of occurrence and the value of the parameter from say 0.1 to 1.0. Plot the probabilities of the following observer counts of the number of vehicles passing a point of observation:

Count	0	1	2	3	4	5	6
Frequency	221	95	24	12	5	2	1

Is the Poisson a reasonable model? What is the estimated parameter from the plot?

5.37. Tensile strengths on Weibull probability paper. The original work of W. Weibull (1951), "A statistical distribution function of wide applicability," *J. Appl. Mech.*,

Vol. 18, pp. 293–297, on the strengths of materials suggests that the breaking tensile stress of concrete has a Weibull distribution. The following results of tensile strengths, in newtons per square millimeter, were obtained from 12 tests conducted in a laboratory:

14.8 15.7 15.1 13.8 14.3 16.6 14.1 16.4 16.1 13.7 13.9 14.6

Plot these results on Weibull probability paper. Fit a straight line by eye and comment on the results.

5.38. Hanging histogram of the Tevere flood flows. For the annual maximum flows of the Tevere River of Problem 5.16, draw a hanging histogram using the lognormal distribution. Comment on the results.

5.39. Accommodation of outliers. Consider the annual maximum flows in the North Fork Sun River listed in Table E.5.7, series 1. It is noted that there is one suspected outlier in the series of 25 annual maximum flows.

(*a*) Plot the data on normal probability paper.
(*b*) Fit a straight line by eye without considering the outlier.
(*c*) Excluding the outlier, calculate the mean $\bar{x}$ and unbiased variance $\hat{s}^2$.

For incorporating outliers, the procedure adopted by Water Resources Council (1981), *Guidelines for Determining Flood Flow Frequency*, Bulletin 17B, Resources Council, Washington, DC, p. 222, is to empirically increase the database using historical evidence, if available, and then revise the mean and variance. Suppose l outliers are identified in a record of n_R years; and from past information, such as marks on bridge piers, it is found that the highest recorded flood level has not been exceeded in n_T years. Then the revised mean $\tilde{x}$ and revised variance $\tilde{s}^2$ of the extended database, are calculated as follows:

$$\tilde{x} = \frac{\left[(n_T - l)\bar{x} + \sum_{i=n_R-l+1}^{i=n_R} x_{(i)}\right]}{n_T}$$

$$(n_T - 1)\tilde{s}^2 = \left[\hat{s}^2(n_T - l) + (n_T - l)(\bar{x} - \tilde{x})^2 + \sum_{i=n_R-l+1}^{i=n_R} (x_{(i)} - \tilde{x})^2\right].$$

[The first equation is a direct adjustment of the mean and the second equation follows from the ANOVA methods of Section 5.7.]

If the magnitude of the outlier has not been exceeded for 500 years, calculate the annual maximum flow with a return period of 500 years assuming a normal distribution.

Chapter 6
Methods of Regression and Multivariate Analysis

In previous chapters we discussed cases and encountered problems in which two or more variables are related. These ranged from the scatter plots of Chapter 1 to the bivariate and multivariate distributions of Chapters 3 and 4 and the ANOVA methods of Chapter 5. We now examine methods of exploiting such relationships in order to make them useful in engineering assessments and predictions. Regression analysis is the most popular of these methods. Basically, it involves the use mathematical functions to model and investigate the dependence of one variable, say, Y, called the *response* variable on one or more other observable variables, say, X, known as the *explanatory* variables. Let it be clear that the purpose of regression is not to search for a cause-and-effect relationship without prior knowledge of the behavior and interactions between variables.The way such a search should be done was discussed with respect to the design of experiments in Section 5.7. In regression we start with our physical knowledge of the processes involved, limited though it may be, and then formulate, fit, evaluate, and validate the relationship through a mathematical model. The procedure is iterative.

As a first step we can say, for example, that the compressive strength of a concrete at 28 days is dependent on the water-cement ratio measured by the slump value of a cone of the mixture. If the mixture had excess or insufficient water, one would not expect the concrete to be strong. Subsequently, there should be a relationship with the methods of curing the concrete (which should enhance its strength if properly done) and also the ultimate density.

The model we propose is of a probabilistic rather than deterministic; that is, there is a certain amount of error in predicting one variable that cannot be explained in terms of other variables. Even if we incorporate all conceivable variables (and eliminate any instrumental, observational, and recording errors), there will still be some modeling error, called the *measurement error*; this is attributable to our incomplete knowledge of the physical processes involved or to factors beyond our control. We proceed from simple to multiple regression in a systematic search for an optimal relationship with a minimum error of prediction. For this purpose we use correlation analysis and the analysis of variance introduced in Chapter 5.

The assumptions we make concern linearity in the model equations and some properties in the series of modeling errors such as normality in distribution which is essential for the statistical tests, homogeneity of variance, and independence—serial and with respect to the response variable. We also assume that the model is fully specified by including all relevant variables and excluding those that are irrelevant. An important part of the analysis is the verification that these assumptions are satisfied, by using graphical methods and appropriate tests. There are other issues we deal with such as outliers in regression, influential observations, and high leverage points.

Cases arise, however, in which strong dependencies between the exploratory variables, called *multicollinearity*, influence the estimates of the regression coefficients with adverse effects upon application of the model. *Ridge regression* is devised to counteract this problem. Our discussion extends to alternative methods in other situations where the assumptions of a linear regression model are not met such as transformations to linearity, general and generalized linear models, nonlinear regression, and nonparametric methods.

There are instances when we consider several variables simultaneously without initially identifying one variable as the response variable. In such situations we are interested in the interdependence between a number (or groups) of variables rather than the dependence of one variable on others, which we investigate in regression. This interdependence is what we consider when we apply multivariate analysis. We discuss principal component and factor analyses, which reduce the variables to a few components or factors, thus minimizing the parameters and the dimensions of the representative matrix. There are similarities and common objectives in the approaches, as seen, for instance, when we find linear combinations of the original variables. We then consider cluster analysis which finds, wherever possible, different groups or clusters, with similar individual items, from the given observations of multivariate phenomena.

The effects of *correlation* over distances are often important to civil and environmental engineers. An example is the relationship between storm rainfall measured at different locations in a catchment, which influences the design of a storm-sewer system. We use spatial correlation and the complementary technique of the *variogram* for the purpose.

6.1 SIMPLE LINEAR REGRESSION

In common with workers in other fields, scientists and engineers relate one variable to two or more other variables for purposes of prediction, optimization, and control. Such relationships are expressed in the form of mathematical equations. *Regression*, as it is called, is indeed the most commonly used technique in statistics. If applied with care and correct interpretation, it can be a boon for a wide spectrum of users. It may, however, be misused at times. This section is confined to the simple case of linear regression of two variables.

Our objective is to investigate the relationship between a random variable and another variable. One starts with an initial assumption of a straight-line relationship. This takes the form

$$Y = \beta_0 + \beta_1 x + \varepsilon, \tag{6.1.1}$$

where Y is an observable random variable, x is an observable nonrandom variable, and β_0 and β_1 are (fixed and) unknown parameters, also called *regression coefficients*; β_0 is the intercept and β_1 is the slope. The regression is that of Y on x. We call Y the *response variable* and x, which we have observed, the *explanatory variable*. These are sometimes referred to as the *dependent and independent variables*, respectively, but then we need to remind ourselves that such an interpretation is not applicable in the probabilistic sense; also the x variable is termed alternatively as the *input, regressor*, or *predictor variable*. The unobservable random variable ε represents the difference between Y and the deterministic component, $\beta_0 + \beta_1 x$; it is usually called the *error* term. If we assume that x is a precise observation, the error accounts for extraneous or unknown or unmeasured factors that influence Y. It is then called the *measurement* error. The model is formulated so that $E[\varepsilon] = 0$. We also assume that the errors have a constant variance for all x and they are independent. Other assumptions and conditions, which we shall come to later, are needed to specify the model completely.

As an example of regression, let us reconsider the data of Table E.1.2, with the compressive strength of concrete at 28 days as the response variable Y and the density of concrete as the explanatory variable x, both of which have been observed. We discussed the relationship in Examples 1.28 and 1.29 and referred to the scatter diagram of Fig. 1.4.1. This is a plot of corresponding pairs of observations (x_i, y_i), $i = 1, 2, \ldots, n$, in a two-dimensional coordinate system. Such a graph is helpful as an indicator of the type of association between the variables. For the data cited, the relationship is not a particularly strong one but is seen to be linear, as noted in Chapter 1. That is, there is much scatter, but the points can be considered to be located randomly above and below a straight line (as defined shortly) that passes through the cluster. In cases such as this, we find that a density of concrete which is higher (lower) than the average density tends to be associated with a compressive strength that is higher (lower) than the average strength; however, it is also seen that some high values of one kind are paired with average values of the other kind.[1] If the locations of the points tend on average to follow a curve, a nonlinear relationship may be a feasible alternative, but, on the other hand, if they are spread over the entire plot it indicates the lack of any relationship. Graphs such as Figs. 1.4.1 and 1.4.2 are an essential part of the procedure adopted by engineers. They are indeed complementary to the regression analysis that follows. In this aspect there is a striking similarity to the probability plots and the goodness-of-fit criteria of Chapter 5.

In situations that conform to Eq. (6.1.1), it is reasonable to write the following *linear conditional relationship*:

$$E[Y \mid X = x] = \beta_0 + \beta_1 x. \tag{6.1.2}$$

This conditional expectation denotes a regression that is linear. In practice a linear relationship may be an approximation. For the data cited—concrete strengths and densities—we have statistical justification for such an assumption. This arises from our postulation of a bivariate normal distribution in Example 4.32, which we confirmed from the normal marginal distributions as in Examples 5.31 and 5.37 and the estimated correlation coefficient of 0.44 in Example 1.29.

Note that the linearity in the context of a statistical regression model for the mean of Y (corresponding to a particular x) is defined with respect to the unknown parameters β_0 and β_1. For instance, $E[Y \mid X = x] = \beta_0 + \beta_1 x^2$ is a linear function of β_0 and β_1; we can also see that $E[Y \mid X = x]$ is a linear function of x^2 but not of x, in this case, unlike in Eq. (6.1.2).

It is clear that the expected values of Y fall on a straight line as defined by the conditional relationship of Eq. (6.1.2) with respect to the observations x; that is, the mean of Y is $\beta_0 + \beta_1 x$ for a particular x. For this purpose it does not matter whether x is the observed outcome of a random process or is known and fixed by the experimenter. The slope β_1 is the change in the mean of Y for a unit increase in x. How Y fluctuates at a particular value of X is governed by the variance of the random variable ε. This is illustrated in Fig. 6.1.1 for a general case.

6.1.1 Estimates of the parameters

Our first task is to estimate the unknown parameters from observations of the two variables Y and x. So far, we have not said anything about the probability distribution of ε. Although distributions are sketched in Fig. 6.1.1, these are not important for our initial objective.

[1] In 1885, Francis Galton, an anthropologist and meteorologist born in Birmingham, England, called this behavior a regression (which means "stepping back") toward mediocrity (or toward the mean, as we now call it) in reference to the relationship between the heights of sons and fathers, and it led to the current usage of regression.

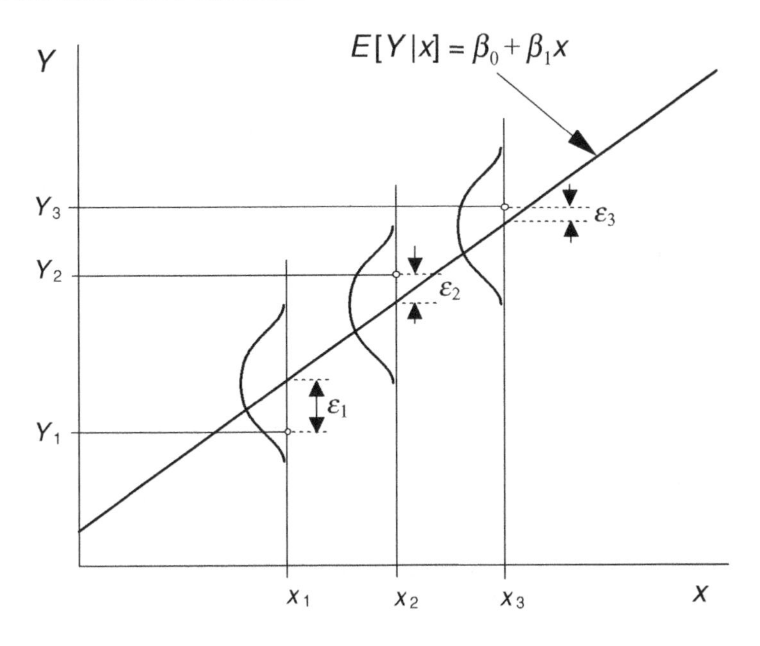

Fig. 6.1.1 Graphical representation of the simple regression model.

As shown in Fig. 6.1.2, for the strengths and densities of concrete listed in Table E.1.2 and plotted originally in Fig. 1.4.1, we can draw a straight line to pass through the cluster of points to provide a "best fit" by eye. But "eyeball fitting" seldom provides the best estimates of the parameters, as indicated by the probability plots of Chapter 5. We could minimize the absolute deviations from the straight line, but their sum is not a convenient quantity from the mathematical point of view.[2] Instead we can minimize the sum of squared deviations from the mean to solve for the parameters, as suggested by Karl Gauss

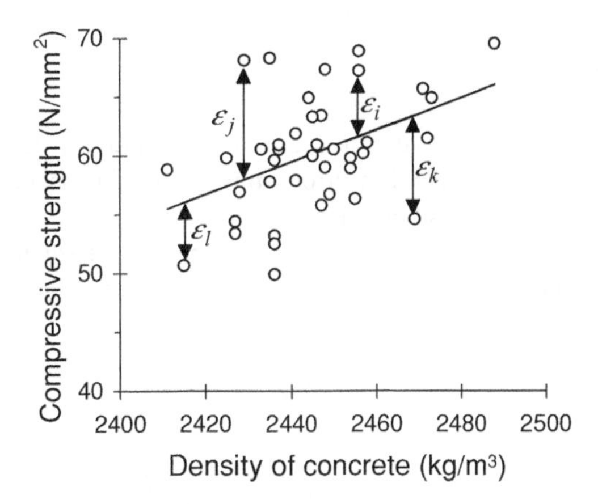

Fig. 6.1.2 Simple linear regression plot of concrete test data; four examples of estimated errors (residuals) are shown.

[2] Likewise, we recall that the mean absolute deviation was not preferred as a measure of dispersion in Chapter 1 for theoretical reasons [Eqs. (1.2.4) and (1.2.5)].

of Germany about 200 years ago and independently by others such as A. M. Legendre of France, after Euler of Switzerland. Then we follow what is known as the *method of least squares*. This is the most commonly used procedure.

Suppose we have n pairs of observations $(x_i, y_i), i = 1, 2, \ldots, n$. For the model specified by Eq. (6.1.1), the sum of the squared deviations (or vertical distances on the plot) from the population regression line with intercept β_0 and slope β_1 is given by

$$S^2 = \sum_{i=1}^{n} \varepsilon_i^2 = \sum_{i=1}^{n} (y_i - \beta_0 - \beta_1 x_i)^2. \tag{6.1.3}$$

The least squares solutions for the unknown parameters are found by minimizing S^2. Accordingly, these are obtained from the solutions to $\partial S^2/\partial\beta_0 = 0$ and $\partial S^2/\partial\beta_1 = 0$. Hence,

$$-2\sum_{i=1}^{n} (y_i - \hat{\beta}_0 - \hat{\beta}_1 x_i) = 0 \qquad \text{and} \qquad -2\sum_{i=1}^{n} (y_i - \hat{\beta}_0 - \hat{\beta}_1 x_i)x_i = 0.$$

These least squares equations can be solved simultaneously because they are linear with respect to β_0 and β_1. Hence,

$$\hat{\beta}_1 = \frac{\sum_{i=1}^{n} y_i x_i - \left(\sum_{i=1}^{n} x_i\right)\left(\sum_{i=1}^{n} y_i\right)/n}{\sum_{i=1}^{n} x_i^2 - \left(\sum_{i=1}^{n} x_i\right)^2/n} = \frac{\sum_{i=1}^{n} (y_i - \bar{y})(x_i - \bar{x})}{\sum_{i=1}^{n} (x_i - \bar{x})^2} \tag{6.1.4}$$

and

$$\hat{\beta}_0 = \frac{\sum_{i=1}^{n} y_i}{n} - \hat{\beta}_1 \frac{\sum_{i=1}^{n} x_i}{n} = \bar{y} - \hat{\beta}_1 \bar{x}, \tag{6.1.5}$$

where

$$\bar{y} = \frac{\sum_{i=1}^{n} y_i}{n}$$

$$\bar{x} = \frac{\sum_{i=1}^{n} x_i}{n}.$$

In Eq. (6.1.4), the first expression seems to be relatively accurate but if the numbers are large and n is large roundoff errors may occur. The second expression is therefore recommended under such situations.

The definitions of the following terms for the sum of squares and cross-products are useful for computations in regression:

$$S_{xx} = \sum_{i=1}^{n} (x_i - \bar{x})^2 = \sum_{i=1}^{n} x_i^2 - \left(\sum_{i=1}^{n} (x_i)\right)^2 \Big/ n, \tag{6.1.6a}$$

$$S_{yy} = \sum_{i=1}^{n} (y_i - \bar{y})^2 = \sum_{i=1}^{n} y_i^2 - \left(\sum_{i=1}^{n} (y_i)\right)^2 \Big/ n, \tag{6.1.6b}$$

and

$$S_{xy} = \sum_{i=1}^{n} (x_i - \bar{x})(y_i - \bar{y}) = \sum_{i=1}^{n} x_i y_i - \left(\sum_{i=1}^{n} x_i\right)\left(\sum_{i=1}^{n} y_i\right) \Big/ n. \tag{6.1.6c}$$

One can, for instance, write the second expression in Eq. (6.1.4) for the slope parameter as

$$\hat{\beta}_1 = \frac{S_{xy}}{S_{xx}}. \tag{6.1.7}$$

Definition and properties: A simple linear regression model that relates a random response Y to an explanatory variable x takes the form

$$Y = \beta_0 + \beta_1 x + \varepsilon,$$

where β_0 and β_1 are the intercept and slope parameters, respectively, and ε is the random error term. Least squares estimates of the parameters are obtained from n pairs of observations (x_i, y_i), $i = 1, 2, \ldots, n$, as

$$\hat{\beta}_1 = \frac{\sum_{i=1}^{n} y_i x_i - \left(\sum_{i=1}^{n} x_i\right)\left(\sum_{i=1}^{n} y_i\right) \Big/ n}{\sum_{i=1}^{n} x_i^2 - \left(\sum_{i=1}^{n} x_i^2\right) \Big/ n} = \frac{\sum_{i=1}^{n} (y_i - \bar{y})(x_i - \bar{x})}{\sum_{i=1}^{n} (x_i - \bar{x})^2} = \frac{S_{xy}}{S_{xx}}$$

and

$$\hat{\beta}_0 = \left(\sum_{i=1}^{n} y_i\right) \Bigg/ n - \hat{\beta}_1 \left(\sum_{i=1}^{n} x_i\right) \Bigg/ n = \bar{y} - \hat{\beta}_1 \bar{x},$$

where $\bar{y} = (\sum_{i=1}^{n} y_i)/n$,

$$\bar{x} = \left(\sum_{i=1}^{n} x_i\right) \Bigg/ n,$$

$$S_{xx} = \sum_{i=1}^{n} (x_i - \bar{x})^2 = \sum_{i=1}^{n} x_i^2 - \left(\sum_{i=1}^{n} (x_i)^2\right) \Bigg/ n.$$

Similarly, S_{yy} is defined and

$$S_{xy} = \sum_{i=1}^{n} (x_i - \bar{x})(y_i - \bar{y}) = \sum_{i=1}^{n} x_i y_i - \left(\sum_{i=1}^{n} x_i\right)\left(\sum_{i=1}^{n} y_i\right) \Bigg/ n.$$

Example 6.1. Simple linear regression model for concrete strengths. From Table E.1.2 with observations x and y for concrete density and strength, respectively, the following summaries are obtained:

$$n = 40; \quad \bar{x} = 2445 \text{ kg/m}^3; \quad \bar{y} = 60.14 \text{ N/mm}^2;$$
$$S_{xx} = 9977; \quad S_{yy} = 980.8; \quad S_{xy} = 1365;$$

[using the first expressions of Eq. (6.1.6)]. Hence,

$$\hat{\beta}_1 = \frac{S_{xy}}{S_{xx}} = \frac{1365}{9977} = 0.1368$$

and

$$\hat{\beta}_0 = \bar{y} - \hat{\beta}_1 \bar{x} = 60.14 - \frac{1365}{9977} \times 2445 = -274.4.$$

The negative value of the intercept is linked to the low value of the coefficient of correlation in the regression (+0.44); it is a reflection of the inadequacy of the simple regression model. The fitted model, used to obtain a mean response, is as follows:

$$\hat{Y} = -274.4 + 0.1368x.$$

This is shown in Fig. 6.1.2 together with the sample data points.

6.1.2 Properties of the estimators and errors

Recall that Eqs. (6.1.1) and (6.1.2) hold when x is a constant (or the known outcome of a random variable X) and Y is a random variable. Equation (6.1.1) is mainly centered around the so-called errors ε, which are postulated to have a zero mean and a constant variance σ^2 and are also independent of each other. For inferential purposes we must also assume that they are normally distributed, that is, $\varepsilon \sim N(0, \sigma^2)$; but this assumption may not be critical for large samples because of the Central Limit Theorem.[3] We also assume that the errors are independent of the x values. It follows that the Y are also independent of each other. The conditional variance of Y is therefore

$$\text{Var}[Y \mid X = x] = \text{Var}[\beta_0 + \beta_1 x + \varepsilon] = \sigma^2 \tag{6.1.8}$$

and from Eq. (6.1.2), $Y \sim N(\beta_0 + \beta_1 x, \sigma^2)$.

6.1.2.1 Properties of the estimators of the regression parameters

The least squares estimators $\hat{\beta}_0$ and $\hat{\beta}_1$ given by Eqs. (6.1.5) and (6.1.4), respectively, are by their nature random variables. We noted that the relationships are linear. The properties of these estimators are of practical use in making inferences. For example, it can be shown as follows [using Eq. (6.1.3) and observations (x_i, y_i), $i = 1, 2, \ldots, n$] that $\hat{\beta}_1$ is an unbiased estimator of the slope parameter $\hat{\beta}_1$. Treating the responses Y_i as random variables and the x_i as constants, as before, we have from Eq. (6.1.4),

$$\beta_1 = \frac{\sum_{i=1}^{n}(x_i - \bar{x})(Y_i - \bar{Y})}{\sum_{i=1}^{n}(x_i - \bar{x})^2} = \frac{\sum_{i=1}^{n} Y_i(x_i - \bar{x}) - \bar{Y}\sum_{i=1}^{n}(x_i - \bar{x})}{\sum_{i=1}^{n}(x_i - \bar{x})^2}.$$

The second term in the numerator on the right-hand side is zero. Hence from Eq. (6.1.7) and because the errors and x values are assumed to be independent,

$$\begin{aligned} E[\hat{\beta}_1] &= \frac{1}{S_{xx}} E[Y_i(x_i - \bar{x})] = \frac{1}{S_{xx}} E[(\beta_0 + \beta_1 x_i + \varepsilon_i)(x_i - \bar{x})] \\ &= \frac{1}{S_{xx}} \left\{ E\left[\beta_0 \sum_{i=1}^{n}(x_i - \bar{x})\right] + E\left[\beta_1 \sum_{i=1}^{n} x_i(x_i - \bar{x})\right] + E\left[\sum_{i=1}^{n} \varepsilon_i(x_i - \bar{x})\right]\right\} \\ &= \frac{1}{S_{xx}}(0 + \beta_1 S_{xx} + 0) = \beta_1. \end{aligned} \tag{6.1.9a}$$

We can similarly obtain the variance of $\hat{\beta}_1$ as follows recalling that, under the model assumptions, the Y_i are independent and have a constant variance σ^2:

$$\text{Var}[\hat{\beta}_1] = \text{Var}\left[\frac{\sum_{i=1}^{n} Y_i(x_i - \bar{x})}{S_{xx}}\right] = \frac{1}{S_{xx}^2}\sigma^2 \sum_{i=1}^{n}(x_i - \bar{x})^2 = \frac{\sigma^2}{S_{xx}}. \tag{6.1.9b}$$

Also, the estimator $\hat{\beta}_i$ can be shown to have minimum variance, and, because it is linear, it is called the *BLUE* of β_1, that is, the *best* (signifying minimum variance) *linear unbiased estimator.*[4] By similar arguments one can show that $\hat{\beta}_0$ is the *BLUE* of β_0.

Let us now consider the mean response to a given value of X, say, x_0, which from Eq. (6.1.2) becomes $E[Y \mid X = x_0] = \beta_0 + \beta_1 x_0$. An unbiased estimator is given by

$$\hat{\mu}_{Y|X=x_0} = \hat{\beta}_0 + \hat{\beta}_1 x_0. \tag{6.1.10a}$$

---

[3] See Chapter 4 and Appendix A.6.

[4] See, for example, Casella and Berger (2002, pp. 544–548).

It follows from Eq. (6.1.5); we can write this as

$$\hat{\mu}_{Y|X=x_0} = \bar{Y} - \hat{\beta}_1\bar{x} + \hat{\beta}_1 x_0 = \bar{Y} + \hat{\beta}_1(x_0 - \bar{x}). \tag{6.1.10b}$$

The variance of the estimated mean response is given by

$$\text{Var}[\hat{Y} \mid X = x_0] = \text{Var}[\bar{Y}] + (x_0 - \bar{x})^2\text{Var}[\hat{\beta}_1] + 2(x_0 - \bar{x})\text{Cov}[\bar{Y}, \hat{\beta}_1].$$

The Y_i are independent and have a constant variance σ^2. Thus,

$$\text{Var}[\bar{Y}] = \frac{1}{n}\text{Var}[Y_i] = \frac{\sigma^2}{n}.$$

Also, it can be shown that $\text{Cov}[\bar{Y}, \hat{\beta}_1]$ is zero.[5] Therefore,

$$\text{Var}[\hat{Y} \mid X = x_0] = \frac{\sigma^2}{n} + (x_0 - \bar{x})^2\frac{\sigma^2}{S_{xx}}. \tag{6.1.11}$$

Hence, the estimated mean response to a given value of x, say, x_0, is distributed (by the Central Limit Theorem) as

$$\hat{\mu}_{Y|X=x_0} \sim N\left[\beta_0 + \beta_1 x_0, \sigma^2\left(\frac{1}{n} + \frac{(x_0 - \bar{x})^2}{S_{xx}}\right)\right]. \tag{6.1.12}$$

For the special case $x_0 = 0$, the random variable $\hat{\mu}_{Y|X=x_0=0}$ is the estimator of β_0, the intercept parameter, with distribution

$$\hat{\beta}_0 \sim N\left[\beta_0, \sigma^2\left(\frac{1}{n} + \frac{\bar{x}^2}{S_{xx}}\right)\right]. \tag{6.1.13}$$

6.1.2.2 Properties of the errors

To estimate the properties of the errors one uses the differences between the observations and the fitted values, or

$$\hat{\varepsilon}_i = y_i - (\hat{\beta}_0 + \hat{\beta}_1 x_i).$$

These are called the residuals. Geometrically, they are the *vertical* distances from the plotted points to the databased straight line as shown in Fig. 6.1.2. The residuals have a mean of zero. This follows from

$$\begin{aligned}\sum_{i=1}^{n}\hat{\varepsilon}_i &= \sum_{i=1}^{n}[y_i - (\hat{\beta}_0 + \hat{\beta}_1 x_i)] = \sum_{i=1}^{n}[y_i - (\bar{y} - \hat{\beta}_1\bar{x} + \hat{\beta}_1 x_i)]\\ &= \sum_{i=1}^{n}(y_i - \bar{y}) - \hat{\beta}_1\sum_{i=1}^{n}(x_i - \bar{x}) = 0 - \hat{\beta}_1 \times 0 = 0.\end{aligned} \tag{6.1.14a}$$

It can also be shown as follows, using the foregoing results, that the residuals and the values of the explanatory variable are not correlated:

$$\begin{aligned}\sum_{i=1}^{n}(x_i - \bar{x})\hat{\varepsilon}_i &= \sum_{i=1}^{n}(x_i - \bar{x})(y_i - \bar{y}) - \hat{\beta}_1\sum_{i=1}^{n}(x_i - \bar{x})(x_i - \bar{x})\\ &= S_{xy} - \hat{\beta}_1 S_{xx} = S_{xy} - S_{xy} = 0.\end{aligned} \tag{6.1.14b}$$

For estimating the variance of the errors using the residuals, we use the divisor $(n - 2)$ because two degrees of freedom are already lost in estimating β_0 and β_1. To

[5] See, for example, Subsection 3.3.3 and Wackerly et al. (2002, p. 546).

put it differently, after one calculates any $(n-2)$ residuals the remaining two residuals are fixed by the constraints set by Eq. (6.1.14a) and (6.1.14b). Hence

$$\hat{\sigma}^2 = \frac{1}{n-2}\sum_{i=1}^{n}\hat{\varepsilon}_i^2 = \frac{1}{n-2}\sum_{i=1}^{n}(y_i - \hat{\beta}_0 - \hat{\beta}_1 x_i)^2. \tag{6.1.14c}$$

Also, from Eqs. (6.1.4) to (6.1.7),

$$\begin{aligned}\hat{\sigma}^2 &= \frac{1}{n-2}\sum_{i=1}^{n}[(y_i - \bar{y}) - \hat{\beta}_1(x_i - \bar{x})]^2 \\ &= \frac{1}{n-2}\sum_{i=1}^{n}[(y_i - \bar{y})^2 + \hat{\beta}_1^2(x_i - \bar{x})^2 - 2\hat{\beta}_1(y_i - \bar{y})(x_i - \bar{x})] \\ &= \frac{1}{n-2}\left(S_{yy} - \frac{S_{xy}^2}{S_{xx}}\right).\end{aligned} \tag{6.1.14d}$$

Either Eq. (6.1.14c) or (6.1.14d) can be used for estimating the error variance, with consideration for possible roundoff errors.

Definitions and properties: Variances of estimators and error variance. The variances of the slope and intercept parameters are as follows:

$$\text{Var}[\hat{\beta}_1] = \frac{\sigma^2}{S_{xx}}$$

and

$$\text{Var}[\hat{\beta}_0] \sim \sigma^2\left(\frac{1}{n} + \frac{\bar{x}^2}{S_{xx}}\right).$$

An unbiased estimator of the error variance is given by

$$\begin{aligned}\hat{\sigma}^2 &= \frac{1}{n-2}\sum_{i=1}^{n}\hat{\varepsilon}_i^2 = \frac{1}{n-2}\sum_{i=1}^{n}(y_i - \hat{\beta}_0 - \hat{\beta}_0 x_i)^2 \\ &= \frac{1}{n-2}\left(S_{yy} - \frac{S_{xy}^2}{S_{xx}}\right).\end{aligned}$$

6.1.2.3 Graphs and residuals

In our assessment of the fit of the linear model, we make the widest possible use of the residuals, which measure the unknown model errors. We must try to verify whether they meet the assumptions made. The main assumptions are (1) constant variance and normality in distribution of the residuals, (2) independence among the residuals, and (3) independence between the residuals and the values of the explanatory variable, x. Graphical methods usually provide confirmation that there are no shortcomings or systematic defects in our model, and are supplemented by statistical tests when necessary. They begin with a scatter plot of the two variables as we had in Chapter 1 (Figs. 1.4.1 and 1.4.2) or we may relate different explanatory variables. Other graphs are based on the ordinary residuals or some associated variables. The normal probability plot is used for indicating departures from normality in the residuals, as we did in Subsection 5.8.2. This may show, for instance, a heavy-tailed distribution or the presence of some unexpected values called *outliers* or *influential observations*; these outliers tend to violate the assumptions made and unduly influence the fitting of the model. In some cases, the graphs may indicate the need for a transformation of the response variable. For example, an increase of variance of the residuals suggests the need for a square-root transformation. (More about this follows.)

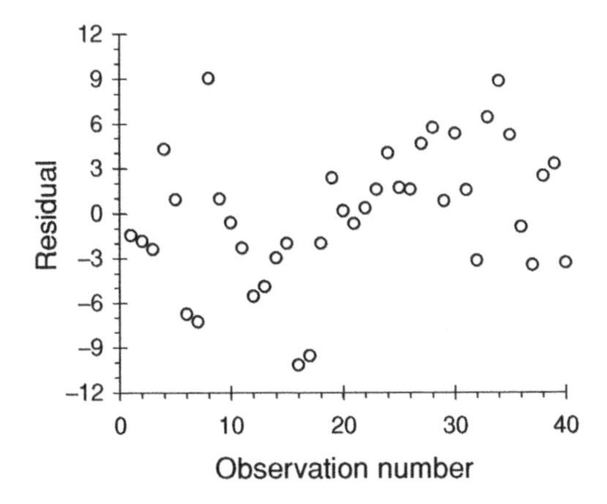

Fig. 6.1.3 Index plot of residuals for regression of concrete strengths and densities.

These aspects of graphical diagnostics, including influential observations and the need for model changes, will be examined in Section 6.2. The discussion here is confined to the basic plots and assumptions.

Example 6.2. Properties of the residuals of linear regression model applied to concrete data. The 40 residuals $\hat{\varepsilon}_i = y_i - (\hat{\beta}_0 + \hat{\beta}_1 x_i)$ are determined from the data of Table 1.2.1 and the parameters estimated in Example 6.1. Here we examine their independence and distributional properties.

To determine whether the residuals are related to each other (in a case like this where there may be time-dependency between the observations), we firstly make an index plot of the residuals (that is residual against observation number). Secondly, we plot the relationship $\hat{\varepsilon}_i$ versus $\hat{\varepsilon}_{i-1}$. This type of plot is relevant when the observations are made at regular intervals of time. These two plots are shown in Figs. 6.1.3 and 6.1.4, respectively. Although the data are by no means perfect (there is evidence of some runs in Fig. 6.1.3), the plots do not show any significant autocorrelation in the residuals.[6] Such a relationship would result in a trend in Fig. 6.1.4.

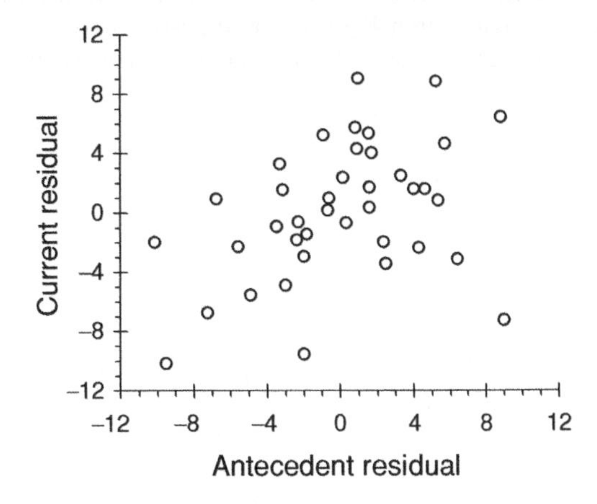

Fig. 6.1.4 Plot of residual versus antecedent residual.

[6] To test whether the residuals are serially related see Durbin and Watson (1951). The test is given by, for example, Draper and Smith (1998, Section 7.2) and Montgomery and Runger (1994). However, this only shows whether there is a particular type of dependence—that of a first-order autoregressive model.

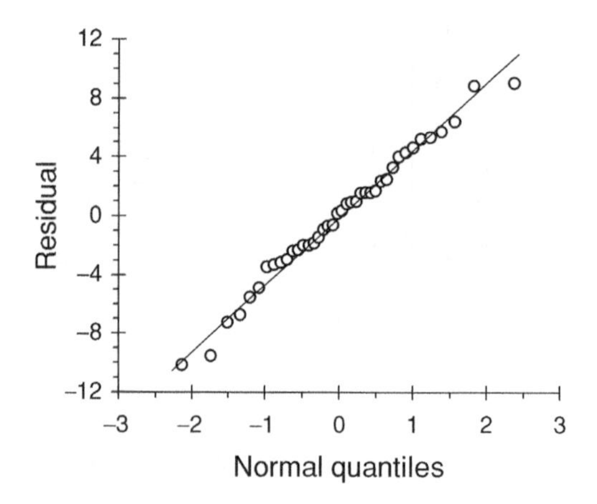

Fig. 6.1.5 Normal probability plot of residuals from regression of concrete strengths and densities.

As given earlier, $S_{xx} = 9977$; $S_{xy} = 1365$; $S_{yy} = 980.8$. From Eq. (6.1.14*d*) the variance of the errors is estimated as

$$\hat{\sigma}^2 = \frac{1}{38}\left(980.8 - \frac{1365^2}{9977}\right) = 20.89.$$

Thus the estimated standard deviation is $\hat{\sigma} = \sqrt{20.89} = 4.57$. Only two of the 40 residuals are outside the range $-2\hat{\sigma}$ to $+2\hat{\sigma}$, that is, -9.14 to $+9.14$. The coefficient of skewness of the residuals is -0.1633, and the coefficient of kurtosis is 2.79; these are in accordance with the normal assumption. A normal probability plot is drawn as shown in Fig. 6.1.5. It does not indicate any outliers or untoward behavior. From the graph and the foregoing statistics, we see that the distribution of the residuals is close to normality. (We may apply the Kolmogorov-Smirnov test as an additional verification.) Figure 6.1.6 is a plot of the residuals $\hat{\varepsilon}_i$ against the densities of concrete x_i.

The dispersion of the points indicates that the errors, as estimated by the residuals, are independent of the explanatory variable, which is one of the assumptions made. Likewise, we can infer the independence by producing a plot of the residuals $\hat{\varepsilon}_i$ against the fitted values $\hat{y}_i$. It is reasonable to assume from Fig. 6.1.6 that the variance is constant; that would not be the case if there were a much larger spread above and below one part of the horizontal axis than another.

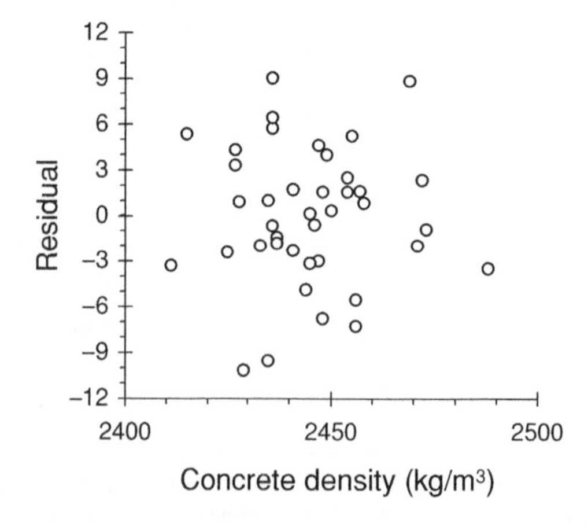

Fig. 6.1.6 Plot of residuals versus concrete densities.

6.1.3 Tests of significance and confidence intervals

In this subsection we apply the methods of Sections 5.3 and 5.4 for inferential purposes. For instance, the significance of the regression can be tested through the slope parameter β_1.

6.1.3.1 Slope parameter

To test whether the slope parameter is equal to a constant, say, β^*, we declare

$$H_0: \beta_1 = \beta^*$$

and

$$H_1: \beta_1 \neq \beta^*$$

as the null and alternative hypotheses, respectively. In this way a two-sided alternative is assumed. Note that the errors are independent and distributed as $\varepsilon \sim N(0, \sigma^2)$ and the Y_i are independent and distributed as $Y_i \sim N(\beta_0 + \beta_1 x_i, \sigma^2)$. As regard the estimator $\hat{\beta}_1$, it is seen from Eq. (6.1.4) that this is linear with respect to the Y_i. Hence, from Eq. (6.1.9a) and (6.1.9b), $\hat{\beta}_1 \sim N(\beta_1, \sigma^2/S_{xx})$. It follows therefore from Eq. (6.1.9) that, under the null hypothesis,

$$\frac{\hat{\beta}_1 - \beta^*}{\sigma/\sqrt{S_{xx}}} \sim N(0, 1).$$

It can also be shown that $(n-2)\hat{\sigma}^2/\sigma^2 \sim \chi^2_{n-2}$, which is a chi-squared distribution with $n-2$ degrees of freedom.[7] In addition, this is independent of $[(\hat{\beta}_1 - \beta^*)/(\sigma/\sqrt{S_{xx}})]$. Thus on consideration of the properties of Student's t distribution as given in Appendix A.7,

$$T = \frac{\hat{\beta}_1 - \beta^*}{\hat{\sigma}/\sqrt{S_{xx}}} \sim t_{n-2}. \tag{6.1.15}$$

This becomes our test statistic. If we set β^* to zero, we can then test the significance of the regression model as will be shown in Example 6.3.

A $100(1-\alpha)$ percent confidence interval for β_1 is found in

$$\Pr\left[\hat{\beta}_1 - t_{n-2,\alpha/2}\frac{\hat{\sigma}}{\sqrt{S_{xx}}} \leq \beta_1 \leq \hat{\beta}_1 + t_{n-2,\alpha/2}\frac{\hat{\sigma}}{\sqrt{S_{xx}}}\right] = 1 - \alpha.$$

Tests and confidence intervals for the intercept β_0 are based on Eq. (6.1.13). Also of practical interest are a confidence interval for the mean value of Y, say, when $X = x_0$, and a prediction interval for a future value of Y. They are based on distributional and independence properties that are similar to those of the estimator $\hat{\beta}_1$.

6.1.3.2 Confidence interval for mean value of Y

For a given value of the explanatory variable, say, $X = x_0$, the population of Y values has an estimated mean $\hat{\beta}_0 + \hat{\beta}_1 x_0$ that is normally distributed as specified by Eq. (6.1.12), under the given assumptions. In usual geometric terms, confidence limits for the (population) mean value of Y, $\beta_0 + \beta_1 x_0$, will be equispaced vertically above and below the regression

[7] See, for example, Mood et al. (1974, pp. 489–491) or Casella and Berger (2002, p. 554).

line. Thus a $100(1-\alpha)$ percent confidence interval for the mean value of Y, say, when $X = x_0$, follows from:

$$\Pr[\hat{\beta}_0 + \hat{\beta}_1 x_0 - t_{n-2,\alpha/2}\hat{\sigma}\sqrt{(1/n) + (x_0 - \bar{x})^2/S_{xx}} \leq \beta_0 + \beta_1 x_0 \\ \leq \hat{\beta}_0 + \hat{\beta}_1 x_0 + t_{n-2,\alpha/2}\hat{\sigma}\sqrt{(1/n) + (x_0 - \bar{x})^2/S_{xx}}] = 1 - \alpha. \quad (6.1.16)$$

We are $100(1-\alpha)$ percent confident that the specified interval includes the unknown mean, $\beta_0 + \beta_1 x_0$.

6.1.3.3 Prediction interval for a future value of Y

We next consider the $100(1-\alpha)$ percent prediction interval for an unobserved or future value of Y, say when $X = x_0$ which is wider than that for the mean value of Y, as shown shortly. From Eq. (6.1.1), the variance of a new observation or predicted value Y_0 of Y (which is independent of previous observations) when $X = x_0$ is obtained from Eq. (3.3.23) as follows:

$$\text{Var}[Y_0] = \text{Var}[\hat{\beta}_0 + \hat{\beta}_1 x_0] + \text{Var}[\varepsilon_0],$$

because $\text{Cov}[\varepsilon_0, \hat{\beta}_0 + \hat{\beta}_1 x_0] = 0$ from the assumptions made. Substituting for the first term on the right from Eq. (6.1.11) and because $\text{Var}[\varepsilon_0] = \sigma^2$, we write

$$\text{Var}[Y_0] = \frac{\sigma^2}{n} + (x_0 - \bar{x})^2 \frac{\sigma^2}{S_{xx}} + \sigma^2 = \sigma^2 \left[1 + \frac{1}{n} + (x_0 - \bar{x})^2 \frac{1}{S_{xx}}\right].$$

The foregoing variance is thus an extension of that for the mean value of Y [Eq. (6.1.11)]; it is greater because of the increased uncertainty. As before, the mean value is given by Eq. (6.1.10a). The $100(1-\alpha)$ percent prediction interval for a future or unknown value, say, Y_0 of Y when $X = x_0$, follows from:

$$\Pr[\hat{\beta}_0 + \hat{\beta}_1 x_0 - t_{n-2,\alpha/2}\hat{\sigma}\sqrt{1 + (1/n) + (x_0 - \bar{x})^2/S_{xx}} \leq Y_0 \\ \leq \hat{\beta}_0 + \hat{\beta}_1 x_0 + t_{n-2,a/2}\hat{\sigma}\sqrt{1 + (1/n) + (x_0 - \bar{x})^2/S_{xx}}] = 1 - \alpha. \quad (6.1.17)$$

We are $100(1-\alpha)$ percent confident that the specified interval includes the unknown random variable Y_0 corresponding to $X = x_0$.

Definitions and properties: Confidence intervals. A $100(1-\alpha)$ percent confidence interval for β_1 follows from

$$\Pr[\hat{\beta}_1 - t_{n-2,\alpha/2}\hat{\sigma}/\sqrt{S_{xx}} \leq \beta_1 \leq \hat{\beta}_1 + t_{n-2,\alpha/2}\hat{\sigma}/\sqrt{S_{xx}}] = 1 - \alpha.$$

A $100(1-\alpha)$ percent confidence interval for the mean value of Y, say, when $X = x_0$, follows from

$$\Pr[\hat{\beta}_0 + \hat{\beta}_1 x_0 - t_{n-2,\alpha/2}\hat{\sigma}\sqrt{(1/n) + (x_0 - \bar{x})^2/S_{xx}} \leq \beta_0 + \beta_1 x_0 \\ \leq \hat{\beta}_0 + \hat{\beta}_1 x_0 + t_{n-2,\alpha/2}\hat{\sigma}\sqrt{(1/n) + (x_0 - \bar{x})^2/S_{xx}}] = 1 - \alpha.$$

A $100(1-\alpha)$ percent prediction interval for a future value of Y, say, Y_0 when $X = x_0$, follows from

$$\Pr[\hat{\beta}_0 + \hat{\beta}_1 x_0 - t_{n-2,\alpha/2}\hat{\sigma}\sqrt{1 + (1/n) + (x_0 - \bar{x})^2/S_{xx}} \leq Y_0 \\ \leq \hat{\beta}_0 + \hat{\beta} x_0 + t_{n-2,a/2}\hat{\sigma}\sqrt{1 + (1/n) + (x_0 - \bar{x})^2/S_{xx}}] = 1 - \alpha.$$

Example 6.3. Test of significance of simple linear regression model applied to concrete data. This is effectively a test on the slope parameter. The null and alternate hypotheses are

$H_0: \beta_1 = 0.$

$H_1: \beta_1 \neq 0.$

Level of significance: $\alpha = 0.05$.

Calculations: $n = 40$, $S_{xx} = 9977$, and from Examples 6.1 and 6.2, $\hat{\beta}_1 = 0.1368$, $\hat{\sigma}^2 = 20.89$, that is, $\hat{\sigma} = 4.57$. Under the null hypothesis it follows from Eq. (6.1.5) that the t statistic for the test is

$$\hat{\beta}_1\sqrt{S_{xx}}/\hat{\sigma} = 0.1368 \times \sqrt{9977}/4.57 = 2.99.$$

From Table C.2, this has a probability of nonexceedance of around 0.999.

Decision: We reject the null hypothesis: $\beta_1 = 0$.

Example 6.4. Confidence interval for the mean value and a prediction interval of a future value of compressive strength. By using the foregoing relationships, a confidence interval can be given to the mean value of compressive strength for a given value of concrete density using the data of Table E.1.2. For the 95% confidence and prediction intervals, we note from Table C.2 that $t_{38,0.025} = 2.026$. The confidence limits for the mean concrete strength, for a given value of density x, is found as follows, using the summaries given in Examples 6.1 to 6.3:

$$-274.4 + x \times 0.1368 \pm 2.026 \times 4.57 \times \sqrt{\left[\frac{1}{40} + \frac{(x - 2445)^2}{9977}\right]}.$$

For example, if $x = 2450$ kg/m^3, the 95% confidence interval for the mean strength is given by

$$(59.22, 62.30).$$

Similarly, one can give the 95% prediction limits for a future value Y_0 of compressive strength from a given value of density x:

$$-274.4 + x \times 0.1368 \pm 2.026 \times 4.57 \times \sqrt{\left[1 + \frac{1}{40} + \frac{(x - 2445)^2}{9977}\right]}.$$

For example, if $x = 2450$ kg/m^3, a 95% prediction interval is given by

$$(51.38, 70.14).$$

The 95% confidence intervals are shown in Fig. 6.1.7. We see that the intervals widen on either side of the mean value $\bar{x} = 2445$ kg/m^3 (arising from the last term under the square-root sign) and also, as expected, the prediction limits are throughout wider than the confidence limits.

6.1.4 The bivariate normal model and correlation

In previous sections the regression analysis was based on the assumption that the random variable Y is a linear function of x, which is fixed by the experimenter or, alternatively, is the known outcome of a random variable. In many applications, however, in which both X and Y are treated as random variables, we do not differentiate between the predictor and response variables. In such cases we consider that the observations come from a population with a bivariate pdf $f(x, y)$. It is convenient to assume that the joint distribution

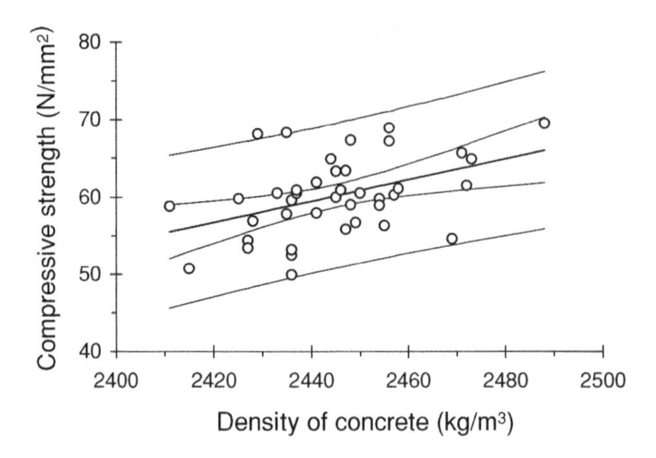

Fig. 6.1.7 Scatter diagram of concrete data with fitted regression line, 95% confidence limits (inner lines) and prediction limits (outer lines).

is bivariate normal as discussed in Subsection 4.3.1, a condition that is sometimes reached by transformation of the variables.

We recall from Eq. (3.3.27) that the coefficient of linear correlation is defined as

$$\rho = \frac{\mathrm{Cov}[X, Y]}{\sigma_X \sigma_Y},$$

where $-1 \leq \rho \leq +1$. The conditional distribution of Y when X takes a value$x$ is normal with conditional expectation

$$E[Y \mid X = x] = \mu_Y + \rho \frac{\sigma_Y}{\sigma_X}(x - \mu_X). \tag{6.1.18}$$

Note that in the bivariate normal model the regression is linear, whereas in the simple regression model linearity was assumed. The intercept and slope parameters [corresponding to Eq. 6.1.2] are, respectively,

$$\beta_0 = \mu_Y - \rho \frac{\sigma_Y}{\sigma_X} \mu_X \qquad \text{and} \qquad \beta_1 = \rho \frac{\sigma_Y}{\sigma_X}.$$

The conditional variance of Y [corresponding to Eq. (6.1.8)] is

$$\mathrm{Var}[Y \mid X = x] = \sigma_Y^2(1 - \rho^2), \tag{6.1.19}$$

which is independent of x.

For example, investigating whether X and Y are independent is equivalent to testing whether $\rho = 0$ (under the normality assumption). For a set of paired data, $x_i, y_i, i = 1, 2, \ldots, n$, the maximum likelihood estimator of the coefficient of linear correlation is given by

$$r = \frac{\sum_{i=1}^{n} (x_i - \bar{x})(y_i - \bar{y})}{\sqrt{\sum_{i=1}^{n} (x_i - \bar{x})^2 \sum_{i=1}^{n} (y_i - \bar{y})^2}} \equiv \frac{S_{xy}}{\sqrt{S_{xx} \times S_{yy}}}, \tag{6.1.20}$$

which is called the *sample correlation coefficient* (or *Pearson's product-moment correlation coefficient*). If we compare this with the estimator for the slope parameter in simple regression, as given by Eq. (6.1.7),

$$\hat{\beta}_1 = r\sqrt{\frac{S_{yy}}{S_{xx}}}. \tag{6.1.21}$$

Thus the slope and correlation coefficient have the same sign.

To test whether $\rho = 0$, we use the following approximation[8]:

$$\frac{1}{2}\ln\frac{1+r}{1-r} \sim N\left(\frac{1}{2}\ln\frac{1+\rho}{1-\rho}, \frac{1}{n-3}\right). \tag{6.1.22}$$

This expression may suffice even for moderate sample sizes, say, $n = 30$. Thus the random variable

$$Z = \frac{\sqrt{n-3}}{2}\ln\frac{(1+r)(1-\rho)}{(1-r)(1+\rho)}$$

is approximately $N(0, 1)$ distributed.

Definition and properties: Significance of correlation. The test statistic used is $Z = \sqrt{n-3}/2\ln[(1+r)(1-\rho)/(1-r)(1+\rho)]$ which is approximately $N(0, 1)$ distributed, where n is the number of observations; also, r and ρ are the sample and population correlation coefficients.

Example 6.5. Testing for positive correlation of concrete densities and strengths.

The null hypothesis H_0: $\rho = 0$.

The alternate hypothesis H_1: $\rho > 0$.

Level of significance: $\alpha = 0.05$.

Calculations: For the data of Table E.1.2, $S_{xx} = 9977$, $S_{yy} = 980.8$, $S_{xy} = 1365$, and $n = 40$. Hence, from Eq. (6.1.18)

$$r = \frac{S_{xy}}{\sqrt{S_{xx} \times S_{yy}}} = \frac{1365}{\sqrt{9977 \times 980.8}} = 0.436$$

and under the null hypothesis

$$z = \frac{\sqrt{37}}{2}\ln\frac{1.436}{0.564} = 2.85.$$

The critical region for a one-tailed test is from Table C.1: $z > z_{0.01} = 2.33$.

Decision: The null hypothesis of no correlation is rejected. We conclude that there is a positive relationship between the concrete densities and strengths.

When there is significant positive correlation between two variables as in Example 6.5, it does not necessarily imply causality. We can only conclude that there is a linear trend between the variables; a positive change in one causes a positive change in the other when $\rho > 0$. If a significant test shows that the null hypothesis of $\rho = 0$ cannot be rejected, then we conclude that the variables are independent with reference to the linear model; there may, however, be some other type of association. On the other hand, for variables with no association, $\rho = 0$.

[8] The transformation was originated by R. A. Fisher; see, for example, Stuart and Ord (1987, pp. 532–533).

6.1.5 Summary of Section 6.1

In this Section we have examined the basic aspects of simple linear regression. From our physical evidence we called x the known predictor variable and Y the unknown response variable. The data we have examined have a linear relationship, which justifies our assumption of linearity. There is another point. If for some physical reason the predictor and response variables are interchanged, the regression changes from one of Y on x to one of X on y; the straight line then has a different slope and intercept, which we can easily define from the foregoing relationships. Furthermore, our discussion has not included other methods of model specification when our basic assumptions are not met, such as weighted least squares.[9] We shall deal with such problems and their possible solutions in Section 6.2.

6.2 MULTIPLE LINEAR REGRESSION

In Section 6.1 we discussed simple linear regression between two variables. We found, for instance, that the compressive strength of a concrete at 28 days is related to its density. If we carry our investigations further, we find that the strength is related to the water-cement ratio and also to other variables, such as the methods of curing the concrete. Likewise, an engineer planning the water resources of a region finds it useful to establish a relationship between the runoff from a catchment and the rainfall input. In addition, one includes catchment area, altitude, length of mainstream, and other factors as explanatory variables. Elsewhere, a soils engineer establishes trends in the strengths of soils over different strata. These are examples where the equation contains more than one explanatory variable. The procedure is known as *multiple regression.*

In considering the problem, we cannot view all the variables simultaneously as we do in a simple (two-dimensional) graph. When there are three variables, a three-dimensional surface can be drawn to illustrate the variability and the fit. With more variables, however, we need to adopt mathematical modeling for the purpose. The analysis is through linear algebra, which provides a compact notation.

One should bear in mind that the exact relationship between variables is usually of a very complex nature. With our linear model and the ensuing least squares fit, as in the case of simple regression, we try to find a simplified but best possible solution on the basis of certain assumptions.

Our strategy begins with formulating the problem. An appropriate set of the explanatory (regressor) variables are chosen and arranged in descending order of considered physical importance. The model assumptions are specified. We then fit the model by least squares. This is followed by the validation of the model and the assumptions made. The procedure includes hypothesis testing made on parameters both individually and collectively. Another important aspect is the use of numerous graphical diagnostics, which can be highly revealing. In addition, if there are one or more sets of observations that exert a strong influence on the parameters or predictions, further tests are made prior to discarding them. These steps constitute an iterative procedure that can lead to a partial systematic elimination or possible changes in the chosen set of explanatory variables if some of them are not contributing significantly to the regression. In addition, one may see a need for a transformation of the response variable. If the validation, which includes the linearity assumption, is acceptable we evaluate the fit. A poor fit will require further iterations and modifications or perhaps a different model.

[9] See, for example, Draper and Smith (1998, pp. 223–229); see also Problem 6.10.

Strong dependencies in the explanatory variables can cause instability in the parameters. *Ridge regression* is devised to find solutions. Our discussion extends to alternative methods in other situations where the assumptions of a linear regression model are not met. These include transformations to linearity, as already mentioned, general and generalized linear models, accommodation of outliers through robust statistics and other means, nonlinear regression and nonparametric methods.

6.2.1 Formulation of the model

A multiple linear regression model takes the form

$$Y = \beta_0 + \beta_1 x_1 + \beta_2 x_2 + \cdots + \beta_{p-1} x_{p-1} + \varepsilon, \tag{6.2.1}$$

where Y is the response variable and there are $p-1$ explanatory variables $x_1, x_2, \ldots, x_{p-1}$, with p parameters (regression coefficients or constants) $\beta_0, \beta_1, \beta_2, \ldots, \beta_{p-1}$. The error (innovation) term ε plays the same role as in the simple linear model. It is assumed to be independently and identically distributed with mean 0 and variance σ^2. For hypothesis testing and the setting of confidence limits, we also assume that ε is normally distributed; alternative distributions of ε are discussed in Subsection 6.2.11. The model is thus represented in the p-dimensional hyperspace of the variables. The parameters $\beta_1, \beta_2, \ldots, \beta_{p-1}$ are sometimes called *partial regression coefficients* because β_j, for example, represents the mean change in Y per unit change in x_j while all the other x variables are held constant. As in the simple model, the linearity of a multiple regression model is defined with respect to the regression coefficients. As before, we assume that the explanatory variables are known and are error-free but the response variable is treated as a random variable.

6.2.2 Linear least squares solutions using the matrix method

The first objective is to estimate the p unknown parameters $\beta_0, \beta_1, \beta_2, \ldots, \beta_{p-1}$. We represent these by a $(p \times 1)$ vector $\boldsymbol{\beta}$ (that is, a column vector). The random errors and the response variables are represented by $(n \times 1)$ vectors, denoted by $\boldsymbol{\varepsilon}$ and $\mathbf{Y}$, respectively. Estimation of the unknown error variance σ^2 follows in the next subsection. The observations x_{ij} are contained in the $(n \times p)$ matrix:

$$\mathbf{X} = \begin{bmatrix} 1 & x_{11} & x_{12} & \cdots & x_{1p-1} \\ 1 & x_{21} & x_{22} & \cdots & x_{2p-1} \\ \vdots & \vdots & \vdots & \ddots & \vdots \\ 1 & x_{n1} & x_{n2} & \cdots & x_{np-1} \end{bmatrix}. \tag{6.2.2}$$

This is sometimes called a *carrier matrix* because it includes the $p-1$ explanatory variables and, according to the specification given by Eq. (6.2.1), has a column of 1s to cater to the constant β_0.

Thus the multiple regression model is written as

$$\mathbf{Y} = \mathbf{X}\boldsymbol{\beta} + \boldsymbol{\varepsilon}. \tag{6.2.3}$$

The vector of mean values $E[\mathbf{Y}]$ of Y is

$$E[\mathbf{Y}] = \mathbf{X}\boldsymbol{\beta}. \tag{6.2.4}$$

For estimation purposes let us revert briefly to the procedure adopted in Subsection 6.1.1 [see Eq. (6.1.3)]. Suppose we have n sets of observations of all the variables

$(y_i, x_{i1}, x_{i2}, \ldots, x_{ip-1}), i = 1, 2, \ldots, n$. The least squares solution is obtained by minimizing

$$S^2 = \sum_{i=1}^{n} \varepsilon_i^2 = \sum_{i=1}^{n} (y_i - \beta_0 - \beta_1 x_{i1} \cdots - \beta_{p-1} x_{ip-1})^2 \quad (6.2.5)$$

with respect to the unknown parameters $\beta_0, \ldots, \beta_{p-1}$. Thus, as before, the sum of squared errors is partially differentiated with respect to each of the unknown parameters and equated to zero. This gives the following p linear equations:

$$n\hat{\beta}_0 + \hat{\beta}_1 \sum_{i=1}^{n} x_{i1} + \hat{\beta}_2 \sum_{i=1}^{n} x_{i2} + \cdots + \hat{\beta}_{p-1} \sum_{i=1}^{n} x_{ip-1} = \sum_{i=1}^{n} y_i$$

$$\hat{\beta}_0 \sum_{i=1}^{n} x_{i1} + \hat{\beta}_1 \sum_{i=1}^{n} x_{i1} x_{i1} + \hat{\beta}_2 \sum_{i=1}^{n} x_{i1} x_{i2} + \cdots + \hat{\beta}_{p-1} \sum_{i=1}^{n} x_{i1} x_{ip-1} = \sum_{i=1}^{n} y_i x_{i1}$$

$$\vdots$$

$$\hat{\beta}_0 \sum_{i=1}^{n} x_{ip-1} + \hat{\beta}_1 \sum_{i=1}^{n} x_{i1} x_{ip-1} + \hat{\beta}_2 \sum_{i=1}^{n} x_{i2} x_{ip-1} + \cdots + \hat{\beta}_{p-1} \sum_{i=1}^{n} x_{ip-1} x_{ip-1} = \sum_{i=1}^{n} y_i x_{ip-1}. \quad (6.2.6)$$

These are called the *normal equations* and provide the p estimators, $\hat{\beta}_0, \hat{\beta}_1, \hat{\beta}_2, \ldots, \hat{\beta}_{p-1}$ of the parameters.

The matrix representation of the Eq. (6.2.6) is $\mathbf{X}^T\mathbf{X}\hat{\boldsymbol{\beta}} = \mathbf{X}^T\mathbf{y}$, where T denotes transpose and $\mathbf{y}$ is a $(n \times 1)$ vector of observed Y values. If the $(p \times p)$ matrix $\mathbf{X}^T\mathbf{X}$, which is symmetric, can be inverted (see the next illustration) the least squares solution to the unknown parameters is

$$\hat{\boldsymbol{\beta}} = (\mathbf{X}^T\mathbf{X})^{-1}\mathbf{X}^T\mathbf{y}, \quad (6.2.7)$$

which is a $(p \times 1)$ vector of fitted parameters.

Thus the vector of estimated mean values of Y is given by

$$\hat{\mathbf{y}} = \mathbf{X}\hat{\boldsymbol{\beta}}. \quad (6.2.8)$$

As in the simple linear regression model, the residuals

$$\hat{\boldsymbol{\varepsilon}} = \mathbf{y} - \mathbf{X}\hat{\boldsymbol{\beta}}, \quad (6.2.9a)$$

which constitute a $(n \times 1)$ vector of differences between the observed and estimated mean values of Y are taken as estimators of the errors

$$\boldsymbol{\varepsilon} = \mathbf{Y} - \mathbf{X}\boldsymbol{\beta} \quad (6.2.9b)$$

and are used in the assessment of the model.

Definitions and properties: Multiple regression. The model is written as $\mathbf{Y} = \mathbf{X}\boldsymbol{\beta} + \boldsymbol{\varepsilon}$, where the vector $\mathbf{Y}$ denotes the response variable, $\mathbf{X}$ is a matrix of explanatory variables, β is a vector of parameters, and $\boldsymbol{\varepsilon}$ is the vector of errors.

Least squares estimates of parameters are given by the $(p \times 1)$ vector $\hat{\boldsymbol{\beta}} = (\mathbf{X}^T\mathbf{X})^{-1}\mathbf{X}^T\mathbf{y}$.

Example 6.6. Multiple regression on stream basin characteristics. Table E.6.1 gives some characteristics of 20 stream basins in the Valtellina region of northern Italy. Physical evidence supports the hypothesis that in this area mean annual runoff is related to the mean annual rainfall and also to the mean elevation of the basin. The statistical significance of these relationships is considered in subsequent subsections. To formulate the model, therefore, we can treat mean annual runoff as the response variable Y and the mean annual rainfall and the mean elevation as explanatory variables X_1 and X_2, respectively. Observed values are listed in columns 2–4 of Table 6.2.1; some of the entries in columns 5–12 are used here and in subsequent examples. Thus,

$$\mathbf{X} = \begin{bmatrix} 1 & 1350 & 2329 \\ 1 & 1621 & 1593 \\ \vdots & \vdots & \vdots \\ 1 & 1283 & 2206 \end{bmatrix} \quad \text{and} \quad \mathbf{y} = \begin{bmatrix} 1654 \\ 1374 \\ \vdots \\ 1023 \end{bmatrix}.$$

$$\mathbf{X}^{\mathrm{T}}\mathbf{X} = \begin{bmatrix} 1 & 1 & \dots & 1 \\ 1350 & 1621 & \dots & 1283 \\ 2329 & 1593 & \dots & 2206 \end{bmatrix} \begin{bmatrix} 1 & 1350 & 2329 \\ 1 & 1621 & 1593 \\ \vdots & \vdots & \vdots \\ 1 & 1283 & 2206 \end{bmatrix}$$

$$= \begin{bmatrix} 20 & 29{,}596 & 33{,}724 \\ 29{,}596 & 45{,}361{,}666 & 48{,}105{,}718 \\ 33{,}724 & 48{,}105{,}718 & 65{,}828{,}584 \end{bmatrix} = (\text{say}) \begin{bmatrix} a & b & c \\ b & e & f \\ c & f & g \end{bmatrix},$$

taking into account the symmetry of the square ($p \times p$) matrix ($\mathbf{X}^{\mathrm{T}}\mathbf{X}$).

The determinant of this matrix is $d = aeg + 2bcf - af^2 - b^2g - c^2e$. If the determinant is zero, the matrix is called *singular* and it does not have an inverse. If ($\mathbf{X}^{\mathrm{T}}\mathbf{X}$) is nonsingular, its inverse is the transpose of the matrix of which the elements are the signed cofactors divided by the determinant. The inverse is thus given by

$$(\mathbf{X}^{\mathrm{T}}\mathbf{X})^{-1} = \begin{bmatrix} (eg - f^2)/d & (cf - bg)/d & (bf - ce)/d \\ (cf - bg)/d & (ag - c^2)/d & (bc - af)/d \\ (bf - ce)/d & (bc - af)/d & (ae - b^2)/d \end{bmatrix}$$

$$= \begin{bmatrix} 3.1121632701 & -0.0015096654229 & -0.00049114006481 \\ -0.0015096654229 & 0.00000083028568222 & 0.0000001666520403 \\ -0.00049114006481 & 0.00000016665204030 & 0.00000014501742117 \end{bmatrix}.$$

Also,

$$\mathbf{X}^{\mathrm{T}}\mathbf{y} = \begin{bmatrix} 1 & 1 & \dots & 1 \\ 1350 & 1621 & \dots & 1283 \\ 2329 & 1593 & \dots & 2206 \end{bmatrix} \begin{bmatrix} 1654 \\ 1374 \\ \vdots \\ 1023 \end{bmatrix} = \begin{bmatrix} 25{,}661 \\ 38{,}852{,}792 \\ 45{,}285{,}738 \end{bmatrix}.$$

Therefore,

$$\hat{\boldsymbol{\beta}} = (\mathbf{X}^{\mathrm{T}}\mathbf{X})^{-1}\mathbf{X}^{\mathrm{T}}\mathbf{y} = \begin{bmatrix} -1035.1 \\ 1.0664 \\ 0.4390 \end{bmatrix}.$$

Hence the fitted model, used to obtain a mean response, is as follows:

$$\hat{Y} = -1035.1 + 1.0664x_1 + 0.4390x_2.$$

Note that roundoff errors will cause differences in solutions to Examples 6.6 and through 6.16. Many software programs are available to find solutions that account for roundoff errors and nonsingularities.[10]

[10] See the discussion in Example 6.6. For matrixes in statistics and linear algebra, see, for example, Nicholson (1990), Graybill (1983), or Strang (1980). Incidentally, note that the dimensions of the matrix $(\mathbf{X}^{\mathrm{T}}\mathbf{X})^{-1}$ are reduced by 1 if the transformations $y_i - \bar{y}$, $x_{i1} - \bar{x}_1$, and so on, are made (see Problems 6.8 and 6.12).

6.2.3 Properties of least squares estimators and error variance

6.2.3.1 Expectations of the least squares estimators

The least squares estimators $\hat{\beta}_0, \hat{\beta}_1, \hat{\beta}_2, \ldots, \hat{\beta}_{p-1}$ are unbiased estimators of the multiple regression parameters $\beta_0, \beta_1, \beta_2, \ldots, \beta_{p-1}$. This can be shown as follows using the assumption that the errors are independent of the explanatory variables. From Eq. (6.2.6), $E[\hat{\boldsymbol{\beta}}] = E[(\mathbf{X}^T\mathbf{X})^{-1}\mathbf{X}^T\mathbf{Y}]$. As in the simple regression model, we treat the Y values as the random variables and the X values as known or fixed. Therefore,

$$E[\hat{\boldsymbol{\beta}}] = (\mathbf{X}^T\mathbf{X})^{-1}\mathbf{X}^T E[\mathbf{Y}]. \tag{6.2.10}$$

Hence by using Eq. (6.2.4), $E[\hat{\boldsymbol{\beta}}] = (\mathbf{X}^T\mathbf{X})^{-1}\mathbf{X}^T\mathbf{X}\boldsymbol{\beta}$. We note that $(\mathbf{X}^T\mathbf{X})^{-1}(\mathbf{X}^T\mathbf{X}) = \mathbf{I}$, a $(p \times p)$ identity matrix (with 1s in the leading diagonal and 0s elsewhere). Therefore,

$$E[\hat{\boldsymbol{\beta}}] = \boldsymbol{\beta}. \tag{6.2.11}$$

6.2.3.2 Covariance matrix of the least squares estimators

The covariances of the least squares estimators can be expressed as the elements of a matrix $\mathbf{C}$ as follows:

$$\mathbf{C} = \begin{bmatrix} \mathrm{Var}[\hat{\beta}_0] & \mathrm{Cov}[\hat{\beta}_0, \hat{\beta}_1] & \ldots & \mathrm{Cov}[\hat{\beta}_0, \hat{\beta}_{p-1}] \\ \mathrm{Cov}[\hat{\beta}_0, \hat{\beta}_1] & \mathrm{Var}[\hat{\beta}_1] & \ldots & \mathrm{Cov}[\hat{\beta}_1, \hat{\beta}_{p-1}] \\ \vdots & \vdots & \ldots & \vdots \\ \mathrm{Cov}[\hat{\beta}_0, \hat{\beta}_{p-1}] & \mathrm{Cov}[\hat{\beta}_1, \hat{\beta}_{p-1}] & \ldots & \mathrm{Var}[\hat{\beta}_{p-1}] \end{bmatrix}. \tag{6.2.12a}$$

By definition [see Eq. (3.3.24)], $\mathbf{C} = E[(\hat{\boldsymbol{\beta}} - \boldsymbol{\beta})(\hat{\boldsymbol{\beta}} - \boldsymbol{\beta})^T]$. From Eqs. (6.2.7), (6.2.10), and (6.2.11), and because $(\mathbf{X}^T\mathbf{X})^{-1}$ is symmetric,

$$\mathbf{C} = E[(\mathbf{X}^T\mathbf{X})^{-1}\mathbf{X}^T(\mathbf{Y} - E[\mathbf{Y}])(\mathbf{Y} - E[\mathbf{Y}])^T\mathbf{X}(\mathbf{X}^T\mathbf{X})^{-1}].$$

The errors represented by $\boldsymbol{\varepsilon} = \mathbf{Y} - E[\mathbf{Y}]$ [see Eqs. (6.2.3), (6.2.4), and (6.2.9*b*)] are assumed to have a zero expectation and a common variance σ^2 as stipulated earlier. Also, because the errors are mutually independent,

$$E[(\mathbf{Y} - E[\mathbf{Y}])(\mathbf{Y} - E[\mathbf{Y}])^T] = \sigma^2\mathbf{I},$$

which is a $(p \times p)$ matrix with the diagonal elements equal to σ^2 and the off-diagonal elements equal to zero. It follows that

$$\mathbf{C} = \sigma^2(\mathbf{X}^T\mathbf{X})^{-1}\mathbf{X}^T\mathbf{X}(\mathbf{X}^T\mathbf{X})^{-1} = \sigma^2(\mathbf{X}^T\mathbf{X})^{-1}. \tag{6.2.12b}$$

The quantities $\sigma^2 c'_{ii}$—where $c'_{ii}, i = 0, 1, \ldots, p-1$, are the diagonal elements of the $(\mathbf{X}^T\mathbf{X})^{-1}$ matrix—are the variances of the estimators of the regression parameters; they are used in making inferences on the parameters and for setting confidence limits, as will be shown.

6.2.3.3 The error variance

Because the error variance σ^2 is unknown, we use the residuals for estimation as given by Eq. (6.2.9*a*). The residual sum of squares is estimated as

$$SS_E = \sum_{i=1}^{n} (y_i - \hat{y}_i)^2.$$

In matrix notation this is represented as

$$\begin{aligned} SS_E &= (\mathbf{y} - \mathbf{X}\hat{\boldsymbol{\beta}})^T(\mathbf{y} - \mathbf{X}\hat{\boldsymbol{\beta}}) \\ &= \mathbf{y}^T\mathbf{y} - \hat{\boldsymbol{\beta}}^T\mathbf{X}^T\mathbf{y} - \mathbf{y}^T\mathbf{X}\hat{\boldsymbol{\beta}} + \hat{\boldsymbol{\beta}}^T\mathbf{X}^T\mathbf{X}\hat{\boldsymbol{\beta}}. \end{aligned}$$

Prior to Eq. (6.2.7) we noted that $\mathbf{X}^T\mathbf{X}\hat{\boldsymbol{\beta}} = \mathbf{X}^T\mathbf{y}$. Furthermore, the scalar quantity $\mathbf{y}^T\mathbf{X}\hat{\boldsymbol{\beta}}$ is equivalent to its transpose $\hat{\boldsymbol{\beta}}^T\mathbf{X}^T\mathbf{y}$. Therefore, the residual sum of squares is

$$SS_E = \mathbf{y}^T\mathbf{y} - \hat{\boldsymbol{\beta}}^T\mathbf{X}^T\mathbf{y}. \tag{6.2.13a}$$

Because p parameters need to be estimated, an unbiased estimator of σ^2 is

$$\hat{\sigma}^2 = \frac{SS_E}{n-p} = \frac{\mathbf{y}^T\mathbf{y} - \hat{\boldsymbol{\beta}}\mathbf{X}^T\mathbf{y}}{n-p}. \tag{6.2.13b}$$

Confidence limits can be obtained on σ^2 because the variable $(n-p)\hat{\sigma}^2/\sigma^2$ is χ^2_{n-p} distributed on consideration of the assumptions of independence and normality (see Subsection 5.3.5).

Definition and properties: Error variance. The residual sum of squares is

$$SS_E = \mathbf{y}^T\mathbf{y} - \hat{\boldsymbol{\beta}}\mathbf{X}^T\mathbf{y},$$

after estimating p parameters from n sets of observations. An unbiased estimator of σ^2 is

$$\hat{\sigma}^2 = \frac{\mathbf{y}^T\mathbf{y} - \hat{\boldsymbol{\beta}}\mathbf{X}^T\mathbf{y}}{n-p}.$$

Example 6.7. The error variance and confidence limits. From Eq. (6.2.13*a*), the residual sum of squares is $SS_E = \mathbf{y}^T\mathbf{y} - \hat{\boldsymbol{\beta}}\mathbf{X}^T\mathbf{y}$. As given in Table 6.2.1,

$$\mathbf{y}^T\mathbf{y} = \sum_{i=1}^{n} y_i^2 = 34{,}986{,}383.$$

In addition, from Example 6.6

$$\hat{\boldsymbol{\beta}}^T\mathbf{X}^T\mathbf{y} = [-1{,}035.1 \cdots 1.0664 \cdots 0.4390]\begin{bmatrix} 25{,}661 \\ 38{,}852{,}792 \\ 45{,}285{,}738 \end{bmatrix} = 34{,}747{,}164.$$

Hence,

$$SS_E = 34{,}986{,}383 - 34{,}747{,}164 = 239{,}219.$$

From Eq. (6.2.13*b*) the residual variance (which estimates σ^2) is

$$\hat{\sigma}^2 = \frac{239{,}219}{17} = 14{,}071.7.$$

As stated, $(n-p)\hat{\sigma}^2/\sigma^2$ has a χ^2_{n-p} distribution.

To establish 95% confidence limits for σ^2, we note that from Table C.3 $\chi^2_{17,0.975} = 7.56$ and $\chi^2_{17,0.025} = 30.2$. Therefore,

$$\Pr\left[\frac{239{,}219}{30.2} \le \sigma^2 \le \frac{239{,}219}{7.56}\right] = 0.95.$$

Hence,

$$\Pr[7921 \le \sigma^2 \le 31{,}643] = 0.95.$$

We are therefore 95% confident that the interval (7921, 31,643) includes the variance σ^2.

Table 6.2.1 Multiple regression of basin characteristics

Station i	y	x_1	x_2	y^2	$x_1^2 x_1^2$	$x_2^2 x_2^2$	yx_1	yx_2	$x_1 x_2$	$\hat{Y}$	$\hat{\varepsilon}$
1	1654	1350	2329	2,735,716	1,822,500	5,424,241	2,232,900	3,852,166	3,144,150	1427	227
2	1374	1621	1593							1393	−19
3	910	1263	1479	1,887,876	2,627,641	2,537,649	2,227,254	2,188,782	2,582,253	961	−51
4	1189	1293	1857							1159	30
5	1453	1666	1335	828,100	1,595,169	2,187,441	1,149,330	1,345,890	1,867,977	1327	126
6	1278	1593	1140	1,413,721						1164	114
7	818	932	2136		1,671,849	3,448,449	1,537,377	2,207,973	2,401,101	896	−78
8	1047	1121	1844	2,111,209						970	77
9	589	1398	144		2,775,556	1,782,225	2,420,698	1,939,755	2,224,110	519	70
10	769	1615	472	1,633,284						894	−125
11	1730	2113	1230		2,537,649	1,299,600	2,035,854	1,456,920	1,816,020	1758	−28
12	1571	1457	2146	669,124						1461	0.110
13	1382	1519	1641	1,096,209	868,624	4,562,496	762,376	1,747,248	1,990,752	1305	77
14	1600	1936	1350		1,256,641		1,173,687			1622	−22
15	1295	1427	2120	346,921		3,400,336		1,930,668	2,067,124	1417	−122
16	1428	1735	1480	591,361	1,954,404		823,422			1465	−37
17	1461	1803	1495	2,992,900		20,736	1,241,935	84,816	201,312	1544	−83
18	1733	1280	3112		2,608,225	222,784		362,968	762,280	1696	37
19	1357	1191	2615	2,468,041		1,512,900	3,655,490	2,127,900	2,598,990	1383	−26
20	1023	1283	2206		4,464,769					1301	−278
				1,909,924		4,605,316	2,288,947	3,371,366	3,126,722		
					2,122,849						
				2,560,00		2,692,88	2,099,25	2,267,86	2,492,67		

				0	2,307,36	1	8	2	9	6-278.37
				1,677,02	1	1,822,50	3,097,60	2,160,00	2,613,60	27
				5	3,748,09	0	0	0	0	125.558
				2,039,18	6	4,494,40	1,847,96	2,745,40	3,025,24	114.003
				4	2,036,32	0	5	0	0	-78.350
				2,134,52	9	2,190,40	2,477,58	2,113,44	2,567,80	77.288
				1	3,010,22	0	0	0	0	70.163
				3,003,28	5	2,235,02	2,634,18	2,184,19	2,695,48	–
				9	3,250,80	5	3	5	5	125.222
				1,841,44	9	9,684,54	2,218,24	5,393,09	3,983,36	–28.014
				9	1,638,40	4	0	6	0	
				1,046,52	0	6,838,22	1,616,18	3,548,55	3,114,46	110.419
				9	1,418,48	5	7	5	5	76.987
					1	4,866,43	1,312,50	2,256,73	2,830,29	–21.945
					1,646,08	6	9	8	8	–
					9					122.177
										–36.672
										–82.770
										37.115
										–25.808
										–
										278.372
Sum	25,661	29,596	33,724	34,986,3	45,361,6	65,828,5	38,852,7	45,285,7	48,105,7	
				83	66	84	92	38	18	
Mean	1283.05	1479.80	1686.20							
Std. Dev.	329.44	287.05	686.84							

y: runoff (mm); x_1: rainfall (mm); x_2: elevation (m); i: basin; $\hat{Y}$: estimated y; $\hat{\varepsilon}$: residual.

6.2.4 Model testing

As stated in the introduction, multiple regression is an iterative procedure. One starts with a chosen set of explanatory variables arranged in a decreasing order of physical importance. Then if we follow the commonly used *backward elimination procedure*, we test the significance of these variables starting with the last. Changes are made, where necessary, following the results of the tests. The changes may involve the exclusion of some variables and the inclusion of others.

Significance tests applied to the multiple regression model range from the application of the F test to the model with a chosen number $(p-1)$ of explanatory variables to individual tests on the model parameters. As in the case of the simple linear model, the assumptions made concern the errors represented by the term ε, in Eq. (6.2.1). We have assumed that the errors are mutually independent with a common $N(0, \sigma^2)$ distribution and also that the errors are independent of the explanatory variables. It is part of the test procedure to verify the assumptions made.

6.2.4.1 Initial significance tests on the regression

After estimating the parameters of the model, we should try to find evidence of a linear relationship between the response and a subset of the explanatory variables, as already mentioned, which we can consequently use in forecasting. For the initial significance test, the hypotheses are

$$\textit{Null hypothesis } H_0\text{: } \beta_i = 0, \qquad \text{for all } i, i = 1, 2, \ldots, p-1$$

and

$$\textit{Alternate hypothesis } H_1\text{: } \beta_i \neq 0, \qquad \text{for one or more } i, i = 1, 2, \ldots, p-1.$$

The total sum of squares of the observations of the response variable is the sum of squared deviations from the mean:

$$\begin{aligned} S_{yy} &= \sum_{i=1}^{n} (y_i - \bar{y})^2 \\ &= \mathbf{y}^{\mathrm{T}}\mathbf{y} - \frac{\left(\sum_{i=1}^{n} y_i\right)^2}{n}. \end{aligned} \tag{6.2.14}$$

This can be separated into two parts, $S_{yy} = SS_R + SS_E$ which are respectively the sum of squares due to the regression and the sum of squares due to the errors. From Eq. (6.2.13*a*), $SS_E = \mathbf{y}^{\mathrm{T}}\mathbf{y} - \hat{\boldsymbol{\beta}}\mathbf{X}^{\mathrm{T}}\mathbf{y}$. Therefore,

$$SS_R = \hat{\boldsymbol{\beta}}\mathbf{X}^{\mathrm{T}}\mathbf{y} - \frac{\left(\sum_{i=1}^{n} y_i\right)^2}{n}. \tag{6.2.15}$$

Under the null hypothesis, $SS_R/\sigma^2 \sim \chi^2_{p-1}$, where σ^2 is the common variance of the errors and $p-1$ is the number of explanatory variables (that is, there are p parameters including β_0); also, $SS_E/\sigma^2 \sim \chi^2_{n-p}$. From the F distribution of Appendix A.8 and Subsection 5.4.5, on the assumption that the Y and X variables have a multivariate normal distribution,

$$\frac{SS_R/p-1}{SS_E/(n-p)} \sim F_{p-1,n-p}. \tag{6.2.16}$$

The expression on the left, denoted F, is called *the ratio of the means of the two respective sums of squares* (as in Table 5.7.1 for ANOVA). The null hypothesis is rejected if $F > F_{p-1,n-p,\alpha}$ which is the F value with numerator and denominator degrees of freedom

Table 6.2.2 ANOVA table for testing significance in multiple linear regression with p parameters including β_0 in vector $\boldsymbol{\beta}$ using n observations.

Source of variation	Degrees of freedom	Sum of squares	Mean square	F value
Regression	$p-1$	$SS_R = \hat{\boldsymbol{\beta}}^T\mathbf{X}^T\mathbf{y} - (\sum_{i=1}^n y_i)^2/n$	$MS_R = SS_R/(p-1)$	$F = MS_R/MS_E$
Residual	$n-p$	$SS_E = \mathbf{y}^T\mathbf{y} - \hat{\boldsymbol{\beta}}^T\mathbf{X}^T\mathbf{y}$	$MS_E = SS_E/(n-p)$	
Total	$n-1$	$SS_{yy} = \mathbf{y}^T\mathbf{y} - (\sum_{i=1}^n y_i)^2/n$		

of $p-1$ and $n-p$, respectively, and probability of exceedance α. A summary of the procedure is given in Table 6.2.2.

Definitions and properties: Sums of squares and ANOVA. The total sum of squares from n observations is

$$S_{yy} = \mathbf{y}^T\mathbf{y} - \frac{\left(\sum_{i=1}^n y_i\right)^2}{n}.$$

The estimated regression and error sums of squares are, respectively,

$$SS_R = \hat{\boldsymbol{\beta}}\mathbf{X}^T\mathbf{y} - \frac{\left(\sum_{i=1}^n y_i\right)^2}{n} \quad \text{and} \quad SS_E = \mathbf{y}^T\mathbf{y} - \hat{\boldsymbol{\beta}}\mathbf{X}^T\mathbf{y};$$

with ratio of means

$$\frac{SS_R/p-1}{SS_E/(n-p)} \sim F_{p-1,n-p},$$

where $p-1$ is the number of explanatory variables.

Example 6.8. Significance of regression of basin characteristics as given in Table E.6.1. A test is made on the multiple regression for which the parameters were estimated in Example 6.7.

Null hypothesis H_0: $\beta_i = 0$, for $i = 1, 2$.
Alternative hypothesis H_1: $\beta_i \neq 0$, for $i = 1, 2$ or for $i = 1$ or for $i = 2$.
Level of significance: $\alpha = 0.05$.

Calculations: The residual sum of squares as computed in Example 6.7 is $SS_E = 239{,}219$. Also, the residual mean square, which is the estimated error variance (or residual variance) is

$$\hat{\sigma}^2\ MS_E = \frac{239{,}219}{17} = 14{,}071.7.$$

From Table (6.2.1) and Eq. (6.2.14), the total sum of squares is

$$SS_{yy} = \sum_{i=1}^{n}(y_i - \bar{y})^2 = \sum_{i=1}^{n} y_i^2 - \frac{\left(\sum_{i=1}^n y_i\right)^2}{n}.$$

$$= 34{,}986{,}383 - \frac{25{,}661^2}{20} = 2{,}062{,}037.$$

Hence, the sum of squares due to the regression is

$$SS_R = SS_{yy} - SS_E$$
$$= 2{,}062{,}037 - 239{,}219 = 1{,}822{,}818.$$

Table 6.2.3 ANOVA table for multiple regression of runoff with rainfall and elevation.

Source of variation	Sum of squares	Degrees of freedom	Mean square	F value
Regression	1,822,818	2	911,409	64.8
Residual	239,219	17	14,072	
Total	2,062,037	19		

The mean square due to regression is

$$\text{MS}_R = \frac{1{,}822{,}818}{2} = 911{,}409.$$

The F value is thus $911{,}409/14{,}072 = 64.8$. Results are summarized in Table 6.2.3.

From Table C.4 the significant value is $F_{2,17,0.05} = 3.85$ (approximately). The rejection region is $F > F_{2,17,0.05} = 3.85$.

Decision: We reject the null hypothesis because the computed F value is in the rejection region. We decide that the mean annual runoff is linearly related to the mean annual rainfall or the mean elevation of the basin or to both variables.

6.2.4.2 Significance tests and confidence limits on a regression parameter

It is also useful to make hypothesis tests on each of the regression parameters. This enables us to eliminate one or more of the chosen explanatory variables if they do not make a significant contribution to the regression sum of squares.

Let us follow arguments similar to those in Subsection 6.1.3 leading to Eq. (6.1.15) and use Eq. (6.2.12*b*) on the basis of the assumptions made. We see that the statistic

$$T = \frac{(\hat{\beta}_i - \beta_i)}{\sqrt{\hat{\sigma}^2 c'_{ii}}}, \tag{6.2.17}$$

where $c'_{ii}, i = 0, 1, \ldots, p-1$, are the diagonal elements of the $(\mathbf{X}^{\mathrm{T}}\mathbf{X})^{-1}$ matrix—has a t distribution with $n - p$ degrees of freedom. This equation is used for a significance test on a regression parameter $\beta_i, i = 0, 1, \ldots, p-1$. However, this is only a partial test on the parameter itself (which is a partial regression coefficient for a particular explanatory variable) because the estimate of the parameter depends on all the explanatory variables used in the model as seen from Eq. (6.2.7).

Following Eq. (6.2.14) and as in Subsection 6.1.3, the $100(1-\alpha)$ percent confidence interval for a regression parameter $\beta_i, i = 0, 1, \ldots, p-1$, is found from the following relationship:

$$\Pr\left[\hat{\beta}_i - t_{n-p,\alpha/2}\sqrt{\hat{\sigma}^2 c'_{ii}} \leq \beta_i \leq \hat{\beta}_i + t_{n-p,\alpha/2}\sqrt{\hat{\sigma}^2 c'_{ii}}\right] = 1 - \alpha. \tag{6.2.18}$$

If the confidence interval includes zero, it indicates that the corresponding variable can be eliminated from the equation.

Definitions and properties: The $100(1-\alpha)$ percent confidence interval for regression parameters $\beta_i, i = 0, 1, \ldots, p-1$ is found from the relationship:

$$\Pr\left[\hat{\beta}_i - t_{n-p,\alpha/2}\sqrt{\hat{\sigma}^2 c'_{ii}} \leq \beta_i \leq \hat{\beta}_i + t_{n-p,\alpha/2}\sqrt{\hat{\sigma}^2 c'_{ii}}\right] = 1 - \alpha,$$

where $c'_{ii}, i = 0, \ldots, p-1$, are the diagonal elements of the $(\mathbf{X}^{\mathrm{T}}\mathbf{X})^{-1}$ matrix.

Example 6.9. Significance tests on regression parameters. Firstly, we test the parameter β_1 as follows:

Null hypothesis H_0: $\beta_1 = 0$.
Alternative hypothesis H_1: $\beta_1 \neq 0$.
Level of significance: $\alpha = 0.05$.
Calculations: From the null hypothesis using Eq. (6.2.17) and substituting from Examples 6.6 and 6.7,

$$t = \frac{\hat{\beta}_1}{\sqrt{\hat{\sigma}^2 c'_{11}}} = \frac{1.0664}{\sqrt{14071.7 \times 0.0000008303}} = 9.868.$$

From Table C.2, $t_{17,0.025} = 2.110$.

Decision: We reject the null hypothesis. The conclusion is that the first explanatory variable, mean annual rainfall, contributes significantly to the model for mean annual runoff.

Secondly, the parameter β_2 is tested as follows:

Null hypothesis H_0: $\beta_2 = 0$.
Alternative hypothesis H_1: $\beta_2 \neq 0$.
Level of significance: $\alpha = 0.05$.

Calculations: From the null hypothesis, using Eq. (6.2.17) and substituting from Examples 6.6 and 6.7, the test statistic is computed as

$$t = \frac{\hat{\beta}_2}{\sqrt{\hat{\sigma}^2 c'_{22}}} = \frac{0.4390}{\sqrt{14071.7 \times 0.0000001450}} = 9.717.$$

From Table C.2, $t_{17,0.025} = 2.110$. The rejection region is therefore $t > t_{17,0.025} = 2.110$.

Decision: We reject the null hypothesis. The conclusion is that the second explanatory variable, mean elevation above sea level, contributes significantly to the model for mean annual runoff.

Example 6.10. Construction of confidence intervals on regression parameters. Using Eq. (6.2.18) and substituting from Example 6.6, we can construct 95% confidence intervals on β_0, β_1, and β_2, respectively, as follows:

$$\left(-1035 - t_{17,0.025}\sqrt{14{,}071 \times 3.112163},\ -1035 + t_{17,0.025}\sqrt{14{,}071 \times 3.112163}\right);$$

$$\left(1.0664 - t_{17,0.025}\sqrt{14{,}071.7 \times 0.0000008303},\ 1.0664 + t_{17,0.025}\sqrt{14{,}071.7 \times 0.0000008303}\right);$$

and

$$\left(0.4390 - t_{17,0.025}\sqrt{14{,}071.7 \times 0.0000001450},\ 0.4390 + t_{17,0.025}\sqrt{14{,}071.7 \times 0.0000001450}\right).$$

From Table C.2, $t_{17,0.025} = 2.110$. Hence the 95% confidence intervals are $(-1477, -593)$, $(0.8383, 1.2945)$, and $(0.3437, 0.5343)$ for β_0, β_1, and β_2, respectively.

6.2.4.3 Significance tests on a set of parameters

As noted, a significance test on each of the parameters is an approximate procedure. An alternative procedure is to use a statistic that has the F distribution, as in the case represented by Eq. (6.2.16) to test a set of parameters; note that this too is based on assumptions such as the multivariate normal distribution of the variables. In the modification, the denominator remains the same. The numerator in the F ratio is, however, changed so that it represents the difference between

- the sum of squares due to the regression when a full set of variables is included and
- the sum of squares when a chosen partial set of variables is eliminated from the regression.

Let the original model contain $p - 1$ explanatory variables (that is there are p parameters, including β_0) arranged in descending order of importance. As stated, we make the choice through physical considerations. Suppose we wish to test that the last m variables do not make a significant contribution to the regression. Then the two hypotheses are

Null hypothesis H_0: $\beta_{p-m} = \beta_{p-m+1} = \cdots = \beta_{p-1} = 0$.
Alternate hypothesis H_1: $\beta_i \neq 0$, for at least one i, $i = p - m, p - m + 1, \ldots, p - 1$.

Also, let

$SS_{R,p-1}$ be the sum of squares due to the regression using all $p - 1$ explanatory variables,
$SS_{R,p-m-1}$ be the sum of squares due to the regression using the first $p - m - 1$ explanatory variables, and
$SS_{E,p-1}$ be the sum of squared residuals using all $p - 1$ explanatory variables.

We have m and $n - p$ numerator and denominator degrees of freedom, respectively. Then

$$\frac{(SS_{R,p-1} - SS_{R,p-m-1})/m}{SS_{E,p-1}/(n-p)} \sim F_{m,n-p}. \tag{6.2.19}$$

Definitions and properties: F test on a set of regression parameters. The test statistic is

$$\frac{(SS_{R,p-1} - SS_{R,p-m-1})/m}{SS_{E,p-1}/(n-p)} \sim F_{m,n-p}.$$

Here $SS_{R,p-1}$ and $SS_{R,p-m-1}$ are the sums of squares due to the regression using all $p - 1$ and the first $p - m - 1$ explanatory variables, respectively. Also $SS_{E,p-1}$ is the sum of squared residuals using all $p - 1$ explanatory variables. We have m and $n - p$ numerator and denominator degrees of freedom, respectively.

Example 6.11. Significance of contribution of a partial set of explanatory variables to the regression. A test is made on the multiple regression for which the parameters were estimated in Example 6.6. Suppose we wish to determine whether the second explanatory variable, mean basin elevation, makes a significant additional contribution to the linear regression of mean annual runoff using initially the first explanatory variable, mean annual rainfall.

Null hypothesis H_0: $\beta_2 = 0$.
Alternate hypothesis H_1: $\beta_2 \neq 0$.
Level of significance: $\alpha = 0.05$.

Calculations: The denominator of the left-hand side of Eq. (6.2.19) is the residual sum of squares; as computed in Example 6.7 it is $SS_E = 239{,}219$ with 17 degrees of freedom. The numerator is $(\boldsymbol{\beta}^T\mathbf{X}^T\mathbf{y})_2 - (\boldsymbol{\beta}^T\mathbf{X}^T\mathbf{y})_1$ [which follows from Eq. (6.2.15)], which is the difference in this case between

(a) a scalar quantity obtained using both explanatory variables and hence three parameters in the $\boldsymbol{\beta}$ vector, and

(b) a scalar quantity obtained using the simple linear regression model for runoff based only on mean rainfall and hence two parameters in the $\boldsymbol{\beta}$ vector.

From Example 6.7,

$$(\hat{\boldsymbol{\beta}}^{\mathrm{T}}\mathbf{X}^{\mathrm{T}}\mathbf{y})_2 = 34{,}747{,}164.$$

For the simple regression model, we estimate the slope-parameter from Eq. (6.1.4) and Table 6.2.1 as

$$\hat{\beta}_1 = \frac{n\sum_{i=1}^{n} y_i x_i - \left(\sum_{i=1}^{n} x_i\right)\left(\sum_{i=1}^{n} y_i\right)}{n\sum_{i=1}^{n} x_i^2 - \left(\sum_{i=1}^{n} x_i\right)^2}$$

$$= \frac{20 \times 38{,}852{,}792 - 29{,}596 \times 25{,}661}{20 \times 45{,}361{,}666 - 29{,}596^2} = 0.5619$$

and the intercept from Eq. (6.1.5) as

$$\hat{\beta}_0 = \bar{y} - \hat{\beta}_1\bar{x} = 1{,}283.05 - 0.5619 \times 1{,}479.8 = 451.55$$

Thus,

$$(\boldsymbol{\beta}^{\mathrm{T}}\mathbf{X}^{\mathrm{T}}\mathbf{y})_1 = \hat{\beta}_0\sum_{i=1}^{n} y_i + \hat{\beta}_1\sum_{i=1}^{n} y_i x_i = 451.55 \times 25{,}661 + 0.5619 \times 38{,}852{,}792$$

$$= 33{,}418{,}608.$$

Hence, from Eq. (6.2.19), the sample F value is

$$\frac{34{,}747{,}164 - 33{,}418{,}608}{14{,}017.7} = 94.4.$$

From Table C.4 the significant value is $F_{1,17,0.05} = 4.50$ (approximately). The rejection region is therefore $F > F_{1,17,0.05} = 4.50$.

Decision: We reject the null hypothesis because the computed F value is in the rejection region. We decide that the mean elevation makes a significant contribution to the multiple linear regression of mean annual runoff.

It is interesting to verify a corollary given at the end of Appendix A.8, that is, if $X \sim t_{n-p}$, $X^2 \sim F_{1,n-p}$. From Example 6.9, $t = 9.717$ for the test of significance of the parameter β_2, the partial regression coefficient for the second explanatory variable. We note that $t^2 = 9.717^2 = 94.4$ is equivalent to the F value calculated above which is also a test on the significance of the second explanatory variable.

The iterative model testing procedure extends to Subsections 6.2.5 (model adequacy), 6.2.6 (residual plots), and beyond as necessary.

6.2.5 Model adequacy

6.2.5.1 Coefficient of determination

From the sums of squares defined in ANOVA, Table 6.2.2, one can define a measure of model adequacy by the statistic

$$R^2 = \frac{\mathrm{SS}_R}{\mathrm{SS}_{yy}}. \tag{6.2.20}$$

This is the ratio of the sum of squares due to regression to the total sum of squares; it is sometimes called the *coefficient of multiple correlation*; or simply, R^2. It gives the proportion (or fraction) of the variability of the response variable that is accounted by the explanatory variables. Tests of hypotheses, however, should be used to determine the explanatory variables to be included in the regression. In simple regression the coefficient is equivalent to the square of the correlation coefficient. High values of R^2 obtained by transformation may not indicate the best approach if one does a comparison for different transformations.[11]

Example 6.12. Coefficient of determination. From Table 6.2.3, the sample coefficient of determination for the multiple regression model is obtained as

$$r^2 = \frac{1{,}822{,}818}{2{,}062{,}037} = 0.884.$$

This shows that 11.6% of the variation is not accounted for by the regression.

For the simple regression model using only the mean annual rainfall, the numerator term in the above coefficient is obtained as follows from Table 6.2.2, the results of Example 6.11, and Table 6.2.1.

$$\begin{aligned} SS_R &= \hat{\boldsymbol{\beta}}^{\mathrm{T}}\mathbf{X}^{\mathrm{T}}\mathbf{y} - \frac{(\sum_{i=1}^{n} y_i)^2}{n} \\ &= 33{,}418{,}608 - \frac{25{,}661^2}{20} = 494{,}262. \end{aligned}$$

For the simple linear regression model, therefore,

$$r^2 = \frac{494{,}262}{2{,}062{,}037} = 0.2397.$$

This shows that only about 24% of the variability of the regression is explained by the mean annual rainfall. The corresponding product-moment correlation coefficient is $r = \sqrt{0.2397} = 0.49$.

6.2.6 Residual plots

The formulation of the regression model $\mathbf{Y} = \mathbf{X}\boldsymbol{\beta} + \boldsymbol{\varepsilon}$ and consequent tests of hypotheses are dependent on the assumptions made. In all of this the error term ε plays a central role. We assume that it is independently, identically, and normally distributed. Also, it is assumed to be independent of **X**. However, the only practical way of measuring the model errors is to use the residuals represented by the vector $\hat{\boldsymbol{\varepsilon}} = \mathbf{y} - \mathbf{X}\hat{\boldsymbol{\beta}}$ [as given by Eq. (6.2.9*a*)] to estimate the errors. This is a reiteration of our discussion in Subsection 6.1.2.

Graphical diagnostics form an essential part of the verification of a multiple regression model. They enable summary inferences to be made. Apart from the scatter plots of the variables that we have seen in Chapter 1 and in this chapter, these inferences concern the residuals. As in the case of simple regression, graphical methods provide evidence that no deficiencies or systematic defects are present in the multiple linear regression model. If the observations are recorded in time, an index plot (time sequence) of the residuals is made. The normal probability plot is used as before for indicating departures from normality.[12]

[11] In fact R^2 is misleading when we compare regressions involving different numbers of explanatory variables (Healy, 1984). There are other measures of model adequacy. Akaike's criterion, for instance, is sometimes used to decide on the order of a regression by choosing p, the number of parameters, to minimize the prediction error, based on the residual mean square, given a natural sequence for the introduction of successive predictors. See Draper and Smith (1998, pp. 138–140) or, in water resources, Helsel and Hirsch (1992).

[12] The half-normal (also called the *folded-normal*) plot is sometimes drawn as an aid for detecting unusual observations. The method is useful when the signs of the residual are not important. The variable takes only positive values (see Subsection 4.2.6, truncated normal) and has cdf

$$F(x) = \int_0^x \sqrt{\frac{2}{\pi}} \exp(-y^2/2)dy.$$

Accordingly, the absolute values of the residuals are plotted. Normal and half-normal plots are compared by Draper and Smith (1998). Sparks (1970) gives an algorithm for the distribution, and graphical displays are shown by Atkinson (1981).

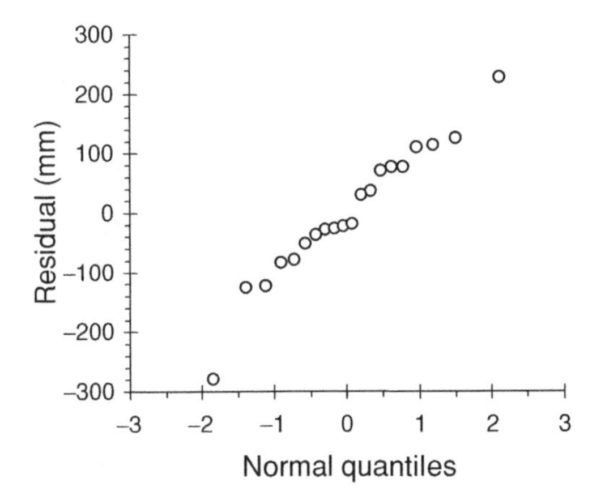

Fig. 6.2.1 Normal probability plot of residuals from multiple regression.

With at least 30 items, we would also draw a stem-and-leaf plot or, with more observations, a histogram. Plots of the residuals against the y and x variables, including explanatory variables that are not used in the regression, bring out possible shortcomings in the model and would indicate methods of overcoming them. For instance, we should know whether all the explanatory variables are suitable and whether, by extending our search, there are other variables that could be included. A particularly important aspect to be studied is the presence of any influential observations that unduly influence the estimation of the parameters. This follows shortly.

Example 6.13. Properties of the residuals of linear regression model applied to basin characteristics. The residuals $\hat{\varepsilon}_i = y_i - \hat{y}_i$ are determined after fitting the model and are shown in the last column of Table 6.2.1. We examine their independence and distributional properties.

In order to determine whether the residuals are normally distributed, a normal probability plot of residuals is drawn as shown in Fig. 6.2.1.

It is seen that the residuals are close to normality in distribution. (Curvatures at the ends which may indicate a uniform or other nonnormal distribution are not indicated here.)

Figure 6.2.2 shows a plot of the residuals $\hat{\varepsilon}_i$ against the fitted runoff $\hat{y}_i$.

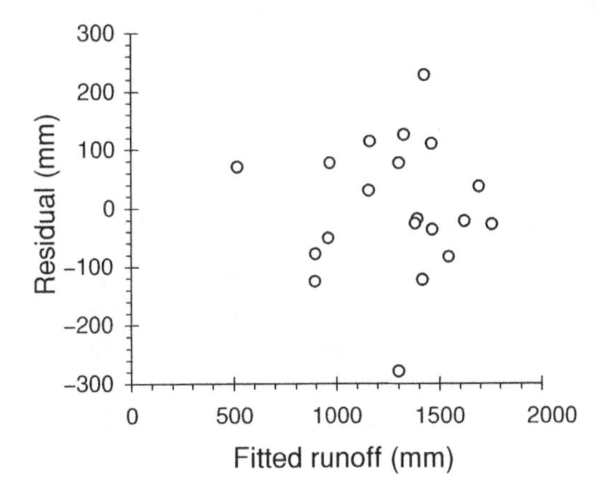

Fig. 6.2.2 Plot of residuals against fitted runoff.

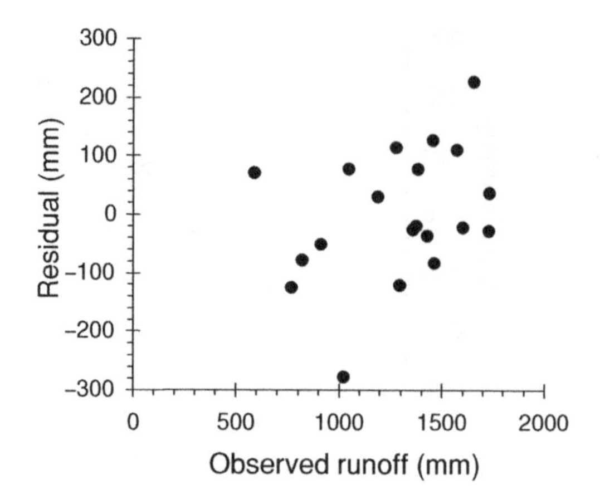

Fig. 6.2.3 Plot of residuals against observed runoff.

A curvature in the general structure of the residuals would mean that the linearity assumption does not hold. There is sufficient indication here that the residuals are random because of the horizontal spread. It is reasonable to assume that the variance is constant. This would not be the case if there was a much larger spread above and below one part of the horizontal axis (through the zero residual) than another, such as the left and right parts or the middle and end parts; alternatively the variance could have varied from one side of the plot to the other. In such situations a possible remedy is to transform the response variable. More about this follows.

The evidence, however, is inconclusive because even if some of the assumptions are incorrect this plot may seem to represent random behavior.[13] We should therefore pursue with additional graphical diagnostics. Figure 6.2.3 shows a plot of the residuals $\hat{\varepsilon}_i$ against the observed runoff y_i. In addition, Fig. 6.2.4 is a plot of the residuals $\hat{\varepsilon}_i$ against mean annual rainfall x_{i1}. These confirm the random behavior seen in Fig. 6.2.2.

Figure 6.2.5 shows the residuals from the simple linear regression of mean annual runoff y and mean annual rainfall x_1 plotted against mean elevation x_2, as in Table 6.2.1.

This figure clearly shows a trend line indicating the need to include mean elevation x_2 as an explanatory variable. After inclusion of this variable in the multiple regression, Fig. 6.2.6 is drawn to show that the residuals $\hat{\varepsilon}_i$ (which are numerically the same as in Figs. 6.2.1 to 6.2.4) have no relationship also with mean elevation x_2.

In Figs. 6.2.1 to 6.2.4 and Fig. 6.2.6, the lowest point appears to be separated from the rest of the plotted points. This corresponds to the last set of observations (station 20), and as shown in Table 6.2.1 the residual has a value of −278.372. Whether this is an influential observation, and whether there are any other influential observations or outliers that are latent will be examined in the next subsection.

6.2.7 Influential observations and outliers in regression

In Section 5.9 we discussed how unexpectedly high or low values, called outliers, can unduly influence the estimation of the parameters of a probability model unless one identifies and deals appropriately with them. Likewise in regression analysis there can be some uncharacteristic observations that can have an excessive influence on the estimates of the parameters and the tests of hypotheses [as we had thought when we viewed the position of

[13] See, for example, Ghosh (1987).

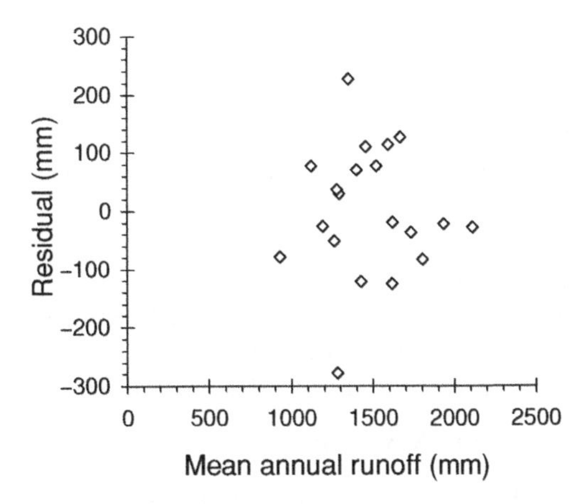

Fig. 6.2.4 Plot of residuals against mean annual rainfall.

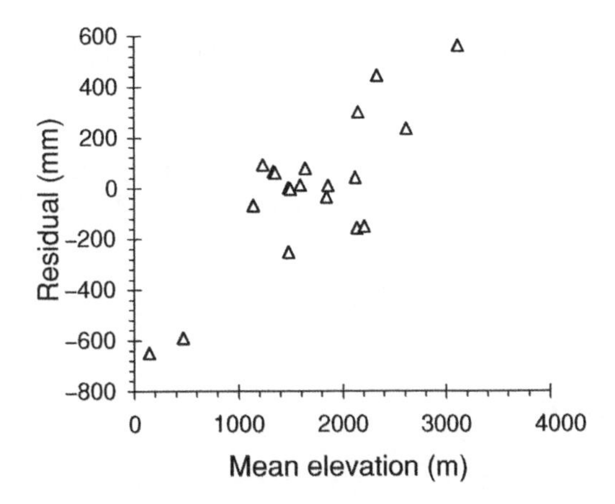

Fig. 6.2.5 Plot of residuals from simple regression of mean annual runoff and mean annual rainfall against mean elevation.

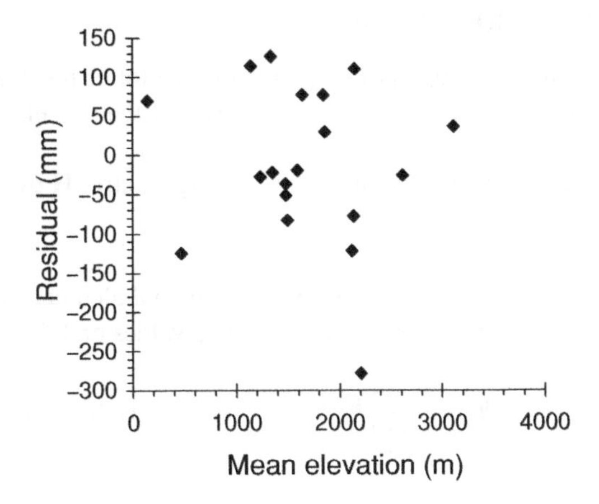

Fig. 6.2.6 Plot of residuals from multiple regression against mean elevation.

the last row of observations (station 20) in Table 6.2.1 which seems to be a remotely placed point in Figs. 6.2.1 to 6.2.4 and Fig. 6.2.6]. This section deals with such observations. We examine whether they can be classed as influential, which means that by removing them there will be a significant change in the estimates of the parameters. Our diagnosis is based on the leverage matrix, the use of standardized residuals, and a measure of influence called *Cook's distance.*[14]

6.2.7.1 The leverage matrix

For a set of n observations, we define the $n \times n$ leverage matrix $\mathbf{H}$ by substituting the solution of the vector of estimated parameters [Eq. (6.2.7)] in the vector of estimated expected values of the response variable [Eq. (6.2.8)]. That is,

$$\hat{\mathbf{y}} = \mathbf{X}\hat{\boldsymbol{\beta}} = \mathbf{X}(\mathbf{X}^{\mathrm{T}}\mathbf{X})^{-1}\mathbf{X}^{\mathrm{T}}\mathbf{y} = \mathbf{H}\mathbf{y}. \tag{6.2.21}$$

Sometimes $\mathbf{H}$ is called the *hat matrix* because it puts a "hat" (circumflex) on $\mathbf{y}$. Note that $\mathbf{H}$ is formed solely by the X values. Thus, when one premultiplies the vector of observed Y values by the leverage matrix $\mathbf{H}$, one obtains the vector of fitted values of Y estimated by the least squares method.

From Eq. (6.2.9*a*) the residuals $\hat{\boldsymbol{\varepsilon}}$ are related to $\mathbf{H}$ as follows:

$$\hat{\boldsymbol{\varepsilon}} = \mathbf{y} - \mathbf{X}\hat{\boldsymbol{\beta}} = (\mathbf{I}-\mathbf{H})\mathbf{y}, \tag{6.2.22}$$

where $\mathbf{I}$ is an $n \times n$ identity matrix. Also because the leverage matrix $\mathbf{H}$ and the residuals matrix $\mathbf{I} - \mathbf{H}$ are symmetrical and idempotent, that is, $\mathbf{H}^2 = \mathbf{H}$, the following relationships hold:

$$\hat{\sigma}^2 = \frac{\hat{\boldsymbol{\varepsilon}}^{\mathrm{T}}\hat{\boldsymbol{\varepsilon}}}{n-p} = \frac{\mathbf{y}^{\mathrm{T}}(\mathbf{I}-\mathbf{H})\mathbf{y}}{n-p}, \tag{6.2.23a}$$

$$\mathrm{Var}[\hat{\mathbf{y}}] = \sigma^2\mathbf{H}, \tag{6.2.23b}$$

$$\mathrm{Var}[\hat{\boldsymbol{\varepsilon}}] = \sigma^2(\mathbf{I}-\mathbf{H}), \tag{6.2.23c}$$

and

$$\mathrm{Cov}[\hat{\boldsymbol{\varepsilon}}, \hat{\mathbf{y}}] = \sigma^2\mathbf{H}(\mathbf{I}-\mathbf{H}) = 0. \tag{6.2.23d}$$

It is because of the last property that points in the plot of residuals against fitted values, shown in Fig. 6.2.2, appear as a horizontal spread if the model assumptions are met (see also Fig. 6.1.6).

We denote the diagonal elements of the leverage matrix $\mathbf{H}$ by

$$h_i = \mathbf{x}_i(\mathbf{X}^{\mathrm{T}}\mathbf{X})^{-1}\mathbf{x}_i^{\mathrm{T}}, \tag{6.2.23e}$$

where $\mathbf{x}_i$ is the ith row of $\mathbf{X}$. Then if the off-diagonal elements are denoted by h_{ij}, from Eq. (6.2.21) the ith fitted value and the observed values of Y have the relationship

$$\hat{y}_i = \sum_{j=1}^{n} h_{ij}y_j = h_i y_i + \sum_{j \neq i} h_{ij}y_j, \qquad \text{for } i = 1, 2, \ldots, n.$$

[14] Originated by Cook (1977).

Hence,

$$\frac{\partial \hat{y}_i}{\partial y_i} = h_i, \quad \text{for } i = 1, 2, \ldots, n.$$

We can therefore say that h_i is a measure of the effect each value y_i has on its own prediction, that is, the determination of $\hat{y}_i$. Furthermore, it is seen from Eq. (6.2.23*c*) that a point with a high value of h_i has low variance and at the maximum value, $h_i = 1$, the residual is zero and the fitted model becomes irrelevant. Similarly, we can interpret h_{ij} as the effect of the jth observation on the prediction of $\hat{y}_i$. For these reasons $\mathbf{H}$ is termed the *leverage matrix* and h_i is termed the *leverage.*

The elements of $\mathbf{H}$ are constrained as follows:

$$0 \leq h_i \leq 1 \tag{6.2.24a}$$

and

$$-0.5 \leq h_{ij} \leq 0.5. \tag{6.2.24b}$$

It can also be shown that the average value of h_i is p/n, where p is the number of parameters.[15] Even moderately high values, say,

$$h_i \geq \frac{2.5p}{n},$$

signify that the ith point has high leverage. The constant in the numerator of this fraction becomes 2 or 1.5 if p exceeds 5 and 14, respectively.[16] If one or more of the h_i are around or greater than the suggested critical value, we may also evaluate the leverage measure

$$h_i' = \frac{h_i}{1 - h_i}, \tag{6.2.25}$$

which is unbounded and is hence more sensitive to highly leveraged observations. This is the ratio of the variances of the fitted value of the response variable and the residual [as seen from Eq. (6.2.23*b*) and (6.2.23*c*]. It represents the distance of the ith point in the regression from the centroid of the other points.

6.2.7.2 Standardized residuals

For comparative purposes in the assessment of the magnitudes of residuals, it is useful to compute the standardized residuals

$$r_i = \frac{\hat{\varepsilon}_i}{\sqrt{\hat{\sigma}^2(1 - h_i)}}, \tag{6.2.26a}$$

$i = 1, 2, \ldots, n$, obtained by dividing the residuals $\hat{\varepsilon}_i$ by the square root of the estimated variance arising from Eq. (6.2.23*c*). The r_i are also called the *internally Studentized residuals.*

The alternate term *externally Studentized residuals* denotes residuals obtained by using the estimated error variance $\hat{\sigma}^2_{(i)}$ computed after deletion of the contribution of $\hat{\varepsilon}_i$. That is,

$$t_i = \frac{\hat{\varepsilon}_i}{\sqrt{\hat{\sigma}^2_{(i)}(1 - h_i)}} \tag{6.2.26b}$$

[15] See, for example, Stuart and Ord (1991, p. 1080).

[16] See, for example, Chatterjee and Hadi (1988, p. 101) and Atkinson (1985, p. 18); however, Draper and Smith (1998, p. 207) "de-emphasizes" the use of leverages because of conflicting opinions on their role.

where

$$\hat{\sigma}_{(i)}^2 = \frac{(n-p)\hat{\sigma}^2 - \hat{\varepsilon}_i^2/(1-h_i)}{n-p-1}. \tag{6.2.26c}$$

For normally distributed errors, the t_i have a t_{n-p} distribution.

The r_i are sometimes used in place of the ordinary residuals $\hat{\varepsilon}_i$ in plots such as Figs. 6.2.1 to 6.2.6. The standardization can be advantageous and will enable influential observations to be more easily seen. They are highlighted somewhat more in the case of externally Studentized residuals t_i.

6.2.7.3 Cook's distance

We noted that the h_i are based solely on the **X** matrix as seen from Eq. (6.2.21). A more representative measure of detecting influential observations is provided by Cook's distance

$$C_i = \frac{(\hat{\boldsymbol{\beta}}_{(i)} - \hat{\boldsymbol{\beta}})^{\mathrm{T}}\mathbf{X}^{\mathrm{T}}\mathbf{X}(\hat{\boldsymbol{\beta}}_{(i)} - \hat{\boldsymbol{\beta}})}{p\hat{\sigma}^2}, \tag{6.2.27}$$

where $\hat{\boldsymbol{\beta}}_{(i)}$ represents the vector of p estimated parameters after deleting the ith case. A large value of C_i signifies a corresponding influence exerted by the ith case. This measure can be shown to be equal to

$$C_i = \frac{r_i^2 h_i}{p(1-h_i)}. \tag{6.2.28a}$$

It is seen that Cook's distance incorporates the leverage measure $h_i' = h_i/(1-h_i)$ of Eq. (6.2.25), and the internally Studentized residual r_i of Eq. (6.2.26a). The divisor $(1-h_i)$ in Eq. (6.2.28a) aids calibration by means of a confidence region.[17] In practice, however, values of C_i greater than 1 are considered to be large, as a rule of thumb.

6.2.7.4 Other statistics related to Cook's distance

DFFITS, for instance, can be obtained from Cook's distance C_i as follows:

$$\text{DFFITS}_i = \sqrt{\frac{C_i p\hat{\sigma}^2}{\hat{\sigma}_{(i)}^2}} \tag{6.2.28b}$$

where $\hat{\sigma}_{(i)}^2$ is defined by Eq. (6.2.26c). The modification A_i to DFFITS by Atkinson (1985) is another example:

$$A_i = \text{DFFITS}_i\sqrt{\frac{(n-p)}{n}}. \tag{6.2.28c}$$

There are other modifications such as the one by Chatterjee and Hadi (1988). As in the case of Cook's distance, one considers values greater than 1 to be large.

[17] See, for example, Chatterjee and Hadi (1988, pp. 118–119).

Table 6.2.4 Residuals, leverages, and Cook's distances of basin characteristics

Index	$\hat{e}_i$	r_i	h_i	$h_i/(1-h_i)$	Cook's d
1	227.1876	2.0158	0.0961	0.1063	0.1440
2	−18.7111	−0.1631	0.0634	0.0677	0.0006
3	−50.9097	−0.4553	0.1102	0.1239	0.0086
4	30.1667	0.2642	0.0726	0.0782	0.0018
5	125.5582	1.1012	0.0749	0.0809	0.0327
6	114.0029	1.0045	0.0833	0.0909	0.0306
7	−78.3501	−0.7614	0.2464	0.3269	0.0632
8	77.2879	0.7037	0.1416	0.1650	0.0272
9	70.1633	0.7927	0.4425	0.7938	0.1663
10	−125.222	−1.1994	0.2243	0.2891	0.1386
11	−28.0140	−0.2859	0.3168	0.4637	0.0126
12	110.4192	0.9699	0.0776	0.0841	0.0264
13	76.9872	0.6667	0.0510	0.0537	0.0080
14	−21.9446	−0.2054	0.1881	0.2316	0.0033
15	−122.177	−1.0699	0.0720	0.0776	0.0296
16	−36.6723	−0.3248	0.0927	0.1022	0.0036
17	−82.7697	−0.7449	0.1214	0.1382	0.0256
18	37.1154	0.3698	0.2830	0.3947	0.0180
19	−25.8075	−0.2368	0.1549	0.1834	0.0034
20	−278.372	−2.4580	0.0872	0.0956	0.1925

6.2.7.5 Outliers in regression

An approximate method of detecting a single outlier in regression is obtained by means of the highest absolute internally Studentized residual, $r_{\max}$, from Eq. (6.2.26a) following a method used in simple regression involving simulation.[18] The test statistic is

$$T(x_i, y_i) = \max_i |r_i| \equiv r_{\max} \qquad (6.2.29a)$$

and one finds C_α such that, conditional to the presence of no more than one outlier,

$$\Pr[T(x_i, y_i) > C_\alpha] \leq \alpha. \qquad (6.2.29b)$$

Then the ith observation is declared an outlier if $T(x_i, y_i) > C_\alpha$. An approximation to C_α is given by

$$C_\alpha = \sqrt{\frac{(n-p)F}{n-p-1+F}}, \qquad (6.2.29c)$$

where $F \equiv F_{1,n-p-1,\alpha/n}$, with n being the number of observations and p the number of parameters. (The F value has 1 and $n-p-1$ degrees of freedom in the numerator and denominator, respectively, and is exceeded with probability α/n.)

Example 6.14. High leverage and influential observations. For the basin characteristics of Table E.6.1 (excluding the last two columns) and listed in Table 6.2.1, the residuals $\hat{\varepsilon}_i$, internally Studentized residuals r_i, the leverage h_i, the leverage measure $h_i' = h_i/(1-h_i)$, and Cook's distances C_i are given in Table 6.2.4.

The leverages and leverage measures are also shown in the index plot, Fig. 6.2.7.

[18] See Tietjen et al. (1973).

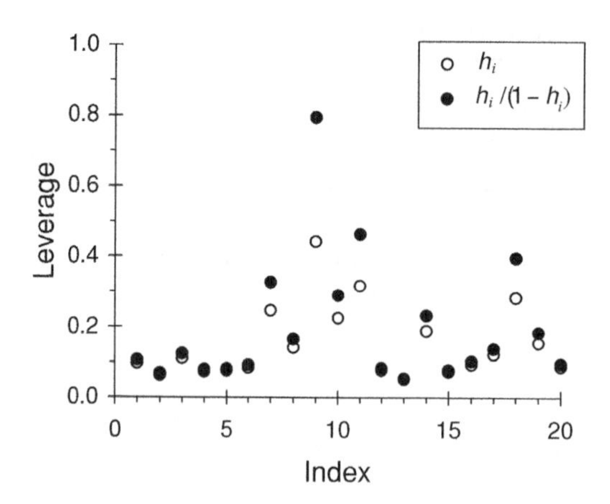

Fig. 6.2.7 Index plot of leverages from residuals of basin characteristics. h_i is the ith diagonal element of the leverage hat matrix.

The 9th point has $h_i = 0.4425$, which is the highest leverage in Table 6.2.4. This exceeds the approximate 95% probability point for high leverage, which is $2.5p/n = 7.5/20 = 0.375$. In the next column, the leverage measure $h_i' = h_i/(1 - h_i) = 0.7938$ for the 9th point which is more than double the value for the second ranked point (18th point) of this column. On examination of the data presented in Table 6.2.1, the high leverage measure for the 9th point is attributable to the very low value of mean elevation, 144 m, which is remote from the other values of elevation.

In order to verify the possible influence exerted by the 9th point, we reestimate the parameters of the regression model after deleting the 9th row from the data set of Table 6.2.1. Hence we obtain

$$\beta_{0(9)} = -1152, \qquad \beta_{1(9)} = 1.1073, \qquad \beta_{2(9)} = 0.4688.$$

These are within the 95% confidence limits computed from the full sample in Example 6.10 as $(-1477, -593)$, $(0.8383, 1.2945)$, and $(0.3437, 0.5343)$, respectively.

Also, let us delete the 9th row and calculate the predicted value of mean annual runoff for mean annual rainfall of 1000 mm and an elevation of 2000 m, for example. We thus obtain $\hat{y}_{1000,2000(9)} = 892.6$ mm (as we shall see in Example 6.16 which follows shortly, this is very close to the value of 909.2 obtained without deleting the 9th row).

We can also compare the variances of the residuals and the variances of the estimated parameters to find that the differences are not large. We conclude that the 9th point has no significant influence on the regression.

Finally, we have confirmation from Cook's statistic, which takes values of 0.1663 and 0.1925 for the 9th and 20th points, respectively; these are much lower than the (rule-of-thumb) critical value of 1.0. Furthermore, calculations of the modified statistics DFFITS and A_i, as given by Eq. (6.2.28b) and (6.2.28c), respectively, for the 20th point give values of 0.5229 and 0.4961, respectively, which are well below the limit of 1.0.

In the following example we shall check for outliers using the internally standardized residuals in column 3 of Table 6.2.4:

Example 6.15. Outliers in regression. In Figs. 6.2.1 to 6.2.4 and Fig. 6.2.6, it is seen that the lowest point is a possible outlier. We noted from Table 6.2.1 that this is the 20th point, with the highest absolute residual $\hat{\varepsilon}_i$ of 278.372. This point has also the highest absolute internally standardized residual r_i, that is, $r_{\max} = 2.458$, as seen in Table 6.2.4. A hypothesis test is made with respect to this point.

Null hypothesis H_0: $(x_{1,20}, x_{2,20}, y_{20}) \neq$ *outlier.*
Alternative hypothesis H_1: $(x_{1,20}, x_{2,20}, y_{20}) =$ *outlier.*
Level of significance: $\alpha = 0.10$.

Partly because the test is not accurate and partly because we need to give ourselves an extra chance of declaring an outlier without incorrectly rejecting it, we use a higher level of significance α than is usual. (This increases the chance of a Type I error, that is, an incorrect rejection, noting that we thereby reduce β, the probability of making a Type II error, that is an incorrect acceptance.)

Calculations: The rejection region is $r_{\max} > C_{0.10}$, where

$$C_{0.10} = \sqrt{\frac{(n-p)F}{n-p-1+F}},$$

from Eq. (6.2.29c), in which $F \equiv F_{1,16,0.10/20} \approx 11.5$ after substituting $n = 20$, $p = 3$, and interpolating from Table C.4. Hence,

$$C_{0.10} = \sqrt{\frac{17 \times 11.5}{16 + 11.5}} = 2.67.$$

Decision: We do not reject the null hypothesis because $r_{\max} = 2.458 < C_{0.10}$.

6.2.7.6 Discussion

From Example 6.14 we note that a high leverage point, such as the 9th point examined here, need not be influential, although one suspects that a high leverage point is usually influential. Influence (as meant here) is measured with respect to the changes, caused by the omission of this point, to the error variances, the variances and estimates of the regression parameters, and the predictions of future observations. Not all of these are significantly affected in some cases. As given in the last column of Table 6.2.4, Cook's distance is a very useful measure for influential observations. High leverage means that a point appears as an outlier in the **X** space but not necessarily in the total **X, Y** space where the regression may, in some instances, cause it to have a low residual or otherwise reduce its influence. Thus, the 9th point in Fig. 6.2.7 has an internally Studentized residual of 0.7927 compared to $r_{\max} = 2.458$ for the 20th point as seen in Table 6.2.4. The 20th point appears as an outlier in Figs. 6.2.1 to 6.2.4 and 6.2.6 but in Example 6.15 we rejected the hypothesis, albeit by a small margin. Proceeding further, we find that the externally Studentized residual $t_{20} = -2.927$, which is significantly high. On the other hand, the 20th point has very low leverage as seen in Table 6.2.4. This tells us that an outlier (or one which seems to be of the type as in this case) need not be a high leverage point or an influential one. A row of observations (as in Table E.6.1) can of course have all three characteristics. With a wide range of diagnostics to consider, users can decide which rows of observations require further scrutiny.

6.2.8 Transformations

The iterative procedure may not result in an acceptable model with respect to the assumptions made as verified by means of residual analysis and various tests of hypotheses. These assumptions include constant variance and normality in distribution of the residuals and a linear structure in the residuals. We emphasized before that a key role is played by residual plots. If there is a violation, the next step would be to consider a transformation of the response variable. As a first step, logarithmic, reciprocal, square root, or cube root transformations can be tried. A particular method is to use the Box-Cox parametric family of power transformations, given here in the more general form:

$$
\begin{aligned}
y(\lambda, \alpha) &= (y+\alpha)^{\lambda} - 1/\lambda, && \text{for } \lambda \neq 0,\\
&= \ln(y+\alpha), && \text{for } \lambda = 0.
\end{aligned} \qquad (6.2.30)
$$

As discussed here, suitable values of the parameters λ and α are found which produces normality in Y and hence in the residuals.

In the case of a single parameter, λ, one plots a normal probability plot for various values of the parameter and selects the value for which a straight line is obtained; simultaneously one seeks confirmation by searching for the maximum correlation between the vertical and horizontal variables of the normal probability plot. In the general case just given, a suggested procedure is to plot log likelihoods for different values of the power parameter λ and repeat the procedure for the shift parameter α to find the maximum likelihood values and hence the optimal values of the parameters; in case α is close to zero one may decide to have only a single parameter, λ.[19]

One may also investigate the extension of the X variables by taking squares and cross products (thereby fitting a conic surface rather than a plane when $p = 3$) and the effects on the residual plots. If outliers are present, we must examine the data carefully and look for possible errors. If they are not erroneous observations, it may be possible to accommodate the outliers by means of a suitable transformation. The assumption of linearity, however, could be violated as evidenced by a curvature in the plot of residuals against fitted values of the response. In such cases a new approach is required. More about this follows.

6.2.9 Confidence intervals on mean response and prediction

6.2.9.1 Mean response

The confidence interval on the mean response when the explanatory variable in a simple linear regression takes a value, say, $X = x_0$ was defined by Eq. (6.1.16) using the estimated mean, its variance, and the t statistic. Likewise, we can determine the mean response when the $p-1$ explanatory variables take the following values, for instance: $x_1 = a_1, x_2 = a_2, \ldots, x_{p-1} = a_{p-1}$. For this set of values, the mean response is $E[Y \mid \mathbf{x} = \mathbf{a}] = \mathbf{a}\boldsymbol{\beta}$ where $\mathbf{a} = [1 \ldots a_1 \ldots a_2 \ldots a_{p-1}]$ [in which the unit value on the left, as defined by Eq. (6.2.2), corresponds to the intercept]. This is estimated by, say,

$$
\hat{\mu}_{Y \mid \mathrm{x=a}} = \mathbf{a}\hat{\boldsymbol{\beta}}. \qquad (6.2.31)
$$

The estimator is unbiased because

$$
\hat{E}[Y \mid \mathbf{x} = \mathbf{a}] = E[\hat{\beta}_0] + a_1 E[\hat{\beta}_1] + \cdots + a_{p-1} E[\hat{\beta}_{p-1}] = \mathbf{a}\boldsymbol{\beta} = E[Y \mid \mathbf{x} = \mathbf{a}].
$$

The variance of the mean response is estimated [using Eq. (3.4.30)] as follows:

$$
\mathrm{Var}[Y \mid \mathbf{X} = \mathbf{a}] = \sum_{i=0}^{p-1} a_i^2 \mathrm{Var}[\hat{\beta}_i] + 2 \sum_{i=1}^{p-1} \sum_{j=1, j\neq i}^{p-1} a_i a_j \,\mathrm{Cov}[\hat{\beta}_i, \hat{\beta}_j],
$$

where $a_0 = 1$. Then from Eq. (6.2.12a) and (6.2.12b),

$$
\mathrm{Var}[Y \mid \mathbf{x} = \mathbf{a}] = [\mathbf{a}(\mathbf{X}^{\mathrm{T}}\mathbf{X})^{-1}\mathbf{a}^{\mathrm{T}}]\sigma^2. \qquad (6.2.32)
$$

[19] See also Atkinson (1985, Chapters 6–9) and Dolby (1963).

Hence, by substituting the residual variance $\hat{\sigma}^2$ for the unknown error variance σ^2, we have the Student's t variable

$$T = \frac{\mathbf{a}\hat{\boldsymbol{\beta}} - \mathbf{a}\hat{\boldsymbol{\beta}}}{\sqrt{\hat{\sigma}^2 \mathbf{a}\,(\mathbf{X}^{\mathrm{T}}\mathbf{X})^{-1}\mathbf{a}^{\mathrm{T}}}} \sim t_{n-p}.$$

This can be used in hypothesis testing and also to specify a $100(1-\alpha)$ percent confidence interval on the mean response. The confidence limits are found from

$$\Pr\left[\mathbf{a}\hat{\boldsymbol{\beta}} - t_{n-p,\alpha/2}\sqrt{\hat{\sigma}^2\mathbf{a}(\mathbf{X}^{\mathrm{T}}\mathbf{X})^{-1}\mathbf{a}^{\mathrm{T}}} \le E[Y \mid x\mathbf{a}] \le \mathbf{a}\hat{\beta} + t_{n-p,\alpha/2}\sqrt{\hat{\sigma}^2\mathbf{a}(\mathbf{X}^{\mathrm{T}}\mathbf{X})^{-1}\mathbf{a}^{\mathrm{T}}}\right] = 1-\alpha. \tag{6.2.33}$$

6.2.9.2 Prediction interval for a future value of Y

Let the explanatory variables take values as just given. From Eqs. (6.2.4) and (6.2.31), the expectation of Y_0 is given by

$$E[Y_0 \mid \mathbf{x} = \mathbf{a}] = \mathbf{a}\boldsymbol{\beta}.$$

The variance of Y_0 [obtained from Eq. (3.4.30)] is as follows:

$$\mathrm{Var}[Y_0 \mid \mathbf{x} = \mathbf{a}] = \mathrm{Var}[\mathbf{a}\hat{\boldsymbol{\beta}}] + \mathrm{Var}[\varepsilon_0],$$

because $\mathrm{Cov}[\mathbf{a}\hat{\boldsymbol{\beta}}, \varepsilon_0] = 0$ by the assumptions made. Substituting from Eq. (6.2.32) for the first term on the right-hand side and because $\mathrm{Var}[\varepsilon_0] = \sigma^2$,

$$\mathrm{Var}[Y_0 \mid \mathbf{x} = \mathbf{a}] = \sigma^2[1 + \mathbf{a}(\mathbf{X}^{\mathrm{T}}\mathbf{X})^{-1}\mathbf{a}^{\mathrm{T}}].$$

Hence, we can construct a $100(1-\alpha)$ percent prediction interval on the future or unknown value Y_0. This follows from

$$\Pr[\mathbf{a}\hat{\boldsymbol{\beta}} - t_{n-p,\alpha/2}\sqrt{\hat{\sigma}^2[1 + \mathbf{a}(\mathbf{X}^{\mathrm{T}}\mathbf{X})^{-1}\mathbf{a}^{\mathrm{T}}]} \le Y_0 \le \mathbf{a}\hat{\boldsymbol{\beta}} + t_{n-p,\alpha/2}\sqrt{\hat{\sigma}^2[1 + \mathbf{a}(\mathbf{X}^{\mathrm{T}}\mathbf{X})^{-1}\mathbf{a}^{\mathrm{T}}]}] = 1-\alpha. \tag{6.2.34}$$

Definition and properties: ***(a)*** **100(1 − α) percent confidence interval on the mean response** when the $p-1$ explanatory variables take values given by $\mathbf{a} = [1 \ldots a_1 \ldots a_2 \ldots a_{p-1}]$, respectively [with an initial unit value representing the intercept, as defined by Eq. (6.2.2)]. This is found from

$$\Pr\left[\mathbf{a}\hat{\boldsymbol{\beta}} - t_{n-p,\alpha/2}\sqrt{\hat{\sigma}^2\mathbf{a}(\mathbf{X}^{\mathrm{T}}\mathbf{X})^{-1}\mathbf{a}^{\mathrm{T}}} \le E[Y \mid \mathbf{x} = \mathbf{a}] \le \mathbf{a}\hat{\boldsymbol{\beta}} + t_{n-p,\alpha/2}\sqrt{\hat{\sigma}^2\mathbf{a}(\mathbf{X}^{\mathrm{T}}\mathbf{X})^{-1}\mathbf{a}^{\mathrm{T}}}\right] = 1-\alpha.$$

(b) **100(1 − α) percent prediction interval on** Y_0 is found from

$$\Pr\left[\mathbf{a}\hat{\boldsymbol{\beta}} - t_{n-p,\alpha/2}\sqrt{\hat{\sigma}^2[1 + \mathbf{a}(\mathbf{X}^{\mathrm{T}}\mathbf{X})^{-1}\mathbf{a}^{\mathrm{T}}]} \le Y_0 \le \mathbf{a}\hat{\boldsymbol{\beta}} + t_{n-p,\alpha/2}\sqrt{\hat{\sigma}^2[1 + \mathbf{a}(\mathbf{X}^{\mathrm{T}}\mathbf{X})^{-1}\mathbf{a}^{\mathrm{T}}]}\right] = 1-\alpha.$$

Example 6.16. Confidence intervals on mean response and predicted value. Suppose we are planning to utilize the resources of a basin with mean annual rainfall of 1000 mm at a mean elevation of 2000 m, and without any prior knowledge of the runoff, from the region in which the basin characteristics are given in Table 6.2.1. We proceed as follows to construct confidence limits on the mean response.

From Eq. (6.2.31),

$$\mathbf{a}\hat{\boldsymbol{\beta}} = [1 \quad 1000 \quad 2000]\begin{bmatrix} -1035.14 \\ 1.0664 \\ 0.4390 \end{bmatrix} = 909.2.$$

For the estimation of the variance of the mean response

$$\mathbf{a}(\mathbf{X}^T\mathbf{X})^{-1} = [1 \quad 1000 \quad 2000]$$

$$\times \begin{bmatrix} 3.1121632701 & -0.0015096654229 & -0.00049114006481 \\ -0.0015096654229 & 0.00000083028568222 & 0.000000166520403 \\ -0.00049114006481 & 0.000000166520403 & 0.00000014501742117 \end{bmatrix}$$

$$= [0.6202177176 \quad -0.0003460757 \quad -0.0000344531].$$

From Eq. (6.2.32) therefore, the variance of the mean response is estimated as

$$\hat{\sigma}^2[\mathbf{a}(\mathbf{X}^T\mathbf{X})^{-1}\mathbf{a}^T]$$

$$= 14071.7[0.6202177176 \quad -0.0003460757 \quad -0.0000344531] \begin{bmatrix} 1 \\ 1000 \\ 2000 \end{bmatrix}$$

$$= 14071.7 \times 0.2052358 = 2888.0.$$

From Table C.2 of Appendix C, $t_{17,0.025} = 2.11$. Therefore, the 95% confidence interval on the mean response, with $\mathbf{a} = [1 \quad 1000 \quad 2000]$ representing the vector of explanatory variables, is found from the relationship

$$\Pr[909.2 - 2.11\sqrt{2888.0} \le E[Y \mid \mathbf{x} = \mathbf{a}] \le 909.2 + 2.11\sqrt{2888.0}] = 1 - \alpha.$$

Thus the confidence limits are (796, 1022).

Similarly, one can give a prediction interval to a value Y_0 of mean annual runoff with $\mathbf{a} = [1 \quad 1000 \quad 2000]$ representing the vector of explanatory variables.
The 95% prediction interval is found from the relationship

$$\Pr\Big[909.2 - 2.11\sqrt{14071.7\,(1 + 0.2052358)} \le Y_0 \le 909.2 + 2.11\sqrt{14071.7\,(1 + 0.2052358)}\Big] = 1 - \alpha.$$

Thus the prediction limits are (634, 1184).

6.2.10 Ridge regression

The procedure called ridge regression was suggested by A. E. Hoerl in the 1960s to overcome problems caused by strong dependencies between explanatory variables used in multiple regression; such a situation is referred to as a *multicollinearity*. This yields a singular $\mathbf{X}^T\mathbf{X}$ matrix and, consequently, estimates of parameters become unstable. The remedy is to increment the diagonal elements of the matrix by some constant, and the problem lies with the choice of the constant. The minimum increment is chosen through graphical or other means so that the set of parameters is stabilized. The technique is an alternative to the least squares procedure, and as practiced it is empirical. It arose from Hoerl's work on ridge analysis of higher dimensional quadratic response surfaces. This was devised to find an engineering solution to the optimization of industrial processes involving more than two or three explanatory variables; it is desirable to a numerical optimization of the estimated function, which has drawbacks. There is a parallel between the illuminating graphics of the effects of all factors viewed simultaneously in ridge analysis and the behavior of the graphs of each of the regression constants in ridge regression as the diagonal elements of $\mathbf{X}^T\mathbf{X}$ are incremented by equal amounts (as shown later).

Let us assume that the response and the explanatory variables are standardized by subtracting their individual means and dividing by the individual square root of the sum of squared differences from the mean. It will be easily seen that the off-diagonal elements of $\mathbf{X}^T\mathbf{X}$ will then be correlation coefficients and the diagonal elements will be 1s. The presence of multicollinearity can be detected firstly from the determinant of the $\mathbf{X}^T\mathbf{X}$ matrix in its correlation form. If the X variables are totally unrelated, the determinant is 1 and, at the other extreme, a value of 0 signifies full dependency, a property that the elements of the matrix will show. Secondly, a diagnostic tool is provided by the eigenvalues of $\mathbf{X}^T\mathbf{X}$ which are the roots of

$$|\mathbf{X}^T\mathbf{X} - \lambda\mathbf{I}| = 0 \tag{6.2.35}$$

(where $|\cdot|$ denotes the determinant) and given by $\lambda_1, \lambda_2, \ldots, \lambda_k$, where k is the number of explanatory variables. An eigenvalue which is nearly zero is an indicator of multicollinearity, and so will a high ratio of the highest and lowest eigenvalues, say, greater than 99.

Based on the results of the preceding tests, we may decide to eliminate one or more of the explanatory variables and choose others. However, quite often we find that data are limited and we need all the available information. So we keep all the variables and make some adjustment to the parameters if their estimates are in an unstable region, through ridge regression. As discussed, the application of ridge regression is an extension of Eq. (6.2.7). The algorithm is given by

$$\hat{\boldsymbol{\beta}}(\theta) = (\mathbf{X}^T\mathbf{X} + \theta\mathbf{I})^{-1}\mathbf{X}^T\mathbf{y}. \tag{6.2.36}$$

This is the basic form, and usually $0 < \theta < 1$. We try to keep θ as low as possible in order to minimize the bias caused by its introduction. The method is thus highly subjective and should be used with care, when the situation demands.[20]

Example 6.17. Ridge regression of some basin characteristics. Table E.6.1 gives some characteristics of stream basins on the left bank of the river Po, in northern Italy. Our objective is to apply a multiple regression model for estimating the mean annual runoff in stream basins in the area, as in previous examples. However, for the sake of this exercise, let us assume that the only available data are, apart from the runoff in column 3, the area of basin measured above the observation station, the length of the longest flow path, and the mean elevation as listed in the last three columns (in reverse order). We therefore use the last three data columns as values of three explanatory variables to form the $\mathbf{X}$ matrix in the regression. As discussed above, the data sets that include the annual runoff in column 3, which we treat as the response variable $\mathbf{y}$, are standardized as follows:

$$x_{ij} = \frac{b_{ij} - \bar{b}_j}{\sqrt{\sum_{i=1}^{n}(b_{ij} - \bar{b}_j)^2}}, \qquad \text{for } i = 1, 2, \ldots, n \qquad \text{and} \qquad j = 1, 2, 3.$$

In this equation, b_{ij} denotes the ith value of the jth basin characteristic, which has a sampling mean $\bar{b}_j$. Also, $n = 20$. The 3×3 $\mathbf{X}^T\mathbf{X}$ matrix gives the correlations of the three basin characteristics in the order stated earlier as follows:

$$\mathbf{X}^T\mathbf{X} = \begin{bmatrix} 1.0000 & 0.9788 & -0.0214 \\ 0.9788 & 1.0000 & -0.0708 \\ -0.0214 & -0.0708 & 1.0000 \end{bmatrix}.$$

It is seen that the variable area has a correlation of 0.9788 with the longest flow path but its correlation with the altitude is −0.0214. In addition, the correlation between the longest flow path and altitude is −0.0708.

[20] Examples are given by Montgomery and Ringer (1994, Section 10.13). See Marquardt and Snee (1975) and Hoerl and Kennard (1970, *a* and *b*). Draper and Smith (1998, Chapter 17) give additional references including different methods of estimating θ.

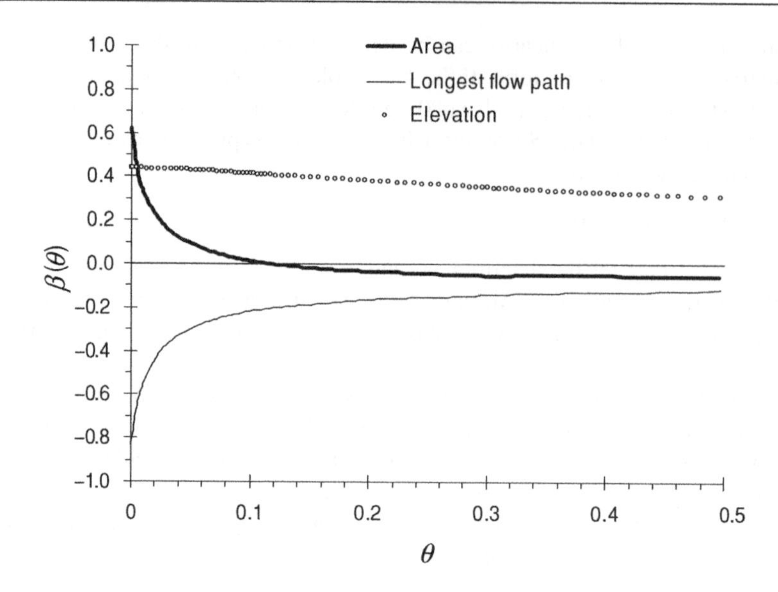

Fig. 6.2.8 Ridge trace of basin characteristics.

The determinant of the matrix is 0.0395, obtained as discussed in Example 6.6. The low determinant and the high value of one of the correlation coefficients indicate unstable estimates of the parameters. This is confirmed by the eigenvalues of the $\mathbf{X}^T\mathbf{X}$ matrix as defined by Eq. (6.2.35), which are as follows: 1.9830, 0.0200, and 0.9970. (It is seen that these sum to 3: the number of explanatory variables; there is more about eigenvalues in the next section.) The ratio between the highest and lowest eigenvalues is 99.15, which is high. Hence (by our rule of thumb), there is justification for a ridge regression.

The inverse of the matrix is given by

$$(\mathbf{X}^T\mathbf{X})^{-1} = \begin{bmatrix} 25.16 & -24.72 & -1.21 \\ -24.72 & 25.28 & 1.26 \\ -1.21 & 1.26 & 1.06 \end{bmatrix}.$$

We applied Eq. (6.2.36) for $0 < \theta < 0.50$ and the results are given in Table 6.2.5. This also shows the rate of change (derivative) of the variables within the limits studied. Figure 6.2.8 shows the ridge trace.

We find that the parameter for elevation undergoes little change, whereas the parameters for area and longest flow path, the highly correlated variables, undergo rapid changes initially.

In choosing an optimum solution, we try to find a sufficiently low value of θ for which the solutions are not within the unstable region. With the help of Fig. 6.2.8 and Table 6.2.5, we make $\theta = 0.05$. Hence the fitted model, used to obtain a mean response, is as follows:

$$\hat{Y} = 0.08\,x_1 - 0.30\,x_2 + 0.43\,x_3.$$

AU: Is the edit OK? [See Section 6.3.4 on p. 385]

6.2.11 Other methods and discussion of Section 6.2

Regression methods are used in prediction, interpolation, and data fitting. It is important to note that such uses are valid only within the limits of the data used in the calibration of the model. One should critically examine any spurious relationships that may arise in an analysis. As noted in Chapter 1, one may find that two variables are correlated because they are associated with a third variable and not on account of any physical relationship between the first two.

Table 6.2.5 Estimates for ridge regression.

θ	β (area)	Rate of change	β (longest flow path)	Rate of change	β (elevation)	Rate of change
0	0.527		−0.756		0.427	
0.005	0.401	−25.137	−0.63	25.291	0.431	0.822
0.01	0.318	−16.746	−0.545	16.88	0.433	0.402
0.015	0.258	−11.951	−0.485	12.073	0.434	0.164
0.02	0.213	−8.955	−0.44	9.069	0.434	0.017
0.035	0.128	−4.541	−0.353	4.644	0.432	−0.194
0.05	0.079	−2.731	−0.303	2.829	0.428	−0.274
0.075	0.033	−1.437	−0.254	1.53	0.42	−0.32
0.1	0.006	−0.877	−0.225	0.968	0.412	−0.331
0.15	−0.023	−0.414	−0.191	0.501	0.396	−0.322
0.2	−0.038	−0.232	−0.172	0.316	0.38	−0.303
0.25	−0.047	−0.143	−0.159	0.224	0.365	−0.283
0.3	−0.053	−0.093	−0.149	0.171	0.352	−0.264
0.4	−0.059	−0.042	−0.135	0.115	0.327	−0.229
0.5	−0.062	−0.018	−0.126	0.086	0.306	−0.201

We demonstrated that residual plots play a major role in the diagnostics, such as in the verification of model assumptions. Residuals should be normally distributed. Plots of residuals against predicted values of the response variable should show constant variance and a linear pattern; possible remedies are weighted least squares and nonlinear regression. Transformations of the explanatory variables and inclusion or exclusion of some of them can be judged on the basis of residuals against response variables.

We have confined this section to linear multiple regression in which least squares solutions are found for the unknown parameters. Influential observations and outliers have been examined. It should be noted that the presence of more than one outlier will cause *masking*, and this means a distortion in the residuals caused by the outliers left in the sample of observations; however, more than one notion of masking has emerged.[21] Robust statistics are expected to provide the answer to the problem of accommodation of outliers. The performances of such methods are not generally affected when outliers are present.[22] Robust methods can also be used to improve the performance of the diagnostic procedures. One method, which is highly robust but inefficient, is the least median squares (LMS) method of Rousseeuw; the fitting poses some difficulties.[23] The estimation of $\hat{\boldsymbol{\beta}}$ is made through the minimization of median squared residuals.

Robust methods include weighted least squares estimators[24] called *M-estimators for regression coefficients.*[25] This method is a recommended alternative to the transformation of variables when the variance of the residuals increases or decreases in a horizontal direction on a residual plot; the property is called *heteroscedasticity* (which is not evident in the case studied: see Fig. 6.2.2 or 6.2.4) as opposed to the classical property of *homoscedasticity*.

---

[21] Lawrance (1995) discusses the subject and also the role of the h_{ij} (off diagonal) elements of the leverage matrix.

[22] Huber (1981).

[23] See Rousseeuw and Leroy (1987, Chapters 3 and 4); see also cautionary note by Hettmansperger and Sheather (1992). The procedure is available in many statistical computer packages.

[24] See Draper and Smith (1998, Chapters 9 and 25).

[25] See Mason et al. (1989, Section 28.2).

In recent years there has been more interest in the general linear model which dates back to the work of Gauss and others in the nineteenth century. Originally, the theory of algebraic invariants sought to identify those quantities in systems of equations that are unaffected by linear transformations of the variables in the system. The general linear model is an extension of the linear multiple regression model for a *single* response (Y) variable. It goes beyond the basic model because linear transformations or linear combinations of multiple-dependent variables are made possible. Thus multivariate tests of significance can be used. Another advantage is that one can find a solution to the normal equations when the explanatory (X) variables are not linearly independent and the inverse of the $\mathbf{X}^{\mathrm{T}}\mathbf{X}$ does not exist (because of which we used ridge regression, somewhat subjectively). Another extension is the generalized linear model. This model is based on the generalization of normality in the assumptions made by using the exponential family, of which the normal distribution is a member. Also, the homoscedasticity assumption, that is, the assumption on the equality of the variances of the individual observations, is relaxed. One uses the generalized least squares regression function (when the disturbances are said to be nonspherical) to provide the best linear unbiased estimator of the expected value of the response. We noted that in the standard regression model (with spherical disturbances), one uses the classical least squares regression function as the estimator.[26]

In some practical situations, the conditions for a linear regression model may not hold, and diagnostics such as a curvature in the plot of residuals against fitted values may indicate the need for a nonlinear model of the type:

$$Y = h(x, \theta) + \varepsilon,$$

based on multidimensional data where h is some nonlinear function with respect to unknown parameters θ. Some examples of nonlinear models are as follows:

(1) $Y = \alpha \ \exp(\beta \ \mathrm{x} + \gamma \ \mathrm{x}^2 + \varepsilon)$,
(2) $Y = \alpha \ x^{\beta}\varepsilon$,
(3) $Y = \frac{1}{1+\exp(\alpha+\beta x)}$,
(4) $Y = \frac{\alpha}{\alpha-\beta}(\exp^{-\beta x} - \exp^{-\alpha x})$.

In the first two examples, the models can be transformed to polynomial and linear types, respectively, by taking natural logarithms. We can apply the transformation $\ln[(1/Y) - 1]$ in the third. These are called *intrinsically nonlinear* because of possible transformations. In the fourth case a transformation is not possible and the model is called *intrinsically nonlinear.*[27]

Some of the least squares iterative methods for nonlinear models are based on the Gauss-Newton, steepest descent, and Marquardt algorithms.[28]

At a somewhat less sophisticated level, nonparametric methods of regression have been adopted in recent years. They are not restricted in application by assumptions such as linearity and normality in distribution. Also, one can treat the relationship between the explanatory variables and the response variable as unknown. This approach allows a

[26] See, for example, McCullagh and Nelder (1989), Elian (2000), Goldberger (1962), and the introduction by Stuart and Ord (1994). The GLIM package was developed by the Working Party on Statistical Computing of the Royal Statistical Society to provide a framework for the fitting of generalized linear models.

[27] For an introduction see Draper and Smith (1998, Chapter 24), Mason et al. (1989, Section 26.1), and Ratkowsky (1983).

[28] See Dutter and Huber (1981), Stuart and Ord (1991, pp. 1089–1090), Hougaard (1988), Atkinson (1985, p. 228), and Kennedy and Gentle (1980, Chapter 10). Box (1966) discusses some abuses of regression.

wider exploration and may reveal hidden structures in the data than are otherwise possible. However, for higher dimensions, the complexity of the possible structure will increase at faster rates, a property which R. E. Bellman called "the curse of dimensionality," and it arises because data are sparse and multicollinearity prevails.[29]

6.3 MULTIVARIATE ANALYSIS

Multiple regression, as we have seen in Section 6.2, is concerned with the variations of one (response) variable and how they can be explained by means of other (explanatory) variables which are related to it physically or in any other plausible way. In multivariate analysis, on the other hand, one considers the relationships between three or more variables that are initially treated as equally important; data analysis involves several observations or measurements of each variable (or individual). Electronic computers have accelerated the use of such methods, which require extensive data handling and mathematical techniques—more than other aspects of statistics. The approach involves looking for simple methods of representing a complex set of variables while retaining most of the information contained in them. For instance, we can verify whether two so-called components suffice to represent three or more variables. The investigation includes ways of classifying the variables into groups or clusters and the relationships between groups. In this section we introduce principal components, factor analysis, and cluster analysis.

6.3.1 Principal components analysis

The aim of principal component analysis is to explain the variance-covariance structure in multiple data sets using a few linear combinations of the original variables. The main objectives are data reduction and interpretation.

Principal components analysis came into practical use about 60 years ago in education psychology. Since then it has found favor in many areas, including some of the pure and applied sciences. The technique is devised to transform, say, p correlated X variables, which are known or observed, into an equal number of uncorrelated (orthogonal) Z indices. These are linear functions of the original variables. The first index accounts for as much of the variance of the original variables as possible, subject to conditions stipulated in this subsection. The second index retains as much of the remaining variance as possible. This continues to the pth index, which has the smallest fraction of the original variance. These Z indices are called *principal components*. The objective is to use a number of components that are less than p to account for most of the variation in the original p variables; thereby, some economy is achieved. One expects this to happen when the correlation structure of the X variables is strong.

Given n measurements in each of the X variables, which are p in number, we can represent the p principal components in matrix notation as

$$\mathbf{Z} = \mathbf{XA} \tag{6.3.1}$$

[29] See, for example, Hettmansperger (1984) for inferences based on rank tests; see, in general, Takezawa (2005) who traces the dependence of a response variable on one of several exploratory variables without specifying initially the model type or structure, using different methods such as smoothing splines and kernel estimation; Mendenhall and Sincich (1995) discuss some aspects.

in which **Z** and **X** are $n \times p$ matrices and **A** is a $p \times p$ matrix of coefficients. For example, if $n = 4$ and $p = 3$ we can write the system as

$$\begin{bmatrix} Z_{11} & Z_{12} & Z_{13} \\ Z_{21} & Z_{22} & Z_{23} \\ Z_{31} & Z_{32} & Z_{33} \\ Z_{41} & Z_{42} & Z_{43} \end{bmatrix} = \begin{bmatrix} X_{11} & X_{12} & X_{13} \\ X_{21} & X_{22} & X_{23} \\ X_{31} & X_{32} & X_{33} \\ X_{41} & X_{42} & X_{43} \end{bmatrix} \begin{bmatrix} a_{11} & a_{12} & a_{13} \\ a_{21} & a_{22} & a_{23} \\ a_{31} & a_{32} & a_{33} \end{bmatrix}. \tag{6.3.2}$$

In general, the jth principal component Z_j is given by

$$\mathbf{z}_j = \mathbf{X}\mathbf{a}_j, \quad \text{for } j = 1, 2, \ldots, p. \tag{6.3.3}$$

in which $\mathbf{z}_j$ is an $n \times 1$ (column) vector and $\mathbf{a}_j$ is a $p \times 1$ (column) vector of coefficients. For example, if $p = 3$, as in Eq. (6.3.2), the system is represented by

$$[\mathbf{z}_1\ \mathbf{z}_2\ \mathbf{z}_3] = \mathbf{X}[\mathbf{a}_1\ \mathbf{a}_2\ \mathbf{a}_3].$$

Then the second principal component Z_2, as given by the second column of the **Z** matrix of Eq. (6.3.2), for instance, is related to the X variables with $n = 4$ through the second vector of coefficients $\mathbf{a}_2$ as follows:

$$\mathbf{z}_2 = \begin{bmatrix} Z_{12} \\ Z_{22} \\ Z_{32} \\ Z_{42} \end{bmatrix} = \begin{bmatrix} X_{11}a_{12} + X_{12}a_{22} + X_{13}a_{32} \\ X_{21}a_{12} + X_{22}a_{22} + X_{23}a_{32} \\ X_{31}a_{12} + X_{32}a_{22} + X_{33}a_{32} \\ X_{41}a_{12} + X_{42}a_{22} + X_{43}a_{32} \end{bmatrix}.$$

In the same way the kth principal component is related to the $n \times p$ matrix of X variables through the kth (column) vector of coefficients $\mathbf{a}_k$.

Let us assume that the X variables are deviations from their respective means. The total variance of **X** is defined as the sum of the variances of the p individual X variables. The variance-covariance matrix of **X** can be defined by Σ, a $p \times p$ matrix of coefficients $\sigma_{ij}; i, j = 1, 2, \ldots, p$. In practice Σ is estimated by the sample covariance matrix **C**:

$$\mathbf{C} = \mathbf{X}^{\mathrm{T}}\mathbf{X}/(n-1) = \begin{bmatrix} c_{11} & c_{12} & \cdots & c_{1p} \\ c_{21} & c_{22} & \cdots & c_{2p} \\ \vdots & \vdots & \cdots & \vdots \\ c_{p1} & c_{p2} & \cdots & c_{pp} \end{bmatrix} \tag{6.3.4}$$

in which the elements are defined as follows:

$$c_{ij} = \frac{1}{n-1} \sum_{k=1}^{n} X_{kj} X_{ki}. \tag{6.3.5}$$

In order to make the principal components independent of the units of the X variables, let us modify the **X** matrix so that the X variables which are deviations from the respective means are divided by the respective standard deviations [considering each column separately as in Eq. (6.3.2)]. Hence **C** becomes the sample correlation matrix. However, because the new X variables have unit variance only asymptotically, these operations are not recommended for small samples (Press, 1972, p. 294).

From Eq. (6.3.3), the variance of the jth principal component is given by

$$\mathrm{Var}[\mathbf{z}_j] = \mathrm{Var}[\mathbf{X}\mathbf{a}_j] = \mathbf{a}_j^{\mathrm{T}}\mathrm{Var}[\mathbf{X}]\mathbf{a}_j, \qquad \text{for } j = 1, 2, \ldots, p.$$

This is estimated by $\mathbf{a}_j^{\mathrm{T}}\ \mathbf{C}\mathbf{a}_j$. Hence, for the first principal component,

$$\mathrm{Var}[\mathbf{z}_1] = \mathbf{a}_1^{\mathrm{T}}\mathbf{C}\mathbf{a}_1. \tag{6.3.6}$$

We proceed as follows by maximizing the variance of the first principal component, and then the second, and so on. In this way we can have, depending on the correlations between the Z and the X variables, fewer than p significant principal components. Accordingly, we estimate the A coefficients by maximizing the right-hand side of Eq. (6.3.6). This is subject to the orthogonality condition

$$\mathbf{a}_1^{\mathrm{T}}\mathbf{a}_1 = 1, \tag{6.3.7}$$

that is, $a_{11}^2 + a_{12}^2 + \cdots + a_{1p}^2 = 1$. (It is seen that without the condition one can increase the variance by increasing one or more of the coefficients.)

Using the Lagrange multiplier λ_1 with the constraint of Eq. (6.3.7), let us maximize the following:

$$P = \mathbf{a}_1^{\mathrm{T}}\mathbf{C}\mathbf{a}_1 + \lambda_1(1 - \mathbf{a}_1^{\mathrm{T}}\mathbf{I}\mathbf{a}_1)$$

(in which we have incorporated the $p \times p$ identity matrix $\mathbf{I}$, which has 1s in the leading diagonal and 0s elsewhere). Thus,

$$\frac{\partial P}{\partial \mathbf{a}_1} = 2\mathbf{C}\mathbf{a}_1 - 2\lambda_1\mathbf{I}\mathbf{a}_1 = \mathbf{0}.$$

(This result follows from the multiplication of the vectors and matrices taking account of the symmetry of the $\mathbf{C}$ matrix.) Hence,

$$(\mathbf{C} - \lambda_1\mathbf{I})\mathbf{a}_1 = \mathbf{0}. \tag{6.3.8}$$

A nontrivial solution to λ_1 is obtained from

$$|\mathbf{C} - \lambda_1\mathbf{I}| = 0 \tag{6.3.9}$$

(where $|\cdot|$ denotes the determinant). Also, if we premultiply Eq. (6.3.8) by $\mathbf{a}_1^{\mathrm{T}}$ it follows from Eq. (6.3.6) that

$$\mathbf{a}_1^{\mathrm{T}}\mathbf{C}\mathbf{a}_1 = \lambda_1 = \mathrm{Var}[\mathbf{z}_1]. \tag{6.3.10}$$

We continue the maximization with respect to the vector $\mathbf{a}_2$ subject to the constraint $\mathbf{a}_1^{\mathrm{T}}\mathbf{a}_1 = 1$. There is an additional constraint $\mathbf{a}_1^{\mathrm{T}}\mathbf{a}_2 = 0$, imposed because of the condition that the Z components are uncorrelated (orthogonal); that is, Z_1 is uncorrelated with Z_2, and in the same way Z_3 has no correlation with Z_1 and Z_2, and so on. Similarly, the maximization is repeated for the third and other principal components. It is seen that (regardless of the number of additional constraints) solutions correspond directly with Eq. (6.3.9). Thus, we can extend the result from the first principal component to all principal components. Accordingly, the λ 's are obtained from the roots of

$$|\mathbf{C} - \lambda\mathbf{I}| = 0. \tag{6.3.11}$$

The λ's are called the eigenvalues (also called *latent* or *characteristic roots*) of the matrix $\mathbf{C}$. From Eq. (6.3.10) and so on, these are shown to be the estimated variances of the respective principal components. The analysis of principal components is mainly concerned with the eigenvalues of the sample covariance or correlation matrix. The corresponding $\mathbf{a}$ vectors are called the *eigenvectors* (also called *latent* or *characteristic vectors*) of matrix $\mathbf{C}$.

We can generalize the result of Eq. (6.3.10) as follows:

$$\mathbf{A}^{\mathrm{T}}\mathbf{C}\mathbf{A} = \mathbf{D}, \tag{6.3.12}$$

Table 6.3.1 Correlation matrix of water pollution data.

	BOD	NO_3-N	NH_3-N
BOD	1	0.65	0.515
NO_3-N	0.65	1	0.415
NH_3-N	0.515	0.415	1

where **D** is a diagonal matrix the diagonal elements of which are the eigenvalues of **C**. Also, from Eq. (6.3.3), the covariance between the standardized X variables and the jth principal component is

$$\text{Cov}[\mathbf{X}, \mathbf{z}_j] = \text{Cov}[\mathbf{X}, \mathbf{X}\mathbf{a}_j] = \mathbf{C}\mathbf{a}_j$$

(which is a $p \times 1$ vector). It follows from the foregoing that this covariance is equivalent to $\lambda_j \mathbf{a}_j$. Hence, if we consider the ith row of X variables and the jth principal component,

$$\text{Cov}[\mathbf{x}_i, \mathbf{z}_j] = \lambda_j a_{ij}.$$

The correlation between the ith row of X variables and the jth principal component is given by

$$\text{Cor}[\mathbf{x_i}, \mathbf{z_j}] = \frac{\text{Cov}[\mathbf{x_i}, \mathbf{z_j}]}{\sqrt{\text{Var}(\mathbf{z_j})}} = a_{ij}\sqrt{\lambda_j}. \tag{6.3.13}$$

This follows from Eq. (6.3.10) and also because the X variables are standardized.[30]

Example 6.18. Principal components analysis of water quality data. Table E.6.2 gives three sets of water quality measurements at 38 stations, separated by distances of 1 km, on the Blackwater River, England. The respective means and standard deviations as given at the bottoms of the columns are used to transform the three columns of data sets into standardized units. The correlation matrix is given in Table 6.3.1.

The variables are arranged according to the magnitudes of the correlation coefficients. If X, Y, and Z represent BOD, NO_3-N (nitrates), and NH_3-N (ammonia), respectively, the eigenvalues are the solutions of the following equation (based on the correlations c_{xy} and so on) obtained from Eq. (6.3.11):

$$\begin{aligned} |\mathbf{C} - \lambda\mathbf{I}| &= (1-\lambda)^3 - (1-\lambda)\left(c_{xy}^2 + c_{xz}^2 + c_{yz}^2\right) + 2c_{xy}c_{xz}c_{yz} = 0 \\ &= (1-\lambda)^3 - 0.86001(1-\lambda) + 0.27808. \end{aligned}$$

After substituting from Table 6.3.1, this is solved by the Newton-Raphson method.[31] The eigenvalues are ranked in descending order and given in the top row of Table 6.3.2.

It is seen from the eigenvalues that the first principal component accounts for $2.059/3.0 = 69\%$ of the total variance in the three variables. The first two components account for $(2.059 + 0.605)/3.0 = 89\%$ of the total variation. The eigenvectors, which are the A coefficients for each vector $\mathbf{a}_j$ corresponding to the ranked eigenvalue λ_j, are obtained from the solutions of Eq. (6.3.8), applied similarly to all components, for each pair of vector, and eigenvalue. For the first vector,

$$\begin{bmatrix} 1-\lambda_1 & c_{xy} & c_{xz} \\ c_{xy} & 1-\lambda_1 & c_{yz} \\ c_{xz} & c_{yz} & 1-\lambda_1 \end{bmatrix} \begin{bmatrix} a_{11} \\ a_{21} \\ a_{31} \end{bmatrix} = \begin{bmatrix} 0 \\ 0 \\ 0 \end{bmatrix}.$$

[30] Equation (6.3.13) is used, in the relationship between **A** and matrix **R** of correlations between the X variables and the principal components Z, in factor analysis of Subsection 6.3.2.

[31] See also Harris (1985, pp. 381–385) for solutions to a cubic equation.

Table 6.3.2 Eigenvalues and eigenvectors of water pollution data.

$\lambda_j, j = 1, 2, 3$	2.059	0.605	0.336
$a_{1j}, j = 1, 2, 3$	0.616	0.205	0.761
$a_{2j}, j = 1, 2, 3$	0.585	0.529	−0.615
$a_{3j}, j = 1, 2, 3$	0.529	−0.824	−0.205

These equations are solved simultaneously for the first eigenvector substituting appropriate correlation coefficients from Table 6.3.1, and similarly the second and third vectors are found using corresponding eigenvalues from the first row of Table 6.3.2. The resulting **a** vectors are shown in Table 6.3.2, by the respective columns below the top row of eigenvalues.

The first two principal components are found using Eq. (6.3.3), noting that each column of X values is initially transformed to a mean of zero and a standard deviation of 1 (see statistics at the bottom of Table E.6.2). The two principal components are given in Table 6.3.3.

(As a matter of interest we can verify that the sample correlation coefficient between the first and second principal components is as low as 0.000008. Similarly, the correlation between the first and second and the second and third components are −0.000009 and 0.000010, respectively.)

Although Z_1 and Z_2 are uncorrelated, there are some local features which can be seen in Fig. 6.3.1.

About 50% of the stations, namely, 16–35, have low negative values of Z_2, above −0.43, but Z_1 varies from −1 to +0.9, approximately. The group 8–14 has values of Z_1 from 0.6 to 2.1 but Z_2 is low, varying from −0.9 to −0.65, approximately. Stations 2, 3, 4, and 5 have positive values of Z_1 and Z_2.

There are several general-purpose computer packages that can be used in principal component analysis; these packages require the data matrix and some simple instructions. As seen in the preceding example, it is possible to reduce the number of variables to

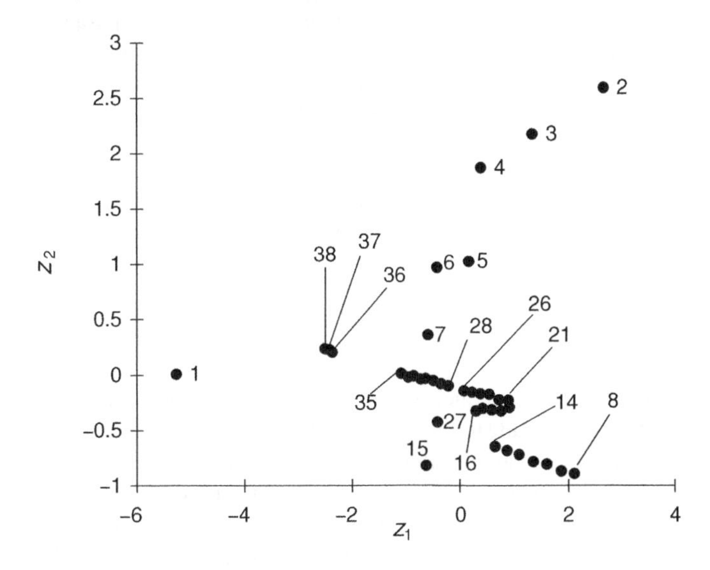

Fig. 6.3.1 Plot of first two principal components of river quality data. Station numbers are given beside the points.

Table 6.3.3 First and second principal components of water pollution data.

Station	Z_1	Z_2
1	−5.2711	0.0082
2	2.6453	2.5931
3	1.3385	2.1698
4	0.3795	1.8678
5	0.1618	1.0231
6	−0.4163	0.9664
7	−0.5868	0.3585
8	2.1124	−0.8958
9	1.8720	−0.8730
10	1.6105	−0.8083
11	1.3650	−0.7830
12	1.0948	−0.7190
13	0.8692	−0.6829
14	0.6436	−0.6468
15	−0.6209	−0.8199
16	0.9162	−0.2939
17	0.7563	−0.3277
18	0.5841	−0.3187
19	0.4155	−0.3063
20	0.2879	−0.3251
21	0.8912	−0.2288
22	0.7202	−0.2257
23	0.5406	−0.1763
24	0.3734	−0.1698
25	0.2222	−0.1559
26	0.0711	−0.1419
27	−0.4093	−0.4257
28	−0.2076	−0.0997
29	−0.3513	−0.0791
30	−0.4826	−0.0542
31	−0.6263	−0.0336
32	−0.7254	−0.0407
33	−0.8530	−0.0126
34	−0.9521	−0.0198
35	−1.0760	0.0117
36	−2.3635	0.2050
37	−2.4242	0.2254
38	−2.5048	0.2349

a fewer number of components that have a high proportion of the original variance. Attempts have been made to remove the subjective element from the choice of the number of components. For example, the scree test is a plot of eigenvalues against the order of the principal components; one decides to curtail the number components from the point at which the plot levels off on the right; in geology, scree pertains to the debris collected on the lower part of a rocky slope. However, interpretation of principal components is often difficult. Using factor analysis, which follows, we extend the case study and attempt to find interpretations by rotating the principal components.

6.3.2 Factor analysis

Factor analysis is a procedure similar to principal components analysis. They are both powerful explorative tools. It was devised by Charles Spearman around 1904 following a study of the correlations among various school examination results. The objective in both cases is to describe a set of, say, p variables $X_1, X_2, \ldots, X_p$ through a smaller number of m factors or components, which are related by a set of coefficients. In factor analysis the variables are transformed using a particular algorithm. This is equivalent to a rotation of the axes of reference. We aim to find factors that are highly related to one or more of the X variables but not to the others and in this way a set of factors (smaller in number than the original variables) are found that accounts for most of the variation in the array of data. The rotation can also be applied to the principal components, as shown here.

The basic relationship in factor analysis is as follows:

$$X_i = \sum_{j=1}^{m} b_{ij} F_j + \varepsilon_i, \tag{6.3.14}$$

where X_i is the ith variable standardized so that the mean is zero and the standard deviation is unity; where b_{ij}, $j = 1, 2, \ldots, m$ are called the factor loadings, $-1 < b_{ij} < +1$; where F_j, $j = 1, 2, \ldots, m$ are the mutually uncorrelated common factors, which also have zero mean and unit variance; and where ε_i is a factor specific to the ith variable, which has a mean of zero and is uncorrelated with the other factors.

It follows that

$$\mathrm{Var}[X_i] = \sum_{j=1}^{m} b_{ij}^2 + \mathrm{Var}[\varepsilon_i].$$

The sum of the squares of the factor loadings and the second term on the right-hand side, which has no relationship with the common factors, are called the *communality* and *specificity* (or *uniqueness*), respectively, of the variables X_i. Further, it can be easily verified that

$$\mathrm{Cor}[X_i, X_j] = \sum_{k=1}^{m} b_{ik} b_{jk}.$$

Thus, the correlation between the variables is related to the corresponding factor loadings.

In factor analysis, a common practice is to ignore the last term in Eq. (6.3.14)—which concerns the specific or unique factors—and to consider only the common factors. In this way, we can write the relationship between the factors and X variables as

$$\mathbf{F} = \mathbf{XH}, \tag{6.3.15}$$

where $\mathbf{F}$ is $n \times p$ matrix of factor scores, $\mathbf{X}$ is an $n \times p$ matrix of observations standardized for each of the p variables (as in principal components analysis), and $\mathbf{H}$ is a $p \times p$ matrix of factor score coefficients.

For $\mathbf{F}$ to be orthogonal, it can be shown as follows that

$$\mathbf{H} = \mathbf{AD}^{-1/2}, \tag{6.3.16}$$

where $\mathbf{A}$ is defined by Eq. (6.3.1) and $\mathbf{D}^{-1/2}$ is diagonal matrix in which the nonzero elements are the reciprocals of the square roots of the eigenvalues of the covariance matrix $\mathbf{C}$, defined by Eq. (6.3.4). Thus from Eqs. (6.3.15) and (6.3.16),

$$\mathbf{F}^{\mathrm{T}}\mathbf{F} = \mathbf{D}^{-1/2}\mathbf{A}^{\mathrm{T}}\mathbf{X}^{\mathrm{T}}\mathbf{XAD}^{-1/2}.$$

Hence, from Eqs. (6.3.4) and (6.3.12), this simplifies to

$$\mathbf{F}^{\mathrm{T}}\mathbf{F} = (n-1)\mathbf{I},$$

thus showing the orthogonality of **F**.

We can incorporate the principal components by relating Eqs. (6.3.1) and (6.3.15) using (6.3.16) as follows:

$$\mathbf{F} = \mathbf{XAD}^{-1/2} = \mathbf{ZD}^{-1/2}. \tag{6.3.17}$$

Because of this relationship, which simply involves a set of p constants, the correlations between the X variables and the factors F are the same as those between the variables X and the principal components Z. It follows from Eq. (6.3.13) that we can write this correlation matrix as

$$\mathbf{R} = \mathbf{AD}^{1/2} \tag{6.3.18}$$

(where $\mathbf{D}^{1/2}$ is a diagonal matrix in which the nonzero elements are the square roots of the eigenvalues of the correlation matrix **C**). The matrix **R**—which gives the correlations between the principal components (columns) and the standardized X variables (rows)—is sometimes called the *factor loading matrix*. It has the property that

$$\mathbf{RR}^{\mathrm{T}} = \mathbf{AD}^{1/2}\mathbf{D}^{1/2}\mathbf{A}^{\mathrm{T}} = \mathbf{ADA}^{\mathrm{T}} = \mathbf{C}$$

[which can be obtained by premultiplying Eq. (6.3.12) by **A** and postmultiplying it by $\mathbf{A}^{\mathrm{T}}$, noting the orthogonality of the **A** matrix, which follows from Eq. (6.3.7), that is, $\mathbf{A}^{\mathrm{T}}\mathbf{A} = \mathbf{AA}^{\mathrm{T}} = \mathbf{I}$]. Also, if **R** is subject to an orthogonal rotation such as **RP** (where **P** is orthogonal, that is, $\mathbf{P}^{\mathrm{T}}\mathbf{P} = \mathbf{I}$), the result is the same.

From Eq. (6.3.18) and by premultiplying Eq. (6.3.12) by **A** and because of the orthogonality of the **A** matrix,

$$\mathbf{R} = \mathbf{ADD}^{-1/2} = \mathbf{CAD}^{-1/2} = \mathbf{CH}$$

[using Eq. (6.3.16)]. That is,

$$\mathbf{H} = \mathbf{C}^{-1}\mathbf{R}. \tag{6.3.19}$$

On account the correspondence between the factors F and the principal components Z, as in Eq. (6.3.17), the effect of an orthogonal rotation **RP** on the principal components (which changes the correlations with the X variables) and the corresponding factors are the same. We may therefore investigate changes to the principal components. This will enable us to make comparisons with the unrotated components (as in Example 6.18). The rotation should be such that meaningful physical interpretations are possible from the resulting components. As in the following example, we aim to have some high correlations between the X variables and the rotated principal components and some low correlations between others.

The original principal components can be written as

$$\mathbf{Z} = \mathbf{XA} = \mathbf{XHD}^{1/2} = \mathbf{XC}^{-1}\mathbf{RD}^{1/2},$$

obtained from Eqs. (6.3.1), (6.3.16), and (6.3.19). An orthogonal rotation will change **R** to **RP**, as stated. Thus the rotated components are given by

$$\mathbf{Z}_r = \mathbf{XC}^{-1}\mathbf{RPD}^{1/2}. \tag{6.3.20}$$

It follows from Eqs. (6.3.4) and (6.3.18) and then (6.3.16) and (6.3.19) that

$$\begin{aligned}\mathbf{Z}_r^{\mathrm{T}}\mathbf{Z}_r &= \mathbf{D}^{1/2}\mathbf{P}^{\mathrm{T}}\mathbf{R}^{\mathrm{T}}\mathbf{C}^{-1}\mathbf{X}^{\mathrm{T}}\mathbf{X}\mathbf{C}^{-1}\mathbf{RPD}^{1/2}\\ &= (n-1)\mathbf{D}^{1/2}\mathbf{P}^{\mathrm{T}}\mathbf{D}^{1/2}\mathbf{A}^{\mathrm{T}}\mathbf{C}^{-1}\mathbf{RPD}^{1/2}\\ &= (n-1)\mathbf{D}^{1/2}\mathbf{P}^{\mathrm{T}}\mathbf{D}^{1/2}\mathbf{A}^{\mathrm{T}}\mathbf{A}\mathbf{D}^{-1/2}\mathbf{PD}^{1/2} = (n-1)\mathbf{D},\end{aligned}$$

which shows that the rotated principal components are uncorrelated.

Example 6.19. Rotation of principal components analysis (factor analysis). In Example 6.18, a principal components analysis was applied to the water quality measurements at 38 stations on the Blackwater River, England, given in Table E.6.2. In this example we apply the following orthogonal rotation to the principal components

$$\mathbf{P} = \begin{bmatrix} 1/\sqrt{2} & 1/\sqrt{2} & 0 \\ -1/\sqrt{2} & 1/\sqrt{2} & 0 \\ 0 & 0 & 1 \end{bmatrix}.$$

(It is seen that, as necessary, $\mathbf{P}^{\mathrm{T}}\mathbf{P} = \mathbf{I}$.) As stated, the purpose is to make the correlations between some of the variables and the components high and the others low, so that we can make a few of the new components accountable for most of the variation in the data sets.

The elements of matrix **A** are given in the three lower rows of Table 6.3.2. From Eq. (6.3.18) and the eigenvalues of the upper row of Table 6.3.2, the factor-loading matrix is obtained as

$$\mathbf{R} = \mathbf{AD}^{1/2} = \begin{bmatrix} 0.883 & 0.160 & 0.441 \\ 0.839 & 0.411 & -0.356 \\ 0.759 & -0.641 & -0.119 \end{bmatrix}.$$

For the rotation specified earlier, this becomes

$$\mathbf{RP} = \begin{bmatrix} 0.512 & 0.737 & 0.441 \\ 0.302 & 0.884 & -0.356 \\ 0.989 & 0.083 & -0.119 \end{bmatrix}.$$

The correlation matrix **C** of the water pollution data is given in Table 6.3.1. Its inverse is (following the procedure in Example 6.6)

$$\mathbf{C}^{-1} = \begin{bmatrix} 1.979 & -1.042 & -0.587 \\ -1.042 & 1.757 & -0.193 \\ -0.587 & -0.193 & 1.383 \end{bmatrix}.$$

Hence,

$$\mathbf{C}^{-1}\mathbf{RPD}^{1/2} = \begin{bmatrix} 0.168 & 0.381 & 0.761 \\ -0.276 & 0.598 & -0.615 \\ 1.448 & -0.380 & -0.205 \end{bmatrix}.$$

From Eq. (6.3.20), the rotated principal components are given by $\mathbf{Z}_r = \mathbf{XC}^{-1}\mathbf{RPD}^{1/2}$. These are (partly) presented in Table 6.3.4. The key to the interpretation of the rotated principal components is the matrix **RP**, which gives the correlations between the rotated components (columns) and the X variables (rows). It is seen that the first rotated component accounts for the variation of the third variable, ammonia (NH_3-N), of Table E.6.2. The second rotated component has no relationship with the ammonia but has fairly close relationships with the other two, BOD and nitrates. The third rotated component is poorly correlated with any of the variables and can therefore be ignored. These findings can be verified by comparing, for example, the high, low, and intermediate values in the columns of Table E.6.2 with those of the appropriate columns of Table 6.3.4.

The correlations between the first two standardized Z variables and the three X variables are indicated in Fig. 6.3.2; see the first two columns of the preceding matrix **R**. The rotation

Table 6.3.4 Rotated first and second principal components of water pollution data

Station	Z_{1r}	Z_{2r}
1	−3.738	−2.014
2	−1.513	2.847
3	−1.884	2.047
4	−2.169	1.466
5	−1.220	0.785
6	−1.555	0.524
7	−0.883	0.029
8	2.662	0.176
9	2.463	0.100
10	2.193	0.046
11	1.987	−0.031
12	1.712	−0.089
13	1.506	−0.150
14	1.299	−0.211
15	0.631	−0.818
16	1.031	0.143
17	0.962	0.058
18	0.829	−0.001
19	0.693	−0.057
20	0.628	−0.120
21	0.929	0.180
22	0.804	0.116
23	0.612	0.082
24	0.486	0.023
25	0.360	−0.025
26	0.235	−0.073
27	0.266	−0.458
28	−0.017	−0.150
29	−0.145	−0.191
30	−0.271	−0.223
31	−0.399	−0.264
32	−0.460	−0.307
33	−0.587	−0.336
34	−0.647	−0.379
35	−0.776	−0.404
36	−1.939	−0.761
37	−2.008	−0.770
38	−2.078	−0.794

RP is equivalent to a rotation of the axes by 45°, and the rotated axes are shown. We can note (from matrix **RP** and Fig. 6.3.2) the very close relationship between X_3, ammonia (NH_3-N), and the first rotated principal component Z_{1r}. We also see the poor association of this variable with the second rotated principal component Z_{2r}, which is associated in varying degrees with nitrates (NO_3-N) and biological oxygen demand (BOD) of the two preceding columns of Table E.6.2, representing X_2 and X_1, respectively.

Numerous types of algorithms with orthogonal factor rotations have been used to obtain the desired final results (see, for example, Cooper, 1983, and in particular the varimax method originally devised by Kaiser, 1958, which we have used here). The basic idea is

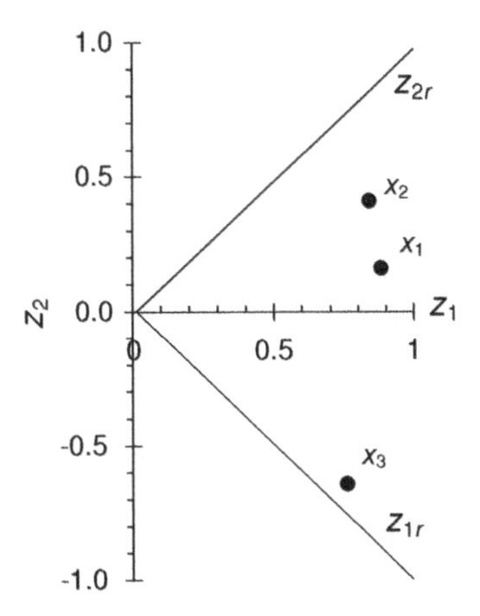

Fig. 6.3.2 Rotation for factor analysis; Z_1, Z_2 are original axes; Z_{1r} and Z_{2r} are rotated axes.

to standardize the factor loadings and maximize the sum of their squares separately for each factor. For the conditions of maximum variance, the factor loadings (which are the correlations between the rotated factors and the X variables) tend to be close to unity or zero with a minimum of intermediate values.

Under the so-called oblique rotation, on the other hand, correlations are permitted between factors to obtain loadings close to unity or zero. However, this approach is apparently less common in practice.

The varimax and other methods are described by Harman (1976), and computer software is widely available.

6.3.3 Cluster analysis

In principal component and factor analysis, the aim was to find linear combinations of a set of variables such that one ends with a few components or factors that represent most of the variation in the original set. There is another branch of multivariate analysis associated with data reduction, in which one groups objects into classes so that there is some similarity between the objects in a given class. Cluster analysis (originated in 1939 by R. Tryon) is the simplest procedure of this kind. There are many advantages in cluster analysis. Most commonly it can be applied to define groups objectively, without using strict mathematical definitions.

Cluster analysis is a step-by-step fusion of individuals in which firstly one group is formed, then another, and so on; with the gradual merging of groups already formed one ends with only one group. The result is termed a *dendogram* (from the Greek word dendron—tree), but this is the end result of only one approach. There are many algorithms proposed for cluster analysis. We shall deal with only the *nearest neighbor* method.

We start with the calculation of distances from each individual to all the other individuals in the set. Initially there are groups of one throughout. The closest pair is joined; if the distance is the same as that to another individual, then all three are merged simultaneously and this is applicable to more than three individuals separated by equal distances. The next smallest pair or group is formed and so on. Groups gradually merge until there is only one united family. This procedure is called an *agglomeration* in order to differentiate from the other method of cluster analysis, called *division*.

In practice, an individual may have more than one attribute, and thus distances must take account of all these attributes. The measure of distance adopted is the *generalized*

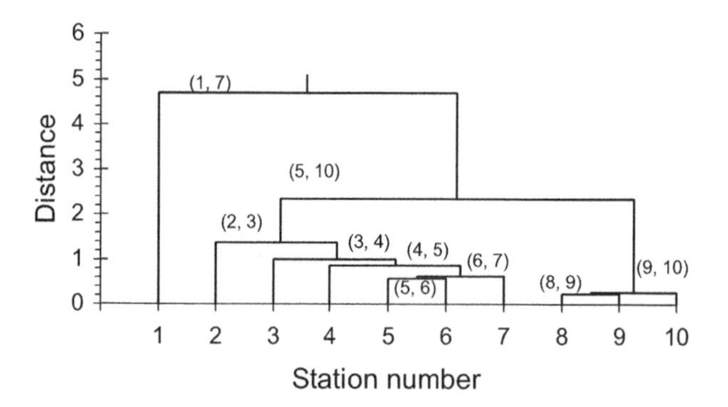

Fig. 6.3.3 Dendrogram for stations 1–10, measuring water quality data, based on first two principal components.

Euclidean distance. For example, if there are p variables, $Z_1, Z_2, \ldots, Z_p$ that quantify the attributes, this distance between individuals i and j is given by

$$d_{ij} = \sqrt{\sum_{k=1}^{p} (z_{ik} - z_{jk})^2}. \tag{6.3.21}$$

If $p = 2$, we have the distance resulting from the use of Pythagoras' theorem. This is the most commonly used measure; there are alternatives, such as the *squared Euclidean distance*, given in the references cited at the end of this chapter.

In order to do a cluster analysis, one possible approach is to start with a principal component analysis of the set. If two principal components can explain most of the variation in the data sets, we can then see how the individuals sort themselves into groups. This is, however, adopted only as a guide, and one should be warned that it may produce results different from a direct analysis of the data. The principal components may not be fully representative of the properties to be considered.

Example 6.20. Cluster analysis of principal components. In Example 6.18, a principal component analysis was applied to the water quality observations at 38 stations on the Blackwater River, England, given in Table E.6.2. The first two principal components are given in Table 6.3.3 and plotted in Fig. 6.3.1, which also shows the individual stations. In this example, we initially make a cluster analysis of the first 10 and the last 11 stations. This is based on the first two principal components and the nearest neighbor method [Eq. (6.3.21)].

Taking only the first ten stations and if only two clusters are to be considered, station 1 is on its own and the second cluster is formed by the remaining nine stations (see Fig. 6.3.3). Proceeding further, we can divide this second cluster into two so that stations 8, 9, and 10 form a cluster on their own, and this will give us three clusters. If six clusters are required, then stations 1–4 will form four different clusters (see Fig. 6.3.1) and the remaining six are subdivided into two as shown.

Figure 6.3.4 gives the dendogram for stations 28–38. It shows that stations 36–38 form one cluster because of their close proximity (as evident from Fig. 6.3.1) and the remaining eight stations will form a second cluster if only two clusters are to be considered. On the other hand, if five clusters are required then we have stations 28–30 forming the first cluster. The other four clusters are formed as follows: stations 31 and 32; stations 33 and 34; station 35 on its own; and stations 36, 37, and 38 in the last cluster.

Finally, let us do a cluster analysis of all 38 stations measuring water quality (using the first two principal components listed in Table 6.3.3). The dendogram is shown in Fig. 6.3.5.

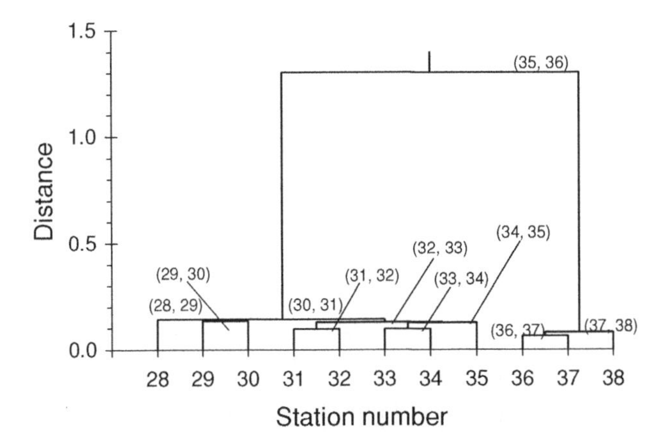

Fig. 6.3.4 Dendrogram for stations 28–38, measuring water quality data, based on first two principal components.

The results are quite close to those obtained from Figs. 6.3.3 and 6.3.4. The small differences arising can be attributed to the data from stations 11 to 27, not taken into account before. If we are forming only two clusters, then station 1 is on its own and the other stations act as one with an amalgamation distance of 1.383. Near the other extreme, we have 13 clusters at an amalgamation distance of 0.283: The largest cluster has stations from 21 to 35 with the exception of 27. (See Fig. 6.3.1, with respect to the principal components.) The second largest group is formed by stations 8–14 (with a maximum distance of 0.280). Stations 16–20 form the next group (with a maximum distance of 0.174). Stations 36, 37, and 38 form the closest group (with a maximum distance of 0.083). The other nine stations 1–7, 15, and 27 are each on its own.

Several potential benefits can result from this particular study. For instance, we may consider that station 1 needs scrutiny to determine why it gives different results. These may arise from errors of instrumentation or measurement. Although the distances between the stations are equal, some of them such as 36–38 may be combined as seen originally in Fig. 6.3.1.

Because methods of cluster analysis are not based on mathematical definitions, they are not strictly inferential. As mentioned before, principal components should only be used as a preliminary assessment of the similarities or dissimilarities of the individuals. Nevertheless, this case study has highlighted additional advantages of principal component analysis, because their configuration, seen with respect to the measurement stations in Fig. 6.3.1, has been beneficial for other methods of multivariate analysis, such as, factor and cluster analyses. Furthermore, principal component analysis can be used as a supplementary method of eliminating explanatory variables in regression analysis that do not sufficiently explain the variation present.

6.3.4 Other methods and summary of Section 6.3

Our coverage of multivariate methods is by no means exhaustive. We have not, for instance, discussed discriminant function analysis. This is a method of separating into two or more groups, wherever possible, using available data. In this respect it is similar to cluster analysis. However, as in principal component analysis, we find linear combinations of the original variables for our purpose. This is also the approach in canonical correlation in which one divides the individuals into two groups and then finds the correlation between them. Furthermore, multidimensional scaling is a method of finding distances between

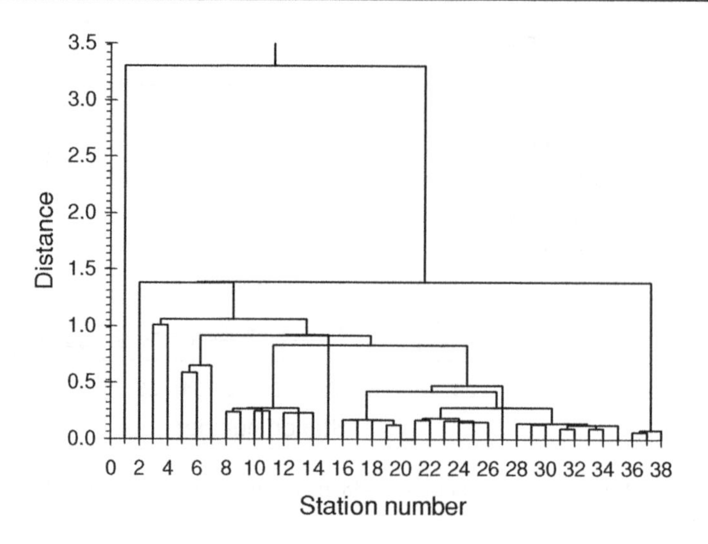

Fig. 6.3.5 Dendrogram for 38 stations measuring water quality.

individuals and is thus an alternative to cluster analysis without going into geometrical relationships.[32]

Of the three methods discussed here, principal component analysis is probably the most widely used procedure. However, by itself such a method may sometimes give results that are not straightforward for interpretation. We have seen that factor analysis, which is an alternative form of principal component analysis, can be useful for interpretation. Cluster analysis, on the other hand, shows similarities between individuals or indicates how some are different from others. Its practical benefits through the use of the dendogram were indicated.

6.4 SPATIAL CORRELATION

In previous sections we considered the correlation between random variables that are applicable in regression or in multivariate analysis. In this section we discuss some aspects of spatial correlation. This enables us to see how variables such as pollutant loads measured at different points in space are related. More importantly from the practical viewpoint, such relationships can be used for estimating values at sites where no measurements are taken.

A sample of data may consist of observations taken in one, two, or three dimensions. Measurements of water quality along a river are taken in one dimension. On the other hand, rainfall and other meteorological variables are measured at particular points, but collectively they comprise a two-dimensional random field. This also applies to remote sensing where picture elements (pixels) are used in areal surveys (as discussed in Chapter 2; see, for example, Fig. 2.1.5) for observing hydrological data over a region. Pollution in groundwater is sometimes treated as a three-dimensional problem because it can occur over different levels.

[32] For more details of these methods see, for example, Sharma (1996) and Krzanowski and Marriott (1994).

Spatial correlation as discussed in this section concerns a field of two or three dimensions. The tools that we use for the measurements of spatial characteristics are the semivariogram and the spatial correlation coefficient.

6.4.1 The estimation problem

In spatial estimation, a basic question is that, given a set of observations of a variable Z_i and the coordinates of the points of observation, what are the values taken by the variable at a point, say, k, where data are unavailable? The approach differs from regression in that local features can affect the solution. On the global scale, however, all measurements should be considered. Thus the first approximation would be to take the arithmetic mean of observations: $z_1, z_2, \ldots, z_n$:

$$\bar{z}_k = \frac{1}{n}\sum_{i=1}^{n} z_i.$$

However, as in most infilling and prediction problems, some measurements (in the vicinity of the point investigated, or sometimes elsewhere) are more closely related to the true value at point k than others. Accordingly, we take a weighted mean

$$\hat{z}_k = \sum_{i=1}^{n} \lambda_i z_i,$$

where $\lambda_i, i = 1, 2, \ldots, n$; the so-called weights are such that $\sum_{i=1}^{n} \lambda_i = 1$. The method of Kriging is used to find a solution to the optimal values of the weights. We shall return to this procedure later but initially we define spatial correlation and the variogram.

6.4.2 Spatial correlation and the semivariogram

The mean of an integrated stochastic process $Z(\mathbf{u})$ such as rainfall or some water quality aspect, $E[Z(\mathbf{u})]$, is the mean of all possible realizations of the process at points $\mathbf{u} \equiv (x, y)$. The process is said to act over a random field. Additionally, the differenced random process $Z(\mathbf{u}) - E[Z(\mathbf{u})]$ represents departures of the original process from the mean at the points considered. The study of such processes is based on the identification of appropriate characteristics of regularity; one of the main characteristics is termed *stationarity*, in the context of stochastic processes, which is defined shortly.

The association of the values taken by the process $Z(\mathbf{u})$ at two points $\mathbf{u}_1$ and $\mathbf{u}_2$ within a given area is represented by the spatial covariance function,

$$\begin{aligned}\mathrm{Cov}\left[\mathbf{u}_1, \mathbf{u}_2\right] &= E\left\{(Z(\mathbf{u}_1) - E\left[Z(\mathbf{u}_1)\right])(Z(\mathbf{u}_2) - E\left[Z(\mathbf{u}_2)\right])\right\} \\ &= E\left[Z(\mathbf{u}_1)Z(\mathbf{u}_2)\right] - E\left[Z(\mathbf{u}_1)\right]E\left[Z(\mathbf{u}_1)\right].\end{aligned} \tag{6.4.1}$$

If $\mathbf{u}_1 \equiv \mathbf{u}_2 \equiv \mathbf{u}_i$ (say), this becomes the variance, the square root of which

$$\sigma(\mathbf{u}_i) = (\mathrm{Var}[Z(\mathbf{u}_i)])^{1/2} = \{E[(Z(\mathbf{u}_i) - E[Z(\mathbf{u}_i)])^2]\}^{1/2} \tag{6.4.2}$$

is the standard deviation of the process at the given point.

The alternative definition, which is commonly used in the estimation problems referred to earlier, is that of the semivariogram,

$$\begin{aligned}\Gamma[\mathbf{u}_1, \mathbf{u}_2] &= \frac{1}{2}\mathrm{Var}[Z(\mathbf{u}_1) - Z(\mathbf{u}_2)] \\ &= \frac{1}{2}E\{(Z(\mathbf{u}_1) - Z(\mathbf{u}_2) - E[Z(\mathbf{u}_1)] + E[Z(\mathbf{u}_2)])^2\}. \end{aligned} \quad (6.4.3)$$

It follows from Eqs. (6.4.1) and (6.4.3) that

$$\mathrm{Cov}[\mathbf{u}_1, \mathbf{u}_2] = \mathrm{Cov}[\mathbf{u}_2, \mathbf{u}_1] \quad \text{and} \quad \Gamma[\mathbf{u}_1, \mathbf{u}_2] = \Gamma[\mathbf{u}_2, \mathbf{u}_1], \text{respectively.}$$

The difference between these functions is that the covariance is a direct function of the *association* between two variables, whereas the semivariogram measures the *disassociation.*

Under *first-order stationary* conditions, one obtains

$$E[Z(\mathbf{u}_1)] = E[Z(\mathbf{u}_2)] = \text{constant.} \quad (6.4.4a)$$

With the additional conditions of *second-order stationarity*,

$$\mathrm{Var}[Z(\mathbf{u}_1)] = \mathrm{Var}[Z(\mathbf{u}_2)], \quad (6.4.4b)$$

$$\mathrm{Cov}[\mathbf{u}_1, \mathbf{u}_2] = \mathrm{Cov}[\mathbf{u}_1 - \mathbf{u}_2], \quad (6.4.4c)$$

and

$$\Gamma[\mathbf{u}_1, \mathbf{u}_2] = \Gamma[\mathbf{u}_1 - \mathbf{u}_2] \quad (6.4.4d)$$

for any two points $\mathbf{u}_1$ and $\mathbf{u}_2$ within the given area. If the first three of the above equations for stationarity hold, we say there is *homogeneity of the mean and the variance-covariance function.* The second property of a random field is that of *isotropy*. It means that the spatial relationships given in terms of the covariance [Eq. (6.4.1)] or the semivariogram [Eq. (6.4.3)] are not conditioned by the vector of distance $\mathbf{h} = \mathbf{u}_1 - \mathbf{u}_2$ but depends only on the absolute value of the distance. Thus,

$$\mathrm{Cov}(\mathbf{h}) = \mathrm{Cov}(h) \quad (6.4.5)$$

and

$$\Gamma(\mathbf{h}) = \Gamma(h). \quad (6.4.6)$$

There is also the theoretical hypothesis of *ergodicity*, which is required for the estimation of the characteristics of the process (and hence of the random field) based on its realizations. *This is applicable if the estimates of its moments taken from the available realizations converge in probability to the theoretical moments, when the available sample increases.* Hence, under ergodicity one realization will suffice for the estimation of these moments. In practice we assume that this property exists (except in some particular cases).

Proceeding further, the correlation function is defined as the normalized covariance function:

$$R[\mathbf{u}_1, \mathbf{u}_2] = \mathrm{Cov}[\mathbf{u}_1, \mathbf{u}_2] / [\sigma(\mathbf{u}_1)\sigma(\mathbf{u}_2)], \quad (6.4.7)$$

which is bounded in the interval $[-1, +1]$.

6.4.3 Some semivariogram models and physical aspects

We assume that the conditions of stationarity and isotropy hold. Two characteristics of the empirical semivariogram should be considered. First, as the separation distance h increases, the variogram tends to approach a constant value. This limiting state is called the *sill*. It is reached at a distance, say, a, known as the *radius of influence*; as h increases beyond $h = a$, the theoretical variogram is constant. Second, it is seen from Eqs. (6.4.3) and (6.4.6) that $\Gamma(0) = 0$; that is, at very close distances the disassociation between values of the variable approaches zero. In practice, however, at very small separation distances the empirical semivariogram can be significantly different from zero, reflecting some local effects. This is called the *nugget effect* (arising from original applications in the gold-mining industry). In the case of rainfall this effect is more evident if one considers hourly data and measurements at shorter time intervals.

The simplest type of semivariogram model is the linear model

$$\begin{aligned} \Gamma(h) &= A_0\delta(h) + A_1 h, && \text{for } 0 \le h < a, \\ &= A_0 + A_1 a, && \text{for } h \ge a, \end{aligned} \tag{6.4.8}$$

where,

$$\begin{aligned} \delta(h) &= 1, && \text{for } h > 0, \\ &= 0, && \text{for } h = 0; \end{aligned}$$

a the radius of influence or correlation distance; A_0 models the discontinuity at the origin called the *nugget effect*; and A_1 is the slope in the linear model.

A closer fit to empirical semivariograms may be obtained using the exponential model:

$$\begin{aligned} \Gamma(h) &= A_0\delta(h) + w\left[1 - \exp\left(-\frac{h}{a}\right)\right], && \text{for } 0 \le h < a \\ &= A_0 + w(1 - 1/e), && \text{for } h \ge a, \end{aligned} \tag{6.4.9}$$

where $e = 0, 2.71828\ldots$, and w is a constant.

The spherical model can provide a better approximation:

$$\begin{aligned} \Gamma(h) &= A_0\partial(h) + \frac{w}{2}\left[3\frac{h}{a} - \left(\frac{h}{a}\right)^3\right], && \text{for } 0 \le h < a, \\ &= A_0 + w, && \text{for } h \ge a. \end{aligned} \tag{6.4.10}$$

The empirical semivariogram is defined, under isotropic conditions, as

$$\hat{\gamma}(h) = \frac{1}{2n}\sum_{i=1}^{n}[z_i(\mathbf{u}) - z_i(\mathbf{u} + \mathbf{h})]^2 \tag{6.4.11}$$

for n observations of $Z(\cdot)$ at two points separated by a distance h. Each observation may represent a cumulative total over a time period, say T. Likewise, spatial correlation between the two points is estimated from field measurements by

$$r(h) = \frac{\sum_{i=1}^{n}[z_i(\mathbf{u}) - \hat{m}(\mathbf{u})][z_i(\mathbf{u}+\mathbf{h}) - \hat{m}(\mathbf{u}+\mathbf{h})]}{\sum_{i=1}^{n}[z_i(\mathbf{u}) - \hat{m}(\mathbf{u})]^2 \sum_{i=1}^{n}[z_i(\mathbf{u}+\mathbf{h}) - \hat{m}(\mathbf{u}+\mathbf{h})]^2}, \tag{6.4.12}$$

where $\hat{m}(\mathbf{u})$ and $\hat{m}(\mathbf{u} + \mathbf{h})$ are the sample means of the observations at the two points.

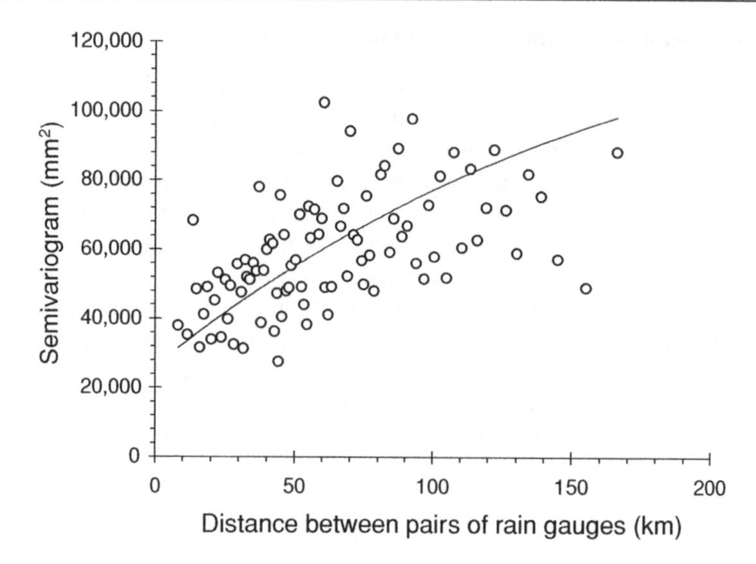

Fig. 6.4.1 Semivariogram versus distance for total annual rainfall from 43 rain gauges around Milan, Italy.

Example 6.21. Spatial correlation of rainfall data. A study was made of the spatial correlation in annual rainfall measured at 43 gauges in an area of approximately 150,000 km^2 in the Lombardia region, around Milan, in northern Italy. The data span a period of 58 years. The area is relatively flat. Figure 6.4.1 shows the semivariogram fitted by an exponential model.

Although some scatter is shown, as commonly found, the variability is constant throughout the measured region. The spatial correlation is shown in Fig. 6.4.2.

An exponential function is fitted to these points after excluding negative values. From the above pairs of stations, ten groups were formed depending on the proximities of the separating

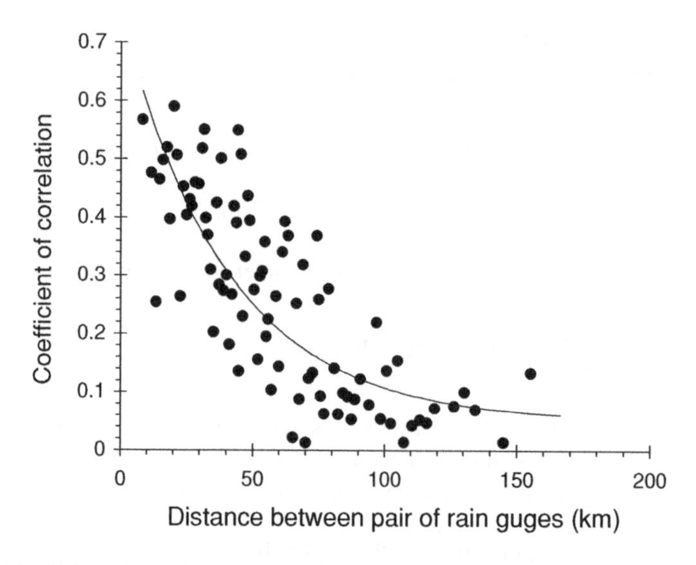

Fig. 6.4.2 Coefficient of correlation versus distance for annual total rainfall from 43 gauges around Milan, Italy.

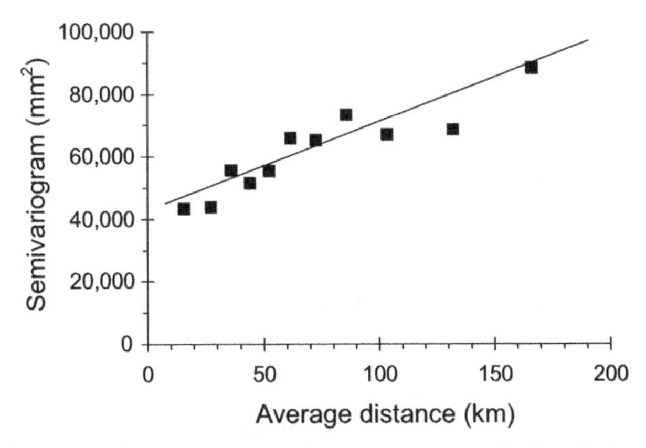

Fig. 6.4.3 Semivariogram for grouped pairs of rain gauges around Milan, Italy.

distances. Calculations for the semivariograms and spatial correlations were made on grouped data, taking mean distances between groups. The results are shown in Figs. 6.4.3 and 6.4.4. It is seen that the empirical semivariogram and spatial correlation functions for grouped data can be approximated by straight lines, which is expected from the nature of Figs. 6.4.1 and 6.4.2.

The study was extended to the Valtellina subcatchment of the Po basin, situated in the mountainous part of the Lombardia region in northern Italy. There were 34 items of data for each of 28 rain gauges in the area. These data were extreme values represented by seven consecutive days of rainfall during the highest flood per year for the period 1927–1967. As in the case of Figs. 6.4.3 and 6.4.4, divisions were made on the basis of distances apart. However, for these data sets, four further subdivisions were made with respect to the directions of the vectors joining the pairs of stations. The purpose was to verify whether the isotropy assumption holds. The results are shown in Figs. 6.4.5 and 6.4.6.

One can say that, considering the large fluctuations that empirical semivariograms generally show, the assumption of isotropy is not unreasonable.

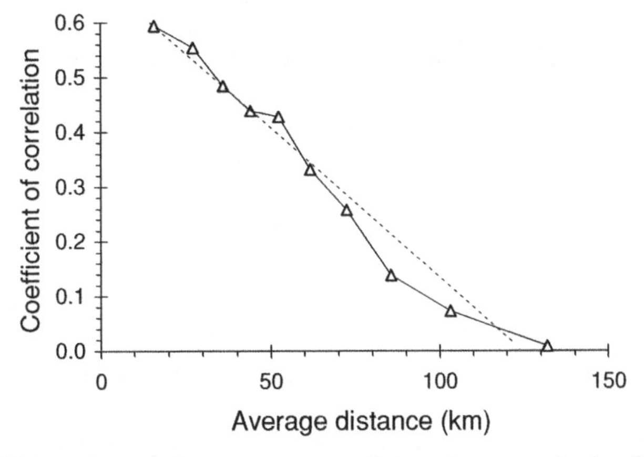

Fig. 6.4.4 Coefficient of correlation versus average distance for grouped pairs of rain gauges around Milan, Italy.

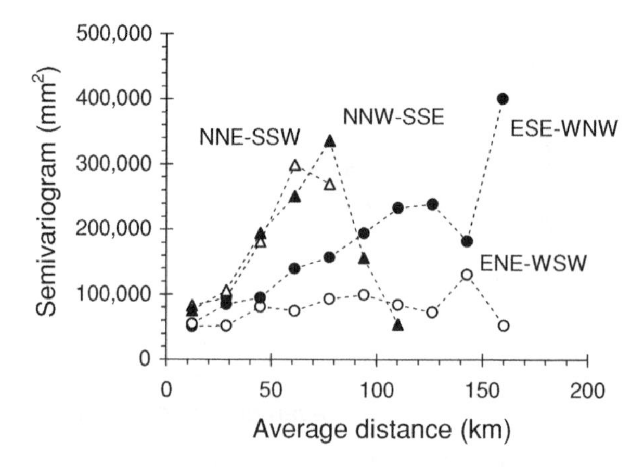

Fig. 6.4.5 Semivariogram versus average distance for grouped pairs of rain gauges from the Valtellina region of northern Italy; four different directions, ten groups of pairs.

6.4.4 Spatial interpolations and Kriging

The methods just described above can be extended to spatial interpolations where data are not available. Given, for instance, n realizations of the process $Z(\mathbf{u})$ at points $\mathbf{u}_i, i = 1, 2, \ldots, n$, we need to obtain a linear estimate $\hat{Z}(\mathbf{u}_k)$ at the point $\mathbf{u}_k$. The linear combination

$$\hat{Z}(\mathbf{u}_k) = \sum_{i=1}^{n} \lambda_i Z(\mathbf{u}_i) \tag{6.4.13}$$

is considered to be an optimum estimate if the coefficients $\lambda_i, i = 1, 2, \ldots, n$ are such that they sum to one and the estimator is unbiased and has minimum variance. This best linear unbiased estimator is given the acronym BLUE. Only data within the radius of influence are considered. Furthermore, it is a common practice to estimate the coefficients in Eq. (6.4.13) using the semivariogram after fitting a theoretical function, such as those specified

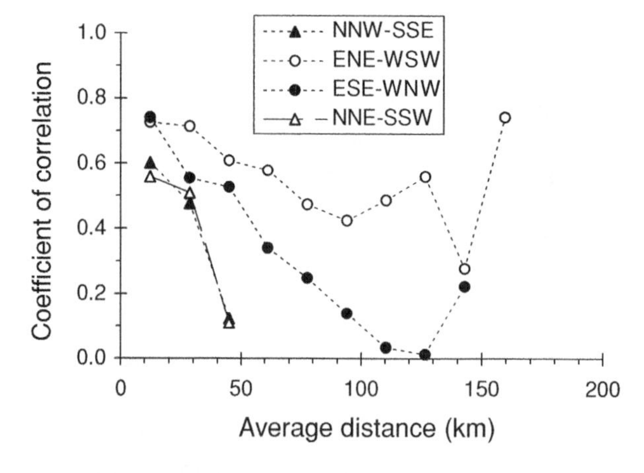

Fig. 6.4.6 Coefficient of correlation versus average distance for grouped pairs of rain gauges from the Valtellina region of northern Italy; four different directions, ten groups of pairs.

by Eqs. (6.4.8) to (6.4.10). The method is called *Kriging*, named after D. G. Krige who worked in the South African mining industry.[33]

The optimum solution is found as in the case of principal components (Subsection 6.3.1) by using the Lagrange multiplier. The results are summarized here. In an isotropic field with estimated semivariogram values $\hat{\gamma}(h_{ij})$ between points i and j at distances h_{ij}, the estimated weights, $\hat{\lambda}_j$, $j = 1, 2, \ldots, n$, are found by solving the following $n + 1$ simultaneous equations:

$$\sum_{j=1}^{n} \hat{\lambda}_i \hat{\gamma}(h_{ij}) + \lambda = \hat{\gamma}(h_{ik}), \qquad i = 1, 2, \ldots, n,$$

$$\sum_{j=1}^{n} \hat{\lambda}_i = 1, \tag{6.4.14}$$

where, clearly, $\hat{\gamma}(h_{ij}) = 0$ for $i = 1, 2, \ldots, n$; k denotes the point where there is no data; and λ is the Lagrange multiplier. One of the models described in Subsection 6.4.3, or a suitable alternative, is used for estimating the semivariogram at stated distances h_{ij}.

Example 6.22. Groundwater. The quality of groundwater is measured by an indicator at three wells, A, B, and C, which are defined in a two-dimensional field as follows:

	East 1 km	East 1.5 km
North 1 km	A	B
North 2 km	C	K

An estimate is required at point K.

The separation distances are calculated and are given in the following table with the water quality indicator Z as measured at the three wells:

	A	B	C	K	z (mg/L)
A	0	0.5	1.0	1.12	12
B	0.5	0	1.12	1.00	15
C	1.0	1.12	0	0.50	10
K	1.12	1.00	0.5	0	–

The semivariogram has a sill value of 4.5 (mg/L)^2 at a radius of influence of 2.0 km and a low point of 0.5 (mg/L)^2. Using this data, the following linear model [Eq. (6.4.8)] is fitted:

$$\hat{\gamma}(h) = 0.5 + 2h.$$

This gives the following semivariogram values in (mg/L)^2 for the respective pairs at distances:

	A	B	C	K
A	0	1.5	2.5	2.74
B	1.5	0	2.74	2.5
C	2.5	2.74	0	1.5
K	2.74	2.5	1.5	0

[33] See also Cressie (1991) and Journel and Huijbregts (1978) who give the original theory. Details of this procedure and theory are found in the work of Bacchi and Kottegoda (1995). Applications in groundwater and surface water hydrology are given by Virdee and Kottegoda (1984) and Kassim and Kottegoda (1991), respectively.

From Eq. (6.4.14) the following equations are obtained:

$$\lambda + 0\hat{\lambda}_1 + 1.5\hat{\lambda}_2 + 2.5\hat{\lambda}_3 = 2.74,$$
$$\lambda + 1.5\hat{\lambda}_1 + 0\hat{\lambda}_2 + 2.74\hat{\lambda}_3 = 2.5,$$
$$\lambda + 2.5\hat{\lambda}_1 + 2.74\hat{\lambda}_2 + 0\hat{\lambda}_3 = 1.5,$$

and

$$\hat{\lambda}_1 + \hat{\lambda}_2 + \hat{\lambda}_3 = 1.$$

The solutions are $\lambda = 0.611$, $\hat{\lambda}_1 = 0.030$, $\hat{\lambda}_2 = 0.297$, and $\hat{\lambda}_3 = 0.673$. Hence, at point K,

$$\hat{Z}_k = 0.030 \times 12 + 0.297 \times 15 + 0.673 \times 10 = 11.5 \text{ mg/L}.$$

The method of estimation described in this section is called *ordinary Kriging*, which is based on certain conditions. When the assumption of stationarity is not justifiable, the method of *universal Kriging* is commonly adopted. If there is a sloping surface, that is, a spatial trend, a local flat region or a similar type of nonstationarity, as seen in the scatter of data points, a transformation to stationarity is possible. A polynomial function is generally used to model the average values of the scatter points. The function is called the drift term. Kriging is then applied to the residuals as given by the differences between the drift term and observed scatter points. Estimated values are obtained from the sum of the interpolated residuals and the drift term.

6.4.5 Summary of Section 6.4

In this section we have discussed methods of estimating spatial association and disassociation. These are the correlation function and the semivariogram, respectively. Some examples have been provided. The scatter in the points is a common feature and this can be much more than in the scatter diagrams used in regression. Hence confidence intervals can be quite wide. It is usual practice to fit theoretical functions for purposes of applications. The semivariogram is commonly used for estimation purposes by the methods of Kriging.

6.5 SUMMARY OF CHAPTER 6

The focus of this chapter, to a large extent, is on methods of regression. Simple regression includes estimation and plotting methods, which are an essential part. Also included are testing of hypotheses and establishment of confidence limits. Linear multiple regression is covered in detail with many aspects of model testing, verification, and revalidation. In the methods of simple and multiple regression, we have emphasized the importance of graphical diagnostics which supplement statistical tests. The assumptions are given at each stage, and it is essential to verify whether these are met in order to validate the model. This is done with the aid of plots, analysis of variance, and numerous other tests. A particular feature in regression is the investigation of influential observations, such as points of high leverage, and outliers. Regarding the last aspect, the approach adopted here is necessarily different from that in probability plotting, covered in Chapter 5, because of the regression relationships. Ridge regression can be helpful, as shown, when estimates of parameters are unstable. However, as indicated, it is a subjective procedure; it should be used in special cases where there is no alternative and with an awareness of different solutions to the problem.

In multivariate analysis, we have highlighted principal component analysis which is apparently used more often than the other methods. However, direct interpretation of the results may sometimes be difficult. The associated procedure of factor analysis can be helpful as indicated. The methods are similar in many respects, and thus in our water quality example we applied factor rotation to principal components. In grouping methods, we have featured cluster analysis and the application of the dendogram. The usefulness of principal components as a guide to clustering is demonstrated.

The estimation of spatial association and disassociation measured by the correlation function and the semivariogram, respectively, are presented in the last section. The examples cited may seem to show some uncertainties in the plots as seen sometimes in the scatter diagrams of regression. However, methods of grouping are suggested and assumptions such as isotropy are verified graphically. The methods of Kriging are used for estimating missing values.

REFERENCES

General. The following references are given for further reading:

Alt, M. (1990). *Exploring Hyperspace—A Non-mathematical Explanation of Multivariate Analysis*, McGraw-Hill, New York. Written for absolute beginners.

Chatterjee, S., and B. Price (1991). *Regression Analysis by Example*, 2nd ed., John Wiley and Sons, New York. Readers familiar with regression will find this book useful.

Comrey, A. L., and H. B. Lee (1992). *A First Course in Factor Analysis*, 2nd ed., Lawrence Erbaum Associates, Hillsdale, NJ. Recommended reading for factor analysis.

Cook, R. D., and S. Weisberg (1994). *An Introduction to Repression Graphics*, John Wiley and Sons, New York. Well written book with computer software; does not require much training in statistics.

Draper, N. R., and H. Smith (1998). *Applied Regression Analysis*, 3rd. ed., John Wiley and Sons, New York. A standard reference for applications of regression analysis; includes ridge regression (Chapter 17), generalized linear models (Chapter 18), nonlinear regression (Chapter 24), and robust regression (Chapter 25).

Everitt, B., S. Landau, and M. Leese (2001). *Cluster Analysis*, 4th ed., Arnold, London. Popular reference on cluster analysis, includes Ward's method. Also gives details of software but says none are ideal.

Fruchter, B. (1954). *Introduction to Factor Analysis*, Van Nostrand-Reinhold, New York. Recommended as an easy introduction.

Haan, C. T. (1977). *Statistical Methods in Hydrology*, Iowa University Press, Ames, IA. Reference for applications of principal components and factor analysis.

Harman, H. H. (1976). *Modern Factor Analysis*, 3rd ed., University of Chicago Press, Chicago. A comprehensive treatment with a mathematical emphasis; reference for the varimax method.

Harris, R. J. (1985). *A Primer in Multivariate Statistics*, 2nd ed., Academic, New York. A useful, practical introduction with some math.

Helsel, D. R., and R. M. Hirsch (1992). *Statistical Methods in Water Resources*, Elsevier, New York. Reference book for exploratory data analysis in water resources engineering.

Jackson, J. E. (1991). *A User's Guide to Principal Components*, John Wiley and Sons, New York. Ample guidance (569 pages) for users of all levels.

Johnson, C., and D. Wichern (1998). *Applied Multivariate Statistical Analysis*, 4th ed., Prentice Hall, Englewood Cliffs, NJ. Reference for multivariate methods.

Kaufman, L., and P. J. Rousseau (1990). *Finding Groups in Data—An Introduction to Cluster Analysis*, John Wiley and Sons, New York. Mathematical level is not high. Ward's algorithm and others are included.

Krzanowski, W. J., and F. H. C. Marriott (1994). *Multivariate Analysis, Part I, Distributions, Ordination and Inference*, Edward Arnold, London. Comprehensive text on multivariate analysis includes outliers, transformations, and nonlinear principal components.

Manly, B. F. J. (1994). *Multivariate Statistical Methods—A Primer*, 2nd ed., Chapman and Hall, London. Very good introduction—a wide coverage of topics.

Mason, R. L., R. F. Gunst, and J. L. Hess (1989). *Statistical Design and Analysis of Experiments with Applications to Engineering and Science*, John Wiley and Sons, New York. See Section 26 on regression.

Ratkowsky, D. A. (1983). *Nonlinear Regression: A Unified Approach*, Marcel Dekker, New York. Nonlinear regression at an introductory level.

Rice, J. A. (1995). *Mathematical Statistics with Data Analysis*, 2nd. ed, Duxbury Press, Belmont, CA. Mathematical statistics at a comprehensible level.

Seber, G. A. F., and C. J. Wild (1989). *Nonlinear Regression*, John Wiley and Sons, New York. Nonlinear regression at an advanced level.

Sharma, S. (1996). *Applied Multivariate Techniques*, John Wiley and Sons, New York. See Chapters 8, 9, and 13, in particular.

Additional references quoted in text

Atkinson, A. C. (1981). "Two graphical displays for outlying and influential observations in regression," *Biometrika*, Vol. 68, pp. 13–20. Reference for outliers and influential observations.

Atkinson, A. C. (1985). *Plots, Transformations and Regression*, Clarendon Press, Oxford. Leverage measures p. 18, pp. 228–233 including discussion on Gauss-Newton algorithm, Chapters 6–9. Transformations in regression Chapters 6–9.

Bacchi, B., and N. T. Kottegoda (1995). "Identification and calibration of spatial correlation patterns of rainfall," *J. Hydrol.*, Vol. 165, pp. 311–348. Reference for Kriging.

Box, G. E. P. (1966). "Use and abuse of regression," *Technometrics*, Vol. 8, pp. 625–629. On regression.

Casella, G., and Berger, R. L. (2002). *Statistical Inference*, 2nd. ed., Duxbury, Pacific Grove, CA. Chi-squared distribution in regression, p. 554.

Chatterjee, S., and A. S. Hadi (1988). *Sensitivity Analysis on Linear Regression*, John Wiley and Sons, New York. Reference for plotting methods in regression and detection of influential observations; Cook's distance, pp. 118–119.

Cook, R. D. (1977). "Detection of influential observations in linear regression," *Technometrics*, Vol. 19, pp. 15–18. Reference for plotting methods in regression and detection of influential observations.

Cooper, J. C. B. (1983). "Factor analysis: An overview," *Am. Stat.*, Vol. 37, No. 2, pp. 141–147. Gives salient features without using complex mathematics.

Cressie, N. A. C. (1991). *Statistics for Spatial Data*, John Wiley and Sons, New York. See Chapters 2 and 3 for variogram and Kriging.

Dolby, J. L. (1963). "A quick method choosing a transformation," *Technometrics*, Vol. 5, pp. 317–325. On transformations.

Durbin, J., and G. S. Watson (1951). "Testing for serial correlation in least squares regression II," *Biometrika*, Vol. 38, pp. 159–178. Test for (only) first-order Markov relationship in residuals.

Dutter, R., and P. J. Huber (1981). "Numerical methods for the nonlinear robust regression problem," *J. Stat. Comput. Simul.*, Vol. 13, pp. 79–114. Robust regression.

Elian, S. N. (2000). "Simple forms of the best linear unbiased predictor in the general linear regression model, " *Am. Stat.*, Vol. 54, No. 1, pp. 25–48. Advanced reference.

Ghosh, S. (1987). "Note on a common error in regression diagnostics using residual plots," *Am. Stat.*, Vol. 41, No. 4, p. 338. Graphical dignostics in regression.

Goldberger, A. S. (1962). "Best linear unbiased prediction in the generalized linear regression model," *J. Am. Statist. Assoc.*, Vol. 57, No. 298, pp. 360–375. Advanced reference.

Graybill, F. A. (1983). *Matrices with Applications in Statistics*, 2nd ed., Wadsworth, Belmont, CA. More advanced than Strang (1980).

Hald, A. (1952). *Statistical Theory with Engineering Applications*, John Wiley and Sons, New York. Data for Problem 6.14.

Healy, M. J. E. (1984). "The use of R^2 as a measure of goodness of fit," *J. R. Stat. Soc.* A, Vol. 147, No. 4, pp. 608–609. Possible misuses of the R^2 statistic.

Hettmansperger, T. P. (1984). *Statistical Inference Based on Ranks*, John Wiley and Sons, New York. Nonparametric regression.

Hettmansperger, T. P., and S. J. Sheather (1992). "A cautionary note on the method of least median squares," *Am. Stat.*, Vol. 46, No. 2, pp. 79–83. Points to possible local instability in the LMS method.

Hoerl, A. E., and R. W. Kennard (1970*a*). "Ridge regression: biased regression for nonorthogonal problems," *Technometrics*, Vol. 12, pp. 55–67. Ridge regression.

Hoerl, A. E., and R. W. Kennard (1970*b*). "Ridge regression: applications to nonorthogonal problems," *Technometrics*, Vol. 12, pp. 69–82. Ridge regression.

Hougaard, P. (1988). "The asymptotic distribution of nonlinear regression parameter estimates: Improving the approximation," *Int. Stat. Rev.*, Vol. 56, p. 221. Nonlinear regression.

Huber, P. J. (1981). *Robust Statistics*, John Wiley and Sons, New York. Robust methods in regression, Chapter 7.

Journel, A. G., and C. J. Huijbregts (1978). *Mining Geostatistics*, Academic Press, New York. Reference for spatial statistics.

Kaiser, H. F. (1958). "The varimax criterion for analytic rotation in factor analysis," *Psychometrika*, Vol. 23, pp. 187–200. Algorithm for varimax rotation.

Kassim, A. H. M., and N. T. Kottegoda (1991). "Rainfall network design through comparative Kriging methods," *Hydrol. Sci. J.*, Vol. 36, No. 3, pp. 223–240. Reference for spatial interpolation in hydrology.

Kennedy, W. J., and J. E. Gentle (1980). *Statistical Computing*, Marcel Dekker, New York. Nonlinear least squares, Chapter 10.

Lawrance, A. J. (1995). "Deletion influence and masking in regression," *J. R. Stat. Soc. B*, Vol. 57, No. 1, pp. 181–189. Regression diagnostics with reference to Cook's distance and further work on the off-diagonal terms of the leverage matrix.

Marquardt, D. W., and R. D. Snee (1975). "Ridge regression in practice," *Am. Stat.*, Vol. 29, pp. 3–20. On ridge regression.

McCullagh, P., and J. Nelder (1989). *Generalized Linear Models*, 2nd ed., Chapman and Hall, London. The theory of generalized linear models.

Mendenhall, W., and T. Sincich (1995). *Statistics for Engineers and the Sciences*, Prentice Hall, Englewood Cliffs, NJ. Nonparametric regression, p. 957; Theil test for significance of slope parameter (distribution-free) p. 960.

Montgomery, A. M., and G. C. Runger (1994). *Applied Statistics and Probability for Engineers*, 5th ed., John Wiley and Sons, New York. Ridge regression, Section 10.13.

Mood, A. M., F. A. Graybill, and D. C. Boes (1974). *Introduction to the Theory of Statistics*, 3rd ed., McGraw-Hill, New York. Chi-squared distribution in regression (pp. 489–491).

Nicholson, W. K. (1990). *Linear Algebra with Applications*, 3rd ed., PWS Publ. Co., Boston. Textbook for matrices—an alternative to the book by Strang.

Press, S. J. (1972). *Applied Multivariate Analysis*, Holt, Rinehart and Winston, New York. On principal component analysis.

Rousseeuw, P. J., and A. M. Leroy (1987). *Robust Regression and Outlier Detection*, John Wiley and Sons, New York. Plotting methods in regression, detection of influential observations, and coping with unusual types of real data.

Ruppert, D., and R. J. Carroll (1980). "Trimmed least squares estimation in the linear model," *J. Am. Stat. Assoc.*, Vol. 75, pp. 828–838. Data for Problem 6.13.

Sparks, D. N. (1970). "Half-normal plotting Algorithm AS 30," *Appl. Stat.*, Vol. 19, pp. 192–196. Computer algorithm for half-normal plot.

Strang, S. (1980). *Linear Algebra and Its Applications*, 2nd ed., Academic Press, New York. Matrix algebra for regression.

Stuart, A., and J. K. Ord (1987). *Kendall's Advanced Theory of Statistics, Vol. 1: Distribution Theory*, 5th ed., Edward Arnold, London. See Section 16.33 for transformation of sample correlation coefficient.

Stuart, A., and J. K. Ord (1991). *Kendall's Advanced Theory of Statistics*, Vol. 2, 5th ed., Edward Arnold, London. Leverage measures p. 1080, pp. 1089–1090.

Stuart, A., and J. K. Ord (1994). *Kendall's Advanced Theory of Statistics, Vol. 2A: Classical Inference and Linear Model*, 6th ed., Edward Arnold, London. See Chapter 29 on the general linear model.

Takezawa, K. (2005). *Introduction to Nonparametric Regression*, John Wiley and Sons, New York. Comprehensive methods of nonparametric regression.

Tietjen, G. L., R. H. Moore, and R. J. Beckman (1973). "Testing for a single outlier in simple linear regression," *Technometrics*, Vol. 15, pp. 717–721. Pioneering work on outliers in regression.

Virdee, T. S., and N. T. Kottegoda (1984). "A brief review of Kriging and its application to optimal interpolation and observation well selection," *Hydrol. Sci. J.*, Vol. 29, No. 4, pp. 367–387. Reference for Kriging in groundwater.

Wackerly, D. D., W. Mendenhall, and R. L. Scheaffer (2002). *Mathematical Statistics with Applications*, 6th ed., Duxbury, Pacific Grove, CA. Covariance properties in regression, p. 546.

PROBLEMS

6.1. Dissolved oxygen. The following observations of dissolved oxygen (DO) were made with respect to time of travel downstream from a point of regulation in a river:

Time of travel (days)	0	0.6	1.1	1.7	1.9	2.4	2.8	3.3	3.7
DO (ppm)	0.39	0.37	0.31	0.28	0.27	0.25	0.20	0.17	0.16

Fit a linear regression, using DO as the response variable, and estimate the parameters. Calculate the coefficient of determination. Does a straight line provide a reasonable fit?

6.2. Population growth. A small city has doubled in population in 9 years. The following approximate counts have been made during the period:

Year	1	2	3	4	5
Population in 1000s	100	107	115	124	135
Year	6	7	8	9	10
Population in 1000s	146	158	171	185	200

Plot the data. Determine the sample correlation coefficient. Decide whether a linear model provides a good fit or whether there should be a transformation of the response variable (population). Give 95% confidence limits for the mean or expected population in year 15 if growth patterns do not change.

6.3. Water quality. For the water quality measurements on the River Ouse at Clapham, England, data given in Problem 1.15 (Chapter 1) determine the linear regression equation by least squares using phosphate as the explanatory variable. Comment on the model and the results.

6.4. Correlation of low flows. The lowest annual flows measured, in cubic meters per second, at stations X and Y on the Jackson and Cowpasture rivers, respectively, in the United States are to be correlated in order to extend the shorter record at station Y. A simple linear regression model is to be used. The following summary statistics have been computed over a 12-year period:

Sum $X = 28.77$; Sum $Y = 28.23$; Sum $XX = 73.14$; Sum $YY = 71.20$; Sum $XY =$ 71.53.

(*a*) Find least squares estimates of the parameters.
(*b*) What is the standard error of the residuals?
(*c*) Estimate the coefficient of correlation.
(*d*) Find approximate 95% confidence limits for the population correlation coefficient.

6.5. Extension of steel wires. Ten steel wires of diameter 0.5 mm and length 2.5 m were extended in a laboratory by applying vertical forces of varying magnitudes. Results are as follows:

Force (kg)	15	19	25	35	42	48	53	56	62	65
Increase in length (mm)	1.7	2.1	2.5	3.4	3.9	4.9	5.4	5.7	6.6	7.2

(*a*) Estimate the parameters of a simple linear regression model with force as the explanatory variable.
(*b*) Find 95% confidence limits for the two parameters.
(*c*) Test the hypothesis that the intercept is zero.
(*d*) What are the conclusions?

6.6. Rainfall-runoff relationship. Table E.7.2 gives 61 years of rainfall and runoff (see columns 6 and 7) at Pontelagoscuro, on the Po River, in northeast Italy. Fit a simple regression model. Test the hypothesis that the slope is zero. Comment on the results. Suggest methods of forming a multiple regression model and the inclusion of other measurements that can enhance the relationship.

6.7. Asbestos concentrations. M. J. Keifer, R. M. Buchan, T. J. Keefe, and K. D. Blehm (1987), "A predictive model for determining asbestos concentrations for fibres less than five millimeters in length," *Environ. Res.*, Vol. 43, pp. 31–38, give the following data for PCM (phase contrast microscopy) and SEM (scanning electron microscopy) concentrations:

Filter	PCM	SEM	Filter	PCM	SEM
1	3.14	7.79	16	0.41	1.86
2	2.61	6.85	17	0.77	2.90
3	3.03	7.60	18	1.63	4.92
4	4.03	9.29	19	3.99	9.22
5	7.82	14.8	20	2.94	7.44
6	5.61	11.72	21	1.02	3.54
7	4.23	9.61	22	1.67	5.00
8	0.62	2.49	23	6.33	12.76
9	1.09	3.71	24	2.38	6.42
10	0.73	2.79	25	1.93	5.34
11	0.70	2.71	26	6.29	12.70
12	7.92	14.94	27	3.77	8.86
13	2.50	6.64	28	4.50	10.04
14	1.91	5.50	29	4.54	12.10
15	4.98	10.78	30	0.48	2.08

Plot the data with PCM as the explanatory variable. Estimate the parameters of a simple linear regression model and show the straight line and the 95% confidence limits. Comment on the model. For a future value of PCM = 8.5, give the predicted value of SEM and the 95% prediction interval. Data used with permission from the Academic Press, Orlando, FL and the authors.

6.8. Alternative least squares. Rewrite the multiple regression model with two explanatory variables subtracting the sample means from each of the variables. Hence write equations for the matrix and parameters.

6.9. Road rutting. The rate of cutting of road ruts was measured with properties of asphalt and road materials in 31 experiments by J. W. Gorman and R. J. Toman (1966), "Selection of variables for fitting variables to data," *Technometrics*, Vol. 8, pp. 27–51. The following is a modified and reduced form of the equation with specified variables and residual sums of squares (RSS):

$$y = \beta_0 + \beta_1 x_1 + \beta_2 x_2 + \beta_3 x_3 + \beta_4 x_4 + \varepsilon,$$

where y = log (change of rut depth per million wheel passes)
x_1 = log (viscosity of asphalt)
x_2 = percent asphalt in surface course
x_3 = percent asphalt in base course
x_4 = percent fines in surface course

Explanatory variables used in equation	RSS
0	11.058
1	0.607
2	0.499
3	0.498
4	0.475

Determine the "best" form of equation to use. (Many more variables are used in the original work.)

6.10. Weighted least squares. The simple linear regression model

$$Y = \beta_0 + \beta_1 x + \varepsilon$$

is modified so that the variance of Y depends in the magnitude of the x as

$$\sigma^2(Y_i \mid x_i) = \sigma_i^2, \qquad i = 1, 2, \ldots, n.$$

Rewrite the least squares equations.

6.11. Trend in precipitation. Annual precipitation in millimeters at Saracay in the Puyango Basin, Ecuador, are given in the last column of Tables E.10.1. By using an appropriate regression equation, test the hypothesis that there is a decreasing trend in the precipitation.

6.12. Extending flow records. The River Oba in western Nigeria has been gauged near Imo and a 5-year record is available. Also, 60-year records are available for monthly rainfalls measured in the cities of Illorin and Ibadan with estimates of evaporation losses during the same period. It is proposed to extend the Imo flow record Q by correlating with the Illorin (current R, antecedent RA) and the Ibadan (current S)

residual rainfalls over the 5-year period of flow observations. A multiple regression model is to be used. The following statistics are provided for the four *monthly* variables (in Imperial units):

Variable	Mean	Standard deviation
R	1.58	2.73
RA	1.58	2.73
S	1.45	2.40
Q	55.75	111.68

The following are the sums of squares and cross-products of deviations from the mean:

	R	RA	S	Q
R	441.49	160.04	197.80	9,426.61
RA	160.04	441.49	176.21	13,005.22
S	197.80	176.21	340.16	10,642.07
Q	9,426.61	13,005.22	10,642.07	735,890.35

(1) Write the four normal equations from which the parameters are estimated.
(2) If the variable RA is not taken into account,

(*a*) estimate the parameters,
(*b*) estimate the standard error of estimate of Q from R and S, and
(*c*) estimate the coefficient of determination.

6.13. Salinity data. The following are part of the salinity data reported by D. Ruppert and R. J. Carroll (1980), "Trimmed least squares estimation in the linear model," *J. Am. Stat. Assoc.*, Vol. 75, pp. 828–838, for water during the spring season in Pamlico Sound, NC USA:

Index	1	2	3	4	5	6	7	8	9
Salinity	7.6	7.7	4.3	5.9	5.0	6.5	8.3	8.2	13.2
Lagged salinity	8.2	7.6	4.6	4.3	5.9	5.0	6.5	8.3	10.1
Discharge	23.01	23.87	26.42	24.87	29.90	24.20	23.22	21.86	22.27
Trend	4	5	0	1	2	3	4	5	0

Index	10	11	12	13	14	15	16	17	18
Salinity	12.6	10.4	10.8	13.1	12.3	10.4	10.5	7.7	9.5
Lagged salinity	13.2	12.6	10.4	10.8	13.1	13.3	10.4	10.5	7.7
Discharge	23.83	25.14	22.43	21.79	22.38	23.93	33.49	24.86	22.69
Trend	1	2	3	4	5	0	1	2	3

Index	19	20	21	22	23	24	25	26	27	28
Salinity	12.0	12.6	13.6	14.1	13.5	11.5	12.0	13.0	14.1	15.1
Lagged salinity	10.0	12.0	12.1	13.6	15.0	13.5	11.5	12.0	13.0	14.4
Discharge	21.79	22.04	21.03	21.01	25.87	26.29	22.93	21.31	20.77	21.39
Trend	0	1	4	4	0	1	2	3	4	5

The biweekly average salinity is given in milligrams per liter with salinity lagged 2 weeks, discharge in cubic millimeters per second, and trend as a dummy variable for the time period.

(*a*) Estimate the parameters for a linear model with salinity as the response variable.
(*b*) Determine the "best" form of model.
(*c*) Determine Cook's distances and leverage measures.
(*d*) Test for any outliers.
(*e*) Comment on the foregoing results (*c* and *d*).

Data used with permission from the *Journal of the American Statistical Association*,

6.14. Hald cement data. The following is a part of the data reported by A. Hald (1952), *Statistical Theory with Engineering Applications*, John Wiley and Sons, New York, p. 647, for the heat-generated H in calories per gram, during hardening, for a type of cement as a function of four additives. The table gives H and four additives A_1, A_2, A_3, and A_4.

Item	H	A_1	A_2	A_3	A_4
1	78.5	7	26	6	60
2	74.3	1	29	15	52
3	104.3	11	56	8	20
4	87.6	11	31	8	47
5	95.9	7	52	6	33
6	109.2	11	55	9	22
7	102.7	3	71	17	6
8	72.5	1	31	22	44
9	93.1	2	54	18	22
10	115.9	21	47	4	26
11	83.8	1	40	23	34
12	113.3	11	66	9	12
13	109.4	10	68	8	12

Complete a ridge analysis and write a predictive equation for H.
(Data used with the kind permission of the author.)

6.15. Principal components. The following covariance matrix $\mathbf{C}$ was computed in a study of catchment characteristics of a river basin:

$$\begin{bmatrix} 3.67 & -4.93 & 2.08 \\ -4.93 & 98.1 & -3.01 \\ 2.08 & -3.01 & 2.01 \end{bmatrix}.$$

Determine the eigenvalues and the eigenvectors, and comment on the results.

6.16. Nitrates in river. For the data given in Table E.6.2, where water quality is given at 1-km intervals, draw a semivariogram of the nitrate values, $h = 1, 2, \ldots, 30$. What type of model is suggested?

6.17. Salinity of groundwater. The following salinity observations were recorded in 25 wells in a coastal aquifer, in milligrams per liter. The wells are spaced at distances of approximately 1 km in the NS and EW directions.

10.5	9.3	10.4	9.1	10.0
9.2	10.1	11.1	10.2	10.3
11.2	10.8	10.2	11.5	11.5
10.9	9.5	11.5	11.0	12.0
11.1	10.5	11.0	10.7	12.5

(a) Determine the semivariogram under conditions of isotropy.
(b) Determine the semivariograms in the NE-SW and NW-SE directions.
(c) Comment on the results.

6.18. Linear semivariogram model. The following data are used for the semivariogram of the grouped data of Fig. 6.4.3. Fit a linear model by regression.

Item	km	Semivariogram
1	15.8	43,423.3
2	27.2	43,886.0
3	36.0	55,604.7
4	44.0	51,576.5
5	52.2	55,397.1
6	61.6	65,899.6
7	72.4	65,089.6
8	85.6	73,292.7
9	103.2	67,044.6
10	131.9	68,610.7
11	166.1	88,218.9

6.19. Exponential semivariogram model. The following data are used for the semivariogram of the annual rainfall data of Fig. 6.4.1. Fit an exponential model stating any assumptions made. If one takes account of the results from Example 6.21, what can be said about the properties of the annual rainfall in the region?

Item	km	Semivariogram	Item	km	Semivariogram
1	8.3	37,851.9	47	58.6	64,391.7
2	11.7	35,326.8	48	59.8	68,915.5
3	13.6	68,326.4	49	60.4	102,468.4
4	14.9	48,446.3	50	61.0	49,045.3
5	16.1	31,541.6	51	62.1	41,186.5
6	17.5	41,053.2	52	63.3	49,151.1
7	18.7	49,047.8	53	65.2	79,739.4
8	20.2	33,951.9	54	66.5	66,704.0
9	21.4	45,263.4	55	67.6	71,922.7
10	22.5	53,088.1	56	69.0	52,255.0
11	23.8	34,547.4	57	70.0	94,147.4
12	25.1	51,042.2	58	71.1	64,181.3
13	26.1	39,733.1	59	72.6	62,752.7
14	27.0	49,519.7	60	74.0	56,775.6
15	28.2	32,489.5	61	74.9	49,858.3
16	29.5	55,755.0	62	75.6	75,493.2

Item	km	Semivariogram	Item	km	Semivariogram
17	30.8	47,523.9	63	77.0	58,419.7
18	31.6	31,275.0	64	78.6	48,015.3
19	32.2	56,937.5	65	80.8	81,717.9
20	32.9	51,891.9	66	82.4	84,134.3
21	33.9	51,271.0	67	84.2	59,251.8
22	35.0	56,113.3	68	85.8	68,939.1
23	36.2	53,660.4	69	87.3	89,176.8
24	37.1	77,947.4	70	88.6	63,794.1
25	37.9	38,722.0	71	90.6	66,877.4
26	38.8	53,830.4	72	92.3	97,726.8
27	39.9	60,068.5	73	94.1	56,009.0
28	41.0	62,769.7	74	96.9	51,560.0
29	41.9	61,731.6	75	98.4	72,857.7
30	42.7	36,311.1	76	100.6	58,023.1
31	43.6	47,268.9	77	102.4	81,231.9
32	44.1	27,706.2	78	104.9	51,767.3
33	44.6	75,630.6	79	107.5	88,162.3
34	45.3	40,477.0	80	110.6	60,517.2
35	46.1	64,279.0	81	113.5	83,272.4
36	47.0	48,014.5	82	116.1	62,780.6
37	47.9	48,837.3	83	119.1	72,207.9
38	48.7	55,357.7	84	122.0	88,796.7
39	50.3	56,927.1	85	126.3	71,400.8
40	51.8	70,026.0	86	130.2	58,951.1
41	52.5	49,162.2	87	134.3	81,787.8
42	53.4	43,975.7	88	138.9	75,341.3
43	54.4	38,393.8	89	144.9	57,169.7
44	55.0	72,487.2	90	155.1	49,060.9
45	55.7	63,407.0	91	166.1	88,218.9
46	57.1	71,495.0			

6.20. Exponential model. Repeat the Kriging analysis of the groundwater quality example of Example 6.22 replacing the linear model by an exponential model stating any assumption made. Estimate the value at K. Comment on the models and the results.

Chapter 7

Frequency Analysis of Extreme Events

Civil and environmental engineers are often concerned with natural hazards. Extreme events, such as floods, hurricanes, and earthquakes, can take many human lives and cause billions of dollars in damages. Paradoxically, one also needs to address the consequences of unusually low streamflows, which can result in high pollutant concentrations. Additionally, engineers must design buildings to withstand high winds and maritime structures to cater for abnormally high sea waves. The survival of a given system depends on its capability to resist those extreme conditions it can be subject to—and not simply the typical values. Therefore, the management and design of civil and environmental systems should account for the likelihood of rare events.

Given an adequately large sample of river flows, pollutant loadings, sea levels, earthquake intensities, or wind velocities, for instance, a probability distribution could be determined for a particular site to some level of precision. However, the availability of data of extreme values is usually restricted to less than 100 observations. Thus, one is unable to estimate directly the frequencies and magnitudes of extreme events, such as large floods or destructive earthquakes, the design values of which may have exceedance probabilities of less than 0.001. Nevertheless, an engineer is expected to use his or her practical knowledge of the processes involved to obtain the best estimate of the risks involved.

Engineers are aware of the uncertainties associated with the probability distributions of natural phenomena or those involving human beings. Even if one knows the parent distribution, in a hypothetical case, its functional representation remains a problem. One can select an appropriate probability model in a practical situation, considering the data available, to describe the phenomenon of interest, and then estimate the parameters and assess the risks involved. But how does one cope with the nonstationarities that often prevail and the associated outliers discussed in Chapters 1, 5, and 6?

The largest value of an event, such as a flood, tornado, or earthquake, can be critical to a given system, as already mentioned. The capacity of a system may likewise depend on extremes, such as the strength of the weakest of many elementary components. A failure may occur when a supply system is incapable of meeting the target demand as a result of low availability of the supplied resource or because the quality of this resource is insufficient to meet a standard. Processes that produce low rainfalls, river flows, lake levels, or groundwater levels involve variables of which the smallest value in a sequence may be critical. Therefore, the search for the frequency distribution of extreme events must also include the distribution of the smallest values.

The first section of this chapter deals with the statistics that represent the frequencies of the largest or smallest random variables. In the second section, probability models commonly used to represent extreme events are presented and discussed. Engineering applications of extreme-value analysis are introduced in the third section.

7.1 ORDER STATISTICS

In Section 1.2, some numerical summaries of data were introduced. Statistics of random variables were discussed in Section 3.2; it was noted therein that the sample moments are physically meaningful statistics. The concept of *order statistics* was introduced in Subsection 5.8.1, and some related properties are studied in this section. These order statistics play an important role in statistical inference and are linked to population quantiles as sample moments are to population moments. Moreover, order statistics which represent data often contain a meaningful experimental content.

7.1.1 Definitions and distributions

Consider a random sample of size n of a variate X with known distribution $F_X(x)$. The sample can be viewed as a sequence of random variables, $X_1, X_2, \ldots, X_n$. If these variates are arranged in order of increasing magnitude, a new sequence is obtained, say, $X_{(1)}, X_{(2)}, \ldots, X_{(n)}$, where $X_{(1)} = \min(X_1, X_2, \ldots, X_n)$, and $X_{(n)} = \max(X_1, X_2, \ldots, X_n)$. Then $X_{(1)}, X_{(2)}, \ldots, X_{(n)}$ are defined as *order statistics*. It must be noted that the order statistics are clearly not independent, for if $X_{(i)} \geq x$, then clearly $X_{(i+1)} \geq x$. In this section, we seek the marginal and joint distribution and some functions of order statistics.

Definition: Order statistics. Let $X_1, X_2, \ldots, X_n$ denote a random sample of size n from a cumulative distribution function $F_X(x)$. Then $X_{(1)} \leq X_{(2)}, \ldots \leq X_{(n)}$—where the $X_{(i)}$s are the X_is arranged in ascending order—are the order statistics corresponding to that random sample. The $X_{(i)}$s are statistics because they are functions of the random sample and are in a stated order.

For given x, let $L_i = I_{(-\infty,x]}(X_i)$; that is, $L_i = 1$ if $X_i \leq x$ and 0 otherwise. Thus, the random variable $Z = \sum_{i=1}^{n} L_i$ represents the number of variates X_i not exceeding x, and Z has a binomial distribution with parameters n and $F_X(x)$. There is equivalence between the two events $\{X_{(k)} \leq x\}$ and $\{Z \geq k\}$; that is, if the kth-order statistic is less than or equal to x, then the number of X_i less than or equal to x is greater than or equal to k, and the converse follows. Therefore, the marginal cumulative distribution of an arbitrary order statistic, say, $X_{(k)}$, with $k = 1, \ldots, n$, is given by

$$F_{X_{(k)}}(x) = \Pr[X_{(k)} \leq x] = \Pr[Z \geq k] = \sum_{j=k}^{n} \binom{n}{j} [F_X(x)]^j [1 - F_X(x)]^{n-j}. \quad (7.1.1)$$

This result provides the marginal distribution for many applications, such as the maximum, $X_{\max} = X_{(n)}$, and the minimum, $X_{\min} = X_{(1)}$, of a random sample $X_1, X_2, \ldots, X_n$. These are

$$F_{X_{(n)}}(x) = \sum_{j=n}^{n} \binom{n}{j} [F_X(x)]^j [1 - F_X(x)]^{n-j} = [F(x)]^n, \quad (7.1.2)$$

and

$$F_{X_{(1)}}(x) = \sum_{j=1}^{n} \binom{n}{j} [F_X(x)]^j [1 - F_X(x)]^{n-j} = 1 - [1 - F_X(x)]^n, \qquad (7.1.3)$$

respectively.[1]

Example 7.1. Urban storm drainage. Following an ancient practice, the design of storm drains in Italy is sometimes based on the third annual maximum value of rainfall intensity recorded in a standard period of n years. For instance, the outlet channel of the municipal sewer system of a town in southern Italy was designed to drain a storm of average intensity 38 mm/h lasting 35 minutes. This was the third highest value recorded over a period of 25 years. Further data analysis indicated that annual maximum storm intensity at that site, X, is exponentially distributed with mean 20 mm/h for this duration. The engineer is interested in estimating the probability that the third highest storm in a period of 25 years exceeds the selected design value. That is, the cdf of $X_{(25-3+1)} = X_{(23)}$ is required. From Eq. (7.1.1),

$$\begin{aligned} F_{X_{(23)}}(x) &= \sum_{j=23}^{25} \binom{25}{j} [F_X(x)]^j [1 - F_X(x)]^{25-j} \\ &= \sum_{j=23}^{25} \binom{25}{j} [1 - e^{-0.05x}]^j [e^{-0.05x}]^{25-j}, \end{aligned}$$

so that,

$$F_{X_{(23)}}(38) = 300 \times 0.024 \times 0.022 + 25 \times 0.020 \times 0.150 + 1 \times 0.017 \times 1 = 0.256,$$

and

$$\Pr[X_{(23)} > 38] = 1 - F_{X_{(23)}}(38) = 1 - 0.256 = .744$$

is the required probability.

If one assumes that the X_i of the random sample $X_1, X_2, \ldots, X_n$ are continuous and come from a pdf $f_X(\cdot)$, the pdf of $X_{(k)}$ is obtained by differentiating the right-hand side of Eq. (7.1.1). This yields

$$f_{X_{(k)}}(x) = \frac{n!}{(k-1)!(n-k)!} f_X(x)[F_X(x)]^{k-1}[1 - F_X(x)]^{n-k}. \qquad (7.1.4)$$

The marginal density of the maximum, $X_{\max} = X_{(n)}$, and the minimum, $X_{\min} = X_{(1)}$, of a random sample $X_1, X_2, \ldots, X_n$ are thus given by

$$f_{X_{\max}}(x) = n f_X(x)[F_X(x)]^{n-1} \quad \text{and} \quad f_{X_{\min}}(x) = n f_X(x)[1 - F_X(x)]^{n-1},$$

respectively. Similarly, one can determine the joint density of $X_{(k)}$ and $X_{(h)}$ for $1 \le k < h \le n$. This pdf is given by

$$f_{X_{(k)},X_{(h)}}(y, x) = \frac{n!\,[F_X(y)]^{k-1}[F_X(x) - F_X(y)]^{h-k-1}[1 - F_X(x)]^{n-h}}{(k-1)!(h-k-1)!(n-h)!} f_X(y)\, f_X(x) \qquad (7.1.5)$$

[1] Note that Eqs. (7.1.2) and (7.1.3) are also determined in Chapter 3 by using the concept of a function of random variables. See Eqs. (3.4.12) and (3.4.15).

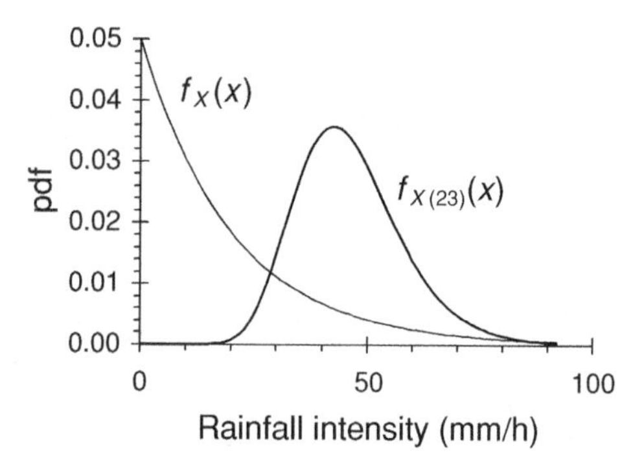

Fig. 7.1.1 Pdfs of annual maximum 35-minute rainfall intensity and the third highest annual maximum in 25 years.

for $y < x$, and, for $y \geq x$, it is zero.[2] In general, the joint pdf of order statistics, $X_{(1)}, X_{(2)}, \ldots, X_{(n)}$, can be written as

$$f_{X_{(1)},\ldots,X_{(n)}}(x_1, \ldots, x_n) = \begin{cases} n!\ f_X(x_1) \cdots f_X(x_n), & \text{for } x_1 < x_2 < \cdots < x_n, \\ 0, & \text{otherwise.} \end{cases} \quad (7.1.6)$$

Any set of marginal densities can be obtained from Eq. (7.1.6) by integrating out the unwanted variables.

Example 7.2. Urban storm drainage. Consider again the third maximum $X_{(23)}$ in 25 years for maximum annual 35-minute average rainfall intensity X of Example 7.1. From Eq. (7.1.4),

$$\begin{aligned} f_{X_{(23)}} &= \frac{25!}{(23-1)!(25-23)!} 0.05e^{-0.05x}[1-e^{-0.05x}]^{23-1}[e^{-0.05x}]^{25-23}, \\ &= 345\, e^{-0.15x}(1-e^{-0.05x})^{22}, \end{aligned}$$

which is shown in Fig. 7.1.1.

During the early years of civil engineering practice, the design variable was often based on the most extreme event among past observations. For example, a spillway designed to pass a flood 50–100% larger than the largest recorded flood in a period of, say, 25–50 years, was considered adequate. Then one should evaluate the probability that such a design will successfully meet its intended function in the future. Let $X_{(n)}$ denote the largest among n subsequent outcomes of a variable, and $X'_{(m)}$ the largest of m previous observations of this variable, say, $X'_1, \ldots, X'_m$. Because

$$\Pr\left[X_{(n)} \leq X'_{(m)}\right] = \int_{z=0}^{\infty} \Pr\left[X_{(n)} \leq z\right] \Pr\left[z < X'_{(m)} \leq z + dz\right],$$

it follows that

$$\Pr\left[X_{(n)} \leq X'_{(m)}\right] = \int_0^{\infty} [F_X(z)]^n m [F_X(z)]^{m-1} f_X(z) dz = m \int_0^{\infty} [F_X(z)]^{n+m-1} dF_X(z).$$

Hence,

$$\Pr\left[X_{(n)} \leq X'_{(m)}\right] = \frac{m}{n+m}. \quad (7.1.7)$$

[2] See Mood et al. (1974, p. 253) and David and Nagaraja (2003, Section 2.2).

The complementary probability, that is, the previously observed largest value will be exceeded in n subsequent observations is given by

$$\Pr\left[X_{(n)} > X'_{(m)}\right] = \frac{n}{n+m}. \tag{7.1.8}$$

Note that as n tends to infinity, the ratio $n/(n+m)$ tends to unity. This shows that in the absence of physical constraints, the probability of exceeding a previously observed largest value approaches 1 if the period of observation is sufficiently long.

This approach can be extended to the more general problem of determining the probability that the kth value $X'_{(k)}$ (counted from the lowest value) will be exceeded r times in n future observations. Denoting this probability by $p_{r|m,k,n}$, one obtains

$$p_{r|m,k,n} = \frac{k}{n+k-r}\frac{\binom{m}{k}\binom{n}{r}}{\binom{n+m}{n+k-r}}. \tag{7.1.9}$$

For example,

$$p_{0|m,m,n} = \frac{m}{n+m},$$

which is equivalent to Eq. (7.1.7). [This is the probability that a previously observed largest value, in m observations, will not be exceeded in n future observations—as obtained by substituting 0 for r and m for k in Eq. (7.1.9).]

Example 7.3. Flood control design. Suppose that a flood control structure is designed to accommodate the largest flood observed over the last 66 years. The engineer must determine the lifetime of this structure for a survival probability of 90%. From Eq. (7.1.7),

$$\Pr\left[X_{(n)} \le X'_{(66)}\right] = \frac{66}{n+66} = .9.$$

Hence $n = (66 - 0.9 \times 66)/0.9 \approx 7$ years.

Equation (7.1.9) also gives the probability that r among n future observations are smaller than the kth smallest value among m previous observations (that is, when k is counted from the top). In this case, Eq. (7.1.8) is also the probability that in n future observations there will be at least one value that is smaller than the previously observed smallest value.

7.1.2 Functions of order statistics

In many engineering applications some functions of order statistics are required. The sample mean, the sample median, and the sample range can provide meaningful information of system behavior in several practical problems.

As defined in Subsection 1.2.1, the sample mean is $1/n \sum_{i=1}^{n} X_i$. The *sample median* is defined as the middle-order statistic if n is odd and the average of the middle two-order statistics if n is even. The *range* is defined as $X_{(n)} - X_{(1)}$.

If the sample size is odd, then the pdf of the sample median is obtained from Eq. (7.1.4); for example, we can let $n = 2l + 1$, where l is some positive integer, and then compute by using Eq. (7.1.4) the density of $X_{(l+1)}$, the sample median. If the sample size is even, say, $n = 2l$, then the sample median is $(X_{(l)} + X_{(l+1)})/2$, the distribution of which can be obtained by a transformation, starting with the joint density of $X_{(l)}$ and $X_{(l+1)}$ which is obtained from Eq. (7.1.5).

The distribution of $R_n = X_{(n)} - X_{(1)}$, the sample range, can be derived by observing that Eq. (7.1.5) yields the joint density of $X_{(1)}$ and $X_{(n)}$ given by

$$f_{X_{(1)},X_{(n)}}(y,x) = \frac{n![F_X(x) - F_X(y)]^{n-2}}{(n-2)!} f_X(y) f_X(x),$$

with $y < x$. Therefore, one obtains the density of R_n as the difference between two random variables.[3] This can be done by putting $r = x - y$ and then integrating; that is,

$$f_{R_n}(r) = \int_{-\infty}^{+\infty} f_{X_{(1)},X_{(n)}}(x - r, x)dx$$

$$= \int_{-\infty}^{+\infty} \frac{n![F_X(x) - F_X(x-r)]^{n-2}}{(n-2)!} f_X(x-r) f_X(x)dx.$$

Hence,

$$f_{R_n}(r) = n(n-1) \int_{-\infty}^{+\infty} f_X(y)[F_X(y+r) - F_X(y)]^{n-2} f_X(y+r)dy.$$

The cdf of R_n can be found by integrating the above density over the appropriate range $R_n \geq 0$; that is,

$$F_{R_n}(r) = n \int_{-\infty}^{+\infty} f_X(y)[F_X(y+r) - F_X(y)]^{n-1} dy. \tag{7.1.10}$$

This result may also be obtained by noting that

$$n f_X(y)dy[F_X(y+r) - F_X(y)]^{n-1}$$

is the binomial probability conditional to y; that is, the probability that one of the X_is is in the interval $(y, y + dy)$ and all of the $n - 1$ remaining X_i are in $(y, y + r)$. Equation (7.1.10) has several applications, such as sizing and operation of reservoirs.

Example 7.4. Tank capacity. It is observed that the hydraulic head X in a tank can vary uniformly each day from $a = 0$ to $b = 5$ units, that is, from empty to full capacity of the tank. For a uniform(a, b) distribution, one obtains from Eq. (7.1.10) after integration (see, for some guidance, Wilks, 1962, p. 236)

$$F_{R_n}(r) = n\left(\frac{r}{b-a}\right)^{n-1} - (n-1)\left(\frac{r}{b-a}\right)^{n-1}.$$

Hence for a uniform(0, 5) distribution,

$$F_{R_n}(r) = n\left(\frac{r}{5}\right)^{n-1} - (n-1)\left(\frac{r}{5}\right)^{n-1}.$$

This is plotted in Fig. 7.1.2 for different values of n.

It is seen that for large n, say, greater than 100, the cdf of the range tends to be a step function of the upper bound, which is the hydraulic head corresponding to full capacity of the tank.

[3] See Section 3.4.

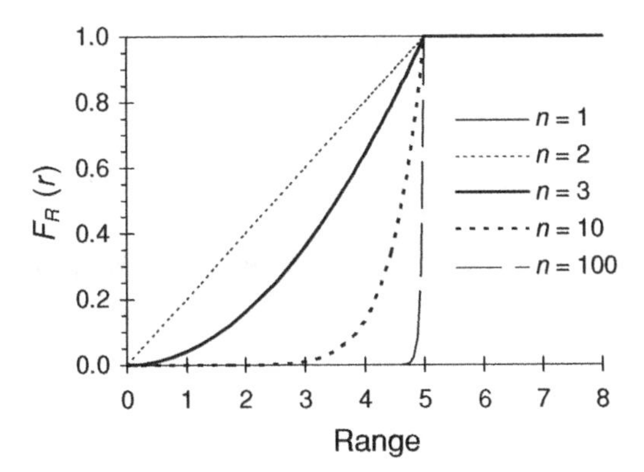

Fig. 7.1.2 Cdfs of sample range for increasing number of sampling variates with common uniform distribution.

Certain functions of order statistics are also statistics and may be used to make statistical inferences. For example, both the sample median and the midrange, $(X_{(1)} + X_{(n)})/2$, can be used to estimate the mean of the population.

7.1.3 Expected value and variance of order statistics

In certain engineering applications the expected value and the variance can provide useful information of system behavior. If $f_X(\cdot)$ is continuous, the means or expected values of Y_k, the kth-order statistics, can be determined from

$$E\left[X_{(k)}\right] = \int_{-\infty}^{+\infty} x f_{X_{(k)}}(x)dx = n\binom{n-1}{k-1}\int_{-\infty}^{+\infty} y[F_X(y)]^{k-1}[1 - F(x)]^{n-k}dF_X(y), \tag{7.1.11}$$

which is obtained by substituting Eq. (7.1.4) for the marginal density of the kth-order statistic.

Example 7.5. Expected frequency of observations. The relative frequencies of a set of random variables arranged in ascending order can be interpreted as a set of random uniform (0, 1) variates. From Eq. (7.1.11),

$$E[X_{(k)}] = n\binom{n-1}{k-1}\int_{-\infty}^{+\infty} xx^{k-1}(1-x)^{n-k}dx = \frac{n\binom{n-1}{k-1}}{(n+1)\binom{n}{k}} = \frac{k}{n+1}$$

is the expectation of the frequency of the kth-ordered variable. This result implies that the order statistics divide the area under the curve $f_X(\ )$ into $n+1$ parts, each with expected value $1/(n+1)$, where $f_X(x)$ is standard uniform.[4] The above sampling frequency is often referred to as *Weibull plotting position* (which refers to the probability at which each $x_{(i)}$ should be plotted in a graph of x versus $\Pr[X > x]$).[5] Note that the corresponding return period is $(n+1)/(n+1-k)$.

[4] This is in view of the probability integral transformation discussed in Chapter 8, on simulation.

[5] Plotting positions are discussed in Chapter 5.

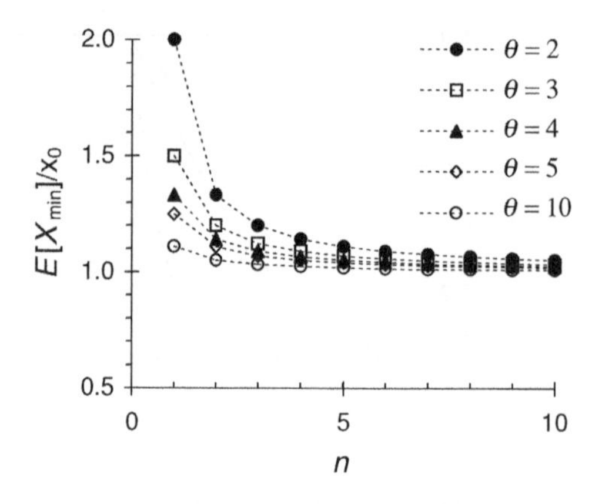

Fig. 7.1.3 Decrease of dimensionless expected minimum delay with the number of daily arrivals.

The expectation of the maximum, $X_{\max} \equiv X_{(n)}$, of a random sample $X_1, X_2, \ldots, X_n$ can be found by substituting n for k in Eq. (7.1.11), thus obtaining

$$E[X_{\max}] = n \int_{-\infty}^{+\infty} x[F_X(x)]^{n-1} f_X(x)dx. \tag{7.1.12}$$

Similarly, for the minimum, $X_{\min} \equiv X_{(1)}$, one gets

$$E[X_{\min}] = n \int_{-\infty}^{+\infty} x[1 - F_X(x)]^{n-1} f_X(x)dx. \tag{7.1.13}$$

Example 7.6. Minimum flight delay. An airport is designed to receive a given daily number of flights, say, n. Let the interarrival time between two successive flights be Pareto distributed with parameter $\theta > 1$, and lower bound x_0. Thus,

$$F_X(x) = 1 - \left(\frac{x_0}{x}\right)^{\theta}.$$

The expected minimum delay is given by Eq. (7.1.13); that is,

$$E[X_{\min}] = n \int_{x_0}^{+\infty} x \left[1 - 1 + \left(\frac{x_0}{x}\right)^{\theta}\right]^{n-1} \frac{\theta}{x_0} \left(\frac{x_0}{x}\right)^{\theta+1} dx$$

$$= n \int_{x_0}^{+\infty} \theta \left(\frac{x}{x_0}\right)^{1-\theta(n-1)-\theta-1} dx = \frac{n\theta}{n\theta - 1} x_0.$$

This is shown in Fig. 7.1.3, where $E[X_{\min}]/x_0$ is plotted against n for different values of θ.

It can be observed that increasing the number of daily arrivals, n, yields the expected delay that defines the lower bound.

The variance of $X_{(k)}$, the kth-order statistic, can be determined from

$$\text{Var}[X_{(k)}] = \int_{-\infty}^{+\infty} \left(x - E\left[X_{(k)}\right]\right)^2 f_{X_{(k)}}(x)\, dx, \tag{7.1.14}$$

and, for $k < h$, the lag$(h-k)$ covariance from

$$\text{Cov}\left[X_{(k)}, X_{(h)}\right] = \int_{-\infty}^{+\infty}\int_{-\infty}^{+\infty} \left(u - E\left[X_{(k)}\right]\right)\left(v - E\left[X_{(h)}\right]\right) f_{X_k,X_h}(u,v)\,du\,dv, \quad (7.1.15)$$

which can be computed by substituting Eqs. (7.1.4) and (7.1.5) for the marginal and the joint density of the kth-order statistics, respectively.

The variance of the maximum, $X_{\max} \equiv X_{(n)}$, of a random sample, $X_1, X_2, \ldots, X_n$, can be found by substituting n for k in Eq. (7.3.17), thus obtaining

$$\text{Var}[X_{\max}] = \int_{-\infty}^{+\infty} \left(x - n \int_{-\infty}^{+\infty} u[F_X(u)]^{n-1} f_X(u)du \right)^2 f_X(x)n[F_X(x)]^{n-1}\,dx. \quad (7.1.16)$$

Similarly, for the minimum, $X_{\min} \equiv X_{(1)}$, one gets

$$\text{Var}[X_{\min}] = \int_{-\infty}^{+\infty} \left(x - n \int_{-\infty}^{+\infty} u[1 - F_X(u)]^{n-1} f_X(u)du \right)^2 f_X(x)n[1 - F_X(x)]^{n-1}dx. \quad (7.1.17)$$

Example 7.7. Minimum flight delay. Consider again Pareto-distributed interarrival time of flight arrivals of Example 7.6. The variance of the minimum delay is given by Eq. (7.1.17); that is,

$$\begin{aligned}\text{Var}[X_{\min}] &= \int_{x_0}^{+\infty} \left\{ x - n \int_{x_0}^{+\infty} u\left[1 - 1 + \left(\frac{x_0}{u}\right)^{\theta}\right]^{n-1} \frac{\theta}{x_0}\left(\frac{x_0}{u}\right)^{\theta+1} du \right\}^2 \frac{n\theta}{x_0}\left(\frac{x_0}{x}\right)^{n\theta+1} dx \\ &= \left[\frac{n\theta}{n\theta - 2} - \left(\frac{n\theta}{n\theta - 1}\right)^2\right] x_0^2.\end{aligned}$$

Note that $\text{Var}[X_{\min}]$ exists only for $\theta > 2/n$.

Because the range is a linear function of the order statistics, its expectation can be simply determined as

$$\begin{aligned}E[R_n] &= E[X_{\max}] - E[X_{\min}] \\ &= n\left(\int_{-\infty}^{+\infty} x[F_X(x)]^{n-1} f_X(x)dx - \int_{-\infty}^{+\infty} x[1 - F_X(x)]^{n-1} f_X(x)dx \right).\end{aligned} \quad (7.1.18)$$

Note that

$$E[X] = \int_{-\infty}^{0} x\,dF_X(x) - \int_{0}^{+\infty} x\,d[1 - F_X(x)] = \int_{0}^{+\infty} [1 - F_X(x)]dx - \int_{-\infty}^{0} F_X(x)dx.$$

Equation (7.1.18) can also be written as

$$E[R_n] = \int_{-\infty}^{+\infty} \{1 - [F_X(x)]^n - [1 - F_X(x)]^n\} dx. \tag{7.1.19}$$

The variance of the range can be similarly found.

Example 7.8. Hurst phenomenon. In range analysis it is advantageous for practical purposes to calculate the difference between the maximum D_n^+ and minimum D_n^- of the accumulated departures from the sampling mean $\bar{X}_n$ of observations $X_1, X_2, \ldots, X_n$. This gives the adjusted range as

$$R_n^* = \max_{1 \leq i \leq n} \left(\sum_{j=1}^{i} X_j - i\bar{X}_n \right) - \min_{1 \leq i \leq n} \left(\sum_{j=1}^{i} X_j - i\bar{X}_n \right) = D_n^+ + |D_n^-|.$$

In order to compare the results from different observed sequences, the range is divided by the standard deviation S_n estimated by

$$S_n = \sqrt{\frac{\sum_{i=1}^{n} (X_i - \bar{X}_n)^2}{n}}$$

to give the adjusted rescaled range

$$R_n^{**} = \frac{R_n^*}{S_n}.$$

Of greater significance is the relationship

$$E[R_n^{**}] = cn^H,$$

where H is termed the *Hurst exponent*. This is deemed to be a constant for an observed sequence, and c is another constant, which was taken as $(1/2)^H$ in the original work by Hurst (1951) on long-term capacity of reservoirs. For natural sequences observed by Hurst, the value of H varies from 0.46 to 0.96, with a mean of 0.73 and a standard deviation of 0.09. However, for a sequence of independent random variables with finite variance, the asymptotic result

$$E[R_n^{**}] = \left(\frac{\pi}{2}\right)^{0.5} n^{0.5}$$

holds regardless of the distribution. The asymptotic value of $H = 0.5$ can also be proved to hold for various types of mutually dependent sequences of random variables. The fact that values of $H \neq 0.5$ are found in observed natural sequences—whereas for a normal (or gamma) sequence the asymptotic value of H is 0.5—is referred to as the *Hurst phenomenon*. Sometimes this has been interpreted as a transient effect, because for finite sequences the asymptotic value of 0.5 is reached only when n is much larger than the sample sizes usually found in practice. However, analysis of some exceedingly long geophysical time series has yielded values of H in the range 0.7–0.9. Consequently, this phenomenon was attributed to the effect of long-range dependence or memory, and some time series were modeled accordingly. In current practice, it is more plausible to attribute the Hurst effect to nonstationarity (by which we mean that there is variability in time of the mean and other statistical properties) in a physical process. See the application of nonstationary modeling using Gibbs sampling in Example 10.13.

7.1.4 Summary of Section 7.1

It is considered that a detailed discussion of order statistics is a necessary prelude to the study of extreme values. We have shown some of the important properties of order statistics in this section. Functions, expected value, and variance are discussed. Examples are given here on the practical use of order statistics.

7.2 EXTREME VALUE DISTRIBUTIONS

Extreme value theory, which is used in storm, flood, wind, sea waves, and earthquake estimation, dates back to the pioneering works by Fréchet (1927) and Fisher and Tippett (1928). This theory was extensively developed by Gumbel (1958) following the extremal type theorem originated by Gnedenko (1943).

7.2.1 Basic concepts of extreme value theory

According to the theory of extreme values, the largest or smallest value from a set of independent identically distributed random variables tends to an asymptotic distribution that only depends on the *tail* of the distribution of the basic variable. Let $X_1, X_2, \ldots, X_n$ denote a set of independent random variables with a common distribution $F_X(x)$, where x is an observed value and n is the number of equispaced data points within a fixed period of 1 year. Also, let $X_{(1)}, X_{(2)}, \ldots, X_{(n)}$ represent the ordered set of the same variables, with $X_{(1)} \leq X_{(2)}, \ldots \leq X_{(n)}$. From Eq. (3.4.12) or (7.1.2), the distribution of $X_{\max} = X_{(n)}$ is given by

$$F_{X_{\max}}(x) = [F_X(x)]^n.$$

As n increases indefinitely, this distribution approaches zero for every finite x in the domain of X. Therefore, standardization is necessary for the derivation of the limiting distribution. If $Y_n = (X_{(n)} - b_n)/a_n$, where $a_n > 0$ denotes a scaling constant and b_n is a location constant, then this limiting distribution must be one of the three following types:

$$\text{Type I:} \quad F_Y(y) = \exp(-e^{-y}), \qquad -\infty < y < +\infty. \tag{7.2.1a}$$

$$\text{Type II:} \quad F_Y(y) = \begin{cases} \exp(-y^{-\gamma}), & y > 0; \\ 0, & y \leq 0. \end{cases} \tag{7.2.1b}$$

$$\text{Type III:} \quad F_Y(y) = \begin{cases} \exp[-(-y)^{\gamma}], & y < 0; \\ 1, & y \geq 0. \end{cases} \tag{7.2.1c}$$

Here $\gamma > 0$ denotes a positive constant, and Y is the asymptote of Y_n (that is, taken when n tends to infinity). The existence of these three asymptotic forms of the distribution of extremes relies on the *stability postulate*, which states that if X has an extreme value distribution, the maximum of n independent observations of X has the same distribution, but with different location and scale parameters. Thus, the solution of

$$[F_X(x)]^n = F_X\left(\frac{x - b_n}{a_n}\right),$$

where a_n and b_n are functions of n, yields all the possible limiting forms of $F_X(x)$ as n tends to infinity.

Necessary and sufficient conditions that a set of random variables belongs to each of the three domains of attraction were given by Gnedenko (1943) and in simpler form by

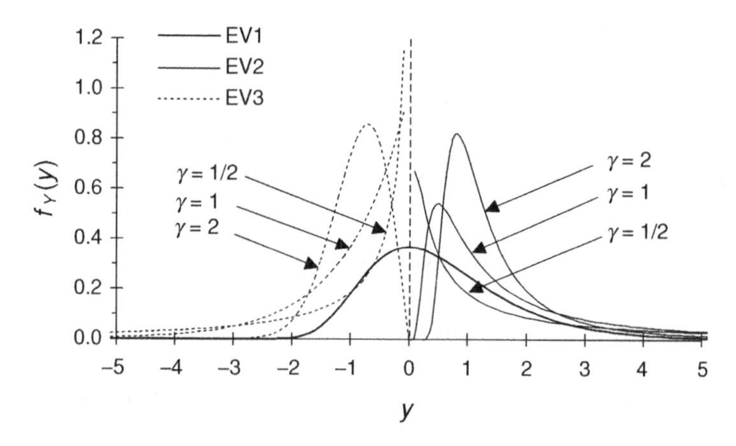

Fig. 7.2.1 Pdfs of asymptotic distributions for largest extreme values.

de Haan (1976). Further, the assumption that the X_is are independent can be relaxed.[6] If $F_X(x)$ is strictly monotonic and continuous, a sufficient condition for convergence to the Type I asymptotic form of the largest value is given by

$$\lim_{x \to \omega} \frac{d}{dx}\left[\frac{1 - F_X(x)}{f_X(x)}\right] = 0, \tag{7.2.2}$$

where ω denotes the upper bound[7] of X, and $f_X(x)/[1 - F_X(x)]$ is called the *hazard function*.[8] The pdfs and the cdfs of the extreme value distributions are shown in Figs. 7.2.1 and 7.2.2, respectively. Henceforth, we refer to these asymptotic distributions as EV1, EV2, and EV3.

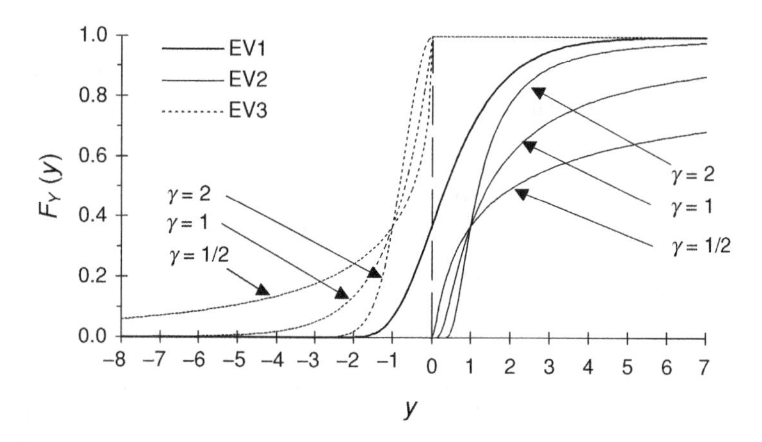

Fig. 7.2.2 cdfs of asymptotic distributions for largest extreme values.

[6] For a sequence of identically distributed random variables, the Type I, II, and III distributions are the only limiting distributions, provided that there is no long-range dependence of high-level exceedances (Leadbetter, 1991).

[7] If X is not bounded, the limit must be obviously taken for x tending to infinity. However, Eq. (7.2.2) can be useful for physically bounded variables (see, for example, Eliasson, 1994).

[8] This is extensively used in reliability theory (see Chapter 9).

Example 7.9. Limiting case for an exponentially distributed variate. Let X be an exponentially distributed variate with cdf $F_X(x) = 1 - e^{-\lambda x}$, for $x \geq 0$. From Eq. (7.2.2),

$$\lim_{x\to+\infty} \frac{d}{dx}\left[\frac{1-F_X(x)}{f_X(x)}\right] = \lim_{x\to+\infty} \frac{d}{dx}\left[\frac{e^{-\lambda x}}{\lambda e^{-\lambda x}}\right] = \lim_{x\to+\infty} \frac{d}{dx}\left[\frac{1}{\lambda}\right] = 0.$$

Therefore, the exponential distribution has the EV1 as its corresponding limiting extreme value distribution.

If $F_X(x)$ is strictly monotonous and continuous, the sufficient condition for convergence to the EV2 distribution is

$$\lim_{x\to+\infty} x\frac{f_X(x)}{1-F_X(x)} = \gamma, \tag{7.2.3}$$

with $\gamma > 0$ denoting a constant. The pdf and cdf of the EV2 distribution are also shown in Figs. 7.2.1 and 7.2.2, respectively.

Example 7.10. Limiting case for a Pareto distributed variate. Let

$$F_X(x) = 1 - \left(\frac{a}{x}\right)^{\theta},$$

for $x > x_0 > 0$, with $\theta > 0$. From Eq. (7.2.3),

$$\lim_{x\to+\infty} x\frac{f_X(x)}{1-F_X(x)} = x\frac{\theta a^{\theta}\, x^{-\theta-1}}{a^{\theta}x^{-\theta}} = \theta.$$

Therefore, X_{max} converges to the EV2 distribution.

The sufficient condition for convergence of a bounded random variable X to the EV3 distribution is

$$\lim_{x\to+\omega} (\omega - x)\frac{f_X(x)}{1-F_X(x)} = \gamma, \tag{7.2.4}$$

where $\gamma > 0$ denotes a constant, ω is the upper bound of X, and $F_X(\omega) = 1$. The pdf and cdf of the EV3 distribution are also shown in Figs. 7.2.1 and 7.2.2, respectively.

Example 7.11. Limiting case for a reflected power-distributed variate. Let $F_X(x) = 1 - a(b-x)^{\gamma}$, with $x \leq b$, $a > 0$, and $\gamma > 0$. Examples include the uniform distribution, where $\gamma = 1$ and the triangular pdf, where $\gamma = 2$. Since $F_X(b) = 1$, and $F_X(x) < 1$ for every $x < b$, b is the upper bound of X. Thus,

$$\lim_{x\to\infty} (\omega - x)\frac{f_X(x)}{1-F_X(x)} = \lim_{x\to b} (b-x)\frac{\gamma a(b-x)^{\gamma-1}}{a(b-x)^{\gamma}} = \gamma.$$

Therefore, the reflected power distribution has the EV3 as its corresponding limiting extreme value distribution.

As noted above, the limiting distribution depends on the general way in which the appropriate tail of the underlying distribution, $F_X(x)$, behaves. Examples of distributions yielding EV1 extreme variates are exponential, gamma, Weibull, normal, lognormal, logistic, and EV1 itself. Examples of distributions yielding EV2 variates are Pareto, Student's t, Cauchy, log-gamma, and EV2 itself. Finally, examples of distributions yielding EV3 variates are reflected-power, uniform, beta, and EV3 itself. However, the convergence to a limiting case can be quite slow in some cases.

In many applications, the probability model is used in the inverse form, because a specified probability of nonexceedance q is selected and the design value is evaluated as the qth quantile of the design variate. For the standard EV1 distribution,

$$y = -\ln(-\ln q), \tag{7.2.5}$$

from Eq. (7.2.1a) with $F_Y(y) = q$. To compare with other types, we can write the qth quantile of the standard EV1 variate as

$$\xi_{q,\text{EV1}} = -\ln(-\ln q) = y. \tag{7.2.6}$$

For the EV2 distribution, one gets $\xi_{q,\text{EV2}} = \exp(y/\gamma)$, and for the EV3 distribution, $\xi_{q,\text{EV3}} = \exp(-y/\gamma)$. Accordingly, the general formulation of extreme value distributions in the inverse form can be written as

$$\xi_q = \frac{1 - \exp(-ky)}{k}, \tag{7.2.7}$$

where y is often referred to as a *reduced*, or *standard Gumbel variate*; it is a surrogate of the probability of nonexceedance, q. By substituting the series expansion of $\exp(-ky)$ and then dividing by k, the special case $k = 0$ leads to the linear relationship for ξ against y that characterizes the EV1 distribution as given by Eq. (7.2.6). The EV2 distribution is applicable when $k < 0$ and, if $k > 0$, the EV3 distribution is signified. Therefore, the probability distribution given in inverse form by Eq. (7.2.7) is called the *general extreme value* (GEV) *distribution*, which can be written as

$$F_Y(y) = \exp[-(1 - ky)^{1/k}]. \tag{7.2.8}$$

The GEV distribution is sometimes referred as the von Mises-Jenkinson distribution in the determination of the probability model of the unknown distribution of the largest value data within a generalized framework; see von Mises (1936) and Jenkinson (1969). We shall return to the GEV distribution in Subsection 7.2.5.

Example 7.12. Quantiles of extreme value distributions. Since y is a surrogate of the probability of nonexceedance, one can compare the quantiles arising from the three different types of the extreme value distributions as represented by Eq. (7.2.8). For example, if $q = 0.9$ is the required design level, the corresponding reduced variate is, from Eq. (7.2.5),

$$y = -\ln[-\ln(0.9)] = 2.250,$$

which is also the quantile for the EV1 distribution. If, for example, $k = -0.25$, the corresponding quantile for the EV2 distribution is found from Eq. (7.2.7), resulting in

$$\xi_{0.9,\text{EV2}} = \frac{1 - \exp(0.25 \times 2.250)}{(-0.25)} = 3.021 > y.$$

Because of the positive exponential form, ξ_{EV2} increases faster than for the EV1 distribution. Therefore, the EV2 distribution can be represented by a curve that is *concave upward* on the (y, ξ) plane. For $k = 0.25$, one has, from Eq. (7.2.7),

$$\xi_{0.9,\text{EV3}} = \frac{1 - \exp(-0.25 \times 2.250)}{0.25} = 1.721 < y.$$

Because of the negative exponential form, the EV3 distribution can be represented by a curve that is *concave downward* on the (y, ξ) plane, as shown in Fig. 7.2.3.

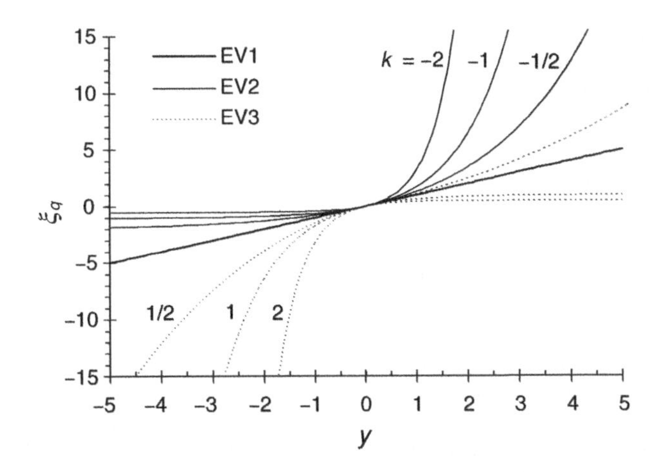

Fig. 7.2.3 Relationship between the qth quantile, ξ_q, and the reduced variate, y, for the three asymptotic extreme value distributions.

7.2.1.1 Smallest value distributions

In some applications, the asymptotic distribution of interest is that of the smallest value. To this effect, one notes that the distributions of the largest and smallest values are related by the *principle of symmetry* first introduced by Gumbel (1958). If X denotes a variate with pdf $f_X(x)$, the variate X^*, whose pdf is the mirror image of $f_X(x)$, has the property $1 - F_X(x) = F_{X^*}(-x)$, as shown in Fig. 7.2.4. Thus, $[1 - F_X(x)]^n = [F_{X^*}(-x)]^n$. From Eq. (7.1.3),

$$[1 - F_X(x)]^n = 1 - F_{X_{(1)}}(x) = 1 - F_{X_{\min}}(x)$$

is the probability of exceedance of the smallest value of X, and $[F_{X^*}(-x)]^n$ is the probability of nonexceedance of the largest value of X^*; that is,

$$1 - F_{X_{\min}}(x) = F_{X^*_{\max}}(-x). \qquad (7.2.9)$$

The corresponding pdfs are also related by

$$f_{X_{\min}}(x) = f_{X^*_{\max}}(-x). \qquad (7.2.10)$$

Using the principle of symmetry, the asymptotic distribution of the smallest value of a random variable can be determined from the distribution of its largest value by reversing the sign and taking the complementary probabilities.

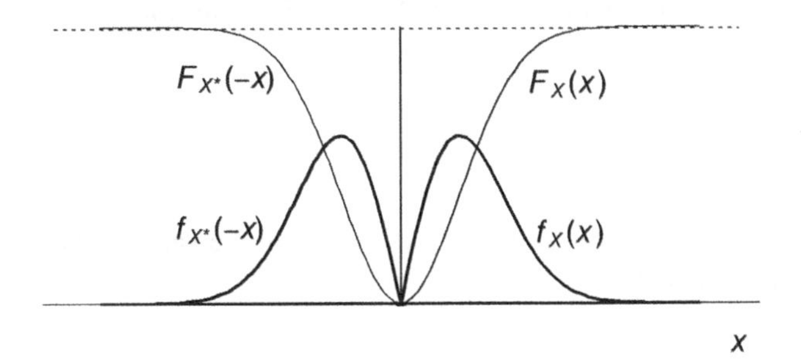

Fig. 7.2.4 Principle of symmetry.

Example 7.13. Smallest value distribution from an exponentially distributed variate. Consider an exponentially distributed variate X with $f_X(x) = \lambda\exp(-\lambda x)$, for $x \geq 0$. The pdf of the corresponding mirror image of X, say, X^* is

$$f_{X^*}(y) = f_X(-y) = \lambda\exp[-\lambda(-y)] = \lambda e^{\lambda y},$$

for $y \leq 0$, and its cdf is given by

$$F_{X^*}(y) = 1 - F_X(-y) = 1 - 1 + e^{\lambda y} = e^{\lambda y},$$

for $-x = y \geq 0$. For a sample of size n of X, one gets

$$1 - F_{X_{\min}}(x) = F_{X^*_{\max}}(-x) = [F_{X^*}(-x)]^n = e^{-n\lambda x},$$

for $x \geq 0$. It follows that the probability of exceedance of the smallest value is the same as the cdf of the largest value of X^*.

Consider, for example, the smallest value from a variate W with pdf $f_W(w) = \lambda\exp(\lambda w)$, for $w \leq 0$ and $\lambda > 0$. The largest value of the mirror image of W, say, W^*, with pdf $f_{W^*}(w) = f_W(-w) = \lambda\exp(-\lambda w)$, has the EV1 distribution. Therefore, the cdf of the smallest value of W will be given by

$$F_{W_{\min}}(w) = 1 - F_{W^*_{\max}}(-w) = 1 - \exp(e^{\lambda w}),$$

for $w \leq 0$; this is defined as the EV1 distribution of the smallest value.

Through the principle of symmetry the following three asymptotic distributions of the smallest value are found:

$$\text{EV1:} \quad F_Z(z) = 1 - \exp(-e^z), \qquad -\infty < z < +\infty. \tag{7.2.11a}$$

$$\text{EV2:} \quad F_Z(z) = \begin{cases} 1 - \exp[-(-z)^{-\gamma}], & z < 0; \\ 1, & z \geq 0. \end{cases} \tag{7.2.11b}$$

$$\text{EV3:} \quad F_Z(z) = \begin{cases} 1 - \exp(-z^{\gamma}), & z > 0; \\ 0, & z \leq 0. \end{cases} \tag{7.2.11c}$$

Here $\gamma > 0$ is a constant, and Z denotes the linearly standardized smallest value of a sample of size n from a random variable under a linear transformation, for n tending to infinity. The pdfs and cdfs of the three asymptotic smallest value distributions are shown in Figs. 7.2.5 and 7.2.6, respectively.

The criteria of convergence to one of the three types of asymptotic distributions of the smallest value are similar to those applicable to the determination of that of the largest value. If $F_X(x)$ is strictly monotonous and continuous, the sufficient condition for convergence to the EV1 distribution of the smallest value is

$$\lim_{x\to+\infty} \frac{d}{dx}\left[\frac{F_X(x)}{f_X(x)}\right] = 0, \tag{7.2.12}$$

where $f_X(x)$ and $F_X(x)$ denote the pdf and cdf of the variable. For the EV2 distribution of the smallest value,

$$\lim_{x\to+\infty} -x\frac{f_X(x)}{F_X(x)} = \gamma, \tag{7.2.13}$$

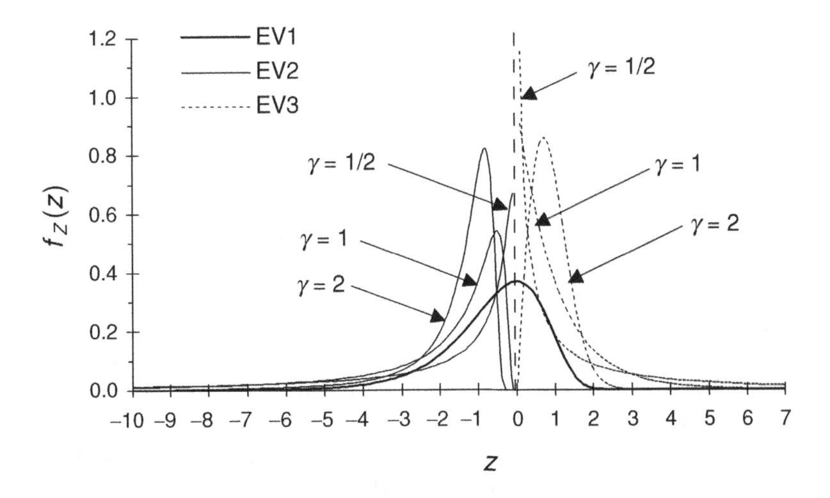

Fig. 7.2.5 Pdfs of asymptotic distributions for smallest extreme values.

with $\gamma > 0$ denoting a constant. For the EV3 distribution of the smallest value,

$$\lim_{x \to \varepsilon^+} (x - \varepsilon) \frac{f_X(x)}{F_X(x)} = \gamma, \tag{7.2.14}$$

where ε denotes the lower bound of X. Note that it is possible for the largest value distribution of a variable to belong to one asymptotic form, whereas its smallest value distribution can belong to another asymptotic form.

Example 7.14. Limiting cases for a Rayleigh distributed variate. Consider a random variable X with pdf $f_X(x) = (x/\sigma^2)\exp[-x^2/(2\sigma^2)]$, for $x \geq 0$. Thus, $F_X(x) = 1 - \exp[-x^2/(2\sigma^2)]$.

The application of the convergence criterion of Eq. (7.2.2) for the EV1 largest value distribution yields

$$\lim_{x \to +\infty} \frac{d}{dx} \left[\frac{1 - 1 + e^{-x^2/(2\sigma^2)}}{(x/\sigma^2)e^{-x^2/(2\sigma^2)}} \right] = \lim_{x \to +\infty} \frac{d}{dx} \left(\frac{\sigma^2}{x} \right) = - \lim_{x \to +\infty} \frac{\sigma^2}{x^2} = 0.$$

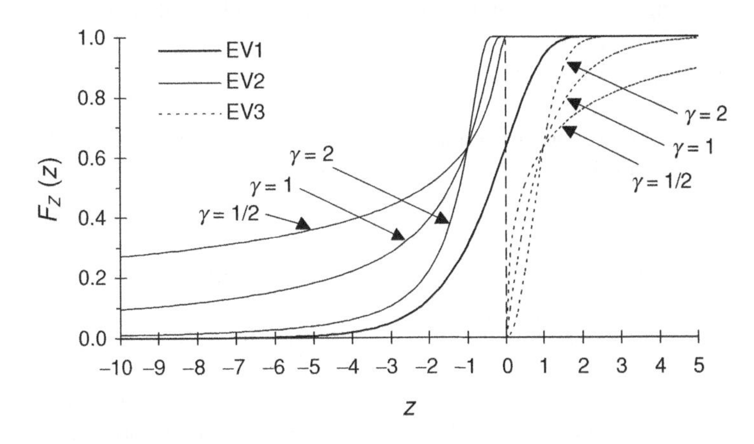

Fig. 7.2.6 Cdfs of asymptotic distributions for smallest extreme values.

This shows that the largest value of X follows the EV1 distribution. The application of the convergence criterion of Eq. (7.2.14) yields

$$\lim_{x\to\varepsilon^+}(x-\varepsilon)\frac{f_X(x)}{F_X(x)}=\lim_{x\to 0^+}(x-0)\frac{(x/\sigma^2)e^{-x^2/(2\sigma^2)}}{1-e^{-x^2/(2\sigma^2)}}=\lim_{x\to 0^+}\frac{(x^2/\sigma^2)}{e^{x^2/(2\sigma^2)}-1}$$
$$=\lim_{x\to 0^+}\frac{2}{e^{x^2/(2\sigma^2)}}=2,$$

which indicates that the smallest value distribution from a Rayleigh variate converges to the EV3 smallest value distribution.

Note that the derivation of the extreme value distribution requires that the underlying distribution of the variable is known. In many engineering applications, one finds that only extreme value data are available. The determination of the appropriate type of the extreme value distribution is then made without any information on the underlying variable, based on some prior knowledge on the physical process or inferred from the data. Although the GEV approach provides a framework to determine the appropriate model of extreme value data, it is often difficult to infer this model from the small samples usually available in applications. Furthermore, the convergence to a limiting case can be quite slow. Finally, the three asymptotic forms presented here are not exhaustive, because distributions exist whose largest value or smallest value distributions do not converge to one of the three types discussed here. Nevertheless, these distributions are useful to analyze extreme events from observed data and provide a framework to predict future outcomes.

7.2.2 Gumbel distribution

The EV1 distribution was extensively developed and applied to extreme values by Gumbel (1935, 1941); therefore, it is often referred to as the *Gumbel distribution*. This distribution results from any underlying distribution of the X_is of the exponential type. The exponential is the obvious candidate, but also the upper tail of other distributions (say, the gamma, the Weibull, the normal, the lognormal, the logistic, and the EV1 itself) converge to the exponential form for large values of the variable. Accordingly, the initial density, $f_X(x)=dF_X(x)/dx$, can be approximated to the form $\lambda e^{-\lambda x}$; this leads to the probability of nonexceedance of $1-e^{-\lambda x}$, which can be substituted for $F_X(x)$ in Eq. (7.1.2) to obtain

$$F_{X_{\max}}(x)=[F(x)]^n=(1-e^{-\lambda x})^n. \tag{7.2.15}$$

By changing the location and scale, Eq. (7.2.15) can be written as

$$F_{X_{\max}}(x)=\left[1-\frac{1}{n}\exp\left(-\frac{x-b}{\alpha}\right)\right]^n, \tag{7.2.16}$$

where α and b are the dispersion (or scale) and location parameters. Taking the limit of the right-hand side of Eq. (7.3.16) as n tends to infinity,

$$F_{X_{\max}}(x)=\exp[-e^{-(x-b)/\alpha}], \tag{7.2.17}$$

which is the Gumbel distribution function. The corresponding pdf is

$$f_{X_{\max}}(x)=\frac{1}{\alpha}\exp\left[-\frac{x-b}{\alpha}-e^{-(x-b)/\alpha}\right],\quad \text{for } -\infty<x<+\infty. \tag{7.2.18}$$

The above result is asymptotic, being approximately true for any large value of n. The engineer may argue, for instance, that the annual maximum flow in a river has such an extreme value distribution because it is the largest daily flow or that the annual maximum height of waves in a harbor represents the largest of some unknown large number of storms in a year. This distribution also arises from the theory of point processes if one assumes that storms occur as Poisson arrivals and that an independent exponentially distributed random variable is associated with each such arrival.[9]

The location parameter b is the mode of the distribution, because $df(x)/dx = 0$ for $x = b$. The parameter α is a measure of dispersion, as already stated, and it only depends on the variance of X_{max}. The parameter b is a measure of location that depends on both the variance and the mean. The moment-generating function is found to be $M_{X_{max}}(t) = \exp(bt)\Gamma(1 - \alpha t)$, for $t < 1/\alpha$. Hence, the mean and the variance of X_{max} are

$$E[X_{max}] = \mu = b + n_e \alpha, \tag{7.2.19}$$

and

$$\text{Var}[X_{max}] = \sigma^2 = \frac{\pi^2 a}{6}, \tag{7.2.20}$$

where n_e denotes the Euler constant, approximately equal to 0.5772. The skewness coefficient is 1.1396, and the kurtosis coefficient is 5.400.

If the first two moments of X_{max} are known, the values of the parameters α and b can be determined by the method of moments from the mean, μ, and the variance, σ^2, of the extreme value population. From Eqs. (7.2.19) and (7.2.20), one obtains

$$\alpha = \frac{\sqrt{6}}{\pi}\sigma, \tag{7.2.21}$$

and

$$b = \mu - n_e \alpha = \mu - \frac{n_e \sqrt{6}}{\pi}\sigma. \tag{7.2.22}$$

This method can be used if a finite sample of the values taken by X_{max} is available: such as the annual maximum river flows for a period of n years. One may compute the values of the parameters α and b by estimating the mean and the variance of the population based on this sample.

Example 7.15. Estimation of Gumbel distribution for storm rainfall data. The mean and standard deviation of annual maximum hourly rainfalls estimated from a 58-year record available from 1931 to 1988 at Genoa University, Italy, listed in Table E.7.1, are 48.16 and 23.76 mm, respectively. (The collected data are correct to the nearest tenth of a millimeter; we have made these statistics more accurate merely for subsequent comparisons.) From Eq. (7.2.21),

$$\hat{a} = \frac{\sqrt{6}}{\pi}\hat{\sigma} = 0.780 \times 23.76 = 18.52 \text{ mm}$$

is the estimated scale parameter of the Gumbel distribution by the method of moments. From Eq. (7.2.22), the location parameter is estimated as

$$\hat{b} = \hat{\mu} - n_e \hat{\alpha} = 48.16 - 0.5772 \times 18.52 = 37.47 \text{ mm}.$$

In Fig. 7.2.7, the cdf of the EV1 distribution with the above estimates of α and b is shown, where the Gumbel or reduced variate is used to represent the frequency level. This distribution is also compared with that of the observed data using the Weibull plotting position (see Example 7.5).

[9] See renewal and point processes in Subsection 4.2.2.5.

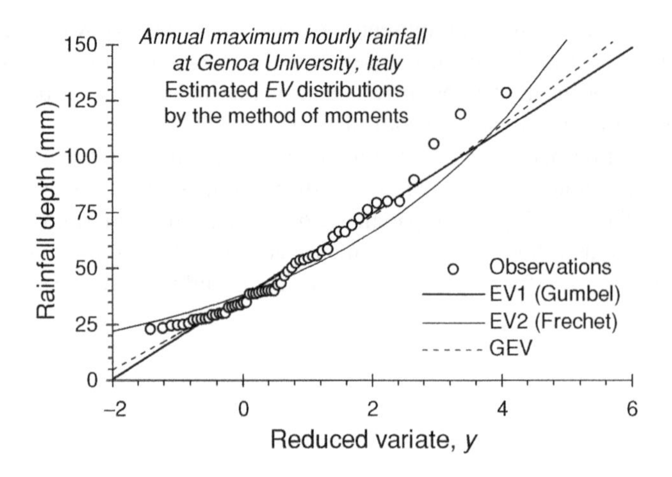

Fig. 7.2.7 Gumbel probability plots of extreme value distributions of annual maximum hourly rainfall observed at Genoa University, Italy.

Quantiles of the Gumbel distribution can be easily obtained by introducing the new (dimensionless) variable $Y = (X_{\max} - b)/\alpha$, which is a standardized EV1 variate with $\alpha = 1$ and $b = 0$, because $F_Y(y) = \exp(-e^{-y})$. Therefore, the qth quantile of $X_{\max}$ can be evaluated as

$$\xi_q = b - \alpha \ln\left[\ln\left(\frac{1}{q}\right)\right] = b + \alpha y, \tag{7.2.23}$$

where y is the reduced variate of Eq. (7.2.5). The cartesian plane with coordinates y and ξ is useful to represent the behavior of the EV1 distribution. Because this distribution plots as a straight line, it is often referred to as a *Gumbel probability plot*, as shown in Fig. 7.2.8.

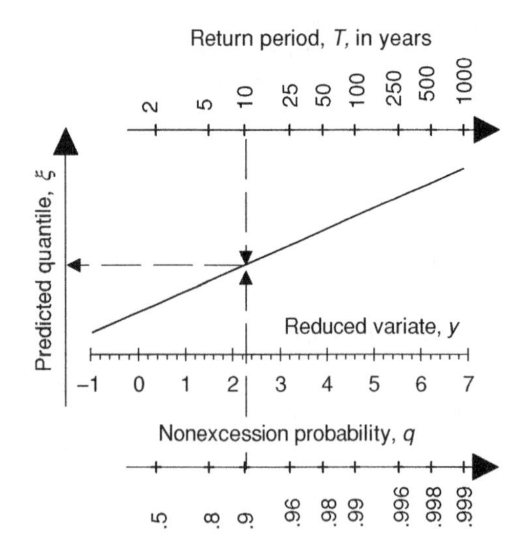

Fig. 7.2.8 Gumbel probability plot.

Another type of standardization can be introduced by combining Eq. (7.2.23) with Eqs. (7.2.19) and (7.2.20) to obtain an inverse formulation similar to that usually adopted for the normal distribution. Thus,

$$\xi_q = \mu - \sigma\left\{\frac{\sqrt{6}}{\pi}\left[n_e + \ln\left(\ln\frac{1}{q}\right)\right]\right\} = \mu + \sigma K, \tag{7.2.24}$$

where K is called the *frequency factor* (Chow, 1951); this plays the same role as the standard normal deviate, because it is a (nonlinear) function of the probability of nonexceedance. The design value of a given EV1 distributed largest value variate has to be selected on the basis of a specified return period, say, T. As shown in Subsection 4.1.4, the return period is related to the probability of exceedance of the variate as $T = 1/\Pr[X_{\max} > x]$. Therefore, the T-year event will be given by the qth quantile of $X_{\max}$ for $q = 1 - 1/T$. When the parameters α and b are known, Eq. (7.2.23) yields

$$x_{\max}(T) = b - \alpha\ln[\ln(1/q)] = b - \alpha\ln\left[\ln\left(\frac{T}{T-1}\right)\right], \tag{7.2.25}$$

where $x_{\max}(T)$ denotes the required design value. Also, from Eq. (7.2.24),

$$x_{\max}(T) = \xi_q = \mu - \sigma\left\{\frac{\sqrt{6}}{\pi}\left[n_e + \ln\left(\ln\frac{1}{q}\right)\right]\right\}$$

$$= \mu - \sigma\left\{\frac{\sqrt{6}}{\pi}\left[n_e + \ln\left(\ln\frac{T}{T-1}\right)\right]\right\}, \tag{7.2.26}$$

in which the mean, μ, and the standard deviation, σ, are used to predict the design value, and, as already stated, $n_e \approx 0.5772$ is the Euler constant.

Example 7.16. Storm rainfall prediction using Gumbel distribution. Consider again the annual maximum hourly rainfall totals recorded at Genoa University. Substituting the estimates of Example 7.15 for α and b in Eq. (7.2.25) gives the 10-year hourly design storm as

$$x_{\max}(10) = \xi_{0.9} = 37.47 - 18.52 \times \ln\left(\ln\frac{1}{0.9}\right) = 79.1\,\text{mm}.$$

Figure 7.2.9 shows how the required quantile of $X_{\max}$ can be predicted as a function of y or T on a Gumbel probability plot.

Note that the probability of nonexceedance q, the return period T, the reduced variate y, and the frequency factor K, are mutually related. For instance, the Gumbel variate can be written as a function of the return period, T, as follows:

$$y = -\ln\left[-\ln\left(1 - \frac{1}{T}\right)\right],$$

which can be approximated by using MacLaurin's theorem[10] as

$$y = -\ln\left[\frac{1}{T} + \frac{1}{2}\left(\frac{1}{T}\right)^2 + \cdots\right] \approx \ln\left(\frac{T^2}{T + 1/2}\right),$$

if only two terms are used. By dividing T^2 by $(T + 1/2)$ one gets

$$y \approx \ln(T - 0.5),$$

[10] Note that $\ln(1 - x) = -\sum_{k=0}^{\infty}(x^{k+1})/(k+1)$, as shown by Eq. (4.1.6*b*).

if T is large. For $T > 10$ years, the error in this approximation is less than 0.5%. For instance, if the 100-year hourly design storm is required,

$$x_{\max}(100) = \xi_{0.99} = 37.47 - 18.52 \times \ln\left(\ln \frac{1}{0.99}\right) = 122.6648\,\text{mm}.$$

And by the approximate method

$$x_{\max}(100) = \xi_{0.99} \approx 37.47 + 18.52 \times \ln(100 - 0.5) = 122.6649\,\text{mm},$$

which is very close to the exact method.

The method of moments is a convenient procedure for providing estimates of parameters from a sample of extreme value data. The maximum likelihood (ML) procedure yields, on the other hand, asymptotically minimum variance estimates which are asymptotically unbiased, but it requires iterative computations. Alternatively, the method of probability weighted moments provides satisfactory parameter estimates for the Gumbel distribution (see Example 3.21).

Example 7.17. Parameter estimation for the Gumbel distribution by the methods of maximum likelihood and probability weighted moments. Substituting Eq. (7.2.18) for $f_X(x)$ in Eq. (3.2.25) the log-likelihood function becomes

$$\ln L = -\sum \frac{x_i - b}{\alpha} - \sum \exp\left(-\frac{x_i - b}{\alpha}\right) - n \ln \alpha,$$

where $x_i, i = 1, \ldots, n$ is the sample of the largest value data, and all summations are taken for $i = 1, \ldots, n$, the size of sample. The partial derivatives of $\ln L$ are

$$\frac{\partial \ln L}{\partial \alpha} = \sum \frac{x_i - b}{\alpha^2} - \sum \frac{x_i - b}{\alpha^2} \exp\left(-\frac{x_i - b}{\alpha}\right) - \frac{n}{\alpha},$$

and

$$\frac{\partial \ln L}{\partial b} = \frac{n}{\alpha} - \sum \frac{1}{\alpha} \exp\left(-\frac{x_i - b}{\alpha}\right).$$

The ML estimators, $\tilde{\alpha}$ and $\tilde{b}$, of the parameters are obtained by setting $\partial \ln L/\partial \alpha = 0$ and $\partial \ln L/\partial b = 0$. From the second equation,

$$\exp\left(\frac{\tilde{b}}{\tilde{\alpha}}\right) = \frac{n}{\sum \exp(-x_i/\tilde{\alpha})},$$

which is used in the first equation to obtain, after simplifying,

$$\tilde{\alpha} = \bar{x} - \frac{\sum x_i \exp(-x_i/\tilde{\alpha})}{\sum \exp(-x_i/\tilde{\alpha})},$$

where $\bar{x}$ denotes the arithmetic mean of the sample. This equation has α as the only unknown, and it must be solved by using numerical iteration. A very simple method (alternative to Newton-Raphson, for instance) is to estimate an initial value of α by the method of moments, and then to substitute it in the right-hand side of this equation, to obtain the next trial value. The third value of α can be made equal to the weighted average of the first and second, and the equation is used again to obtain a fourth value; here, the most recent value is given a greater weight. The procedure is repeated until there is no significant difference, and the final value is substituted for $\tilde{\alpha}$ in

$$\tilde{b} = \tilde{\alpha} \ln\left[\frac{n}{\sum \exp(-x_i/\tilde{\alpha})}\right].$$

Let us consider the 58-year record of annual maximum hourly rainfall at Genoa University, Italy (see Table E.7.1). The sample mean is 48.16 mm, and the scale and location parameters estimated by using the method of moments are 18.52 and 37.47 mm, respectively (see Example 7.15). The first iteration is made by using $\alpha = 18.52$ mm, which yields a second value of $\alpha = 14.30$ mm. The new initial value is obtained by using weights of 0.25 for the first, and 0.75 for the second, so obtaining

$$\alpha = 18.52 \times 0.25 + 14.30 \times 0.75 = 15.36 \text{ mm},$$

which yields $\alpha = 15.47$ mm. The new initial value is computed as

$$\alpha = 15.36 \times 0.25 + 15.47 \times 0.75 = 15.44 \text{ mm}.$$

It is seen that the value obtained from the right-hand side of the equation in α is again 15.44 mm, which is thus taken as the ML estimate of the scale parameter. The ML estimate of the location parameter is

$$\tilde{b} = 15.44 \times \ln\left(\frac{58}{4.89}\right) = 38.37 \text{ mm}.$$

Following the procedure shown in Example 3.21, the values of the pwms are found to be $M_0 = 48.16$ mm (equal to the arithmetic mean) and $M_1 = 30.32$ mm, respectively, The corresponding parameter estimates are

$$\hat{\alpha} = \frac{2M_1 - M_0}{\ln 2} = 18.01 \text{ mm},$$

and

$$\hat{b} = M_0 - n_e\hat{\alpha} = 48.16 - 0.5772 \times 18.01 = 37.77 \text{ mm}.$$

It is seen that the pwm estimates are very close to those estimated by the method of moments ($\alpha = 18.52$ mm and $b = 37.47$ mm; see Example 7.15). The similarities are seen in Fig. 7.2.9. On the contrary, the ML estimates of α and b are substantially different. For small samples, it can be argued that the ML procedure gives undue weight to the smaller values; although this may not be a fair criticism, it should be noted that engineers looking for practical means of extrapolation tend to give more attention to the larger values in the data.

The Gumbel distribution of the smallest value arising from an initial variate with an exponential tail can be determined by using the principle of symmetry. The resulting cdf

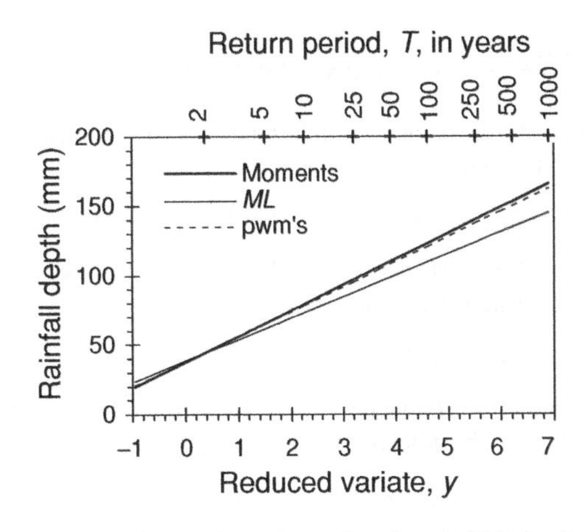

Fig. 7.2.9 Gumbel predictions of annual maximum hourly raindfall depth at Genoa University, Italy.

is given by

$$F_{X_{\max}}(x) = 1 - \exp[-e^{x-b/\alpha}], \tag{7.2.27}$$

and the corresponding pdf is

$$f_{X_{\max}}(x) = \frac{1}{\alpha}\exp\left[\frac{x-b}{\alpha} - e^{(x-b)/\alpha}\right], \quad \text{for } -\infty < x < +\infty, \tag{7.2.28}$$

where α and b are the scale and location parameters, respectively. The mean of $X_{\min}$ is given by

$$E[X_{\min}] = b - n_e\alpha, \tag{7.2.29}$$

where $n_e \approx 0.5772$ denotes the Euler constant. The variance of $X_{\min}$ is the same as that for $X_{\max}$, given by Eq. (7.2.20). The skewness and the kurtosis coefficients are again two constants, equal to 1.1396 and 5.400, respectively, as for $X_{\max}$.

If the first two moments of $X_{\min}$ are known, the values of the parameters α and b can be determined from the mean, μ, and the variance, σ^2, of the extreme value population. From Eq. (7.2.20) one gets

$$\alpha = \frac{\sqrt{6}}{\pi}\sigma. \tag{7.2.30}$$

And from Eq. (7.2.29)

$$b = \mu + n_e\alpha = \mu + \frac{n_e\sqrt{6}}{\pi}\sigma. \tag{7.2.31}$$

This method can be used if a finite sample of the values taken by $X_{\min}$ is available, such as the annual minimum river flow for a period of n years at a particular site. One can compute the values of the parameters α and b by estimating the mean and the variance of the population based on this sample. Alternatively, the methods of maximum likelihood and probability weighted moments can provide satisfactory results.

Example 7.18. Low-flow analysis using Gumbel distribution. The mean and standard deviation of annual minimum flow in the Po River at Pontelagoscuro, Italy, recorded from 1918 to 1978 (see Table E.7.2, column 3) are 554.6 and 190.5 m³/s, respectively. From Eq. (7.2.30),

$$\hat{\alpha} = \frac{\sqrt{6}}{\pi}\hat{\sigma} = 0.780 \times 190.5 = 148.5\,\text{m}^3/\text{s}$$

is the estimated scale parameter of the Gumbel distribution as fitted to the data by the method of moments. From Eq. (7.2.31),

$$\hat{b} = \hat{\mu} + n_e\hat{\alpha} = 554.6 + 0.5772 \times 148.5 = 640.3\,\text{m}^3/\text{s}.$$

The cdf of the EV1 distribution with these values of α and b is shown in Fig. 7.2.10, where the Gumbel variate for the smallest value,

$$y^* = \ln[-\ln(1-q)],$$

is used to represent the frequency level. This distribution is also compared with the distribution of observed data. The probability that $X_{\min} \leq 0$ is given by

$$F_{X_{\min}}(0) = 1 - \exp(-e^{-\hat{b}/\hat{\alpha}}) = 1 - \exp(-e^{-4.313}) = 1 - 0.987 = 0.013.$$

This is of course not possible for the Po River. Therefore, this distribution is not generally suitable to model river flows and similar variables.

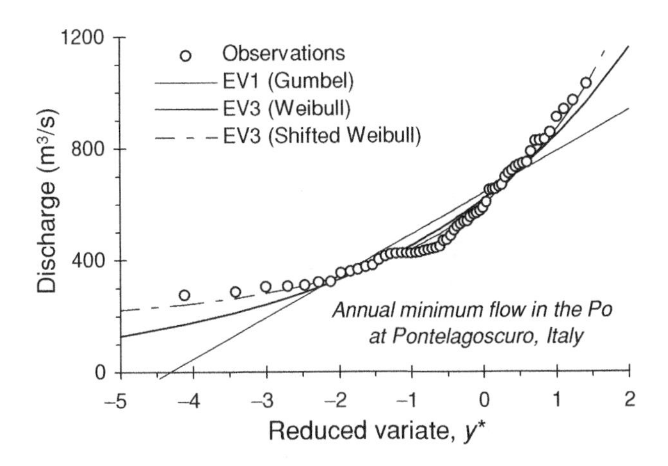

Fig. 7.2.10 Gumbel probability plots of extreme value distributions of annual minimum flow in the Po River at Pontelagoscuro, Italy.

7.2.3 Fréchet distribution

The EV2 distribution was first developed and applied by Fréchet (1927). Hence it is often referred to as the *Fréchet distribution.* This distribution results from any underlying distribution of the Pareto type, that is, in power form. Because the upper tail of other distributions (for example, the Student's t, the Cauchy, the log-gamma, and the EV2 itself) converges to the power form for large values of the variable, these distributions have also the EV1 as limiting distribution. The cdf of the Fréchet distribution is of the form

$$F_{X_{\max}}(x) = \exp\left[-\left(\frac{x_0}{x}\right)^{\theta}\right], \tag{7.2.32}$$

for $x > 0$, and the corresponding pdf is

$$f_{X_{\max}}(x) = \frac{\theta}{x_0}\left(\frac{x_0}{x}\right)^{\theta+1}\exp\left[-\left(\frac{x_0}{x}\right)^{\theta}\right], \tag{7.2.33}$$

where $x_0 > 0$ denotes a scale parameter and $\theta > 0$ is a shape parameter.

The moments of order r, which exist only for $r < \theta$, are given by

$$E\left[X^r_{\max}\right] = x_0^r\Gamma(1 - r/\theta); \tag{7.2.34}$$

consequently,

$$E[X_{\max}] = x_0\Gamma(1 - 1/\theta), \quad \text{for } \theta > 1, \tag{7.2.35}$$

$$\text{Var}[X_{\max}] = x_0^2[\Gamma(1 - 2/\theta) - \Gamma^2(1 - 1/\theta)], \quad \text{for } \theta > 2. \tag{7.2.36}$$

Since

$$V^2_{X_{\max}} = \frac{\Gamma(1 - 2/\theta)}{\Gamma^2(1 - 1/\theta)} - 1, \quad \text{for } \theta > 2, \tag{7.2.37}$$

the shape parameter only depends on the coefficient of variation (see Fig. 7.2.11).

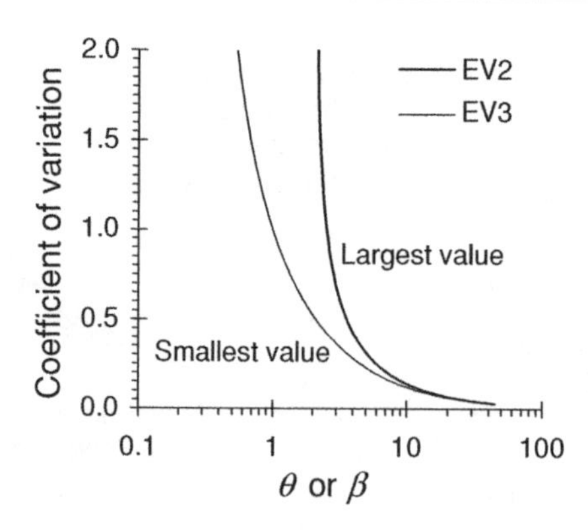

Fig. 7.2.11 Coefficient of variation versus the exponent (shape parameter) of the largest value Fréchet distribution and of the smallest value Weibull distribution.

If the first two moments of X_{max} exist and are known, the values of the parameters x_0 and θ can be determined from the mean, μ, and the variance, σ^2, of the extreme value population. However, using the mean and the coefficient of variation V is relatively straightforward, because Eq. (7.2.37) indicates that the shape parameter θ depends only on V. After V is estimated (as the ratio of the sample standard deviation to the sample mean), Eq. (7.2.37) must be solved via numerical iteration to find θ. Then, using Eq. (7.2.35), one can estimate the value of the scale parameter,

$$x_0 = \frac{\mu}{\Gamma(1 - 1/\theta)}. \tag{7.2.38}$$

If a finite sample of the values taken by X_{max} is available—such as the annual maximum wind velocity for a period of n years at a particular site—one can compute the values of the parameters from those of the sampling mean and coefficient of variation.

Example 7.19. Estimation of Fréchet distribution for storm rainfall. Consider again the annual maximum hourly rainfall of Table E.7.1. The sampling mean and coefficient of variation are 48.16 and 0.493 mm, respectively, as in Example 7.15. The shape parameter of the Fréchet distribution is found by using the method of moments. From Eq. (7.2.37)

$$\frac{\Gamma(1 - 2/\hat{\theta})}{\Gamma^2(1 - 1/\hat{\theta})} = 1 + 0.493^2,$$

which is solved by numerical iterations to obtain

$$\hat{\theta} = 3.62.$$

Then, from Eq. (7.2.38),

$$\hat{x}_0 = \frac{\hat{\mu}}{\Gamma(1 - 1/\hat{\theta})} = \frac{48.16}{\Gamma(1 - 1/3.62)} = 38.15 \text{ mm}.$$

Figure 7.2.7 shows the cdf of the Fréchet distribution with the above values of x_0 and θ on a Gumbel probability plot. An alternative graphical representation is shown in Fig. 7.2.12, where the quantiles of the variate are plotted on the logarithmic scale.

It is seen that the Fréchet distribution plots as a straight line on the $(y, \ln \xi)$ plane. This gives a Weibull probability plot, as discussed shortly.

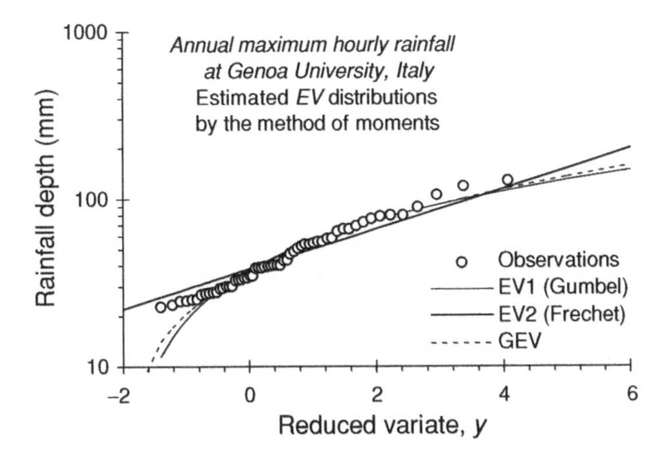

Fig. 7.2.12 Log-Gumbel probability plots of extreme value distributions of annual maximum hourly rainfall at Genoa University, Italy.

The inverse form of the Fréchet distribution is given by the quantile

$$\xi_q = x_0(-\ln q)^{-1/\theta} = x_0 \exp\left(\frac{y}{\theta}\right), \tag{7.2.39}$$

where q is the probability of nonexceedance. Because of the positive exponential form for $\theta > 0$, ξ increases faster than for the Gumbel distribution, when the reduced variate y is increased. Therefore, the distribution can be represented by a curve that is concave upward on a Gumbel probability plot. The quantile estimates of the Fréchet distribution can be the design value for a given return period. Thus,

$$x_{\max}(T) = \xi_q = x_0(-\ln q)^{-1/\theta} = x_0\left(\ln\frac{T}{T-1}\right)^{-1/\theta}, \tag{7.2.40}$$

where $x_{\max}(T)$ denotes the required design value.

Example 7.20. Storm rainfall prediction using Fréchet distribution. Substituting the values of 38.15 mm and 3.62 estimated in Example 7.19 for x_0 and θ, respectively, in Eq. (7.2.40) gives the 10-year hourly design storm:

$$\hat{x}_{\max}(10) = \hat{\xi}_{0.9} = 38.15 \times \left(\ln\frac{1}{0.9}\right)^{-0.27} = 70.0 \text{ mm}.$$

Note that this value is smaller than that predicted by the Gumbel distribution in Example 7.16. Conversely, the 100-year hourly design storm,

$$\hat{x}_{\max}(100) = \hat{\xi}_{0.99} = 38.15 \times \left(\ln\frac{1}{0.99}\right)^{-0.27} = 132.1 \text{ mm},$$

is larger than that predicted by the Gumbel distribution. This is because the Fréchet distribution is concave upward on a Gumbel probability plot.

Note that the Gumbel and Fréchet distributions are mutually related through the logarithmic transformation. If $X_{\max}$ is a Fréchet-distributed variate with parameters x_0 and θ, the logarithmic transformation of $X_{\max}$ will be a Gumbel variate with scale parameter equal to $1/\theta$, and location parameter equal to $\ln x_0$. Because of this relationship, the Fréchet distribution plots as a straight line on a Log-Gumbel probability plot. The Fréchet distribution of the largest value accounts for a lower bound of X_{max}, say, $\varepsilon < x_0$. The corresponding cdf is written as

$$F_{X_{max}}(x) = \exp\left[-\left(\frac{x_0 - \varepsilon}{x - \varepsilon}\right)^{\theta}\right], \tag{7.2.41}$$

with $x > \varepsilon$, and $x_0 > \varepsilon$. Since the smallest value Fréchet distribution is defined for negative values of the variable, this distribution is of little practical interest. Nevertheless, its properties can be determined by using the principle of symmetry.

7.2.4 Weibull distribution as an extreme value model

The most useful applications of the EV3 (third asymptotic) distribution to practical problems deal with the smallest values. The cdf of this distribution [as given by Eq. (4.2.16)] is

$$F_{X_{min}}(x) = 1 - \exp\left[-\left(\frac{x}{\lambda}\right)^{\beta}\right], \quad x \geq 0, \tag{7.2.42}$$

where the scale parameter $\lambda \geq 0$ is sometimes called the characteristic smallest value, and the shape parameter $\beta > 0$ gives a measure of dispersion. The corresponding pdf is

$$f_{X_{min}}(x) = \frac{\beta}{\lambda}\left(\frac{x}{\lambda}\right)^{\beta-1} \exp\left[-\left(\frac{x}{\lambda}\right)^{\beta}\right], \quad x \geq 0. \tag{7.2.43}$$

The rth-order moments of X_{min} are given by

$$E\left[X_{min}^{r}\right] = \lambda^{r}\Gamma(1 + r/\beta), \tag{7.2.44}$$

which yields

$$E[X_{min}] = \lambda\Gamma(1 + 1/\beta), \tag{7.2.45}$$

and

$$\mathrm{Var}[X_{min}] = \lambda^{2}[\Gamma(1 + 2/\beta) - \Gamma^{2}(1 + 1/\beta)]. \tag{7.2.46}$$

The coefficient of variation depends only on the shape parameter, that is,

$$V_{X_{min}}^{2} = \frac{\Gamma(1 + 2/\beta)}{\Gamma^{2}(1 + 1/\beta)} - 1; \tag{7.2.47}$$

this relationship is plotted in Fig. 7.2.11.[11]

If the first two moments of X_{min} are known, the values of the parameters λ and β can be determined from the mean, μ, and the coefficient of variation, V, of the extreme value population by solving Eq. (7.2.47) for β via numerical iteration. Then, using Eq. (7.2.45), one can estimate the value of the scale parameter as

$$\lambda = \frac{\mu}{\Gamma(1 + 1/\beta)}. \tag{7.2.48}$$

[11] See also Table C.6. Table C.5 gives the gamma function.

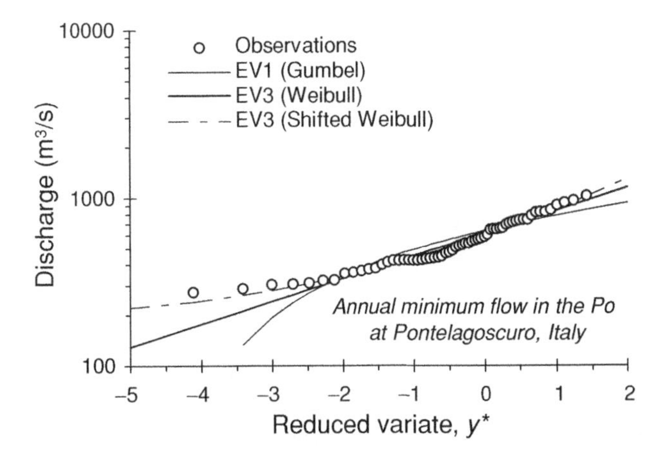

Fig. 7.2.13 Weibull probability plots of extreme value distributions of annual minimum flow in the Po River at Pontelagoscuro, Italy.

Example 7.21. Low-flow analysis using Weibull distribution. The mean and coefficient of variation of annual minimum flow data of Table E.7.2 are 554.6 m^3/s and 0.343, respectively. From Eq. (7.2.47),

$$\frac{\Gamma(1+2/\beta)}{\Gamma^2(1+1/\beta)} = 1 + 0.343^2,$$

the numerical solution of which yields $\hat{\beta} \approx 3.195$. Thus, $\hat{\lambda} = 619.2$ m^3/s. The cdf of the Weibull distribution with these values of λ and β is shown in Fig. 7.2.13, where the Gumbel variate for the smallest value, $y^* = \ln[-\ln(1-q)]$, is used to represent the frequency level (Table C.6 gives $\hat{\beta} \approx 3.17$).

Note that the two-parameter Weibull distribution plots as a straight line on the $(y^*, \ln\xi)$ plane. It is also seen that the Weibull distribution provides a better fit of the observed distribution than the Gumbel distribution.

In many applications of the smallest value distribution of a variate, one may be interested in its lower limit. Introducing the lower bound ε of $X_{\min}$ as a parameter, the cdf of the Weibull distribution is modified as

$$F_{X_{\min}}(x) = 1 - \exp\left[-\left(\frac{x-\varepsilon}{\lambda-\varepsilon}\right)^{\beta}\right], \tag{7.2.49}$$

with $x \geq \varepsilon$ and $\lambda > \varepsilon$. The pdf of the *shifted Weibull distribution* is

$$f_{X_{\min}}(x) = \frac{\beta}{\lambda-\varepsilon}\left(\frac{x-\varepsilon}{\lambda-\varepsilon}\right)^{\beta-1} \exp\left[-\left(\frac{x-\varepsilon}{\lambda-\varepsilon}\right)^{\beta}\right], \tag{7.2.50}$$

and the rth-order moments of $(X_{\min} - \varepsilon)$ are given by

$$E[(X_{\min}-\varepsilon)^r] = (\lambda-\varepsilon)^r\Gamma(1+r/\beta). \tag{7.2.51}$$

From Eq. (7.2.51), the mean and variance of $X_{\min}$ are found as

$$E[X_{\min}] = \varepsilon + (\lambda-\varepsilon)\,\Gamma(1+1/\beta), \tag{7.2.52}$$

and

$$\mathrm{Var}[X_{\min}] = (\lambda-\varepsilon)^2[\Gamma(1+2/\beta) - \Gamma^2(1+1/\beta)]. \tag{7.2.53}$$

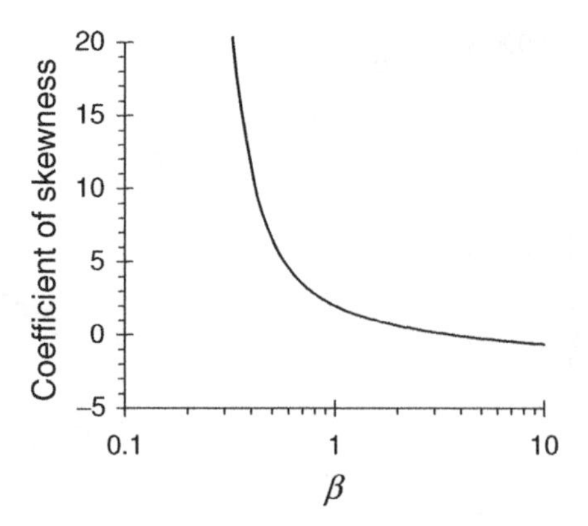

Fig. 7.2.14 Coefficient of skewness versus the exponent (shape parameter) of the smallest value Weibull distribution.

The skewness coefficient, which is given by

$$\gamma_{1,X_{\min}} = \frac{[\Gamma(1+3/\beta) - 3\Gamma(1+2/\beta)\Gamma(1+1/\beta) + 2\Gamma^3(1+1/\beta)]}{[\Gamma(1+2/\beta) - \Gamma^2(1+1/\beta)]^{3/2}}, \tag{7.2.54}$$

only depends on the shape parameter, β (see Fig. 7.2.14).

In his studies on material strength in fatigue, Weibull (1939) assumed lower bounds, denoted by ε, on strength and on the number of cycles before which no failure occurs.[12]

Example 7.22. Fatigue failure analysis using shifted Weibull distribution. After repeated experiments, the observed number of cycles at failure of steel specimens under reversed torsion for a given stress is found to be 477.1 on average, with a standard deviation of 36.4 and a skewness coefficient of 0.328. An engineer needs to estimate the lower bound for this experiment by using the Weibull distribution of the smallest value. Therefore, the estimated value of the skewness coefficient is used in Eq. (7.2.54) to determine the shape parameter of the distribution. Accordingly, the numerical solution of

$$\frac{[\Gamma(1+3/\hat{\beta}) - 3\Gamma(1+2/\hat{\beta})\Gamma(1+1/\hat{\beta}) + 2\Gamma^3(1+1/\hat{\beta})]}{[\Gamma(1+2/\hat{\beta}) - \Gamma^2(1+1/\hat{\beta})]^{3/2}} = 0.328$$

yields $\hat{\beta} = 2.57$. From Eq. (7.2.53),

$$(\hat{\lambda} - \hat{\varepsilon}) = \frac{\hat{\sigma}}{[\Gamma(1+2/\hat{\beta}) - \Gamma^2(1+1/\hat{\beta})]^{1/2}}$$
$$= \frac{36.4}{[\Gamma(1+2/2.57) - \Gamma^2(1+1/2.57)]^{1/2}} = 98.2,$$

which is substituted for $(\lambda - \varepsilon)$ in Eq. (7.2.52) to obtain

$$\hat{\varepsilon} = \hat{\mu} - (\hat{\lambda} - \hat{\varepsilon})\Gamma(1+1/\hat{\beta}) = 477.1 - 98.2 \times \Gamma(1+1/2.57) = 389.9.$$

That is, the estimate of the lower bound or smallest value of the number of cycles at failure for the steel specimens is 389.9.

Also, the scale parameter or characteristic smallest value is estimated as,

$$\hat{\lambda} = 98.2 + 389.9 = 488.1.$$

[12] Since this distribution was also applied by Goodrich (1927) to hydrological data, it is sometimes referred to as the *Goodrich distribution*.

The inverse form of the Weibull distribution is given by

$$\xi_q = \varepsilon + (\lambda - \varepsilon)[-\ln(1-q)]^{1/\beta} = \varepsilon + (\lambda - \varepsilon)\ \exp\left(\frac{y^*}{\beta}\right). \tag{7.2.55}$$

Because of the positive exponential form for $\beta > 0, \xi_q$ increases faster than for the Gumbel distribution, when y^* is increased. Therefore, the distribution can be represented by a curve which is concave upward on a Gumbel probability plot for $\varepsilon > 0$. If $\varepsilon = 0$, this becomes a straight line on a Weibull probability plot. The application of the Weibull distribution with the lower bound to smallest value data of physical variates must be carefully considered. For example, if the Weibull distribution is fitted to the low-flow data of Example 7.21, with a skewness coefficient is 0.617, the parameters estimates obtained by the method of moments are $\hat{\beta} = 2.02$, $\hat{\lambda} = 601.9$ m^3/s, and $\hat{\varepsilon} = 186.8$ m^3/s. Although the lower bound (186.8 m^3/s) is smaller than the smallest observed value of the 61-year sample (275 m^3/s), the existence of this bound should be supported by physical arguments. Nevertheless, as expected, the introduction of a lower bound to the Weibull distribution gives a better fit (see Figs. 7.2.10 and 7.2.13).

For completeness, let us consider also the cdf of the Weibull distribution of the largest value. This can be written as

$$F_{X_{\max}}(x) = \exp\left[-\left(\frac{\lambda - x}{\lambda - \varepsilon}\right)^{\beta}\right], \tag{7.2.56}$$

for $x \leq \lambda$, with λ denoting the upper bound of $X_{\max}$, ε a location parameter, and $\beta > 0$ a shape parameter. The corresponding pdf is

$$f_{X_{\max}}(x) = \frac{\beta}{\lambda - \varepsilon}\left(\frac{\lambda - x}{\lambda - \varepsilon}\right)^{\beta - 1} \exp\left[-\left(\frac{\lambda - x}{\lambda - \varepsilon}\right)^{\beta}\right], \tag{7.2.57}$$

and the rth-order moments of $(\lambda - X_{\max})$ are given by

$$E[(\lambda - X_{\max})^r] = (\lambda - \varepsilon)^r\, \Gamma(1 + r/\beta), \tag{7.2.58}$$

which can be used to derive the moments and the other statistics of $X_{\max}$. Because of the upper bound λ, this distribution is seldom applied in practice.

7.2.5 General extreme value distribution

The general extreme value (GEV) distribution was applied by Jenkinson (1955 and 1969) to identify the frequency distribution of the largest values of meteorological data when the limiting form of the extreme value distribution is unknown. The basic form was introduced in Eq. (7.2.8). Using three parameters, the cdf of the GEV distribution is given by

$$F_{X_{\max}}(x) = \exp\left\{-\left[1 - \frac{k(x - \varepsilon)}{\alpha}\right]^{1/k}\right\}, \tag{7.2.59}$$

where α denotes a scale parameter, ε a location parameter, and k is the shape parameter introduced in Subsection 7.2.1, which determines the type of the asymptotic tail. We discussed previously that for $k < 0$, the GEV represents an EV2 distribution and it is defined only for $x > (\varepsilon + \alpha/k)$; for $k > 0$, this model becomes the EV3 distribution, and it is defined only for $x < (\varepsilon + \alpha/k)$; the case of $k = 0$ corresponds to the Gumbel distribution of Eq. (7.2.17) with scale parameter α and location parameter b.

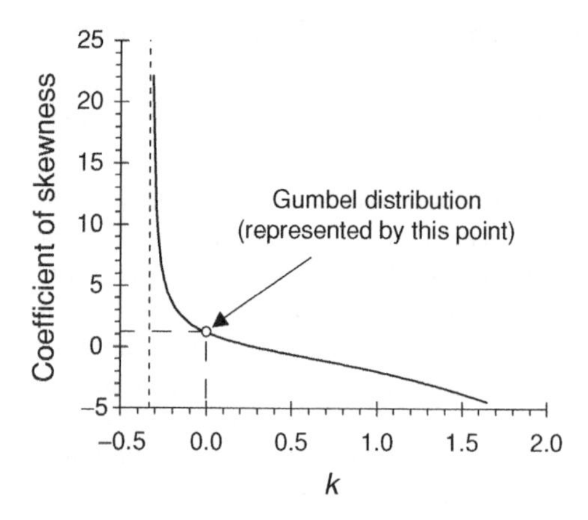

Fig. 7.2.15 Coefficient of skewness versus the exponent (shape parameter) of the GEV distribution.

As in the case of the EV2 distribution, the rth-order moments exist only if $k > -1/r$. The mean and variance of the GEV distribution are given by

$$E[X_{\max}] = \varepsilon + \frac{\alpha}{k}[1 - \Gamma(1+k)], \quad \text{for } k > -1, \tag{7.2.60}$$

and

$$\text{Var}[X_{\max}] = \left(\frac{\alpha}{k}\right)^2 [\Gamma(1+2k) - \Gamma^2(1+k)], \quad \text{for } k > -0.5, \tag{7.2.61}$$

respectively. Therefore, the mean is not defined for $k < -1$, and the variance for $k < -1/2$. The coefficient of skewness is given by

$$\gamma_{1,X_{\max}} = \text{sign}(k)\frac{-\Gamma(1+3k) + 3\Gamma(1+k)\Gamma(1+2k) - 2\Gamma^3(1+k)}{[\Gamma(1+2k) - \Gamma^2(1+k)]^{3/2}}, \quad \text{for } k > -1/3, \tag{7.2.62}$$

where $\text{sign}(k) = +1$ for $k > 0$ and $\text{sign}(k) = -1$ for $k < 0$, while it is not defined for $k < -1/3$; therefore, one notes that the shape parameter only depends on the coefficient of skewness if the third moment exists (see Fig. 7.2.15).

If the first three moments of $X_{\max}$ exist and are known, the values of the three parameters ε, α, and k can be determined from the mean, the variance, and the skewness coefficient of the data. Since Eq. (7.2.62) indicates that k only depends on the coefficient of skewness for $k > -1/3$, one can solve this equation in k by substituting the sampling skewness coefficient. Then, from Eq. (7.2.61) the scale parameter is found as

$$\alpha = \sqrt{\frac{k^2\sigma^2}{\Gamma(1+2k) - \Gamma^2(1+k)}}, \tag{7.2.63}$$

where the sample variance is substituted for $\sigma^2 = \text{Var}[X_{\max}]$. Finally, the location parameter is computed from

$$\varepsilon = \mu - \frac{\alpha}{k}[1 - \Gamma(1+k)], \tag{7.2.64}$$

where the sample mean is substituted for μ.

Example 7.23. Estimation of GEV distribution for storm rainfall data. The mean, variance, and coefficient of skewness for the annual maximum of hourly rainfall total which are estimated from the 58-year record at Genoa University, Italy (as given in Table E.7.1 and used in Example 7.15) are 48.16 mm, 564.33 mm^2, and 1.501, respectively. To estimate the shape parameter of the GEV distribution by the method of moments, the sampling skewness is used in Eq. (7.2.62), which is then solved for k by numerical iteration. Thus, $\hat{k} = -0.05$. Then, from Eq. (7.2.63),

$$\hat{\alpha} = \sqrt{\frac{\hat{k}^2\sigma^2}{\Gamma(1+2\hat{k}) - \Gamma^2(1+\hat{k})}} = \sqrt{\frac{0.0025 \times 564.33}{\Gamma(1-0.10) - \Gamma^2(1-0.05)}} = 17.27\,\text{mm},$$

and, from Eq. (7.2.64),

$$\hat{\varepsilon} = \hat{\mu} - \frac{\hat{\alpha}}{\hat{k}}[1 - \Gamma(1+\hat{k})] = 48.16 - \frac{17.27}{(-0.05)}[1 - \Gamma(1-0.05)] = 37.30\,\text{mm}.$$

Note that the GEV distribution, as applied here, corresponds to the EV2 type. However, the estimated k is very different from the value of $-1/\theta$ found in Example 7.19 where the Fréchet distribution was used. This is because of the introduction of a location parameter ε. Figure 7.2.7 shows the cdf of the GEV distribution with the above values of α, ε, and k as plotted on a Gumbel probability plot; the same distribution is shown in Fig. 7.2.12 which is a Weibull probability plot.

Example 7.24. Estimation of probability distribution for flood flow data. The GEV distribution for storm rainfall data of Example 7.23 has a negative exponent, signifying that it has a heavier tail than the exponential. This often occurs in the largest value analysis of several physical variables, such as storm rainfall, temperature, flood discharge, and snow cover. However, there are exceptions to the general rule when the skewness coefficient is not significant. For example, consider the data of annual maximum flow in the Po River observed at Pontelagoscuro, Italy, from 1918 to 1978 (see Table E.7.2). The mean, standard deviation, and skewness coefficient of the data are 5408 m^3/s, 4627 m^3/s, and 0.0882, respectively. If we fit a GEV distribution to this data the k value is 0.008. Because this is very close to $k = 0$, a Gumbel distribution was fitted with $\alpha = 353.0$ and $b = 4626.9$. The Gumbel probability plot is shown in Fig. 7.2.16 for which the Weibull plotting position was used as in Example 7.15.

The GEV distribution provides, as intended, a flexible model for extreme value data when compared to the Gumbel and the Fréchet distributions. Note that wind velocities and sea wave heights often display a lighter tail than the exponential.

The inverse form of the GEV distribution is given by

$$\xi_q = \varepsilon + \frac{\alpha}{k}[1 - (-\ln q)^k] = \varepsilon + \frac{\alpha}{k}[1 - \exp(-ky)], \tag{7.2.65}$$

providing the predicted qth quantile, which is required to determine the design value for a specified probability of nonexceedance or return period. Figure 7.2.17 shows a plot of $(\xi_q - \varepsilon)/\alpha$ against y for different values of k. As already emphasized, it is seen that the curve is concave downward for $k > 0$, it is linear for $k = 0$, and it is concave upward for $k < 0$. Note that, for $k < 0$, the rate of increase of $(\xi_q - \varepsilon)/\alpha$ with y is very sensitive to the value of k.

The design value of a given GEV distributed random variable is associated with a given return period of exceedance as follows:

$$x_{\max}(T) = \xi_{1-1/T} = \varepsilon + \frac{\alpha}{k}\left[1 - \left(\ln\frac{T}{T-1}\right)^k\right], \tag{7.2.66}$$

where $x_{\max}(T)$ denotes the required design value.

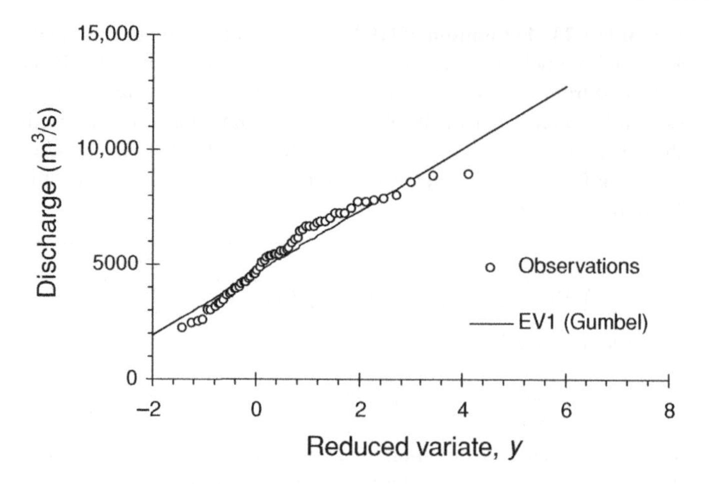

Fig. 7.2.16 Gumbel probability plot of annual maximum flow in the Po River at Pontelagoscuro, Italy.

Example 7.25. Storm rainfall prediction using the GEV distribution. Substituting the parameter estimates of Example 7.23 ($\bar{k} = -0.05$, $\hat{\alpha} = 17.27$ mm, and $\hat{\varepsilon} = 37.30$ mm) for k, α, and ε in Eq. (7.2.66) gives the estimated 10-year hourly design storm as

$$\hat{x}_{\max}(10) = \xi_{0.9} = 37.30 + \frac{17.27}{(-0.05)}\left[1 - \left(\ln\frac{10}{9}\right)^{-0.05}\right] = 78.4 \text{ mm}.$$

This value is higher than that predicted by the Fréchet distribution (see Example 7.20), and it is very close to that predicted by the Gumbel distribution (see Example 7.16), both of which exclude the location parameter ε. Conversely, the 100-year hourly design storm,

$$\hat{x}_{\max}(100) = \xi_{0.99} = 37.30 + \frac{17.17}{(-0.05)}\left[1 - \left(\ln\frac{100}{99}\right)^{-0.05}\right] = 126.1 \text{ mm}.$$

This is higher than that predicted through the Gumbel distribution, and it is lower than that predicted by the Fréchet distribution, both of which exclude the location parameter ε.

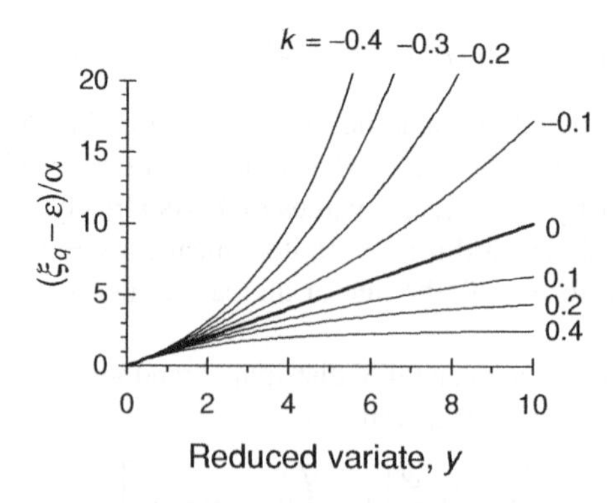

Fig. 7.2.17 Values of $(\xi_q - \varepsilon)/\alpha$ predicted by the GEV distribution for different values of exponent k.

In Examples 7.18 to 7.25 we used the method of moments. A consistent estimator for the parameters of the GEV distribution is given by the method of maximum likelihood.[13] Also, the ML estimates are unbiased in large samples, and experience has shown that they do very well with recorded observations. However, ML estimators cannot always be reduced to simple formulas, so estimates must be calculated using numerical methods. Moreover, ML estimators sometimes perform poorly when the distribution of the observations deviates significantly from the fitted distribution. An alternative method is given by the *method of* L-*moments*, introduced in Section 3.2. The parameters of the GEV distribution are related to the first three L-moments as

$$L_1 = \varepsilon + \frac{\alpha}{k}[1 - \Gamma(1 + k)], \tag{7.2.67a}$$

$$L_2 = \frac{\alpha}{k}(1 - 2^{-k})\Gamma(1 + k), \tag{7.2.67b}$$

$$\frac{L_3}{L_2} = \frac{2(1 - 3^{-k})}{(1 - 2^{-k})} - 3. \tag{7.2.67c}$$

After the values of L_1, L_2, and L_3 are estimated from the data, one can solve for k first from Eq. (7.2.67c). An approximate explicit solution for $-0.5 \le k \le 0.5$ is given by Hosking et al. (1985) as

$$k = 7.8590\left(\frac{2L_2}{L_3 + 3L_2} - \frac{\ln 2}{\ln 3}\right) + 2.9554\left(\frac{2L_2}{L_3 + 3L_2} - \frac{\ln 2}{\ln 3}\right)^2.$$

Then, the estimate of α is obtained from Eq. (7.2.67b) as

$$\alpha = \frac{kL_2}{(1 - 2^{-k})\Gamma(1 + k)}.$$

Finally, the location parameter is found using Eq. (7.2.67a); that is,

$$\varepsilon = L_1 - \frac{\alpha}{k}[1 - \Gamma(1 + k)].$$

7.2.6 Contagious extreme value distributions

Some extreme events can be described by the maximum (or minimum) value taken by a sequence of a random number of random variables. Such a concept was introduced in Section 3.4 (see Example 3.61) by using the concept of contagious distributions. In extreme value theory, a number of equispaced data points is assumed. The approach adopted here, however, is somewhat different, because a finite but random number of occurrences is considered. Let $X_1, X_2, \ldots, X_N$ denote a set of independent random variables with a common distribution $F_X(x)$, where x is an observed value and N is the random number of data points occurring within a fixed period of, say, 1 year. Examples are the number of floods occurring at a site in a year, the annual number of earthquakes in a region, and the number of high wind speeds. To define such events, a lower threshold must be often introduced considering that, for instance, a flood event is defined by the peak flow exceeding a given level. If $p_N(n)$ denotes the pmf of N, the cdf of X_{max} can be derived

[13] However, the maximum likelihood estimator may not exist for $k > 1$ (Smith, 1985).

from that of the X_is by weighting $[F_X(x)]^n$ by $p_N(n)$ for all possible values n of N. Thus,

$$F_{X_{max}}(x) = \sum_{n=0}^{+\infty} [F_X(x)]^n p_N(n). \tag{7.2.68}$$

A probabilistic model useful for counting the occurrences of extreme events is the Poisson distribution. Under the assumption that N is Poisson distributed with pdf given by Eq. (4.1.7), one has

$$F_{X_{max}}(x) = \sum_{n=0}^{+\infty} [F_X(x)]^n \frac{v^n e^{-v}}{n!} = \sum_{n=0}^{+\infty} \frac{[v F_X(x)]^n e^{-v}}{n!}$$

$$= e^{-v[1-F_X(x)]} \sum_{n=0}^{+\infty} \frac{[v F_X(x)]^n e^{-v F_X(x)}}{n!},$$

where v denotes the mean number of occurrences in a given time period, such as 1 year. The sum of the series on the right-hand side of the above equation is unity because the pdf of a Poisson variate with parameter $v F_X(x)$ sums to unity over all its possible values. Therefore,

$$F_{X_{max}}(x) = e^{-v[1-F_X(x)]}, \tag{7.2.69}$$

where $1 - F_X(x) = \Pr[X > x]$ is the probability of exceedance of the underlying variable. For example, if X is a shifted exponentially distributed variate, say,

$$F_X(x) = 1 - \exp[-\lambda(x - \varepsilon)], \tag{7.2.70}$$

with $\lambda > 0$ denoting the scale parameter, ε the location parameter and $x \geq \varepsilon$, the distribution of X_{max} is obtained from Eq. (7.2.69) as

$$F_{X_{max}}(x) = \exp[-v e^{-\lambda(x-\varepsilon)}] = \exp[-e^{-\lambda(x-\varepsilon-\lambda^{-1}\ln v)}]. \tag{7.2.71}$$

This is a Gumbel distribution with scale parameter $1/\lambda$, and location parameter equal to $\varepsilon + \lambda^{-1} \ln v$.

Example 7.26. Flood occurrence and peak discharge. In Table 1.1.1 the number of flood occurrences a year recorded from 1939 to 1972 at the gauging station of Calamazza on the Magra River, Italy, is reported. A flood occurrence is defined as the event of river discharge exceeding 300 m^3/s. Because 133 floods were observed in a period of 34 years, the average number of floods per year is estimated as 133/34, that is, 3.91. Thus, the Poisson distribution with $v = 3.91$ is taken to describe the count of annual flood occurrences at that site (see Fig. 7.2.18).

The average discharge of the recorded peak flows exceeding 300 m^3/s is 925 m^3/s, and flood discharge is assumed to be an exponentially distributed variate with scale parameter λ; that is,

$$F_X(x) = 1 - \exp[-\lambda(x - 300)],$$

for $x \geq 300$ m^3/s, and $F_X(x) = 0$ for $x < 300$ m^3/s. Since $E[X - 300] = 1/\lambda$, the scale parameter is estimated from

$$\hat{\lambda} = \frac{1}{(925 - 300)} = \frac{1}{625} = 0.0016\ (\text{m}^3/\text{s})^{-1}.$$

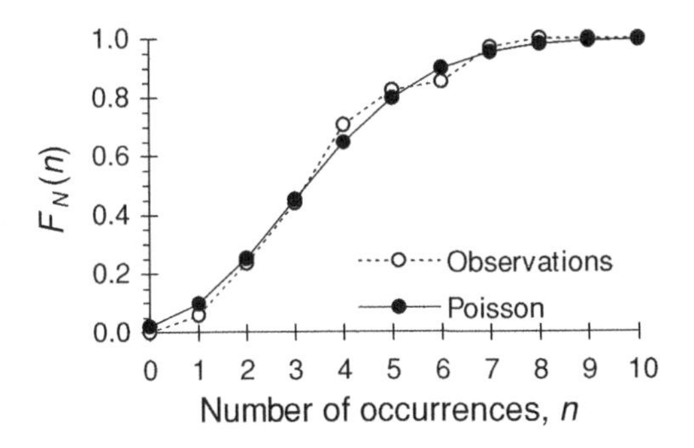

Fig. 7.2.18 Cdf of the number of annual occurrences of flood flows exceeding 300 m^3/s in the Magra River at Calamazza between Pisa and Genoa, Italy.

Thus,

$$F_X(x) = 1 - \exp[-0.0016(x - 300)].$$

From Eq. (7.2.71), the cdf of maximum annual flood discharge is found as

$$F_{X_{\max}}(x) = \exp\{-\exp[-0.0016(x - 300) + \ln 3.91]\} = \exp\left(-e^{-(x-1152)/625}\right),$$

with $x > 300$ m^3/s. The observed flood flows exceeding a specified level are sometimes referred to as a *peaks over threshold* (POT) *series* in Europe, and as a *partial duration series* (PDS) in the United States. Data analysis shows that the accuracy of the Poisson assumption to model flood occurrences increases with increasing threshold level.

Depending on the common distribution $F_X(x)$ of the X_is, the other asymptotic types of the largest value distribution can also be derived. If the X_is have a common Pareto distribution, say,

$$F_X(x) = 1 - \left(\frac{a}{x}\right)^{\theta}, \tag{7.2.72}$$

with $x > a$, where $\theta > 0$ is a shape parameter and $a > 0$ a scale parameter, Eq. (7.2.69) yields

$$F_{X_{\max}}(x) = \exp\left[-\nu\left(\frac{a}{x}\right)^{\theta}\right] = \exp\left[-\left(\frac{a\nu^{1/\theta}}{x}\right)^{\theta}\right], \tag{7.2.73}$$

for $x > a$. This is the Fréchet distribution of Eq. (7.2.32) with shape parameter θ, and scale parameter $x_0 = a\nu^{1/\theta}$, where $x_0 > 0$. Similarly, the GEV distribution is found for X_is distributed according to a generalized Pareto distribution, say,

$$F_X(x) = 1 - \left[1 - k\left(\frac{x - b}{c}\right)\right]^{1/k}, \tag{7.2.74}$$

with $x > b$ for $k < 0$, and $b \leq x < b + c/k$ for $k > 0$, where k is a shape parameter, c a scale parameter, and b a location parameter. Substituting Eq. (7.2.74) for $F_X(x)$ in Eq. (7.2.69) yields Eq. (7.2.59) after some manipulations, where $\alpha = c\nu^{-k}$, and $\varepsilon = b + (1 - \nu^{-k})c/k$. Note that for $k = 0$ the generalized Pareto distribution of Eq. (7.2.74) is a shifted exponential distribution, which leads to a Gumbel distribution with scale and location parameters equal to c and $b + c \ln \nu$, respectively.

Example 7.27. Hurricane winds. Suppose that a 24-year record of wind speeds is available for a given area affected by hurricanes. The record contains the maximum wind speeds of nine hurricanes, the average of which is 34 m/s and the coefficient of variation is 0.298. An engineer wishes to determine the wind speed corresponding to a 50-year return period, and assumes that the number of hurricanes per year is a Poisson-distributed variate. Thus, $\hat{\nu} = 9/24 = 0.375$. If the wind speed is a Pareto-distributed variate, the mean and coefficient of variation are related to the parameters of Eq. (7.2.72) as $\mu = a\theta/(\theta - 1)$ and $V^2 = 1/[\theta(\theta - 2)]$. Hence, for $\theta > 0$,

$$\hat{\theta} = 1 + (1 + 1/\hat{V}^2)^{1/2} = 1 + (1 + 1/0.298^2)^{1/2} = 4.5;$$

thus,

$$\hat{a} = \hat{\mu}(\hat{\theta} - 1)/\hat{\theta} = 34 \times (4.5 - 1)/4.5 = 26.4 \text{ m/s}.$$

Therefore, the estimated scale parameter of the (truncated) Fréchet distribution of annual maximum wind speed is given by

$$\hat{x}_0 = \hat{a}\hat{\nu}^{1/\theta} = 26.4 \times 0.375^{1/4.5} = 21.2 \text{ m/s}.$$

For $q = 1 - 1/T = 1 - 1/50 = 0.98$, $y = 3.902$. From Eq. (7.2.39),

$$\hat{x}_{\max}(50) = \hat{x}_0 \exp\left(\frac{y}{\theta}\right) = 21.2 \exp\left(\frac{3.902}{4.5}\right) = 50.5 \text{ m/s}$$

is the required wind velocity.

This approach has the advantage of accounting for the complete sequence of extreme events associated with a given physical variable. An example is the highest wind velocity in a thunderstorm event, which can be modeled as a Weibull-distributed variate (see Example 4.26). If thunderstorms occur as a Poisson process, substituting Eq. (4.2.16) for $F_X(x)$ in Eq. (7.2.69) gives the extreme value distribution of annual maximum thunderstorm wind speed as

$$F_{X_{\max}}(x) = \exp\left\{-\exp\left[-\left(\frac{x}{\lambda}\right)^{\beta} + \ln\nu\right]\right\}, \tag{7.2.75}$$

where $\lambda > 0$ and $\beta > 0$ are the scale and shape parameters of the underlying Weibull-distributed maximum wind speed in a thunderstorm, respectively, and ν is the average number of thunderstorms in a year. It is seen that $(X_{\max})^{\beta}$ is a Gumbel-distributed variate with scale parameter λ^{β}, and location parameter equal to $\lambda^{\beta}\ln\nu$. This distribution is represented on a Gumbel probability plot by a curve which is concave upward for $\beta < 1$, linear for $\beta = 1$, and concave downward for $\beta > 1$ (see Fig. 7.2.19).

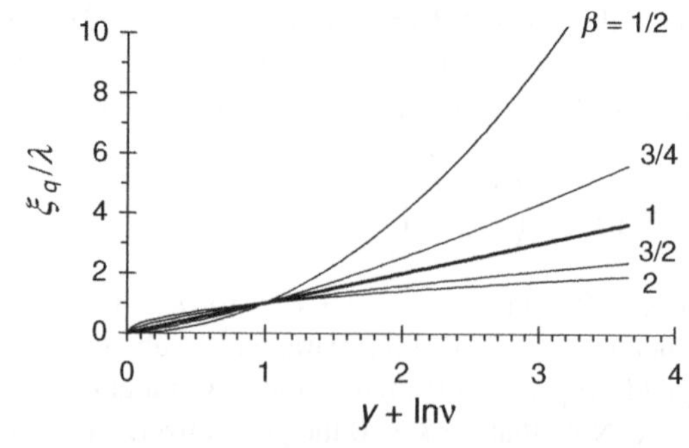

Fig. 7.2.19 Values of ξ_q/λ versus $y + \ln\nu$ predicted by the Poisson-Weibull largest value contagious distribution for different values of exponent β.

Extreme values of a design variable can also arise from the combined effect of more than one physical variable. For example, the maximum annual flood in a river can be caused by high runoff and snowmelt; and the maximum height of sea waves may be due to the combined effect of astronomical tides and storm surges. Mixture distributions are needed to model, for example, extreme wind speeds, if one must distinguish thunderstorm winds from hurricane and tornado winds, which have different probability distributions. Assuming m independent sequences of random variables, each of which is associated with a Poisson-distributed occurrence process with parameter ν_j, $j = 1, \ldots, m$, gives

$$F_{X_{\max}}(x) = \prod_{j=1}^{m} e^{-\nu_j[1-F_{X_j}(x)]}, \tag{7.2.76}$$

where $F_{X_i}(x)$ denotes the common cdf of the jth component. For example, if two components, say, the X_{1i}s and X_{2i}s, are considered that are shifted exponentially distributed variates, the extreme value distribution is

$$\begin{aligned} F_{X_{\max}}(x) &= \exp\left[-e^{-\lambda_1(x-\varepsilon_1-\lambda_1^{-1}\ln\nu_1)} - e^{-\lambda_2(x-\varepsilon_2-\lambda_2^{-1}\ln\nu_2)}\right] \\ &= \exp\left(-e^{-\frac{x-b_1}{\alpha_1}} - e^{-\frac{x-b_2}{\alpha_2}}\right), \end{aligned} \tag{7.2.77}$$

where ν_1 and ν_2 denote the mean number of occurrences per year of the X_{1i}s and X_{2i}s, respectively. This distribution is sometimes referred to as the *two-component extreme value* (TCEV) distribution (see Rossi et al., 1984).

Example 7.28. Highest sea waves. An 11-year record of sea storms is available for the Adriatic Sea in Venice, Italy. The annual maximum height of sea waves for eastern storms (direction from 45° to 100° North) is a Gumbel-distributed variate with scale and location parameters of 0.34 and 2.83 m, respectively. The same distribution is applied to sea wave heights for northern storms (direction from 100° to 160° North), and the scale and location parameters are estimated as 0.51 and 3.15 m, respectively. Denoting the annual maximum height independent of storm direction by $X_{\max}$, Eq. (7.2.77) yields

$$F_{X_{\max}}(x) = \exp\left(-e^{-(x-2.83)/0.34} - e^{-(x-3.15)/0.51}\right).$$

This distribution is shown in Fig. 7.2.20 using a Gumbel probability plot.

Assuming that the highest record in a year independent of direction to be a Gumbel variate, values of 0.43 and 3.62 m are estimated for the scale and location parameters, respectively.

Another type of contagious extreme value distribution is obtained for the extreme value of variables with model parameters that are themselves random variables. (More about this concept of parameters follows in Chapter 10.) For example, let the X_is be Weibull-distributed variates with common parameters β and λ; we model λ as a random variate Λ with known pdf, say, $f_\Lambda(\lambda)$. Accordingly, the cdf of X is found using Eq. (3.4.41), and the extreme value distribution is then determined from Eq. (7.2.68). Let $f_\Lambda(\lambda) = \eta e^{-\eta\lambda}$. From Eq. (3.4.41)

$$F_X(x) = \int_0^\infty \left\{1 - \exp\left[-\left(\frac{x}{\lambda}\right)^\beta\right]\right\} \eta e^{-\eta\lambda} d\lambda. \tag{7.2.78}$$

For Poisson-distributed occurrences with parameter ν, one obtains

$$F_{X_{\max}}(x) = \exp\left[-\nu \int_0^{+\infty} \eta \exp\left(-\frac{x^\beta + \eta\lambda^{\beta+1}}{\lambda^\beta}\right) d\lambda\right], \tag{7.2.79}$$

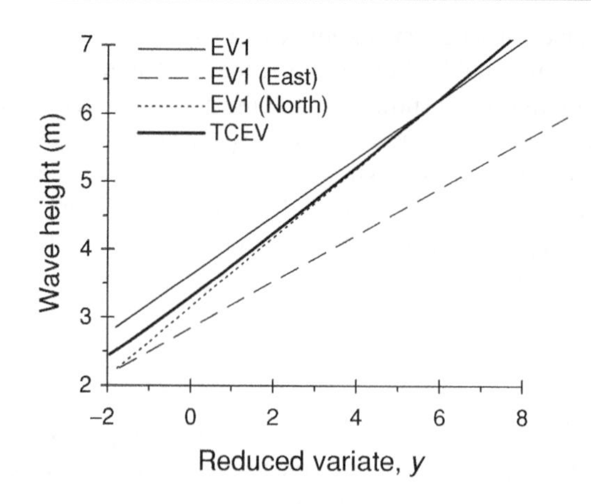

Fig. 7.2.20 Gumbel probability plot of annual maximum heights of sea waves in Venice.

which must be solved numerically to compute the probabilities of nonexceedance of extreme values. This model might be applied, for instance, to highest sea waves, because the height of the highest wave in a sea storm can be considered to be a Rayleigh variate ($\beta = 2$) with parameter λ depending on the spectral properties of the storm. Assuming that these properties vary randomly from one storm to another, one can take λ to be outcome of an exponentially distributed variate Λ with scale parameter η, and use Eq. (7.2.79) to evaluate the probabilities of nonexceedance of annual maximum sea wave heights.

Example 7.29. Avalanche size. Snow has some complex properties such as density, cohesion, and angle of internal friction. If snow falls continuously and accumulates over a long slope, at some critical depth of snow, the frictional resistance of the sloping surface will be overcome and movement of the snow mass will commence. An avalanche is a large mass of snow that moves on a mountain slope causing destruction in its wake.

It is assumed that avalanches in a given mountain area occur as Poisson events and their size (volume of detached snow) is an exponentially distributed variate X with parameter λ. This parameter depends on local topography. Let us assume that it varies uniformly in a particular study area from $a = 0.001\ \text{m}^{-3}$ to $b = 0.01\ \text{m}^{-3}$. Thus,

$$F_X(x) = \int_a^b \frac{1 - e^{-\lambda x}}{b - a} d\lambda = 1 - \frac{e^{-ax} - e^{-bx}}{(b - a)x},$$

and the associated extreme value cdf is given by

$$F_{X_{\max}}(x) = \exp\left[-v \frac{e^{-ax} + e^{-bx}}{(b - a)x}\right],$$

where v is the average number of avalanches occurring in the area in a year. If $v = 9$, one has

$$F_{X_{\max}}(x) = \exp\left[-9 \frac{e^{-0.001x} + e^{-0.01x}}{(0.01 - 0.001)x}\right] = \exp\left(-\frac{e^{-0.001x} + e^{-0.01x}}{0.001x}\right),$$

which is shown in Fig. 7.2.21 on a Gumbel probability plot. Note that this distribution plots as a concave upward curve.

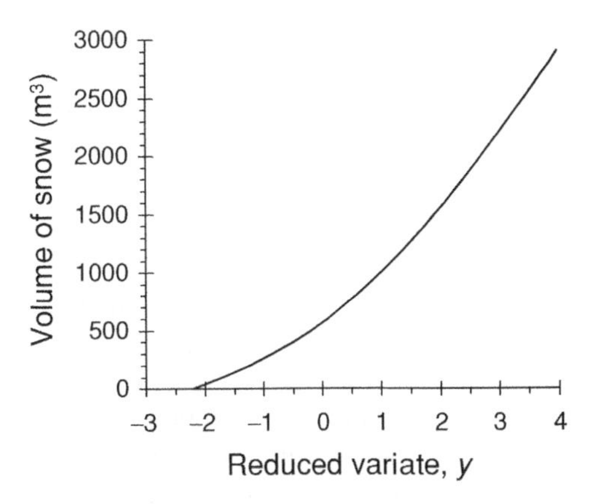

Fig. 7.2.21 Gumbel probability plot of avalanche volume.

7.2.7 Use of other distributions as extreme value models

In some engineering applications no information is available on an underlying variable, the largest or smallest value distribution of which may be required. Because a physical variable can be the result of complex physical processes, the derivation of its extreme value distribution should account for manifold factors affecting extreme events. In most cases, it remains undetermined due to inadequate knowledge of these factors, so that an engineer must infer it from the available extreme value data. Because the three asymptotic forms are not exhaustive, and the convergence to a limiting case can be quite slow, other parametric distributions should also be considered. Here the problem may be one of selecting a distribution from among a number of contending mathematical models when no single one is preferred on the basis of the physical characteristics of the phenomena. In practice, lognormal, gamma, and Pareto distributions are among those that have been applied to extreme value data.

The lognormal distribution, which is discussed in Subsection 4.2.7, has been successfully applied to largest values of hydrologic, meteorologic, and climatic data. One could assume that these extremes result from the joint multiplicative action of a vast number of meteorological and geographical effects. If this number is infinitely large, $\ln X_{\max}$ is the sum of an infinite number of variables, and, accordingly, it is normally distributed by the Central Limit Theorem and its extensions to the sum of dependent variables. However, it is likely that the interactions of the contributory effects are manifold, such as of additive, multiplicative, exponential, and power types. Therefore, the lognormal distribution can provide only an approximation to real-world situations, just as when other theoretical distributions are applied to extreme events.

The pdf of the lognormal distribution is given in Eq. (4.2.27). In practice, the *shifted lognormal distribution* is often adopted as an extreme value model, where a location parameter is introduced to account for the presence of an inner cutoff of the variate. The cdf of the shifted lognormal distribution is given by

$$F_{X_{\max}}(x) = \int_{\varepsilon}^{x} \frac{1}{u\sigma_{\ln(X-\varepsilon)}\sqrt{2\pi}} \exp\left\{-\frac{1}{2}\left[\frac{\ln(u-\varepsilon)-\mu_{\ln(X-\varepsilon)}}{\sigma_{\ln(X-\varepsilon)}}\right]^2\right\} du, \qquad (7.2.80)$$

with $x > \varepsilon$, where ε is the location parameter, $\mu_{\ln(X-\varepsilon)}$ is the mean of $\ln(X_{\max} - \varepsilon)$, and $\sigma_{\ln(X-\varepsilon)}$ is the standard deviation of $\ln(X_{\max} - \varepsilon)$. For $\varepsilon = 0$, the lognormal distribution is signified. This distribution is sometimes referred to as *Gibrat-Galton distribution* because it was first introduced by Galton (1875) and it was applied to storm data by Gibrat (1932). The inverse form is not explicit, so that numerical methods are used to find design values for a specified probability of nonexceedance. By introducing the standard normal variate, we can evaluate the qth quantile as

$$\xi_q = \varepsilon + \exp\left(\mu_{\ln(X-\varepsilon)} + \Phi_q^{-1}\sigma_{\ln(X-\varepsilon)}\right), \tag{7.2.81}$$

where Φ_q^{-1} is the qth quantile of the standard normal variate. If the inner cutoff ε is known from physical reasoning, the estimation of $\mu_{\ln(X-\varepsilon)}$ and $\sigma_{\ln(X-\varepsilon)}$ by the method of moments can be made by using Eq. (4.2.28). When ε too must be estimated from the data, the method of moments can be developed as follows. Since the standard deviation and the skewness coefficient of $X_{\max} - \varepsilon$ equal those of $X_{\max}$, these are in turn estimated by the corresponding sampling statistics. The skewness coefficient of $X_{\max}$ can be thus written as

$$\gamma_1 = 3\frac{\sigma_{X-\varepsilon}}{\mu_{X-\varepsilon}} + \left(\frac{\sigma_{X-\varepsilon}}{\mu_{X-\varepsilon}}\right)^3 = 3\frac{\sigma_X}{\mu_{X-\varepsilon}} + \left(\frac{\sigma_X}{\mu_{X-\varepsilon}}\right)^3 = 3\frac{\sigma}{\mu-\varepsilon} + \left(\frac{\sigma}{\mu-\varepsilon}\right)^3,$$

where μ, σ, and γ_1 denote the mean, the standard deviation, and the skewness coefficient of $X_{\max}$, respectively. The numerical solution of

$$\gamma_1(\mu-\varepsilon)^3 - 3(\mu-\varepsilon)^2\sigma - \sigma^3 = 0, \tag{7.2.82}$$

for ε yields the required estimate of the location parameter. From Eq. (4.2.28*b*),

$$\sigma_{\ln(X_{\max}-\varepsilon)} = \sqrt{\ln\left[1 + \left(\frac{\sigma}{\mu-\varepsilon}\right)^2\right]} \tag{7.2.83}$$

and

$$\mu_{\ln(X_{\max}-\varepsilon)} = \ln(\mu-\varepsilon) - \frac{1}{2}\ln\left[1 + \left(\frac{\sigma}{\mu-\varepsilon}\right)^2\right]. \tag{7.2.84}$$

For $\varepsilon = 0$, Eqs. (7.2.83) and (7.2.84) provide the parameter estimates for the lognormal distribution.

Example 7.30. Storm rainfall analysis using lognormal and shifted lognormal distributions. An engineer wishes to compute the 100-year maximum hourly rainfall depth from the data of Table E.7.1, for Genoa University, Italy, using the lognormal and shifted lognormal models. The sample mean, standard deviation, coefficient of variation, and skewness coefficient of the data are 48.16 mm, 23.76 mm, 0.493, and 1.501, respectively, as in Example 7.23. For $\varepsilon = 0$, Eqs. (7.2.83) and (7.2.84) yield

$$\hat{\sigma}_{\ln(X_{\max})} = \sqrt{\ln(1 + 0.493^2)} = 0.467$$

and

$$\hat{\mu}_{\ln(X_{\max})} = \ln(48.16) - 0.5\ln(1 + 0.493^2) = 3.766,$$

respectively. From Eq. (7.2.81), the estimated 100-year rainfall depth

$$\hat{x}_{\max}(100) = \exp\left(3.766 + \Phi_{0.99}^{-1} \times 0.467\right) = \exp(3.766 + 2.326 \times 0.467) = 127.9\text{ mm}.$$

If the shifted lognormal distribution is used, Eq. (7.2.82) is written as

$$1.501(48.16 - \hat{\varepsilon})^3 - 3 \times 23.76 \times (48.16 - \hat{\varepsilon})^2 - 23.76^3 = 0,$$

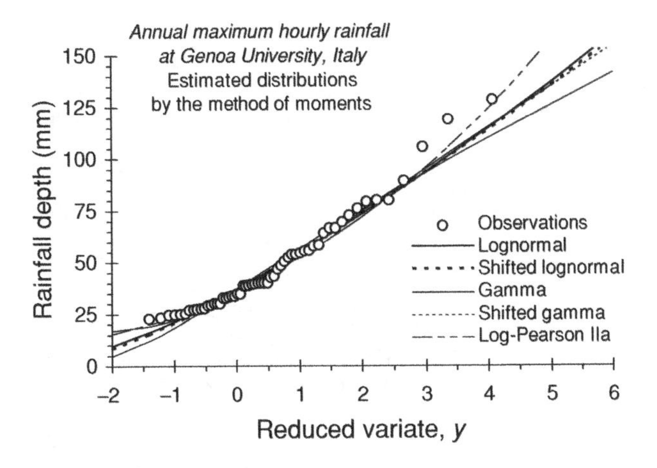

Fig. 7.2.22 Gumbel probability plots of annual maximum hourly rainfall at Genoa University, Italy.

the numerical solution of which yields $\hat{\varepsilon} = -2.75$ mm. This value is substituted for ε in Eqs. (7.2.83) and (7.2.84) to obtain

$$\hat{\sigma}_{\ln(X_{\max}-\varepsilon)} = \sqrt{\ln\left[1 + \left(\frac{23.76}{48.16 + 2.75}\right)^2\right]} = 0.444,$$

and

$$\hat{\mu}_{\ln(X_{\max}-\varepsilon)} = \ln(48.16 + 2.75) - \frac{1}{2}\ln\left[1 + \left(\frac{23.76}{48.16 + 2.75}\right)^2\right] = 3.832.$$

Note that the estimated negative value of the location parameter is physically unrealizable and should be set to zero (but is included here merely for comparative purposes). From Eq. (7.2.81),

$$\hat{x}_{\max}(100) = -2.75 + \exp(3.832 + 2.326 \times 0.444) = 126.8\,\text{mm}.$$

The cdfs of the lognormal and the shifted lognormal models are shown in Fig. 7.2.22. Both distributions are concave upward on the Gumbel probability plots shown.

The gamma distribution[14] is sometimes used in the analysis of observed extreme value data. Introducing an inner cutoff level usually improves the fit to the data. The *shifted gamma distribution*, which is also referred to as *Pearson Type III distribution*, is determined by using the same procedure adopted above to derive the shifted lognormal distribution. If ε denotes a location parameter, $(X_{\max} - \varepsilon)$ is a gamma-distributed variate with cdf written as

$$F_{X_{\max}}(x) = \int_{\varepsilon}^{x} \frac{\lambda^r (u - \varepsilon)^{r-1}}{\Gamma(r)} \exp[-\lambda(u - \varepsilon)]du \tag{7.2.85}$$

[14] See Subsection 4.2.3.

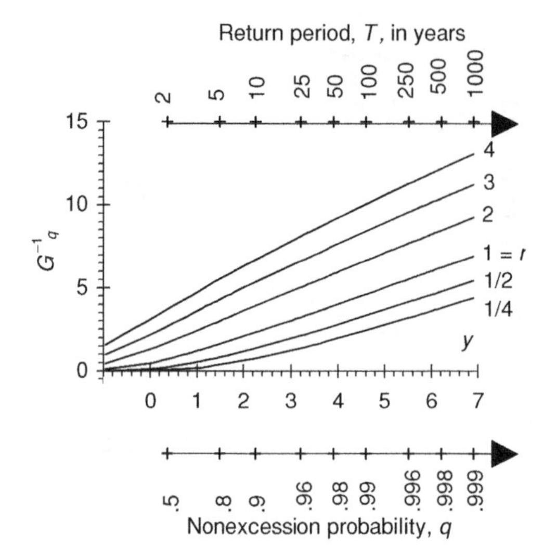

Fig. 7.2.23 Values of the standard gamma quantiles versus Gumbel variate, y, probability of nonexceedance, q, and return period, T, for different values of exponent r.

with $x > \varepsilon$, where λ is a scale parameter, r is a shape parameter, and ε is a location parameter.[15] To obtain the qth quantile, numerical computations are required. This is done by using the standard gamma distribution with cdf

$$F_W(w) = G(w) = \int_0^w \Gamma^{-1}(r) z^{r-1} \exp(-z) dz, \tag{7.2.86}$$

which is widely tabulated.[16] Thus,

$$\xi_q = \varepsilon + \frac{G_q^{-1}(r)}{\lambda}, \tag{7.2.87}$$

where $G_q^{-1}(r)$ denotes the qth quantile of the standard gamma variate (see Fig. 7.2.23).

If ε is known from physical reasoning, Eqs. (4.2.12) provide the shape and scale parameters of the shifted gamma distribution from the mean and standard deviation of $(X_{\max} - \varepsilon)$. When ε must be estimated from the data, the estimators by the method of moments are

$$r = 4\gamma_1^{-2}, \tag{7.2.88}$$

$$\lambda = \frac{2}{\sigma \gamma_1}, \tag{7.2.89}$$

and

$$\varepsilon = \mu - \frac{r}{\lambda} = \mu - \frac{2\sigma}{\gamma_1}, \tag{7.2.90}$$

respectively.

---

[15] The pdf is given by Eq. (4.2.11) but a $(x - \varepsilon)$ modification is necessary.

[16] As shown in Subsection 4.2.3, one can use the tables of the chi-squared distribution (see Table C.3) or a computer algorithm for the purpose.

In applications to extreme value data, the logarithmic transformation of the variate is often taken. The resulting distribution is known as the *log-Pearson Type III distribution*, which was recommended previously for flood frequency analysis by the American Water Resources Council (however, some recent publications by the U.S. Geological Survey are based on the GEV distribution). The cdf of the log-Pearson Type III distribution is written as,

$$F_{X_{\max}}(x) = \int_{\varepsilon}^{x} \frac{\lambda^r (\ln u - \varepsilon)^{r-1}}{u\Gamma(r)} \exp[-\lambda(\ln u - \varepsilon)]du, \tag{7.2.91}$$

with $x > \varepsilon$, where λ denotes a scale parameter, r a shape parameter, and ε a location parameter. The qth quantile is obtained from the corresponding quantile of the standard gamma variate as

$$\xi_q = \exp\left[\varepsilon + \frac{\lambda G_q^{-1}(r)}{\lambda}\right]. \tag{7.2.92}$$

When ε is known, the estimation of λ and r by the method of moments can be easily performed by computing the mean and standard deviation of $\ln(X_{\max} - \varepsilon)$. If it must be estimated from the data, the parameter estimates by the method of moments can be computed by substituting the sample mean, standard deviation, and skewness coefficient of $\ln(X_{\max})$ for μ, σ, and γ_1, respectively, in Eqs. (7.2.88) to (7.2.90).

Example 7.31. Storm rainfall analysis using gamma, shifted gamma, and log-Pearson Type III distributions. Consider again the data sample of Table E.7.1, for Genoa University, Italy. The mean, standard deviation, and skewness coefficient of these data are 48.16 mm, 23.76 mm, and 1.501, respectively, as in Example 7.30. An engineer wishes to compute the 100-year maximum hourly rainfall depth using the gamma, shifted gamma, and log-Pearson Type III models. The shape and scale parameters of the gamma distribution are estimated by the method of moments by substituting the sample mean and standard deviation of $X_{\max}$, respectively, for μ and σ in Eq. (4.2.12). Thus,

$$\hat{r} = \frac{\hat{\mu}_X^2}{\hat{\sigma}_X^2} = \frac{48.16^2}{23.76^2} = 4.11,$$

and

$$\hat{\lambda} = \frac{\hat{\mu}_X}{\hat{\sigma}_X^2} = \frac{48.16}{23.76^2} = 0.0853\,\text{mm}^{-1},$$

respectively. For $q = 0.99$, the 99% quantile of the standard gamma variate is $G_{0.99}^{-1}(4.11) = 10.220$; this can be approximated from Table C.3 (as shown in Example 4.23). This value is used to estimate the 100-year rainfall depth for the gamma model. From Eq. (7.2.87),

$$\hat{x}_{\max}(100) = \frac{G_q^{-1}(r)}{\hat{\lambda}} = \frac{10.220}{0.0853} = 119.8\,\text{mm}.$$

From Eq. (7.2.90), the location parameter ε of the shifted gamma distribution is estimated using the mean, standard deviation, and skewness coefficient of $X_{\max}$. Thus,

$$\hat{\varepsilon} = \hat{\mu} - \frac{2\hat{\sigma}}{\hat{\gamma}_1} = 48.16 - 2 \times \frac{23.76}{1.501} = 16.52\,\text{mm}.$$

From Eq. (7.2.88),

$$\hat{r} = \frac{4}{\hat{\gamma}_1^2} = \frac{4}{1.501^2} = 1.77,$$

and, from Eq. (7.2.89)

$$\hat{\lambda} = \frac{2}{\hat{\sigma}\hat{\gamma}_1} = \frac{2}{23.76 \times 1.501} = 0.0561\,\text{mm}^{-1}.$$

The estimated 100-year rainfall depth for the shifted gamma model is given by

$$\hat{x}_{max}(100) = 16.52 + \frac{G_{0.99}^{-1}(1.77)}{0.0561} = 16.52 + \frac{6.212}{0.0561} = 127.3\,\text{mm}.$$

The mean, standard deviation, and skewness coefficient of $\ln(X_{max})$ are 3.775, 0.436, and 0.570, respectively, which are used to estimate the parameters of the log-Pearson Type III distribution. From Eq. (7.2.88),

$$\hat{r} = \frac{4}{\hat{\gamma}_{1,\ln(X_{max})}^2} = \frac{4}{0.570^2} = 12.31;$$

from Eq. (7.2.89)

$$\hat{\lambda} = \frac{2}{\hat{\sigma}_{\ln(X_{max})}\hat{\gamma}_{1,\ln(X_{max})}} = \frac{2.0}{0.436 \times 0.570} = 8.048;$$

and, from Eq. (2.7.90),

$$\hat{\varepsilon} = \hat{\mu}_{\ln(X_{max})} - \frac{2\hat{\sigma}_{\ln(X_{max})}}{\hat{\gamma}_{1,\ln(X_{max})}} = 3.775 - \frac{2 \times 0.436}{0.570} = 2.247.$$

Thus,

$$\hat{x}_{max}(100) = \exp\left[2.247 + \frac{G_{0.99}^{-1}(12.31)}{8.048}\right] = e^{2.247+21.899/8.048} = 143.4\,\text{mm}.$$

The cdfs of these models are shown in Fig. 7.2.22. It is seen that these distributions are represented by curves that are concave upward on Gumbel probability plots.

Example 7.32. Goodness-of-fit in storm rainfall analysis. If the cdf of a random variable is inferred from the available extreme value data, the comparison between observed and theoretical frequencies on a probability plot can help in evaluating the goodness-of-fit. Goodness-of-fit testing procedures are used (as shown in Chapter 5) to compare observed and fitted cumulative frequencies. Consider again the maximum annual storm rainfall depths of Table E.7.1. Table 7.2.1 shows the results of the chi-squared, Kolgomorov-Smirnov, and Anderson-Darling tests for eight extreme value models fitted to this data by the method of moments.

The tests reject the null hypothesis only in the case of one out of eight models. Although goodness-of-fit testing procedures can help in discriminating among different mathematical models when no single one is preferred on the basis of the physical characteristics of the phenomena, their capability to discriminate among extreme value models is often poor.

7.2.8 Summary of Section 7.2

In this section we have provided a detailed discussion of distributions applicable to extreme values. These range from the Gumbel to other extreme value models, such as Frechét and Weibull. We have also introduced contagious distributions. We show how specific distributions can produce extremes that conform to different types. We extend the use of extreme value distributions to various natural hazards in the next section, in addition to those discussed here. A summary of extreme value distributions is given in Table 7.2.2.

Table 7.2.1 Chi-squared, Kolgomorov-Smirnov, and Anderson-Darling tests for annual maximum hourly storm depth model at Genoa University, Italy

		Distribution							
Class width (mm)	Observed absolute freqency	Gumbel	Fréchet	GEV	Lognormal	Gamma	Shifted lognormal	Shifted gamma	Log-Pearson III
0–27.5	9	0.195	20.931	0.055	0.046	0.427	0.083	0.142	0.127
27.6–33.5	10	2.110	0.027	1.392	0.973	2.305	1.171	0.540	0.121
33.6–40.0	10	0.879	0.805	0.480	0.388	1.227	0.486	0.455	0.016
40.1–53.5	9	1.689	4.683	1.949	1.810	1.306	1.759	1.293	2.057
53.6–69.0	10	0.007	0.316	0.001	0.017	0.007	0.008	0.080	0.240
>69.0	10	0.012	1.997	0.077	0.079	0.079	0.057	0.024	0.296
Σ or χ^2	58	4.891	<u>28.76</u>	3.953	3.313	5.274	3.564	2.535	2.856
Degrees of freedom		3[(1)]	3[(1)]	2[(2)]	3[(1)]	3[(1)]	2[(2)]	2[(2)]	2[(2)]
$d_n^{(3)}$		0.130	<u>0.179</u>	0.122	0.113	0.126	0.116	0.096	0.089
$A^{2(4)}$		1.005	<u>4.005</u>	0.788	0.660	1.117	0.725	0.519	0.397

Sample values of normalized square deviations, χ^2, d_n, and A^2 are given. Underlined values indicate that the distribution can be rejected with $\alpha = 0.05$. [(1)] $\chi^2_{3,0.05} = 7.81$; [(2)] $\chi^2_{2,0.05} = 5.99$; [(3)] $D_{58,0.05} = 0.131$; [(4)] $A^2_{0.05} = 2.492$; as shown in Subsection 5.6.4.

Table 7.2.2 Summary of extreme value distributions

Distribution		cdf	Mean, $E[X]$	Variance, $\mathrm{Var}[X]$	Quantile, ξ_q
Gumbel (EV1)	Largest	$\exp\{-\exp[-(x-b)/\alpha]\}$, $-\infty < x < +\infty$, $-\infty < b < +\infty, \alpha > 0$	$b + n_e\alpha \approx b + 0.577\alpha$	$\pi^2\alpha^2/6 \approx 1.645\alpha^2$	$b - \alpha\ln(-\ln q)$
	Smallest	$1-\exp\{-\exp[(x-b)/\alpha]\}$, $-\infty < x < +\infty$, $-\infty < b < +\infty, \alpha > 0$	$b - n_e\alpha \approx b - 0.577\alpha$	$\pi^2\alpha^2/6 \approx 1.645\alpha^2$	$b + \alpha\ln[-\ln(-\ln q)]$
Fréchet (EV2)	Largest	$\exp[-(x_0/x)^\theta]\, x, x_0 > 0$, $\theta > 0$	$x_0\Gamma(1-1/\theta)$, for $\theta > 1$	$x_0^2[\Gamma(1-2/\theta) - \Gamma^2(1-1/\theta)]$, for $\theta > 2$	$x_0(-\ln q)^{-1/\theta}$
	Smallest	$1-\exp[-(x_0/x)^\theta]\, x < 0, x_0 > 0$, $\theta > 0$	$x_0\Gamma(1-1/\theta)$, for $\theta > 1$	$x_0^2[\Gamma(1-2/\theta) - \Gamma^2(1-1/\theta)]$, for $\theta > 2$	$x_0[-\ln(1-q)]^{-1/\theta}$
Shifted Weibull	Largest	$\exp\{-[(\lambda - x)/(\lambda - \varepsilon)]^\beta\}\, x \le \lambda$, $\varepsilon < \lambda, \beta > 0$	$\lambda - (\lambda - \varepsilon)\Gamma(1 + 1/\beta)$	$(\lambda - \varepsilon)^2[\Gamma(1 + 2/\beta) - \Gamma^2(1 + 1/\beta)]$	$\lambda - (\lambda - \varepsilon)(-\ln q)^{1/\beta}$
	Smallest	$1-\exp\{-[(x-\varepsilon)/(\lambda-\varepsilon)]^\beta\}\, x \ge \varepsilon$, $\lambda > \varepsilon, \beta > 0$	$\varepsilon + (\lambda - \varepsilon)\Gamma(1 + 1/\beta)$	$(\lambda - \varepsilon)^2[\Gamma(1 + 2/\beta) - \Gamma^2(1 + 1/\beta)]$	$\varepsilon + (\lambda - \varepsilon)[-\ln(1-q)]^{1/\beta}$
GEV	Largest	$\exp\{-[1-k(x-\varepsilon)/\alpha]^{1/k}\}\alpha > 0$; for $k < 0, x > (\varepsilon + \alpha/k)$; for $k > 0, x < (\varepsilon + \alpha/k)$	$\varepsilon + (\alpha/k)[1-\Gamma(1+k)]$, for $k > -1$	$(\alpha/k)^2[\Gamma(1+2k) - \Gamma^2(1+k)]$, for $k > -1/2$	$\varepsilon + (\alpha/k)[1-(-\ln q)^k]$

7.3 ANALYSIS OF NATURAL HAZARDS

7.3.1 Floods, storms, and droughts

Extreme value analysis of hydrological processes plays a relevant role in water resources planning and analysis, because random variables describing storm rainfall, flood, and low flows are essential to predict design values in engineering projects.[17]

The study of extreme values related to floods has a long history. Initially, ancient agricultural nations which depended heavily on water flows realized the economic significance of floods. Currently, the importance of this natural phenomenon has increased in modern industrialized countries. Water is a renewable and inexpensive source of energy. Impounded in reservoirs or diverted from streams, it is essential for irrigating field crops. Furthermore, one must have a sufficient knowledge of the quantity of water flow to control erosion. It is widely accepted that life and property need to be protected against the effects of floods. In most societies a high price is paid to reduce the possibilities of damages arising from future floods. Indeed, the failure of a dam caused by overtopping is a serious national disaster. To safeguard against such an event, engineers need to cater for the safe passages of possible extremely rare floods.

The initial problem is that the historical record of flow data at any particular site does not extend over an adequately long period. Therefore, some realistic way of extrapolation is sought. One method is to use paleoflood data. However, such records are seldom available at the site of interest. The study of outliers, which are unexpectedly high values, provides clues of the tail behavior, so that the model can be applied to the type of events that a reservoir spillway must be designed to accommodate. Unfortunately these items are too few in many cases to estimate the parameters or to give a representation of the underlying diverse population. Accordingly, they indicate possible upper limits and do not usually provide probabilities of exceedance. The term *regionalization* is used to describe regional methods that are applied using regional data. This approach provides a means of extrapolating in space, where space is substituted for time. For this purpose, one defines a homogeneous region, including the site in question, with many historical data series. There are drawbacks firstly in defining a homogeneous region and secondly in accounting for possible correlation between events at different sites. Solutions are often sought through empirical methods. Faced with the difficulties encountered, the engineer should make use of all information available, including single-site streamflow and rain gage data, regional streamflow data, and any historic and paleoflood data.

It is common practice to choose a parent distribution such as the log-Pearson Type III, or the general extreme value distribution (which have been used in the United States, and the United Kingdom, respectively). As an aid to the choice, a histogram or probability plot can be prepared for each site or a pooled plot for the region in order to perform a goodness-of-fit analysis.[18] Thus a regional curve can be obtained for extrapolating beyond available sample sizes. Furthermore, one needs to make assumptions about the parameters of the distribution of maximum flows at different sites.[19]

---

[17] Extensive reviews of the methods used for analyzing extreme value hydrologic data are given by Kite (1988) and Stedinger et al. (1992) among others.

[18] Large measurement errors can occur in extreme flow data thus affecting the choice of an appropriate cdf of maximum flows (Rosso, 1985). More importantly, there is often evidence of nonstationarity in the flow regime, thus giving rise to outliers (Kottegoda, 1984) as already mentioned in the text; in Italy, the two-component extreme value distribution (Rossi et al., 1984) has been used to resolve this problem.

[19] For instance, we might consider the bootstrap applied in Chapter 3 with respect to bias reduction. However, in relation to distributional assumptions, the procedure is not straightforward.

To address the problems encountered, three principles have been suggested for improving extreme flood estimation (Committee on Techniques for Estimating Probabilities of Extreme Floods, 1988). These principles can be summarized as: (1) substitution of space for time; (2) introduction of a so-called structure into the model by incorporating spatial stochastic dependence and relationships between regional parameters; and (3) focus on the extreme right tails of empirical distributions.

In addition, extreme flood analysis should be accompanied by explicit uncertainty analysis. This includes assessment of model assumptions and errors arising from incorrect models. Some aspects of uncertainty analysis, such as errors of Type I and Type II, are discussed in Chapter 5. An assessment of the effects of alternative distributions may be needed. Types of uncertainty analysis should include Bayesian methods as shown in Chapter 10.

Example 7.33. Flood frequency estimation for rivers of Italy. The National Research Council of Italy (2000) proposed regional curves for flood frequency analysis of the form

$$\frac{Q(T)}{Q_{\text{index}}} = g(y; \Xi),$$

where $Q(T)$ denotes the T-year annual maximum flood flow, y is the standard Gumbel reduced variate, and Q_{index} is the index flood, which is taken as the mean annual flood. The analytical form of function $g(y; \Xi)$ depends on the extreme value distribution adopted to fit the regional random variable $X = Q(T)/Q_{\text{index}}$. This approach was pioneered by Natural Environmental Research Council (1975) for rivers in the United Kingdom and is currently adopted in several countries; see also Robson and Reed (1999).

Bocchiola et al. (2003) used the GEV distribution to accommodate flood data for Italian rivers. Accordingly,

$$\frac{Q(T)}{Q_{\text{index}}} = \varepsilon + \frac{\alpha}{k}(1 - e^{-ky}),$$

where the reduced variate y corresponds with $Q(T)$. This is based on the assumption that annual maximum floods are GEV-distributed with common coefficients of skewness and variation, which are estimated from all rivers in the region (see Fig. 7.3.1 for North-West Italy).

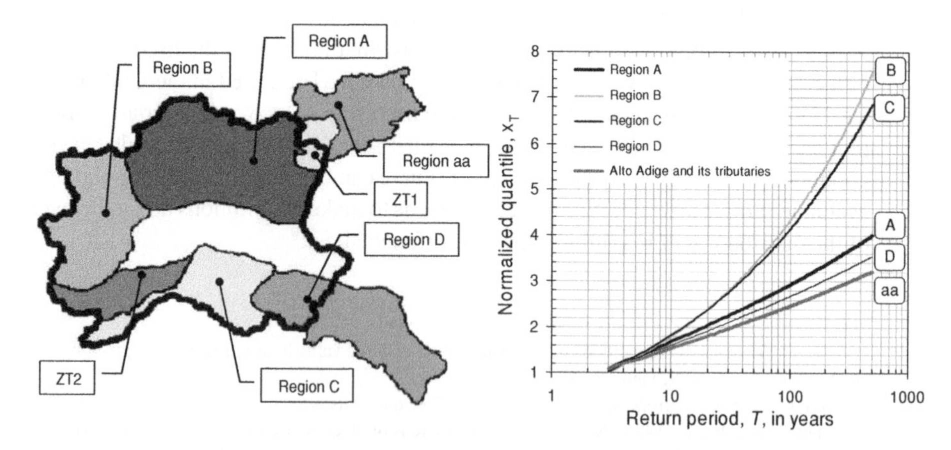

Fig. 7.3.1 Normalized annual maximum flood versus return period as obtained from the GEV distribution estimated from regional data in north-west Italy. Areas denoted with ZT are transition zones.

If all records in Region C are used, these including rivers in the regions of Thyrrenian Liguria and Northern Appennines, the parameter estimates are $k = -0.276$, $\alpha = 0.377$, and $\varepsilon = 0.643$.

Thus, if the 100-year flood is required,

$$y(100) = -\ln\left(\ln\frac{T}{T-1}\right) = -\ln\left(\ln\frac{100}{99}\right) = 4.600,$$

so that

$$\frac{Q(100)}{Q_{\text{index}}} = 0.643 - \frac{0.377}{0.276}\left(1 - e^{0.276\times 4.600}\right) = 4.14.$$

The suggestion was to determine the index flood from basin characteristics (as in previous studies in the United States and United Kingdom). For this purpose, one can use regression analysis of Q_{index} against a number of independent variables representing these characteristics, such as basin area, geomorphologic parameters, and basin soil index. For instance, if one takes basin area in km^2 (AREA) as the only explanatory variable describing catchment characteristics, the following relationship was found with the flood index:

$$Q_{\text{index}} = 5.2\ \text{AREA}^{0.75}.$$

For the River Bisagno at la Presa in Liguria, for example, with a catchment area of 34.2 km^2,

$$Q_{\text{index}} = 5.2 \times 34.2^{0.75} = 73.5\ \text{m}^3/\text{s},$$

so that the 100-year flood can be estimated as

$$Q(100) = 5.17 \times 73.5 = 304\ \text{m}^3/s.$$

This simple equation is inaccurate if used in this way. In fact, the estimated Q_{index} is 94.8 m^3/s from a record of 48 years. Alternatively, one can use the four-variable regression equation determined for Region C:

$$Q_{\text{index}} = 0.21\ \text{AREA}^{0.897}\text{RAIN}^{0.678}\,\text{ELEV}^{-0.686}\,\text{SHAP}^{0.285}.$$

In this equation, RAIN is the expected maximum annual hourly rainfall depth in mm, ELEV is the average elevation of the drainage basin (from a 1:10,000 map) expressed in kilometers; and SHAP is the dimensionless ratio of basin area to the squared mainstream length. Hence, after substituting from available data,

$$Q_{\text{index}} = 0.21 \times 34.2^{0.897} \times 41.18^{0.678} \times 0.413^{-0.686} \times 0.489^{0.285} = 92.9\ \text{m}^3/\text{s},$$

so that $Q(100) = 4.14 \times 92.9 = 385$ m^3/s. Although this regression provides here a rather accurate estimate of Q_{index}, regression methods are usually affected by large uncertainties. Therefore, methods involving hydrologic models are often used to estimate Q_{index} in order to account for runoff generation processes occurring in the catchment.[20]

The frequencies of precipitation of various intensities and durations are used in the hydrologic design of structures aimed at controlling storm runoff and floods, such as road culverts, storm sewers, and small dams, and at preventing erosion and mass movements on hill slopes. Rainfall frequency analysis provides values of cumulated rainfall at a point in space for a specified return period and a continuous range of durations. Basin average rainfall values are usually developed from point rainfall by introducing a reduction factor for basin areas larger than a typical size of storm cells, say, 10 km^2. For a site where rainfall data are available, frequency analysis is performed by using an extreme value probability model to represent the annual maximum of rainfall depth recorded over a

[20] See, for example, Bocchiola et al. (2003).

specified duration, say, t. The Gumbel, Fréchet, lognormal, log-Pearson Type III, and GEV distribution with $k < 0$ are most frequently adopted.

Because some regularities arise for rainfall depths recorded over various durations, the variability of storm depth x with duration t for a specified frequency level is often represented by an empirical relationship in the form

$$x_{\max}(T;t) = \xi_q(t) = \frac{ct}{t^{\kappa} + f}, \tag{7.3.1}$$

where $x_{\max}(T;t)$ is the design rainfall depth associated with a return period T and duration t, and c, κ, and f are parameters that depend on location and probability of nonexceedance or return period.[21] To estimate the parameters of the *depth-duration-frequency curve* of Eq. (7.3.1), a common procedure is to initially determine the T-year quantiles of annual maximum storm depth for a given number of durations; then a nonlinear regression procedure is applied to relate these quantile values with the corresponding durations according to Eq. (7.3.1). Because this method might yield contradictory results, κ and f are sometimes constrained as constants for a given location, while c is allowed to vary with the return period. However, this procedure is somewhat cumbersome, and it is affected by a certain degree of subjectivity because of the selection of the durations and return periods to be considered.

An alternative approach is based on the concept of *scale invariance*, which describes a physical feature of rainfall fields. Accordingly, annual maximum rainfall depth $X_{\max}$ is parameterized by duration t, and the properties of the random variable $X_{\max}(t)$ are studied. Denoting by $\lambda > 0$ a scaling factor, scale invariance holds if the random variables $X_{\max}(\lambda t)$ and $\lambda^n X_{\max}(t)$ have the same probability distribution for $t_{\text{in}} \leq t \leq t_{\text{out}}$, and $t_{\text{in}} \leq \lambda t \leq t_{\text{out}}$.[22] Here, the values of t_{in} and t_{out} are the inner and outer cutoffs representing the physical bounds for which scale invariance holds, and $n > 0$ is a scaling exponent. Thus,

$$\xi_q(\lambda t) = \xi_q(t)\,\lambda^n. \tag{7.3.2}$$

Without loss of generality, measuring time in units of t_{in}, so that $t_{\text{in}} = 1$, we can write

$$\xi_q(t) = \xi_q(1)\,t^n,$$

for $1 \leq t \leq t_{\text{out}}$, since $\xi_q(\lambda t) = \xi_q(1)(\lambda t)^n = [\xi_q(1)t^n]\lambda = \xi_q(t)\lambda^n$. Because statistical scale invariance also leads to the following scaling relationship for the rth-order moments,

$$E\left[X^r_{\max}(\lambda t)\right] = \lambda^{rn} E\left[X^r_{\max}(t)\right], \tag{7.3.3}$$

the coefficients of variation, skewness, and kurtosis of $X_{\max}(t)$ are independent of the temporal scale, so that the variability of $X_{\max}(t)$ with duration can be expressed by a power law relating the mean annual maximum storm depth to the corresponding duration. The resulting depth-duration-frequency curves of station precipitation can be expressed as

$$\xi_q(t) = \xi'_q\,\mu_1\,t^n, \tag{7.3.4}$$

where ξ'_q denotes the qth quantile of annual maximum storm depth normalized by its mean for any duration in the range t_{in} to t_{out}, μ_1 is the mean annual maximum storm depth for

---

[21] For example, values of $c = 0.55$ mm/minute, $\kappa = 0.82$, and $f = 6.57$ minute$^{0.82}$ are given by Wenzel (1982) for a return period of 10 years in New York, NY.

[22] Statistical scale invariance is defined as $\Pr[X(\lambda t) \leq x] = \Pr[\lambda^n X(t) \leq x]$ for any x of X, and it is denoted as $X(\lambda t) \underline{\underline{d}} \lambda^n X(t)$ in compact form.

Table 7.3.1 Estimated rth-order moments and scaling exponents with duration for the observed annual maximum storm depth at Lanzada in Valtellina, Italy

	Values of $E[X^r_{\max}(t)]$, in mm						
t, in hours	$r = 1$	$r = 3$	$r = 6$	$r = 12$	$r = 24$	$\eta(r)$	$r\eta(1)$
1	14.00	21.45	30.35	42.10	59.13	0.457	0.457
2	211.21	484.65	961.85	1886.67	3704.01	0.910	0.913
3	3423.4	11660.0	31851.4	90313.4	245427.3	1.359	1.370
4	59589	301627	1103476	4624875	17168900	1.803	1.827
5	1113150	8442009	40033446	2.53E + 08	1.26E + 09	2.240	2.284

unit duration, and n is a scaling exponent that depends on location. Note that the curves determined from Eq. (7.3.4) for different values of q plot as parallel straight lines on the plane (t, ξ) when using logarithmic scales. For example, if $X_{\max}(t)$ is a GEV-distributed variate,[23] one obtains

$$x_{\max}(T; t) = \xi_q(t) = \left[\varepsilon + \frac{\alpha}{k}(1 - e^{-ky})\right] \mu_1 \, t^n, \tag{7.3.5}$$

where ε, α, and k are the location, scale, and shape parameters of the GEV distribution estimated for the normalized annual maximum storm depth, μ_1 is the mean value of annual maximum storm depth for unit duration, and n is the scaling exponent. This can be determined from the two values of the means of annual maximum storm depth for t_{in} and t_{out}, the inner and outer cutoff levels.[24]

Example 7.34. Scaling depth-duration-frequency curves. Table 7.3.1 shows the sample statistics for annual maximum rainfall totals recorded at Lanzada station in Valtellina, Italy, for durations from 1 to 24 hours in a period of 50 years. To analyze the scaling properties of these data, one must search for the exponent $\eta(r)$ in the relationship $E[X^r_{\max}(t)] \propto t^{\eta(r)}$, as shown in Fig. 7.3.2*a*.

Linear regression of $\ln E[X^r_{\max}(t)]$ against $\ln t$ is performed for $r = 1, \ldots, 5$, and the resulting values of $\eta(r)$ are given in Table 7.3.1. Since $\eta(r)$ does not differ significantly from nr, with $n = \eta(1) = 0.457$, one can assume Eq. (7.3.3) to hold exactly. The scaling model of rainfall depth with duration is then developed by fitting the Gumbel distribution to extreme value data for various durations divided by the corresponding mean for the specified duration (see Fig. 7.3.2*b*). The resulting estimated values of the scale and location parameters are $\hat{\alpha} = 0.191$ and $\hat{b} = 0.890$, respectively. The scaling model of depth-duration-frequency curves is thus written as

$$x_{\max}(T; t) = (b + \alpha y)\, \mu_1 t^n, \quad \text{or} \quad x_{\max}(T; t) = (b_1 + \alpha_1 y)\, t^n,$$

where $\alpha_1 = \mu_1 \alpha$, and $b_1 = \mu_1 b$, the estimated value of μ_1 from the above regression procedure is 14.0 mm/h$^{0.457}$. Thus,

$$x_{\max}(T; t) = (12.06 + 2.59y) t^{0.457}$$

[23] See, for example, Buishand (1989).

[24] Burlando and Rosso (1996) suggest alternative methods for estimating μ_1 and n, and provide an extension of this method to multiscaling rainfall.

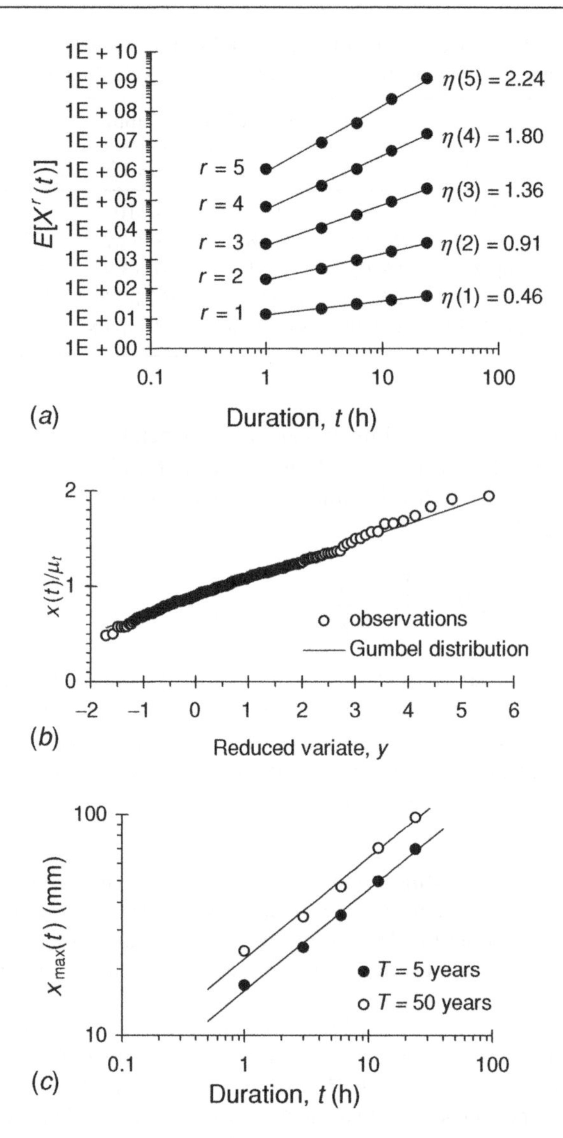

Fig. 7.3.2 Annual maximum storm depth for Lanzada station, northern Italy: scaling of rth-order moments with duration (a); distribution of normalized data (b), and scaling depth-duration-frequency curves where points show the predicted quantiles for the specified durations (c).

is the predictive equation required to estimate design rainfall depth for a specified duration. For example, if $T = 5$ years is the design return period, $q = 0.8$ and $y = 1.500$. The predicted storm depth for a duration of 6 hours is then found as

$$x_{\text{max}}(5; 6) = (12.06 + 2.59 \times 1.500)\, 6^{0.457} = 36.36 \text{ mm}.$$

This value is close to that of 35.0 mm, which is obtained by fitting the Gumbel distribution to 6-hour duration data (see Fig. 7.3.2c).

The scaling concept can be useful in other hydrologic applications because it can represent complex variability using parsimonious statistical models. For instance, scaling with drainage basin area can be used for flood frequency regionalization using the flood index method.[25] Also, multiparametric scaling can be introduced to investigate the joint variability of a random variable in space and time. For example, annual maximum precipitation over a drainage basin depends on both storm duration and the area covered by the storm. Rainfall intensity over a given area generally decreases with increasing duration; rainfall intensity averaged over a given duration generally decreases with increasing area. Therefore, duration and area play a combined role in determining the outcomes of the investigated variable. This can be modeled using the self-affine approach[26] to represent the *intensity-duration-area-frequency* curves of areal precipitation.[27]

Droughts and associated water shortages play a fundamental role in human life and their analysis is needed in environmental and civil engineering. Low-flow statistics are used in water supply planning to determine allowable water transfers and withdrawals, and they are needed in allocating waste loads and in siting treatment plants and sanitary landfills. Frequency analysis of low flows is necessary to determine minimum downstream release requirements from hydropower, water supply, cooling plants, and other facilities.

Although a single variable is often sufficient to characterize maximum flood flows, the definition of drought and low flows in rivers often involve more than one variable, such as the minimum flow level, the duration of flows which are less than that level, and the cumulated water deficit.[28] To overcome the problem of evaluating the joint probability of mutually related variates, a low-flow index can be used, such as the annual minimum d-day consecutive average discharge with probability of nonexceedance q, say, $\xi_q(d)$. For instance, the 10-year 7-day average low-flow index, $\xi_{0.1}(7)$ is widely used with droughts in the United States (d is 7 and q is taken to be 0.1). A preliminary step in performing low-flow frequency analyses is the "deregulation" of the low-flow series to obtain "natural" streamflows. Also, low-flow series should be subjected to trend analysis so that identified trends can be reflected in frequency analyses. This includes accounting for the impact of large withdrawals and diversions from water and wastewater treatment facilities, as well as lake regulation, urbanization, and other factors modifying flow regime.

To estimate $\xi_q(d)$ from streamflow records, one generally fits a parametric probability distribution to the annual minimum d-day consecutive average low-flow data series. The Gumbel distribution of the smallest value and the EV3 or Weibull distribution is theoretically plausible for low flows, as shown in Example 7.21. Studies in the United States and Canada have recommended the shifted Weibull, the log-Pearson Type III, lognormal, and shifted lognormal distributions based on apparent goodness-of-fit. Moreover, low-flow data can contain zero values, such as in some arid areas, where zero flows are recorded more often than nonzero flows. Accordingly, the cdf of a low-flow index, say, X has a probability mass at the origin, p_0, and a continuous distribution for nonzero values of X, which can be interpreted as the conditional cdf of nonzero values, say, $F_X(x|x > 0)$. Thus,

$$F_X(x) = p_0 + (1 - p_0)F_X(x|x > 0).$$

The parameters of $F_X(x|x > 0)$ can be estimated by any procedure appropriate for complete samples using only nonzero data, whereas the extra parameter p_0 denotes the

[25] See, for example, De Michele and Rosso (2002).

[26] Statistical affinity is defined as $X(\lambda^a t, \lambda^b A) \underline{\underline{d}} \lambda^{-H} X(t, A)$ with a, b, and H denoting the scaling exponents. Here $X(t, A)$ is, for example, precipitation intensity delivered by a storm in t hours as averaged over an area of A square kilometers (De Michele et al., 2001).

[27] See, for example, De Michele et al. (2002).

[28] Some flood problems, however, require one to evaluate the joint probability distribution of peak flow and volume for the design flood (see, for example, Bacchi et al., 1992 and De Michele et al., 2005).

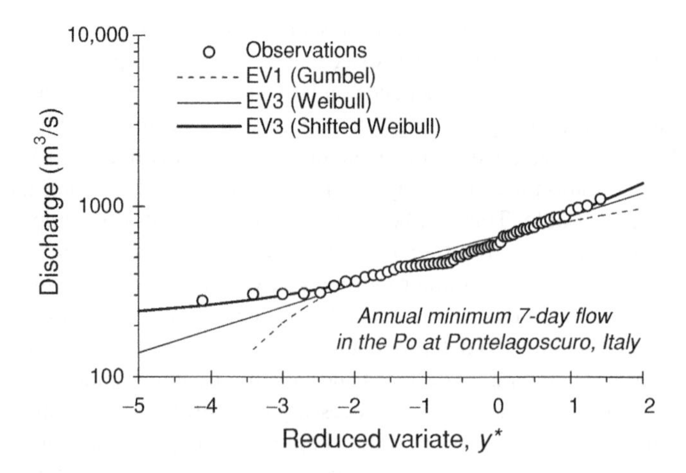

Fig. 7.3.3 Weibull probability plots of extreme value distributions of annual 7-day minimum flow in the Po River at Pontelagoscuro, Italy.

probability that an observation is zero. If r nonzero values are observed in a sample of n values of data, the natural estimator of the exceedance probability $q_0 = (1 - p_0)$ of the zero value or perception threshold is r/n, and $p_0 = 1 - r/n$.

Example 7.35. Annual minimum 7-day flow. The mean and standard deviation of the 7-day annual minimum flows in the Po River at Pontelagoscuro, Italy, recorded from 1918 to 1978 (see Table E.7.2, column 4) are 579.2 and 196.0 m^3/s, respectively, and the skewness coefficient is 0.338. The estimated values of the parameters of the Gumbel and Weibull distributions of the smallest value fitted to this data by the method of moments are

$$\hat{\alpha} = 152.8\,\text{m}^3/\text{s}, \quad \hat{b} = 667.4\,\text{m}^3/\text{s}, \quad \text{and} \quad \hat{\beta} = 3.26, \hat{\lambda} = 646.1\,\text{m}^3/\text{s},$$

respectively, where the same notation of Section 7.2 is used. If the shifted Weibull distribution is used, one gets

$$\hat{\beta} = 2.556, \quad \hat{\lambda} = 1007\,\text{m}^3/\text{s}, \quad \text{and} \quad \hat{\varepsilon} = -2803\,\text{m}^3/\text{s}$$

(which is unacceptable but given here for comparison).

These three cdfs are shown in Fig. 7.3.3 on a Weibull probability plot for the smallest value, where the reduced variate $y^* = \ln[-\ln(1-q)]$ is used to represent the frequency level. Note that the fitted distributions are close to those fitted to the annual minimum daily flow. This occurs because of the large drainage area of more than 70×10^3 km^2.

For the Gumbel distribution, the estimated 10-year 7-day average low flow, $\xi_{0.1}(7)$, is given by

$$\xi_{0.1}(7) = \ln[-\ln(1-0.1)] \times 152.8 + 667.4 = 323.5\,\text{m}^3/\text{s}.$$

For the Weibull distribution,

$$\xi_{0.1}(7) = 646.1 \times \exp\left\{\ln\frac{[-\ln(1-0.1)]}{3.25}\right\} = 323.3\,\text{m}^3/\text{s},$$

However, for the shifted Weibull distribution we get the unacceptable result:

$$\xi_{0.1}(7) = -2803 + (1007 + 2803) \times \exp\left\{\ln\frac{[-\ln(1-0.1)]}{2.556}\right\} = -1223\,\text{m}^3/\text{s},$$

given here for completeness.

Regional regression procedures are employed at ungauged sites to estimate low-flow statistics by using basin characteristics. In the simple drainage area method, for example, one estimates a low-flow quantile for an ungauged river site draining an area of A as $(A/A_x)\xi_q(d)$, where $\xi_q(d)$ is the corresponding low-flow quantile for a gauging station in the vicinity which drains an area of A_x. This may be modified by a scaling factor $(A/A_x)^b$, $b < 1$, which is estimated by regional regression of quantiles for several gauged sites.

An alternative to d-day averages is the flow duration curve introduced in Subsection 1.1.6 and shown in Example 1.6. This gives the proportions of the time over the whole record in which different daily flow levels are exceeded, but, unlike $\xi_q(d)$, it cannot be interpreted on an annual event basis.

An alternative to d-day averages is the flow duration curve introduced in Subsection 1.1.6 and shown in Example 1.6. This gives the proportions of the time over the whole record in which different daily flow levels are exceeded, but, unlike $\xi_q(d)$, it cannot be interpreted on an annual event basis.

7.3.2 Earthquakes and volcanic eruptions

Earthquakes pose a severe threat to the built environment. There can be more than 50 potentially destructive earthquakes annually. When they occur close to urban areas, the consequences are catastrophic. Many types of improperly designed buildings collapse and, further away, dams, bridges, and transport systems are destroyed.

The *magnitude* of an earthquake is a measure based on ground motion amplitude obtained from seismographs at a specified distance from the rupture of the crust. A seismograph is a recorded time series of the displacement, velocity, and acceleration experienced by a particle at the site of the instrument. The associated science is termed seismology; its field is mainly observational. A well-known measure of the strength of an earthquake is the Richter scale, which takes a logarithmic basis and classifies earthquakes essentially on a scale from 0 to 10.[29]

Alternative definitions of magnitude include the local magnitude and that determined by seismic waves. The global seismic network can monitor earthquakes occurring anywhere in the world with a magnitude larger than 4, and many regions in the world have dense seismic networks that are capable of monitoring earthquakes as small as magnitude 2 or less. The *intensity* scale provides a primarily qualitative description of earthquake effects, including human perception and effects on buildings, infrastructures, landscape, and natural surroundings. For example, the Mercalli–Cancani–Sieberg (MCS) intensity scale is used in southern Europe, the Medvedev–Sponheuer–Karnik (MSK) intensity scale in central and eastern Europe, and the modified Mercalli (MMI) intensity scale in the United States. Also, the European Macroseismic Scale (EMS), developed after significant contributions by Italian seismologists, is probably a superior tool for describing intensity. The epicentral intensity data are converted to magnitude using a linear relationship; for example, epicentral intensities of 6 and 9 as measured in MMI units are associated with magnitudes of 5 and 7, respectively. However, field data indicate that this relationship is affected by a certain degree of uncertainty.

Statistical methods play an important role in seismology. Original work was done by Jeffreys (1967), especially with regard to travel timetables. Later Bolt (1999) pioneered the use of data from sensors along fault lines. Seismologists are involved with a number of types of massive data sets with large variability; these include first arrival times, signal duration, maximum amplitude, and oscillation periods. Their main concern is the risk.

[29] No earthquakes with magnitudes greater than 9.5 have been observed.

Although various statistical relationships have been used to relate the frequency of occurrence of earthquakes to their magnitudes, the log-linear relationship by Gutenberg and Richter (1944) is the most generally accepted method. This is written as $\log(\nu) = \log(a) - bx$; that is,

$$\nu = a10^{-bx}, \tag{7.3.6}$$

where ν denotes the mean number of occurrences of earthquakes in a unit of time, say, a year, with magnitude greater than x, and the values taken by parameters a and b depend on the particular area. For example, worldwide data analysis yields $a = 10^8$, and $b = 1$, for surface-wave magnitude based on surface waves with a period of 20 seconds (Turcotte, 1992). (Note that b is modified into $b' = b\ln(10) = 2.3b$ when taking the natural logarithms instead of logarithms to base 10.) Thus, about ten earthquakes exceeding magnitude 7 are expected each year around the world. A lower bound, say, $X_{\min}$, is often introduced to represent the level of earthquakes below which there is no engineering interest or data are insufficient, and an upper bound, say, $X_{\max}$, to represent the largest possible earthquake considered for a particular zone. This modifies the Guttenberg-Richter law of Eq. (7.3.6) to the truncated exponential recurrence relationship, which is written in the form

$$\nu = \nu_0 \frac{e^{-b'(x-x_{\min})} - e^{-b'(x_{\max}-x_{\min})}}{1 - e^{-b'(x_{\max}-x_{\min})}}, \tag{7.3.7}$$

where ν_0 is the number of earthquakes equal to the lower bound or larger.

Crustal deformation takes place at the boundaries between major surface plates, and relative displacements would occur on well-defined faults, which are considered to have memory. Because the Poisson model is memoryless, one could argue that this model is inappropriate to account for the occurrences of earthquakes. However, the Poisson model can be applied in many situations where fault memory actually exists; it is inappropriate only if the elapsed time between significant events, with memory, exceeds the average recurrence time between such events (Cornell and Winterstein, 1988). Assuming the annual number of earthquakes in the study area to be a Poisson variate with mean ν, the probability that no earthquake occurs in a year with magnitude greater than x is $e^{-\nu}$. This is also the probability that the annual maximum magnitude does not exceed a value of X. Thus, the cdf of annual maximum earthquake magnitude is

$$F_{X_{\max}}(x) = e^{-\nu} = \exp(-a\, e^{-b'x}). \tag{7.3.8}$$

This is the Gumbel distribution of Eq. (7.2.11) with scale parameter $1/b'$, and location parameter $b'^{-1}\ln(a)$. For worldwide data, $b = 1$ and $a = 10^8$, the scale and location parameters of the Gumbel distribution are 0.43 and 8, respectively (see Fig. 7.3.4).

This distribution is widely used to predict the annual maximum magnitude of earthquakes in a region. For example, values of $b = 0.90$ and $a = 7.73 \times 10^4$ are estimated from the catalog of earthquakes exceeding magnitude 6 in southern California for the period 1850–1994 as reported by the Working Group on California Earthquakes Probabilities (1995). Thus, the annual maximum of earthquake magnitude in this region can be estimated using the Gumbel distribution with scale and location parameters of 0.48 and 5.4, respectively.[30]

[30] These estimates are close to the estimated scale and location parameters of 0.49 and 5.8, respectively, by Lomnitz (1974), from records of annual maximums of earthquake magnitude in southern California for the period 1932–1962 (see Problem 7.16).

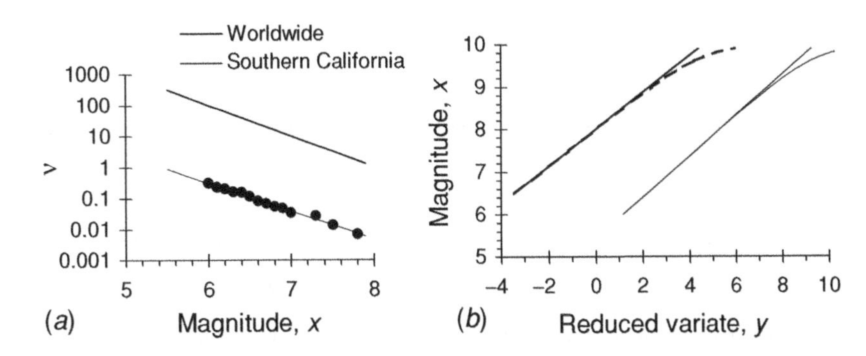

Fig. 7.3.4 The Gutenberg-Richter law (a) and the corresponding probability distribution of annual maximum earthquake magnitude (b). The dotted lines show the effect of truncation with $X_{\min} = 6$ and $X_{\max} = 10$.

Example 7.36. Earthquake intensity in Rome. The *Catalog of Italian earthquakes from year 1000 to 1980* contains all the available historical information on earthquakes that occurred in Italy during the past 1000 years. It also includes values of earthquake intensity in terms of the Mercalli–Canconi–Sieber (MCS) index. Using the data of MCS intensity X for the metropolitan area of Rome which are reported in Problem 1.22, one estimates the values of b' and a in Eq. (7.3.6) by performing the regression of $\ln \nu$ against x. Here ν denotes the observed number of earthquakes with intensity not less than x divided by the number of years of observation, that is, 980. Accordingly, $\hat{b}' = 1.10$ and $\hat{a} = 4.21$ are found. Note that only intensities greater than 2 are considered (see Fig. 7.3.5).

Accordingly, the Gumbel distribution for the annual maximum earthquake MCS intensity of Eq. (7.3.8) has parameters $1/\hat{b}' = 0.91$, and $\hat{b}'^{-1}\ln(\hat{a}) = 1.3$ for this area. Because of linearity between intensity and magnitude, the corresponding cdf of annual maximum magnitude can be easily determined.

The total energy in the seismic waves generated by an earthquake can be related to its magnitude by a log-linear relationship, such as $\log(E) = 1.44X + 5.24$, for E in Joules. The strain release during an earthquake is proportional to the moment of the earthquake, which can also be related to its magnitude by using either a heuristic linear relationship

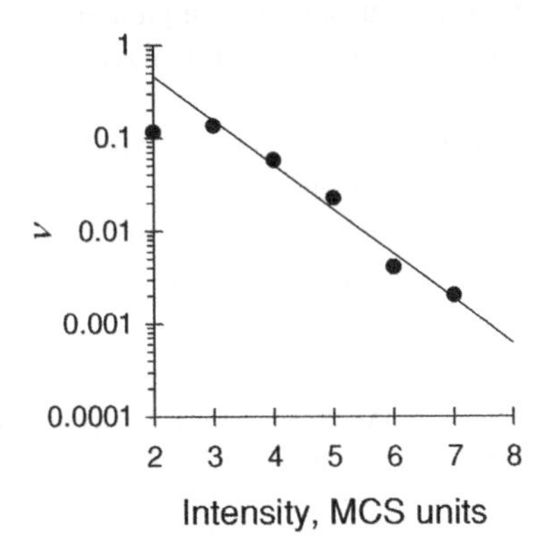

Fig. 7.3.5 Mean number of earthquakes per year for the metropolitan area of Rome with an MCS intensity greater than a specified value.

or a theoretical log-linear law. The area of the rupture is also related to the moment by a power law.[31]

Example 7.37. Earthquake area of rupture. Assume that the moment z of an earthquake is related to its magnitude x by

$$\log(z) = d + cx \text{ and by } \log(z) = \log(\alpha) + \varepsilon \log(w)$$

to the area of rupture w, where c, d, α, and ε are known parameters. Thus,

$$x = \log(z)/c - d/c = \varepsilon \log(w)/c + \log(\alpha)/c - d/c,$$

which is substituted for x in the Guttenberg-Richter law to obtain

$$\log(v) = \log(a) - b\varepsilon \log(w)/c - b \log(\alpha)/c + bd/c, \quad \text{that is, } v = kw^{-\theta},$$

where $\theta = b\varepsilon/c$ is a scaling exponent, and $\log(k) = \log(a) - b\log(\alpha)/c + bd/c$ is a constant. Using a value of 1.5 for both c and ε as indicated, for instance, by Kanamori and Anderson (1975), the value of θ is found to be unity for worldwide data with $b = 1$. Therefore, the annual number of worldwide earthquakes with area of rupture larger than w is inversely proportional to the area of rupture. In zoning seismic risk, the value of the scaling exponent equals the scale factor of the corresponding Guttenberg-Richter law, that is, $\theta = b$. For instance, one gets $\theta = 0.9$ for southern California, and $\theta = 1.1/2.3 = 0.48$ for the metropolitan area of Rome.[32] The corresponding cdf of the annual maximum area of rupture is

$$F_{W_{\max}}(w) = \exp(-kw^{-\theta}),$$

which is the Fréchet distribution with shape parameter θ, and scale parameter equal to $k^{1/\theta}$.

The assessment of seismic hazard for a given site requires the evaluation of ground motion acceleration at that site (see Problem 1.1). This can be determined by combining intensity or magnitude of earthquakes in the region with the attenuation of epicentral magnitude or intensity for a specified probability distribution of the distance from the epicenter. Therefore, one must determine the spatial distribution of epicentral distance for the earthquakes of the region, which is dictated by active faults or point sources if they are present in the region. Alternatively, a uniform or other probability distribution is used.

Volcanoes are the manifestation of a thermal process situated deep inside the earth from which heat is not easily emitted by conduction and radiation. A volcano is formed as part of the heat eviction process, where the earth's crust opens and magma, a molten rock material that results in igneous rocks on cooling, reaches out from enormous pressure chambers. The magma comes out as lava usually accompanied by an avalanche of hot gases, steam, ash, and rock debris. Some volcanoes seem to have erupted only once, whereas others are known to have had several eruptions.

Because various types of volcanic eruptions require particular quantitative approaches, it is considerably more difficult to quantify a volcanic eruption than an earthquake. Volcanoes have a wide spectrum of sizes, and eruptions from a single volcano have many variations. Some eruptions produce mainly ash, or tephra, whereas others yield primarily liquid rock,

---

[31] The works by Cornell (1968) and McGuire (2005 are of seminal importance. Furthermore, the books by Gutenberg and Richter (1954), Lomnitz (1974), Lomnitz and Rosenblueth (1976), Reiter (1990), Turcotte (1992) and Vere-Jones (1992), among others, provide excellent reviews of probabilistic seismic analysis, and give exhaustive references on these topics. These relationships can be used to determine the distribution of these variables from that of magnitude using the concept of a function of a random variable. Also, Lomnitz (1994) deals with earthquake predictions and Naeim (1989) provide improved seismic designs.

[32] This is because of linearity between intensity and magnitude.

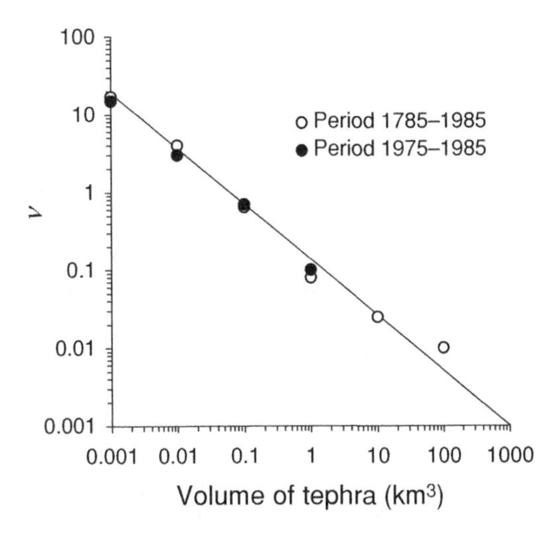

Fig. 7.3.6 Mean number of volcanic eruptions per year with a volume of tephra larger than a specified value.

or magma. The volume of tephra or magma in an eruption depends on circumstances that are poorly understood. Using the volume of tephra as a measure of size, McLelland et al. (1989) give frequency-volume statistics for volcanic eruptions during the period from 1975 to 1985 and for historic eruptions of the last 200 years. Using this data, Turcotte (1992) shows that the mean number ν of eruptions per year, with a volume of tephra larger than x, varies according to a power law; that is,

$$\nu = cx^{-d}, \tag{7.3.9}$$

where $d = 0.71$, and $c = 0.14$ for the volume of tephra measured in cubic kilometers (see Fig. 7.3.6).

Assuming that the number of eruptions in a year is a Poisson variate with mean ν, the probability that no eruptions occur in a year with tephra volume larger than x is $e^{-\nu}$. This probability is also the probability that the maximum tephra volume of a volcanic eruption in a year does not exceed a value of x. Thus, the cdf of annual maximum tephra volume is found as

$$F_{X_{\max}}(x) = e^{-\nu} = \exp(-cx^{-d}). \tag{7.3.10}$$

This is the Fréchet distribution with shape parameter d and scale parameter $c^{1/d}$.

7.3.3 Winds

About 70% of total claims by insurers concern windstorms. Hurricane Andrew, for instance, devastated the Gulf Coast of United States in 2002 and led to a record sum of \$17 billion in insurance losses. Consequent to different natural hazards, earthquakes and windstorms account for most human fatalities. In the design of tall structures and long-span bridges, engineers provide resistance to counter the effects of high-speed winds.

Structural engineers use the extreme or fastest values of wind speed with return periods such as 50 years for most permanent structures, 25 years for structures having no human occupants or where there is a negligible risk to human life, and 100 years for structures with an unusually high degree of hazard of life and property in case of failure. Because the probability law describing extreme wind speed applies to homogeneous micrometeorological conditions, one must consider initially the averaging time, the height above ground, and the roughness of the surrounding terrain before using a specified probability law to represent wind data. If different sampling frequencies were used to collect the data, the whole sample must be adjusted to a unique averaging time, such as a period of 5 minutes. If the anemometer elevation changed during the recording period, the data must be adjusted to a common value (for example, 10 m above ground) using a logarithmic law to represent vertical profile of wind speed. Similarity models must be used to adjust wind data from different nearby locations to a common uniform roughness over a distance of about 100 times the elevation of the instrument; in addition, sheltering effects and small-scale wind obstacles must be properly considered. Finally, in modeling extreme wind speeds the engineer must also distinguish cyclonic winds from hurricane and tornado winds because they follow different probability laws.[33]

In well-behaved climates (with a stationary distribution of extreme winds) the annual maximum wind speed is often represented by the Gumbel distribution (ANSI/ASCE, 1988). Accordingly, Eq. (7.2.26) is used to predict design wind speeds based on the estimated mean and standard deviation. For stations with very short records, the analysis based on the annual maxima often yields poor design estimates, and the maximum wind in each month is used instead. Thus, the design wind is predicted as

$$x_{\max}(T) = \mu_m - \sigma_m \left\{ \frac{\sqrt{6}}{\pi} \left[n_e + \ln \left(\ln \frac{12T}{12T - 1} \right) \right] \right\}, \tag{7.3.11}$$

where μ_m and σ_m denote the mean and standard deviation of the sample of monthly maxima, respectively (assuming that there is no significant seasonal variation). An alternative to the Gumbel distribution is the Fréchet distribution, although appreciable differences are found only for large return periods, say, greater than 100 years. Because the Weibull distribution was found to accommodate wind speed data for Europe (Troen and Petersen, 1989), one might also determine the extreme value distribution using either the contagious distribution approach of Subsection 7.2.6 with a specified threshold or the probability distribution of continuous wind speed. Such data are usually available as a sequence of time averages, for example, as 10-minute average wind speeds, and the cdf of the average wind speed X is taken as

$$F_X(x) = p_0 + (1 - p_0)\{1 - \exp[-(x/\lambda)^{\beta}]\} \tag{7.3.12}$$

for $x \geq 0$, where λ and β denote the scale and shape parameters of the Weibull distribution used to model nonzero values, and p_0 is the probability of zero values. The cdf of annual maximum wind speed $X_{\max}$ is usually determined using one of two methods.[34] The first one assumes that there are m independent data in a year, so that

$$F_{X_{\max}}(x) = [F_X(x)]^m. \tag{7.3.13}$$

Alternatively,

$$F_{X_{\max}}(x) = \exp[-\eta f_X(x)] \tag{7.3.14}$$

[33] Extensive reviews of methods for determining design wind speed can be found in the books by Simiu and Scanlan (1986) and Liu (1991).

[34] See, for example, Solari (1966).

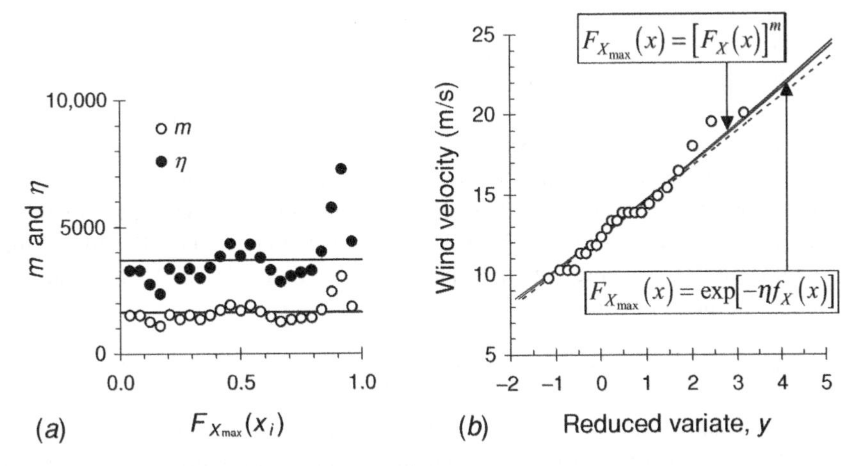

Fig. 7.3.7 Annual maximum wind velocity at Forlanini Airport in Milan, Italy. (*a*) Estimates of parameters m and η. (*b*) Gumbel probability plot where the dotted line shows the EV1 distribution fitted to the annual maximum values by the method of moments.

is used under the assumption that the data in the series are mutually independent.[35] The values of m and η are found by combining the observed frequencies of annual maxima with the values of $F_X(x)$ and $f_X(x)$ computed for the $x_{\max(i)}$s.[36]

Example 7.38. Extreme wind speed in Milan. The analysis of 10-minute average wind speed data recorded at the Forlanini Airport in Milan, Italy, from 1951 to 1973 yields $\hat{p}_0 = 0.511$ and $\hat{\lambda} = 1.284$ m/s, and $\hat{\beta} = 0.836$ (Lagomarsino and Solari, 1995). Using the maximum annual data shown in Table E.7.4, one fits the values of m and η of Eqs. (7.3.13) and (7.3.14) by computing the observed frequencies of the $x_{\max(i)}$'s. From Eq. (7.3.13)

$$m = \frac{\ln[F_{X_{\max}}(x)]}{\ln[F_X(x)]};$$

then

$$\hat{m} = \frac{1}{n}\sum_{i=1}^{n} \frac{\ln[i/(n+1)]}{\ln[p_0 + (1-p_0)\{1 - \exp[-(x_{\max(i)}/\lambda)^{\beta}]\}]},$$

which yields a value of $\hat{m} = 1627$ (see Fig. 7.3.7*a*).

From Eq. (7.3.14),

$$\eta = -\frac{\ln[F_{X_{\max}}(x)]}{f_X(x)};$$

then

$$\hat{\eta} = \frac{1}{n}\sum_{i=1}^{n} \frac{\ln[i/(n+1)]}{p_0 + (1-p_0)(\beta/\lambda)(x_{\max(i)}/\lambda)^{\beta-1}\exp[-(x_{\max(i)}/\lambda)^{\beta}]}.$$

This yields a value of $\hat{\eta} = 3693$ (see Fig. 7.3.7*a*). The Gumbel probability plot of Fig. 7.3.7*b* shows the above probability models and $X_{\max} \sim$ Gumbel (2.21 m/s, 12.34 m/s) which one obtains when fitting the EV1 distribution to the annual maximum data by the method of moments.

[35] Equation (7.3.14) descends from the crossing properties of a series of mutually independent identically distributed variables.

[36] Refined methods for estimation of m and η are discussed in Lagomarsino et al. (1992).

In hurricane regions, the data are a mixture of hurricane and cyclonic winds, so that a single probability law cannot adequately model these data.[37] The extreme value distribution results as a mixture between the two underlying distributions, such as the two-component extreme value distribution of Eq. (7.2.77). For stations having a relatively long record of hurricanes, say, longer than 50 years, it is often found that some years have more than one value listed whereas in other years there are none.[38] Therefore, one can apply the contagious distribution approach to model extreme hurricane winds, where the number of occurrences of hurricanes is a Poisson-distributed variate; and the Pareto distribution, for example, is used to represent wind speed data (see Example 7.26). The resulting extreme value distribution of annual maximum winds is

$$F_{X_{\max}}(x) = \exp\left[-e^{-\frac{x-b_1}{\alpha_1}} - \nu_2\left(\frac{a_2}{x}\right)^{\theta}\right], \qquad (7.3.15)$$

where α_1 and b_1 are the scale and location parameters of Gumbel-distributed cyclonic wind speed as estimated from the annual maxima of cyclonic winds, a_2 and θ are the scale and shape parameters of Pareto-distributed hurricane wind speed, and ν_2 is the mean number of annual occurrences of hurricanes.

Example 7.39. Design wind speed. The number of annual occurrences of hurricanes in a given location is a Poisson-distributed variate with mean 0.375, and the associated wind speed is a Pareto-distributed variate with shape and scale parameters of 4.5 and 26.4 m/s, respectively (see Example 7.27). The annual maximum cyclonic (thunderstorm) wind speed is a Gumbel-distributed variate with mean and standard deviation of 27.5 and 4.3 m/s, respectively. Thus, the corresponding scale and location parameters are 3.4 and 25.6 m/s, respectively. The cdfs of extreme thunderstorm and hurricane winds are shown in Fig. 7.3.8 using Gumbel probability plots.

From Eq. (7.3.15) the cdf of the annual fastest wind speed is found to be

$$F_{X_{\max}}(x) = \exp\left[-e^{-\frac{x-25.6}{3.4}} - 0.375\left(\frac{26.4}{x}\right)^{4.5}\right],$$

which is also plotted in Fig. 7.3.8. Note that this equation cannot be inverted to compute a required design value, for example, the 50-year design wind speed. A numerical solution is therefore found for x with probability of nonexceedance taken as 0.98, so obtaining $\hat{x}_{\max}(50) =$ 50.8 m/s. Note that the corresponding quantiles for thunderstorm and hurricane winds are 38.7 and 50.5 m/s, respectively.

The estimation of the probability distribution of the annual maximum tornado wind speed is affected by larger uncertainties than that achievable for well-behaved climates and hurricane-prone regions because of the lack of tornado wind speed records.[39] To find

[37] See, for example, Gomes and Vickery (1977).

[38] Because in most cases adequate data are not available, the procedure for performing statistical analysis of hurricane wind speed is (1) to identify the geographical areas in which hurricanes that are capable of striking the location (station) considered occur; (2) to set up a physically based simulation model with random variables described through probability distributions estimated from the data gathered in the area of interest; (3) to generate series of hurricanes using Monte Carlo simulation (see Chapter 8) in order to produce sufficient data for the location considered; and (4) to fit an extreme value distribution of hurricane winds to the generated data. Further developments of this procedure, first introduced by Russel (1971), are given by Batts et al. (1980), Delaunay (1987), and Sánchez-Sesma et al. (1988) among others.

[39] Because the area struck by an individual tornado is very limited, simulation of tornado winds is much more difficult than that of hurricane winds, due to the length of time required (perhaps centuries) to get sufficient

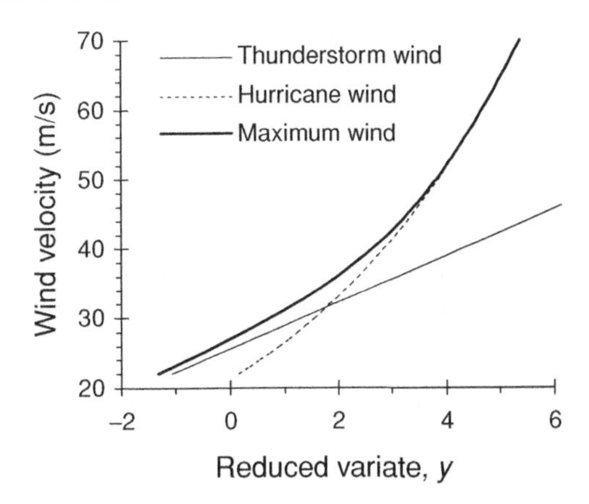

Fig. 7.3.8 Annual maximum thunderstorm and hurricane wind speeds.

this distribution, the probability Pr[A] that a tornado strikes a particular location in a given year is considered jointly with the conditional probability Pr[$W|A$] that this tornado has a wind speed higher than x. Thus, Pr[A] Pr[$W|A$] is the probability that the annual maximum tornado wind exceeds the value of x. Since areas of tornado paths are independent of the tornado intensity, the cdf of the annual fastest tornado wind is usually estimated as Pr[A] Pr[W]; that is,

$$F_{X_{\max}}(x) = 1 - p_A p_W, \tag{7.3.16}$$

where p_A denotes the *strike probability* Pr[A] and p_W the *speed probability* Pr[W]. The strike probability is generally evaluated from $\nu_0 a/A_0$, where a is the average damage area of a tornado, A_0 is a reference area taken as a one-degree longitude-latitude square, and ν_0 is the average number of tornadoes per year in the area A_0. For example, the estimated values of p_A in the United States are reported over a five-degree longitude-latitude square grid by Markee et al. (1974), where $\hat{a}$ is taken as 7.3 km^2. The estimation of p_W is generally based on tornado classification associating wind speed with damage potential and observed effects. For example, the wind speed of tornado gusts in the United States is approximately a lognormal variate with a mean of about 43 m/s and a coefficient of variation of 0.38 (Markee et al., 1974). Note that the annual fastest wind speed used in the analysis of extratropical and hurricane winds must be adjusted by a multiplicative factor of about 1.2 to account for wind speeds of tornado gusts.

An alternative approach is to consider the occurrence of a tornado in the location of interest as a Poisson event with parameter $\nu = \nu_0 a/A_0$. This gives

$$F_{X_{\max}}(x) = \exp\left\{-\nu_0 a A_0^{-1}[1 - F_X(x)]\right\}, \tag{7.3.17}$$

where $F_X(x)$ denotes the cdf of tornado wind speed, and ν_0, a, and A_0 are regional values.[40]

data for a given location. Moreover, there is a lack of anemometric data as a result of gauge failures at those few stations that experience a tornado; the anemometer is often destroyed by such an event. Therefore, the observations currently available are derived from scales of intensity based on structural damage in the areas hit (see, for example, Wolde-Tinsae et al., 1985 and Fujita, 1985).

[40] Developments of tornado wind-risk analysis are provided by Garson et al. (1975) and Twisdale and Dunn (1978), among others.

Example 7.40. Tornado wind speed in the United States. Power plants in the United States must be designed against tornadoes with exceedance probability p of 10^{-7}. The engineer must determine the corresponding wind speed for power plants. In the state of Florida, the estimated tornado strike probability is 1.5×10^{-3}. Accordingly, the corresponding design intensity has a probability of

$$p_B = p/p_A = 10^{-7}/1.5 \times 10^{-3} = 6.67 \times 10^{-5}.$$

Assuming that gust wind speed is a lognormal variate X with mean of 43 km/h and coefficient of variation of 0.38, from Eqs. (7.2.83) and (7.2.84) one obtains $\hat{\sigma}_{\ln(X)} = 0.367$, and $\hat{\mu}_{\ln(X)} = 3.694$. The required design quantile is then obtained as

$$\hat{\xi} = \exp\left[\hat{\mu}_{\ln(X)} + \Phi^{-1}_{1-p_1}\hat{\sigma}_{\ln(X)}\right] = \exp[3.694 + 3.821 \times 0.367] = 164\,\text{m/s}.$$

Note that this value is close to the design value of 160 km/h recommended by the U.S. Nuclear Regulatory Commission for Region I (Eastern and Central United States) including Florida.

7.3.4 Sea levels and highest sea waves

Sea levels change continuously, taking a variety of scales. In particular, tides, arising mainly from the gravitational pull of the moon, cause periodic variations. Tides generally have periods of 12 hours but there can be 24-hour tides of smaller amplitudes. Waves cause instantaneous quasi-cyclical changes in levees with amplitudes exceeding 10 m in open seas. Wave action increases with storms. For example, the storm surge caused by hurricane Katrina on 29 August 2005 in the southern United States reached a record height of over 9 m. Such effects have to be accounted for in designs.

In offshore and coastal engineering, one should select an appropriate wave height for design purposes. This requires a statistical analysis of extreme waves. Measurements of heights of ocean waves have been taken since the 1960s, and a common procedure was to fit a lognormal distribution to the observed data (Draper, 1963). Subsequently, digital computers made it possible to simulate the largest storms that occurred at a particular site using meteorological data. In such cases, the parts of the wave heights attributed to the storm were estimated, and the peaks over a threshold method were used (Petrauskas and Aagaard, 1971). The steps in the statistical procedure for extreme wave heights are (1) to select appropriate data for, (2) to fit a suitable probability distribution to the observed data, (3) to compute (extreme) values from the fitted distribution, and (4) to compute their confidence intervals.[41]

Data selection is a primary issue in this type of analysis. Wave heights vary in time as a continuous nonstationary process because of periodic (seasonal) variation. Waves resulting from different physical processes should be modeled separately. In midlatitude areas, for instance, high waves may arise from different causes, such as tropical and extratropical storms. In other areas, there may also be differences, mostly because of fetch (usually concerning maximum distance from land) limitations. One uses mixed distributions in these conditions. Because of nonstationarity and mixed distributions, the statistical analysis becomes difficult. Considering such uncertainties, extreme wave heights from different seasons and different causes should be analyzed separately.

The peaks over a threshold method is generally used to select the sample from the full data set of wave heights, because the number of annual maxima is most often too small

[41] Extensive reviews of the methods used for analyzing extreme sea wave height data are reported by Muir and El-Shaarawi (1986), Tawn (1993), and Mathiesen et al. (1994), among others.

to make a reasonable extrapolation of extremes. The choice of the threshold is usually made on a physical or meteorological basis. The use of weather charts is one method of determining the number n_a of significant storm events per year; if there is a seasonal variation in storms, the threshold is often obtained by stipulating that an average of less than two storms (which could be a noninteger) are included for the season with only a few storms in the storm peak data set.

The three-parameter Weibull distribution seems to provide an acceptable fit to significant wave heights for most oceans and seas. The truncated distribution for storm peaks above the threshold x_0 is given by

$$F_{X|X>x_0}(x) = \frac{F_X(x) - F_X(x_0)}{1 - F_X(x_0)} = 1 - \exp\left[-\left(\frac{x-b}{a}\right)^c + \left(\frac{x_0-b}{a}\right)^c\right], \quad (7.3.18)$$

where a, b, and c are parameters to be estimated from the data. The values of c which are most frequently found from data analysis range from 1 to 2 and many are close to 1. Therefore, a truncated exponential distribution can be used as an approximation.[42] The T-year wave height is then computed as the value x satisfying

$$F_{X|X>x_0}(x) = 1 - \frac{\tau}{T}, \quad (7.3.19)$$

where τ is the average time between two subsequent storms.

Alternatively, one must consider the annual number of storm occurrences to be a random variable. If N_a is a Poisson-distributed variate with mean ν, from Eq. (7.2.69) we obtain the cdf of annual maximum wave heights as

$$\begin{aligned} F_{X_{\max}}(x) &= \exp\{-\nu[1 - F_{X|X>x_0}(x)]\} \\ &= \exp\left\{-\nu \exp\left[-\left(\frac{x-b}{a}\right)^c + \left(\frac{x_0-b}{a}\right)^c\right]\right\}. \end{aligned} \quad (7.3.20)$$

Example 7.41. Highest sea waves in the Adriatic Sea. Consider the data set of observed highest sea waves above a threshold x_0 of 2 m in the upper Adriatic Sea, given in Problem 1.23. This data include 18 independent storms recorded in a period of 13 months. One notes that the truncated distribution of Eq. (7.3.18) can be written as

$$(\kappa - \ln p)^{1/c} = (x - b)/a,$$

where p is the probability of exceedance, and $\kappa = [(x_0 - b)/a]^c$ is the normalization factor for truncation. For a selected value of c, one plots the ordered observations against $(\kappa - \ln p)^{1/c}$ to fit the straight line interpolating these points, so determining the values of a and b by a trial-and-error procedure (see Fig. 7.3.9a). Assuming $c = 2$, one obtains $\hat{a} = 1.7$ m, $\hat{b} = 1.3$ m, with $\hat{\kappa} = [(2 - 1.3)/1.7]^2 = 0.170$. The fitted distribution is also shown in Fig. 7.3.9b as an exponential probability plot.

The T-year wave height is then computed as the value of x satisfying Eq. (7.3.19); that is,

$$x(T) = b + a[\kappa - \ln(\tau/T)]^{1/c},$$

where $\hat{\tau} = (13/18)/12 = 0.06$ year. For $T = 10$ years, one gets

$$\hat{x}(10) = 1.3 + 1.7\left[0.170 - \ln\left(\frac{0.06}{10}\right)\right]^{1/2} = 5.21 \text{ m}.$$

[42] See Subsection 4.2.5 and Eq. (4.2.17d). As regard the peaks over a threshold method (particularly with respect to Pareto-distributed wind speeds of Subsection 7.3.3), Harris (2005) provides an interesting discussion.

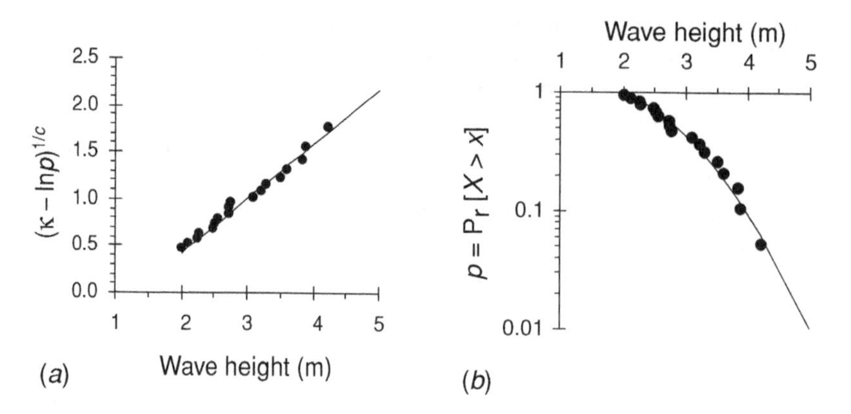

Fig. 7.3.9 Highest sea waves in the upper Adriatic sea in a period of 13 months: (a) observations versus $(\kappa - \ln p)^{1/c}$, and (b) exponential probability plot.

Alternatively, the contagious Poisson model of Eq. (7.3.20) is considered. From this model the design value is found from

$$\hat{x}_{\max}(T) = b + a\left\{\kappa - \ln\left[-\nu^{-1}\ln\left(1 - \frac{1}{T}\right)\right]\right\}^{1/c}.$$

The mean number of storm occurrences in a year is estimated as $\hat{\nu} = 1/\hat{\tau} = 16.2$. Thus, for $T = 10$,

$$\hat{x}_{\max}(10) = 1.3 + 1.7\left\{0.170 - \ln\left[-0.06 \times \ln\left(1 - \frac{1}{10}\right)\right]\right\}^{1/2} = 5.19\,\mathrm{m},$$

which practically equals the previous estimate from the peaks over a threshold model.

More generally, one can consider the wave height X at a given time to comprise three unobserved additive components: mean sea level U, tidal level W, and surge level S. The mean sea level, which is a measure of the variability in the data of frequencies longer than a year, varies as a result of changes in land and global water levels. For example, 100 years of data show that the mean sea level increases with a rate of 1–2 mm/year on the global scale; also, the presence of interannual variations due to the Southern Oscillation means that nonstationarity (of the mean) can no longer be modeled by a simple linear trend in the Pacific Ocean. The deterministic astronomical tidal component, generated by changing forces on the ocean produced by planetary motion, can be predicted from a cyclic equation including global and local (site) constants. The random surge component, generated by short-term climatic behavior, is identified as the $X-U-W$ residual.[43] Therefore, the probability distribution of $X_{\max}$, the annual maximum sea wave height, must account for nonstationarity. Also, the extreme values of S may cluster around the highest values of W, because extreme sea levels typically arise in storms that happen to produce large surges at or around the time of a high tide. However, it is often assumed that the astronomical tide does not affect the magnitude of a storm surge (as in the

[43] For example, around the coast of Great Britain, Y appears to be linearly increasing (Woodworth, 1987), the dominant cycle in the tide has a period of 12 hours and 26 minutes, and extreme sea levels typically arise in storms which happen to produce large surges at or around the time of a high tide (Tawn, 1993).

Example 7.42). It is then unlikely that the highest values of S coincide with the highest values of W.

Example 7.42. Extreme sea levels. To account for the clustering of large values of the underlying variables determining extreme sea levels, the family of limiting distributions of Eqs. (7.2.1) is taken as $[F_Y(y)]^\theta$, with $0 \le \theta \le 1$ denoting the extremal index for the process. This parameter is estimated by the amount of clustering among the extreme values of the multicomponent process, and $1/\theta$ is the limiting mean cluster size. The limiting distribution is the generalized extreme value distribution of Eq. (7.2.59) with the same shape parameter of the stationary case, whereas the scale and location parameters are modified by the dependence components. Tawn (1992) suggested the evaluation of the cdf of the annual maximum surge $S_{\max}$ as

$$F_{S_{\max}}(s) = F_S^{\tau\theta_s}(s) = \exp\left\{-\left[1 - \frac{k_s(s - s_0)}{\alpha_s}\right]^{1/k_s}\right\},$$

for $s > s_0$, where τ is the number of observations per year of hourly surges above a threshold of s_0, and θ_s is the extremal index for the hourly surge sequence. Thus,

$$F_S(s) = \exp\left\{-\frac{1}{\tau\theta_s}\left[1 - \frac{k_s(s - s_0)}{\alpha_s}\right]^{1/k_s}\right\}.$$

The estimator of θ_s is the inverse of the sample mean of independent clusters in the surge sequence, where the cluster maxima exceeds the threshold of s_0. The cdf of $X_{\max}$ is then found as

$$F_{X_{\max}}(x) = \Pr[S_t \le x - u_t - w_t \quad \text{for } t \in \tau] = \left[\prod_{t\in\tau} F_S(x - u_t - w_t)\right]^\theta,$$

where $\theta_s \le \theta \le 1$ is the hourly sea-level extremal index. (Note that U and W are as previously defined and t is a point in time.) Estimation of θ is via the same approach as for θ_s but is based on the hourly sea-level series. The design value ξ_q is found to satisfy

$$-\frac{\theta_s}{\theta}\log q = \frac{1}{\tau}\sum_{t\in\tau_i}\left[1 - \frac{k_s(\xi - s_0 - u_t - w_t)}{\alpha_s}\right]^{1/k_s},$$

which is solved numerically for ξ_q.

Multisite analysis generally yields more robust parameter and quantile estimates. For example, Tawn (1993) analyzed a 34-year record of hourly sea levels for Port Adelaide in south Australia, and estimated $\hat{\xi}_{0.99} = 2.48$ m from site analysis, with a standard error of 0.12 m; from multisite analysis a corresponding value of 2.59 m was obtained with a standard error of 0.07 m.

7.3.5 Summary of Section 7.3

In this section we have considered diverse types of natural hazards, of importance in civil and environmental engineering, and demonstrated the practical use of extreme value distributions. Applications range from floods and droughts to earthquakes, volcanic eruptions, high winds, and sea waves and levels. For further elaborations and discussions of other types of natural hazards, such as tsunamis, landslides, avalanches, and forest fires, see Salvadori et al. (2007).

7.4 SUMMARY OF CHAPTER 7

Civil and environmental engineers are often concerned with extreme events. Because of the short length of observations usually available as a random sample of extreme values, large uncertainties affect the properties of extremes. Order statistics provide the statistical properties of the arranged outcomes of a variable, such as its maximum, minimum, and range. Extreme value theory shows that the maximum or minimum of a sequence of variates with a common distribution is distributed asymptotically according to a general form. This is called the *general extreme value distribution*. It includes three asymptotic types, which depend on the distribution of the underlying variables. Alternative models for extreme values include contagious distributions based on random counting of extreme events, and other parametric distributions, such as the lognormal, the gamma, and the log-Pearson Type III, which also provide a satisfactory fit to extreme value data in some cases. Theoretical developments are followed by 42 practical examples. As an aid to engineering judgments, numerous probability plots are shown. However, we have not emphasized the exact choice of a plotting position, because of uncertainties regarding the probability distribution, and differences tend to be small as discussed in Chapter 5. Besides, there are the effects of nonstationarity to consider, such as outliers (Section 5.9). As stated before, engineers looking for practical means of extrapolation tend to give more attention to the larger values in the data. See the application of the two-component extreme value distribution in Example 7.28 and the case of extreme sea levels of Example 7.42; see also nonstationary modeling in Example 10.13.

Current practices for analyzing extreme values of hydrological, seismic, volcanic, wind, and sea-level variables concerning natural hazards are briefly outlined and discussed, taking account of nonstationarity where necessary. Continuing efforts toward better understanding of natural hazards should result in more realistic assessments.

REFERENCES

General. The following references are given for further reading:

Bolt, B. (1999) *Earthquakes*, WH Freeman & Co., New York, 366 pp.

Coles, S. (2001). *An Introduction to Statistical Modeling of Extreme Values*, Springer-Verlag, New York. Medium-level text.

Committee on Techniques for Estimating Probabilities of Extreme Floods (1988). *Techniques for Estimating Probabilities of Extreme Floods. Methods and Recommended Research*, National Academy Press, Washington, DC. A survey of problems encountered in evaluating extreme flood probabilities.

David, H. A., and H. N. Nagaraja (2003). *Order Statistics*, 3rd ed., John Wiley and Sons, New York. Advanced reference for order statistics.

Galambos, J. (1978). *The Asymptotic Theory of Extremes*, John Wiley and Sons, New York. A Comprehensive advanced reference on the subject.

Gumbel, E. J. (1958) *Statistics of Extremes*, Columbia University Press, New York. A classical treatise on the statistical properties and models of extremes.

Kite, G. W. (1988). *Frequency and Risk Analysis in Hydrology*, Water Resources Publications, Littleton, CO. A handbook of probability distributions used in extreme value analysis of hydrologic processes.

Kotz, S., and S. Nadaraja (2000). *Extreme Value Distributions: Theory and Applications*, Imperial College Press, London. Suitable for engineers.

Leadbetter, M. R., G. Lindgren, and H. Rootzén (1983). *Extremes and Related Properties of Random Sequences and Processes*, Springer-Verlag, New York. An advanced book on statistical extremes.

Liu, H. (1991). *Wind Engineering: A Handbook for Structural Engineers*, Prentice Hall, Englewood Cliffs, NJ. A handbook for structural engineers including a short survey of current practice in estimating design winds (Chapter 2) and some building codes and standards (Chapter 7).

Lomnitz, C. (1974). *Global Tectonics and Earthquake Risk*, Elsevier Scientific Publishing Company, New York. An introduction on stochastic modeling of earthquake processes and the risks involved.

Lomnitz, C., and E. Rosenblueth (eds.) (1976). *Seismic Risk and Engineering Decisions*, Elsevier Scientific Publishing Company, New York. A well-edited engineering-oriented book on seismic hazards, including tsunamis.

Reiter, L. (1990). *Earthquake Hazard Analysis: Issues and Insights*, Columbia University Press, New York. A comprehensive reference on earthquake hazard analysis.

Rutenberg, A. (ed.) (1994). *Earthquake Engineering*, A. A. Balkema, Rotterdam. It includes seismic risk and Bayesian analysis.

Simiu, E., and R. H. Scanlan (1986). *Wind Effects on Structures: An Introduction to Wind Engineering*, 2nd ed., John Wiley and Sons, New York. A comprehensive reference in wind engineering including current practices for design wind evaluation.

Smith, R. L. (1990). "Extreme value theory," in *Handbook of Applicable Mathematics*, edited by W. Ledermann, Vol. 7, Chapter 14, pp. 437–472, John Wiley and Sons, New York. Advanced theory.

Turcotte, D. L. (1992). *Fractals and Chaos in Geology and Geophysics*, Cambridge University Press, Cambridge, UK. This text includes an alternative approach to the analysis of extremes of some natural processes.

Additional references quoted in the text

ANSI/ASCE 7 (1988). *Minimum Design Loads for Buildings and Other Structures*, American Society of Civil Engineers, New York. Gumbel fitting of wind speeds.

Bacchi, B., A. Brath, and N. T. Kottegoda (1992). "Analysis of the relationship between flood peaks and flood volumes based on crossing properties of river flow processes," *Water Resour. Res.*, Vol. 10, pp. 2773–2782. Distribution of flood volumes.

Batts, M. E., L. R. Russel, and E. Simiu (1980). "Hurricane wind speeds in the United States," *J. Struct. Eng. Div., ASCE*, Vol. 106, pp. 2001–2016. Extreme value distributions.

Bocchiola, D., C. De Michele, and R. Rosso (2003). "Review of recent advances in index flood estimation," *Hydrol. Earth Syst. Sci.*, Vol. 7, No. 3, pp. 283–296. Regional floods.

Bocchiola, D., C. De Michele, and R. Rosso (2004). "L'appplicazione della legge generalizzata del valore estremo GEV all'analisi regionale delle piene in Italia," *L'Acqua*, Vol. 1/2004, pp. 35–52 (in Italian). Regional floods.

Buishand, T. A. (1989). "Statistics of extremes in climatology," *Stat. Neerl.*, Vol. 43, pp. 1–30. Meteorological applications.

Burlando, P., and R. Rosso (1996). "Scaling and multiscaling depth-duration-frequency curves of storm precipitation," *J. Hydrol.*, Vol. 187, pp. 45–64. Multiscaling of rainfall.

Chow, V. T. (1951). "A general formula for hydrologic frequency analysis," *Trans. Am. Geophys. Union*, Vol. 32, pp. 231–237. Frequency factors for extreme values.

Cornell, C. A. (1968). "Temporal and magnitude dependence in earthquake recurrence models," *Bull. Seismologic. Soc. Am.*, Vol. 58, pp. 1583–1606. Work of seminal importance.

Cornell, C. A., and S. R. Winterstein (1988). "Temporal and magnitude dependence in earthquake recurrence models," *Bull. Seismologic. Soc. Am.*, Vol. 78, pp. 1522–1537. Poisson model for earthquakes.

de Haan, L. (1976). "Sample extremes: An elementary introduction," *Stat. Neerl.*, Vol. 30, pp. 161–172. Asymptotic distributions.

De Michele, C., and R. Rosso (2002). "A multi-level approach to flood frequency regionalization," *Hydrol. Earth Syst. Sci.*, Vol. 6, No. 2, pp. 185–194. Regional studies.

De Michele, C., N. T. Kottegoda, and R. Rosso (2001). "The derivation of areal reduction factor of storm rainfall from its scaling properties," *Water Resour. Res.*, Vol. 37, pp. 3247–3252. On scaling.

De Michele, C., N. T. Kottegoda, and R. Rosso (2002). "IDAF curves of extreme storm rainfall: A scaling approach," *Water Sci. Techol.*, Vol. 45, No. 2, pp. 83–90. On scaling.

De Michele, C., G. Salvadori, A. Petaccia, and R. Rosso (2005). "Bivariate statistical approach to spillway design flood," *J. Hydrol. Eng. ASCE*, Vol. 10, No. 1, pp. 50–57. Bivariate medeling.

Delaunay, D. (1987). "Extreme wind speed distribution for tropical cyclones," *J. Wind Eng. Ind. Aerodyn*, Vol. 28, pp. 61–68. Extreme value distributions.

Draper, L. (1963). "Derivation of a "design wave" from instrumental records of sea waves," *Proc. Inst. Civ. Eng.*, Vol. 26, pp. 291–304. Application of lognormal distributions.

Eliasson, J. (1994). "Statistical estimates of PMP values," *Nor. Hydrol.*, Vol. 25, pp. 301–312. Extreme value distributions associated with deterministic procedures.

Fisher, R. A., and H. C. Tippett (1928). "Limiting forms of the frequency distribution of the largest or smallest member of a sample," *Proc. Camb. Phil. Soc.*, Vol. 24, pp. 180–190. Origin of extreme value theory.

Fréchet, M. (1927). "Sur la loi de probabilité de l'écart maximum," *Ann. de la Societé Pol. de Math. (Cracow)*, Vol. 6, pp. 93–117. Origin of extreme value theory.

Fujita, T. T. (1985). "The Downburst," SMRP Research Paper no. 210, p. 122, University of Chicago, Chicago. On structural damage.

Galton, F. (1875). "Statistics by intercomparison with remarks on the law of frequency levels," *Phil. Mag.*, 4th Series, Vol. 49, pp. 33–46. Origin of lognormal distribution.

Garson, R. C., J. Morla-Catalan, and C. A. Cornell (1975). "Tornado risk evaluation using wind speed profiles," *J. Struct. Eng. Div.,ASCE*, Vol. 101, pp. 1167–1171. Risk analysis.

Gibrat, R. (1932). "Amenagement hydroélectriques des cours d'eau: Statistique mathématique et calcul dés probabilités," *Rev. Générale de l'Electricité*, Vol. 32, pp. 15–16. Original application of the lognormal distribution.

Gnedenko, B. V. (1943). "Sur la distribution limite du terme maximum d'une série aléatoire," *Ann. Math.*, Vol. 44, pp. 423–453. Extremal type theorem.

Gomes, L., and B. J. Vickery (1977). "Extreme wind speeds in mixed climates," *J. Ind. Aerodyn*, Vol. 2, pp. 331–344. Modeling of hurricanes.

Goodrich, R. D. (1927). "Straight line plotting of skew frequency data," *Trans. Am. Soc. Civ. Eng.*, Vol. 91, pp. 1–118. Special distribution in hydrology.

Gumbel, E. J. (1935). "Les valeurs extrèmes des distributions statistiques," *Ann. Inst. Henri Poincaré*, Vol. 4, pp. 115–158. Origin of Gumbel distribution.

Gumbel, E. J. (1941). "The return period of flood flows," *Ann. Math. Stat.*, Vol. 12, pp. 163–190. Application of Gumbel distribution.

Gutenberg, B., and C. F. Richter (1944). "Frequency of earthquakes in California," *Bull. Seismologic. Soc. Am.*, Vol. 34, pp. 185–188. Probabilistic seismic analysis.

Gutenberg, B., and C. F. Richter (1954). *Seismicity of the Earth and Associated Phenomenon*, 2nd ed., Princeton University Press, Princeton, NJ. Earthquake frequencies.

Harris, I. (2005). "Generalized Pareto methods for wind extremes. Useful tool or mathematical mirage?" *J. Wind Eng. Ind. Aerodyn., ASCE*, Vol. 93, pp. 341–360. Comments on the use of the Pareto distribution in wind engineering.

Hosking, J. R. M., J. R. Wallis, and E. F. Wood (1985). "Estimation of the generalized extreme-value distribution by the method of probability-weighted moments," *Technometrics*, Vol. 27, pp. 251–261. Application of *L*-moments.

Hurst, H. E. (1951). "Long-term storage capacity of reservoirs (with discussion)," *Trans. Am. Soc. Civ. Eng.* Vol. 116, paper 2447, pp. 770–808. Reference for Example 7.8.

Jeffreys, H. (1967). "Statistical methods in seismology," in *International Dictionary of Geophysics*, edited by K. Runcorn, Pergamon, London. Pioneering work by a versatile scientist.

Jenkinson, A. F. (1955). "The frequency distribution of the annual maximum (or minimum) value of meteorological elements," *Q. J. R. Meteorol. Soc.*, Vol. 81, pp. 158–171. General extreme value distribution.

Jenkinson, A. F. (1969). "Estimation of maximum floods," World Meteorological Organization, Technical Note, no. 98, Chapter 5, pp. 183–257. General extreme value distribution.

Kanamori, H., and D. L. Anderson (1975). "Theoretical basis of some empirical relations in seismology," *Bull. Seismolc. Soc. Am.*, Vol. 65, pp. 1073–1096. Reference for Example 7.37.

Kottegoda, N. T. (1984). "Investigation of outliers in annual maximum flow series," *J. Hydrol.*, Vol. 72, No. 1, pp. 105–137. Nonstationarity and methods of detecting outliers.

Lagomarsino, S., and G. Solari (1995). "The wind induced dynamic behaviour of the South-Milan telecommunication tower," *Studi e Ricerche, Scuola di Specializzazione in Costruzioni in Cemento Armato Fratelli Pesenti*, Politecnico di Milano, Vol. 16, pp. 231–266. Reference for Example 7.38.

Lagomarsino, S., G. Piccardo, and G. Solari (1992). "Statistical analysis of high return period wind speeds," *J. Wind Eng. Ind. Aerodyn., ASCE*, Vol. 41–44, pp. 485–496. Annual maximum wind speeds.

Leadbetter, M. R. (1991). "On a basis for peaks over threshold modeling," *Stat. Probab. Lett.*, Vol. 12, pp. 357–362. Asymptotic distributions.

Lomnitz, C. (1994). *Fundamentals of Earthquake Prediction*, John Wiley and Sons, New York. On earthquake predictions.

Markee, E. H., J. R. Beckerley, and K. E. Sansders (1974). "Technical basis for interim regional tornado criteria," WASH-1300 (UC-11), U.S. Atomic Energy Commission, Office of Regulation, Washington, DC. Maximum wind speeds.

Mathiesen, M., Y. Goda, P. J. Hawkes, E. Mansard, M. J. Martin, E. Pelthier, E. F. Thomson, and G. Van Vledder (1994). "Recommended practice for extreme wave analysis," *J. Hydraul. Res.*, Vol. 32, No. 6, pp. 803–814. Analysis of wave heights.

McGuire, R. K. (2005). "Seismic hazard and risk analysis," Earthquake Engineering Research Institute Publication No. MNO-10, Oakland, CA. Work of seminal importance.

McLelland, L., T. Simkin, M. Summers, E. Nielson, and T. C. Stein (1989). *Global Volcanism 1975–1985*, Prentice Hall, Englewood Cliffs, NJ. Volcanic eruptions.

Mood, A. M., F. A. Graybill, and D. C. Boes (1974). *Introduction to the Theory of Statistics*, 3rd ed., McGraw-Hill, New York. Joint density of order statistics.

Muir, L. R., and A. H. El-Shaarawi (1986). "On the calculation of extreme wave heights," *Ocean Eng.*, Vol. 13, pp. 93–118. Extensive reviews.

Naeim, F. (ed.) (1989). *The Seismic Design Handbook*, Van Nostrand, New York. On improved designs related to earthquakes.

Natural Environmental Research Council (1975). *Flood Studies Report*, NERC Publication, London. Extensive analysis of flood data; Example 7.33.

National Research Council, National Group for Prevention of Hydrogeological Hazard (2000). The VAPI Project. General Report, edited by S. Gabriele and F. Rossi, CNR GNDCI Publication, Rome (in Italian).

Petrauskas, C., and P. M. Aagaard (1971). "Extrapolation of historical storm data for estimating design-wave heights," *J. Soc. Pet. Eng.*, Vol. 11, pp. 23–37. Peaks over a threshold method.

Robson, A., and D. Reed (1999). "Statistical procedures for flood frequency estimation," in *Flood Estimation Handbook*, Vol. 3, Institute of Hydrology, Wallingford, UK. Pooling of data in regional flood studies.

Rossi, F., M. Fiorentino, and P. Versace (1984). "Two component extreme value distribution for flood frequency analysis," *Water Resour. Res.*, Vol. 20, pp. 847–856. Application in Example 7.28.

Rosso, R. (1985). "A linear approach to the influence of discharge measurement error on flood estimates," *J. Hydrol. Sci.*, Vol. 20, pp. 137–149. Observational errors.

Russel, L. R. (1971). "Probability distribution of hurricane effects," *J. Waterways, Harbors Coastal. Eng.. Div., ASCE*, Vol. 97, pp. 139–154. Extreme value model for hurricanes.

Salvadori, G., C. De Michele, N. T. Kottegoda, and R. Rosso (2007). *Extremes in Nature: An Approach Using Copulas*, Springer. Statistics for natural hazards.

Sánchez-Sesma, J., J. Aguirre, and M. Sen (1988). "Simple modeling procedure for estimation of cyclonic wind speeds," *J. Struct. Eng. Div., ASCE*, Vol. 114, pp. 352–370. Extreme value distributions of hurricane winds.

Smith, R. L. (1985). "Maximum likelihood estimation in a class of nonregular cases," *Biometrika*, Vol. 72, pp. 67–90. Bounds for maximum likelihood estimation.

Solari, G. (1996). "Statistical analysis of extreme wind speed," in *Modeling of Atmosphere Flow Fields*, edited by D. P. Lalas and C. F. Ratto, World Scientific, Singapore. Annual maximum wind speed.

Stedinger, J. R., R. M. Vogel, and E. Foufula-Georgiu (1992). "Frequency analysis of extreme events," in *Handbook of Hydrology*, Chapter 18, edited by D. R. Maidment, McGraw-Hill, New York. Extreme value analysis in hydrology.

Tawn, J. A. (1992). "Estimating probabilities of extreme sea levels," *Appl. Stat.*, Vol. 41, pp. 77–93. Extreme value analysis of sea levels.

Tawn, J. A. (1993). "Extreme sea-levels," in *Statistics for the Environment*, edited by V. Barnett and K. Feridum Turkman, John Wiley and Sons, Chichester, UK. Reviews in analyzing sea-wave heights.

Troen, I., and E. L. Petersen (1989). "European wind atlas," Commission of the European Communities, Directorate-General for Science, Research and Development, Brussels, Belgium. Wind speeds in Europe.

Twisdale, L. A., and W. L. Dunn (1978). "Tornado data characterization and wind speed risk," *J. Struct. Eng. Div.,ASCE*, Vol. 104, pp. 1611–1630. Risk analysis.

Vere-Jones, D. (1992). "Statistical methods for the description and display of earthquake catalogs," in *Statistics in Environmental and Earth Sciences*, edited by A. T. Walden and P. Guttorp, pp. 220–246, Halstead, New York. Presents recent work.

von Mises, R. (1936). "La distribution de la plus grande de n valeurs," *Rev. Math. de L'Union Interbalkaniqee*, Vol. 1, pp. 141–160. Origin of GEV distribution.

Weibull, W. (1939). "A statistical theory of the strength of materials," *Proc. Roy. Soc. Swed. Inst. Eng. Res.*, No. 151, pp. 1–45. Origin of Weibull distribution.

Wenzel, H. G. (1982). "Rainfall for urban stormwater design," in *Urban Stormwater Hydrology*, edited by D. Kibler, Water Resources Monograph 7, American Geophysical Union, Washington, DC. Depth-duration-frequency curves.

Wilks, S. S. (1962). *Mathematical Statistics*, John Wiley and Sons, New York. cdf of range for Example 7.4.

Wolde-Tinsae, A. M., M. L. Porter, and D. I. McKeown (1985). "Wind speed analysis of tornadoes based on structural damage," *J. Clim. Appl. Meteorol.*, Vol. 24, pp. 699–710. Damage assessment.

Woodworth, P. L. (1987). "Trends in UK mean sea level," *Mar. Geod.*, Vol. 11, pp. 57–87. Research on coast of Great Britain.

Working Group on California Earthquakes Probabilities (1995). "Seismic hazards in southern California: probable earthquakes, 1994 to 2024," *Bull. Seismologic. Soc. Am.*, Vol. 85, pp. 379–439. Special report.

PROBLEMS

7.1. Observed frequency of maximum annual storm rainfall data. Consider the 58-year data of maximum annual hourly storm depth reported in Table E.7.1.

(*a*) Find the expected frequency of nonexceedance of the largest recorded value.

(*b*) Find the theoretical probability of nonexceedance of this value resulting from the fitted GEV distribution.

(*c*) Compute the plotting positions of the observations using the equation $p_i = (i - 0.35)/n$ where i is the rank in increasing order and n is the number of items of data. Compare the observed frequency estimates with these expected frequencies and with the theoretical ones.

7.2. Hurst effect in hydrologic data. Using rescaled range analysis shown in Example 7.8 compute the Hurst exponents for annual rainfall and runoff in the Po River at Pontelagoscuro, Italy (see data in Table E.7.2). Use equispaced values of ln n.

7.3. Minimum flight delay. An airport is designed to receive n daily flight arrivals. Find the mean and variance of the expected minimum delay if the interarrival time X

is a shifted exponentially distributed variate with scale parameter λ and location parameter x_0.

7.4. Flood Discharge. Consider the data of maximum annual flood flows in the Tevere River at Ripetta, Italy, reported in Table E.5.8. Compute the 100-year flood discharge using (*a*) the Gumbel distribution, (*b*) the GEV distribution, (*c*) the lognormal distribution, and (*d*) the gamma distribution. Perform a goodness-of-fit test using the chi-squared, Kolgomorov-Smirnov and Anderson-Darling tests. Consider $\alpha = .10$

7.5. Depth-duration-frequency curves of storm rainfall. Consider the statistical summaries of Table 7.3.1 for the annual maximum storm depth for various durations observed at Lanzada, Italy. The estimated L-moments for the normalized annual maximum storm depth (extreme value data divided for various durations divided by the corresponding mean for the specified duration) are $L_1 = 1$, $L_2 = 0.1330$, and $L_3 = 0.0182$, respectively.

(*a*) Compute the parameters of the Gumbel distribution by the method of L-moments and compare this distribution with that estimated in Example 7.34 by the method of moments.

(*b*) Compute the parameters of the GEV distribution by the methods of moments and L-moments, and compare these distributions with the Gumbel model on a Gumbel probability plot.

(*c*) Find the depth-duration-frequency curve for a return period of 100 years using the GEV model estimated by the method of L-moments.

7.6. Dry spells. A period of days on which no rainfall is experienced continuously is called a *dry run* if preceded and succeeded by one or more wet days. As shown in Example 4.16, the run length X of a dry spell can be modeled as a log-series distributed variate. Suppose that X has a mean of 5 days, so the estimated p is .07, and the number of dry spells in a year is a Poisson variate with a mean of 40. Find the cdf of the annual maximum run length of a dry spell. Compute the return period of a dry spell 60 days long.

7.7. Highest sea wave in a storm. The highest sea wave X in a storm is modeled as a Rayleigh-distributed variate with pdf $f_X(x) = (x/\lambda^2)\exp[-(x/\lambda)^2/2]$. Suppose that parameter λ varies randomly from one storm to another, and assume that the number of storms in a year with $\lambda \geq \lambda_0$ is a Poisson-distributed variate with mean ν. Find the cdf of the annual maximum sea wave height, $X_{\max}$, if $\lambda - \lambda_0$ is an exponentially distributed variate with parameter α.

7.8. Overflooding. The annual maximum flood discharge $X_{\max}$ at a given river site is a Gumbel-distributed variate with parameters $\alpha = 625$ m^3/s and $b = 1152$ m^3/s. The overflooding volume Y during a flood with peak discharge X exceeding a value of $\varepsilon = 2000$ m^3/s is modeled as an exponentially distributed variate with mean $c(X/\varepsilon)^\beta$, where $c = 5 \times 10^6$ m^3 and $\beta = 0.5$. Assume that $N \sim$ Poisson(ν) is the number flood events in a year with peak discharge exceeding ε with $\nu = 2.1$. Find the cdf of the annual maximum overflooding volume, $Y_{\max}$, and compute the 100-year overflooding volume.

7.9. Sea waves. Consider the data set of simulated highest sea waves above a threshold in the upper Adriatic Sea of Problem 1.23. Find the 10-year design wave height resulting from the data set obtained by using calibration strategy no. 1, and that for calibration strategy no. 2 (assume Poisson events and the shifted exponential distribution to fit the data). Compare these values with that obtained from the analysis of observed data reported in Example 7.41.

7.10. Maximum annual wind speed predictions. Find the probability distribution of maximum annual wind speed and the 50-year wind velocity for (*a*) Cagliari and (*b*) Pantelleria, Italy, by fitting the Gumbel distribution to the extreme value data shown in Table E.7.3 using the method of moments.

7.11. Rescaling of wind speed estimates. The maximum annual wind speed $X(t, z)$ averaged over a period of length t that is recorded at ground elevation z in a particular site with roughness length z_0 scales as

$$X(\lambda t, \eta z) = X(t, z)\left[1 + \frac{0.98c(\lambda)}{\ln(z/z_0)}\right]\frac{\ln(\eta z/z_0)}{\ln(z/z_0)},$$

where $t < 1$, $0 < \lambda < 1$, and η are two scaling factors, with $c(\lambda)$ denoting a scaling function, $c(1) = 0$. Suppose that the available records are averaged for 5 minutes, say, $\lambda = 1/12$, for t in hours, with $c(1/12) = 0.54$, the gauging station (site A) is located at elevation of $z = 8$ m in open terrain with $z_0 = 0.01$ m, and the sample mean and standard deviation of the annual maximum data are 15 and 3 m/s, respectively. Reference wind speed X_A is taken as the 10-minute average wind speed at a standard elevation of 10 m, that is, $X_A = X(1/6 \text{ hour}, 10 \text{ m})$, with $\eta = 1.25$, $\lambda = 1/6$, and $c(1/6) = 0.36$.

(*a*) Find the cdf of annual maximum reference wind speed for site A using the Gumbel distribution and the method of moments. To design a building located in the downtown area, one must determine the 10-minute average wind speed at a ground elevation of 50 m knowing that the roughness length for this location (site B) is 0.3 m. The vertical profile of wind velocity is given by $2.5\,u^* \ln(z/z_0)$, where u^* denotes the friction velocity, which scales as $u^*_B/u^*_A = (z_{0B}/z_{0A})^\gamma$ for two sites A and B with different roughness length (see Fig. 7.P1). Therefore, one must rescale the reference wind speed as

$$X_B = X(t, z_B) = \frac{X(t, z_A)}{\ln(z_A/z_{0A})}\left(\frac{z_{0B}}{z_{0A}}\right)^\gamma \ln\left(\frac{z_B}{z_{0B}}\right)$$

with $t = 1/6$ hour and $\gamma = 0.07$.

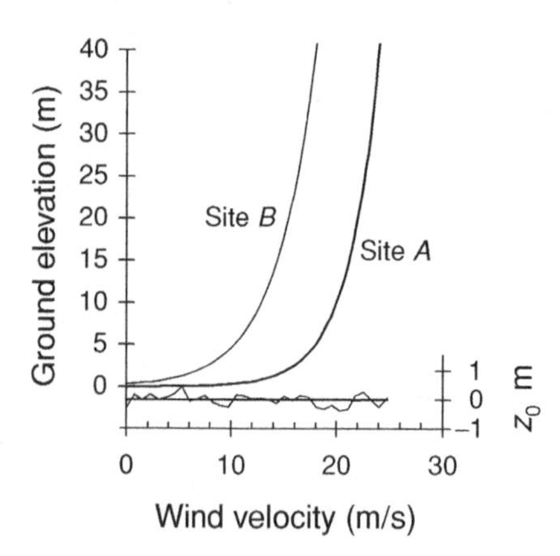

Fig. 7.P1 Comparison of wind velocity profiles at different sites.

(*b*) Find the cdf of annual maximum wind speed at $z_B = 50$ m for site B using the Gumbel distribution and the method of moments.
(*c*) Compute the 50-year design wind speed.

7.12. Wind speed prediction for Milan Park Tower. The 110-m Park Tower in Milan, Italy, is a steel tower built in 1933 on the occasion of the Fifth Triennial Decorative Arts Exhibition. From 1951 to 1973 the following twelve thunderstorms were recorded by an anemometer located at the elevation of 108 m with 10-minute average wind speed X exceeding the critical mean velocity of 20.18 m/s.

Date	Wind direction (°N)	Average velocity (m/s)
29/04/53	15	20.86
08/01/58	135	20.98
05/01/59	315	25.06
08/01/59	315	25.06
09/01/59	315	22.19
20/04/59	45	22.17
28/07/59	345	20.26
10/02/61	315	21.48
12/02/61	315	27.21
03/04/71	315	21.48
20/11/71	315	21.48
15/12/73	345	26.34

Find the cdf of the annual maximum 10-minute average wind speed if the probability distribution of X is (*a*) exponential, and (*b*) Pareto, and compare the corresponding extreme values for a return period of 50 years. Use a Gumbel probability plot to compare these results with $X_{max} \sim$ Gumbel(2.89 m/s, 15.79 m/s) which is obtained by rescaling the extreme value data recorded at the Forlanini Airport station [G. Ballio, S. Lagomarsino, G. Piccardo, and G. Solari (1999). "Probabilistic analysis of Italian extreme winds: Reference velocity and return criterion," *Wind Struct*, Vol. 2, no. 1, pp. 51–68].

7.13. Annual maximum wind speed in Pisa. Consider the following data set of 41 annual maximum 10-minute average wind speeds at Pisa Airport, Italy:

year	1951	1952	1953	1954	1955	1956	1957	1958	1959	1960	1961
m/s	15.43	15.43	15.37	15.43	22.03	18.52	16.46	18.00	19.55	19.03	18.00

year	1962	1963	1964	1965	1966	1967	1968	1969	1970	1971	1972
m/s	19.41	18.52	13.89	18	14.40	16.46	14.40	16.46	13.43	21.50	13.37

year	1973	1974	1975	1976	1977	1978	1979	1980	1981	1982	1983
m/s	15.43	20.58	11.32	15.43	11.32	14.40	16.46	16.98	13.59	22.63	15.95

year	1984	1985	1986	1987	1988	1989	1990	1991
m/s	13.89	13.89	19.05	13.89	13.89	16.98	19.95	12.33

Use the Kolmogorov-Smirnov and Anderson-Darling goodness-of-fit tests to compare the observed and theoretical cumulative frequencies as predicted by the (*a*) Gumbel, (*b*) Fréchet, (*c*) lognormal, (*d*) gamma, (*e*) GEV, (*f*) shifted lognormal, and (g) shifted gamma distributions. Discuss the decision of rejecting the null hypothesis when applicable. Consider $\alpha = .01$ and $.05$.

7.14. Design return period tornado wind speed. For the case study of Example 7.40, compute (a) the return period of a design tornado wind speed of 160 m/s and (b) the associated probability of exceedance using the contagious model of Eq. (7.3.17) with $X \sim$ lognormal(43 m/s, 16.34 m/s) and $\nu = 1.5 \times 10^{-3}$.

7.15. Confidence limits of design values. Consider the Gumbel distribution given in inverse form by Eq. (7.2.26) where the sample mean and standard deviation $\hat{\mu}_X$ and $\hat{\sigma}_X$ are used to estimate the population mean and standard deviation, μ and σ, respectively. Assuming that $\hat{\mu}_X$ and $\hat{\sigma}_X$ are asymptotically normally distributed show that

$$\text{Var}[\hat{\xi}_q] = \text{Var}\left[\frac{\pi^2}{6} + 1.1396(y - n_e) + 1.1(y - n_e)^2\right]\frac{\sigma^2}{n},$$

where y denotes the reduced variate. This expression can be used to determine the confidence interval by approximating the sample distribution of ξ_q as $N \sim (\hat{\xi}_q, \text{Var}[\hat{\xi}_q])$ and substituting the sampling variance for σ^2. Using this procedure, compute the 95% confidence interval for the annual maximum hourly storm rainfall predicted in Example 7.16. It can be shown that the variance of any estimator of a parameter is larger than, or at least equal to, a theoretically specified variance known as the *Cramer-Rao lower bound*, which makes use of the Cramer-Rao inequality of Subsection 5.2.3. This method may be used to derive the variance of quantile estimates from a given extreme value distribution.

7.16. Southern California earthquakes. Consider the ordered sample of magnitudes of southern California annual maximum earthquakes from 1932 to 1962 reported by C. Lomnitz (1974), *Global Tectonics and Earthquake Risk*, Elsevier Scientific Publishing Company, New York.

4.9 5.3 5.3 5.5 5.5 5.5 5.5 5.6 5.6 5.6 5.8 5.8 5.8 5.9 6.0 6.0
6.0 6.0 6.0 6.0 6.2 6.2 6.3 6.3 6.4 6.4 6.5 6.5 6.5 7.1 7.7

Originally, the Gumbel distribution was fitted to these data, but other potential candidates are the Fréchet and lognormal distributions. Use the Anderson-Darling goodness-of-fit test to compare the observed and theoretical cumulative frequencies as predicted by the (1) Gumbel, (2) Fréchet, (3) lognormal distributions. Consider $\alpha = .05$. Compare the theoretical and observed cdfs on a Gumbel probability plot.

7.17. Historical records in extreme value analysis. One wishes to supplement the information available from an s-year sample of observed extreme value data with a historical record of h years given that a perception threshold or detection limit was exceeded l times. This occurs, for example, for paleoflood data, and also for water quality data exceeding a prescribed level, for sea wave heights estimated by sailors, and for earthquake intensity estimated from earthquake effects on landscapes. If e denotes the number of observations that exceeded the threshold in the s-year

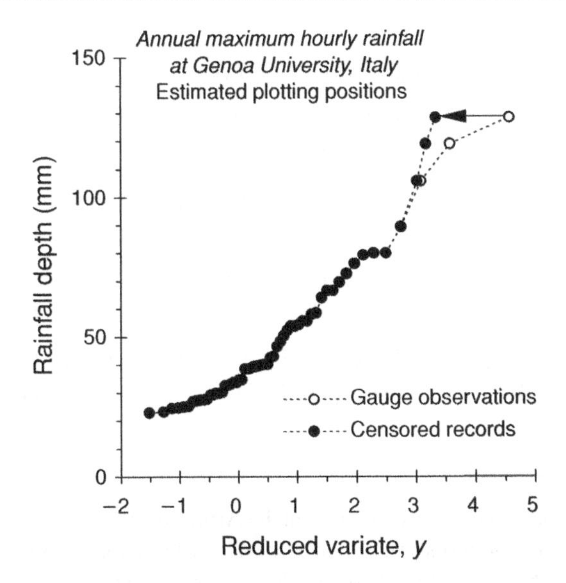

Fig. 7.P2 Comparison of Gumbel probability plots of annual maximum hourly rainfall at Genoa University, Italy, for gauge observations and censored records.

sample, a total of $r = l + e$ observations exceeded this threshold for the $n = s + h$ years of record, which is referred to as a *censored sample*. The natural estimator of the probability of exceedance of the detection threshold is r/n. If these r values are indexed by $j = 1, \ldots, r$, the reasonable plotting positions accommodating the probabilities of exceedance within the interval $(0, r/n)$ are

$$p_j = \frac{r}{n}\left(\frac{j - \eta}{r + 1 - 2\eta}\right),$$

where p_j is the probability of exceedance of the jth observation arranged in descending order, and η is a value depending on the underlying distribution, say, $\eta = 0.4$. Note that e observations that exceeded the threshold are counted among the r exceedances of that threshold. Plotting positions within $(r/n, 1)$ for the remaining $(s - e)$ data in the s-year sample are

$$p_j = \frac{r}{n} + \left(1 - \frac{r}{n}\right)\left(\frac{j - \eta}{s - e + 1 - 2\eta}\right),$$

for $j = 1, \ldots, s - e$. For instance, consider the $s = 58$ years of data of maximum annual hourly storm depth shown in Table E.7.1. Suppose that during a supplementary historical period of $h = 98$ years, the maximum annual hourly storm depth in Genoa exceeded 100 mm in $l = 5$ years. The total length of the record is $s + h = 156$ years, and $r = l + e = 5 + 3 = 8$. The observed frequencies are thus modified as shown in Fig. 7.P2, assuming that all historical storm depths exceeded the largest observed value.

Fit the GEV distribution to the censored sample of maximum annual hourly storm depth at Genoa University using L-moments.

Consider the following data of annual maximum flood flows in m^3/s for the Arno River in Florence, Italy, with $s = 40$ years:

year	1929	1930	1931	1932	1933	1934	1935	1936	1937	1938	1939
x	1642	—	1264	1130	1220	1780	1520	1100	1490	633	1350
year	1940	1941	1942	1943	1944	1945	1946	1947	1948	1949	1950
x	1250	1345	1079	—	2068	na	978	1594	1206	1425	922
year	1951	1952	1953	1954	1955	1956	1957	1958	1959	1960	1961
x	1780	937	1760	901	820	776	899	1600	1670	2070	1390
year	1962	1963	1964	1965	1966	1967	1968	1969	1970	1971	1972
x	1000	—	—	1120	3540	—	1430	1120	738	540	428
year	1973	1974									
x	385	1060									

It is reported that the discharge in the Arno exceeded $l = 3$ times a threshold of about 2400 m^3/s in a historical period of $h = 145$ years. None of these floods exceeded the 1966 flood, which had a peak discharge of 3540 m^3/s. Fit the GEV distribution to the censored sample by the method of L-moments. Find the return period of a flood with peak discharge exceeding 3000 m^3/s.

7.18. Maximum local earthquake intensity and ground motion. Using the epicentral intensity data in the Charleston area, South Carolina, from 1893 to 1984 the following recurrence relationship is found:

$$\log(\nu) = 1.02 - Y,$$

where ν is the number of earthquakes with intensity larger than Y in a year [see D. Amick, and P. Talwani (1986), "Earthquake recurrence rates and probability estimates for the occurrence of significant seismic activity in the Charleston area: The next 100 years," in *Proceedings of the Third Annual Conference on Earthquake Engineering*, Charleston, South Carolina, Vol. 1, pp. 55–64]. Find the cdf of annual maximum earthquake intensity. Suppose that peak ground motion is related to local site intensity as

$$\log(Z) = 0.3Y + 0.014,$$

where Z denotes the average horizontal peak acceleration in m/s^2 peak [see M. D. Trifunac, and A. G. Brady (1975), "On the correlation of seismic intensity scales with peaks of recorded strong ground motion," *Bull. Seismol. Soc. Am.*, Vol. 65, pp. 139–162]. Find the cdf of the annual maximum average horizontal component of epicentral peak acceleration.

7.19. Ground motion acceleration in earthquakes. The horizontal peak ground motion acceleration Z is a basic quantity in seismic hazard analysis at a particular site (as discussed in Problem 1.1). It depends on different factors, because it increases with the epicentral intensity Y of an earthquake and with its magnitude X, and it decreases with the epicentral distance r of the site. The relationship between these variables also depends on the geographic region, and it is determined using multiple regression. Suppose that

$$\log Z = 0.14Y + 0.24X - 0.68\log(r) + 0.60,$$

where Y is the epicentral intensity as measured in the modified Mercalli scale, r is the epicentral distance in km, and Z is measured in meters per square second. This relationship provides a good fit for data from the western United States [see J. R. Murphy and L. J. O'Brian (1977), "The correlation of the peak ground acceleration amplitude with seismic intensity and other physical parameters," *Bull. Seismol. Soc. Am.*, Vol. 67, pp. 877–915]. Suppose that X is a Gumbel-distributed variate with parameters $\alpha = 0.49$ and $b = 1.4 \times 10^5$, and $X = 2Y/3 + 1$. Evaluate the cdf of Z at an epicentral distance of 100 km.

7.20. Ground motion acceleration in earthquakes. The horizontal peak ground motion acceleration z is a basic quantity in seismic hazard analysis at a particular site. For a magnitude-x earthquake that occurred at a distance u from a given site, an estimate of z can be obtained as

$$z = \frac{Ae^{Bx}}{(u + C)^2}.$$

Typical values are $A = 1230$, $B = 0.8$, $C = 25$ km for u in kilometers and z in centimeters per square second [see, for example, N. M. Newmark and E. Rosenblueth (1971), *Fundamentals of Earthquake Engineering*, Prentice Hall, Englewood Cliffs, NJ]. Suppose that, in a homogeneous area, the annual number of earthquakes exceeding a given threshold x_0 is a Poisson variate with mean ν, and $X - x_0$ has an exponential distribution with scale parameter λ. If there are no recognized point sources or active faults, one can assume the epicentral distance as $U \sim \text{uniform}(0, l)$, with l denoting the maximum distance between two points in the given area. Show that the annual maximum of Z follows the Fréchet distribution. This distribution can be used to predict design values of horizontal peak ground motion acceleration in this area.

7.21. Southern California earthquakes. Consider the magnitude data listed in Table E.7.3.

(*a*) Check the Poisson assumption for the occurrence of earthquakes exceeding magnitude 6 by fitting the exponential distribution to the interarrival time W. Compare the observed and fitted cdf of W on an exponential probability plot.

(*b*) Compute the parameters of the Gutenberg-Richter law for type A zones, and find the return period of a magnitude-7 earthquake assuming that magnitude is bounded by $X_{\text{min}} = 6$ and $X_{\text{max}} = 8.22$.

7.22. Design return period of snow load. Snow load is evaluated as the product $Z = XW$, with X denoting the depth of snow cover, and W its specific weight. Based on a long record of observations of snow cover in the Italian Apennines, one models the depth X of snow delivered by a snowstorm as $X \sim \text{lognormal}(0.32\,\text{m}, 0.29\,\text{m})$. The specific weight of snow W depends on weather and season, and one should model snow pack dynamics to achieve accurate estimates of Z. However, measurements of density and temperature of snow yield $W \sim \text{lognormal}(3500\,\text{N/m}^3, 800\,\text{N/m}^3)$. Also, X and W are positively correlated with $\rho_{X,W} = 0.60$. If 4.7 snowstorms are expected to occur in a year on average, show that the cdf of maximum annual snow load (see Fig. 7.P3) can be written as

$$F_{Z_{\text{max}}}(z) = \exp\left\{-\nu\left[1 - \Phi\left(\frac{z - \mu_{\ln(Z)}}{\sigma_{\ln(Z)}}\right)\right]\right\},$$

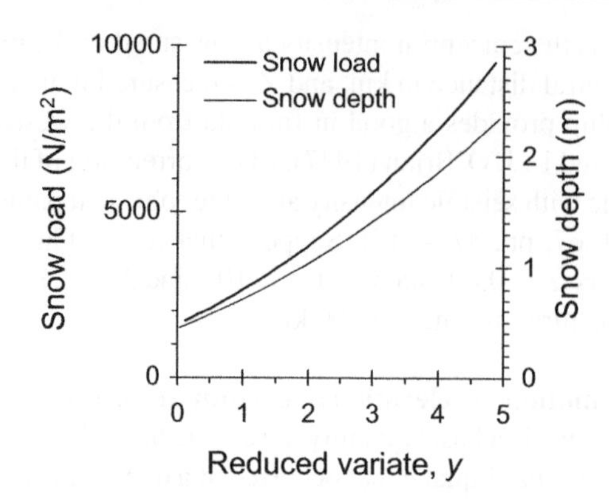

Fig. 7.P3 Gumbel probability plots of annual maximum snow depth and load.

with $v = 4.7$, $\mu_{\ln(Z)} = 6.696$, and $\sigma_{\ln(Z)} = 0.837$. Note that

$$\rho_{\ln(X),\ln(Y)}\sigma_{\ln(X)}\sigma_{\ln(Y)} = \ln(1 + V_X V_Y \rho_{X,Y})$$

if $\ln(X)$ and $\ln(Y)$ have a bivariate normal distribution.

Find the return period for design values of (*a*) $z_{max} = 8000$ N/m^2, and (*b*) $x_{max} = 2.15$ m.

Chapter 8
Simulation Techniques for Design

An engineering system can be studied by physical experiments using replications that reproduce its essential features. For instance, wind tunnel experiments can be implemented by loading the small-scale replication of a tower with specified design winds. Laboratory channel experiments, performed by subjecting the hydraulic model of a harbor breakwater to specified design waves, constitute another example. Such experiments with models are assumed to *simulate*, that is, to reproduce the behavior of the (real-world) prototype. Accordingly, one applies the physical laws of continuum and fluid mechanics to evaluate the response of the prototype by rescaling the response from these experiments. This approach can be extended to theoretical and numerical experiments. Thus, *simulation* is generally defined as the process of replicating the real-world prototype based on a set of assumptions and conceived models of reality. In practice, theoretical simulation is performed numerically, and numerical experiments have become an increasingly popular method to analyze engineering systems since the advent of digital computers; they are also substituted for physical experiments in many applications. This is because numerical experiments can allow a more detailed representation of the investigated system than that achievable through a physical model, and they are often much cheaper. On the other hand, numerical and physical experiments can be coupled; for example, laboratory channel experiments can be performed by loading a hydraulic model with wave characteristics generated from mathematical models of sea wave motion. Similar procedures can be adopted in structural and geotechnical engineering.

The simulation process predicts the response or performance of a system using a prescribed set of values for the system parameters or design variables. After repeated simulations, one can assess the sensitivity of the system response to variations in the parameters or variables. Alternative designs are thereby evaluated, and the optimal design is determined.

The method of *Monte Carlo simulation* is used when dealing with random variables.[1] The procedure is usually repeated to generate a different set of values of the variables in accordance with a specified probability distribution. In this way, a series of solutions is obtained corresponding to different sets of values of the random variables. Such samples are (statistically) similar to samples of experimental observations. The methods of statistical estimation and inference can then be applied. For the type of finite samples that are commonly used, the results of Monte Carlo simulations have sampling variability. Therefore, sampling theory should be utilized in designing a Monte Carlo experiment.

To perform a Monte Carlo simulation one needs to specify the probability distribution of the variables involved, which must be known or assumed. Therefore, the generation of outcomes from a prescribed probability distribution is a fundamental task in the simulation process. This process is deterministic for a given set of generated variables; it describes the relationships among system variables and parameters which define the response or

[1] The term *Monte Carlo* is said to have been introduced by the physicist John Von Neuman as a code word connected with his secret work on the atomic bomb at Los Alamos during World War II (Hammersley and Handscomb, 1964).

performance of the system. The uncertainties affecting these relationships and the random relationships that may eventually occur are usually represented by additional or dummy variables with known distributions.

We conclude with a discussion on sensitivity analysis and uncertainty analysis, as related to simulation. The distinction between aleatory and epistemic uncertainty is made with implications in model formulation, implementation and assessment, and in decision-making.

8.1 MONTE CARLO SIMULATION

8.1.1 Statistical experiments

Monte Carlo experiments have been known for many years. For instance, Buffon's needle problem, which may be used to compute the circumference π of a circle of unit diameter (which is also the area of a circle of unit radius), dates back more than 200 years, and its probabilistic generalization was made by Laplace in 1812 (Beckman, 1971). We commence with an illustration of this problem in order to demonstrate the advantages of simulation.[2]

Example 8.1. Computation of π. Consider a horizontal floor on which parallel lines are drawn at equal distances a. A needle of length b, where $b < a$, is dropped at random on the floor. Our initial problem is to find the probability that the needle will intersect a line. Let X be a random variable that gives the distance of the midpoint of the needle to the nearest line, with $0 < x \leq a/2$, and let Y be the variable which gives the acute angle between the needle (or its extension) and the line.

The outcomes of X and Y are bounded as $0 < x \leq a/2$ and $0 < y \leq \pi/2$. Since $\Pr[x < X \leq x + dx] = (2/a)dx$, and $\Pr[y < Y \leq y + dy] = (2/\pi)dy$, one obtains $f_X(x) = 2/a$, and $f_Y(y) = 2/\pi$. Noting that X and Y are independent, the joint pdf is the product of the marginals, that is, $f_{X,Y}(x, y) = 4/(a\pi)$. From Fig. 8.1.1a, it is seen that the needle actually crosses a line when $X \leq (b/2)\sin Y$.

The probability of this event is given by

$$p = \frac{4}{a\pi} \int_0^{\pi/2} \int_0^{(b/2)\sin y} dx\, dy = \frac{2b}{a\pi}.$$

When this expression is equated to the frequency of hits (or crossings) observed in actual (physical) experiments, accurate values of π can be obtained. First, we specify a and $b < a$ and assume an appropriate value of π. Then m independent pairs of X and Y that follow the foregoing uniform distributions are generated numerically. Second, p is estimated as the ratio between the random number N of those pairs (x, y) that satisfy $x \leq (b/2)\sin y$ and the number of trials m. Finally, π is computed as $2b/(ap)$, that is, $2bm/(aN)$. The accuracy increases with the number m of trials, as shown in Fig. 8.1.1b. It is seen that when $m > 100$, a stable value of π is reached. Such experiments are called *urn extractions* and are used for generation.

Statistical simulation is a conceptualization of a trial-and-error procedure, as just demonstrated, in terms of probability. It combines the notions of prior and posterior probabilities.

---

[2] The value of $\pi (= 3.14159\ldots$, and currently known to more than 5 billion places) has fascinated mankind since Babylonian times, notably Archimedes who evaluated the interval: $223/71 < \pi < 22/7$, with the limits averaging to about 3.1419. It can also be empirically measured by drawing a large circle and measuring its diameter D and circumference πD; π is the first letter of the Greek words for periphery and parameter, that is, circumference. The conventional method of calculating the constant is to use a mathematical equation such as the symmetric formula of Sandow: $\pi = \frac{\prod_{n=1}^{\infty}(1+(1/4n^2-1))}{\sum_{n=1}^{\infty} 1/(4n^2-1)}$ (among equations by many others); subsequently in 2005, he produced a faster product equation for $\pi/2$.

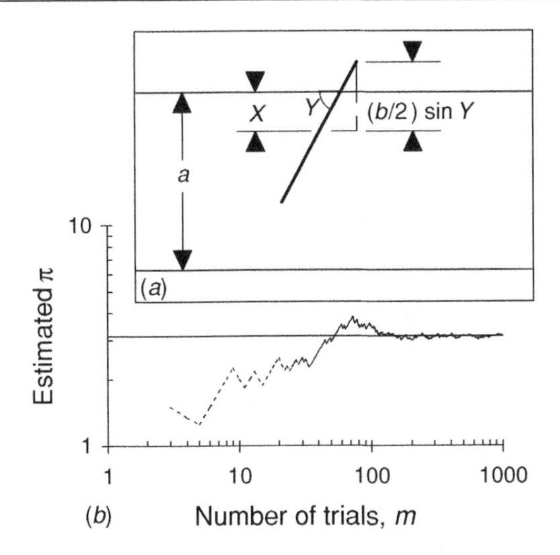

Fig. 8.1.1 Buffon's needle problem: (*a*) sketch of the experiment and (*b*) results of simulation.

For instance, the relationship of π to the probability that Buffon's needle intersects a line on the floor is known a priori, that is, from system assessment. From subsequent experiments using this system, the value of π is determined from the relative frequency of hits observed in actual experiments, or the posterior probability. In some practical applications, the prior probability is unknown or vaguely defined and thus one performs statistical experiments to estimate its value as is done, for instance, in Monte Carlo integration.

Example 8.2. Monte Carlo integration. The definite integral of a function $g(u) > 0$ from a to b—that is,

$$I = \int_a^b g(u)du,$$

is the area bounded by the curve $g(u)$ within the interval $[a, b]$, as shown in Fig. 8.1.2.

Consider a rectangle embedding this area, and suppose that one were to throw darts at the rectangle of area $A = c(b - a$, where $c \geq g(u)$ for $a \leq u \leq b$. Let n denote the (large) number of darts thrown uniformly against this target. If N is the number of darts falling below

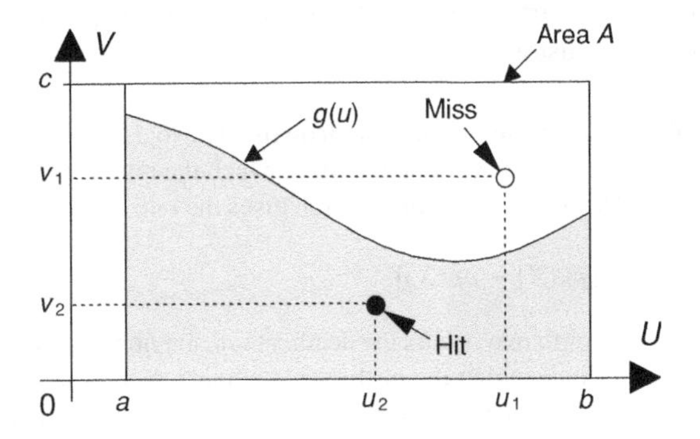

Fig. 8.1.2 Monte Carlo method of integration. Random points are chosen within the area A. The integral of the function $g(\cdot)$ is estimated as the area of A rescaled by the fraction of random points falling below the curve g.

the curve $g(u)$, the integral may be estimated as the area A multiplied by the fraction N/n of random points that fall below $g(u)$; that is,

$$I = \int_a^b g(u)du \approx c(b-a)\frac{N}{n},$$

where $p = N/n$ is the probability of a hit. Instead of throwing darts, one might generate n pairs u and v of two independent uniformly distributed variates U and V, with $a \leq u \leq b$ and $0 \leq v \leq c$, respectively, and count the number N of pairs with $v \leq g(u)$. For an increasing number of generated pairs, one expects that the value of the estimated integral approaches its theoretical value. This method can also be used in the multidimensional case by picking n random points, say, $x_1, \ldots, x_n$, uniformly distributed in a multidimensional volume Ω. Then, the basic theorem of Monte Carlo integration estimates the integral of a function g over Ω as

$$I = \int_\Omega g d\Omega = \Omega\langle g\rangle \pm Z,$$

where Z is a random variable representing the error in the estimated integral, with zero mean and standard deviation

$$\sigma_Z \sqrt{\frac{\langle g^2\rangle - \langle g\rangle^2}{n}}.$$

Here the angle brackets denote taking the arithmetic mean over the n sample points; that is,

$$\langle g\rangle = \frac{1}{n}\sum_{i=1}^{n} g(x_i) \quad \text{and} \quad \langle g^2\rangle = \frac{1}{n}\sum_{i=1}^{n} [g(x_i)]^2.$$

There is no guarantee that the error is distributed as normal, so that the error term should be taken only as a rough indication of probable error. Note that the implementation of this method requires the generation of uniform random numbers in a specified domain, say, the rectangle A or the hypervolume Ω.

The preceding examples show that statistical experiments can be performed to solve problems that are not probabilistic by using random numbers generated from a parent uniformly distributed variate, such as X and Y in Buffon's needle problem, and U and V in the Monte Carlo integration method for a function of a single variable. One can extend this approach to systems described by probabilistic models. For example, to evaluate the probability distribution of times spent waiting for a taxicab on a particular street, one can perform numerical or physical experiments, such as urn extractions or on-the-road trials. In numerical experiments, one can assume that the interarrival time of two subsequent cabs is a uniformly distributed variate and can assume a specified probability that such a cab will stop at the customer's call.

Example 8.3. Logistic population growth. The logistic model is used in applied ecology to represent the growth of a population in which the rate of growth is a net balance of births and deaths. The logistic growth equation gives the rate of increase as

$$\frac{dX}{dt} = X[\rho_B(X) - \rho_D(X)],$$

where both the birth rate ρ_B and the death rate ρ_D are nonnegative functions of the population size X. Let us assume that the birth rate is a linear decreasing function of X, say, $\rho_B(X) = a_1 - b_1 X$, and the death rate is a linear increasing function of X, say, $\rho_D(X) = a_2 + b_2 X$, with $a_1, a_2 > 0$, and $b_1, b_2 \geq 0$. The growth rate dX/dt is thus given by

$$\frac{dX}{dt} = X[(a_1 - a_2) - (b_1 + b_2)X] = X(r - sX),$$

which is the Verhulst-Pearl logistic equation with intrinsic rate of increase $r = a_1 - a_2$, and the so-called saturation parameter $s = b_1 + b_2$. The events of interest constitute a sequence of births and deaths. If we disregard the time elapsing between each event, there are two possibilities for the next event: it may be a birth, event B, or a death, event D. If B occurs, the population will increase from its present value x to $x + 1$; conversely, the population will decrease from x to $x - 1$ if D occurs. The corresponding probabilities are proportional to the birth and death rates as

$$\Pr[B] \propto X\rho_B(X) = a_1 X - b_1 X^2,$$
$$\Pr[D] \propto X\rho_D(X) = a_2 X + b_2 X^2.$$

Since these events are mutually exclusive and collectively exhaustive,

$$\Pr[B] = p = \frac{(a_1 X - b_1 X^2)}{[(a_1 + a_2)X - (b_1 - b_2)X^2]},$$

$$\Pr[D] = 1 - p = \frac{(a_2 X + b_2 X^2)}{[(a_1 + a_2)X - (b_1 - b_2)X^2]}.$$

Suppose $a_1 = 0.7, a_2 = 0.2, b_1 = 0.0045$, and $b_2 = 0.0005$; this gives $r = 0.5$ and $s = 0.005$. The deterministic equation of growth rate becomes

$$\frac{dX}{dt} = 0.5X - 0.005X^2,$$

so that the equilibrium size of the population (when $dx/dt = 0$) is $r/s = 0.5/0.005 = 100$. To simulate the probabilistic growth of the population, that is, to generate a sequence of births and deaths starting with a population of given size, say, $x = 69$, one must first calculate the foregoing probabilities:

$$p = \frac{(0.7 \times 69 - 0.0045 \times 69^2)}{[(0.7 + 0.2) \times 69 - (0.0045 - 0.0005) \times 69^2]} = 0.624,$$

$$1 - p = 1 - 0.624 = 0.376.$$

From a random number generator (or by other means), one then picks a number u uniformly distributed in the range (0, 1]. If $u \leq p$, the next event is a birth, so that the population increases in size to 70; if $u > p$, the next event is a death, so that the population decreases to 68. Once the event has happened and the population size is adjusted accordingly, one calculates the new probabilities for the next event and proceeds as before. Table 8.1.1 lists a short sequence of events generated using this procedure.

After the initial value of 69 is specified, the value of X in each row is obtained by altering the value in the row above, according to the previous event, which may be a birth, B, or death, D.

Example 8.4. Seepage under a dike wall. Figure 8.1.3 shows a river valley resting on a homogeneous alluvial layer of porous material bounded by impervious rock. Supposing the level of the groundwater table BL is constant at h_g, the total river head is constant at h_r so that there is a head loss of $h_l = h_r - h_g$. Also, the boundaries $ABCDEFGHI$ and LMN are impervious and there are no sinks in the flow. One superimposes a square grid, or lattice, on the aquifer to study the movement of a particle of water at an interior point under a two-dimensional random walk; in other words, this particle is made to move horizontally and vertically from one point to another through the lattice.[3] Whenever the particle meets one

[3] In general, the path traversed during a random walk by a particle that moves in steps is determined by chance either with respect to direction or with respect to distance or both direction and distance (like the steps of a drunken sailor). Quite often, movement is assumed from one point to any of the nearest neighboring points, as in the lattice assumed here, on an equally likely basis.

Table 8.1.1 Simulation of logistic population growth

Step	Population size, X	Pr[B]	Pr[D]	u	Event
1	69	0.624	0.376	0.730	D
2	68	0.627	0.373	0.170	B
3	69	0.624	0.376	0.824	D
4	68	0.627	0.373	0.689	D
5	67	0.631	0.369	0.386	B
6	68	0.627	0.373	0.872	D
7	67	0.631	0.369	0.595	B
8	68	0.627	0.373	0.606	B
9	69	0.624	0.376	0.648	D
10	68	0.627	0.373	...	...
	...	...	...	...	...

of the aforementioned impervious boundaries on its random walk in the limited zone, it is reflected back. On the other hand, when it reaches the groundwater table or the river bed, the path is terminated and values of h_g and h_r, respectively, are given to it. After n such random walks from an interior point S (where n is large, say, greater than 400), suppose that n_g paths are assigned h_g and n_r paths are assigned h_r, where $n = n_g + n_r$. We estimate the head at point S:

$$h_S = \frac{n_g h_g + n_r h_r}{n_g + n_r}.$$

Consider an interior node 0 and a particle at any of its neighboring grid points 1, 2, 3, or 4 in Fig. 8.1.3. The probability of a particle arriving at point 0 is obtained by weighting the probabilities that it arrives at the four neighboring points. For the random walk, we assign equal weights of 1/4, as shown. If the point 0 is on a boundary, such as point Q, then the

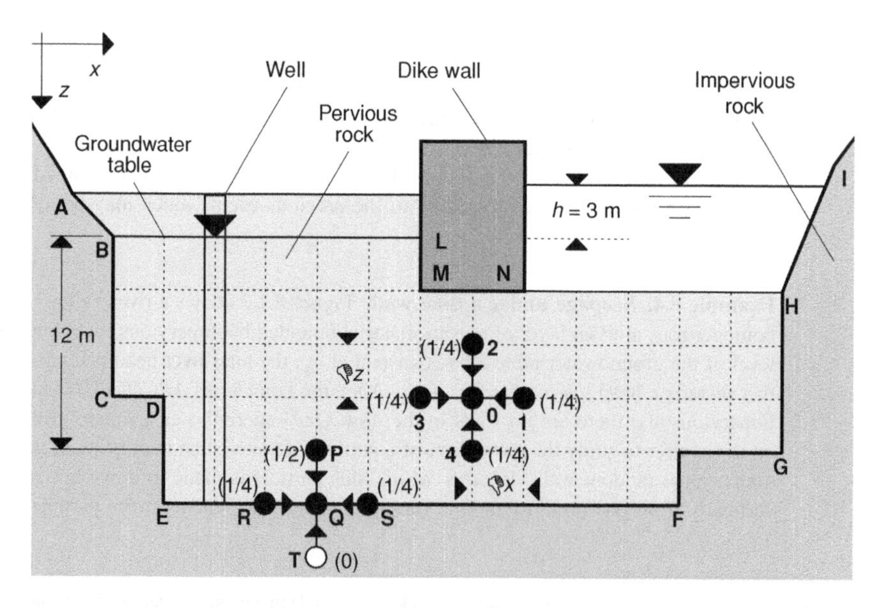

Fig. 8.1.3 Groundwater head under a dike wall in a river valley transect. The values in parentheses are the next-step probabilities in the traced direction.

weighting is adjusted to 1/4, 1/2, and 1/4, as shown. A Monte Carlo simulation of the random walks from point 0 in Fig. 8.1.3 produced 423 paths that terminated at the groundwater table boundary, and 577 paths that terminated at the river bed boundary. The groundwater table boundary represents a total head of $h_g = 0$ m, and the total head at the riverbed boundary is $h_r = 3$ m. Hence, the pressure head at 0 is

$$h_0 = \frac{n_g h_g + n_r h_r}{n} = \frac{423 \times 0 + 577 \times 3}{423 + 577} = 1.73 \text{ m}.$$

The elevation head at point 0, relative to the groundwater table, is 9 m. Hence, the pressure in the water at point 0 is $9.81 \times (9 + 1.73) \approx 105$ kPa.

Numerical experiments using random numbers can be used to simulate processes such as population growth for births and deaths considered as chance occurrences, or the growth of a random pattern on a lattice. Roulette wheels, such as those used at casinos in Monte Carlo and elsewhere, were originally used in obtaining random numbers. Since then investigators have tried coin tossing, urn extraction, numbers in a telephone directory, and subsequently tables of random numbers (Rand Corporation, 1955). The modern method is, of course, to use a computer routine; some techniques for generating random numbers are presented in the following section. However, most engineering systems are modeled by random variables with distributions different from the uniform. Then the unknown distribution of a design variable must be determined from known distributions of other variates. The probability integral transform provides the theoretical basis for stochastic simulation by generating random numbers from a specified probability distribution.

8.1.2 Probability integral transform

Since the cdf of a continuous random variable X is a monotonic and continuous function of x, $F_X(x)$ is a candidate for $g(\cdot)$ in the one-to-one transformation $u = g(x)$ studied in Section 3.4. Because $u = g(x) = F_X(x)$ is a nondecreasing function, the inverse function $x = \xi(u)$ can be defined for any value of u between 0 and 1 as the smallest x satisfying $F_X(x) \geq u$ (according to the definition of quantiles in Sub-subsection 3.2.1.6). Thus, defining the random variable $U = F_X(X)$,

$$F_U(u) = \Pr[U \leq u] = \Pr[F_X(X) \leq u] = \Pr[X \leq \xi(u)] = F_X(\xi(u)) = u,$$

for $0 < u < 1$, $F_U(u) = 0$ for $u \leq 0$, and $F_U(u) = 1$ for $u \geq 1$. The pdf of U is thus given by

$$f_U(u) = \frac{dF_U(u)}{du} = 1,$$

for $0 < u < 1$ and 0 elsewhere, signifying that $U \sim$ uniform (0, 1). The transformation $U = F_X(X)$ is called the *probability integral transform*. Figure 8.1.4 shows the relationship between U and X.

Property: Probability integral transform. If a random variable X has continuous cdf $F_X(x)$, the transformation $U = F_X(X)$ yields $U \sim$ uniform (0, 1). Conversely, if $U \sim$ uniform (0, 1), then $X = \xi(U)$ has cumulative distribution function $F_X(x)$, if $\xi(\cdot)$ maps any value of u into the smallest x satisfying $F_X(x) \geq u$.

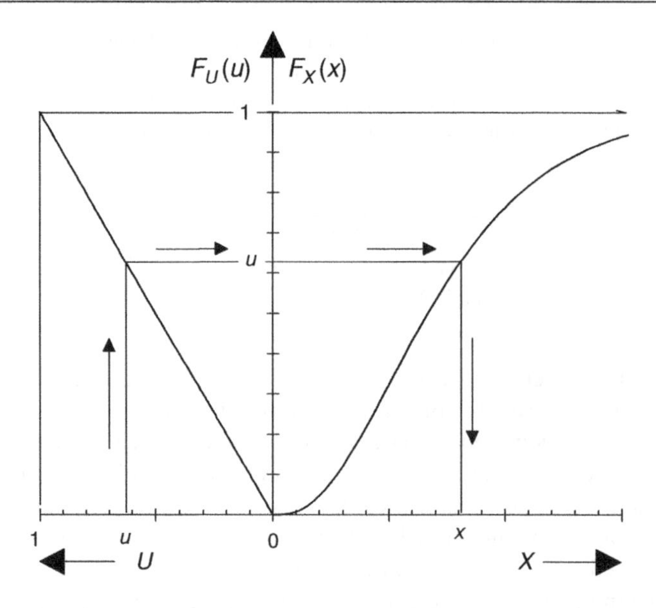

Fig. 8.1.4 Relationship between a uniform variate U and a variate X from an another distribution.

Engineering applications of statistical simulation methods involve the generation of values of random variables. To obtain an outcome x of a variate X with continuous cdf $F_X(x)$, one can generate a value u of a $(0, 1)$ uniform random variate U. Then, the required value of X is found by using the inverse cdf as

$$x = \xi_u, \tag{8.1.1}$$

the uth quantile of the variate. For example, if X is an exponentially distributed variate with parameter λ, the cdf

$$F_X(x) = 1 - \exp(-\lambda x)$$

can be inverted to obtain

$$x = \xi_u = -\lambda^{-1} \ln(1 - u) \equiv -\lambda^{-1} \ln(u),$$

where $u = F_X(x)$; the last passage from $\ln(1 - u)$ to $\ln(u)$ follows from the fact that $U \sim 1 - U$ when $U \sim$ uniform $(0, 1)$. Accordingly, one generates a value u from a uniform $(0, 1)$ distribution and then computes the corresponding value x of X as the logarithmic transformation of u rescaled by $-1/\lambda$.

Obviously, this method is straightforward when the cdf of X can be inverted.

Example 8.5. Peaks of sea waves. During a severe storm the height X of wave peaks at a site are found to follow the Rayleigh distribution with pdf

$$f_X(x) = (x/\lambda^2)\exp[-(x/\lambda)^2/2],$$

with λ denoting a scale parameter depending on the characteristics of the energy spectrum of the sea storm. To obtain a random outcome of X, its cdf

$$F_X(x) = 1 - \exp[-(x/\lambda)^2/2]$$

is inverted to obtain

$$x = \lambda[-2\ \ln(1-u)]^{1/2} \equiv \lambda[-2\ \ln(u)]^{1/2},$$

where u is an outcome of a standard uniform variate.

The probability integral transform provides a basic concept in the application of statistical simulation methods to engineering problems. In many cases, the engineer needs to study the response of a system that is subject to a random input. However, the complexity of system transformation between input and output does not facilitate the derivation of the statistical properties of the output from those of the given input. If the transformation of the input into the system output is known, the random output of the system is then found by simulation; that is, by generating a sequence of outcomes of the input and then determining the associated output of each outcome. The statistical properties of the system response are studied by investigating the simulated sample of outcomes of the system output.

8.1.3 Sample size and accuracy of Monte Carlo experiments

In Monte Carlo integration it is seen that choosing n points uniformly and randomly distributed in a multidimensional space leads to an error term that decreases as $n^{-1/2}$, because each new point sampled adds linearly to an accumulated sum of squares that will become the variance, and the estimated error comes from the square root of the variance. In designing a Monte Carlo experiment, one must determine how many simulations are required to assess the system behavior. When simulation is used to evaluate the probability p that some event occurs, such as unsatisfactory system performance, one must search for the sample size required to obtain a specified accuracy of the estimated p. If N denotes the observed number of occurrences of the event in a sample of size n, the obvious estimator of p is the proportion N/n. When sequential simulations are independent of each other, N is a binomial variate with parameters n and p. From Eq. (5.3.7), the standard error of the estimated proportion is

$$\sigma_{\hat{p}} = \sqrt{\frac{p(1-p)}{n}}, \tag{8.1.2}$$

and for large n (say, $n > 30$ and $np > 5$), the sampling distribution is very nearly normal with mean np and variance $np(1-p)$. In practice, the sample estimate $\hat{p}$ is substituted for p, and the $100(1-\alpha)$ percent two-sided confidence limits on the true value p given n and an observed value of the estimator $\hat{p}$ are determined as

$$\hat{p} - z_{\alpha/2}\sqrt{\frac{\hat{p}(1-\hat{p})}{n}} \quad \text{and} \quad \hat{p} + z_{\alpha/2}\sqrt{\frac{\hat{p}(1-\hat{p})}{n}} \tag{8.1.3}$$

(where $z_{\alpha/2}$ denotes a standard normal variate that is exceeded with probability $\alpha/2$). The necessary sample size n to ensure that the $100(1-\alpha)$ percent confidence limits are within 100ε percent of the true value of p, where $0 \le \varepsilon \le 1$; that is,

$$z_{\alpha/2}\sqrt{\frac{p(1-p)}{n}} \le \varepsilon p$$

is given by

$$n \ge \frac{z_{\alpha/2}^2(1-p)}{\varepsilon^2 p}. \tag{8.1.4}$$

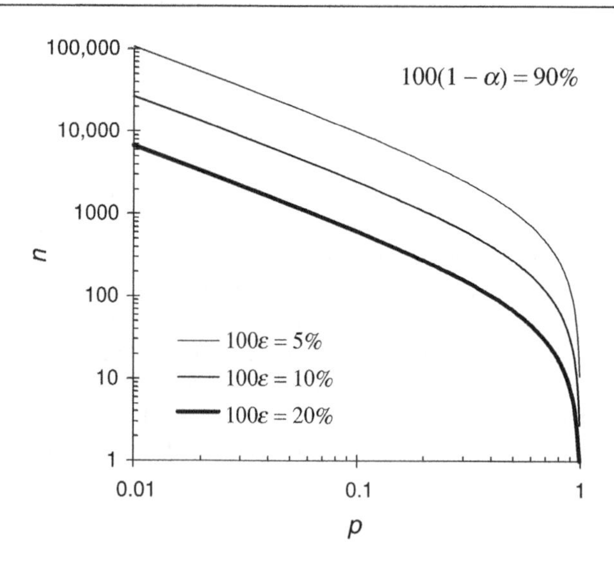

Fig. 8.1.5 Simulation sample size n required to estimate the probability p of the design event within 100ε percent of its true value with $100(1-\alpha)$ percent confidence.

Since n is a function of p, which is unknown before the experiment is performed, one must estimate the value of p before the experiment. Figure 8.1.5 shows the increase of n for decreasing p and different values of acceptable tolerance ε.

8.1.3.1 Antithetic variates

The accuracy of Monte Carlo simulations is closely related to the size of generated samples. However, the variance of simulation results can be reduced without increasing the sample size. This can be done by variance reduction techniques based on the properties of correlated samples. For instance, if X_1 and X_2 denote two unbiased estimators of a variable X, one can combine these estimators to obtain a new estimator $X^* = (X_1 + X_2)/2$—the expectation of which, of course, is still

$$E[X^*] = E\left[\frac{X_1 + X_2}{2}\right] = \frac{E[X_1] + E[X_2]}{2} = (X + X)/2 = X \tag{8.1.5}$$

signifying that X^* is unbiased. The corresponding variance is

$$\mathrm{Var}[X^*] = \mathrm{Var}\left[\frac{X_1 + X_2}{2}\right] = \frac{(\mathrm{Var}[X_1] + \mathrm{Var}[X_2] + 2\mathrm{Cov}[X_1, X_2])}{4}. \tag{8.1.6}$$

If the estimators X_1 and X_2 are negatively correlated, that is, $\mathrm{Cov}[X_1, X_2] < 0$, the variance of X^* will be smaller than that of $(\mathrm{Var}[X_1] + \mathrm{Var}[X_2])/4$ which occurs for independent estimators. The *antithetic variates* method (Hammersley and Morton, 1956) is a simulation procedure to ensure negative correlation between X_1 and X_2. This is done by generating a sequence, say, $u_1, u_2, \ldots, u_n$ of independent standard uniform variates to obtain the size-n estimator X_1. The related sequence $1 - u_1, 1 - u_2, \ldots, 1 - u_n$ is then used to obtain another size-n estimator X_2; the correlation between X_1 and X_2 is negative.[4]

[4] Any improvement will depend on the particular case study. Numerous examples are given by Ang and Tang (1984).

Example 8.6. Storm rainfall total. The total amount of water Z delivered by a storm in a given location was evaluated in Example 3.49 from an independent exponentially distributed duration X and an average rainfall rate Y. Suppose our objective is to estimate the mean rainfall total in a storm $E[Z] = E[XY]$ by Monte Carlo simulation. Two sequences of n standard uniform variates are first considered, say, $u_1, u_2, \ldots, u_n$ and '$v_1, v_2, \ldots, v_n$, respectively. Hence,

$$x_i = -\mu_X \ln(1 - u_i), \quad i = 1, 2, \ldots, n,$$
$$y_i = -\mu_Y \ln(1 - v_i), \quad i = 1, 2, \ldots, n,$$

where μ_X and μ_Y denote the mean of X and Y, respectively. The outcomes of Z are then found as

$$z_{1i} = x_{1i} y_{1i} = \mu_X \mu_Y \ln(1 - u_i) \ln(1 - v_i), \quad i = 1, 2, \ldots, n.$$

With the antithetic uniformly distributed variates one obtains

$$z_{2i} = x_{2i} y_{2i} = \mu_X \mu_Y \ln(u_i) \ln(v_i), \quad i = 1, 2, \ldots, n.$$

By combining the two sets one estimates the mean rainfall as

$$z^* = \frac{1}{2}(z_1 + z_2) = \frac{1}{2n} \sum_{i=1}^{n} (x_{1i} y_{1i} + x_{2i} y_{2i}),$$

which has a variance of

$$\text{Var}[Z^*] = \frac{1}{4n^2} n \text{Var}[X_1 Y_1 + X_2 Y_2].$$

From Eq. (3.4.28),

$$\text{Var}[(X_1 Y_1 + X_2 Y_2)] = \text{Var}[X_1 Y_1] + \text{Var}[X_2 Y_2] + 2\text{Cov}[X_1 Y_1, X_2 Y_2].$$

From Eq. (3.4.35),

$$\text{Var}[X_1 Y_1] = \text{Var}[X_1]\text{Var}[Y_1] + \text{Var}[X_1](E[Y_1])^2 + (E[X_1])^2 \text{Var}[Y_1],$$
$$\text{Var}[X_2 Y_2] = \text{Var}[X_2]\text{Var}[Y_2] + \text{Var}[X_2](E[Y_2])^2 + (E[X_2])^2 \text{Var}[Y_2].$$

Also,

$$E[X_1] = E[-\mu_X \ln(1 - U)] = \mu_X E[-\ln(1 - U)]$$
$$= \mu_X \left[\int_0^1 -\ln(1 - u)\, du \right] = \mu_X (1) = \mu_X.$$

and

$$\text{Var}[X_1] = \text{Var}[-\mu_X \ln(1 - U)] = \mu_X^2 \text{Var}[-\ln(1 - U)]$$
$$= \mu_X^2 \left\{ \int_0^1 [\ln(1 - u)]^2 du - \left[\int_0^1 -\ln(1 - u) du \right]^2 \right\} = \mu_X^2 (2 - 1) = \mu_X^2.$$

Similarly, $E[X_2] = \mu_X, \text{Var}[X_2] = \mu_X^2, E[Y_1] = E[Y_2] = \mu_Y$, and $\text{Var}[Y_1] = \text{Var}[Y_2] = \mu_Y^2$. From Subsection 3.4.3,

$$\text{Cov}[X_1 Y_1, X_2 Y_2] = E[X_1 Y_1 X_2 Y_2] - E[X_1 Y_1] E[X_2 Y_2],$$

where

$$E[X_1 Y_1] = E[X_1] E[Y_1] = \mu_X \mu_Y,$$
$$E[X_2 Y_2] = E[X_2] E[Y_2] = \mu_X \mu_Y,$$

and[5]

$$\begin{aligned}E[X_1Y_1X_2Y_2] &= E[\mu_X \ln(1-U)\mu_Y \ln(1-V)\mu_X \ln(U)\mu_X \ln(V)]\\ &= (\mu_X\mu_Y)^2 E[\ln(1-U)\ln(1-V)\ln(U)\ln(V)]\\ &= (\mu_X\mu_Y)^2 E[\ln(1-U)\ln(U)]E[\ln(1-V)\ln(V)]\\ &= \mu_X^2\mu_Y^2\left\{\int_0^1[\ln(1-u)][\ln(u)]du\right\}\left\{\int_0^1[\ln(1-v)][\ln(v)]dv\right\}\\ &= \mu_X^2\mu_Y^2\left(2-\frac{\pi^2}{6}\right)\left(2-\frac{\pi^2}{6}\right) = \mu_X^2\mu_Y^2\left(2-\frac{\pi^2}{6}\right)^2.\end{aligned}$$

Thus,

$$\text{Cov}[X_1Y_1, X_2Y_2] = (\mu_X\mu_Y)^2\left[\left(2-\frac{\pi^2}{6}\right)^2 - 1\right],$$

and

$$\text{Var}[(X_1Y_1 + X_2Y_2)] = (\mu_X\mu_Y)^2\left\{3+3+2\left[\left(2-\frac{\pi^2}{6}\right)^2-1\right]\right\} = 4.252(\mu_X\mu_Y)^2.$$

The variance of estimator Z^* is then determined as

$$\text{Var}[Z^*] = \frac{1}{4n^2}n\text{Var}[X_1Y_1 + X_2Y_2] = \frac{1.063}{n}\mu_X^2\mu_Y^2.$$

Note that simulating a single sample of length $2n$ yields a variance of

$$\frac{1}{4n^2}(2n)\text{Var}[XY] = \frac{1}{2n}\left(3\mu_X^2\mu_Y^2\right) = \frac{1.500}{n}\mu_X^2\mu_Y^2,$$

which is about 50% greater than that obtained using the antithetic variates technique.

Since Monte Carlo experiments are often used to compare the performance of different design options, this type of experiment can be developed in a combined way to reduce the standard error or variance of simulation results without increasing the sample size. Let $X_A = g(\boldsymbol{A}; Y_1, \ldots, Y_k)$ denote the performance function of a design A, where $\boldsymbol{A}$ is a set of design values, and the Y_i are the random variables to be simulated in order to estimate the system performance. Consider an alternative design B, the performance of which is evaluated as $X_B = g(\boldsymbol{B}; Y_1, \ldots, Y_k)$ for a set $\boldsymbol{B}$ of design values. The difference in performance between the two designs is $X = X_A - X_B$. It is expected that X_A and X_B are highly correlated. To evaluate the mean value of X, the variance of this estimate is given by

$$\text{Var}[\hat{E}[X]] = \text{Var}[\hat{E}[X_A]] + \text{Var}[\hat{E}[X_B]] - 2\text{Cov}[\hat{E}[X_A], \hat{E}[X_B]]. \tag{8.1.7}$$

This variance will be smaller than that corresponding to independent sampling of X_A and X_B (which is the sum of the two individual variances) if $\text{Cov}[X_A, X_B] > 0$. Therefore, one can use the same sequence of the standard uniform random variables to simulate the performances of the two designs.

Example 8.7. Cantilever wood beam. The (maximum) deflection X of a cantilever wood beam can be evaluated as

$$X = \frac{(Y+Z)l^4}{8EI},$$

[5] See, for example, Gradshteyn and Ryzhik (1994, Subsection 4.221, p. 558) for the integral that follows.

with Y and Z denoting the distributed dead and live loads over the span length l of the beam, E the modulus of elasticity, and I the cross-sectional moment of inertia, which, for a rectangular beam, is given by

$$I = \frac{WH^3}{12},$$

where W and H are the width and depth of the cross section. Thus,

$$X = \frac{3(Y + Z)l^4}{2EWH^3}.$$

Let us assume that $Y \sim N[4500\ \text{N/m}, (0.1 \times 4500\ \text{N/m})^2]$, $Z \sim$ lognormal $[5000\ \text{N/m}, (0.5 \times 5000\ \text{N/m})^2]$, and $E \sim N[1.2 \times 10^{10}\ \text{N/m}^2, (0.5 \times 1.2 \times 10^{10}\ \text{N/m}^2)^2]$. An engineer needs to evaluate two alternative designs, say, A and B. A 20-cm-wide and 30-cm-deep rectangular beam is considered in design A, and design B uses a 15-cm-wide and 40-cm-deep rectangular beam. Because the actual dimensions can range within an interval of ± 5 cm of the nominal value in both cases, W and H are assumed to be uniformly distributed random variables. Accordingly, $W_A \sim$ uniform (0.15 m, 0.25 m), $H_A \sim$ uniform (0.25 m, 0.35 m), $W_B \sim$ uniform (0.1 m, 0.2 m), and $H_B \sim$ uniform (0.35 m, 0.45 m).

Let y_{iA}, z_{iA}, and e_{iA} denote the random outcomes of Y, Z, and E required for the ith simulation cycle of design A, and y_{iB}, z_{iB}, e_{iB} the corresponding outcomes for design B. First, to perform correlated simulations for the two designs, the same random outcomes are used in each simulation cycle, that is, $y_{iA} = y_{iB}$, $z_{iA} = z_{iB}$, and $e_{iA} = e_{iB}$. Further,

$$\begin{aligned} w_{iA} &= 0.15 + 0.1u_i, \\ h_{iA} &= 0.25 + 0.1v_i, \\ w_{iB} &= 0.1 + 0.1u_i, \\ h_{iA} &= 0.35 + 0.1v_i, \end{aligned}$$

where u_i and v_i are uniform (0, 1) random variates. The results from 500 simulation runs, each with 100 cycles, are shown in Fig. 8.1.6*a*, with the estimated means of deflection for designs A and B, and the expected differences X^* between X_A and X_B.

The simulation procedure shows the variations in the estimated means. The estimated variances and covariance of the estimates are accordingly computed as

$$\begin{aligned} \text{Var}[\hat{E}[X_A]] &= 3.26 \times 10^{-7}\ \text{m}^2, \\ \text{Var}[\hat{E}[X_B]] &= 9.61 \times 10^{-8}\ \text{m}^2, \end{aligned}$$

and

$$\text{Cov}[\hat{E}[X_A], \hat{E}[X_B]] = 1.73 \times 10^{-7}\ \text{m}^2.$$

These are substituted in Eq. (8.1.7) to give

$$\text{Var}[\hat{E}[X^*]] = 3.26 \times 10^{-7} + 9.61 \times 10^{-8} - 2 \times 1.73 \times 10^{-7} = 7.61 \times 10^{-8}\ \text{m}^2.$$

This value is compared with the sampling variance of

$$\text{Var}[\hat{E}[X_A] - \hat{E}[X_B]] = 7.55 \times 10^{-8}\ \text{m}^2,$$

which is found from the experiment.

Second, uncorrelated simulations are performed with $y_{iA} \neq y_{iB}$, $z_{iA} \neq z_{iB}$, and $e_{iA} \neq e_{iB}$. Also,

$$\begin{aligned} w_{iA} &= 0.15 + 0.1u_{iA}, \\ h_{iA} &= 0.25 + 0.1v_{iA}, \\ w_{iB} &= 0.1 + 0.1u_{iB}, \\ h_{iA} &= 0.35 + 0.1v_{iB}, \end{aligned}$$

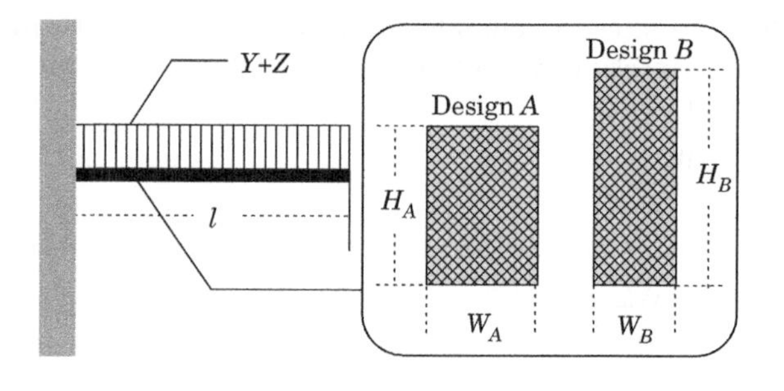

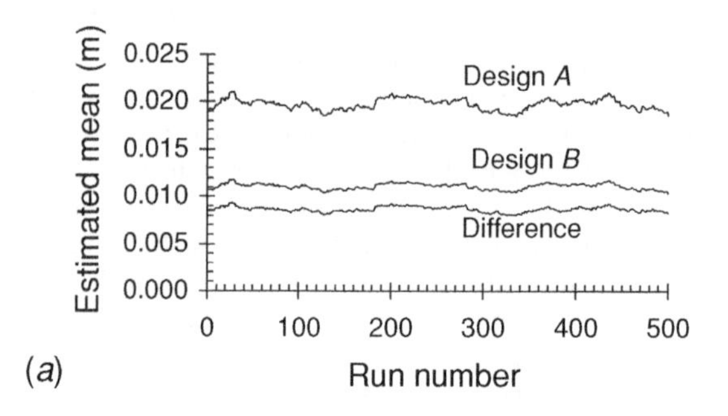

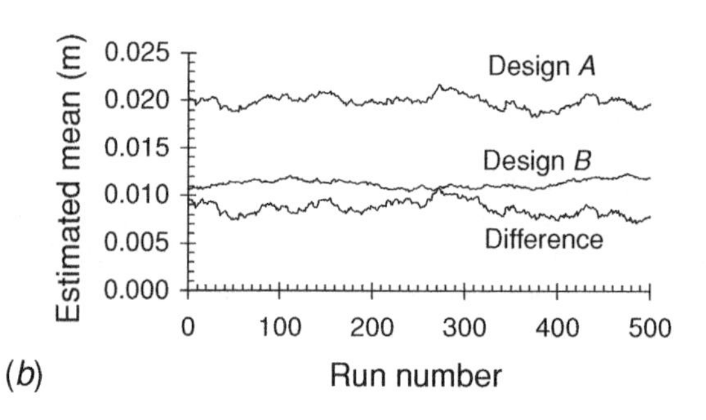

Fig. 8.1.6 Results of 500 simulation runs of expected deflection for a cantilever wood beam by using a size-100 sample for each run: (*a*) correlated and (*b*) independent simulations of the two design alternatives.

where u_{iA}, v_{iA}, u_{iB}, and v_{iB} denote four uniform (0, 1) mutually independent random variates. For this experiment,

$$\text{Cov}[\hat{E}[X_A], \hat{E}[X_B]] = 7.66 \times 10^{-9} \text{ m}^2,$$

and

$$\text{Var}[\hat{E}[X_A] - \hat{E}[X_B]] = 5.78 \times 10^{-7} \text{ m}^2.$$

The results of the uncorrelated simulation are shown in Fig. 8.1.6*b*. The correlated sampling reduces the variance of the estimated difference in design performance by about 87%. Note that from the results of the first experiment

$$\begin{aligned}\text{Var}[\hat{E}[X_A] - \hat{E}[X_B]] &= \text{Var}[\hat{E}[X_A]] + \text{Var}[\hat{E}[X_B]] \\ &= 3.26 \times 10^{-7} + 9.61 \times 10^{-8} = 4.22 \times 10^{-7} \text{ m}^2,\end{aligned}$$

which is somewhat smaller than 5.78×10^{-7} m^2. The second experiment is not needed to derive the result for the case of uncorrelated experiments.

8.1.3.2 Control variates

The accuracy of estimation in Monte Carlo experiments can be increased in some cases by substituting an indirect estimator Y^* for the original estimator X^* of the variate X. This is realized by introducing a *control variate* Z to represent, for instance, the performance function of an approximated model of the system studied. For example, the indirect estimator can be defined as

$$Y^* = X^* - \eta(Z - \mu_Z), \tag{8.1.8}$$

where η is a coefficient and μ_Z denotes the mean of Z, which is a random variable correlated with X. Note that the approximate model must be superceded if possible, by the determination of μ_Z by analytical methods. If X^* is an unbiased estimator of X, then

$$\begin{aligned} &E[X^*] = X, \\ &E[Y^*] = E[X^*] - \eta(E[Z] - \mu_Z) = E[X^*] = X. \end{aligned} \tag{8.1.9}$$

This means that Y^* is also an unbiased estimator; the variance is given by

$$\text{Var}[Y^*] = \text{Var}[X^*] + \eta^2\text{Var}[Z] - 2\eta\text{Cov}[X^*, Z]. \tag{8.1.10}$$

If $2\eta\text{Cov}[X^*, Z] > \eta^2\text{Var}[Z]$, then $\text{Var}[Y^*] < \text{Var}[X^*]$, signifying that the indirect estimator Y^* is more accurate than X^*. To maximize the variance reduction, one can select a value of η such that $\text{Var}[Y^*]$ is minimized; that is,

$$\frac{\partial \text{Var}[Y^*]}{\partial \eta} = 2\eta\text{Var}[Z] - 2\text{Cov}[X^*, Z] = 0,$$

or

$$\eta = \frac{\text{Cov}[X^*, Z]}{\text{Var}[Z]}. \tag{8.1.11}$$

By substituting the right-hand side of Eq. (8.1.11) for η in Eq. (8.1.10), one finds the corresponding minimum $\text{Var}[Y^*]$ as

$$\text{Var}[Y^*] = \left(1 - \rho^2_{X^*,Z}\right)\text{Var}[X^*], \tag{8.1.12}$$

where $\rho_{X^*,Z}$ is the correlation coefficient between X^* and Z. It is seen that the reduction in variance increases as $\rho_{X^*,Z}$ increases; this means that Z should be highly dependent on X^* to ensure an effective reduction. This is usually obtained if the control model provides a good approximation of the system.

8.1.4 Summary for Section 8.1

In this section we have introduced some basic concepts of simulation. Details of the probability integral transform have been given and variance reduction techniques are discussed. The examples have shown how Monte Carlo experiments can be used in practice. We now proceed to the generation of random numbers of specific distributions through the use of computers.

8.2 GENERATION OF RANDOM NUMBERS

8.2.1 Random outcomes from standard uniform variates

The probability integral transform indicates that generation of uniform (0, 1) random numbers is the basic generation process used to derive the outcomes from a variate with known probability distribution. Current methods to generate standard uniform variates are deterministic, in the sense that systematic procedures are used after one or more initial values are randomly selected. For example, system-supplied *random number generators* in most digital computers are almost always *linear congruential generators*. This algorithm is based on recursive calculation of a sequence of integers $k_1, k_2, k_3, \ldots,$ each between 0 and $m - 1$ (a large number) from a linear transformation:

$$k_{i+1} = (ak_i + c)(\text{modulo } m). \tag{8.2.1}$$

Here a and c are positive integers called the *multiplier* and the *increment*, respectively, and the notation (modulo m) signifies that k_{i+1} is the remainder obtained after dividing $(ak_i + c)$ by m, where m denotes a (large) positive integer. Hence, denoting $\eta_i = \text{Int}[(ak_i + c)/m]$, the corresponding residual is defined as

$$k_{i+1} = ak_i + c - m\eta_i. \tag{8.2.2}$$

Hence,

$$u_{i+1} = \frac{k_{i+1}}{m} = \frac{ak_i}{m} + \frac{c}{m} - \text{Int}\left[\left(\frac{ak_i + c}{m}\right)\right], \tag{8.2.3}$$

where the u_i are uniform (0, 1). Because these numbers are repeated with a given period, they are usually called *pseudorandom* numbers. The quality of the results depends on the magnitudes of the constants a, c, and m and their relationships, but the type of computer used will impose constraints. Because the period of the cycle is not greater than m, and it increases with m, the main criterion is that the period after which the original numbers are unavoidably repeated should be as long as possible. In practice, m is set equal to the word length, that is, the number of bits retained as a unit in the computer. Moreover, the constants c and m should not have any common factors, and the value of a should be sufficiently high. Because all possible integers between 0 and $m - 1$ occur after some interval of time, regardless of the generator used, any initial choice of the seed k_0 is as good as any other.

Example 8.8. Linear congruential algorithm. Suppose we assume low values for the constants in Eq. (8.2.1): $a = 5$, $c = 1$, and $m = 8$. Let $k_0 = 1$ be the seed for generating a sequence of random integers $k_i, i = 1, 2, 3, \ldots$. For $i = 1$ one has

$$\begin{aligned} k_1 &= ak_0 + c - m\text{Int}\left[\left(\frac{ak_0 + c}{m}\right)\right] = 5 \times 1 + 1 - 8 \times \text{Int}\left[\left(\frac{5 \times 1 + 1}{8}\right)\right] \\ &= 5 + 1 - 8 \times \text{Int}(0.75) = 5 + 1 - 8 \times 0 = 6, \end{aligned}$$

and, from Eq. (8.2.3),

$$u_1 = \frac{k_1}{m} = \frac{6}{8} = 0.75.$$

The second iteration yields

$$\begin{aligned} k_2 &= ak_1 + c - m\text{Int}\left[\left(\frac{ak_1 + c}{m}\right)\right] = 5 \times 6 + 1 - 8 \times \text{Int}\left[\left(\frac{5 \times 6 + 1}{8}\right)\right] \\ &= 30 + 1 - 8 \times \text{Int}(3.875) = 30 + 1 - 8 \times 3 = 7, \\ u_2 &= \frac{k_2}{m} = \frac{7}{8} = 0.875. \end{aligned}$$

The subsequent iterations yield the following sequence:

0.5, 0.625, 0.25, 0.375, 0, 0.125, <u>0.75, 0.875, 0.5, 0.625, 0.25, 0.375, 0, 0.125</u>, 0.75, 0.875, 0.5, 0.625, 0.25, 0.375, 0, 0.125, <u>0.75, 0.875, 0.5, 0.625, 0.25, 0.375, 0, 0.125</u>, ...

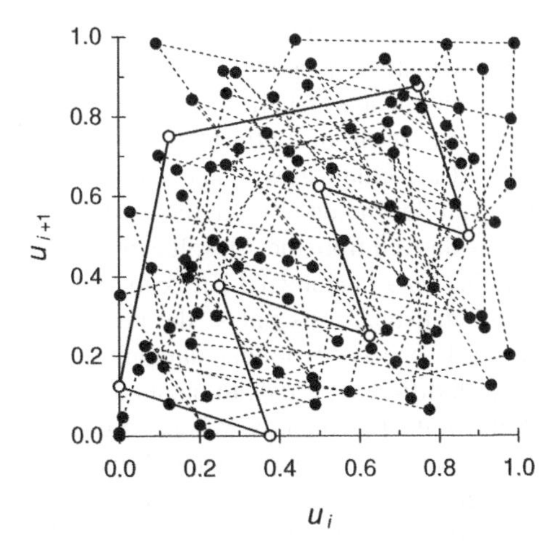

Fig. 8.2.1 Trajectory of 100 sequentially generated standard uniform random numbers with $a = 5, c = 1$, and $m = 8$ (solid line), and with $a = 2^7 + 1, c = 1$, and $m = 2^{35}$ (dotted line).

This is seen to be cyclic with a period of 8, because the underlined sequence of 8 values is repeated indefinitely. It is demonstrated by plotting u_{i+1} against u_i in Fig. 8.2.1.

Also shown in Fig. 8.2.1 are results from the generator $a = 2^7 + 1, c = 1$, and $m = 2^{35}$, which yields a much larger period of cyclicity. This choice gives satisfactory results for binary computers; and $a = 101, c = 1$, and $m = 2^b$ for a decimal computer with a word length b.

The advantage of the linear congruential method when applied through a digital computer is the speed of implementation. Because only a few operations are required each time, its use has become widespread. A disadvantage is that once the seed is specified, the entire series is predictable.

The pseudorandom numbers generated by these procedures may be tested for uniform distribution and for statistical independence. Goodness-of-fit tests, such as the chi-squared and the Kolmogorov-Smirnov tests, may be used to verify that these numbers are uniformly distributed. Both parametric and nonparametric methods, such as the runs test, can be used to check for randomness between successive numbers in a sequence. In spite of the fact that these procedures are essentially deterministic, pseudorandom numbers generated with large m and accurate choices of a and c generally appear to be uniformly distributed, and stochastically independent, so that they can be properly used to perform Monte Carlo simulations. Algorithms to generate pseudorandom numbers, which closely approximate mutually independent standard uniform variates, are a standard feature in statistical software. Standard uniform random numbers are available as a system-supplied function in digital computers, as well as in most customary computational and data management facilities such as spreadsheets and data bases.[6]

By substituting $c = 0$ in Eq. (8.2.1) the *multiplicative congruential generator*

$$k_{i+1} = ak_i \text{ (modulo } m) \tag{8.2.4}$$

[6] Extensive reviews of methods and computer routines for random number generation are given by Knuth (1981), Bratley et al. (1987), and Press et al. (1992), among others.

is obtained. This is a standard algorithm used in pocket calculators and digital computers. The multiplier a and the modulus m should be well chosen as outlined before.[7] Note that for the multiplicative congruential generator the value of 0 is not allowed as the initial seed, because it perpetuates itself, in contrast to any nonzero initial seed.

A more sophisticated algorithm uses a standard generator to compute the required random values, but it shuffles the output to remove low-order serial correlations. In this case, a random variable derived from the jth value in the sequence, I_j, is the output—not on the jth call but rather on a randomized later call, $j + 32$ on average. This algorithm should pass nearly all statistical tests unless the number of calls becomes very large, say, larger than $m/20$. If one needs longer random sequences, one can combine two different sequences with different periods to obtain a sequence with period equal to the least common multiple of the two periods. When these algorithms are implemented on a digital computer, the execution time for the *shuffled multiplicative congruential generator* is about 1.3 times larger than that required by the standard generator, while that required by the *two-component shuffled multiplicative congruential generator* is about twice this time.

It is seen that the error term in Monte Carlo integration decreases as $n^{-1/2}$ when choosing n points uniformly randomly distributed in a multidimensional space. From Eq. (8.1.2) the standard deviation of the estimate of probability of an event of interest decreases as $n^{-1/2}$ when n sequential mutually independent simulations are performed. One might search for a faster decay of the error associated with simulation. For instance, if sample points used for integration lie on a cartesian grid, and one samples each grid point exactly once (in whatever order), the Monte Carlo method thus becomes a deterministic quadrature scheme in which the fractional error decreases at least as fast as n^{-1}, and even faster if the function goes to zero smoothly at the boundaries of the sampled region or it is periodic in the region. However, using a grid one must decide in advance how fine it should be, and it is not convenient to sample until some convergence or termination criterion is met. Therefore, one might search for an intermediate scheme—some way to pick sample points that are random yet spread out in some self-avoiding way, avoiding the chance of clustering that occurs with uniform random points. *Quasi-random*, or *sub-random*, generators, which are based on deterministic algorithms to generate uniformly distributed sequences of variates that maximally avoid each other, can be used for this purpose. An example is shown in Fig. 8.2.2, where 100 and 400 pairs of standard uniform random numbers (u, v) are compared with the same number of pairs generated as gridded uniform random numbers in the unit square using a grid size δ of 0.1 and 0.05, respectively. These points are determined by generating a pair of random numbers from two independent uniformly distributed variates, U and V, for each grid cell, with $(i-1)\delta \leq u \leq i\delta$ and $(j-1)\delta \leq v \leq j\delta$, for $i = 1, \ldots, 10$ and $j = 1, \ldots, 10$. Note that the sample space is covered much more uniformly by the gridded random points than by purely random ones. The statistics of the two sequences (see Table 8.2.1) also show that the two coordinates of a point are independent of each other in both cases. More sophisticated methods can be developed using binary fractions.[8] Although quasi-random sequences provide satisfactory

[7] For example, the "minimal standard" generator of Park and Miller as reported by Press et al. (1992) is based on the choices of $a = 7^5 = 16807$ and $m = 2^{31} - 1 = 2{,}147{,}483{,}647$. The generator has a period of $2^{31} - 2 \approx 2.1 \times 10^9$, and it has seen much successful use. Since the product of a and $m - 1$ exceeds the maximum value for a 32-bit integer, one can use an approximate factorization of m, such as $m = ah + l$, which yields $h = \mathrm{Int}(m/a)$. Thus $l = m$ (modulo a).

[8] See Press et al. (1992) for a review of quasi-random generators and related references.

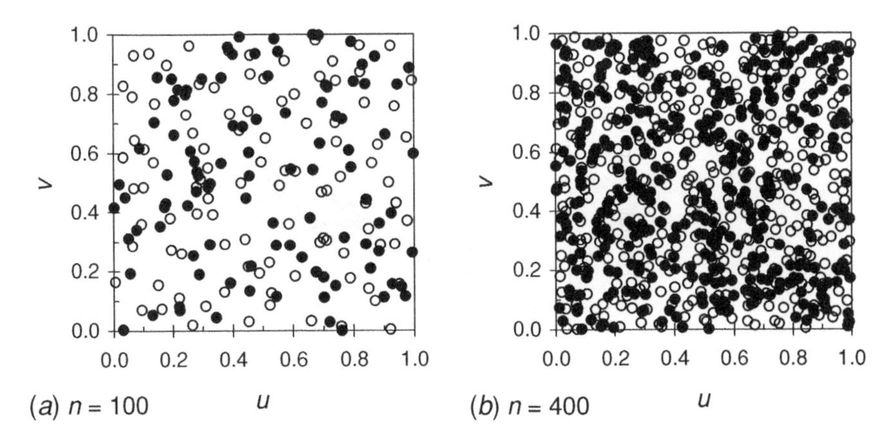

Fig. 8.2.2 Points in a unit square generated as standard uniform random numbers (solid circles) and as uniform random numbers on a uniform grid (empty circles) with size 0.1 (*a*) and 0.05 (*b*).

results when applied to problems of the Monte Carlo type,[9] one should handle these methods with great care and previously test all the statistical properties which are relevant to the simulation of the system.

It seems a paradox that a deterministic machine like a computer is used to produce random numbers.[10] With reference to a digital computer, the program for the random number generator should be supplemented by numerous tests as discussed earlier. An additional test is that if two or more random number generators are used for the input of a simulation, the outputs should not have any significant statistical differences. Furthermore, there should be no relationship between the computer program for the random number generator and that used by the engineer in simulation.

Table 8.2.1 Statistics of n pairs of standard uniform random numbers and of gridded uniform random numbers in the unit square

			Standard uniform		Gridded uniform	
Statistic	Theoretical	n	u	v	u	V
Mean	0.5	100	0.507	0.513	0.502	0.504
		400	0.501	0.492	0.500	0.500
Variance	0.083	100	0.081	0.085	0.082	0.081
		400	0.082	0.083	0.084	0.083
Skewness coefficient	0	100	0.030	−0.005	0.004	−0.033
		400	−0.047	0.063	−0.004	−0.008
Correlation, $\rho_{U,V}$	0	100	0.026		−0.015	
		400	−0.010		0.001	

[9] For example, using the Sobol sequence (not presented here) makes the error in Monte Carlo integration to decrease as n^{-1} if a smooth function is to be integrated, and as $n^{-2/3}$ for a function with step discontinuities (Press et al., 1992, p. 305).

[10] According to von Neumann (1951), "Anyone who considers arithmetical methods of producing random digits is, of course, in a state of sin," for "there is no such thing as a random number. There are only methods to produce random numbers."

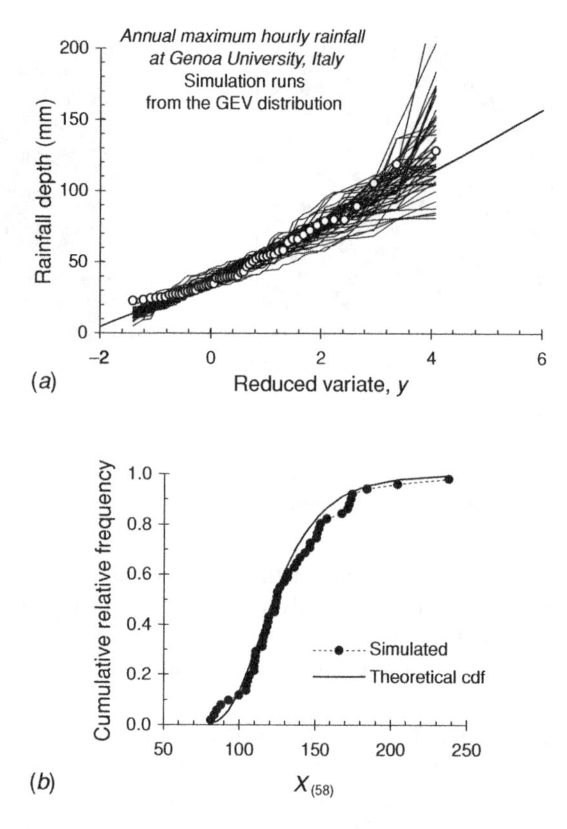

Fig. 8.2.3 Simulation of annual maximum hourly rainfall at Genoa University, Italy, using the GEV distribution: (*a*) Gumbel probability plots of 50 simulations, each with 58 outcomes, as compared with observations; and (*b*) simulated cumulative frequency of the largest value as compared with its theoretical cdf.

8.2.2 Random outcomes from continuous variates

The probability integral transform yields the required outcome x of a random variable X with continuous cdf $F_X(x)$ from a generated value u of the standard uniform variate on 0 to 1. Accordingly, x is determined as the uth quantile of X. Obviously, the inverse transform method is straightforward when the cdf of X can be inverted analytically, that is, the inverse function is available as in the case of an exponential variate.

Example 8.9. GEV-distributed storm depth. The distribution of annual maximum hourly storm depth at Genoa University in northwest Italy is estimated to have a GEV distribution with parameters $k = -0.05$, $\alpha = 17.27$ mm, and $\varepsilon = 37.30$ mm in Example 7.23. An individual outcome from this population is found by substituting a standard uniform random number u for the frequency level in Eq. (7.2.65). Thus,

$$x = \varepsilon + \frac{\alpha}{k}[1 - (-\ln u)^k] = 37.29 - \frac{17.17}{0.05}[1 - (-\ln u)^k].$$

Fifty simulation runs, each with $n = 58$ outcomes, are shown in Fig. 8.2.3*a* on a Gumbel probability plot.

Monte Carlo simulations give an idea of possible sampling variability of a particular variate. For instance, Fig. 8.2.3*b* shows the cumulative relative frequency curve for the largest value of 58 outcomes in 50 runs; this is also compared with the theoretical cdf evaluated from Eq. (7.1.2).

Table 8.2.2 Summary of independent random numbers x generated from selected distributions

Distribution	Parameters	x
Standard normal		$\sqrt{-2\ln u_1}\sin(2\pi u_2)$ and $\sqrt{-2\ln u_1}\cos(2\pi u_2)$
Standard beta	α, β	$u_1^{1/\alpha}\Big/\left[u_1^{1/\alpha}+u_2^{1/\beta}\right]$, provided that $u_1^{1/\alpha}+u_2^{1/\beta}\leq 1$[a]
Standard gamma	r	for $r>1$: $-\ln\left(\prod_{i=1}^{r}u_i\right)=-\sum_{i=1}^{r}\ln u_i$, for integer r; and, in general, $-\sum_{i=4}^{r'+3}\ln u_i+(-\ln u_3)u_1^{1/f}\Big/\left[u_1^{1/f}+u_2^{1/(1-f)}\right]$, with $r'=\mathrm{Int}(r)$, $f=r-r'$, and acceptance region as for beta; for $r<1$: if $u_1(e+r)/e\leq 1$, then $x=[u_1(e+r)/e]^{1/r}$, if $u_2\leq e^{-x}$; if $u_1(e+r)/e>1$, then $x=-\ln[(e+r)(1-u_1)/(er)]$, if $u_2\leq x^{r-1}$; otherwise reject and repeat until accepted[b]
Binomial	n, p	$\sum_{i=1}^{n}k_i$, with $k_i=1$, if $u_i<p$; and $k_i=0$, if $u_i\geq p$
Poisson	v	x such that $\sum_{i=1}^{x}-v^{-1}\ln(u_i)\leq 1$, and $\sum_{i=1}^{x+1}-v^{-1}\ln(u_i)>1$
Geometric	p	$1+\mathrm{Int}[\ln u/\ln(1-p)]$
Negative binomial	r, p	$r+\sum_{i=1}^{m}h_i$, with $h_i=1$, if $u_i\geq p$; $h_i=0$, if $u_i<p$; and m is the smallest integer such that $r=\sum_{i=1}^{m}(1-h_i)$

Note: $u_1, u_2, \ldots, u_i, \ldots$ denote uniform (0, 1) random numbers.
[a] See Jöhnk (1964), also for gamma distribution.
[b] See Ahrens and Dieter (1974).

The inverse transform method is effective if the quantile of the variate has an explicit formulation in terms of probability of nonexceedance. However, many probability models cannot be inverted analytically; examples are the normal, lognormal, beta and gamma distributions, among others. Because numerical computations to obtain the required quantile may be cumbersome, other methods can be developed using the concept of a function of a random variable and the theorem of total probability. A powerful alternative is the rejection method using a procedure similar to that used in Monte Carlo integration. We provide details of this procedure in this section.

When a random variable X can be expressed as a function of other random variates, say, $X=g(Y_1, Y_2, \ldots, Y_k)$ and methods for generating $Y_1, Y_2, \ldots, Y_k$ are available, an outcome of X can be determined as $X=g(y_1, y_2, \ldots, y_k)$, where $(y_1, y_2, \ldots, y_k)$ are random realizations of $Y_1, Y_2, \ldots, Y_k$. This method can be used to derive standard normal-, gamma-, and beta-distributed numbers, among others, as summarized in Table 8.2.2.

Example 8.10. Standard normal random numbers. Let V and W denote two variates defined as $W=-\ln U_1$, and $V=U_2$, where U_1 and U_2 are two independent standard uniform variates. As U_1 and U_2 are independent, the joint pdf of W and V is given by

$$f_{W,V}(w,v)=f_W(w)f_V(v)=f_{U_1}(e^{-w})\left|\frac{de^{-w}}{dw}\right|f_{U_2}(v)=e^{-w},$$

as discussed in Section 3.4. If one writes W and V as $W = [(Z_1)^2 + (Z_2)^2)]/2$, and $V = (2\pi)^{-1} \times \tan^{-1}(Z_2/Z_1)$, the pdf of the new variates Z_1 and Z_2 is written as

$$f_{Z_1,Z_2}(z_1, z_2) = Jf_{W,V}\left(\frac{z_1^2 + z_2^2}{2}, \frac{1}{2\pi}\tan^{-1}\left(\frac{z_2}{z_1}\right)\right),$$

where the Jacobian J is given by

$$J = \begin{vmatrix} \dfrac{\partial w}{\partial z_1} & \dfrac{\partial w}{\partial z_2} \\ \dfrac{\partial v}{\partial z_1} & \dfrac{\partial v}{\partial z_2} \end{vmatrix} = \begin{vmatrix} z_1 & z_2 \\ \dfrac{-z_2}{2\pi\left(z_1^2 + z_2^2\right)} & \dfrac{z_1}{2\pi\left(z_1^2 + z_2^2\right)} \end{vmatrix} = \frac{1}{2\pi}.$$

Thus,

$$f_{Z_1,Z_2}(z_1, z_2) = \frac{1}{2\pi}\exp\left(-\frac{z_1^2 + z_2^2}{2}\right),$$

for $-\infty < z_1, z_2 < +\infty$. This is the joint pdf of two independent standard normal variates, as seen from Eq. (4.3.1) for $\rho = 0$. Inverting the above expressions for Z_1 and Z_2 as functions of W and V, $Z_1 = (2W)^{1/2}\cos(2\pi V)$, and $Z_2 = (2W)^{1/2}\sin(2\pi V)$ are obtained. Hence, substituting $W = -\ln U_1$ and $V = U_2$ for W and V, respectively,

$$Z_1 = (-2\ln U_1)^{1/2}\cos(2\pi U_2),$$

and

$$Z_2 = (-2\ln U_1)^{1/2}\sin(2\pi U_2),$$

showing that a pair of independent standard normal variates can be generated using two independent standard uniform variates. This is called the *Box-Muller method* and is commonly used. Standard normal random numbers are used to generate normal and lognormal outcomes to give a variate with known mean and variance.

8.2.2.1 Decomposition method

Using the theorem of total probability of Eq. (2.2.15), one can express the pdf of a variate X as the weighted sum of a set of other density functions in the form

$$f_X(x) = \sum_{i=1}^{m} f_{X_i}(x)p_i, \tag{8.2.5}$$

where $f_{X_i}(x) = f_X(x|B_i)$, $i = 1, \ldots, m$, is a set of component density functions and $p_i = \Pr[B_i]$ is the probability or relative weight associated with $f_{X_i}(x)$ for the ith component B_i. A complex pdf $f_X(x)$ can be decomposed into a combination of simpler pdfs, whose corresponding cdfs can be inverted analytically. Accordingly, the *decomposition method* first generates a random number for the probability p_i and the corresponding pdf is selected; then, another random number is generated according to the selected pdf. Note that m standard uniform random numbers must be generated to obtain the desired outcome.

Example 8.11. Ferry transportation. Two companies provide ferry transportation across the Strait of Messina in southern Italy. The waiting time X of a ferry is distributed as a mixture of two exponential distributions (see Fig. 8.2.4).

The pdf of this distribution, which is sometimes referred to as *contaminated exponential distribution*, is given by

$$f_X(x) = p\lambda_1\exp(-\lambda_1 x) + (1 - p)\lambda_2\exp(-\lambda_2 x),$$

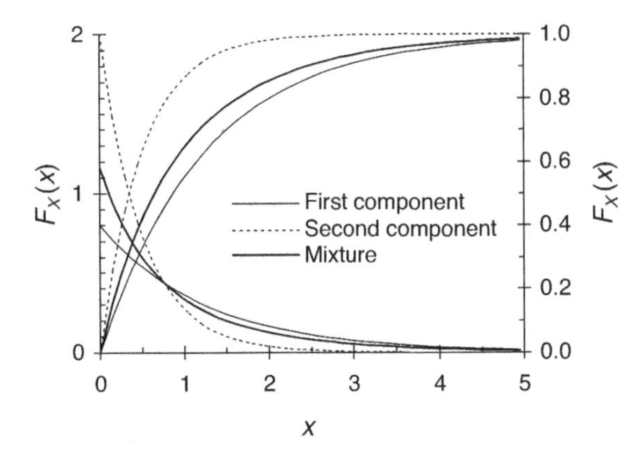

Fig. 8.2.4 Contaminated exponential distribution of waiting time X for ferry arrivals.

and its cdf is

$$F_X(x) = p[1 - \exp(-\lambda_1 x)] + (1 - p)[1 - \exp(-\lambda_2 x)].$$

Since $F_X(x)$ cannot be inverted analytically, but $f_X(x)$ is the sum of two exponential densities weighted by p and $(1 - p)$, respectively, one can use the decomposition method to simulate random waiting times. Accordingly, one first generates a standard uniform random number u_1. If $u_1 < p$, the required outcome x of X is found by applying the inverse transform method to the first component:

$$x = -\frac{1}{\lambda_1}\ln(1 - u_2),$$

where u_2 is another standard uniform random number. Conversely, if $u_1 \geq p$,

$$x = -\frac{1}{\lambda_2}\ln(1 - u_2),$$

which is the random waiting time, found from the u_2th quantile of the second component. For example, let $p = 0.7$, $\lambda_1 = 0.8\ \text{h}^{-1}$, and $\lambda_2 = 1\ \text{h}^{-1}$. Suppose that $u_1 = 0.32$, and $u_2 = 0.44$. Since $0.32 < 0.7$, one gets

$$x = -\frac{1}{0.8}\ln(1 - 0.44) = 0.725 \text{ hour.}$$

8.2.2.2 Rejection method

The *rejection method* provides a general technique for generating outcomes from a variate X with known and computable pdf $f_X(x)$, and it is not essential for the cdf to be computable. This method is based on a simple geometrical concept, similar to that used in Monte Carlo integration. The pdf $f_X(x)$ is shown in Fig. 8.2.5.

On the same graph, one draws another curve $y = g(x)$ that has a finite area under it, and it is such that $g(x) \geq f_X(x)$ for all possible values of X. This is called the *comparison function* and it lies everywhere above the original probability density function. In the standard rejection method, one generates a random variable X from the density $g(x)/\alpha$, where α is the area under $g(x)$. Then a standard uniform variate U is drawn and X is accepted if $U < f_X(x)/g(x)$.

Suppose there is some way of choosing a random point in two dimensions that is uniformly distributed in the area under $g(x)$. If this point lies outside the area under the

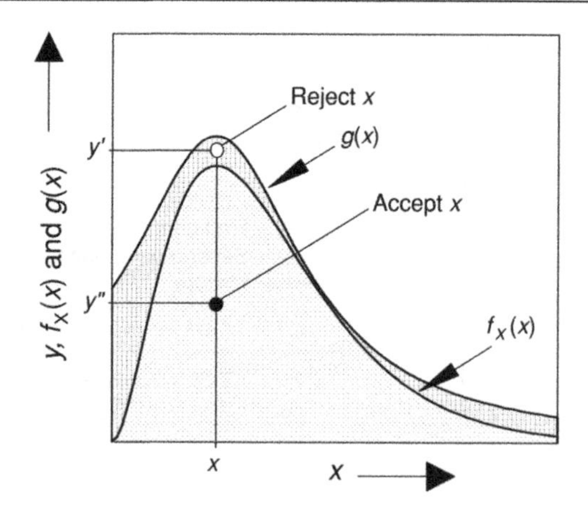

Fig. 8.2.5 Rejection method for generating a variate X from a distribution with known pdf.

original pdf $f_X(x)$, we reject it and proceed to another random chosen point. On the contrary, if the point lies inside the area under $f_X(x)$, we accept it and take x as a random number from X. This is because the accepted points are uniformly distributed in the area of acceptance, which is a subset of the comparison area, so that their values have the desired distribution. The fraction of points rejected depends on the ratio of the area under $g(x)$ to the area under $f_X(x)$, and not on the particular form of either function. If, for instance, the area under $g(x)$ is less than 1.5, less than one-third of the points will be rejected regardless of how well $g(x)$ fits $f_X(x)$.

Example 8.12. Standard gamma random numbers. One wishes to generate random numbers from the standard gamma distribution with pdf

$$f_X(x) = \frac{x^{r-1}e^{-x}}{\Gamma(r)},$$

for a specified value of shape parameter $r > 1$. The form of this pdf is such that it can be embedded using a comparison function derived from the standard Cauchy-distributed variate W with pdf

$$f_W(w) = [\pi(1+w^2)]^{-1},$$

and cdf

$$F_W(w) = 1/2 + \pi^{-1}\text{arc tan}\, w.$$

Thus, one takes

$$g(x) = \frac{c}{[1+(x-b)^2/a^2]},$$

where a, b, and c denote three constants. The values of a, b, and c are selected in such a way that $g(x)$ is everywhere greater than $f_X(x)$, and the area under the curve (which depends on the product ca) is as small as possible. One then applies the inverse transform method to generate a value of W from a standard uniform random number u_1 as $z = \tan[\pi(u_1 - 1/2)]$. This is used to compute a value of x as $x = az + b$; that is,

$$x = a\tan[\pi(u_1 - 1/2)] + b.$$

The ordinate y of the random point on the (x, y) plane is then found as $u_2 g(x)$, where u_2 is another standard uniform random number and is compared with the original pdf evaluated at x. If

$$y = \frac{u_2 c}{1 + \tan^2[\pi(u_1 - 1/2)]} \leq \frac{1}{\Gamma[r]} \left\{ a \tan\left[\pi\left(u_1 - \frac{1}{2}\right)\right] + b \right\}^{r-1} e^{-a\tan[\pi(u_1 - 1/2)] - b} = f_X(x),$$

the value of x is accepted as a standard gamma random number; conversely, if $y > f_X(x)$, it is rejected and another point (x, y) is generated. For example, if $r = 2$, one can take $a = 2.01$, $b = 2$, and $c = 0.3$, as shown in Fig. 8.2.5. Note that to generate n random numbers using this method, one requires more than $2n$ uniform(0, 1) random numbers. This technique can be used for a large variety of variates with a bell-shaped pdf.[11]

8.2.3 Random outcomes from discrete variates

The inverse transform method can also be used to generate a random number from a discrete distribution $F_X(x_i)$ by using a standard uniform random number u. The condition $F_X(x_{(i-1)}) < u \leq F_X(x_{(i)})$ gives the corresponding discrete random number as $x_{(i)}$, the ith-ordered possible value of the random variable X. However, the generation of discrete random numbers by this method requires the calculation of the cdf for all possible values of the random variable or, at least, for many of them. Then, one will search for $x_{(i)}$ each time a number u is generated.

Example 8.13. Binomial random numbers. One wishes to generate random numbers from a binomial distributed variate with cdf

$$F_X(x) = \sum_{k=0}^{x} \binom{n}{k} p^k (1 - p)^{n-k},$$

for $x = 0, 1, \ldots, n$. Also, let $p = 0.2$ and $n = 5$. Suppose that the random variate generated from a uniform(0, 1) distribution is 0.8. Since

$$F_X(1) = \sum_{k=0}^{1} \binom{5}{k} 0.2^k (1 - 0.2)^{5-k} = 0.737,$$

and

$$F_X(2) = \sum_{k=0}^{2} \binom{5}{k} 0.2^k (1 - 0.2)^{5-k} = 0.942,$$

one has $F_X(1) < 0.8 \leq F_X(2)$, and the corresponding value of X is $x = 2$, as shown in Fig. 8.2.6.

For large n, say, $n > 30$ and $np > 5$ the normal approximation can be used to generate binomial random numbers, as given by Eq. (4.2.24a). If z denotes a standard normal random number, one computes

$$x^* = np + [np(1 - p)]^{1/2} z,$$

[11] See the more efficient method of Ahrens and Dieter (1974) in Table 8.2.2, with details by Devroye (1986, pp. 401–428) who gives an evaluation of various techniques for nonuniform random number generation.

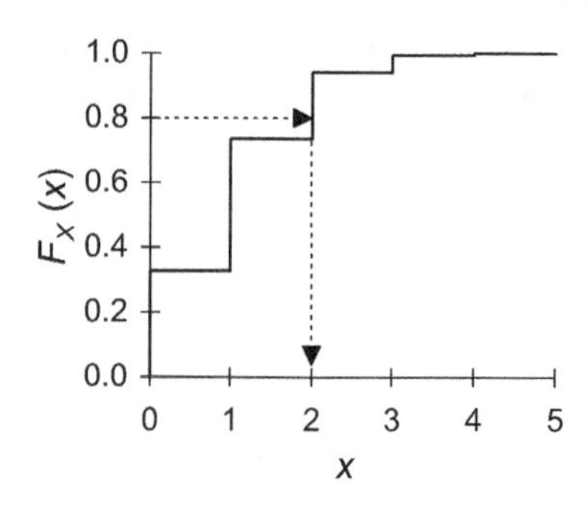

Fig. 8.2.6 Generation of a binomial random number.

to obtain the corresponding value of the binomial variate as

$$x = 0, \quad \text{if } x^* \leq 0,$$
$$x = n, \quad \text{if } x^* \geq n,$$
$$x = x^*, \quad \text{if } 0 < x^* < n,$$

where x^* is rounded off to the nearest integer.

Because the computation of the cdf of a discrete variate may be cumbersome in some cases, other methods can also be used. For instance, one can use the rejection method introduced for continuous variates. The pdf of a discrete variate can be viewed as a sequence of Dirac delta functions of possible outcomes x_i with an area of $p_X(x_i)$, but one can spread the finite area in the spike at x_i into the interval from x_i to x_{i+1}, thus defining a sort of continuous density function $f(x)$ as shown in Fig. 8.2.7. If a uniformly distributed random point in the area upperly bounded by the comparison function also lies inside the area bounded by $f(x)$, the integer part of its abscissa is accepted as a discrete random number; conversely, it is rejected.

Some discrete random variables can be interpreted as a counter of occurrences. One can thus generate values from this variate by using the probability distribution of the distance between occurrences or interarrival time. The discrete outcome is determined as the terminator index of the random sum describing the cumulated distance or interarrival time in a specified distance or period. This is the case, for instance, of the Poisson and negative binomial distributed variates, as summarized in Table 8.2.2.

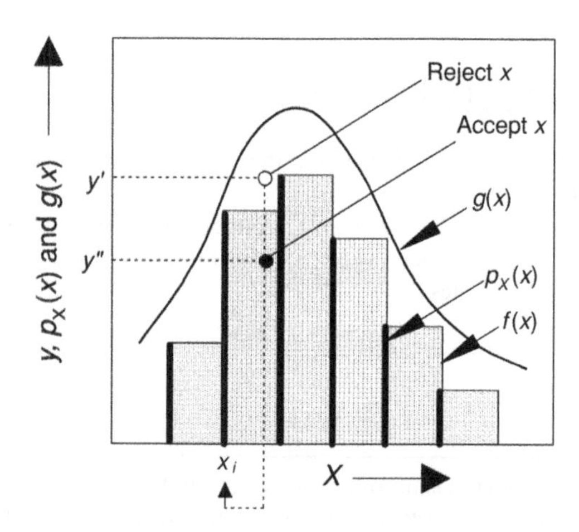

Fig. 8.2.7 Rejection method for generating an outcome from a discrete variate X with known pmf.

Example 8.14. Poisson random numbers. The generation of random numbers from the Poisson distribution with pmf

$$p_X(x) = \frac{\nu^x e^{-\nu}}{x!},$$

is based on the knowledge that the distance between occurrences or interarrival time t is exponentially distributed with mean $1/\lambda$ where $\lambda = \nu/t$ is the rate of occurrence.[12] A sequence of interarrival times $t_i, i = 1, 2, \ldots,$ can thus be generated from the exponential distribution. If

$$t_1 + t_2 + \cdots t_{i-1} + t_i \leq t < t_1 + t_2 + \cdots t_{i-1} + t_i + t_{i+1},$$

one will take i as the appropriate value of the Poisson-distributed variate, because exactly i occurrences are observed within the reference distance or period t. Accordingly, a Poisson random number is generated as the value x such that

$$\sum_{i=1}^{x} -\left(\frac{t}{\nu}\right)\ln(u_i) \leq t, \quad \text{or} \quad \sum_{i=1}^{x} -\nu^{-1}\ln(u_i) \leq 1,$$

and

$$\sum_{i=1}^{x+1} -\left(\frac{t}{\nu}\right)\ln(u_i) > t, \quad \text{or} \quad \sum_{i=1}^{x+1} -\nu^{-1}\ln(u_i) > 1,$$

where the u_i are a sequence of uniform(0, 1) random numbers. Note that for large ν, say, $\nu > 10$, the normal approximation to the Poisson distribution might be used, that is $X \sim N(\nu - 0.5, \nu)$. Hence, x is generated from a standard normal random number z by computing[13]

$$x^* = \nu - 0.5 + \nu^{1/2} z.$$

The corresponding value of the Poisson variate is obtained as

$$x = 0, \quad \text{if } x^* \leq 0;$$
$$x = x^*, \quad \text{if } x^* > 0,$$

where x^* is rounded off to the nearest integer. This approximation is also a useful starting point when using a rejection procedure.

8.2.4 Random outcomes from jointly distributed variates

If a simulation requires the outcome from a set of k stochastically independent variates, say, $X_1, X_2, \ldots, X_k$, the random numbers for each variate can be generated independently of one another using the above methods. This is because the joint pdf is simply the product of the marginals. Conversely, the outcome from a k-dimensional variate with mutually dependent components can be generated in cascade using the concept of conditional probability. From Eq. (2.2.12) and Subsection 3.3.2,

$$F_{X_1,\ldots,X_k}(x_1, \ldots, x_k) = F_{X_1}(x_1) F_{X_2|X_1}(x_2|x_1) \cdots F_{X_k|X_1,\ldots,X_{k-1}}(x_k|x_1, \ldots, x_{k-1});$$

that is, the joint cdf can be written as the product of marginal and conditional cdfs. Accordingly, a value x_1 can be generated independently as the u_1th quantile from the marginal cdf of X_1. With this value of x_1, the conditional cdf of X_2 given X_1 is a function only of x_2, and hence a value x_2 can be determined as the u_2th quantile of the conditional

[12] As discussed in Subsection 4.2.2.

[13] See, also Example 4.30.

cdf of X_1 given X_2. Using the values x_1 and x_2 already obtained, one computes x_3 as the u_3th quantile of the conditional cdf of X_3 given X_1 and X_2. This recursive procedure is then carried out until the required set of dependent random numbers $(x_1, x_2, \ldots, x_k)$ is computed from a set of independent standard uniform random numbers $(u_1, u_2, \ldots, u_k)$. The effectiveness of the method is ensured by the straightforward application of the probability integral transform, which is possible if the initial marginal and its associated conditional cdfs can be inverted analytically.

Example 8.15. Bivariate normal random numbers. In Example 4.32, compressive strength Y and density X of concrete are modeled using the bivariate normal pdf of Eq. (4.3.1). It is seen that the conditional variate Y given X is a normal variate with mean

$$\mu_Y + \rho(x - \mu_X)\left(\frac{\sigma_Y}{\sigma_X}\right),$$

and standard deviation

$$\sigma_Y(1 - \rho^2)^{1/2},$$

where μ_X, μ_Y, σ_X, and σ_Y denote the means and standard deviations of X and Y, respectively, and ρ is the correlation coefficient between the two variates. Therefore, one can generate a value x of X as

$$x = \mu_X + z_1\sigma_X,$$

which is used to compute a value y of Y as

$$y = \mu_Y + \rho(x - \mu_X)\left(\frac{\sigma_Y}{\sigma_X}\right) + z_2\sigma_Y(1 - \rho^2)^{1/2},$$

where z_1 and z_2 are two independent standard normal random numbers.

8.2.5 Summary of Section 8.2

Methods of generating random variates from continuous and discrete distributions are shown in this section. We commenced with different types of uniform random number generators. The Box-Muller technique for normal variates and the generation of other types such as gamma and Poisson variates are given. We demonstrate the use of the decomposition and rejection methods of generation. The next section deals with the use of simulation in design.

8.3 USE OF SIMULATION

Simulation methods can be applied to large and complex systems, which would require large simplifications to be modeled using analytical methods. In such cases, more realistic simulation models can be used. Furthermore, simulations are often the only means of verifying or validating approximate analytical solution methods and of searching for a solution to those statistical problems requiring cumbersome analytical developments or those that are yet unsolved.

8.3.1 Distributions of derived design variates

Design variates of engineering systems are often derived from other variates for which observations are available. For example, the wind load on a tower is usually determined from wind speeds measured in the area; pollution loads in a river are the product of the combined effects of wastewater discharges and natural flows; and flood flows in an ungauged stream can be derived from observed storm and basin characteristics, and the elevation of the levees designed to accommodate these flows is found from the relationship between water depth and discharge. These relationships are deterministic if the transformation of the observed variates and the adaptation in the design is known without uncertainties. Alternatively, they are stochastic. This can occur because of unpredictable system behavior, or it can be caused by uncertainties in modeling and parameterization. Analytical solution methods should be preferred to simulations because they provide general solutions. However, their application is sometimes cumbersome, as previously stated, or one finds that system complexity does not facilitate the development of analytical derivations, so that simulation is the only tool to achieve practical results. A detailed description of a system is possible through simulation, whereas oversimplifications of the system may be required to develop analytical solutions. Also, simulation methods can be used to verify or validate those analytical approximations which are amenable for extrapolations or generalizations.

Example 8.16. Pier scour. Pier foundations of bridges over water can be undermined by local scour. The best-fit scour model for bridge piers proposed by Johnson (1992) gives the scour depth X measured from the average channel bed to the bottom of the scour hole as

$$X = 2.02Y(b/Y)^{0.98} F_r^{0.21} W^{-0.24},$$

where Y is the depth of flow just upstream of the pier, F_r is the upstream Froude number ($F_r = V/(gY)^{1/2}$, V and g denote the approach flow velocity and acceleration due to gravity, respectively), W is sediment gradation (equal to $d_{84\%}/d_{50\%}$, the ratio between the 84% quantile to the median sediment diameter), and b is the pier width. All these quantities are measured in metric units. Using the Manning formula to compute the velocity for a wide rectangular channel cross section,

$$V = (1/n)S^{1/2}Y^{2/3},$$

where n is the roughness coefficient and S is the slope. Hence, the Froude number is

$$F_r = \frac{V}{(gY)^{1/2}} = S^{1/2}Y^{1/6}n^{-1}g^{-1/2},$$

thus,

$$X = 2.02Y(b/Y)^{0.98}(S^{1/2}Y^{1/6}n^{-1}g^{-1/2})^{0.21} W^{-0.24},$$

which, after substituting 9.81 m/s^2 for g, can be written as

$$X = 1.59b^{0.980}Y^{0.055}S^{0.105}n^{-0.210}W^{-0.240}.$$

The estimation of Y, S, n, and W is affected by uncertainties. We propose to model all these quantities as random variables. One can thus determine the probability distribution of X by simulation if the probability distributions of Y, S, n, and W are known. For a pier width of 2.5 m, suppose that sediment gradation $W \sim$ lognormal (4, 1.6^2), the slope $S \sim N(0.002, 0.0004^2)$, the depth $Y \sim N(4.75\,\text{m}, 1.2^2\,\text{m}^2)$, and the roughness coefficient $n \sim$ uniform (0.02, 0.04). Also, one can reasonably assume that Y, S, n, and W are independent of each other. To perform each simulation, one will generate a standard uniform random number, u_i, and three independent standard normal numbers, z_{1i}, z_{2i}, and z_{3i}. The ith outcome of the roughness coefficient n is found by rescaling u_i as

$$n_i = 0.02 + (0.04 - 0.02)u_{1i},$$

and those of W, S, and Y are computed as

$$w_i = \exp(1.312 + 0.385 z_{1i}),$$
$$s_i = 0.002 + 0.0004 z_{2i}, \quad \text{and} \quad y_i = 4.75 + 1.2 z_{3i}$$

[note Eq. (4.2.2*a*) and (4.2.2*b*) for the uniform distribution and Eq. (4.2.28*e*) and the foregoing equation for the lognormal distribution]. Figure 8.3.1*a* shows the sampling cdf $F_X(x)$ of the scour depth resulting from the first 10, 100, and 1000 simulation cycles. It is seen that the size-10 sample provides a rough approximation to the size-1000 sampling cdf, which is a better approximation than that obtained from the size-100 sample. The estimated means and standard deviations of the specified variates are shown in Fig. 8.3.1*b* and 8.3.1*c* for an increasing number of simulation cycles. Note that the sampling means and standard deviations of Y, S, n, and W estimated from 1000 simulation cycles practically overlap with those used as inputs to the simulation procedure.

The 1000-cycle simulated mean and standard deviation of X are 3.39, and 0.36 m, respectively. Note that the estimated mean is very close to the nominal value of 3.32 m determined by substituting the mean values for the corresponding variates in the pier scour model. These results can also be compared with the approximated mean and standard deviation, which are computed by using Taylor's series expansion about the means of independent variates (see Section 3.4). The first and second partial derivatives of X with respect to each independent variate are as follows:

$$\left(\frac{\partial X}{\partial y}\right)_\mu = 1.59 \times 0.0551 b^{0.980} \mu_Y^{-0.945} \mu_S^{0.105} \mu_n^{-0.210} \mu_W^{-0.240} = 0.0325,$$

$$\left(\frac{\partial X}{\partial s}\right)_\mu = 1.59 \times 0.1051 b^{0.980} \mu_Y^{0.055} \mu_S^{-0.895} \mu_n^{-0.210} \mu_W^{-0.240} = 0.0006,$$

$$\left(\frac{\partial X}{\partial n}\right)_\mu = -1.59 \times 0.2100 b^{0.980} \mu_Y^{0.055} \mu_S^{0.105} \mu_n^{-1.210} \mu_W^{-0.240} = -23.21,$$

$$\left(\frac{\partial X}{\partial w}\right)_\mu = -1.59 \times 0.2400 b^{0.980} \mu_Y^{0.055} \mu_S^{0.105} \mu_n^{-0.210} \mu_W^{-1.240} = -0.1989,$$

$$\left(\frac{\partial^2 X}{\partial y^2}\right)_\mu = -1.59 \times 0.0520 b^{0.980} \mu_Y^{-1.945} \mu_S^{0.105} \mu_n^{-0.210} \mu_W^{-0.240} = -0.0065,$$

$$\left(\frac{\partial^2 X}{\partial s^2}\right)_\mu = -1.59 \times 0.0940 b^{0.980} \mu_Y^{0.055} \mu_S^{-1.895} \mu_n^{-0.210} \mu_W^{-0.240} = -17863.8,$$

$$\left(\frac{\partial^2 X}{\partial n^2}\right)_\mu = 1.59 \times 0.2542 b^{0.980} \mu_Y^{0.055} \mu_S^{0.105} \mu_n^{-2.210} \mu_W^{-0.240} = 936.4,$$

$$\left(\frac{\partial^2 X}{\partial w^2}\right)_\mu = 1.59 \times 0.2975 b^{0.980} \mu_Y^{0.055} \mu_S^{0.105} \mu_n^{-0.210} \mu_W^{-2.240} = 0.0616,$$

where μ_Y, μ_S, μ_n, and μ_W denote the means of Y, S, n, and W, respectively. From Eq. (3.4.36),

$$E[X] \approx 3.32 + 0.5(-0.0065 \times 1.2^2 - 17863.8 \times 0.0004^2 + 936.4 \times 0.0058^2 + 0.0616 \times 1.6^2) = 3.41 \text{ m},$$

and, from Eq. (3.4.37),

$$\text{Var}[X] \approx 0.0325^2 \times 1.2^2 + 0.0006^2 \times 0.0004^2 + (-23.21)^2 \times 0.0058^2 + (-0.1989)^2 \times 1.6^2 = 0.1209 \text{ m}^2,$$

which yields an approximate standard deviation of 0.35 m. These approximations provide rather accurate estimates of the mean and standard deviation of scour depth as determined from simulation.

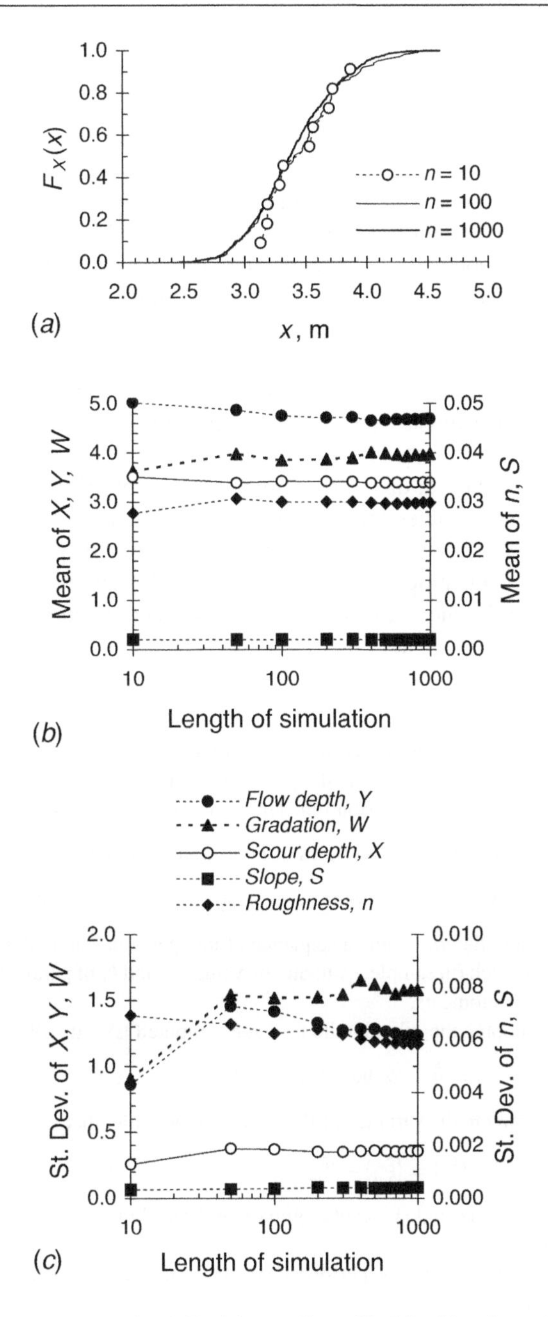

Fig. 8.3.1 Simulation of scour depth X: (a) sampling cdf of X, (b) estimated mean, and (c) standard deviation of involved variates.

8.3.2 Sampling statistics

The analytical solution to certain sampling problems is sometimes unmanageable. For example, a design value is often determined as the qth quantile of the design variate; therefore, one must assess the prediction limits of these values from a sample of observations. The standard error of estimates are not known for some distributions applicable to engineering systems, and the asymptotic results cannot be applied to small samples.

However, Monte Carlo experiments can be performed to determine these statistics, which depend on the probabilistic model used to fit the data, the method used for estimating its parameters, and the sample size. Further, one can use simulation to assess how the presence of either systematic or random errors in observed data can influence the identification and the estimation of the probabilistic model fitted to these data. Other sampling problems are also approached by simulation. For example, the sampling distributions of the coefficients of variation and skewness are used in developing the index-flood method in regionalizing flood flows in rivers of a homogeneous region. Because of the difficulties in obtaining analytical solutions, it must be determined via simulation for most extreme value distributions used in hydrological practice, such as the GEV, the TCEV, and the log-Pearson Type III distributions.[14]

The results of Monte Carlo experiments aimed at determining the sampling properties of statistical estimates are often generalized by introducing best-fit equations to relate the quantities involved. For example, one can search for the relationship of the standard error of the qth quantile estimate to the probability of nonexceedance, sample size, and parameter values for a specified distribution and a given method for parameter estimation. When performing simulation runs for this purpose, one must define the role of different quantities involved and design these runs to explore all possible values jointly taken by these quantities. The results should correspond to the deterministic formulas obtained from field and laboratory experiments on physical and engineering systems.

Example 8.17. Standard error of Gumbel quantile estimates. One can perform Monte Carlo experiments to evaluate the performance of the qth quantile unbiased L-moment estimator for the Gumbel distribution. These involve the following steps:

(1) Generate n size-m samples of EV1 random numbers as

$$x_{ij} = b - \alpha \ln(-\ln u_{ij}), \quad \text{for } j = 1, \ldots, m, \quad \text{and} \quad i = 1, \ldots, n,$$

where the u_{ij} denote a sequence of independent standard uniform random numbers.

(2) For each ith sample, estimate the values $\hat{\alpha}_i$ and $\hat{b}_i$ of parameters α and b by the method of L-moments.

(3) Compute the qth quantile for each estimated EV1 distribution as

$$\hat{\xi}_{qi} = \hat{b}_i - \hat{\alpha}_i \ln(-\ln q), \quad \text{for } i = 1, \ldots, n.$$

(4) Compute the variance of the estimated quantile, that is,

$$\text{Var}[\hat{\xi}_q] = \langle \hat{\xi}_{qi}^2 \rangle - \langle \hat{\xi}_{qi} \rangle^2,$$

where the angle brackets signify that the arithmetic mean is calculated over the sample of n.

(5) Other measures of performance can be evaluated, such as the bias of the estimated quantile,

$$\langle \hat{\xi}_{qi} \rangle - [b - \alpha \ln(-\ln q)]$$

and its root mean square error, that is,

$$\left\langle \{\hat{\xi}_{qi} - [b - \alpha \ln(-\ln q)]\}^2 \right\rangle^{1/2}.$$

[14] See, for example, Chowdhury et al. (1991) and Vogel and McMartin (1991).

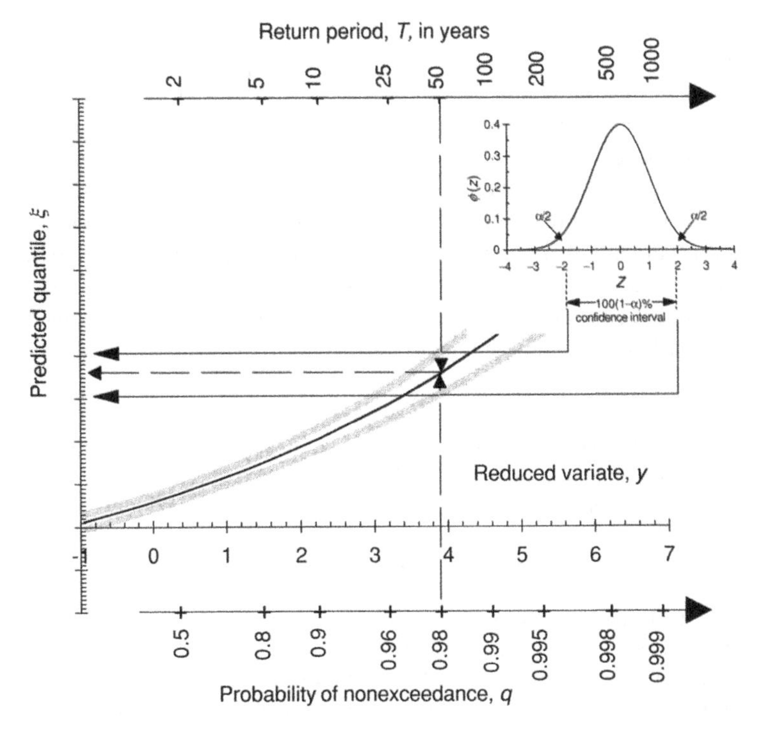

Fig. 8.3.2 Confidence limits of quantile estimates.

The resulting variance of the estimated quantile can be used to check the validity of the formula based on asymptotic theory, that is,

$$\mathrm{Var}\left[\hat{\xi}_q\right] = \frac{\alpha^2}{m-1}\left[\left(1.1128 - \frac{0.9066}{m}\right) - \left(0.4574 - \frac{1.1722}{m}\right)y + \left(0.8046 - \frac{0.1855}{m}\right)y^2\right],$$

where $y = \ln(-\ln q)$ denotes the reduced variate.[15] Note that the variance of the quantile estimates is independent of the location parameter. To evaluate the prediction limits of quantile estimate, one can assume the estimated quantile to be a random variable Z distributed as

$$N(\hat{\xi}_q,\ \mathrm{Var}[\hat{\xi}_q]),$$

as sketched in Fig. 8.3.2.

This method can also be used to approach more sophisticated sampling problems, such as the effects of random errors in data on the bias and standard error of Gumbel quantile estimates.[16]

8.3.3 Simulation of time- or space-varying systems

Simulation is often used to study the dynamics of time- or space-varying systems. A simulation model may be time (space)-sequenced or event-sequenced. For example, when

[15] See Downton (1966) and Phien (1987).

[16] See, for example, Rosso (1985).

simulating the logistic population growth of Example 8.3, one either assumes that births and deaths are a sequence of events occurring at equispaced times, or considers their occurrences to be random points on the time axis. In a time- or space-sequenced model, a fixed time or distance interval, Δt or Δu, is selected, and the model examines the state of the system at successive time or space intervals. Events of interest can sometimes be unnoticed in a sequenced study. For example, consider a reservoir that receives precipitation during Δt, from which the same amount of water is lost by seepage and evaporation. If one examines the initial and terminal reservoir states corresponding to the beginning and end of the interval Δt, there is no evidence of precipitation, or leakage, or evaporation. An event-sequenced simulation considers a sequence of events, like storms, floods, hurricanes, and earthquakes, and the interarrival time of these events is modeled as a random variable.

Example 8.18. Simulation of streamflows. Kottegoda (1970) simulated 20 sets of 25-year sequences of monthly runoff to the Elan Valley reservoirs in Wales using the following recursive equation:

$$\frac{X_t - \mu_\tau}{\sigma_\tau} = \rho_1 \frac{X_{t-1} - \mu_{\tau-1}}{\sigma_{\tau-1}} + (1 - \rho_1)^{1/2}\eta_t,$$

where X_t denotes flow in month t and for the corresponding calendar month τ, μ_τ and σ_τ are the mean and standard deviations.[17] The lag 1 serial correlation coefficient is denoted by ρ_1 and η_t is a generated random variate with a mean of zero and appropriate higher moments. The 12 historical (sample) mean monthly flows, $\bar{x}_\tau$, $\tau = 1, 2, \ldots, 12$, are distributed as

$$\bar{x}_\tau \sim N\left(\mu_\tau, \frac{\sigma_\tau^2}{n}\right)$$

in large samples. Here $n = 50$ is the number of years of data used in the study. Further, the 95% confidence interval for μ_τ is defined by

$$\Pr\left[\bar{x}_\tau - 1.96\frac{\sigma_\tau}{\sqrt{n}} \le \mu_\tau \le \bar{x}_\tau + 1.96\frac{\sigma_\tau}{\sqrt{n}}\right] = .95.$$

These are shown in Fig. 8.3.3.

Also shown are the monthly means estimated from each of the 20 simulated sets. Their large sample distribution is $\bar{s}_\tau \sim N(\mu_\tau, \sigma_\tau^2/n)$. The probability that the means of the simulated sets lie within the above mentioned 95% confidence interval is given by

$$\int_a^b \frac{1}{\sqrt{2\pi}\sqrt{2\sigma_\tau^2/n}} \exp\left(-\frac{x^2}{2 \times 2\sigma_\tau^2/n}\right) dx$$

where $a = -1.96\sigma_\tau/\sqrt{n}$ and $b = 1.96\sigma_\tau/\sqrt{n}$. The probability is the same if standardized units are used, for which the limits are

$$a^* = -\frac{1.96\sigma_\tau/\sqrt{n}}{\sqrt{2\sigma_\tau^2/n}} \approx -\sqrt{2}$$

and (similarly) $b^* \approx \sqrt{2}$. Thus, the proportion of estimated monthly means of simulated flows expected within the 95% confidence limits of the historical monthly means is 0.84. This is the area under the standard normal pdf between $-\sqrt{2}$ and $\sqrt{2}$. However, in these simulations the μ_τs in the recursive equation have been replaced by the $\bar{x}_\tau$s. Then the $(\bar{x}_\tau - \bar{s}_\tau)$s tend to

[17] $\tau = t \pmod{12}$ where $\tau = 1, \ldots, 11, 0$ for January, ..., November, December. Sample estimates were used for all parameters.

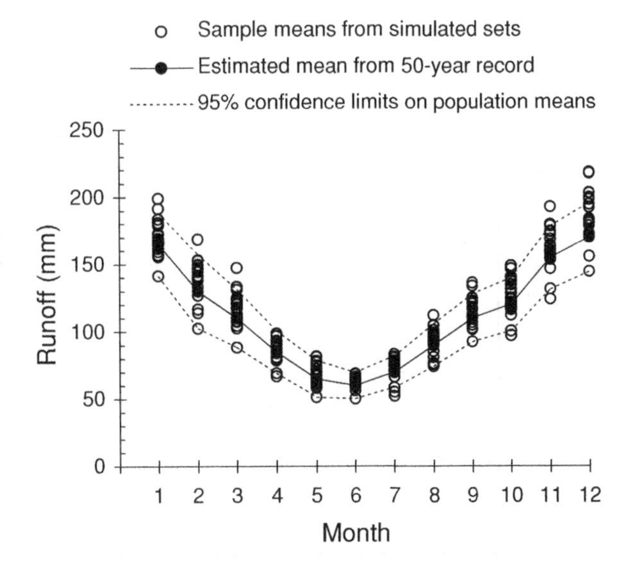

Fig. 8.3.3 Simulation of mean monthly inflows to the Elan Valley reservoirs in Wales, United Kingdom.

have a smaller variance than $N(0, 2\sigma_\tau^2/n)$ variates. The proportion of values within the 95% confidence limits should therefore be larger than 0.84.

In the time-sequenced simulated sets, the observed number of values of monthly means falling within the confidence limits was found to be 0.87. The closeness of these proportions suggests that the model assumptions are validated.

Example 8.19. Simulation of groundwater heads. In the design of structures such as dams and weirs, increasing use is being made of the simulation of seepage and groundwater flow. The hydraulic conductivity of geologic soils depends on many physical factors, including particle size and distribution and the porosity and shapes of particles and their arrangements. Because of the inherent nonuniformity of porous media and because the exact nature of the spatial distribution of hydraulic conductivity is unknown, there is uncertainty in the prediction of hydraulic head. Random variations in hydraulic conductivity and other soil properties and their effects on hydraulic heads can be studied by Monte Carlo methods.

The fundamental equation applicable is

$$\frac{\partial}{\partial x}\left[K(x)\frac{\partial \Phi}{\partial x}\right] = 0,$$

where $\Phi(x)$ is the hydraulic head and $K(x)$ is the hydraulic conductivity at any point x. Kottegoda and Katuuk (1983) applied random walk methods for solving specific boundary value problems, discussed previously. Steps taken by flow particles in a medium represent a random walk between two boundaries. When a particle strikes a boundary, its motion is terminated or it is reflected back; thus its subsequent movement depends on the boundary conditions. Initially, the case of the absorbing barrier was studied.

In the simulation it is assumed that there is simultaneous movement of particles from all interior points in a region for which calculations are required. The probability of movement of a particle in any direction depends partly on the permeability in that direction.

When groundwater flow occurs under steady-state conditions, the potential head at an interior point P in a region is found by making a hypothetical particle wander from one point in the region on a preselected grid until it finally hits a boundary at point j. Once the

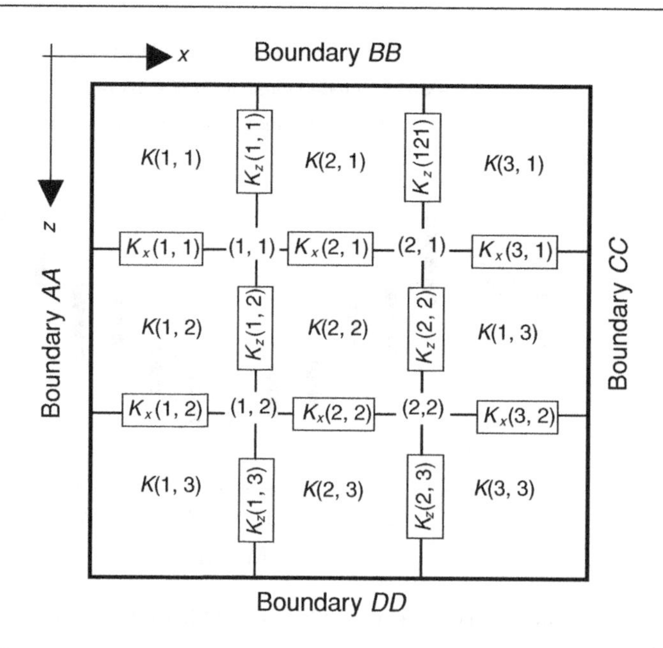

Fig. 8.3.4 Hydraulic conductivity for representation of two-dimensional flow. (x, z) represents a node; $K(x, z)$ is conductivity in block (x, z); $K_x(x, z)$ is conductivity in x direction from node (x, z) ; $K_z(x, z)$ is conductivity in z direction from node (x, z).

random walk of the particle is terminated, a boundary value $H(j)$ is recorded. Here, $H(j)$ is the hydraulic head at point j on the boundary. If n such walks are performed from point P, the Monte Carlo solution for the expected potential head at point P, $\Phi(P)$, is obtained from the sum of the effective boundary values divided by n; that is,

$$\Phi(P) = n^{-1} \sum_{j=1}^{n} H(j).$$

Figure 8.3.4 shows the variations of conductivities $K(x, z)$ between blocks in two dimensions within the study area.

The conductivities in the x- and z-directions are

(1) x-direction: $K_x(1, 1) = 1/2[K(1, 1) + K(1, 2)]$; $K_x(2, 1) = 1/2[K(2, 1) + K(2, 2)]$.
(2) z-direction: $K_z(1, 1) = 1/2[K(1, 1) + K(2, 1)]$; and $K_z(1, 2) = 1/2[K(2, 1) + K(2, 2)]$.

The probabilities of particle movement from a node in each of four directions are calculated as follows. Let

$$\sum K = K_x(1, 1) + K_x(2, 1) + K_z(1, 1) + K_z(1, 2).$$

The probability of moving to the right from point (1, 1), say, is $\Pr[R] = K_x(2, 1)/\sum K$. Probabilities of particles moving to the left, upward, and downward, denoted, respectively, by $\Pr[L]$, $\Pr[U]$, and $\Pr[D]$, are similarly calculated. The actual movement of a particle depends on the random number, U_t, generated at time t from a (0, 1) uniform distribution. If, for instance,

$$0 \leq U_t < \Pr[R],$$

a particle moves to the right; if

$$\Pr[R] \leq U_t < \Pr[R] + \Pr[D],$$

Table 8.3.1 Monte Carlo solution for potential head: Hydraulic heads of two-dimensional flow, matrix 5 × 5 blocks with n_x random numbers; uniform medium ($\sigma_Y = 0$)

zx:	0	1	2	3	4	5
$n_x = 500$						
0	100	100	100	100	100	100
1	100	95	92	82	73	50
2	100	89	84	75	63	50
3	100	86	78	73	59	50
4	100	76	65	59	56	50
5	100	50	50	50	50	50
$n_x = 1200$						
0	100	100	100	100	100	100
1	100	94	92	84	75	50
2	100	90	83	76	65	50
3	100	86	77	68	57	50
4	100	75	67	60	55	50
5	100	50	50	50	50	50

the particle moves downward; if

$$\Pr[R] + \Pr[D] \leq U_t < \Pr[R] + \Pr[D] + \Pr[L],$$

the particle moves to the left; in all other cases, the particle moves upward.

Table 8.3.1 shows the Monte Carlo solutions at each of the nodes for matrix blocks of size 5 × 5. The Monte Carlo solution for an internal node is calculated by weighting each of the four sets of boundary heads by the proportion of particles reaching the boundary and adding the weights. After obtaining the solutions for each of the internal nodes, hydraulic heads are calculated by taking the weighted average of the Monte Carlo solutions at the surrounding points. The weights are in fact the probabilities $\Pr[R]$, $\Pr[L]$, $\Pr[U]$, and $\Pr[D]$ determined from the internal conductivities as shown above.

The first set of results is for $n_x = 500$, where n_x is the total of the random numbers used. In the second set n_x is increased to 1200. It is noted that corresponding values on either side of the leading diagonal become closer to each other as n_x increases. This is due to the symmetrical conditions imposed. However, when the standard deviation of the conductivities is increased there is more divergence in the hydraulic heads. The space-sequenced simulation can be extended to three-dimensional cases with different boundary conditions and lognormally distributed conductivities. Note that the values of hydraulic conductivity depend on the spatial scale, so that appropriate values must be chosen for the size of each block. The concept of scaling introduced in Section 7.3.1 to represent the variability of extreme storm rainfall with duration could be applied for this purpose by substituting space for time.

Example 8.20. Storm clustering. The temporal variability of rainfall intensity during a storm at a point in space depends on the combined effect of two physical processes. One is the movement of the storm field over the area of rain, and the other is the "birth" and "death" (arising and passing) of storm cells. Point rainfall intensity $X(t)$ at time t can be modeled as the sum of a random number N of rectangular pulses of length Y and intensity Z, which are randomly displaced on the time axis with an interarrival time of W (see Fig. 8.3.5a).

Each pulse represents a cell with a lifetime of Y, which delivers an amount of water of YZ. Rainfall intensity is found as

$$X(t) = \sum_{i=1}^{N} K_i(t) Z_i,$$

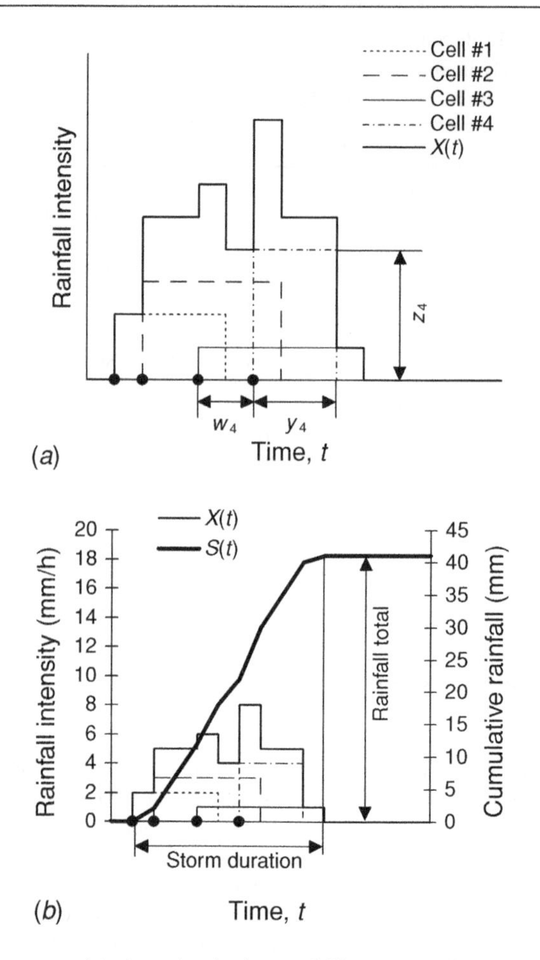

Fig. 8.3.5 Storm structure: (a) clustering in time and (b) storm profile.

with

$$
\begin{aligned}
K_i(t) &= 1, \quad \text{for } \sum_{j=1}^{i} W_j \le t \le Y_i + \sum_{j=1}^{i} W_j, \\
&= 0, \quad \text{elsewhere,}
\end{aligned}
$$

denoting an indicator function for the ith cell with duration Y_i and intensity Z_i to be active at time t. The storm profile $S(t)$ is defined as the temporal evolution of cumulative rainfall and results in a monotonic nondecreasing function shown in Fig. 8.3.5b. In this event-sequenced simulation of storm clustering, one can assume N to be either a Poisson or a geometric variate, whereas the Y's, Z's, and W's are usually assumed to be exponentially distributed random variables.[18] However, to preserve the scaling properties of extreme storms with varying durations (see Subsection 7.3.1) one may consider Pareto-type distributions to model the Ys, Zs, and Ws.

8.3.4 Design alternatives and optimal design

Although Monte Carlo simulations may be limited by the constraint of computational capability, the application of simulation methods is perhaps the most widely used method

[18] See, for example, Burlando and Rosso (1993).

for evaluating alternative designs. The reason for the popularity of the approach is its mathematical simplicity and versatility, which makes it possible to estimate both physical and economic responses of various alternatives. To achieve the optimal design of a system, one begins with a trial design and continues to obtain system responses with different trials until the optimal system configuration is achieved. The continuously improving performance of digital computers and the availability of simulation-oriented software facilities contribute to the wide use of simulation methods.

Example 8.21. Chemical sludge conditioning. Coagulation of the solids dispersed in sludge from a wastewater treatment plant increases the rate of water removal by filtration or air-drying. The coagulating process is called *chemical conditioning*, and a common conditioning chemical for wastewater sludge is ferric chloride or $FeCl_3$. The requirement of this conditioner is expressed as the percentage X of the pure chemical to the weight of the solids fraction on a dry basis. This can be evaluated as

$$X = 1.08 \times 10^{-4} \frac{YZ}{100 - Z} + 1.6 \frac{V}{100 - V},$$

where Y denotes alkalinity of the sludge moisture in mg/L of $CaCO_3$; and Z, V, and $1 - V$ are, respectively, the percentage moisture, volatile matter, and fixed solids in the sludge.[19] To improve the process, one can wash out a share of alkalinity with water of low alkalinity by using repeated washings in multiple tanks. This type of treatment is called *elutriation.* In this case, alkalinity is reduced to

$$Y' = \frac{Y(R - 1) + WR(R^n - 1)}{R^{n+1} - 1},$$

where W and R are, respectively, the alkalinity of washing water and its proportion to wastewater sludge, and n is the number of tanks used by countercurrent operation. One must evaluate the conditioner requirements for unelutriated and elutriated sludge for $Y \sim N(3000 \text{ mg/L}, 1000 \text{ mg}^2/\text{L}^2)$, $Z \sim$ uniform $(85, 95)$, and $V \sim$ uniform $(40, 50)$, $R \sim$ lognormal $(3, 0.5^2)$, and $W \sim$ uniform (15 mg/L, 25 mg/L). The nominal values are computed by considering the mean values of these variates. Thus,

$$x_{\text{nom}} = 1.08 \times 10^{-4} \frac{3000 \times 90}{10} + 1.6 \frac{45}{55} = 4.23\%$$

of $FeCl_3$ on a dry basis for unelutriated sludge. If sludge is elutriated in two tanks by countercurrent operation, alkalinity reduces to

$$y'_{\text{nom}} = \frac{3000(3 - 1) + 20 \times 3(3^2 - 1)}{3^3 - 1} = 249 \text{ mg/L}.$$

Thus,

$$x'_{\text{nom}} = 1.08 \times 10^{-4} \frac{249 \times 90}{10} + 1.6 \frac{45}{55} = 1.55\%$$

of $FeCl_3$ on a dry basis are nominally required for elutriated sludge. Simulations are performed to evaluate the probability that elutriation can effectively reduce the requirement of conditioner, that is, $\Pr[X > X']$. The cdfs of X and X' obtained from 1000 simulations are shown in Fig. 8.3.6 on a normal probability plot, and the corresponding sampling statistics are reported in Table 8.3.2.

The probability that elutriation in two tanks by countercurrent operation reduces the requirement of conditioner is higher than .99.

Example 8.22. Reservoir capacity. Consider a reservoir designed to meet a given minimum monthly demand of water supply, δ_τ, under the operating rule shown in Fig. 8.3.7. Suppose the maximum demand is d_τ.

[19] See Genter (1946).

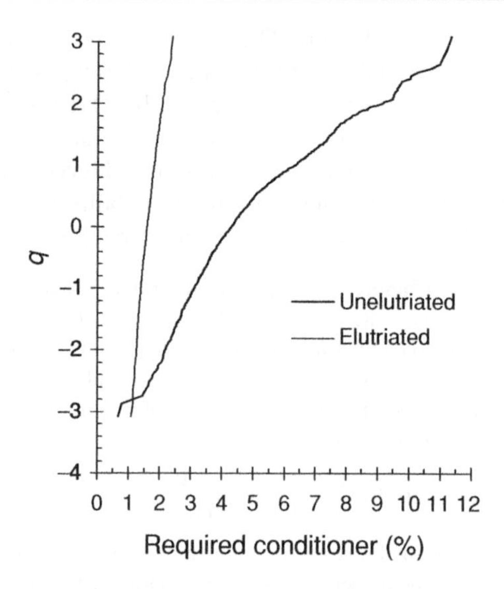

Fig. 8.3.6 Sampling cdfs of required conditioner from 1000 simulations of chemical sludge conditioning on a normal probability plot.

Table 8.3.2 Sampling statistics from 1000 Monte Carlo simulations of chemical sludge conditioning

	Unelutriated sludge	Elutriated sludge	
	Ferrite chloride requirement, X' (%)	Reduced alkalinity, Y' (mg/L)	Ferrite chloride requirement, X' (%)
Mean	4.59	252.9	1.59
Standard deviation	1.67	85.8	0.20
Coefficient of variation	0.36	0.34	0.13
Coefficient of skewness	1.04	0.30	0.49
Nominal values	4.23	249.00	1.55

The total amount of water available in month t is S_t+X_t, with X_t denoting monthly inflow into the reservoir, and S_t initial storage in the reservoir at the beginning of that month.[20] The draft or release R_t is obtained as

$$(a)\ \ R_t = k(S_t + X_t) = \delta_\tau, \quad \text{if } S_t + X_t \le d_\tau,$$
$$(b)\ \ R_t = \delta_\tau + (d_\tau - \delta_\tau)(S_t + X_t - d_\tau)/c, \quad \text{if } d_\tau < S_t + X_t < d_\tau + c,$$

and

$$(c)\ \ R_t = S_t + X_t - c, \quad \text{if } S_t + X_t \ge d_\tau + c,$$

where c denotes the capacity of the reservoir. It is assumed that δ_τ equals d_τ from September to December, $0.85d_\tau$ from May to August, and $0.70d_\tau$ from January to April. Because storage cannot exceed the capacity c of the reservoir, an amount at least equal to $S_t + X_t - c$ must be

[20] $\tau = t \pmod{12}$ where $\tau = 1, \ldots, 11, 0$ for January, ..., November, December.

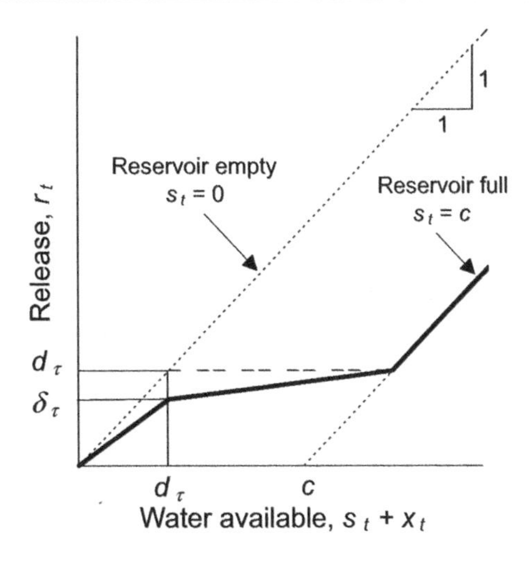

Fig. 8.3.7 Operating rule for a reservoir.

released in that month. Accordingly, one can completely satisfy the demand d_τ and release downstream a surplus of water of $S_t + X_t - c - \delta_\tau$ under rule (*c*) leaving the reservoir full. Conversely, if the minimum demand, δ_τ, exceeds the available water $S_t + X_t$, one delivers all the available water to partially satisfy the demand under rule (*a*) leaving the reservoir empty. Rule (*b*) is applied when the available water exceeds the demand; the release is increased proportionally.

Suppose that inflow X_t is modeled using the recursive stochastic equation of Example 8.18 with $\rho_1 = 0.85$, $\eta_t \sim N(0, 0.42^2)$; also, the mean μ_τ and standard deviation σ_τ of $X(t)$ for the calendar month t are those shown in Fig. 8.3.8, where the demand d_τ of water supply is also indicated.

If a reservoir is to be designed with an active volume c, one can use the above equations to simulate the operation of the reservoir for each t in a period of given length, say, $12n$. For example, a portion of the trajectory resulting from the simulation of $n = 100$ years of operation of a reservoir with capacity of 300 mm is shown in Fig. 8.3.9 with monthly inflow, X_t, storage, S_t, release, R_t, and deficit, Y_t, which is the shortage in a month because insufficient water is available from inflow and initial storage. All data are in equivalent runoff of the drainage basin impounded by the reservoir.

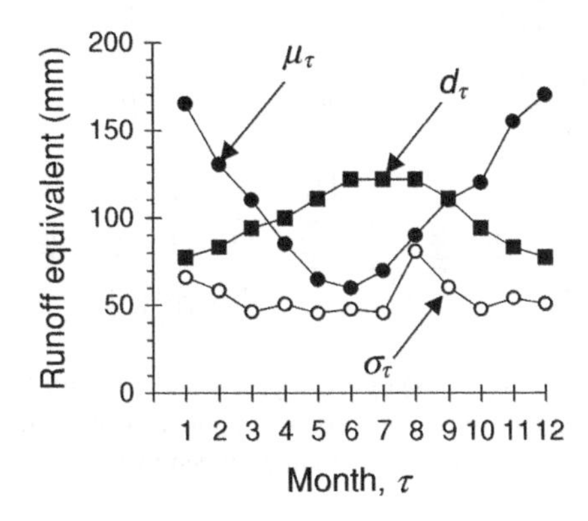

Fig. 8.3.8 Mean μ_τ and standard deviation σ_τ of monthly inflows to a reservoir, and maximum monthly demand d_τ of water supply from the reservoir.

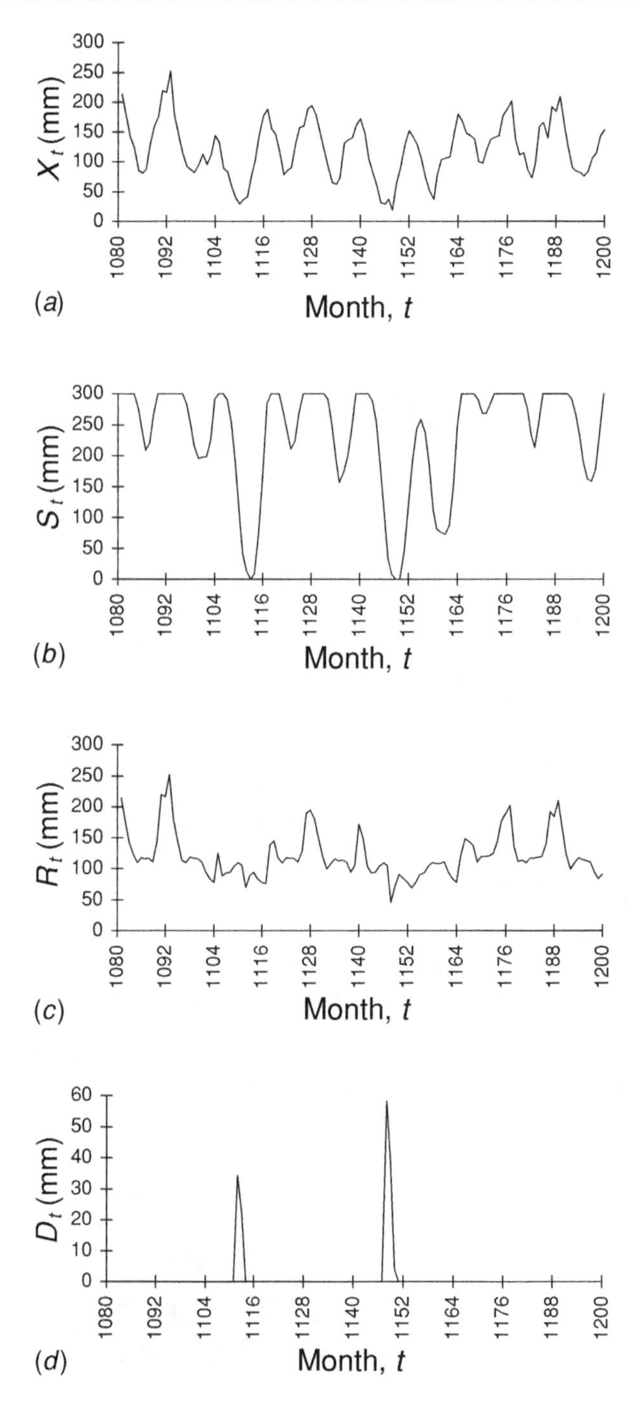

Fig. 8.3.9 Results (last 10 years) of a 100-year simulation run of monthly (*a*) inflow, (*b*) storage, (*c*) release, and (*d*) deficit for the target demand under the normal operating rule. All data are in equivalent runoff from the drainage basin impounded by the reservoir, which has an active volume of 300 mm for this run.

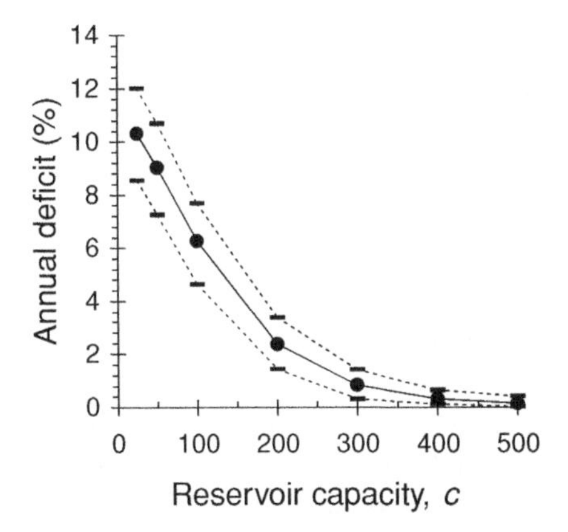

Fig. 8.3.10 Annual deficit as a percentage of annual target demand as averaged from 10 simulation runs of reservoir operation for a period of 100 years. The upper and lower bounds show the extreme values out of 10 runs.

Simulation runs with different values of c are then performed to achieve the optimal reservoir capacity needed to meet the demand. For example, if one wishes to reduce the average annual deficit to a given percentage of the demand, it seems straightforward to increase c. However, the simulation runs shown in Fig. 8.3.10 indicate that the rate of reduction decreases rather rapidly; for example, the amount of reduction that can be achieved by a 20% increase of c, say, from 500 to 600 mm, is only 0.13% on average, resulting in a reduction of the annual deficit from 13.8 to 12.1 mm.

The total length of deficit in a year is shown in Fig. 8.3.11. It is seen that increasing c from 500 to 600 mm reduces the expected number of months of water shortage from about 8 to 7 in 25 years.

The required optimal capacity is generally obtained by introducing economic analysis into the simulation. For a specified lifetime of the structure, the initial costs of construction and operation costs for different reservoir capacities can be compared with the present value of the benefit. Note that operating rules for large reservoirs should take account of commonly applied multipurpose uses.

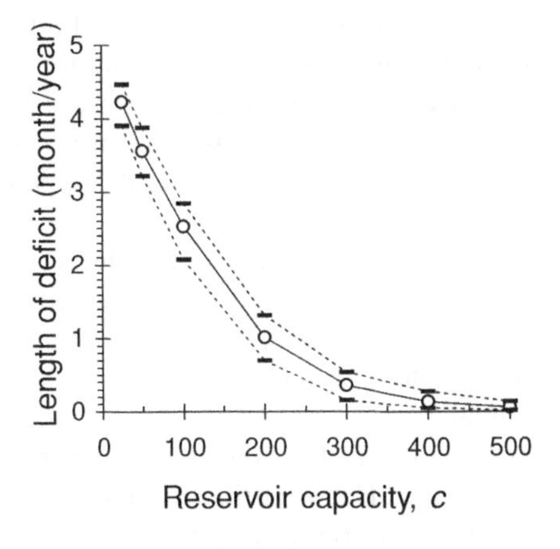

Fig. 8.3.11 Annual length of deficit as averaged from 10 simulation runs of reservoir operation for a period of 100 years. The upper and lower bounds show the extreme values out of 10 runs.

8.3.5 Summary of Section 8.3

The purpose of this section is to highlight practical applications of simulation. In particular, details of how simulation can be beneficial in design and research are given. A wide range of applications is included here. We now discuss sensitivity and uncertainty analysis in relation to simulation.

8.4 SENSITIVITY AND UNCERTAINTY ANALYSIS

Computer-based modeling and simulation are used in many areas of application, particularly in risk assessment and reliability engineering, which is the subject of the next chapter. In this chapter we have provided examples and problems based on mathematical models of a specified physical system. The concept of a mathematical model was initially discussed in Chapter 2. Basically it is represented by a set of equations, input factors, and parameters that characterize the process under investigation. From the perspective of simulation, the model consists of three parts, ordered sequentially as inputs, simulator, and outputs. The inputs can be some parameters that describe the physical characteristics of the system studied or the general description of the system. They are needed for the simulator to produce a sequence of outputs. As already discussed, models and inputs are subject to uncertainties. The sources of uncertainty include errors of measurement, representation of unpredictable or stochastic events, and misconceptions of the system studied.

The main purpose in the analysis of a physical system and its simulation is usually for making decisions related, for example, to the design and optimization of performance or risk and reliability assessment of the system. If modifications and further analysis are required, a commonly used method is *sensitivity analysis*. This activity enables one to investigate whether, for instance, the uncertainties in the input are related to those of the output. The analysis can also be used to determine how the variation in the output of a model can be attributed to different sources or factors of variation. At a basic level, one can vary each model parameter, over a reasonable range of uncertainty, and note the relevant change in the output for a given input. Some examples are changes in flow velocities and hydraulic heads. In this way one can also determine the relative sensitivities of the model parameters. Also, if the data are not sensitive to variations in the parameters, additional investigations are deemed to be necessary. Furthermore, one can ascertain the uncertainties in model structures, specifications and the assumptions made. Electronic spreadsheets, when used with care and acquired expertise, provide a valuable aid in the various types of activity (because sensitivity analysis seems to be an overcharged term). A useful feature in popular packages is a "What if?" type of exercise with Monte Carlo simulation for sensitivity analysis.

Consequently, different levels of acceptance by a decision maker may be associated with different types of uncertainty. It will also lead to increased confidence in the model, which good practice demands. One can also see that sensitivity analysis is closely associated with uncertainty analysis that attempts to quantify the uncertainty associated with the input, simulator, and the output.

In recent years there has been a formal recognition of two types of uncertainty. Firstly, there is *aleatory uncertainty*. This is attributed to the natural or unpredictable variation in the performance of a system. Because aleatory uncertainty is inevitable, one can only quantify it, and hence it is also called irreducible uncertainty, inherent uncertainty, or stochastic uncertainty. Probability theory is most commonly applied in this case although under generalized information theory, alternatives such as fuzzy set theory, as discussed in Chapter 2, exist. It is assumed that sufficient experimental data are available to apply the probabilistic concepts. In a supplementary way, methods of Monte Carlo simulation, described in this chapter, are sometimes used to study the interaction of aleatory variables with changes in time of a dynamic process.

The other type is known as *epistemic uncertainty*. It arises from a lack of knowledge or information regarding one or more aspects of the modeling process. This may refer, for example, to a physical parameter that may be insensitive or may have insufficient data to support it. We assume that the system can be conceptualized, but this may be the main source of error. Subsequently, if there is an increase in knowledge, the uncertainty in the response of the system is reduced. Thus this type is also called reducible uncertainty, subjective uncertainty, and state-of-knowledge uncertainty. It pertains to our level of ignorance. Probability theory has provided a basis to model both types of uncertainty from historical times. One notes, however, that the application of probability theory is not as straightforward as in the case of aleatory uncertainty. Investigations are often made through Monte Carlo simulations.

Without proper consideration of the sources of uncertainty, applications of probability theory may be faulty. Besides, these two types of uncertainty can be used in assessing data requirements and for discussions between experts and users.[21]

8.5 SUMMARY AND DISCUSSION OF CHAPTER 8

In this chapter we have discussed many aspects of simulation and associated design. Simulation provides a method of evaluation of the performance of a system for a given design and operating policies. It can be used as a supplement to physical experiments and sometimes as a replacement. One begins with a trial design and continues to obtain system responses with different trials. Thus, one can have knowledge of anticipated performances resulting from any design or operating rule. In the case of complex systems, such results may be impossible to obtain. However, in these cases, planning and screening of alternatives should be done as a prelude to the definition of boundaries and other details of an experiment. In relation to simulation, we also discuss sensitivity analysis and two types of uncertainty analysis with their implications in modeling and decision-making.

REFERENCES

General. The following references are given for further reading:

Ang, A. H-S., and W. H. Tang (1984). *Probability Concepts in Engineering Planning and Design, Vol. II: Decision, Risk and Reliability*, John Wiley and Sons, New York, pp. 562. See Chapter 5 on simulation with numerous practical examples.

Bratley, P., B. L. Fox, and E. L. Schrage (1987). *A Guide to Simulation*, 2nd ed., Springer-Verlag, New York. Recommended guidance on simulation.

Devroye, L. (1986). *Non-uniform Random Variate Generation*, Springer-Verlag, New York. An excellent evaluation of various techniques for nonuniform random number generation.

---

[21] See for example Oberkampf et al. (2004) for an assessment of the nature of uncertainties with some added challenge problems, and Helton (1999) for the application of uncertainty and sensitivity analysis to a waste isolation pilot plant. See also, Cacuci et al. (2005) for applications of sensitivity and uncertainty analysis, and Saltelli et al. (2004) for a guide to the use of sensitivity analysis.

Evans, M., N. Hastings, and B. Peacock (2000). *Statistical Distributions*, 3rd ed., John Wiley and Sons, New York. Excellent summaries for computer generation of 39 distributions.

Hammersley, J. M., and D. C. Handscomb (1964). *Monte Carlo Methods*, Methuen, London. Origins of simulation.

Knuth, D. E. (1981). *Seminumerical Algorithms, Vol. 2: The Art of Computer Programming*, 2nd ed., Chapter 3, Addison-Wesley, Reading, MA. Standard reference on algorithms.

Press, W. H., S. A. Teukolsky, W. T. Vetterling, and B. P. Flannery (1992) *Numerical Recipes in Fortran: The Art of Scientific Computing*, 2nd ed., Chapter 7, Random Numbers, Cambridge University Press, Cambridge. Essential reading for the programmer.

Additional references quoted in the text

Ahrens, J. H., and U. Dieter (1974). "Computer methods for sampling from gamma, beta, Poisson and binomial distributions," *Computing*, Vol. 12, pp. 223–246. Random number generation.

Beckman, P. (1971). *A History of π, the Monte Carlo Method*, Golden Press, Boulder, CO. Background to the scientific evaluation of the area of a circle of unit radius.

Burlando, P., and R. Rosso (1993). "Stochastic models of temporal rainfall: Reproducibility, estimation and prediction of extreme events," in *Stochastic Hydrology and Its Use in Water Resources Systems Simulation and Optimization*, edited by J. D. Salas, R. Harboe, and J. Marco-Segura, pp. 137-173, Kluwer, Dordrecht. Storm rainfall modeling in continuous time.

Cacuci, D. G., M. Ionescu-Bujor, and M. Navon (2005). *Sensitivity and Uncertainty Analysis, Vol. II: Applications to Large-Scale Systems*, Chapman and Hall, London. Model assessments.

Chowdhury, J. U., J. R. Stedinger, and Li-Hsiung Lu (1991). "Goodness-of-fit tests for regionalized extreme value flood distributions," *Water Resour. Res.*, Vol. 27, pp. 1765–1776. Standard error of estimates from extreme value distributions.

Downton, F. (1966). "Linear estimates of parameters of the extreme value distribution," *Technometrics*, Vol. 8, pp. 3–17. Standard error of quantile estimates for the EV1 distribution.

Genter, A. L. (1946). "Computing coagulant requirements in sludge conditioning," *Trans. Am. Soc. Civ. Eng.*, Vol. 111, p. 641. Sludge engineering.

Gradshteyn, I. S., and I. M. Ryzhik (1994). *Tables of Integrals, Series and Products*, 5th ed. Academic Press, New York. Advanced math: reference for Example 8.6.

Hammersley, J. M., and K. W. Morton (1956). "A new Monte Carlo technique antithetic variates," *Proc. Camb. Phil. Soc.*, Vol. 52, pp. 449–474. Variance reduction in antithetic variables.

Helton, J. C. (1999). "Uncertainty and sensitivity analysis in performance assessment for the Waste Isolation Pilot Plant," *Comput Phys. Commun.*, Vol. 117, pp. 156–180. Useful application.

Johnson, P. A. (1992). "Reliability-based pier scour engineering," *J. Hydraul Eng., Div., ASCE*, Vol. 118, pp. 1344–1358. Risk assessment in pier design.

Jöhnk, M. D. (1964). "Erzeugung von betaverteilten und gammavertleiten zufallszahlen," *Metrika*, Vol. 8, pp. 5–15. Original rejection method for generating beta and gamma variates.

Kottegoda, N. T. (1970). "Statistical methods of river flow synthesis for water resources assessment," *Proc. Inst. Civ. Eng.*, Paper 7339S, Supl. XVIII. Includes simulation techniques for reservoir design.

Kottegoda, N. T., and G. Katuuk (1983). "Effect of spatial variation in hydraulic conductivity on groundwater flow by alternate solution techniques," *J. Hydrol.*, Vol. 65, pp. 349–362. Simulation of groundwater movement.

Oberkampf, W. L., J. C. Helton, C. A. Joslyn, S. F. Wojktiewicz, and S. Ferson (2004). "Challenge problems: Uncertainty in system response given uncertain parameters," *Reliab. Eng. Syst. Saf.*, Vol. 85, pp. 11–19. Discussion on aleatory and epistemic uncertainty and the role of sensitivity analysis.

Phien, H. N. (1987). "A method of parameter estimation for the extreme value type-I distribution," *J. Hydrol.*, Vol. 90, pp. 251–268. Standard error of quantile estimates.

Rand Corporation (1955). *A Million Random Digits with 100,000 Normal Deviates*, Free Press, Glencoe, IL. Table of random numbers

Rosso, R. (1985). "A linear approach to the influence of discharge measurement error on flood estimates," *J. Hydrol. Sci.*, Vol. 30, pp. 137–149. Effect of observational errors on estimated quantiles from the EV1 distribution.

Saltelli, A., S. Tarantola, F. Campolongo, and M. Ratto (2004). *Sensitivity Analysis in Practice: A Guide to Assessing Scientific Models*, John Wiley and Sons, New York. Useful practical guide for sensitivity analysis.

Vogel, R. W., and D. E. McMartin (1991). "Probability plot goodness of fit and skewness estimation procedures for the Pearson Type 3 distribution, *Water Resour. Res.*, Vol. 27, pp. 3149–3158. Standard error of quantile estimates.

von Neumann, J. (1951). "Various techniques used in connection with random digits," *U.S. Natl. Bur. Stand., Appl. Math. Ser.*, Vol. 12, pp. 36–38. Random number generation.

PROBLEMS

8.1. Flood regionalization. Develop an algorithm to generate random numbers to simulate the two-component extreme value distribution of flood flows with cdf

$$F_X(x) = \exp\left(-e^{-(x-b_1/\alpha_1)} - e^{-(x-b_2/\alpha_2)}\right).$$

Generate 100 samples each with 1000 items for given values of parameters and find the sampling probability distribution of the coefficients of variation and skewness. Let $\alpha_1 = 1.15\ \text{m}^3/\text{s}, b_1 = 10\ \text{m}^3/\text{s}, \alpha_2 = 2.20\ \text{m}^3/\text{s}, b_2 = 15\ \text{m}^3/\text{s}$. This method may be used to compare the theoretical and sampling variability of these coefficients as estimated from maximum annual flow data observed at different gauging stations in a region.

8.2. Percolation cluster. A fluid spreading randomly through a medium is represented by particles moving on a square grid, that is, a quadratic lattice, where each node is occupied by a pore with a probability of p and neighboring pores are connected by small capillary channels (see Fig. 8.P1).

A fluid injected into any given pore may only invade another adjacent pore that is directly connected to that pore through a capillary channel. The pores connected to the injection point form a cluster.

(*a*) Find the minimum probability, p_c, that a fluid injected into a site on the left edge of the lattice reaches the right edge for the structure shown with 16×16 nodes. This cluster is called the *spanning cluster* or the *percolation cluster*. Simulations on very large clusters showed that the probability of having a percolation cluster tends to zero as $n \to \infty$ and $p < 0.593$ [from R. M. Ziff

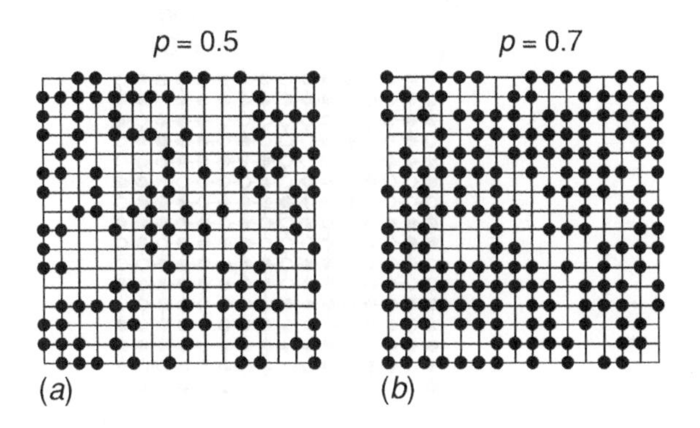

Fig. 8.P1 Quadratic lattice representing a porous medium where the nodes are occupied by pores with a probability of (*a*) $p = 0.5$ and (*b*) $p = 0.7$.

(1986), "Test of scaling exponents for percolation-clusters perimeters," *Phys. Rev. Lett.*, Vol. 56, pp. 545–548].

(*b*) The percolation probability $p_\infty(p)$ is defined as the probability that a fluid injected at a site, chosen at random, will wet infinite number of pores. Then, $p_\infty(p) = 0$ for $p \leq p_c$. Design a Monte Carlo experiment to show that the percolation probability vanishes as a power law near p_c; that is $p_\infty(p) \propto (p - p_c)^\alpha$ for $p > p_c$, and $p \to p_c$. The exponent α is $5/36$ for two-dimensional percolation and about $2/5$ for three-dimensional percolation.

(*c*) Design a Monte Carlo experiment to show that for large n the number of sites of the largest cluster increases as $\ln(n)$ for $p < p_c$; as n^2 for $p > p_c$; and as n^α for $p = p_c$; with a value of α of about 1.89 [from J. Feder (1988), *Fractals*, Plenum Press, New York, 283 p., Section. 7.2, "The infinite cluster at p_c"].

8.3. Invasion percolation. In a porous medium, oil is displaced by water, which is injected very slowly. Invasion percolation occurs when one neglects any pressure drops both in the invading fluid (water) and in the defending fluid (oil) because the capillary forces completely dominate the viscous forces, and the dynamics of the process is determined at the pore level. Simulation of the process on a lattice consists of following the motion of the water particle injected at a given site on the lattice as it advances through the smallest available pore, thus filling the pores with the invading fluid. As the invader advances, it traps regions of the defending fluid by completely surrounding regions of this fluid, that is, by disconnecting finite clusters of the defending fluid from the exit sites of the sample (see Fig. 8.P2).

For an $n \times n$ lattice, the following simulation algorithm describes invasion percolation [from D. J. Wilkinson and J. F. Willemsen (1983), "Invasion percolation: A new form of percolation theory," *J. Phys. A*, Vol. 16, pp. 3365–3376]:

(1) One assigns uniform(0, 1) random numbers to each site of the lattice.

(2) The injection for the invading fluid is assumed to occur at the upper-left corner and extraction for the defending fluid at the lower-right corner.

(3) Growth sites are defined as the sites belonging to the defending fluid and neighbors to the invading fluid.

(4) The invading fluid advances to the growth site that has the lowest random number.

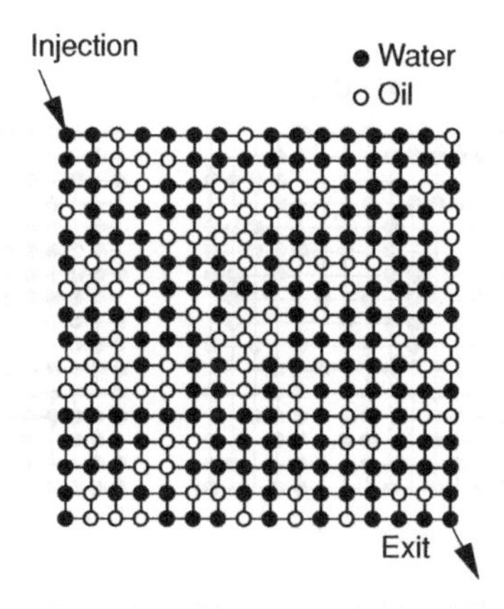

Fig. 8.P2 Quadratic lattice representing invasion percolation of water displacing oil.

(5) Trapping is obtained by eliminating the growth sites in regions completely surrounded by the invading fluid from the list of growth sites.
(6) The invasion process ends when the invading fluid reaches the exit site.

This algorithm is based on the fact that oil is incompressible (thus, water cannot invade trapped regions of oil). Using this simulation algorithm, show that the number of sites in the central $m \times m$ portion of the $n \times n$ lattice (with $m \ll n$) that are occupied by water is proportional to m^α with a value of α of about 1.89 [from M. M. Dias and D. Wilkinson (1986), "Percolation with trapping," *J. Phys. A*, Vol. 19, pp. 3131–3146]. This is, for instance, one origin of the phenomenon of residual oil.

8.4. Water storage. Water storage X in a large reservoir is modeled as a truncated normal variate with pdf

$$f_X(x) = \frac{1}{[\Phi(1) - \Phi(-2)]}\frac{1}{3\sqrt{2\pi}}\exp\left[-\left(\frac{x-6}{3}\right)^2\right]$$

$$= \frac{1.5389}{3\sqrt{2\pi}}\exp\left[-\left(\frac{x-6}{3}\right)^2\right],$$

for $0 \le x \le 9$ units, and zero elsewhere. Find $F_X(7)$ by Monte Carlo integration using 1000 simulation cycles, and compare this result with that obtained using tables of the normal distribution. What is the number of simulation cycles required to achieve a standard error of estimation not larger than 10% of the true value? Assume the mode of X as the maximum ordinate for the rectangular envelope of $f_X(x)$, with $0 \le x \le 7$ units.

8.5. Storm rainfall. The total amount of water Z delivered by a storm in a given location is evaluated as $Z = XY$ from independent duration X and average rainfall rate Y of a storm, with $X \sim$ lognormal (1.2 h, 6 h^2), and $Y \sim$ lognormal (10 mm/h, 100 mm^2/h^2). Assume that the number of storms in a year is a Poisson variate with a mean of 25. Using Monte Carlo simulation find the cdf of the annual maximum hourly storm depth, that is, the maximum amount of rainfall in a year which is delivered in the specified duration of 1 hour.

8.6. Storm rainfall. Solve Problem 8.5 under the assumption that the duration $X \sim$ lognormal (1.2 h, 6 h^2) of a storm and its average intensity $Y \sim$ lognormal (10 mm/h, 100 mm^2/h^2) are negatively correlated variates with $\rho_{X,Y} = -0.3$. Note that if two jointly distributed variates U and W follow the bivariate normal distribution, then the covariance between $X = \exp(U)$ and $Y = \exp(W)$ is given by $\text{Cov}(X, Y) = \mu_X\mu_Y\{\exp[\text{Cov}(U, W)] - 1\}$. One can thus generate correlated values of X and Y from bivariate normal random numbers distributed as $U = \ln(X) \sim N(\mu_{\ln(X)}, \sigma^2_{\ln(X)})$ and $W = \ln(Y) \sim N(\mu_{\ln(Y)}, \sigma^2_{\ln(Y)})$ having a correlation coefficient of

$$\rho_{U,W} = \frac{\ln(1 + V_X V_Y \rho_{X,Y})}{\sqrt{\ln\left(1 + V_X^2\right)\ln\left(1 + V_Y^2\right)}}.$$

8.7. Generation of beta variates. Let $X \sim \text{beta}(a, b)$ with $0 \le x \le 1$. Develop an algorithm to generate beta random numbers based on the rejection method. Compare the cdf resulting from simulation of 100 samples of $X \sim$ beta $(1, 3)$ with its analytical form by using the Kolmogorov-Smirnov test.

8.8. Wastewater treatment plant. An activated-sludge plant includes five serial processes: (1) coarse screening, (2) grit removal, (3) plain sedimentation, (4) contact treatment, and (5) final settling. Let X_i denote the efficiency of the ith treatment, that is, the fraction of remaining pollutant after removal by the ith serial treatment. For example, X_1 is the fraction of pollutant removed by treatment process 1, X_2 is the fraction of the remaining pollutant after removal by treatment process 2, and so on. The amount Q_{out} of pollutant in the effluent is given by

$$Q_{\text{out}} = (1 - X_1)(1 - X_2)(1 - X_3)(1 - X_4)(1 - X_5)Q_{\text{in}},$$

where Q_{in} denotes the amount of pollutant in the untreated inflow. A quality indicator of the performance of the plant is then defined as

$$Y = (1 - X_1)(1 - X_2)(1 - X_3)(1 - X_4)(1 - X_5).$$

Consider a plant with the following single-process mean efficiencies in the removal of the 5-day 20°C biological oxygen demand (BOD):

$$\mu_1 = 0.05, \mu_2 = 0.05, \mu_3 = 0.20, \mu_4 = 0.70, \mu_5 = 0.10,$$

where $\mu_i = E[X_i]$. Suppose that X_1, X_2, X_3, and X_5 are normal variates with common coefficient of variation of 0.2, and $X_4 \sim$ uniform (0.6, 0.8). Find the pdf and cdf of Y by simulation assuming that the five processes are independent of each other. Compare the mean of Y with the nominal value.

8.9. Underground pipeline subject to corrosion. An underground pressured pipeline is subject to stresses caused by external soil pressure and by internal (fluid) pressure. Assuming the radius of pipe r is much larger than the thickness of the pipe wall t, the circumferential stress s_f due to internal pressure is estimated as

$$s_f = \frac{pr}{t},$$

where p is the internal pressure. The bending stress s_s in the circumferential direction produced in the pipe wall by the external soil loading can be estimated from

$$s_s = \frac{6k_m C_d \gamma B_d^2 E t r}{E t^3 + 24 k_d p r^3}.$$

Here C_d is a dimensionless calculation coefficient for soil load, γ is the unit weight of soil backfill, B_d is the width of the ditch at the top of the pipe, E is the modulus of elasticity of the pipe metal, k_m is a bending moment coefficient dependent on the distribution of vertical load and reaction, and k_d is a deflection coefficient dependent on the distribution of vertical load and reaction. The circumferential bending stress s_t produced in the pipe wall due to traffic loads (such as that resulting from roadway, railway, or airplane traffic) may be estimated from

$$s_t = \frac{6k_m I_c C_t F E t r}{A\left(E t^3 + 24 k_d p r^3\right)},$$

where I_c is a dimensionless impact factor, C_t is the dimensionless surface load coefficient, F is the wheel load on surface, and A is the effective length of pipe on which load is computed. If the pipe remains in the elastic range under load, the maximum circumferential stress is given at the critical sections by $s_f + s_s + s_t$. By using simulation, compute the expected maximum circumferential stress and its coefficient of variation. Suppose that the quantities involved have the following distributions [from M. Ahammed and R. E. Melchers (1994), "Reliability of underground pipelines subject to corrosion," *J. Transp. Eng. Div., ASCE*, Vol. 120, pp. 989–1002, reproduced by the permission of the publisher, ASCE]:

Variate	Distribution	Mean	Coefficient of variation
p	Normal	6.205 MPa	0.20
r	Normal	228.6 mm	0.05
t	Normal	8.73 mm	0.05
k_m	Lognormal	0.235	0.20
C_d	Lognormal	1.32	0.20
γ	Normal	18.85×10^{-6} N/mm^3	0.10
B_d	Normal	762 mm	0.15
E	Normal	206,800 MPa	0.05
k_d	Lognormal	0.108	0.20
I_c	Normal	1.5	0.25
C_t	Lognormal	0.12	0.20
F	Normal	267,000 N	0.25
A	Normal	914 mm	0.20

The main effect of corrosion is weight loss. Because we are mainly interested is general corrosion, it is assumed that the loss of wall thickness can be modeled empirically by a power law, $d = k\tau^n$, where τ is the time of exposure in years, k is a multiplying coefficient, and n is a constant. Accordingly, one will substitute $(t - d)$ or $(t - k\tau^n)$ for t in the above equations to account for corrosion. Suppose that both k and n are normal variates with means of 0.3 and 0.6, respectively, and coefficients of variation of 0.3 and 0.2, respectively. Evaluate the expected maximum circumferential stress and its coefficient of variation after an exposure of 30 years.

8.10. Debris flow. Debris flows, also referred as *mudflows*, are a significant hazard in many parts of the world, causing extensive damage to engineering structures such as buildings, bridges, and culverts, as well as causing loss of life. From data analysis in the Los Angeles area, California, the following empirical formula was proposed to estimate the debris volume X in cubic meters:

$$X = 56.56Y^{0.75}a^{1.25}(1 + 80e^{-0.239a-0.537W})^{0.5},$$

where a denotes the watershed area in square kilometers, Y the 72-hour maximum annual rainfall depth in millimeters, and W the time interval between watershed burning in years [from R. H. McCuen, Ayyub, B. M., and T. V. Hromadka (1990), "Risk of Debris-Basin failure," *ASCE J. Water Resour. Plan. Man. Div., ASCE*, Vol. 116, pp. 473–483, reproduced by the permission of the publisher, ASCE]. Assume that Y and W are independent variates, Y is a Gumbel-distributed variate with a mean of 100 mm and a coefficient of variation of 0.444, and W is a lognormal-distributed variate with a mean of 8 years and a coefficient of variation of 1.375. Consider a drainage area a of 2.5 km^2, and find the probability distribution of X by simulation.

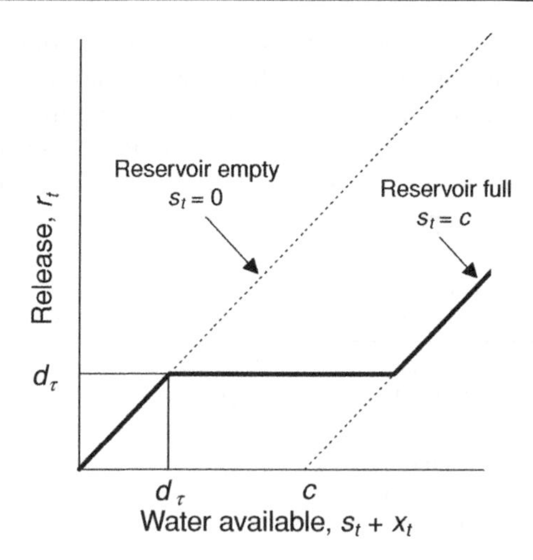

Fig. 8.P3 "Normal" operating rule for a reservoir.

8.11. Reservoir capacity. In determining the optimal capacity of a reservoir, let us assume that the manager will follow the so-called normal operating rule shown in Fig. 8.P3.

In this case, the draft or release R_t is obtained as

(1) $R_t = S_t + X_t$, if $S_t + X_t \leq d_\tau$,
(2) $R_t = d_\tau$, if $d_\tau < S_t + X_t < d_\tau + c$,
(3) $R_t = S_t + X_t - c$, if $S_t + X_t \geq d_\tau + c$,

where c denotes the capacity of the reservoir. The rate of demand of water supply, d_τ, is equal to the mean annual runoff in March, April, November, and December. It is reduced to 85% in May, August, September, and October and to 70% in January, February, June, and July. Using this rule and the other data of Example 8.22, find the optimal capacity of the reservoir for an average annual deficit of 1% of the annual demand. Assume full reservoir as the initial condition. Compare this result with that of Example 8.22.

8.12. Model selection for extreme value data. Let Xdenote a GEV-distributed random variable with parameters $\varepsilon = 0$, $\alpha = 1$, and $k = -0.2$. Perform the following experiment:

(*a*) Generate a sample of 100 outcomes of this variate.
(*b*) Fit the (1) Gumbel, (2) Fréchet, (3) lognormal, (4) gamma, (5) GEV, (6) shifted-lognormal, (7) shifted-gamma, and (8) log-Pearson Type III distributions to the generated sample.
(*c*) Perform a goodness-of-fit testing procedure using the chi-squared, Kolgomorov-Smirnov and Anderson-Darling tests.

Determine the probability models for which the null hypothesis is not rejected. Repeat the experiment for a sample of 10,000 outcomes.

8.13. River network. A river network can be described as a random binary tree, as shown in Fig. 8.P4*a*.

A mathematical tree originates from a root (ancestor) and it grows by subsequent branching, through a bifurcation process. A link is defined as the line segment between two vertices of the tree; external links are those connecting an internal vertex (junction) with an external vertex (source), and internal links are those joining

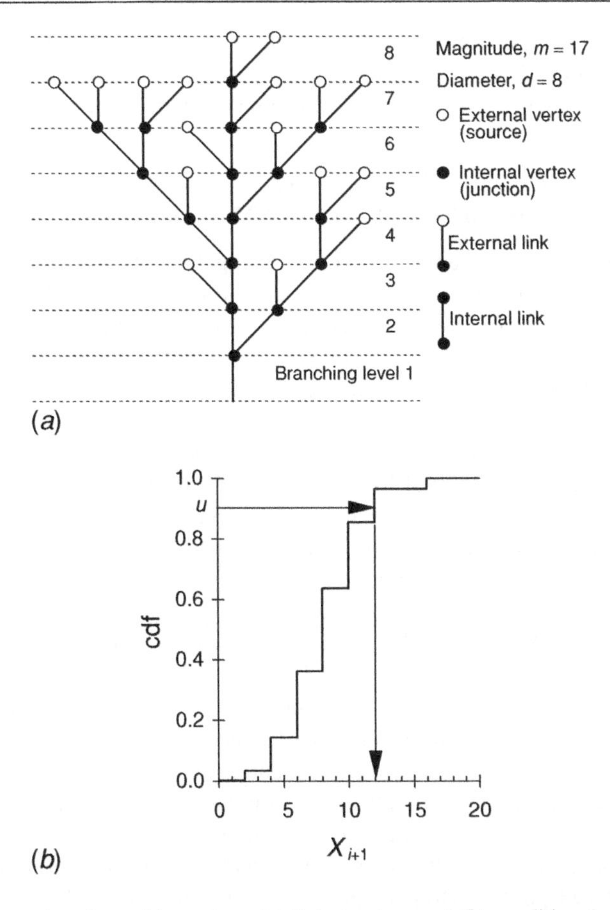

Fig. 8.P4 River network as a binary tree: (*a*) link structure and (*b*) conditional cdf of the number of links at level $i + 1$ given that $x_i = 8$.

two junctions. The total number of external links is called the "*magnitude*" of the tree. A tree of magnitude m has $n = 2m - 1$ links (total progeny). A hierarchical order can be assigned to each element of the tree by indexing a link by its "level" of branching, that is, by progressively numbering the links from 1, which is assigned to the root, to k, which is the level of the source having the highest distance from the root. Let X_i denote the number of links at branching level i. In a standard model of river networks, the tree randomly branches with a constant branching probability p for all the links independently of the bifurcation level. Therefore, the number of links at level $i + 1$, X_{i+1}, depends only on X_i, the number of links at the previous level. The process of branching through upstream growth is called *Markovian*, because each stage of development depends only on the immediately previous one. If $p = 1/2$, the probability that X_{i-1} links at level $i - 1$ will originate X_i links at level i is

$$\Pr[X_i = x_i | X_{i-1} = x_{i-1}] = 2^{-x_{i-1}} \binom{x_{i-1}}{x_i/2},$$

where

$$\binom{x_{i-1}}{x_i/2}$$

denotes the combinations of x_{i-1} links taking $x_i/2$ at a time, and $X_{i-1} \leq 2X_i$. For example, if $X_{i-1} = 8$ at level $i - 1$, the corresponding transition probabilities p_X for X_i are those listed in the following table with the associated conditional cdf F_X:

$X_i =$	0	2	4	6	8	10	12	14	16
$p_X =$	0.0039	0.0313	0.1094	0.2188	0.2734	0.2188	0.1094	0.0313	0.0039
$F_X =$	0.0039	0.0352	0.1445	0.3633	0.6367	0.8555	0.9648	0.9961	1.0000

One can simulate a river network by using the probability integral transform method as shown in Fig. 8.3.P4*b*. The process terminates when $X_k = 0$ (adsorbing state), and level k is called the "diameter" of the river network. Using this model find the probability distribution of the level j for which the number X_j of links is a maximum in trees with diameter of $k = 8$.

8.14. Seismic hazard. In a period of 600 years, about 330 earthquakes occurred in Central Italy having epicentral MCS intensity X exceeding 6. Also, X is modeled as an exponential variate with scale and location parameters of 0.91 and 6, respectively. Seismic hazard in a specific site is represented by MCS intensity Y as evaluated from the following attenuation law:

$$Y = X - \frac{1}{\ln \psi} \ln \left[1 + \frac{\psi - 1}{\psi_0} \left(\frac{Z \varphi^{x_0 - X}}{z_0} - 1 \right) \right],$$

where Z denotes the distance from the epicenter, $z_0 = 9.5$ km is the distance of the isoseismical line for epicentral intensity $x_0 = 10$, and $\psi_0 = 1$, $\psi = 1.5$, and $\varphi = 1.3$ are the estimated values of parameters ψ_0, ψ, and φ for Central Italy [see G. Grandori, A. Drei, F. Perotti, and A. Tagliani (1991), "Macroseismic intensity versus epicentral distance": The case of Cental Italy, in: M. Stucchi, D. Postpischl, and D. Slejko, eds., "Investigations of historical earthquakes in Europe," *Tectonophysics*, Vol. 193, pp. 175–181]. Suppose $Z \sim$ uniform (3 km, 25 km) and find the probability distribution of Y by simulation. Compute the 100-year MCS intensity for this region assuming that Y is a Gumbel variate.

Chapter 9

Risk and Reliability Analysis

In the assessment of civil and environmental engineering systems, one must evaluate the capability of a designed system to respond to project requirements or to meet users' demands. A system can fail to perform its intended function for one or more reasons, such as natural hazards or lower performance than predicted. Failures may even include such rare events as collapse of major structures. For example, a dam break can be caused by a catastrophic flood that exceeds the design value or a structural failure attributable to faulty design. Although the assurance of system performance and safety is primarily a task of engineers, the accepted levels of adequacy or risk are subject to economic and social constraints. Therefore, planning and design of engineering systems require that cost-and-benefit analyses be performed accurately in order to achieve a complete assessment of system performance and safety. Social issues also play an important role in the analysis of civil and environmental engineering systems, because these systems are more directly involved with the public than are other engineering systems.

Defining the inability of a system to perform adequately, such that failure results, is not an easy task. For instance, in the case of a dam break the associated failure is not reversible and it can occur immediately. On the other hand, if an airport suffers occasional traffic congestion, it will be able to work again satisfactorily after the operational conditions causing the congestion are removed; moreover, the airport may be subject to another similar failure in the future. In a wastewater treatment plant, failure may take place gradually; determining the stage at which the system ceases to perform adequately is to a certain degree subjective. All these decisions involve some additional uncertainty in the definition of system performance and safety, as well as the judgments of decision makers, who must balance potential benefits against costs.

To analyze a system's risk of failure, one must clearly identify the input to the system and its consequent response. In the case of a building, structural safety depends on the maximum load that may be imposed over the lifetime of the building, and also on the load-carrying capacity, or strength, of the structure or its components. Because the predictions of maximum load and actual strength of a structure are subject to uncertainties, one cannot ensure its absolute safety, and the engineer must rely on some probabilistic concept indicating the likelihood that the available strength will adequately withstand the maximum load over the lifetime of the building. In a water supply system, one must compare the demand of water from the different users with the available resource. Because the demand undergoes fluctuations, and water resources are subject to natural variability, the probability of available supply relative to demand gives the adequacy of the system's design or operation. Reliability was defined simply as the probability of success in most related examples given previously.[1] A more precise definition of reliability is the probabilistic assessment of the likelihood that a system will perform adequately for a specified period of time under known operating conditions. The acceptance of a given level of reliability must be discussed in the light of possible economic and social costs, and benefits. Risk

[1] See Example 2.19.

and reliability of a system are defined as the probabilities of failure and nonfailure over the specified system lifetime.

> **Definition: Risk and reliability.** The *risk* that a system is incapable of meeting the demand is defined as the probability of failure p_f over the specified system lifetime under specified operating conditions. System *reliability*, denoted by r, is the complementary probability of nonfailure, $r = 1 - p_f$.

The capability of a system to perform under given requirements can be defined using different terms, such as capacity and demand, load and strength, force and resistance. In this chapter, the concept of capacity X and demand Y is used because of its generality. For example, this concept can be applied to describe the landing capacity and the flight arrival rate for an airport, bearing capacity of a terrain and foundation load, allowable and computed stresses, culvert size and stormwater flow depth, structural capacity and earthquake loads, spillway capacity and flood discharge, and allowable and predicted biological oxygen demand.

The objectives of this chapter are to define measures of reliability and to examine different types of failure. We also study the uncertainty aspects of reliability from different perspectives. The final section is devoted to reliability design.

9.1 MEASURES OF RELIABILITY

9.1.1 Factors of safety

The assessments of risk and safety of civil and environmental engineering systems are traditionally based on allowable *factors of safety*; these are estimated from previous experience on the behavior of a particular system or from observed responses of similar systems. A conventional measure of the factor of safety taken by the engineer is the ratio of the assumed nominal values of capacity x^* and demand y^*, as

$$z^* = \frac{x^*}{y^*}. \tag{9.1.1}$$

For example, if the allowable stress in a timber beam is 36 N/mm^2 and the design stress is 24 N/mm^2, as shown in Fig. 9.1.1, the conventional safety factor is 1.5. An engineer

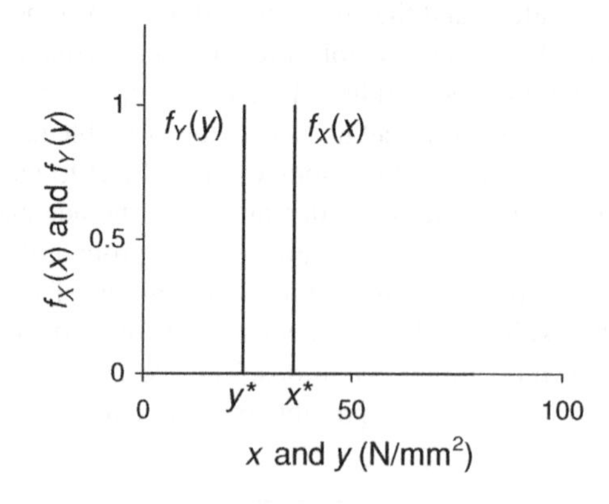

Fig. 9.1.1 Nominal values of capacity and demand, x^* and y^*, respectively.

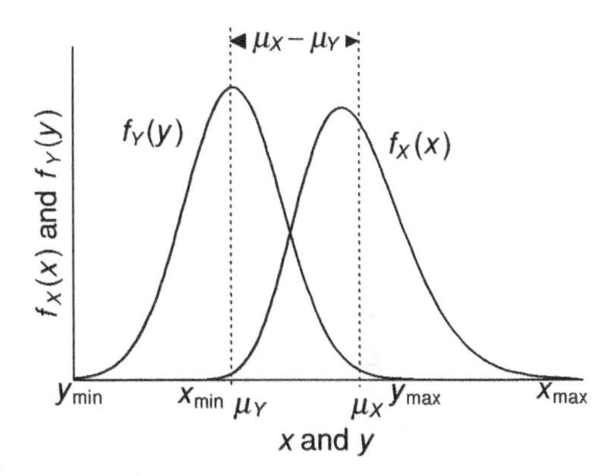

Fig. 9.1.2 Probability density functions of capacity X and demand Y.

would assume that the designed beam is satisfactory if the calculated safety factor is greater than a specified minimum value, which is given by design prescriptions or is based on experience. If a safety factor of 1.5 is considered too low, the engineer should redesign the system to increase the maximum induced stress or to decrease the load.

The demand on a system often results from a number of uncertain components, such as wind loadings, earthquake accelerations, streamflows, water table depths, sea levels, storm intensities, air and water temperatures, and pollutant loads. The capacity usually depends on the variability of system characteristics; these include strengths of materials, construction techniques, testing errors, inspection supervision, and environmental conditions. In order to evaluate the uncertainties affecting both demand and capacity, one designs the prototype system on the hypothesis of a physically based mathematical model; this necessitates a careful scrutiny of the formulas and equations, and their assumptions, that are used to scale model parameters. To take an extreme case, an engineer should not expect a highly empirical formula, developed at a time when knowledge of the system was rather scanty, to model the real world with a reliability close to 100%. This assumption can only be justified after making repeated observations of the performances of a large number of similar systems.

Because the nominal values of both the capacity x^* and the demand y^* cannot be determined with certainty, the capacity and demand functions must be considered as probability distributions, as shown in Fig. 9.1.2. Hence, the safety factor as given by ratio $Z = X/Y$ of two random variables, X and Y, is also a random variable.

Definition: Safety factor. The *safety factor* of a system, treated as a random variable and defined as $Z = X/Y$, is the ratio between capacity X and demand Y of the system.

The inadequacy of the system to meet the demand, as measured by the probability of failure, is associated with that portion of the distribution of the safety ratio wherein it becomes less than unity, that is, the portion in which $Z = X/Y \leq 1$ (see Fig. 9.1.3).

The probability p_f of system failure is thus given by

$$p_f = \Pr[Z \leq 1] = F_Z(1). \tag{9.1.2}$$

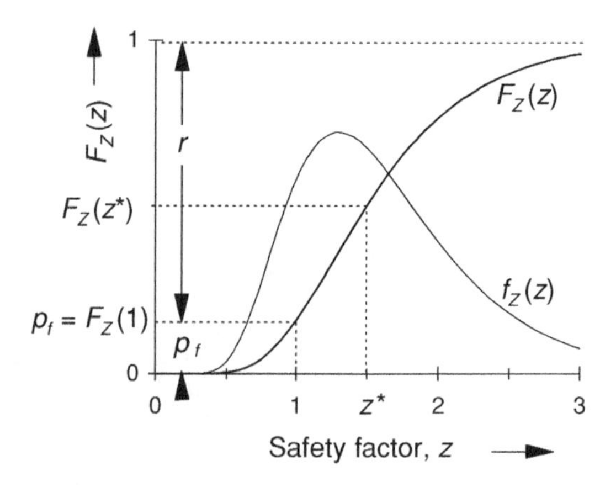

Fig. 9.1.3 cdf and pdf of safety factor $Z = X/Y$.

The corresponding probability of nonfailure is

$$r = 1 - p_f = \Pr[Z > 1] = 1 - F_Z(1), \tag{9.1.3}$$

which can be interpreted as survival probability or simply reliability. When the joint probability distribution of X and Y is known, the reliability of the system can be evaluated by determining the cdf of X/Y. There is a zero probability of failure ($p_f = 0$) and a reliability of 100% ($r = 1$) only if the maximum demand $Y_{\max}$ does not exceed the minimum capacity $X_{\min}$, so that the two distributions do not overlap.

Example 9.1. Structural safety factor for independent lognormally distributed load and strength. Consider a structure whose load-carrying capacity or strength X, and load Y are independent lognormal variates, with means and standard deviations μ_X, μ_Y, and σ_X, σ_Y, respectively. In this case, the safety factor, $Z = X/Y$, is also a lognormal variate. As shown in Eq. (4.2.28)

$$\mu_{\ln(Z)} = \mu_{\ln(X)} - \mu_{\ln(Y)} = \ln(\mu_X) - \frac{1}{2}\ln\left(1 + V_X^2\right) - \ln(\mu_Y) + \frac{1}{2}\ln\left(1 + V_Y^2\right),$$

where $V_X = \sigma_X/\mu_X$ and $V_Y = \sigma_Y/\mu_Y$ are the coefficients of variation of X and Y, respectively, and

$$\sigma_{\ln(Z)}^2 = \sigma_{\ln(X)}^2 + \sigma_{\ln(Y)}^2 = \ln(1 + V_X^2) + \ln(1 + V_Y^2).$$

In terms of the medians, m_X and m_Y, it follows from Eq. (4.2.28d) that

$$\mu_{\ln(Z)} = \ln(m_X) - \ln(m_Y) = \ln(m_X/m_Y),$$

where the ratio (m_X/m_Y) represents the median safety factor. Since $\ln(Z)$ is normally distributed with mean $\mu_{\ln(Z)}$ and standard deviation $\sigma_{\ln(Z)}$, the random variable $[\ln(Z) - \mu_{\ln(Z)}]/\sigma_{\ln(Z)}$ is a standard normal variate. Therefore, the probability of failure is found using Eq. (9.1.2) as

$$p_f = F_Z(1) = \Phi\left(\frac{\ln 1 - \mu_{\ln(z)}}{\sigma_{\ln(z)}}\right) = \Phi\left(-\frac{\mu_{\ln(z)}}{\sigma_{\ln(z)}}\right)$$

$$= 1 - \Phi\left(\frac{\ln(m_X/m_Y)}{\sqrt{\ln\left(1 + V_X^2\right) + \ln\left(1 + V_Y^2\right)}}\right),$$

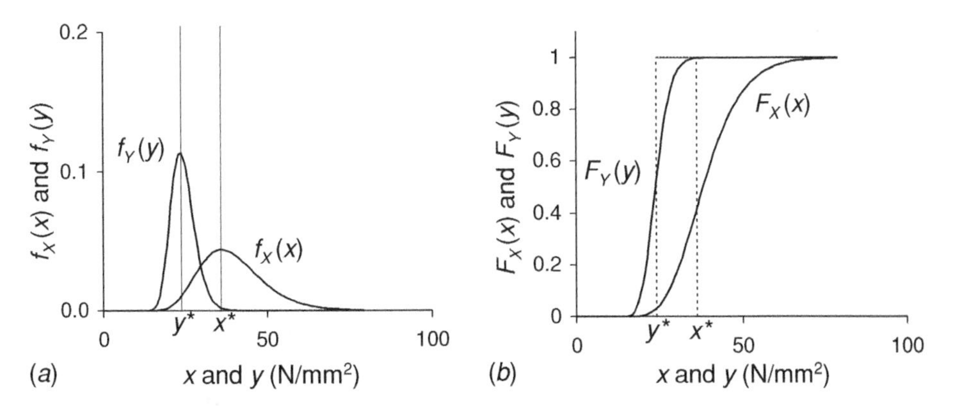

Fig. 9.1.4 Timber strength and load illustration: pdfs (a) and cdfs (b) for independent lognormally distributed X and Y.

where $\Phi(\cdot)$ denotes the cdf of the standard normal distribution. Accordingly, the reliability of the structure, $r = \Pr[Z > 1] = 1 - F_Z(1) = 1 - p_f$, is

$$r = \Phi\left(\frac{\ln(m_X/m_Y)}{\sqrt{\ln\left(1+V_X^2\right)+\ln\left(1+V_Y^2\right)}}\right).$$

Thus, if X and Y are independent and lognormally distributed, the reliability is a function of the median safety factor and the standard deviation $\sigma_{\ln(X/Y)}$.

Consider, for example, a rigid timber beam (see Fig. 2.1.3) with an estimated average strength of 39.1 N/mm², and coefficient of variation of 25% (as in Table 1.2.2). If the beam is designed to carry a load of 24.0 N/mm², with a coefficient of variation of 15%, one can compute the failure probability as follows. Since the means and coefficients of variation of strength X and load Y are $\mu_X = 39.1$ N/mm², $V_X = 0.25$, $\mu_Y = 24.0$ N/mm², $V_Y = 0.15$, respectively, assuming X and Y are independent and lognormally distributed,

$$\sigma_{\ln(Z)} = [\ln(1+0.25^2) + \ln(1+0.15^2)]^{1/2} = 0.288,$$

$$\mu_{\ln(Z)} = \ln(39.1) - \frac{1}{2}\ln(1+0.25^2) - \ln(24.0) + \frac{1}{2}\ln(1+0.15^2) = 0.469.$$

The required probability of failure is thus

$$p_f = F_Z(1) = \Phi\left(\frac{-0.469}{0.288}\right) = \Phi(-1.628) = .052,$$

which indicates that the beam has a reliability of 94.8%. The pdfs of X and Y are shown in Fig. 9.1.4a, and the corresponding cdfs in Fig. 9.1.4b. The pdf and cdf of the safety factor Z are shown in Fig. 9.1.5.

To define a single-valued measure, one may use central measures of capacity and demand. For example, the *central safety factor* is defined as

$$\zeta = \frac{E[X]}{E[Y]} = \frac{\mu_X}{\mu_y}, \tag{9.1.4}$$

where the expected capacity μ_X and demand μ_Y are used.

Definition: Central safety factor. The *central safety factor* of a system, denoted by ζ, is the ratio between expected capacity X and demand Y of the system.

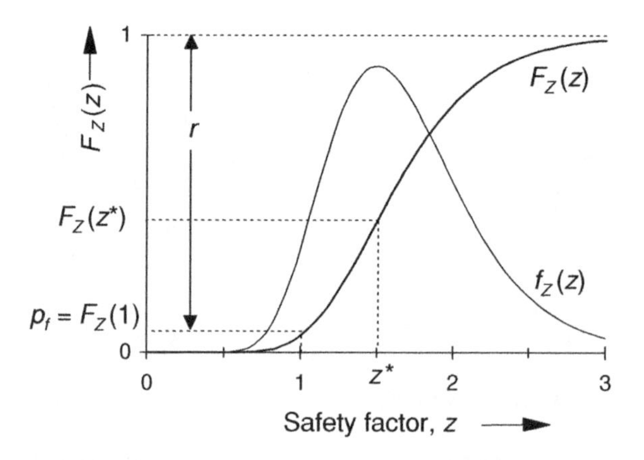

Fig. 9.1.5 Timber strength and load illustration: pdf and cdf of safety factor $Z = X/Y$ for independent lognormally distributed X and Y.

Suppose an engineer assigns a nominal value of capacity less than that of its expected value, say,

$$x^* = \mu_X - h_X\sigma_X, \tag{9.1.5a}$$

and prescribes a nominal value for the demand greater than μ_Y, say,

$$y^* = \mu_Y + h_Y\sigma_Y, \tag{9.1.5b}$$

where h_X and h_Y are sigma units of their respective functions. If these estimates of x^* and y^* are used in Eq. (9.1.1),

$$z^* = \frac{\mu_X - h_X\sigma_X}{\mu_Y + h_Y\sigma_Y}. \tag{9.1.6}$$

Hence, the central factor exceeds the conventional factor of safety, when positive values of h_X and h_Y are taken as usual.

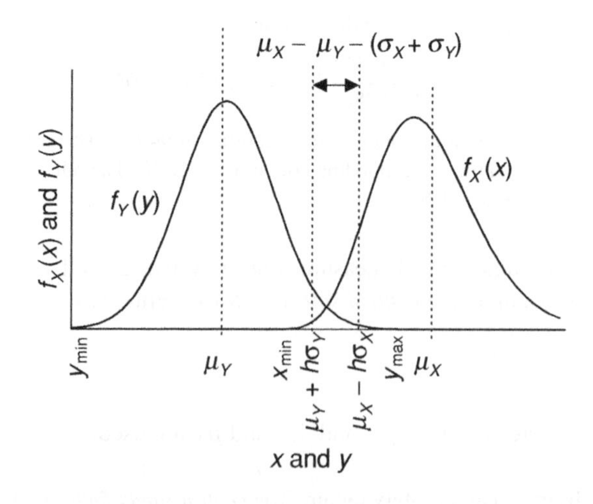

Fig. 9.1.6 Sigma bounds of capacity X and demand Y.

Example 9.2. Central safety factor for a pumping station. A pumping station was designed using a safety factor z^* of 1.8, or 9/5. An engineer has the task of assessing the reliability of the system without any knowledge of possible fluctuations of capacity and demand. Therefore, the coefficients of variation of capacity and demand are assumed to be equal ($V_X = V_Y = V$), as are the sigma bounds, ($h_X = h_Y = h$). (see Fig. 9.1.6). From Eq. (9.1.6),

$$z^* = \frac{\mu_X - h_X V_X \mu_X}{\mu_Y + h_Y V_Y \mu_Y} = \frac{\mu_X}{\mu_Y}\frac{1 - hV}{1 + hV} = \zeta \frac{1 - hV}{1 + hV}.$$

The engineer further assumes that the possible range of V is $0.1 \leq V \leq 0.5$, and $0 \leq h \leq 1$, so that the possible range of hV is $0 \leq hV \leq 0.5$. Since no other information is available regarding the moments of hV, the principle of maximum entropy suggests that hV can be modeled as a uniformly distributed variate with $E[hV] = 1/4$, which yields $\zeta / z^* = 5/3$. Therefore, to improve system reliability in order to achieve a safety factor of ζ the engineer must increase the nominal capacity x^* of the pumping station from $(9/5)y^*$ to $(5/3)(9/5)y^* = 3y^*$.

9.1.2 Safety margin

As shown in Fig. 9.1.2, if the maximum demand $Y_{\max}$ exceeds the minimum capacity $X_{\min}$, the distributions overlap and there is a nonzero probability of failure. To assess this probability one can take the difference between capacity and demand,

$$S = X - Y, \tag{9.1.7}$$

which is usually referred to as the *safety margin*. Because X and Y are random variables, the safety margin is also a random variable (see Fig. 9.1.7).

Definition: Safety margin. The *safety margin* of a system is defined as the random difference $S = X - Y$ between capacity X and demand Y of the system.

The inadequacy of the system to meet the demand, as measured by p_f, is associated with that portion of the distribution of the safety margin wherein S takes negative values, that is, the portion in which $S = X - Y \leq 0$. Thus,

$$p_f = \Pr[(X - Y) \leq 0] = \Pr[S \leq 0], \tag{9.1.8}$$

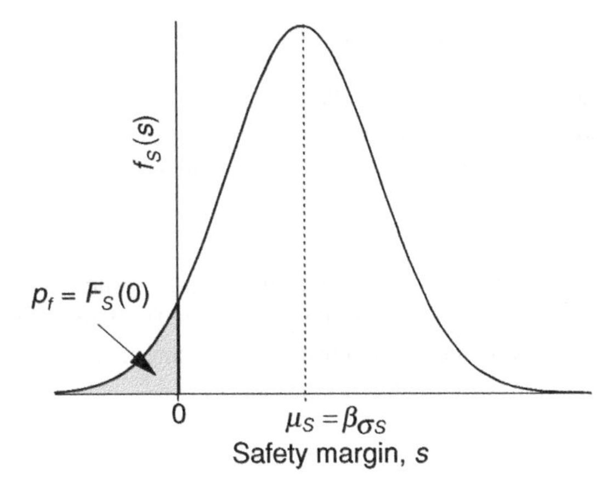

Fig. 9.1.7 pdf of safety margin S.

and the corresponding reliability is

$$r = 1 - p_f = \Pr[(X - Y) > 0] = \Pr[S > 0]. \tag{9.1.9}$$

When the joint probability distribution of X and Y is known, the reliability of the system can be evaluated by determining the cdf of $X - Y$.

Example 9.3. Structural margin of safety for independent normally distributed load and strength. Consider a structure whose load-carrying capacity or strength X and load Y are independent normal variates, with means and standard deviations μ_X, μ_Y and σ_X, σ_Y, respectively. In this case, the safety margin, $S = X - Y$, is shown in Example 3.60 to be also a normal variate with

$$\mu_S = \mu_X - \mu_Y, \quad \sigma_S^2 = \sigma_X^2 + \sigma_Y^2.$$

Since S is normally distributed with mean μ_S and standard deviation σ_S, the random variable $(S - \mu_S)/\sigma_S$ is a standard normal variate, and the reliability of the structure is from Eq. (9.1.9)

$$r = 1 - F_S(0) = 1 - \Phi\left(\frac{0 - \mu_S}{\sigma_S}\right) = 1 - \left[1 - \Phi\left(\frac{\mu_S}{\sigma_S}\right)\right] = \Phi\left(\frac{\mu_X - \mu_Y}{\sqrt{\sigma_X^2 + \sigma_Y^2}}\right),$$

where $\Phi(\cdot)$ denotes the cdf of the standard normal distribution.

For example, consider again the rigid timber beam of Example 9.1, and assume normal and independent strength X and load Y. The probability of failure can be computed as follows. Since

$$\mu_X = 39.1\ \text{N/mm}^2,\ V_X = 0.25,\ \mu_Y = 24.0\ \text{N/mm}^2,\ V_Y = 0.15,$$

one has

$$\sigma_X = 39.1 \times 0.25 = 9.775\ \text{N/mm}^2,\ \sigma_Y = 24.0 \times 0.15 = 3.6\ \text{N/mm}^2,$$

so that

$$\mu_S = 39.1 - 24.0 = 15.1\ \text{N/mm}^2, \quad \text{and} \quad \sigma_S = (9.775^2 + 3.6^2)^{1/2} = 10.42\ \text{N/mm}^2.$$

The required probability of failure is

$$p_f = F_S(0) = \Phi\left(\frac{-15.1}{10.42}\right) = \Phi(-1.450) = .074,$$

which indicates a beam reliability of 92.6%. The pdfs of X, Y, and S are given in Fig. 9.1.8*a*, and the corresponding cdfs are given in Fig. 9.1.8*b*. The estimated reliability from the independent normal model differs from that obtained from the independent lognormal model by only about 2%.

In Example 3.60, it was shown using the moment-generating function that any linear combination of two independent normally distributed variates is itself normally distributed. It follows that, when X and Y are normally distributed, and correlated, the safety margin S is a normal variate with mean and variance

$$\mu_S = \mu_X - \mu_Y, \tag{9.1.10a}$$

$$\sigma_S^2 = \sigma_X^2 - 2\rho_{XY}\sigma_X\sigma_Y + \sigma_Y^2, \tag{9.1.10b}$$

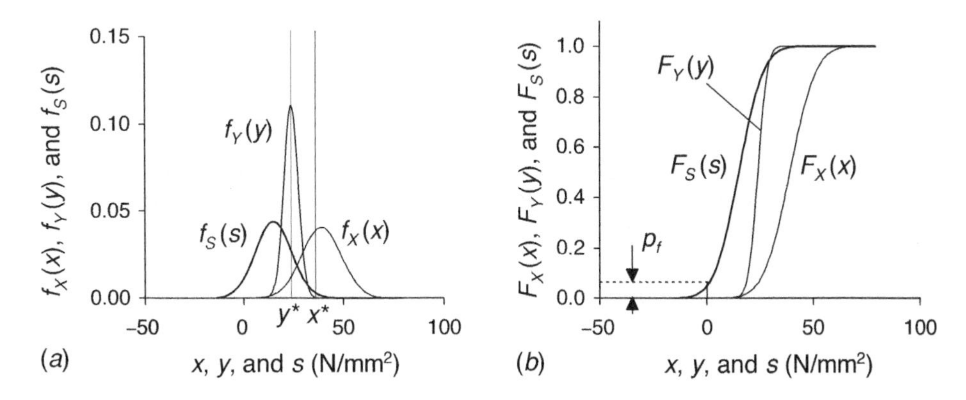

Fig. 9.1.8 Timber strength and load illustration: pdf (*a*) and cdf (*b*) of safety margin S for independent normally distributed X and Y.

respectively, where ρ_{XY} denotes the correlation coefficient between capacity X and demand Y. Hence, the probability of failure is given by

$$p_f = \Phi\left(-\frac{\mu_X - \mu_Y}{\sqrt{\sigma_X^2 - 2\rho_{XY}\sigma_X\sigma_Y + \sigma_Y^2}}\right), \tag{9.1.11}$$

where $\Phi(\cdot)$ denotes the cdf of the standard normal distribution; accordingly,

$$r = \Phi\left(\frac{\mu_X - \mu_Y}{\sqrt{\sigma_X^2 - 2\rho_{XY}\sigma_X\sigma_Y + \sigma_Y^2}}\right) \tag{9.1.12}$$

is the associated reliability.

Example 9.4. Irrigation water supply. During the growing season the expected demand, Y, from an irrigation scheme is 10 units with a coefficient of variation of 50%, which accounts for fluctuations associated with weather variability. The mean available water, X, which is diverted from a river barrage, is 20 units, with a coefficient of variation of 20%, which accounts for fluctuations associated with hydrologic variability in that season. Because of the relationship between hydrology and climate, the natural water availability often tends to decrease when the demand increases, so that the correlation coefficient between X and Y is negative. The estimated value of ρ_{XY} is -0.5. An irrigation engineer needs to estimate the reliability of the system assuming that both capacity X and demand Y are normally distributed variates.

The standard deviations of capacity and demand are

$$\sigma_X = V_X\mu_X = 0.2 \times 20 = 4 \text{ units},$$
$$\sigma_Y = V_Y\mu_Y = 0.5 \times 10 = 5 \text{ units},$$

respectively. The safety margin, $S = X - Y$, is normally distributed with mean

$$\mu_S = \mu_X - \mu_Y = 20 - 10 = 10 \text{ units},$$

and standard deviation

$$\sigma_S = \left(\sigma_X^2 - 2\rho_{XY}\sigma_X\sigma_Y + \sigma_Y^2\right)^{1/2} = (4^2 + 2 \times 0.5 \times 4 \times 5 + 5^2)^{1/2} = 7.81 \text{ units}.$$

The required risk of failure is

$$p_f = F_S(0) = \Phi\left(\frac{-10}{7.81}\right) = 1 - \Phi(1.28) = 1 - 0.9 = .1,$$

and the associated reliability is 90%. In order to increase the reliability of the system to 95%, the diversion of a neighboring stream is considered for the purpose of increasing the mean capacity. Assuming that both V_X and ρ_{XY} do not change, the mean capacity μ_X must be increased by a factor of a, so that

$$r = 1 - F_S(0) = \Phi\left(\frac{a\mu_X - \mu_Y}{\sqrt{a^2\mu_X^2 V_X^2 - 2\rho_{XY} a V_X \mu_X \sigma_Y + \sigma_Y^2}}\right) = 0.95,$$

that is,

$$\frac{a\mu_X - \mu_Y}{\sqrt{a^2\mu_X^2 V_X^2 - 2\rho_{XY} a V_X \mu_X \sigma_Y + \sigma_Y^2}} = 1.65.$$

Hence,

$$\frac{20a - 10}{\sqrt{4^2a^2 + 2 \times 0.5 \times 4 \times 5 \times a + 5^2}} = \frac{20a - 10}{\sqrt{16a^2 + 20a + 25}} = 1.65,$$

which yields $a = 1.20$. This means that the new source must provide a 20% increase in the average water availability in order to increase the reliability of the irrigation system to 95%.

9.1.3 Reliability index

An important measure of the adequacy of an engineering design is the reliability index, defined as

$$\beta = \left(\frac{\mu_S}{\sigma_S}\right). \tag{9.1.13}$$

This can be interpreted as the number h of sigma units (the number of standard deviations σ_S) between the mean value of the safety margin $E[S] = \mu_S$ and its critical value $S = 0$, as shown in Fig. 9.1.7. By definition, the reliability index is also the reciprocal of the coefficient of variation of the safety margin; that is $\beta = 1/V_S$.

Definition: Reliability index. The *reliability index* of a system, denoted by β, is defined as the ratio between the mean and standard deviation of the safety margin of the system.

Example 9.5. Structural reliability index for normally distributed safety margin. Consider again a structure whose load-carrying capacity, or strength, X and load Y are independent normal variates (see Example 9.3). Since $r = \Phi(\mu_S/\sigma_S)$, r is a function of the ratio μ_S/σ_S, which is the safety margin expressed in units of σ_S, that is, the reliability index β. Therefore, system reliability can be written as $r = \Phi(\beta)$, and the corresponding probability of failure is given by $p_f = 1 - r = 1 - \Phi(\beta)$. For normal S, a value of $\beta = 0$ corresponds to $r = 0.5$ (50% reliability). Similarly, $\beta = 1.28$ corresponds to 90% reliability, $\beta = 1.65$ with 95%, $\beta = 2.33$ with 99%, $\beta = 3.10$ with 99.9%, and $\beta = 3.72$ with 99.99%. This illustrates that the level of reliability is a function of both the relative position of $f_X(x)$ and $f_Y(y)$, as measured by the mean safety margin $\mu_S = \mu_X - \mu_Y$, and the degree of dispersion, as measured in terms of the standard deviation $\sigma_S = (\sigma_X^2 + \sigma_Y^2)^{1/2}$. The reliability index β reflects the combined effect of both these factors. A useful approximation of the failure probability is given by

$$p_f \cong 2 \times 10^{-\beta},$$

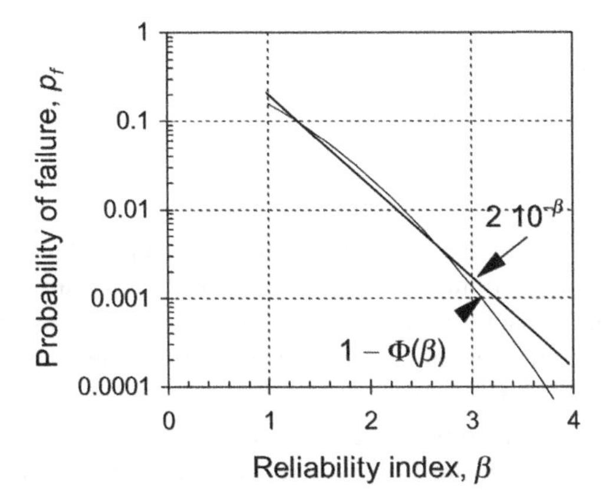

Fig. 9.1.9 Power approximation of the probability of failure as a function of the reliability index.

which can be used for reliability analysis with β taking values from 1 to 2.7, as shown in Fig. 9.1.9.

For example, in the case of the rigid timber beam of Example 9.3, where strength X and load Y are normal and independent variates,

$$\beta = \frac{\mu_S}{\sigma_S} = \frac{15.1}{10.42} = 1.45,$$

so that the reliability index is 1.451 sigma units. The power approximation of the corresponding probability of failure is

$$p_f \cong 2 \times 10^{-1.45} = .071.$$

The previously computed value of $p_f = 1 - \Phi(1.45) = 0.074$, so that the error in the power approximation is only about 3%.

To obtain a general expression for the reliability index in terms of the first two moments of the capacity and the demand functions, one can use Eqs. (3.4.27) and (3.4.28) to write μ_S as $\mu_S = \mu_X - \mu_Y$, and $\sigma_S^2 = \sigma_X^2 - 2\rho_{XY}\sigma_X\sigma_Y + \sigma_Y^2$. Thus,

$$\beta = \frac{\mu_X - \mu_Y}{\sqrt{\sigma_X^2 - 2\rho_{XY}\sigma_X\sigma_Y + \sigma_Y^2}}, \tag{9.1.14}$$

where ρ_{XY} denotes the correlation coefficient between capacity and demand. Accordingly, the reliability index is a maximum if $\rho_{XY} = +1$, and a minimum if $\rho_{XY} = -1$. For normally distributed X and Y, Eq. (9.1.11) gives the probability of failure as

$$p_f = 1 - \Phi(\beta), \tag{9.1.15}$$

and

$$r = \Phi(\beta) \tag{9.1.16}$$

is the associated reliability as obtained from Eq. (9.1.12).

Example 9.6. Irrigation water supply. Consider again the irrigation problem of Example 9.4. The reliability index of this design is

$$\beta = \left(\frac{\mu_S}{\sigma_S}\right) = \left(\frac{10}{7.81}\right) = 1.28,$$

so that the reliability of the system is $r = \Phi(1.28) = 0.9$. If a higher reliability, say, 95% is sought, one must have an index of

$$\beta = \Phi^{-1}(0.95) = 1.65,$$

where $\Phi^{-1}(\xi)$ is the ξth quantile of the standard normal variate.

Although reliability problems are sometimes approached by assuming independent capacity and demand, doing so violates the objective, because engineering structures are designed so that capacity will accommodate the induced demand. That is, to cope with higher loads, structures are made stronger. Therefore, engineering practice often requires a positive correlation to link capacity with demand. For instance, when the capacity is known to be reduced, the demand is restricted, as in the case when heavy traffic is restricted if one or more lanes of a bridge are closed for repairs. To investigate the influence of correlation, one can rearrange Eq. (9.1.14) in terms of the central safety factor, thus obtaining

$$\beta V_Y = \frac{\zeta - 1}{\sqrt{v^2\zeta^2 - 2\rho v\zeta + 1}}, \tag{9.1.17}$$

where $v = V_X/V_Y$, and $\rho = \rho_{XY}$. This relationship is plotted in Fig. 9.1.10 for a range of correlation coefficients, and values of v of 1, 2, and 1/2.

A comparison of Fig. 9.1.10*b* and 9.1.10*c* shows that the influence of correlation between capacity and demand increases with decreasing ratio v between the coefficients of variation of capacity and demand. Thus, increasing the variability of demand relative to that of capacity increases the correlation. For $V_X = V_Y = V$, Eq. (9.1.17) simplifies to

$$\beta V = \frac{\zeta - 1}{\sqrt{\zeta^2 - 2\rho\zeta + 1}}. \tag{9.1.18}$$

This case is shown in Fig. 9.1.10*a*. Consider, for instance, a structural design with $V = 0.5$, and $\zeta = 2$. From Eq. (9.1.18), one has $\beta V \cong 0.45$ for $\rho = 0$, $\beta V \cong 0.33$ for $\rho = -1$, and $\beta V = 1$ for $\rho = 1$. If $\rho = 0$, one obtains $\beta = 0.45/0.5 = 0.9$, meaning that the risk of failure p_f equals $p_f = 1 - \Phi(0.9) = 0.18$. For $\rho = -1$, one gets $\beta = 0.33/0.5 = 0.66$, yielding $p_f = 1 - \Phi(0.66) = 0.25$. For $\rho = 1$, one has $\beta = 1/0.5 = 2$, and $p_f = 1 - \Phi(2) = 0.02$. The magnitude of the difference highlights the importance of recognizing the dependency between capacity and demand.

To measure the correlation coefficient between capacity and demand is not an easy task, because it depends on many factors. For instance, structural design often reflects a positive correlation coefficient of at least $\rho = 0.5$. However, with capacity and demand positively correlated, conservative design estimates of reliability are obtained when disregarding the dependency between the two; this results in high probabilities of failure for structures known to be very safe. On the contrary, nonconservative estimates are obtained if one disregards negatively correlated capacity and demand (see Example 9.4), although this error is substantially lower than that introduced by neglecting positive correlation.

Example 9.7. Irrigation water supply. Consider again the irrigation problem of Example 9.4, and assume that correlation between capacity and demand can be neglected. Assuming that $\rho_{XY} = 0$ yields

$$\sigma_S = \left(\sigma_X^2 + \sigma_Y^2\right)^{1/2} = (4^2 + 5^2)^{1/2} = 6.40 \text{ units},$$

and

$$\beta = \frac{\mu_S}{\sigma_S} = \frac{10}{6.40} = 1.56;$$

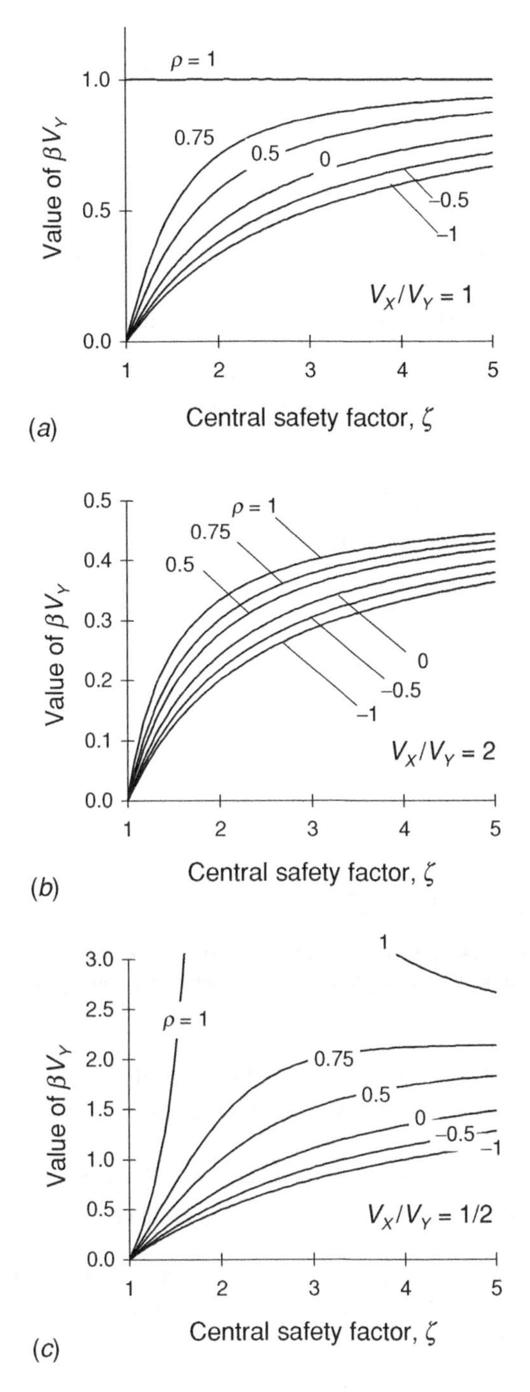

Fig. 9.1.10 βV_Y versus ζ for different correlation coefficients between capacity X and demand Y for (a) $V_X/V_Y = 1$, (b) $V_X/V_Y = 2$, and (c) $V_X/V_Y = 1/2$.

thus, the estimated reliability of the system is $r = \Phi(1.56) = 0.94$. If compared with the original estimate of 90%, this result illustrates that an engineer who disregards the correlation between capacity and demand can come to the misleading conclusion that the goal of 95% reliability can be reached.

One can use Eqs. (9.1.15) and (9.1.16) to compute the failure and nonfailure probabilities also if either X or Y or both are nonnormal. This is a straightforward exercise for two independent lognormal variates, as shown in the following example:

Example 9.8. Structural reliability index for independent lognormally distributed load and strength. Consider again a structure whose load-carrying capacity or strength X and load Y are independent lognormal variates, with means and standard deviations μ_X, μ_Y and σ_X, σ_Y, respectively (see Example 9.1). Because the nonfailure probability of this structure is given by

$$r = \Phi\left(\frac{\ln(m_X/m_Y)}{\sqrt{\ln\left(1+V_X^2\right)+\ln\left(1+V_Y^2\right)}}\right),$$

the reliability index is, from Eq. (9.1.16),

$$\beta = \frac{[\ln(m_X/m_Y)]}{\left[\ln\left(1+V_X^2\right)+\ln\left(1+V_Y^2\right)\right]^{1/2}} = \frac{\mu_{\ln(Z)}}{\sigma_{\ln(Z)}},$$

where m_X and m_Y denote the medians of X and Y, respectively. For the rigid timber beam of Example 9.1,

$$\ln\left(\frac{m_X}{m_Y}\right) = \mu_{\ln(Z)} = 0.469,$$

and

$$\sigma_{\ln(Z)} = \left[\ln\left(1+V_X^2\right)+\ln\left(1+V_Y^2\right)\right]^{1/2} = 0.288,$$

so that $\beta = 0.469/0.288 = 1.628$. This result can be compared with the value of 1.451, which is found under the assumption of independent and normally distributed strength and load (see Example 9.5).

If capacity and demand are lognormal variates, then $\ln X$ and $\ln Y$ are normal variates. A general form of the reliability index is found as

$$\beta = \frac{\ln(m_X/m_Y)}{\sqrt{\ln\left(1+V_X^2\right)+\ln\left(1+V_Y^2\right)-2\rho_{XY}\sqrt{\ln\left(1+V_X^2\right)\ln\left(1+V_Y^2\right)}}}, \tag{9.1.19}$$

where m_X and m_Y denote the medians of X and Y, respectively. Introducing the central safety factor as in Eq. (9.1.17), and assuming equal coefficients of variation $V_X = V_Y = V$, Eq. (9.1.19) reduces to

$$\beta = \frac{\ln\zeta}{\sqrt{2(1-\rho)\ln(1+V^2)}}. \tag{9.1.20}$$

Then, from the series expansion $\ln(1+x) = x - x^2/2 + x^3/3 - \cdots$, one can approximate $\ln(1+V^2)$ with V^2, so

$$V\beta = \frac{\ln\zeta}{\sqrt{2(1-\rho)}}. \tag{9.1.21}$$

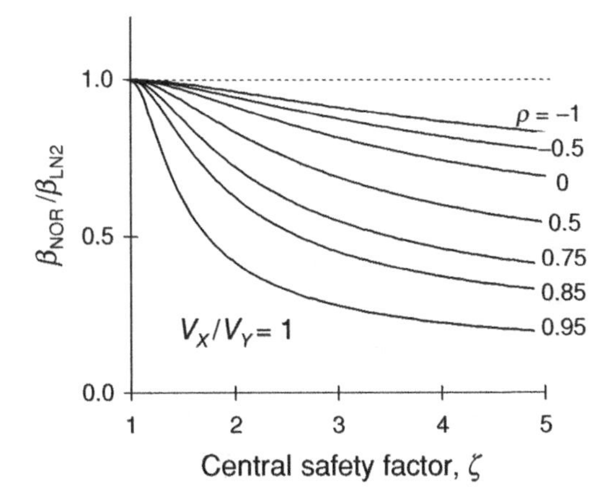

Fig. 9.1.11 $\beta_{NOR}\beta_{LN2}$ versus ζ for different correlation coefficients between capacity X and demand Y.

Figure 9.1.11 shows the ratio β_{NOR}/β_{LN2} as a function of the central safety factor ζ for a range of correlation coefficients relating capacity and demand, where β_{NOR} is computed from Eq. (9.1.18) and β_{LN2} from Eq (9.1.21). One can see that normally distributed variates produce smaller values of the reliability index than lognormal variates. Although this difference is minor for negative correlation and the ratio does not exceed $7/10$ for uncorrelated variates, reliability indices of normal and lognormal variates diverge from each other with rapidly increasing positive correlation. The magnitude of the reliability index is not known in most cases and as seen here the index by itself may not predict very effectively the performance of the investigated system, especially for correlated capacity and demand. However, this index is widely used as a basis for scaling performance in civil engineering practice.

As stated previously, Eqs. (9.1.15) and (9.1.16) can also be used to evaluate the failure and nonfailure probabilities if either X or Y or both are not normally distributed. The usual procedure is to transform these variates into equivalent normal deviates by using an appropriate transformation, such as the Rosenblatt transformation shown in the following example; accordingly, two or more jointly distributed random variables are transformed into another set of two or more normal variates. Thus, Eq. (9.1.14) is used in terms of the means, variances, and correlation coefficient of the equivalent normal variates.[2] However, in engineering practice the test of adequacy for the reliability index concept generally hinges on a comparison of the computed reliability index β with the recommended value.

Example 9.9. Thermal pollution in a river. The discharge Y from the cooling system of a thermal power plant flows into a river. To prevent thermal pollution in the river, it is desirable that Y does not exceed a fraction of natural flow Q in the river, say, $X = Q/a$, where a denotes a constant that depends on the difference in temperature between the two flows. An engineer wishes to evaluate the risk that thermal pollution occurs in the river. Assume that Y is normally distributed with mean 2 m^3/s and coefficient of variation of 20%, as shown in Fig. 9.1.12.

For the period in which the river receives the discharge Y, Q can be approximated by an exponential distribution with mean 40 m^3/s, and $a = 5$. Inflow Y and streamflow Q are

[2] For some additional details see, for example, by Ang and Tang (1984, p. 350).

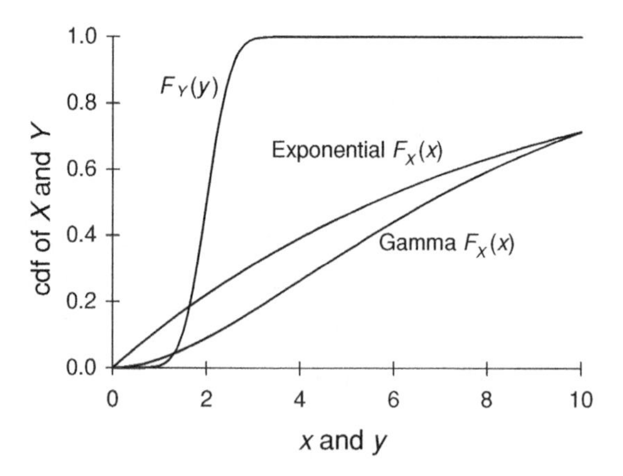

Fig. 9.1.12 cdfs of capacity X and demand Y for illustration of thermal pollution in a river.

further assumed to be independent variates. The problem is approached by using the univariate Rosenblatt transformation in order to determine the equivalent normal distribution for the nonnormal capacity X. Since X is exponentially distributed with mean $40/5 = 8\ \text{m}^3/\text{s}$,

$$f_X(x) = \left(\frac{1}{8}\right) \exp\left(-\frac{x}{8}\right) = 0.125 \exp(-0.125x)$$

and

$$F_X(x) = 1 - \exp(-0.125x).$$

The mean μ_{X^*} and the standard deviation σ_{X^*} of the equivalent normal distribution for the exponential capacity X are found from the assumption that, at the failure point x^*,

$$\Phi\left[\frac{(x^* - \mu_{X^*})}{\sigma_{X^*}}\right] = F_X(x^*),$$

where $\Phi(\cdot)$ denotes the cdf of the standard normal variate. Thus,

$$\mu_{X^*} = x^* - \sigma_{X^*}\Phi^{-1}[F_X(x^*)],$$

where $\Phi^{-1}(\xi)$ denotes the ξth quantile of the standard normal distribution. It also follows from the previous assumption at the failure point that, by equating the corresponding probability densities at the failure point,

$$\left(\frac{1}{\sigma_{X^*}}\right)\phi\left[\frac{(x^* - \mu_{X^*})}{\sigma_{X^*}}\right] = f_X(x^*),$$

where $\phi(\cdot)$ denotes the pdf of the standard normal variate. Hence,

$$\sigma_{X^*} = \frac{\phi\{\Phi^{-1}[F_X(x^*)]\}}{f_X(x^*)}.$$

By substitution,

$$\sigma_{X^*} = \frac{\phi\{\Phi^{-1}[1 - \exp(-0.125x^*)]\}}{[(0.125 \exp(-0.125x^*)]},$$

and

$$\mu_{X^*} = x^* - \sigma_{X^*}\Phi^{-1}[1 - \exp(-0.125x^*)],$$

whereas $\mu_Y = 2\ \text{m}^3/\text{s}$, and $\sigma_Y = 0.2 \times 2 = 0.4\ \text{m}^3/\text{s}$.

Because the failure point is unknown, the problem is solved by iteration. If $x^* = 1\ \text{m}^3/\text{s}$ is taken as the initial value,

$$\sigma_{X^*} = \frac{\phi\{\Phi^{-1}[1 - \exp(-0.125 \times 1)]\}}{[(0.125 \exp(-0.125 \times 1)]}$$
$$= \frac{\phi[\Phi^{-1}(0.118)]}{0.110}$$
$$= \frac{\phi[(-1.188)]}{0.110} = 1.79,$$

and

$$\mu_{X^*} = x^* - \sigma_{X^*}\Phi^{-1}[1 - \exp(-0.125 \times 1)] = 1 - 1.79\Phi^{-1}(0.118) = 3.12;$$

these are used in Eq. (9.1.14) to obtain, for independent capacity and demand,

$$\beta = \frac{(\mu_{X^*} - \mu_Y)}{\left(\sigma_{X^*}^2 + \sigma_Y^2\right)^{1/2}} = \frac{(3.12 - 2)}{(1.79^2 + 0.4^2)^{1/2}} = 0.61.$$

For the second iteration, let us take $x^* = 1.5$, which yields $\beta = 0.74$. As shown in Table 9.1.1, this procedure is then followed until the difference between two subsequent estimates of β is negligible. Accordingly, one obtains $\beta = 0.76$; that is, the reliability of the system $r = \Phi(0.76) \cong 78\%$.

One can also use the same approach for a capacity distribution different from the exponential. For example, if X is gamma distributed with mean $8\ \text{m}^3/\text{s}$, and its coefficient of variation is $1/\sqrt{2}$ (see Fig. 9.1.12), the parameters of the gamma pdf are found to be, by the method of moments,

$$r = \left(\frac{1}{V_X}\right)^2 = \left(\frac{1}{\sqrt{2}}\right)^2 = 2, \quad \lambda = \frac{r}{\mu_X} = \frac{2}{8} = 0.25\ \text{m}^{-3}\text{s}.$$

Thus, from Eq. (4.2.7),

$$f_X(x) = \left[\frac{\lambda^r}{\Gamma(r)}\right] x^{r-1} \exp(-\lambda x) = 0.25^2 x \exp(-0.25x),$$

Table 9.1.1 Risk evaluation for thermal pollution in a river with exponentially distributed streamflow

Exponential capacity, X						
Mean of X =	**8**					
λ =	0.125					
Iteration process						
Point of failure x^* =	1.0	1.5	2.0	2.5	2.1	**1.9**
$F(x^*)$ =	0.1175	0.1710	0.2212	0.2684	0.2309	0.2114
$f(x^*)$ =	0.1103	0.1036	0.0974	0.0915	0.0961	0.0986
$\Phi^{-1}[F(x^*)]$ =	−1.188	−0.950	−0.768	−0.618	−0.736	−0.802
$\phi\{\Phi^{-1}[F(x^*)]\}$ =	0.197	0.254	0.297	0.330	0.304	0.289
Mean of X^* =	3.12	3.83	4.34	4.73	4.43	4.25
Standard deviation of X^* =	1.79	2.45	3.05	3.60	3.17	2.94
Normal demand, Y						
Mean of Y =	**2**					
Standard deviation of Y =	**0.4**					
Evaluation of reliability index, β						
β =	**0.61**	**0.74**	**0.76**	**0.75**	**0.76**	**0.76**
Reliability: $\Phi(\beta)$ =						.777
Risk: $1 - \Phi(\beta)$ =						.223

Table 9.1.2 Risk evaluation for thermal pollution in a river with gamma-distributed streamflow

Gamma capacity, X						
Mean of X =	**8**					
Coefficient of variation of X =	**0.707**					
r =	2					
λ =	0.25					
Iteration process						
Point of failure, x^* =	1.0	1.5	2.0	2.5	1.9	2.1
$F(x^*)$ =	0.0265	0.0550	0.0902	0.1302	0.0827	0.0979
$f(x^*)$ =	0.0487	0.0644	0.0758	0.0836	0.0738	0.0776
$\Phi^{-1}[F(x^*)]$ =	−1.935	−1.598	−1.339	−1.125	−1.387	−1.294
$\phi\{\Phi^{-1}[F(x^*)]\}$ =	0.061	0.111	0.163	0.212	0.152	0.173
Mean of X^* =	3.44	4.26	4.87	5.35	4.76	4.98
Standard deviation of X^* =	1.26	1.73	2.15	2.53	2.06	2.23
Normal demand, Y						
Mean of Y =	**2**					
Standard deviation of Y =	**0.4**					
Evaluation of reliability index, β						
β =	**1.09**	**1.27**	**1.32**	**1.31**	**1.31**	**1.32**
Reliability, $\Phi(\beta)$ =						0.906
Risk, 1 $\Phi(\beta)$ =						0.094

and, for $r = 2$,

$$F_X(x) = \int_0^x \frac{\lambda^2}{\Gamma(2)} z^{2-1} \exp(-\lambda z) dz = 1 - (1 + \lambda x)e^{-\lambda x}$$

$$= 1 - (1 + 0.25x)e^{-0.25x}.$$

Using this procedure, one gets, for the initial value of $x^* = 1$ m^3/s,

$$\sigma_{X^*} = \frac{\phi\{\Phi^{-1}[1 - (1 + 0.25 \times 1) \times \exp(-0.25 \times 1)]\}}{[(0.25^2 \times 1 \times \exp(-0.25 \times 1)]} = 1.26,$$

$$\mu_{X^*} = 1 - 1.26 \times \Phi^{-1}[1 - (1 + 0.25 \times 1) \times \exp(-0.25 \times 1)] = 3.44;$$

and, using these values in Eq. (9.1.14),

$$\beta = \frac{(\mu_{X^*} - \mu_Y)}{\left(\sigma_{X^*}^2 + \sigma_Y^2\right)^{1/2}} = \frac{(3.44 - 2)}{(1.26^2 + 0.4^2)^{1/2}} = 1.09.$$

After some iterations, the reliability index is found to be 1.32. Hence, from Eq. (9.1.16) reliability is about 91%. The procedure is detailed in Table 9.1.2.

9.1.4 Performance function and limiting state

In many engineering applications, the assessment of reliability is made by comparing the calculated reliability index β with that found to be adequate from previous experience for the given system. For this purpose, one must establish a relationship between the capacity (for example, the strength) of the system and the demand (for example, the load) such that if capacity and demand are equal, there is a *limiting state* of interest. The safety factor $Z = X/Y$ is an example of such a relationship, where the *safe state* is represented by

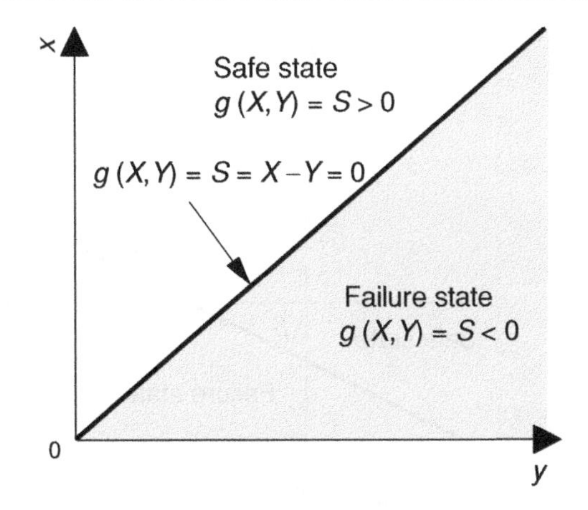

Fig. 9.1.13 Failure state, safe state, and limiting state of interest.

$Z > 1$, the *failure state* by $Z < 1$, and the *limiting state* by $Z = 1$. The margin of safety, defined as $S = g(X, Y) = X - Y$, is another example of this state, where $S > 0$ represents the *safe state*, $S < 0$ the *failure state*, and $S = 0$ the *limiting state*. The major advantage of using S instead of Z is that S is a linear function, whereas Z is highly nonlinear.

More generally, one can define a *performance function* $g(X, Y)$ which gives the limiting state of interest in the form

$$g(X, Y) = 0, \tag{9.1.22}$$

so that safety and failure of the system are represented by two regions in the plane (x, y) as shown in Fig. 9.1.13, separated by $g(X, Y) = X - Y = 0$.

Definition: Performance function. The *performance function* of a system is the random function $g(X, Y)$ of capacity X and demand Y describing system performance, related to its possible failure, or *limiting state of interest*, given by $g(X, Y) = 0$.

Let us define reduced variables as

$$X' = \frac{(X - \mu_X)}{\sigma_X}, \tag{9.1.23a}$$

$$Y' = \frac{(Y - \mu_Y)}{\sigma_Y}, \tag{9.1.23b}$$

for $S = g(X, Y) = X - Y = 0$. Then one obtains

$$g(X', Y') = \sigma_X X' - \sigma_Y Y' + \mu_X - \mu_Y = 0, \tag{9.1.24}$$

which provides an alternative form of the limiting state of interest. If a reduced coordinate system is introduced as shown in Fig. 9.1.14, the straight line generated by this expression is displaced at a distance equal to the reliability index β from the origin. This is because the shortest distance from the origin to a line $ax + by + c = 0$ is $c/(a^2 + b^2)^{1/2}$ (from analytical geometry). That is,

$$\beta = \frac{\mu_X - \mu_Y}{\sqrt{\sigma_X^2 + \sigma_Y^2}},$$

which is the reliability index for uncorrelated variables [from Eq. (9.1.14)]. This concept can be extended to any performance function in linear form. If the limiting state is a

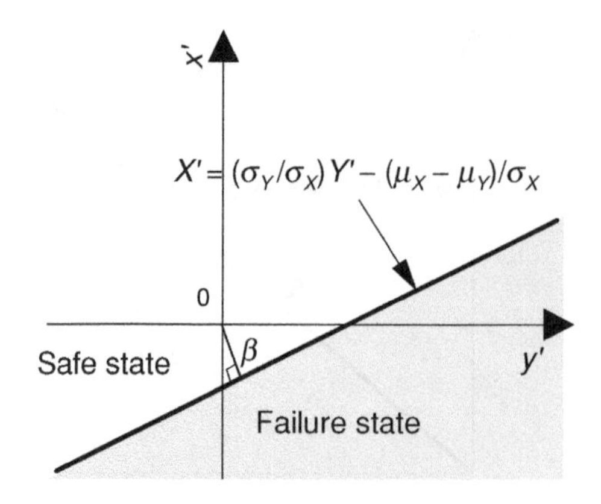

Fig. 9.1.14 Failure state, safe state, and limiting state in a reduced coordinate domain.

nonlinear function of reduced variables, this property does not hold. However, for a strictly monotonic $g(X', Y')$ the reliability index β corresponds to the shortest distance from the origin, as shown in Fig. 9.1.15.

More generally, the limiting state may be a function of a number of capacity and demand variables, $X_1, X_2, \ldots, X_n$, for the operational and environmental conditions of interest. A general limiting state or performance function is thereby introduced:

$$g(X_1, X_2, \ldots, X_n) = 0. \tag{9.1.25}$$

This defines a critical hypersurface in the n-dimensional space, such that $g(\cdot) > 0$ is the safe state and $g(\cdot) < 0$ is the failure state. Accordingly, the probability of failure is given by

$$p_f = \int_{g(x_1,\ldots,x_n)<0} \cdots \int f_{X_1,\ldots,X_n}(x_1, \cdots, x_n)\, dx_1 \cdots dx_n \tag{9.1.26}$$

and the corresponding nonfailure probability is

$$r = \int_{g(x_1,\ldots,x_n)>0} \cdots \int f_{X_1,\ldots,X_n}(x_1, \cdots, x_n) dx_1 \cdots dx_n. \tag{9.1.27}$$

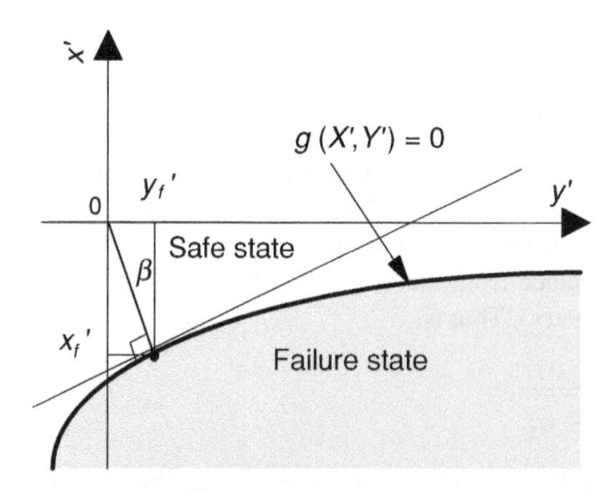

Fig. 9.1.15 Failure state, safe state, and nonlinear limiting state in a reduced coordinate domain.

A measure of the reliability index can be taken as the minimum distance from the critical hypersurface (in the multidimensional coordinate space) to the origin, by extending the concept of the two-dimensional case. Shinozuka (1983) showed that the minimum distance is the most probable failure point. Depending on the form taken by the performance function, the computations needed to obtain the failure point may be rather cumbersome. The solution can be carried out by minimizing the function $(X_1^2 + X_2^2 + \cdots + X_n^2)^{1/2}$, subject to the constraint $g(X_1, X_2, \ldots, X_n) = 0$, using Lagrange multipliers.[3]

If $g(\cdot)$ is expressed in linear form, say,

$$g(X_1, \ldots, X_n) = a_0 + \sum_{i=1}^{n} a_i X_i, \tag{9.1.28}$$

where a_i, $i = 0, \ldots, n$ are known constants, the reliability index defined by the minimum distance is given by

$$\beta = \frac{a_0 + \sum_{i=1}^{n} a_i \mu_i}{\sqrt{\sum_{i=1}^{n} \sum_{j=1}^{n} a_i a_j \rho_{ij} \sigma_i \sigma_j}}, \tag{9.1.29}$$

where μ_i and σ_i denote the mean and standard deviation of X_i, respectively, and ρ_{ij} is the correlation coefficient between X_i and X_j. For mutually independent variates,

$$\beta = \frac{a_0 + \sum_{i=1}^{n} a_i \mu_i}{\sqrt{\sum_{i=1}^{n} a_i^2 \sigma_i^2}}. \tag{9.1.30}$$

Example 9.10. Lake phytoplankton. Climate and water quality are among the factors influencing the quantity of phytoplankton in shallow lakes. Let us assume that the rate of increase of phytoplankton to be expressed as a linear function $g(X_1, X_2, X_3)$ of three variables: the temperature of water X_1, the global radiation X_2, and the concentration of nutrients X_3. The equilibrium corresponds to the limiting state of interest, $g(X_1, X_2, X_3) = 0$, and positive growth rates must be avoided to prevent eutrophication. Field observations indicate that X_1, X_2, and X_3 can be modeled as normal variates with the following means and coefficients of variation:

$$\mu_1 = 16°\text{C}, \quad \mu_2 = 150\ \text{W/m}^2, \quad \mu_3 = 100\ \text{mg/m}^3$$

and

$$V_1 = 0.5, \quad V_2 = 0.3, \quad V_3 = 0.7.$$

Thus,

$$\sigma_1 = 8°\text{C}, \quad \sigma_2 = 45\ \text{W/m}^2, \quad \sigma_3 = 70\ \text{mg/m}^3.$$

Although it is observed that temperature and radiation have no effect on the concentration of nutrients so that $\rho_{13} = \rho_{23} = 0$, mutually they are highly correlated with $\rho_{12} = 0.8$. The equilibrium function is

$$g(X_1, X_2, X_3) = a_0 - (a_1 X_1 + a_2 X_2 + a_3 X_3),$$

with $a_0 = 6.9$ mg/m^3, $a_1 = 0.08$ mg/(m$^3 \times$°C), $a_2 = 0.01$ mg/m $\times$ W, and $a_3 = 0.02$. Other variables should be incorporated, such as those accounting for predation and natural wastage;

[3] The detailed procedure based on the application of Lagrange multipliers is given, for example, by Ang and Tang (1984, p. 343).

these are included in the constant a_0 because of difficulties in estimating them separately. The reliability index is then computed using Eq. (9.1.29). The numerator is given by

$$a_0 - (a_1\mu_1 + a_2\mu_2 + a_3\mu_3) = 6.9 - (0.08 \times 16 + 0.01 \times 150 + 0.02 \times 100) = 2.12,$$

and the argument in the square root of the denominator is

$$\begin{aligned} &a_1^2\sigma_1^2 + a_2^2\sigma_2^2 + a_3^2\sigma_3^2 + 2a_1a_2\rho_{12}\sigma_1\sigma_2 + 2a_1a_3\rho_{13}\sigma_1\sigma_3 + 2a_2a_3\rho_{23}\sigma_2\sigma_3 \\ &\quad = 0.08^2 \times 8^2 + 0.01^2 \times 45^2 + 0.02^2 \times 70^2 + 2 \times 0.08 \times 0.01 \times 0.8 \times 8 \times 45 + 0 + 0 \\ &\quad = 3.03. \end{aligned}$$

Thus,

$$\beta = \frac{2.12}{\sqrt{3.03}} = 1.22,$$

and from Eq. (9.1.16),

$$r = \Phi(1.22) = 0.89.$$

This means there is an 89% chance that positive growth rates are prevented. Hence, the risk that the algal biomass will increase is 11%.

If the limiting state function $g(\cdot)$ is nonlinear, the distance from the failure hypersurface to the origin of the reduced variates may not be unique. Therefore, to find the exact probability of failure one should solve the integral in Eq. (9.1.26), which generally will require multiple numerical quadrature. To obtain an approximate solution one can use the hyperplane tangent to the hypersurface. If the exact nonlinear failure surface is convex toward the origin, the approximation will be conservative, as shown in Fig. 9.1.15 in the two-dimensional case. Conversely, if the surface is concave, it will be nonconservative, because the approximation lies on the unsafe side. By this method the ith component x'_{if} of the failure point $(x'_{1f}, x'_{2f}, \ldots, x'_{nf})$ expressed in reduced variates is given by

$$x'_{if} = -\frac{\left(\frac{\partial g}{\partial X'_i}\right)_f}{\sqrt{\sum_{i=1}^{n}\left(\frac{\partial g}{\partial X'_i}\right)_f^2}}\beta = -\alpha_i\beta, \qquad (9.1.31)$$

where (in geometrical terms) the α_is are the direction cosines of the component axes. The derivates are computed at point $(x'_{1f}, x'_{2f}, \ldots, x'_{nf})$. Thus, as in Eq. (9.1.23*a*),

$$x_{if} = \mu_i + \sigma_i x'_{if} = \mu_i - \alpha_i\sigma_i\beta. \qquad (9.1.32)$$

The required value of β is then found by substituting the right-hand side of Eq. (9.2.32) for the x_is in the performance function $g(x_1, x_2, \ldots, x_n)$ as follows:

$$g(\mu_1 - \alpha_1\sigma_1\beta, \mu_2 - \alpha_2\sigma_2\beta, \ldots, \mu_n - \alpha_n\sigma_n\beta) = 0. \qquad (9.1.33)$$

Solving this limiting state equation for β gives the required value of the reliability index. This usually requires an iterative procedure. One assumes tentative initial values of $x_{1f}, x_{2f}, \ldots, x_{nf}$; then using estimated values of standard deviations, one computes the partial derivates, and the values of the α_i, which are used to solve Eq. (9.1.33) for β, using estimated values of means. At each step one estimates β, and then reevaluates the $x_{if} = \mu_i - \sigma_i\alpha_i\beta$, and the procedure is repeated until there is convergence.

Example 9.11. Low-head run-of-river small hydropower station. The economic performance of the irrigation barrage located at Balcad, Somalia, along the Wabe Shabelle River could be improved by installing a hydropower station to meet the local energy demand. An engineer estimates the power demand X_3 to be 600 kW on average, with a variance of 3600 kW^2. If a standard turboaxial turbine is installed, power output can be estimated as 7.5 X_1X_2, where discharge X_1 is measured in m^3/s and hydraulic head X_2 in m, and 7.5 is a coefficient accounting for gravity, density of water, and the overall efficiency of installed equipment. Accordingly, power is given in units of kW. Although an average discharge of 22 m^3/s and an average head of 5.2 m are available, discharge and head availability depends on variable natural flows; it is also subject to the constraint of barrage handling, which is operated with priority for irrigation demand. Discharge and head can be assumed to be independent normal variates, X_1 and X_2, with coefficients of variation of 0.2 and 0.15, respectively. Assuming that the demand X_3 in kW is also normal and independent of discharge and head, the engineer wishes to evaluate the reliability of the plant. Thus,

$$g(X_1, X_2, X_3) = 7.5X_1X_2 - X_3$$

is the performance function of the hydropower system, with

$$\begin{aligned}
\mu_1 &= 22 \text{ m}^3/\text{s}, & V_1 &= 0.20, & \sigma_1 &= 4.4 \text{ m}^3/\text{s};\\
\mu_2 &= 5.2 \text{ m}, & V_2 &= 0.15, & \sigma_2 &= 0.78 \text{ m};\\
\mu_3 &= 600 \text{ kW}, & V_3 &= 0.10, & \sigma_3 &= 60 \text{ kW}.
\end{aligned}$$

The partial derivates of the performance function with respect to each of the variables evaluated at the failure point are determined as

$$\frac{\partial g}{\partial X_i'} = \left(\frac{\partial g}{\partial X_i}\right)\left(\frac{\partial X_i}{\partial X_i'}\right) = \left(\frac{\partial g}{\partial X_i}\right)\sigma_i,$$

which follows directly from Eq. (9.1.23*a*). Thus,

$$\left(\frac{\partial g}{\partial X_1'}\right)_f = 7.5x_2\sigma_1, \quad \left(\frac{\partial g}{\partial X_2'}\right)_f = 7.5x_1\sigma_2, \quad \left(\frac{\partial g}{\partial X_3'}\right)_f = -\sigma_3 = -60,$$

Taking the means as initial values (that is, $x_1 = 22$ m^3/s and $x_2 = 5.2$ m) one obtains

$$\begin{aligned}
\left(\frac{\partial g}{\partial X_1'}\right)_f &= 7.5 \times 5.2 \times 4.4 = 171.6,\\
\left(\frac{\partial g}{\partial X_2'}\right)_f &= 7.5 \times 22 \times 0.78 = 128.7,\\
\left(\frac{\partial g}{\partial X_3'}\right)_f &= -60.
\end{aligned}$$

Hence, form Eq. (9.1.31),

$$\begin{aligned}
\alpha_1 &= \frac{171.6}{\sqrt{49{,}610}} = 0.770,\\
\alpha_2 &= \frac{128.7}{\sqrt{49{,}610}} = 0.578,\\
\alpha_3 &= -\frac{60}{\sqrt{49610}} = -0.269.
\end{aligned}$$

Thus, the new failure point is given by

$$\begin{aligned}
x_{1(\text{new})} &= \mu_1 - \alpha_1\sigma_1\beta = 22 - (0.770 \times 4.4)\beta = 22 - 3.388\beta,\\
x_{2(\text{new})} &= \mu_2 - \alpha_2\sigma_2\beta = 5.2 - (0.578 \times 0.78)\beta = 5.2 - 0.451\beta,\\
x_{3(\text{new})} &= \mu_3 - \alpha_3\sigma_3\beta = 600 - (-0.269 \times 60)\beta = 600 + 16.14\beta.
\end{aligned}$$

Table 9.1.3 Evaluation of reliability for a low-head run-of-river small hydropower plant

Design data	Unit	Mean	Coefficient of variation	Standard deviation
Normal discharge, X_1	m^3/s	**22**	**0.20**	4.4
Normal hydraulic head, X_2	M	**5.2**	**0.15**	0.78
Normal power demand, X_3	kW	**600**	**0.10**	60
Limiting state of interest is $g(X_1, X_2, X_3) = 7.5X_1X_2 - X_3 = 0$				
Iteration process				
Initial x_{1f} =	22.0	17.8	17.7	17.7
Initial x_{2f} =	5.2	4.6	4.7	4.7
Initial x_{3f} =	600	620	623	623
$(\partial g/\partial X_1')_f$ =	171.6	153.2	154.6	154.8
$(\partial g/\partial X_2')_f$ =	128.7	104.2	103.7	103.6
$(\partial g/\partial X_3')_f$ =	−60.0	−60.0	−60.0	−60.0
$\Sigma(\partial g/\partial X_1')^2_f$ =	49610	37922	38254	38276
α_{1f} =	0.770	0.787	0.790	0.791
α_{2f} =	0.578	0.535	0.530	0.529
α_{3f} =	−0.269	−0.308	−0.307	−0.307
New x_{1f} =	17.8	17.7	17.7	17.7
New x_{2f} =	4.6	4.7	4.7	4.7
New x_{3f} =	620.0	622.8	622.7	622.7
$g(\cdot) = 7.5x_{1f}x_{2f} - x_{3f}$ =	-4.5×10^{-5}	7.5×10^{-6}	1.7×10^{-5}	1.8×10^{-5}
Yields β =	**1.24**	**1.23**	**1.23**	**1.23**
Reliability, $\Phi(\beta)$ =				0.892
Risk, $1 - \Phi(\beta)$ =				0.108

The limiting state equation is given by

$$7.5(22 - 3.388\beta)(5.2 - 0.451\beta) - (600 + 16.14\beta) = 0,$$

which yields $\beta = 1.24$. The second-order algebraic equation can be solved analytically; however, numerical computations are generally required. Results are shown in Table 9.1.3.

To perform the second iteration, one makes use of the new failure point:

$$x_{1(\text{new})} = 22 - 3.388\beta = 22 - 3.388 \times 1.24 = 17.8,$$

$$x_{2(\text{new})} = 5.2 - 0.451\beta = 5.2 - 0.451 \times 1.24 = 4.6,$$

$$x_{3(\text{new})} = 600 + 16.14\beta = 600 + 16.14 \times 1.24 = 620.$$

Then the values of the partial derivates are computed as

$$\left(\frac{\partial g}{\partial X_1'}\right)_f = 153.2, \quad \left(\frac{\partial g}{\partial X_2'}\right)_f = 104.2, \quad \left(\frac{\partial g}{\partial X_3'}\right)_f = -60.$$

Hence, from Eq. (9.1.31),

$$\alpha_1 = \frac{153.2}{\sqrt{37,922}} = 0.787,$$

$$\alpha_2 = \frac{104.2}{\sqrt{37,922}} = 0.535,$$

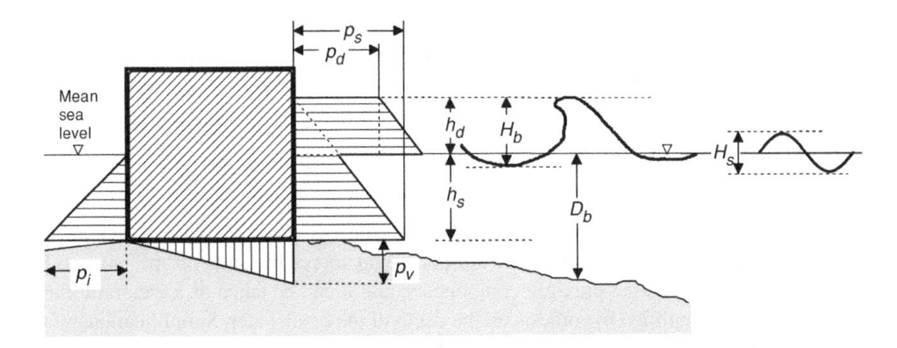

Fig. 9.1.16 Sketch of a vertical wall harbor breakwater.

and

$$\alpha_3 = \frac{-60}{\sqrt{37,922}} = -0.308.$$

This gives a new failure point

$$x_{1\,(\text{new})} = 22 - (0.787 \times 4.4)\beta = 22 - 3.463\beta,$$
$$x_{2\,(\text{new})} = 5.2 - (0.535 \times 0.78)\beta = 5.2 - 0.417\beta,$$
$$x_{3\,(\text{new})} = 600 - (-0.308 \times 60)\beta = 600 + 18.48\beta.$$

Accordingly, the new limiting state equation is

$$7.5(22 - 3.463\beta)(5.2 - 0.417\beta) - (600 + 18.48\beta) = 0,$$

which yields $\beta = 1.23$. Further iterations indicate that $\beta = 1.23$ is the required reliability index, as shown in Table 9.1.3. Therefore, the reliability of the hydropower station is about 89%, and the associated risk of failure is about 11%. The most probable failure point occurs for a demand 623 kW, a discharge of 17.7 m^3/s, and an available head of 4.7 m. System simulations should be implemented to account for the correlation between head and discharge, and for the variability of equipment efficiency with discharge and head (see La Barbera et al., 1983). By this procedure one can evaluate which types of standard turboaxial equipments should be adopted.

This method can also be used to evaluate the failure and nonfailure probabilities if one or more variables are not normal. One must transform these variables into equivalent normal deviates by using an appropriate procedure, such as the Rosenblatt transformation.[4] The usual practice, however, is to compute the reliability index and compare it to the design value.

Example 9.12. Harbor breakwater. Consider a harbor breakwater constructed with massive concrete tanks filled with sand (see Fig. 9.1.16). It is necessary to evaluate the risk that the breakwater will slide under the lateral pressure of a large wave during a major storm. Stability against sliding exists when the ratio of the resultant horizontal force R_h to the resultant of the vertical force R_v does not exceed the coefficient of friction c_f. Therefore, the limiting state of interest can be represented as

$$g(c_f, R_v, R_h) = c_f R_v - Rh = 0,$$

[4] This is introduced briefly in Section 9.4.

where c_f can be interpreted as a random variable, X_1, which represents the inherent uncertainty associated with its field evaluation. The resultant R_v of the vertical forces is given by the algebraic sum of the weight of the tank reduced for buoyancy, X_2, and the vertical component of dynamic uplift pressure due to the breaking wave F_v:

$$R_v = X_2 - F_v,$$

where F_v is proportional to the height of the design wave, H_b, when the slope of sea bottom is known. The resultant R_h of the horizontal forces depends on the balance between the static and dynamic pressure components, and it can be taken as a quadratic function of H_b under a simplified hypothesis on the depth of the breakwater. Simplifications of the shoaling effects indicate that the height H_b of the design wave is proportional to the random deepwater $X_4 = H_s$, which is found from frequency analysis of extreme storms in the area. Finally, the limiting state of interest can be written as

$$g(X_1, X_2, X_3, X_4) = X_1(X_2 - a_1 X_3 X_4) - X_3\left(a_2 X_4^2 + a_3 X_4\right) = 0,$$

where an additional variate X_3 is introduced to represent the uncertainties caused by the simplifications adopted to model the dynamic forces F_v and R_h; the constants a_1, a_2, and a_3 depend on the geometry of the system.

A unit width of a vertical wall located in La Spezia harbor, Italy, is considered, and all variables are assumed to be independent. Suppose X_1, X_2, and X_3 are normal variates with the following means and coefficients of variation:

$$\mu_1 = 0.64, \quad \mu_2 = 3400 \text{ kN/m}, \quad \mu_3 = 1,$$
$$V_1 = 0.15, \quad V_2 = 0.05, \quad V_3 = 0.20.$$

Frequency analysis of severe storms in the study area suggests that X_4 has an extreme value Type I distribution with mean $\mu_4 = 5.16$ m, and standard deviation $\sigma_4 = 0.93$ m. Following methods of Section 7.2, the scale and location parameters of the cdf of X_4 are

$$\alpha = \left(\frac{\sqrt{6}}{\pi}\right)\sigma_4 = 0.78 \times 0.93 = 0.73 \text{ m},$$

and

$$b = \mu_4 - 0.5772\alpha = 5.16 - 0.5772 \times 0.73 = 4.74 \text{ m}.$$

Finally, by accounting for the sea-bottom profile and the geometry of the breakwater wall one estimates that

$$a_1 = 70, a_2 = 17 \text{ m/kN}, a_3 = 145.$$

Accordingly, the limiting state equation becomes

$$g(X_1, X_2, X_3, X_4) = X_1 X_2 - 70 X_1 X_3 X_4 - 17 X_3 X_4^2 - 145 X_3 X_4 = 0.$$

The partial derivatives of the performance function with respect to each of the variables evaluated at the failure point are obtained from

$$\frac{\partial g}{\partial X_i'} = \left(\frac{\partial g}{\partial X_i}\right)\left(\frac{\partial X_i}{\partial X_i'}\right) = \left(\frac{\partial g}{\partial X_i}\right)\sigma_i.$$

Proceeding as in Example 9.11,

$$\left(\frac{\partial g}{\partial X_1'}\right)_f = (x_2 - 70x_3x_4)\sigma_1,$$

$$\left(\frac{\partial g}{\partial X_2'}\right)_f = x_1\sigma_2,$$

$$\left(\frac{\partial g}{\partial X_3'}\right)_f = -(70x_1x_4 + 17x_4^2 + 145x_4)\sigma_3,$$

$$\left(\frac{\partial g}{\partial X_4'}\right)_f = -(70x_1x_3 + 34x_3x_4 + 145x_3)\sigma_4,$$

where x_1,x_2,x_3, and x_4 are the values of the corresponding variables at the failure point. For the first iteration, one takes the means as the initial values. Since X_4 is an EV1 variate, the equivalent normal variate must be determined, as shown in Example 9.9. Thus,

$$\sigma_4^* = \frac{\phi\{\Phi^{-1}[F_{X_4}(x_{4f})]\}}{f_{X_4}(x_{4f})} = \frac{\phi\{\Phi^{-1}[F_{X_4}(5.16)]\}}{f_{X_4}(5.16)}$$

$$= \frac{\phi[\Phi^{-1}(0.570)]}{0.442} = \frac{\phi(0.177)}{0.442} = \frac{0.393}{0.442} = 0.889,$$

and

$$\mu_4^* = x_{4f} - \sigma_4^*\Phi^{-1}[F_{X_4}(x_{4f})] = 5.16 - 0.889 \times 0.177 = 5.002.$$

Also,

$$\left(\frac{\partial g}{\partial X_1'}\right)_f = (3400 - 70 \times 1.00 \times 5.002) \times 0.096 = 292.78,$$

$$\left(\frac{\partial g}{\partial X_2'}\right)_f = 0.64 \times 170 = 108.80,$$

$$\left(\frac{\partial g}{\partial X_3'}\right)_f = -(70 \times 0.64 \times 5.002 + 17 \times 5.002^2 + 145 \times 5.002) \times 0.20 = -274.97.$$

$$\left(\frac{\partial g}{\partial X_4'}\right)_f = -(70 \times 0.64 \times 1.00 + 34 \times 1.00 \times 5.002 + 145 \times 1.00) \times 0.889$$

$$= -320.01.$$

Hence, from Eq. (9.1.31),

$$\alpha_1 = \frac{292.78}{275,575^{1/2}} = 0.558,$$

$$\alpha_2 = \frac{108.80}{275,575^{1/2}} = 0.207,$$

$$\alpha_3 = \frac{-274.97}{275,575^{1/2}} = -0.524,$$

$$\alpha_4 = \frac{-320.01}{275,575^{1/2}} = -0.610.$$

Thus, the new failure point is given by

$$x_{1(\text{new})} = \mu_1 - \alpha_1\sigma_1\beta = 0.64 - (0.558 \times 0.096)\beta = 0.64 - 0.054\beta,$$

$$x_{2(\text{new})} = \mu_2 - \alpha_2\sigma_2\beta = 3400 - (0.207 \times 170)\beta = 3400 - 35.23\beta,$$

$$x_{3(\text{new})} = \mu_3 - \alpha_3\sigma_3\beta = 1 - (-0.524 \times 0.2)\beta = 1 + 0.105\beta,$$

$$x^*_{4(\text{new})} = \mu_4^* - \alpha_4\sigma_4^*\beta = 5.002 - (-0.610 \times 0.889)\beta = 5.002 + 0.542\beta,$$

and the limiting state equation becomes

$$(0.64 - 0.054\beta)(3400 - 35.23\beta) - 70(0.64 - 0.054\beta)(1 + 0.105\beta)(5.002 + 0.542\beta)$$
$$- 17(1 + 0.105\beta)(5.002 + 0.542\beta)^2 - 145(1 + 0.105\beta)(5.002 + 0.542\beta) = 0,$$

which is solved numerically to yield $\beta = 1.446$. Accordingly, the new failure point to be used for the second iteration is given by

$$x_{1\,(\text{new})} = 0.64 - 0.054\beta = 0.563,$$
$$x_{2\,(\text{new})} = 3400 - 35.23\beta = 3349,$$
$$x_{3\,(\text{new})} = 1 + 0.105\beta = 1.152,$$
$$x^*_{4\,(\text{new})} = 5.002 + 0.542\beta = 5.786.$$

The corresponding failure point for X_4 is found as

$$x_{4(\text{new})} = b - \alpha \ln \left\{ -\ln \Phi \left[\frac{x^*_{4\,(\text{new})} - \mu^*_4}{\sigma^*_4} \right] \right\}$$
$$= 4.74 - 0.73 \ln \left\{ -\ln \Phi \left[\frac{5.786 - 5.002}{0.889} \right] \right\} = 5.875.$$

After further iterations, as shown in Table 9.1.4, the value of $\beta = 1.352$ is obtained. Because there is no further change, the reliability of the structure is estimated as $\Phi(1.352) = 0.912$. This means that the sliding risk is about 9%. Simulations can also provide solutions to this problem, as seen in the work of Franco et al. (1986) and Burcharth (1994).

9.1.5 Further practical solutions

Alternative solutions to those presented in the preceding subsections can be used to apply simplified methods for reliability assessment. These include the *first-order, second-moment* (FOSM) method and the approximations obtained by *Taylor series expansion* and *point estimation* to evaluate the first- and second-order moments of the variates or safety indices. Although these solutions have a certain degree of inaccuracy in describing the prototype system, they provide satisfactory results in many cases. These methods are therefore quite attractive for the solution of engineering problems.

9.1.5.1 First-order second-moment (FOSM) method

A simplified reliability model, first introduced in structural steel design, only uses the mean values and coefficients of variation for the resistance X and load Y in a particular limiting state to obtain the reliability index β, which is computed as

$$\beta \approx \frac{\ln(\mu_X/\mu_Y}{\sqrt{V_X^2 + V_Y^2}}, \tag{9.1.34}$$

regardless of the type of distribution of X and Y.[5]

Example 9.13. FOSM reliability index. Consider again the rigid timber beam of Examples 9.3 and 9.8 with

$$\mu_X = 39.1\ \text{N/mm}^2, \quad V_X = 0.25, \quad \mu_Y = 24.0\ \text{N/mm}^2, \quad V_Y = 0.15.$$

[5] See Ravindra and Galambos (1978).

From Eq. (9.1.34),

$$\beta \approx \frac{\ln(39.1/24.0)}{\sqrt{0.25^2 + 0.15^2}} = 1.674.$$

This result is very close to 1.628 found under the assumption of independent and lognormally distributed strength and load (see Example 9.8).

The loss of accuracy due to this approximation may be negligible in some cases. The FOSM approximation is useful when the available information on design variables is not sufficient to evaluate their marginal and joint distributions with a satisfactory degree of accuracy. This method is widely applied to evaluate the reliability of individual system components in structural engineering. For example, the Load and Resistance Factor Design (LRFD) specification for structural steel buildings was developed by assessing the reliability of structural members and connections by the FOSM method.[6]

Example 9.14. LRFD specification for metal structures. The design inequality for metal structures contains partial factors for load effects, γ, and resistances, ϕ, in the form

$$\phi r_n \geq \gamma_d q_d + \sum_i \gamma_{ei} q_{ni},$$

where the subscript n denotes nominal (code-specified) values of resistance r and load effects q, the subscript i denotes different applicable resistance limit states, the subscript d means dead load, and the subscript e defines time-varying load effects due to occupancy, snow, earthquake, wind, and other effects. The nominal resistance r_n and the resistance factor ϕ depend on the limiting state appropriate to each type of structural member or connection. For example, $\phi = 0.85$ applies for the elastic limit state of an axially loaded column, and $\phi = 0.9$ applies for a flexured beam subjected to a bending moment. The LRFD specification gives the load factors γ in order to achieve a targeted value of the reliability index β for a given configuration of loads. For example, if the load combination includes dead, live, or snow loads, typical values of β are 3.0 for structural members (for example, columns and beams) and 4.5 for structural connections (for example, bolts and welds). The targeted β for structural members is 2.5 if a combination of dead, live, and wind loads is considered, and 1.75 for a combination of dead, live, and earthquake loads.

Consider, for example, an axially loaded column subjected to live load. The LRFD limit state equation is

$$0.85 r_n = 1.2 q_d + 1.6 q_{nl},$$

where r_n denotes the nominal capacity, and q_d and q_{nl} denote the dead load and the nominal live load, respectively. Here, the partial load factors are $\gamma_d = 1.2$ for the dead load, and $\gamma_{nl} = 1.6$ for the live load. For a particular case of live load-to-dead-load ratio of 3, the expected total load μ_Y is taken as[7]

$$q_d + q_{nl} = (3 + 1)q_d = 4q_d,$$

and $V_Y = 0.19$. From the limiting state equation, the nominal column resistance r_n is computed as

$$r_n = \frac{q_d(1.2 + 1.6 \times 3)}{0.85} = 7.06 q_d,$$

which is taken to equal the expected column resistance μ_X, with $V_X = 0.05$. From Eq. (9.1.34),

$$\beta \approx \frac{\ln(7.06/4.0)}{\sqrt{0.05^2 + 0.19^2}} = 2.988,$$

which is close to the targeted value of 3.

[6] See American Institute of Steel Construction (1986) for this specification and Smith (1991) for its practical use.

[7] See Ellingwood et al. (1982).

Table 9.1.4 Evaluation of sliding risk for a vertical wall harbor breakwater

Design data	Unit	Mean	Coefficient of variation	Standard deviation	Scale parameter	Location parameter
Normal coefficient of friction, X_1		**0.64**	0.15	0.096		
Normal reduced weight, X_2	kN/m	**3400**	0.05	170		
Normal model error, X_3		**1**	0.20	0.20		
Gumbel wave height, X_4	m	**5.16**	0.18	0.93	0.73	4.74
Limiting state of interest is						
$g(X_1, X_2, X_3, X_4) = X_1X_2 - 70X_1X_3X_4 - 17X_3X_4^2 - 145X_3X_4 = 0$						
Iteration process						
Initial $x_{1f} =$		0.640000	0.562553	0.586312	0.589100	0.589343
Initial $x_{2f} =$		3400.000	3349.036	3367.144	3367.601	3367.589
Initial $x_{3f} =$		1.000000	1.151530	1.132257	1.133355	1.133292
Initial $x_{4f} =$		5.160000	5.875090	6.043751	6.053718	6.055510
$F(x_{4f}) =$		0.570374	0.811049	0.847407	0.848997	0.849340
$f(x_{4f}) =$		0.441644	0.234245	0.193879	0.191666	0.191270
$\Phi^{-1}[F(x_{4f})] =$		0.177326	0.881769	1.023973	1.032142	1.033609
$\phi\{\Phi^{-1}[F(x_{4f})]\} =$		0.392719	0.270442	0.236171	0.234196	0233841
Mean of $X_{4f} =$		5.002318	4.857065	4.796410	4.792546	4.791848
Standard deviation of $X_{4f} =$		0.889221	1.154526	1.218138	1.221897	1.222573
$(\partial g/\partial X_1')_f =$		292.7844	276.7306	277.2945	277.1838	277.1715
$(\partial g/\partial X_2')_f =$		108.8000	95.63393	99.67297	100.1471	100.1883
$(\partial g/\partial X_3')_f =$		−274.967	−327.218	−348.719	−350.086	−350.248

	$(\partial g/\partial X_4')_f =$	−320.012	−506.683	−539.805	−542.945	−543.324
	$\Sigma(\partial g/\partial X_{i'}')^2_f =$	275574.5	449525.4	499820.9	504209.7	504736.7
	$\alpha_{1f} =$	0.557736	0.412743	0.392224	0.390357	0.390136
	$\alpha_{2f} =$	0.207257	0.142638	0.140984	0.141036	0.141021
	$\alpha_{3f} =$	−0.52379	−488046	−0.49325	−0.49302	492996
	$\alpha_{4f} =$	−0.60960	−0.75572	−0.76354	−0.76463	−0.76476
	New $x_{1f} =$	0.562553	0.586312	0.589100	0.589343	0.589372
	New $x_{2f} =$	3349.036	3367.144	3367.601	3367.589	3367.593
	New $x_{3f} =$	1.1515	1.132257	1.133355	1.133292	1.133284
	New $x_{4f}^* =$	5.78640	6.032927	6.053702	6.055509	6.055731
	$g(\cdot) =$	4.0×10^{-5}	6.3×10^{-5}	-3.0×10^{-4}	-2.1×10^{-4}	-2.5×10^{-4}
Yields	$\beta =$	**1.446464**	**1.354967**	**1.351792**	**1.351779**	**1.351779**
	Reliability, $\Phi(\beta) =$					0.911777
	Risk, $1 - \Phi(\beta) =$					0.088223

9.1.5.2 Taylor series expansion method

The *second-order second-moment method* simplifies the implied functional relationships by truncation of the Taylor series expansion of the function. This makes it possible to estimate the mean and variance of a derived variate using Eqs. (3.4.36) and (3.4.37). As implied, inputs and outputs are expressed as expected values and standard deviations. The advantages of this approach are the simple mathematical requirements and low computational needs; the only requirement is the knowledge of the first few moments. However, the mathematical requirements, although simpler than those of exact methods, are generally not elementary.

Example 9.15. Pier scour. Consider again the scour problem in bridge foundation of Example 8.16. The scour depth X measured from the average channel bed to the bottom of the scour hole is evaluated as

$$X = 1.59b^{0.980}Y^{0.055}S^{0.105}n^{-0.210}W^{-0.240},$$

where Y is the depth of flow just upstream of the pier, S is the slope, n is the roughness coefficient, W is sediment gradation, and b is the pier width. All these quantities are measured in metric units. For a pier width of 2.5 m, the sediment gradation $W \sim$ lognormal $(4, 1.6^2)$, the slope $S \sim N(0.002, 0.0004^2)$, the depth $Y \sim N(4.75 \text{ m}, 1.2^2 \text{ m}^2)$, and the roughness coefficient $n \sim$ uniform $(0.02, 0.04)$. Also, Y, S, n, and W are assumed to be independent of each other. Application of the Taylor series expansion (see Example 8.16) gives

$$E[X] = 3.40 \text{ m}, \quad \text{Var}[X] = 0.1208 \text{ m}^2.$$

If $Z \sim N(4 \text{ m}, 0.05^2 \text{ m}^2)$ is the maximum allowable scour, one can compute the reliability index β under the assumption of $X \sim N(3.40 \text{ m}, 0.35^2 \text{ m}^2)$ independent of Z. From Eq. (9.1.14),

$$\beta = \frac{4 - 3.4}{(0.05^2 + 0.35^2)^{1/2}} = 1.71.$$

This means that the reliability of the system is

$$r = \Phi(1.71) = .956.$$

This result can be compared with that obtained by simulation, where a value x of X is determined in each simulation cycle by the same procedure shown in Example 8.17; the corresponding value z of Z is simply generated from the standard normal generator, and a failure occurs if $z < x$. Ten runs each of 1000 simulation cycles yield the following estimates of the risk of failure:

$$0.055, 0.059, 0.054, 0.054, 0.059, 0.060, 0.048, 0.053, 0.051, 0.054,$$

with an average of 0.055, and the corresponding simulated reliability is 0.945. This shows that the Taylor series approximation to second-order moments and the further normal approximation to the cdf of system load X provide rather accurate reliability estimates.

9.1.5.3 Point estimation method

As in the Taylor series method, the point estimate method (Rosenblueth, 1975) provides estimates of the moments of a function of random variables from those of the underlying variables without requiring the specification of their joint probability distribution. It is possible to approach reliability problems by this method, which does not require derivatives of limiting state function. This method is advantageous because it is difficult or impossible to evaluate these derivatives when, for instance, the limiting state of interest is given as an implicit function or in the form of graphs or as finite element solutions. If $Z = g(X_1, X_2, \ldots, X_i, \ldots, X_m)$ denotes a random variable Z that is a function of m random variables $X_1, X_2, \ldots, X_i, \ldots, X_m$, one can obtain 2^m point estimates of Z, say, z_k with $k = 1, \ldots, 2^m$, as

$$z_k = g(\mu_1 + \eta_{1k}\sigma_1, \mu_2 + \eta_{2k}\sigma_2, \ldots, \mu_m + \eta_{mk}\sigma_m), \tag{9.1.35}$$

where the μ_i and σ_i denote the means and standard deviations, respectively, of the X_i; also, the η_{ik} are coefficients which take values of 1 and -1 satisfying

$$k = 1 + \sum_{i=1}^{m} 2^{i-2}(1 + \eta_{ik}). \tag{9.1.36}$$

If $Z = g(X_1)$, for example, one has two-point estimates; four-point estimates are obtained for $Z = g(X_1, X_2)$, and so on. The moments of Z are estimated as

$$E[Z^r] = \sum_{k=1}^{2^m} \pi_k z_k^r, \tag{9.1.37}$$

where π_k denotes the weight for the kth point estimate. This weight is given by

$$\pi_k = 2^{-m}, \tag{9.1.38}$$

for any k if the X_is are mutually independent random variables, and it is computed as

$$\pi_k = 2^{-m} \left(\sum_{i=1}^{m-1} \sum_{j=i+1}^{m} \eta_{ik}\eta_{jk}\rho_{ij} \right), \tag{9.1.39}$$

for correlated X_i, where ρ_{ij} denotes the correlation coefficient between X_i and X_j. Note that this method can also be applied when $g(X_1, X_2, \ldots, X_m)$ is not an explicit function of the X_i, but its value is determined through numerical computations.

For a function of one random variable, $Z = g(X_1)$, $\eta_{11} = -1$ and $\eta_{12} = +1$, so that $z_1 = g(\mu_1 - \sigma_1)$ and $z_2 = g(\mu_2 + \sigma_2)$. Since $\pi_1 = \pi_2 = 1/2$,

$$E[Z] = \sum_{k=1}^{2} \pi_k z_k = 0.5(z_1 + z_2), \tag{9.1.40}$$

and

$$\begin{aligned} \mathrm{Var}[Z] &= E[Z^2] - E^2[Z] = \sum_{k=1}^{2} \pi_k z_k^2 - 0.25(z_1 + z_2)^2 \\ &= 0.5\left(z_1^2 + z_2^2\right) - 0.25(z_1 + z_2)^2 \\ &= 0.25\left(z_1^2 - 2z_1z_2 + z_2^2\right). \end{aligned} \tag{9.1.41}$$

In the one-dimensional case, an insight of this method is given by analogy which can be established between a probability density function and a distributed vertical load on a simply supported horizontal rigid beam. The expected value is the analog of the center of loading, and the standard deviation gives information concerning the central tendency and scatter of the variate. Rosenblueth (1975) suggested that this information could be extracted from the beam analogy with a rigid beam of length b, as shown in Fig. 9.1.17, with reaction π_1 acting at x_1, π_2 acting at x_2, and

$$\pi_1 + \pi_2 = \int_0^b f_X(x)\,dx = 1.$$

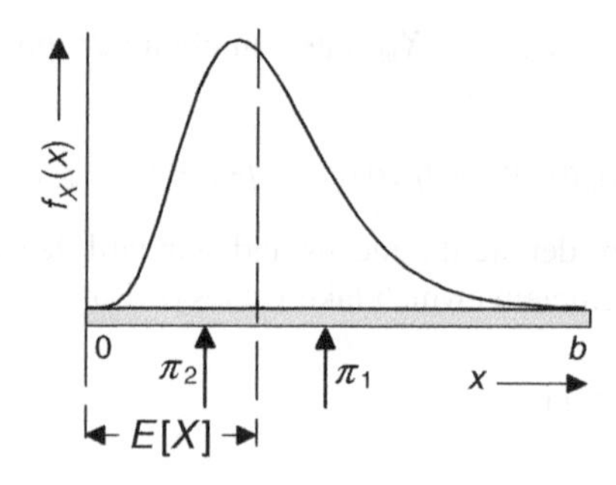

Fig. 9.1.17 Beam analogy with point estimates.

The reactions π_1 and π_2 are said to be *two-point estimates* of the distribution of $f_X(x)$. This analogy indicates that this method should require that the probability density function is symmetric. Harr (1987) introduced modified weights as $\pi_1 = 0.5 + \gamma_1/4$ and $\pi_2 = 0.5 - \gamma_1/4$ to account for skewness coefficient γ_1 from 0 to 1, and studied the accuracy of these weights for varying γ_1, as shown in Table 9.1.5.

Example 9.16. Bearing capacity of soil. In foundation engineering, bearing capacity of a soil depends on

$$Y = \tan^4\left(45 + \frac{X}{2}\right),$$

which is a function of the friction angle X. One wishes to evaluate how the mean and standard deviation of Y are influenced by variability in the friction angle. Application of Eq. (9.1.35) yields

$$y_1 = \tan^4\left(45 + \frac{\mu}{2} - \frac{\sigma}{2}\right),$$
$$y_2 = \tan^4\left(45 + \frac{\mu}{2} + \frac{\sigma}{2}\right),$$

where μ and σ denote the mean and standard deviation of X, respectively. These point estimates are used in Eqs. (9.1.40) and (9.1.41) to obtain

$$\begin{aligned}\mu_Y &= 0.5[\tan^4(45 + \mu/2 - \sigma/2) + \tan^4(45 + \mu/2 + \sigma/2)],\\ \sigma_Y^2 &= 0.25[\tan^8(45 + \mu/2 - \sigma/2) + \tan^8(45 + \mu/2 + \sigma/2)\\ &\quad - 2\tan^4(45 + \mu/2 - \sigma/2)\tan^4(45 + \mu/2 + \sigma/2)].\end{aligned}$$

For example, if $\mu = 25^\circ$ and the coefficient of variation is $V = 0.2$ (that is, $\sigma = 5^\circ$), values of $\mu_Y = 6.580$ and $\sigma_Y^2 = 5.863$ are obtained, so that the estimated coefficient of variation of

Table 9.1.5 Sensitivity of point estimate weight to skewness

$-\gamma_1$	$\pi_1 = 0.5 + \gamma_1/4$	Exact π_1
0.00	0.50	0.50
0.25	0.56	0.56
0.50	0.62	0.62
0.75	0.69	0.68
1.00	0.75	0.72

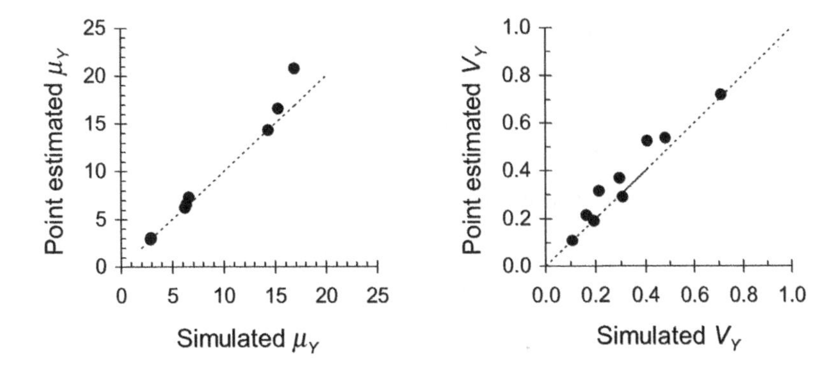

Fig. 9.1.18 Comparison of point estimated and simulated mean and coefficient of variation of bearing capacity factor Y of soil.

Y is $V_Y = 0.368$. Vannucchi (1985) compared the results from the application of the point estimation method with those obtained via simulation for $\mu = 15°, 25°$, and $35°$ and $V = 0.1$, 0.2, and 0.3. These results are shown in Fig. 9.1.18.

Note that the deviation of point estimates from simulations increases for increasing V. However, point estimates provide a satisfactory approximation to simulations.

In the bivariate case with $Z = g(X_1, X_2)$, $\eta_{11} = -1, \eta_{12} = +1, \eta_{13} = -1, \eta_{14} = +1, \eta_{21} = -1, \eta_{22} = -1, \eta_{23} = +1$, and $\eta_{24} = +1$. Thus, $z_1 = g(\mu_1 - \sigma_1, \mu_2 - \sigma_2)$, $z_2 = g(\mu_1 + \sigma_1, \mu_2 - \sigma_2)$, $z_3 = g(\mu_1 - \sigma_1, \mu_2 + \sigma_2)$, and $z_4 = g(\mu_1 + \sigma_1, \mu_2 + \sigma_2)$. From Eq. (9.1.39),

$$\pi_1 = \pi_4 = \rho/4 \quad \text{and} \quad \pi_2 = \pi_3 = -\rho/4,$$

where $\rho = \rho_{12} = \rho_{21}$ is the correlation coefficient between X_1 and X_2. Thus,

$$\begin{aligned} E[Z] &= \sum_{k=1}^{4} \pi_k z_k = 0.25\rho(z_1 + z_4) - 0.25\rho(z_2 + z_3) \\ &= 0.25\rho(z_1 + z_4 - z_2 - z_3) \end{aligned} \tag{9.1.42}$$

and

$$\begin{aligned} \mathrm{Var}[Z] &= 0.25\rho\left[\left(z_1^2 + z_4^2\right) - \left(z_2^2 + z_3^2\right)\right] - E^2[Z] \\ &= 0.25\left(z_1^2 + z_4^2 - z_2^2 - z_3^2\right) - E^2[Z]. \end{aligned} \tag{9.1.43}$$

For mutually independent X_1 and X_2,

$$E[Z] = \sum_{k=1}^{4} \pi_k z_k = 0.25(z_1 + z_2 + z_3 + z_4) \tag{9.1.44}$$

and

$$\mathrm{Var}[Z] = 0.25\left(z_1^2 + z_2^2 + z_3^2 + z_4^2\right) - 0.625(z_1 + z_2 + z_3 + z_4)^2. \tag{9.1.45}$$

Example 9.17. Low-head run-of-river small hydropower station. We consider again the irrigation barrage located at Balcad, Somalia, along the Wabe Shabelle River, where a hydropower station is to be installed to meet the local energy demand (see Example 9.11). Suppose that system demand is a normal variate X_3 independent of system capacity $Z = 7.5X_1X_2$, where X_1 and X_2 are independent normal variates. The mean, coefficient of variation, and standard deviation of these variates are

$$\mu_1 = 22 \text{ m}^3/\text{s}, \quad V_1 = 0.20, \quad \sigma_1 = 4.4 \text{ m}^3/\text{s};$$
$$\mu_2 = 5.2 \text{ m}, \quad V_2 = 0.15, \quad \sigma_2 = 0.78 \text{ m};$$
$$\mu_3 = 600 \text{ kW}, \quad V_3 = 0.10, \quad \sigma_3 = 60 \text{ kW}.$$

Using point estimation to evaluate the mean and variance of Z, one calculates

$$z_1 = 7.5(22 - 4.4)(5.2 - 0.78) = 583.44,$$
$$z_2 = 7.5(22 + 4.4)(5.2 - 0.78) = 875.16,$$
$$z_3 = 7.5(22 - 4.4)(5.2 + 0.78) = 789.36,$$
$$z_4 = 7.5(22 + 4.4)(5.2 + 0.78) = 1184.04,$$

which are used in Eqs. (9.1.44) and (9.1.45) to obtain

$$\bar{z} = 0.25(583.44 + 875.16 + 789.36 + 1184.04) = 858.00,$$

and

$$\hat{\sigma}_z^2 = 0.25(583.44^2 + 875.16^2 + 789.36^2 + 1184.04^2) - 858^2 = 46{,}672.8.$$

Since $X_3 \sim N(600 \text{ kW}, 60^2 \text{ kW}^2)$, assuming $Z \sim N(858 \text{ kW}, 216.0^2 \text{ kW}^2)$ gives the safety margin $S = Z - X_3$ as $S \sim N(258 \text{ kW}, 224.22^2 \text{ kW}^2)$. From Eq. (9.1.13) the reliability index β is estimated as

$$\beta = \frac{\mu_S}{\sigma_S} = \frac{258}{224.22} = 1.15.$$

This value can be compared with that of $\beta = 1.23$ found in Example 9.11 using the analytical approach.

Multivariate problems require the determination of the values of the η_is according to Eq. (9.1.36) as shown in Table 9.1.6.

Table 9.1.6 Coefficients η_{ik} for the point estimate method

	$i =$	1	2	3	4
	$k = 1$	−1	−1	−1	−1
$m = 1$	2	+1	−1	−1	−1
	3	−1	+1	−1	−1
2	4	+1	+1	−1	−1
	5	−1	−1	+1	−1
	6	+1	−1	+1	−1
	7	−1	+1	+1	−1
3	8	+1	+1	+1	−1
	9	−1	−1	−1	+1
	10	+1	−1	−1	+1
	11	−1	+1	−1	+1
	12	+1	+1	−1	+1
	13	−1	−1	+1	+1
	14	+1	−1	+1	+1
	15	−1	+1	+1	+1
4	16	+1	+1	+1	+1

9.1.6 Summary of Section 9.1

In this section we have discussed some fundamental concepts and criteria in the analysis of risk and reliability. These include the factor of safety, safety margin, reliability index, and performance function. As alternative solutions we have introduced the first-order second-moment (FOSM) method for resistance factor design, Taylor series approximation, and the point estimation method.

9.2 MULTIPLE FAILURE MODES

Let us consider the reliability of a system with components that have known reliabilities. A system can have three basic configurations, pertaining to *series*, *parallel* (or *redundant*), and *compound* (*series* and *parallel*) systems (see Fig. 9.2.1).

In general, a system can have one or more subsystems, and each subsystem can be decomposed into components. The engineer initially determines the reliabilities of the components and hence determines the reliabilities of the subsystems. Finally, the system reliability is calculated from the subsystem reliabilities using the axioms of probability and the concept of series and parallel systems.

For example, an earth dam can collapse because of either a destructive flood or a catastrophic earthquake (see, for example, Example 2.19). These two events can be viewed as a series mode producing the failure of the system, because the occurrence of either event

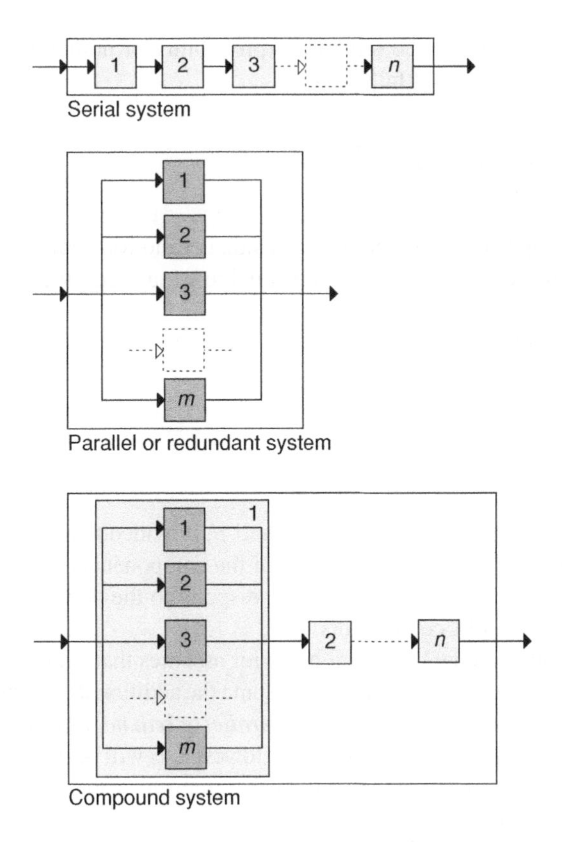

Fig. 9.2.1 Block diagram of series, parallel, and compound system layout.

can cause its collapse. The flood-induced failure due to overtopping occurs because of a heavy storm in the upstream basin. Overtopping of the dam occurs if this storm yields a flood hydrograph causing a flow exceeding spillway capacity; however, even if the spillway can cope with the upstream hydrograph, overtopping can also be caused by the combined effect of high flows and landslides of the slopes adjacent to the reservoir, which in turn can generate additional waves. Because the joint occurrence of these two modes—high flows (lower than spillway capacity) and landslides adjacent to the reservoir—can cause the failure of the dam, this can be interpreted as parallel modes of failure. Generally, the interrelationships among the various components of a system are complex; we will, however, assume initially that they are independent of each other.

9.2.1 Independent failure modes

Let us generalize the different types of system configurations. A *series system* performs only if each and every one of its n components does not fail. If the events $A_i = \{\text{failure of the } i\text{th component}\}$ are independent with probability $\Pr[A_i] = p_i$, the event corresponding to the successful performance of the series system is simply given by $\{A_1^c A_2^c \cdots A_n^c\}$, which describes the event that each and every component does not fail. Therefore, the reliability of the system is

$$r_{ss} = \Pr\left[A_1^c A_2^c \ldots A_n^c\right] = \Pr\left[A_1^c\right] \Pr\left[A_2^c\right] \times \cdots \times \Pr\left[A_n^c\right] = \prod_{i=1}^{n} (1 - p_i), \quad (9.2.1)$$

where $(1 - p_i)$ denotes the reliability (probability of nonfailure) of the ith component. The overall probability of failure is

$$p_f = 1 - \prod_{i=1}^{n} (1 - p_i) \approx \sum_{i=1}^{n} p_i, \quad (9.2.2)$$

in which the approximation on the right side is valid for small p_i. If the individual probabilities of failure p_i, $i = 1, \ldots, n$ are equal to p, say, Eq. (9.2.2) gives

$$p_f = 1 - (1 - p)^n, \quad (9.2.3)$$

and

$$p_f \approx np. \quad (9.2.4)$$

This approximation is quite close for small p, provided that n is not large. In summary, a series system or subsystem fails if any of the components causes the system to fail. That is to say, the adequacy of the system to respond to the demand by the users depends on the adequacy of all its components.

An alternative model of system behavior assumes that if one or more modes of failure occur, the remaining components can assume the additional responsibility to assure system performance. Such a system is called a *parallel* or *redundant system*. It operates if any of its components (or subsystems) functions. However, as will be discussed, we may sometimes require more than one component (or subsystem) to work. For independent components the failure of this system is described by the event $A_1 A_2 \cdots A_m$ that all components fail. The probability of failure is then the product of the individual probabilities. Therefore, the reliability of this system (which is the probability of the event that at least one component works) is

$$r_{rs} = \Pr[(A_1 A_2 \cdots A_m)^c] = 1 - \Pr[A_1 A_2 \cdots A_m] = 1 - \prod_{i=1}^{m} p_i, \tag{9.2.5}$$

and the risk of system failure is the product of the individual component modes, that is, the probability that all components will fail. Thus,

$$p_f = \prod_{i=1}^{m} p_i \tag{9.2.6}$$

for a system with m parallel or redundant components. If all the p_i are equal to p,

$$p_f = p^m. \tag{9.2.7}$$

For example, if there are five parallel modes of system failure, each of which is associated with a 1% chance of failure, the probability of failure of the system is 10^{-10}. For ten possible modes, the probability of failure is decreased to 10^{-20}. The extremely low probability of failure for the parallel model is a consequence of the *redundancy* of components. In other words, a parallel or redundant system fails only when all of its components fail.

The series and parallel models described represent unique situations. In series systems there is a single path connecting the output to the input, so that the removal of any component or link interrupts the path and results in the failure of the entire system. Conversely, there are m paths connecting the output to the input of a redundant system, which fails only if all its components are interrupted. Generally, engineering systems can be represented by a combination of both series and redundant components. With independent series components, total reliability of a system is the product of individual reliabilities; if it has independent redundant components, on the other hand, the total probability of failure equals the product of the individual probabilities of failure. For a combined system, the reliability of a system with $i = 1, 2, \ldots, m$ redundant components each with n_i series components is given by

$$r_s = 1 - \prod_{i=1}^{m} \left(1 - \prod_{j=1}^{n_i} r_{ij} \right). \tag{9.2.8}$$

Here r_{ij} denotes the reliability of the jth series subcomponent of the ith redundant component. The next illustration is devised to compare dependent and independent cases with series and redundant components.

Example 9.18. Pipe network. Consider the part of a pipeline network for urban water supply shown in Fig. 9.2.2.

Knowing the individual probabilities of rupture for each pipe, p_i, and the corresponding reliabilities $r_i = 1 - p_i, i = 1, \ldots, 5$, consider, the reliability of the system with respect to node d. This is the probability that node d does not remain isolated because no water passes through it. From Fig. 9.2.2 it is seen that for this condition to hold at least one of the following routes must work:

$$(1, 3), (2, 4), (1, 5, 4), \text{ and } (2, 5, 3).$$

Let us call these routes A, B, C, and D, respectively. For each route (that is, for each series component), the probability of failure is obtained as shown in the rightmost term

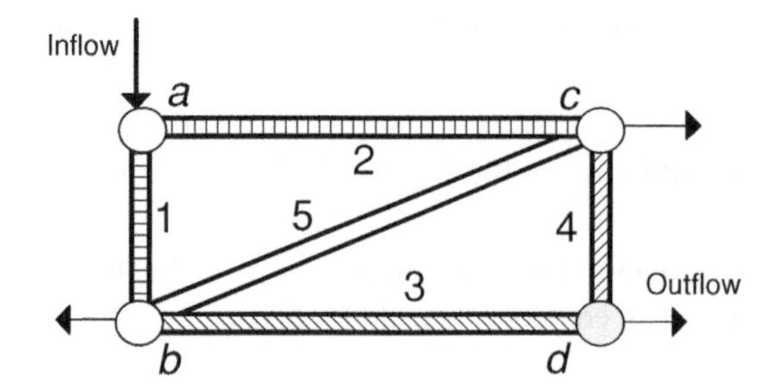

Fig. 9.2.2 Part of pipeline network for urban water supply showing series, without pipe *bc*, and redundant components, with pipe *bc*.

of Eq. (9.2.8). That is,

$$\Pr[A] = 1 - r_1 r_3 \text{ (that is, route A does not work);}$$

$$\Pr[B] = 1 - r_2 r_4;$$

$$\Pr[C] = 1 - r_1 r_4 r_5;$$

$$\Pr[D] = 1 - r_2 r_3 r_5.$$

These routes form four parallel-series (that is, redundant) systems. However, the events described here are *not* independent. For example, routes $A = (1, 3)$ and $D = (2, 5, 3)$ have pipe 3 in common. Therefore, their joint effects must be considered. Routes A and B are independent. Thus from Eq. (9.2.8),

$$\Pr[AB] = (1 - r_1 r_3)(1 - r_2 r_4) = (1 - r_1 r_3 - r_2 r_4 + r_1 r_2 r_3 r_4).$$

On the other hand, routes C and D are *not* independent. From the addition rule of probability [see Eq. (2.2.6)],

$$\begin{aligned}\Pr[CD] &= \Pr[C] + \Pr[D] - \Pr[C + D]\\ &= (1 - r_1 r_4 r_5) + (1 - r_2 r_3 r_5) - (1 - r_1 r_2 r_3 r_4 r_5)\\ &= 1 - r_1 r_4 r_5 - r_2 r_3 r_5 + r_1 r_2 r_3 r_4 r_5.\end{aligned}$$

Proceeding further, the probability of failure of the system is found as follows:

$$\begin{aligned}\Pr[ABCD] = 1 - r_s = 1 &- r_1 r_4 r_5 - r_2 r_3 r_5 + r_1 r_2 r_3 r_4 r_5\\ &- r_1 r_3 + r_1 r_3 r_4 r_5 + r_1 r_2 r_3 r_5 - r_1 r_2 r_3 r_4 r_5\\ &- r_2 r_4 + r_1 r_2 r_4 r_5 + r_2 r_3 r_4 r_5 - r_1 r_2 r_3 r_4 r_5\\ &+ r_1 r_2 r_3 r_4 - r_1 r_2 r_3 r_4 r_5 - r_1 r_2 r_3 r_4 r_5 + r_1 r_2 r_3 r_4 r_5.\end{aligned}$$

Note, incidentally, that there are no squared terms in such multiplications; the effect of any single pipe is considered only once in the same product. Hence,

$$\begin{aligned}1 - r_s = 1 &- r_1 r_3 - r_2 r_4 - r_1 r_4 r_5 - r_2 r_3 r_5 + r_1 r_3 r_4 r_5 + r_1 r_2 r_3 r_5\\ &+ r_1 r_2 r_4 r_5 + r_2 r_3 r_4 r_5 + r_1 r_2 r_3 r_4 - 2 r_1 r_2 r_3 r_4 r_5.\end{aligned}$$

If $r_1 = r_2 = r_3 = r_4 = r_5 = r$,

$$r_s = 2r^2 + 2r^3 - 5r^4 + 2r^5.$$

Let us now suppose that we remove pipe 5. We then apply Eq. (9.2.8) directly (for two independent routes):

$$r_s' = 1 - (1 - r^2)(1 - r^2).$$

Thus,

$$r_s' = 2r^2 - r^4;$$

and

$$r_s - r_s' = 2r^3 - 4r^4 + 2r^5 = 2r^3(1 - 2r + r^2).$$

This expression shows that $0 < r_s - r_s' < 1$ for $0 < r < 1$. For example, if $p = 0.01$, that is, $r = 0.99$,

$$r_s = 2 \times 0.99^2 + 2 \times 0.99^3 - 5 \times 0.99^4 + 2 \times 0.99^5 = 0.9998,$$
$$r_s' = 2 \times 0.99^2 - 0.99^4 = 0.9996.$$

Thus removal of one of the pipes causes only a negligible decrease in the reliability if p is small; even if $p = 0.5$, the decrease is $1/16$.

In designing the layout of an investigated system, one can compare the benefit achievable by decreasing the cost of system failure with the additional cost of adding a redundant component. Engineering designs are generally improved by adding redundant components; for instance, if k components are sufficient to prevent the failure of the system, some additional components are added to increase its reliability. This situation is represented by the k-out-of-m model, describing a system of m components, k of which must be operable for the system to succeed. An example is a pumping station with m parallel pumps, k of which are required to function in order to supply the target flow; in the meantime, maintenance and repairs can be made on the other $m - k$ pumps. If all m components have the same probability of failure p (that is, the same reliability $1 - p$), the binomial distribution gives the probability of success of x out of the m components. As in Eq. (4.1.3),

$$p_X(X = x|m, q) = \binom{m}{x}(1-p)^x p^{m-x}, \tag{9.2.9}$$

with $x = 0, 1, 2, \ldots, m$. Hence, the reliability of a k-out-of-m component system is

$$r = \sum_{x=k}^{m} p_X(X = x|m, 1 - p) = \sum_{x=k}^{m} \binom{m}{x} (1 - p)^x p^{m-x}, \tag{9.2.10}$$

and the probability of failure of the system is

$$p_f = 1 - \sum_{x=k}^{m} \binom{m}{x} (1 - p)^x p^{m-x}. \tag{9.2.11}$$

A block diagram for the k-out-of-m model has $\binom{m}{k}$ reliability paths, each path containing k different elements (which represent one of the combinations of m components taken k at a time). A block diagram for the 3-out-of-4 model is shown in Fig. 9.2.3, where the layout is sketched for a typical pumping station feeding a water treatment facility.

The system will fail only if all four paths fail. For the 3-out-of-4 model, Eq. (9.2.11) gives

$$p_f = 1 - \sum_{x=3}^{4} \binom{4}{x} (1 - p)^x p^{4-x} = 1 - 4(1 - p)^3 p - (1 - p)^4.$$

For example, if $p = 0.03$, then $p_f = 0.005$, so that the estimated reliability will be 99.5%.

Example 9.19. Flight planning. An airline company is planning the types and numbers of carriers to be used between two cities at night. Current estimates are that the minimum number of passengers is 100, the maximum is 250, and the average is 175. Three different options are considered for the aircraft with the following number of seats: 260, 180, and 110, as shown in Fig. 9.2.4.

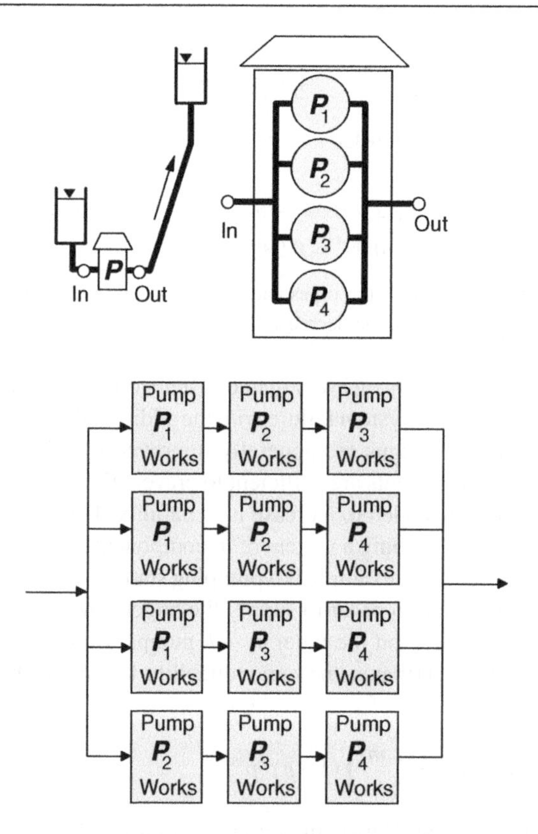

Fig. 9.2.3 Pumping station and pipeline connecting two tanks. The station has four parallel pumps, three of which must operate to ensure the target flow to reach the upper tank. The associated block diagram shows four reliability paths, each of them having three components.

One needs to find the reliability of each demand for each of the three options, assuming a constant risk p that an airplane is out of service. The block diagrams displaying the failure paths and the corresponding reliabilities are also given [see Eqs. (9.2.8) and (9.2.11)]. The curves in Fig. 9.2.4 show how the different reliabilities vary with the individual probability of failure p. As expected, systems with parallel elements are seen to have higher reliability. Thus, the redundant configuration at bottom left of the block diagrams has the smallest risk if $0 < p < 1/2$. The risk is always higher for the series systems. In a practical situation, different individual reliabilities are likely for each aircraft. Costs also need to be accounted for. This method can be applied to water resources management, road traffic, structural foundation, and other problems.

If $m = k$ in Eq. (9.2.11), $p_f = 1 - (1 - p)^m$; that is, Eq. (9.2.3) is obtained. For $k = 1$, $p_f = p^m$; that is, Eq. (9.2.7) is obtained. When all component reliabilities are equal, an m-out-of-m system is equivalent to a series configuration with m independent components; conversely, a 1-out-of-m system is equivalent to a redundant configuration with m independent components. Accordingly, these two configurations give the bounds for the probability of failure, since

$$\prod_{i=1}^{m} p_i \leq p_f \leq \sum_{i=1}^{m} p_i, \tag{9.2.12}$$

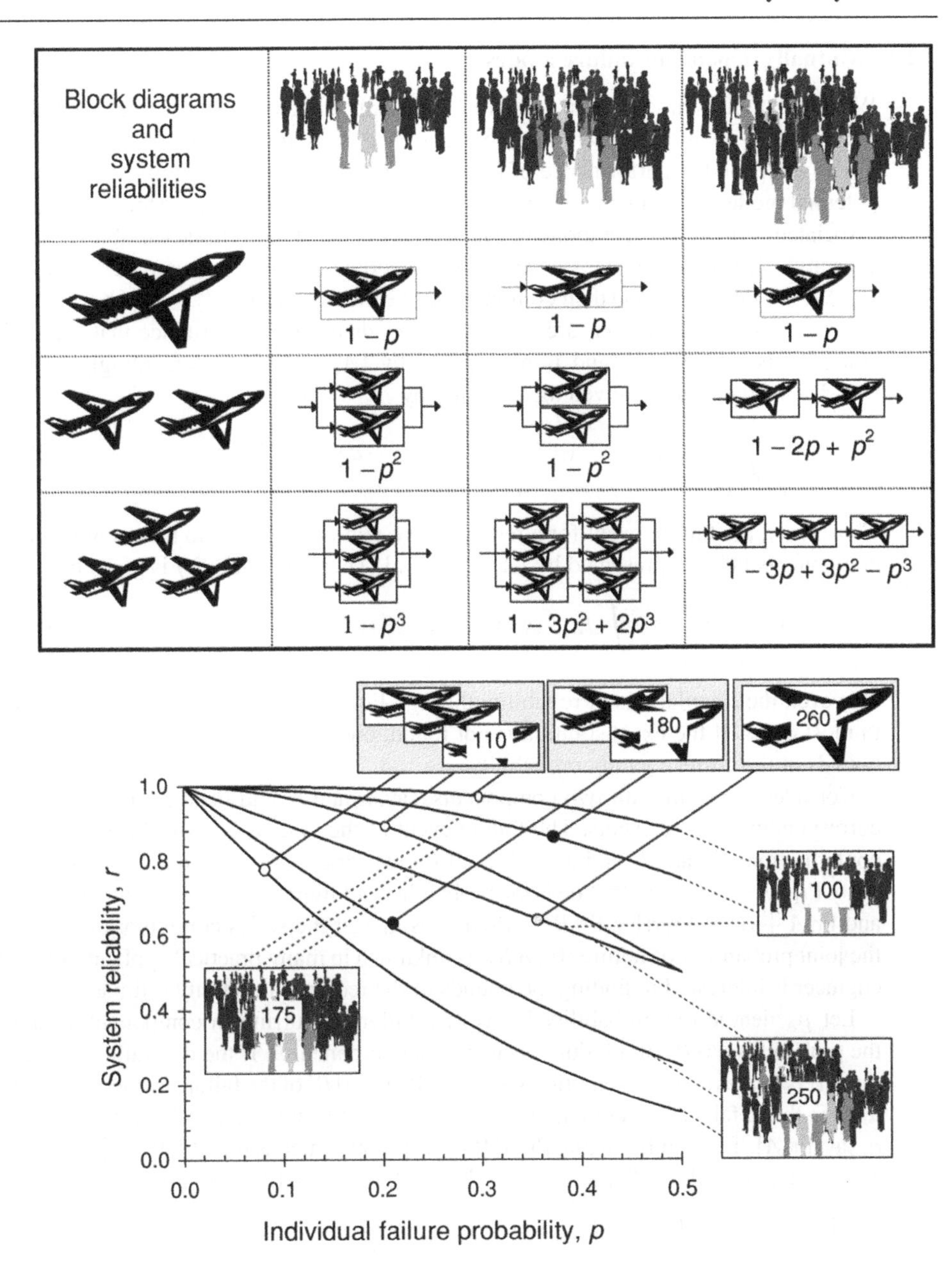

Fig. 9.2.4 Block diagrams and system reliabilities for different configurations of flight transportation, with a plot of reliability r against the individual probability of failure p.

where the p_i are the individual probabilities of failure of the m independent system components. For example, if $m = 10$, and $p_i = p = 10^{-2}$, for $i = 0, 1, 2, \ldots, m$, the lower bound is 10^{-20}, while the upper one is 10^{-1}. This range is rather wide, so some better procedure is necessary to tighten the bounds in order to approach complex systems with a large number of components.[8]

[8] See Harr (1987, p. 149) or Serfling (1974) for a discussion of this subject.

9.2.2 Mutually dependent failure modes

When the failure modes of a system are mutually dependent, the system components are related to each other such that the occurrence of a failure in one of the components has an effect on the performance of the others. Consider a system with n components or potential modes of failure, and denote with $g_i(X_1, X_2, \ldots, X_l)$ the performance function associated with each ith component, where $X_1, X_2, \ldots, X_l$ are l basic variables describing capacity and demand of the systems. The individual failure events are defined as $A_i = \{g_i(X_1, \ldots, X_l) < 0\}$, and their complements are the safe events, $A_i^c = \{g_i(X_1, \ldots, X_l) > 0\}$. For a series system, the safe event corresponds to the circumstance that each of the components proves safe, and is represented by $\{A_1^c A_2^c \cdots A_n^c\}$. Accordingly, from Eq. (9.1.27), its reliability is given theoretically by

$$r = \int\limits_{A_1^c A_2^c \cdots A_n^c} \cdots \int f_{X_1,\ldots,X_l}(x_1, \ldots, x_l) dx_1 \cdots dx_l. \tag{9.2.13}$$

Conversely, the failure event for a redundant system corresponds to the event that all the components fail is represented by $\{A_1 A_2 \cdots A_n\}$. Thus, its reliability is given by

$$r = 1 - \int\limits_{A_1 A_2 \ldots A_n} \cdots \int f_{X_1,\ldots,X_l}(x_1, \ldots, x_l) dx_1 \cdots dx_l. \tag{9.2.14}$$

However, the calculations of reliability through Eq. (9.2.13) or (9.2.14) are cumbersome in most cases. If the exact solution is not found, one can search for the upper and lower bounds of the corresponding probability.

Consider a system with two components, say, a and b, and denote with A and B the corresponding failure events. The failure event of the system is $\{AB\}$ for the redundant configuration of a and b, and it is $\{A + B\}$ for the series configuration. If the joint probability $\Pr[AB]$ is known, $\Pr[AB]$ is the probability of failure of the redundant arrangement, and $\Pr[A + B] = \Pr[A] + \Pr[B] - \Pr[AB]$ is that of the series configuration. However, the joint probability of failure ($\Pr[AB]$) is unknown in many practical applications, so the engineer is interested in finding the bounds of the required reliability estimate.

Let p_A denote the probability $\Pr[A]$ of a failure occurring in component a, and p_B the probability $\Pr[B]$ of a failure occurring in component b. If the two failure modes are *positively correlated*, the conditional probability $\Pr[A|B]$ of the failure event of component a given that b fails is larger than, or at least equal, to the marginal probability of failure of a, say, $\Pr[A]$. The condition that $\Pr[A|B] \geq \Pr[A]$ also means that $\Pr[A^c|B^c] \geq \Pr[A^c] = 1 - p_A$. Because $\Pr[A^c B^c] = \Pr[A^c|B^c]\Pr[B^c]$, from Eq. (2.2.10), it follows that

$$\Pr[A^c B^c] \geq \Pr[A^c]\Pr[B^c] = (1 - p_A)(1 - p_B).$$

This means that the reliability $\Pr[A^c B^c]$ of a series system exceeds (or equals) the product of the individual reliabilities of its components. Noting that the latter product is the reliability for mutually independent failure modes, one can see that positively correlated components increase the reliability of the system compared with the same configuration composed of independent components. Therefore, the product $(1 - p_A)(1 - p_B)$ can be taken as the lower bound of system reliability. Conversely, one notes that $A^c B^c \subset A^c$, and $A^c B^c \subset B^c$, as shown in the Venn diagrams of Fig. 9.2.5. Thus,

$$\Pr[A^c B^c] \leq \min\{\Pr[A^c], \Pr[B^c]\} = \min(1 - p_A, 1 - p_B)$$

provides the upper bound. Thus,

$$(1 - p_A)(1 - p_B) \leq \Pr[A^c B^c] \leq \min(1 - p_A, 1 - p_B).$$

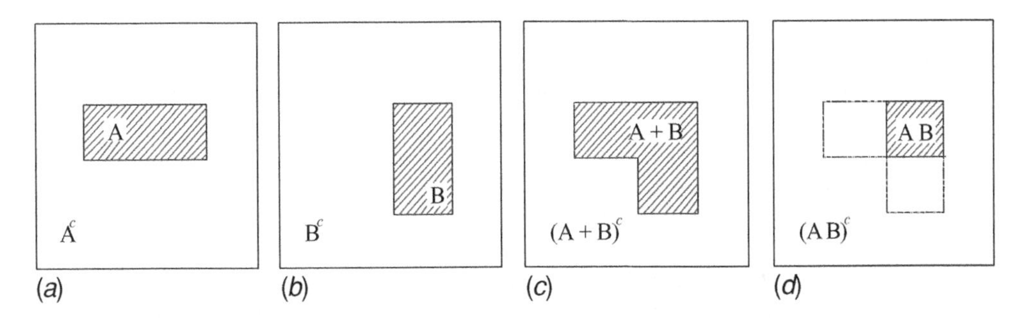

Fig. 9.2.5 Venn diagrams representing the sample space Ω and failure events A and B. The failure event of the series configuration is $A + B$, and $(A + B)^c$ is the safety event (c). Failure of the redundant configuration is AB, and AB^c its safety event (d).

This result can be extended to the general case of n series positively correlated components with individual failure probability p_i; the reliability bounds are thus

$$\prod_{i=1}^{n}(1 - p_i) \le r \le \min_{i=1,\dots,n}(1 - p_i). \tag{9.2.15}$$

With equal failure probability p for each of the components, the bounds become

$$(1 - p)^n \le r \le (1 - p). \tag{9.2.16}$$

For example, if there are five positively correlated serial modes of failure, each of them associated with a 1% chance of failure, the reliability of the system will range from 0.99^5 to 0.99, that is, from 95.1 to 99%. For ten possible modes, the reliability ranges from 90.4 to 99%, and so on. Therefore, the bounds may be widely separated for a large number of potential modes with the same individual risk. However, if the reliability of the system is dominated by a single dominating mode, the bounds will be narrow.

Consider now the redundant configuration of two positively correlated components a and b, where $\{AB\}$ is the failure event for the system. Since $\Pr[A|B] \ge \Pr[A]$, and $\Pr[AB] = \Pr[A|B]\Pr[B]$,

$$\Pr[AB] \ge \Pr[A]\Pr[B] = p_A p_B,$$

which gives the lower bound of the probability of failure. Conversely,

$$\Pr[AB] \le \min\{\Pr[A], \Pr[B]\} = \min(p_A, p_B)$$

provides the upper bound. Thus,

$$p_A p_B \le \Pr[AB] \le \min(p_A, p_B).$$

By extending this result to m redundant positively correlated components with individual failure probabilities p_i, the reliability bounds are

$$\min_{i=1,\dots,m}(1 - p_i) \le r \le 1 - \prod_{i=1}^{m} p_i, \tag{9.2.17}$$

and, for constant component risk p,

$$(1 - p) \le r \le (1 - p^m). \tag{9.2.18}$$

For example, if $m = 5$ and the individual risk is 1%, the reliability of the redundant system will range from 0.99 to $(1 - 10^{-10})$. For $m = 10$ and $p = 0.01$, $0.99 \le r \le (1 - 10^{-20})$.

The separation between these bounds dramatically increases with the number of potential modes of failure.

Using similar reasoning, the upper bound for negatively correlated series modes is found as follows:

$$r \leq \prod_{i=1}^{n} (1 - p_i). \tag{9.2.19}$$

That is, the effect of negatively correlated components is to decrease the reliability of the system as compared with the same configuration with independent components. This shows that the engineer should prevent negative correlations between series modes of failure in designing the configuration of a system. Conversely, the lower bound cannot be found, or it is null, so an analysis of system reliability requires accurate simulations of system behavior. Fortunately, negative correlation is less frequent than positive correlation in civil engineering practice, although it often occurs in many environmental engineering systems.

Example 9.20. Multipurpose reservoir. A reservoir located in a semiarid region is designed to meet two conflicting demands: water supply and flood control. Floods occur randomly with season, owing to variations in the local climate, whereas a water shortage may occur if the impounded water is low at the end of spring and the following summer is dry. Let F denote the event of a catastrophic flood in a year, D the occurrence of a summer drought in that year, and L that of a low reservoir level at the end of spring. From hydrologic analyses, one estimates the associated probabilities $p_F = \Pr[F]$, $p_D = \Pr[D]$, and $p_L = \Pr[L]$. A system failure is said to occur (event E) if the reservoir receives a high flood when it is at a high level or if water supply is insufficient. One can model this system through two series modes, the first one describing flood risk, and the second representing water shortage. The failure event A_1 associated with overflooding is simply $A_1 = F$, whereas that associated with insufficient water supply is $A_2 = DL$, so that the second component is a redundant subsystem. Thus, $E = A_1 + A_2 = F + DL$. Climatic records also indicate that events D and L are positively dependent, but there is a negative correlation between extreme floods and droughts in the region. Hence, the reliability of the system is, from Eq. (9.2.19),

$$r \leq (1 - \Pr[A_1])(1 - \Pr[A_2]) = (1 - p_F)(1 - \Pr[A_2]).$$

From Eq. (9.2.17),

$$\min(1 - p_D, 1 - p_L) \leq \Pr\left[A_2^c\right] \leq (1 - p_D p_L);$$

that is,

$$p_D p_L \leq \Pr[A_2] \leq 1 - \min(1 - p_D, 1 - p_L).$$

Accordingly,

$$r \leq (1 - p_F)(1 - p_D p_L).$$

For example, if $p_F = 0.01$, $p_D = 0.15$, and $p_L = 0.10$,

$$r \leq (1 - 0.01)(1 - 0.15 \times 0.10) = 0.99 \times 0.985 = 0.975,$$

which means that the system has a chance of failure of at least 2.5%.

The bounds obtained in the preceding discussion are generally too wide for obtaining effective reliability estimates, especially for a system with several modes of failure. The search for more effective bounds can be implemented when the underlying joint probability distribution is known. In the case of correlated normal variates $X_1, X_2, \ldots, X_l$, Ditlevsen (1979) found the joint probability of failure $\Pr[AB]$ to be bounded as

$$\max(p_a, p_b) \leq \Pr[AB] \leq p_a + p_b, \tag{9.2.20}$$

with

$$p_a = \Phi(-\beta_A)\Phi\left(-\frac{\beta_B - \rho\beta_A}{\sqrt{1-\rho^2}}\right), \tag{9.2.21a}$$

and

$$p_b = \Phi(-\beta_B)\,\Phi\left(-\frac{\beta_A - \rho\beta_B}{\sqrt{1-\rho^2}}\right), \tag{9.2.21b}$$

where β_A and β_B denote the reliability indexes of component a and b, respectively, and ρ the correlation coefficient between the two. Accordingly, the reliability bounds for a redundant system with two components a and b are

$$1 - p_a - p_b \leq r \leq 1 - \max(p_a, p_b). \tag{9.2.22}$$

For constant component reliability index, $\beta_A = \beta_B = \beta$, substituting Eq. (9.2.21) into (9.2.22) yields

$$1 - 2\Phi(-\beta)\Phi\left(-\beta\frac{1-\rho}{\sqrt{1-\rho^2}}\right) \leq r \leq 1 - \Phi(-\beta)\,\Phi\left(-\beta\frac{1-\rho}{\sqrt{1-\rho^2}}\right). \tag{9.2.23}$$

Note that $\Phi(-\beta)$ is the risk of failure of each individual component. These bounds are shown in Fig. 9.2.6, where the probability of failure $p_f = 1 - r$ is plotted against the reliability index β for three different values of the correlation coefficient. A comparison of the plots of Fig. 9.2.6*a* and 9.2.6*b* for $\rho = 0.5$ and 0.9, respectively, indicates that the reliability of a redundant system decreases, that is, p_f increases, with increasing positive correlation. It is also seen from Fig. 9.2.6*c* that the effect of negative correlation is to increase remarkably the reliability of the system as compared with the same configuration with independent components.

Example 9.21. Road connection. Two mountain resorts are connected by roads a and b. During a snowstorm in the region there is a 20% chance that traffic is suspended in road a, and a corresponding 10% chance for road b. The road between the two resorts can be modeled as a redundant system, with individual probabilities of failure $p_A = 0.2$ and $p_B = 0.1$. Assuming independent failures, the risk p_f that there is no access between the two resorts during a snowstorm is simply

$$p_f = p_A p_B = 0.2 \times 0.1 = 0.02.$$

However, limited facilities in the area delay the removal of snow from the two roads. Accordingly, from Eq. (9.2.17), the system reliability is bounded by

$$\min(1 - p_A, 1 - p_B) \leq r \leq 1 - p_A p_B;$$

that is, $0.8 \leq r \leq 0.98$. Hence, $0.02 \leq p_f \leq 0.2$.

From past experience the failure modes can be assumed to be normally distributed with a positive correlation of $\rho = 0.7$. From Eq. (9.1.15),

$$\beta_A = \Phi^{-1}(1 - p_A) = \Phi^{-1}(0.8) = 0.842$$

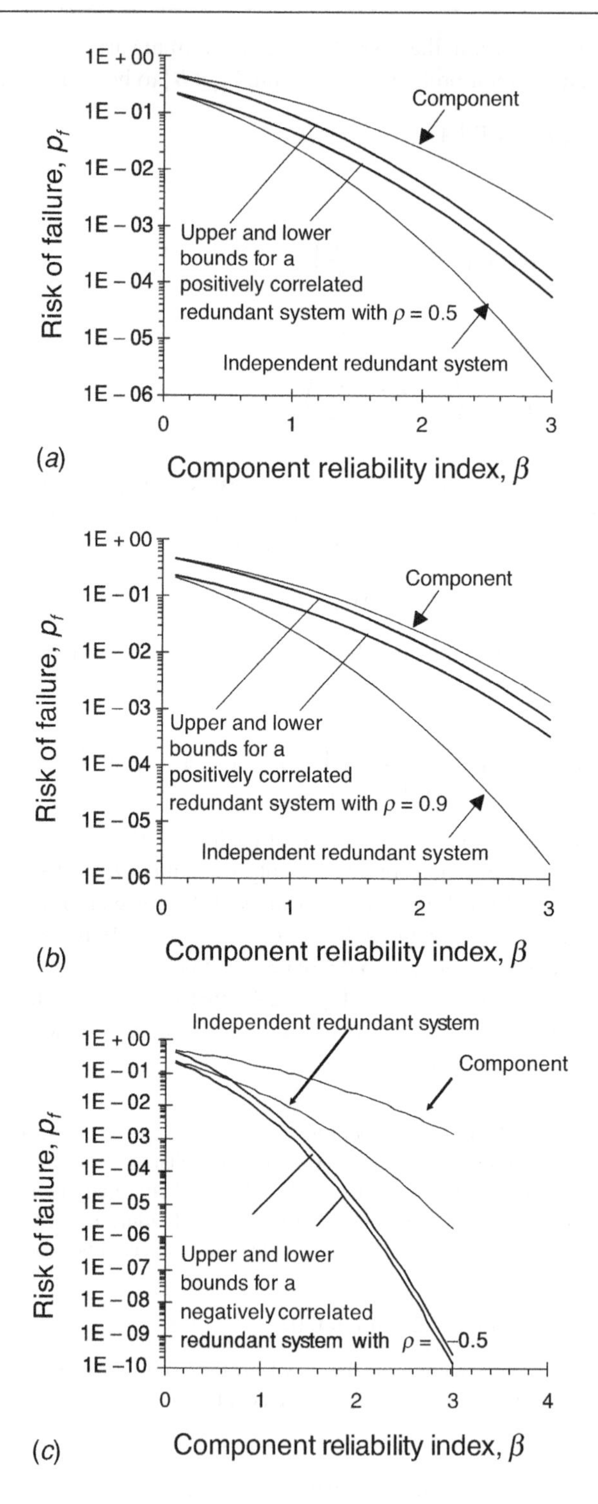

Fig. 9.2.6 Upper and lower bounds for the risk of failure of a correlated redundant system with (*a*) $\rho = 0.5$, (*b*) $\rho = 0.9$, and (*c*) $\rho = -0.5$.

and

$$\beta_B = \Phi^{-1}(1 - p_B) = \Phi^{-1}(0.9) = 1.282.$$

These are substituted in Eq. (9.2.21) to obtain

$$p_a = p_A \Phi\left(-\frac{\beta_B - \rho\beta_A}{\sqrt{1-\rho^2}}\right) = 0.2\Phi\left(-\frac{1.282 - 0.7 \times 0.842}{\sqrt{1 - 0.7^2}}\right) = .0332$$

and

$$p_b = p_B \Phi\left(-\frac{\beta_A - \rho\beta_B}{\sqrt{1-\rho^2}}\right) = 0.1\Phi\left(-\frac{0.842 - 0.7 \times 1.282}{\sqrt{1 - 0.7^2}}\right) = .0531.$$

Hence, from in Eq. (9.2.22),

$$1 - 0.0332 - 0.0531 \leq r \leq 1 - \max(0.0332, 0.0531);$$

that is, $0.914 \leq r \leq 0.947$. Thus, the required risk of failure is bounded as $0.053 \leq p_f \leq 0.086$. These limits are much narrower than those obtained by distribution-free methods, and thus provide an improved assessment of the road system.

For a series system with two components having individual failure probabilities p_A and p_B the failure event is $\{A + B\}$. From the addition rule of probability theory [see Eq. (2.2.6)],

$$\Pr[A + B] = p_A + p_B - \Pr[AB].$$

Thus,

$$p_A + p_B - \max(\Pr[AB]) \leq \Pr[A + B] \leq p_A + p_B - \min(\Pr[AB]),$$

and substituting the lower and upper bounds of Eq. (9.2.20) for $\min(\Pr[AB])$ and $\max(\Pr[AB])$, respectively,

$$p_A + p_B - p_a - p_b \leq \Pr[A + B] \leq p_A + p_B - \max(p_a, p_b), \tag{9.2.24}$$

where p_a and p_b are given in Eq. (9.2.21). Substituting $1 - r$ for the failure probability $\Pr[A + B]$ in Eq. (9.2.24) yields

$$1 - p_A - p_B + \max(p_a, p_b) \leq r \leq 1 - p_A - p_B + p_a + p_b. \tag{9.2.25}$$

For a constant component reliability index β, the individual failure probability is, from Eq. (9.1.15), $p = \Phi(-\beta)$. Substitution into Eq. (9.2.25) yields

$$1 - \Phi(-\beta)\left[2 - \Phi\left(-\beta\frac{1-\rho}{\sqrt{1-\rho^2}}\right)\right] \leq r \leq 1 - 2\Phi(-\beta)\left[1 - \Phi\left(-\beta\frac{1-\rho}{\sqrt{1-\rho^2}}\right)\right]. \tag{9.2.26}$$

Figure 9.2.7 shows these bounds in terms of the probability of failure $p_f = 1 - r$ plotted against the reliability index β for three different values of the correlation coefficient. Comparing these plots with those of Fig. 9.2.6 indicates that the series configuration is much less sensitive to correlation than the redundant one. Only high correlation levels between the components significantly reduce the risk of failure from that corresponding to the independent case, as shown in Fig. 9.2.7*b*. The effect of negative correlation is rather negligible, as shown in Fig. 9.2.7*c*.

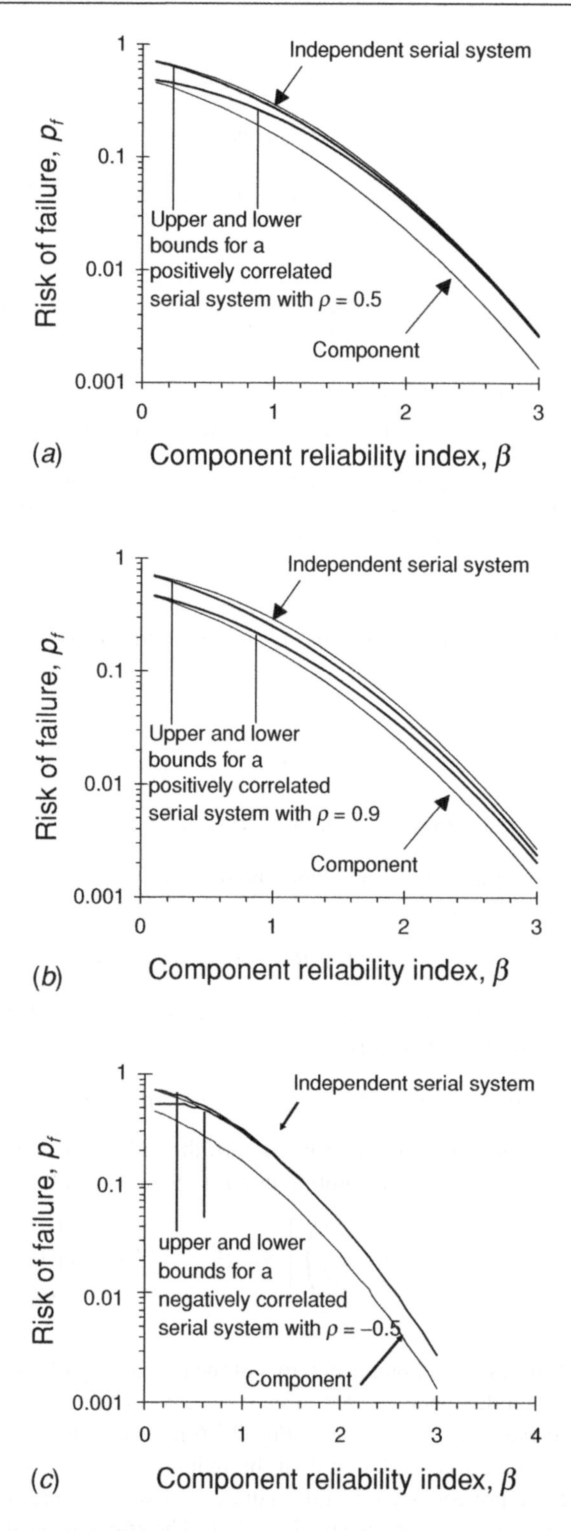

Fig. 9.2.7 Upper and lower bounds for the risk of failure of a correlated series system with (*a*) $\rho = 0.5$, (*b*) $\rho = 0.9$, and (*c*) $\rho = -0.5$.

Example 9.22. Suspension bridge. A suspension bridge relies on twin identical steel wires, the collapse of either of which causes the bridge to collapse. Accordingly, the failure of the bridge is modeled by two series modes. Previous analysis of wire safety has been made assuming independent and normally distributed strength X and load Y, with $\mu_X = 16.3$ N/mm^2, $V_X = 0.2$ N/mm^2, $\mu_Y = 7.2$ N/mm^2, and $V_Y = 0.3$. Thus, from Eq. (9.1.14),

$$\begin{aligned}\beta &= (\mu_X - \mu_Y)/[(V_X\mu_X)^2 + (V_Y\mu_Y)^2]^{1/2} \\ &= (16.3 - 7.2)/[3.26^2 + 2.16^2]^{1/2} = 2.33,\end{aligned}$$

which gives $\Phi(-2.33) = 0.01$, that is, an individual probability of failure of 1%. For independent failure modes, the reliability of the bridge would be

$$r = (1 - 0.01)^2 = 0.98.$$

This means that there is a 2% risk of failure due to wire collapse. However, correlation between the two failure modes must be introduced, because of their identical overall behavior. It can be shown that

$$\rho = \text{Var}[Y]/\text{Var}[X] = 2.16^2/3.26^2 = 0.44,$$

which also yields

$$\beta\frac{1-\rho}{\sqrt{1-\rho^2}} = 2.33\frac{1-0.44}{\sqrt{1-0.44^2}} = 1.45.$$

The reliability of the bridge is then computed from Eq. (9.2.23):

$$\begin{aligned}&1 - \Phi(-2.33)[2 - \Phi(-1.45)] \le r \le 1 - 2\Phi(-2.33)[1 - \Phi(-1.45)] \\ &1 - 0.010 \times (2 - 0.073) \le r \le 1 - 2 \times 0.010 \times (1 - 0.073) \\ &0.9808 \le r \le 0.9815.\end{aligned}$$

The associated risk of failure is $0.0185 \le p_f \le 0.0192$. By neglecting the correlation between the two failure modes, one overestimates the risk of failure up to 8%, although a conservative result is achieved under the assumption of independent modes.

The extension of this method to any multiple failure modes is not trivial. For a redundant system with $A_1, A_2, \ldots, A_m$ correlated normal modes of failure, the solution is obtained by subsequent comparison of the joint probability $\Pr[A_iA_j]$ for each pair of modes. For a series system having $A_1, A_2, \ldots, A_n$ correlated normal modes of failure with individual reliability index β_i the bounds are[9]

$$\begin{aligned}p_f \ge \Phi(-\beta_1) + \max\left(\sum_{i=2}^{n}\left\{p_i - \sum_{j=1}^{i-1}\left[\Phi(-\beta_i)\Phi\left(-\frac{\beta_j - \rho\beta_i}{\sqrt{1-\rho^2}}\right)\right.\right.\right.\\ \left.\left.\left. + \Phi(-\beta_j)\Phi\left(-\frac{\beta_i - \rho\beta_j}{\sqrt{1-\rho^2}}\right)\right]\right\}, 0\right)\end{aligned} \quad (9.2.27a)$$

and

$$\begin{aligned}p_f \le &\sum_{i=1}^{n}\Phi(-\beta_i) \\ &- \sum_{i=2}^{n}\max_{i<j}\left\{\max\left[\Phi(-\beta_i)\Phi\left(-\frac{\beta_j - \rho\beta_i}{\sqrt{1-\rho^2}}\right), \Phi(-\beta_j)\Phi\left(-\frac{\beta_i - \rho\beta_j}{\sqrt{1-\rho^2}}\right)\right]\right\}.\end{aligned} \quad (9.2.27b)$$

[9] Sharper bounds than those given by the following equations are possible. See, for example, Worsley (1982).

9.2.3 Summary of Section 9.2

The application of reliability to multiple failure modes is introduced here. Both independent and dependent cases are discussed. There is a comparison of series and redundant systems. The concepts of k-out-of-m models and reliability bounds are discussed.

9.3 UNCERTAINTY IN RELIABILITY ASSESSMENTS

In this section we discuss two additional approaches to the problem of uncertainty in the estimation of reliability. The first is based on the assumption that after completion of a system its (design) reliability is given by a parameter, say R^*, but it needs to be estimated. We cannot do so directly but we show how to find two bounds or limits that include it. We call these limits credibility limits. To obtain the credibility interval, the beta distribution is postulated.[10] Secondly, Bayes' theorem is used to update the prior estimate of reliability on receipt of additional information on system performance.

9.3.1 Reliability limits

In this subsection we assume that the design reliability of a system is a parameter R^*. This parameter is treated as a random variable in the Bayesian sense, as in Chapter 10. Instead of finding a point estimate, we seek a credibility interval, that is, an interval within which R^* lies at a given level of probability. This credibility interval for the parameter for reliability is analogous to the confidence interval in classical statistics where a parameter is a constant.

As in the classical case, suppose that the upper, r_U, and lower, r_L, limits of the design reliability R^* are those values satisfying $\Pr[R^* \le r_L] = \Pr[R^* > r_U] = \alpha/2$; that is,

$$\Pr[r_L < R^* \le r_U] = 1 - \alpha. \tag{9.3.1}$$

This of course means that, on a frequency basis, there is a $(1 - \alpha)$ percent chance that the credibility interval in Eq. (9.3.1) includes R^*. To resolve the problem, we treat reliability as a standard beta distributed random variate R, which is justifiable because of the imposed limits of 0 and 1. The credibility limits are then determined by solving

$$\alpha = 2F_R(r_L) = 2\int_0^{r_L} \frac{r^a(1-r)^b}{B(a+1, b+1)}dr \tag{9.3.2a}$$

for r_L and

$$\alpha = 2[1 - F_R(r_U)] = 2\int_{r_U}^{1} \frac{r^a(1-r)^b}{B(a+1, b+1)}dr \tag{9.3.2b}$$

for r_U. The solution can be obtained either by using tabulated values of the beta distribution, or more conveniently by numerical methods, for which standard software facilities are available. Suppose that our system is one that experiences a run of successes before a

[10] See Subsections 4.2.4 and 10.2.4.

failure occurs. Thus we set $b = 1$, which follows from the geometric distribution.[11] The pdf of R is then given by

$$f_R(r|a, 1) = (a+1)(a+2)r^a(1-r). \tag{9.3.3}$$

Hence, from Eqs. (9.3.2),

$$(a+2)r_L^{a+1} - (a+1)r_L^{a+2} = \alpha/2 \tag{9.3.4a}$$

and

$$(a+2)r_U^{a+1} - (a+1)r_U^{a+2} = 1 - \alpha/2. \tag{9.3.4b}$$

Example 9.23. Water quality test. Water samples are taken daily from a particular distribution system to test the quality of supply. Suppose that after 20 successful water samples were taken, the twenty-first sample fails the required quality level.[12] Thus,

$$E[R] = \frac{a+1}{a+b+2} = \frac{20+1}{20+1+2} = 0.913$$

is the average estimated reliability. The corresponding standard deviation is

$$\sigma_R = \sqrt{\frac{(a+1)(b+1)}{(a+b+2)^2(a+b+3)}} = \sqrt{\frac{(20+1)(1+1)}{(20+1+2)^2(20+1+3)}} = 0.058.$$

The 90% credibility limits are found by substituting 0.10 for α in Eqs. (9.3.4). Thus,

$$22r_L^{21} - 21r_L^{22} = 0.05,$$

from which $r_L = 0.802$, and

$$22r_U^{21} - 21r_U^{22} = 0.95,$$

which gives $r_U = 0.984$.

9.3.2 Bayesian revision of reliability

In the design of civil and environmental engineering systems, there is often a need to update criteria used after receiving additional data or information. In a statistical assessment, if the reliability of a system is found to be below expectations after it is completed, the estimate of reliability should be revised. As in Subsection 9.3.1 we continue to treat reliability R as a random variable. It may be uncertain in such cases whether the initial estimate is inaccurate or whether a failure arises because of random factors. We need an appropriate procedure to find an answer.

As shown in Chapter 2, Bayes' theorem of Eq. (2.2.16) provides a rational method of updating the prior assessment of the probability of an event on receipt of additional data.[13] In the context of reliability of system reliability, when the introduction of the posterior outcome X alters the prior pdf, the reliability can be interpreted as the outcome of the random variable R in the range $0 \leq R \leq 1$. Within these limits it is reasonable to model R as a beta variate, following our previous postulation, with initially estimated mean and variance. Therefore, from Eq. (4.2.13a) the prior pdf of reliability is

$$f_R(r) = \frac{r^{a_0}(1-r)^{b_0}}{B(a_0+1, b_0+1)}, \tag{9.3.5}$$

[11] See Subsections 4.1.4 and 4.2.4.
[12] See further discussion in Subsection 10.2.4.
[13] Further details are given in Chapter 10.

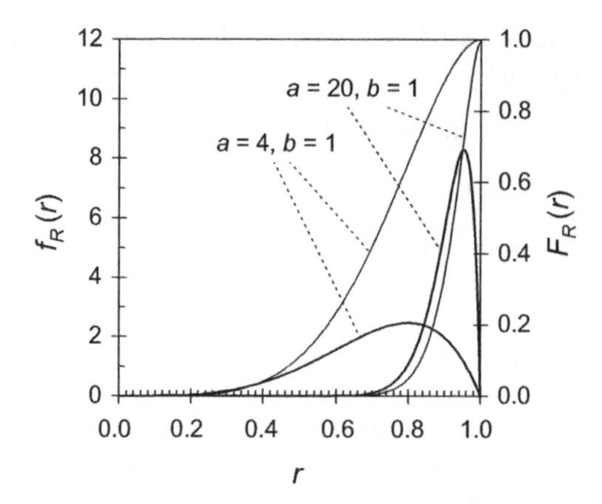

Fig. 9.3.1 Beta-distributed reliability estimates.

where $B(a_0 + 1, b_0 + 1)$ denotes the beta function of Eq. (4.2.13*b*), and $a_0 > 0, b_0 > 0$. The new data, say, X can be interpreted as the number of successes in n trials with r as the probability of a success. Then, as given by Eq. (4.1.3) and subject to independence and other conditions as specified, X is modeled as a binomial variate by the likelihood function

$$f_X(x|r) = \binom{x}{r} r^x(1-r)^{n-x}. \tag{9.3.6}$$

The beta distribution of Eq. (9.3.5) is said to be conjugate prior of the discrete binomial distribution of X given in Eq. (9.3.6).[14] By Bayes' theorem the posterior pdf is given by

$$f_R(r|x) = \frac{f_X(x|r)\, f_R(r)}{\int_0^1 f_X(x|r)\, f_R(r)\, dr}, \tag{9.3.7}$$

in which the role of the denominator is to provide the normalizing constant. Hence,

$$f_R(r|x) = \frac{r^{a_0+x}(1-r)^{b_0+n-x}}{B(a_0+x+1, b_0+n-x+1)}.$$

If we write $a_0 + x = a$, and $b_0 + n - x = b$,

$$f_R(r|x) = \frac{r^a(1-r)^b}{B(a+1, b+1)}. \tag{9.3.8}$$

The pdf of R for selected values of a and b is shown in Fig. 9.3.1, in which the corresponding cdf is also plotted. From Eqs. (4.2.14) the expected value and variance of the posterior reliability R are given by

$$E[R] = \frac{a+1}{a+b+2} \tag{9.3.9a}$$

[14] These conjugate distributions are extensively studied by Raiffa and Schlaifer (1961). The modern practice in applying Bayes' theorem in intractable situations is to simulate using Markov Chain Monte Carlo methods (see Section 10.3).

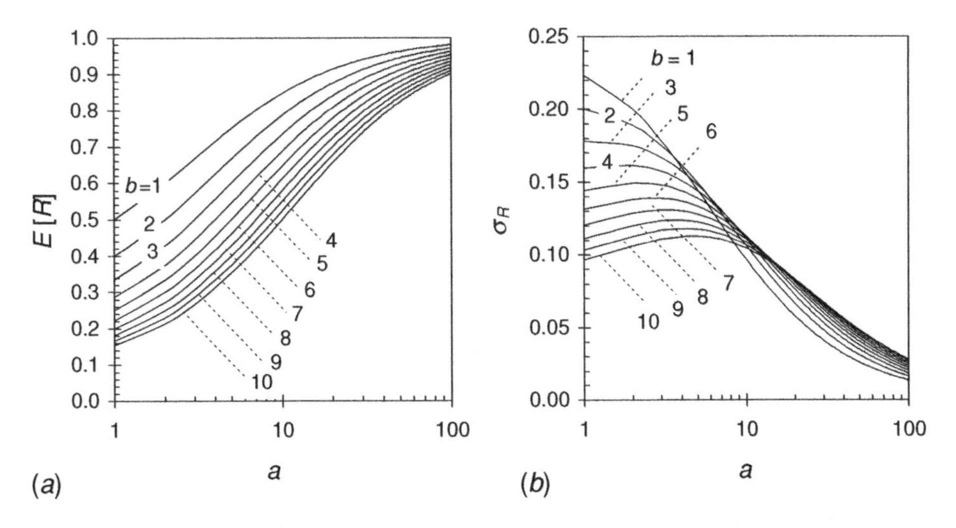

Fig. 9.3.2 Expected value (*a*) and standard deviation (*b*) of beta-distributed reliability estimates.

and

$$\text{Var}[R] = \frac{(a+1)(b+1)}{(a+b+2)^2(a+b+3)}. \tag{9.3.9b}$$

Figure 9.3.2*a* shows a plot of $E[R]$ against a with $1 \leq a \leq 100$, for values of b ranging from 1 to 10. The corresponding plot for the standard deviation of R is shown in Fig. 9.3.2*b*.

Example 9.24. Wastewater treatment. The design reliability of the efficiency in the abatement of water pollutants for a newly constructed wastewater treatment facility is 96%, which has been determined with an error of 2%. Assuming initially $E[R] = 0.95$, and $V_R = 0.02$, one takes R to be a beta-distributed variate with parameter values of $a_0 = 75$ and $b_0 = 2$ (see Fig. 9.3.3).

After the beginning of plant operation, the outflow from the plant is sampled daily and tested in the laboratory to evaluate the actual performance of the system. The engineer wishes to assess the reliability of the system after $n = 90$ samples have been analyzed, and no failures

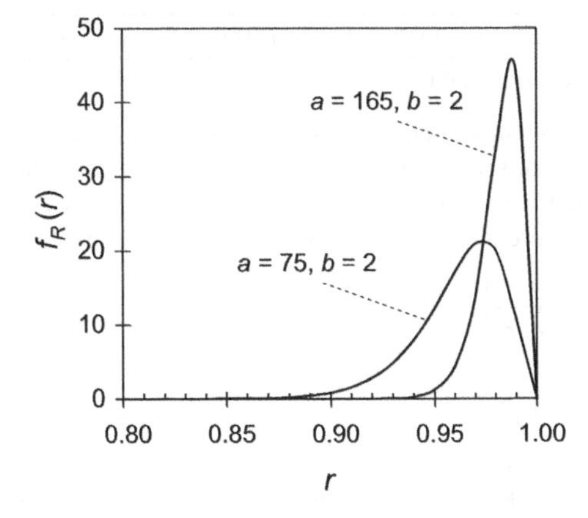

Fig. 9.3.3 Prior and revised pdf of estimated reliability for Example 9.24.

have been observed. Since $x = 90$, and $n - x = 0$, the assessed reliability of outflow quality is beta distributed with parameters

$$a = a_0 + x = 75 + 90 = 165$$

and

$$b = b_0 + n - x = b_0 = 2.$$

Thus,

$$E[R] = \frac{a+1}{a+b+2} = \frac{165+1}{165+2+2} = 0.98$$

is the expected reliability after the assessment. Also,

$$\sigma_R = \sqrt{\frac{(a+1)(b+1)}{(a+b+2)^2(a+b+3)}} = \sqrt{\frac{(165+1)(2+1)}{(165+2+2)^2(165+2+3)}} = 0.010.$$

Note that without prior knowledge of system performance, substituting $a_0 = b_0 = 0$ in Eq. (9.3.5) results in the uniform prior $f_{R|a_{0'},b_0}(r|0,0) = 1$.

Example 9.25. Water quality test. Suppose water samples are taken daily from a distribution system to test the quality of supply. If b out of n samples are found to fail the required quality level, the engineer wishes to determine the expected value and variance of the reliability as a function of n and b assuming a uniform prior distribution. Since the number of successful samples is $a = n - b$,

$$E[R] = \frac{a+1}{a+b+2} = \frac{n-b+1}{n-b+b+2} = \frac{n-b+1}{n+2}.$$

The corresponding variance is

$$\text{Var}[R] = \frac{(a+1)(b+1)}{(a+b+2)^2(a+b+3)} = \frac{(n-b+1)(b+1)}{(n+2)^2(n+3)}.$$

For instance, if 100 samples are taken in a day, 5 of which do not meet the optimal quality requirements,

$$E[R] = \frac{100-5+1}{100+2} = 0.941,$$

and the associated standard deviation of the reliability is

$$\sigma_R = \sqrt{\frac{(n-b+1)(b+1)}{(n+2)^2(n+3)}} = \sqrt{\frac{(100-5+1)(5+1)}{(100+2)^2(100+3)}} = \sqrt{\frac{96 \times 6}{102^2 \times 103}} = 0.023.$$

If the test is aimed at detecting water pollution, it is required that all the samples meet the specified standard. Therefore, if n samples are taken daily, one must obtain $n = a$ continuous successes and $b = 0$ failures. An engineer is thus interested in knowing the number of samples to be taken in order to achieve a given level of reliability. Since, for $n = a$ and $b = 0$,

$$E[R] = \frac{n+1}{n+2},$$

one gets

$$n = \frac{2E[R]-1}{1-E[R]}.$$

Figure 9.3.4 shows a plot of this equation, from which can be determined the number of consecutively successful tested samples required to achieve a given average reliability. For

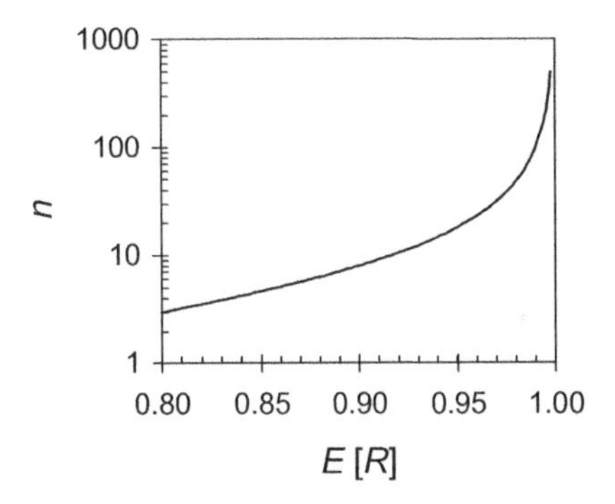

Fig. 9.3.4 Number of continuously successfully test samples required to achieve a given average reliability.

example, to achieve an average reliability of 90%, the required number of consecutively successful test samples is eight. The corresponding standard deviation is 0.028, so the coefficient of variation of the reliability is $0.028/0.9 = 0.03$, or 3%. For an average reliability of 95%, one should have $n = 18$, and the corresponding coefficient of variation of reliability is then 1%. For an average reliability of 99%, one must have $n = 98$, with a 0.1% coefficient of variation of reliability.

Suppose that the reliability of a design is predicted to be 90%. This is based on eight consecutive successes after the system is built (as discussed previously). Then a failure occurs, so that the previous estimate must be revised. For the posterior failure, we have $a = 8$ and $b = 1$. Hence the revised probability of $E[R] = 0.82$. A second failure would make the expected reliability 75%.

9.3.3 Summary of Section 9.3

Two additional methods of estimation of reliability, in the face of uncertainty, are shown here, based on the assumption that reliability is a beta variate. First, credibility limits are obtained for the design value of reliability. Second, Bayesian methods are used in the revision of reliability on receipt of new data.

9.4 TEMPORAL RELIABILITY

As stated in Section 9.1, reliability is the probability that a system performs adequately over a design period if it runs to specifications. A failure is said to occur when the system is incapable of working as specified to perform its intended function. Engineers are aware that most systems fail at some point in time and are therefore interested in evaluating the time to failure, called the *survival time*, particularly in the case of systems such as dams or other structures, which are characterized by irreversible failures. In this section we investigate reliability as a function of time.

9.4.1 Failure process and survival time

Generally, an engineer is interested in evaluating the risk of failure of a system over a specified period of time: the design life t. Because of the inherent randomness, the *survival time* of a system is defined as the random waiting time W to the first failure from the beginning of system operation, usually the time of the project's completion. Accordingly,

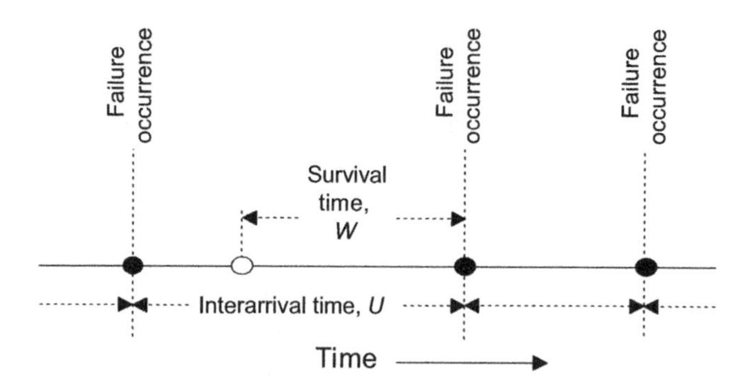

Fig. 9.4.1 Survival and interarrival times.

the risk of failure over the specified lifetime t is given by $p_f(t) = F_W(t)$, where $F_W(t)$ denotes the cdf of the survival time. The probability that the system will survive at least t units of time (for example, t years) from the beginning of system operation is defined as the *reliability function* of the system, denoted $R(t)$. This is the probability that no failure occurs within the specified lifetime t. Accordingly, $R(t)$ is a temporal function indicating the chance of a system to perform its intended function in time.

Definition: Reliability function. The *reliability function* of a system is defined as

$$R(t) = \int_t^\infty f_W(t)\,dt = 1 - F_W(t), \tag{9.4.1}$$

where $f_W(\cdot)$ is the pdf of *survival time*, W, and $F_W(\cdot)$ its cdf.

The probability distribution of survival time depends on the temporal process of failure occurrence (see Fig. 9.4.1). If this process is stationary, it can be represented by the random interarrival time U between two failure occurrences. Accordingly, the return period of failure occurrence is defined as the expected value of U, $\mu_U = E[U]$, and the average rate of failure $\lambda = 1/\mu_U$. To analyze the relationship between survival and interarrival times, W and U respectively, one denotes by U^* a time interval comprising the time origin, which is displaced at random on the time axis. Thus, the pdf of U^* is[15]

$$f_{U^*}(u) = (u/\mu_U)f_U(u) = \lambda u f_U(u),$$

where $f_U(u)$ denotes the pdf of U. The conditional pdf of W given that the failures occur in time with interarrival time U is

$$\begin{aligned} f_W(w|u) &= 1/u, && \text{for } w < u, \\ &= 0, && \text{elsewhere.} \end{aligned}$$

Therefore, the joint pdf of W and U^* is

$$\begin{aligned} f_{W,U^*}(w,u) &= f_W(w|u)f_{U^*}(u) = (1/u)\lambda u f_U(u) = \lambda f_U(u), && \text{for } w < u, \\ &= 0, && \text{elsewhere.} \end{aligned}$$

[15] Cox and Lewis (1961, p. 61) provide a heuristic argument for the derivation of the pdf of U^* (length-based sampling procedure).

The marginal pdf of survival time W is then found by integrating $f_{W,U^*}(w, u)$:

$$f_W(w) = \int_0^\infty f_{W,U^*}(w, u)\,du = \lambda \int_w^\infty f_U(u)\,du = \lambda\,[1 - F_U(w)]. \tag{9.4.2}$$

Also, the probability that at least one failure will occur in a time span of length t is given by

$$p_f(t) = \int_0^t f_W(w)\,dw = \lambda \int_0^t [1 - F_U(w)]\,dw = \lambda t - \lambda \int_0^t F_U(w)\,dw. \tag{9.4.3}$$

Accordingly, the associated reliability function is

$$R(t) = 1 - \lambda \int_0^t [1 - F_U(w)]\,dw = 1 - \lambda t + \lambda \int_0^t F_U(w)\,dw. \tag{9.4.4}$$

Noting that $f_U(u) \geq 0$, $1 - F_U(u) \leq 1$, and

$$\int_0^t [1 - F_U(w)]\,dw \leq t,$$

one finds

$$p_f(t) \leq \lambda t, \quad \text{for } t \leq 1/\lambda, \tag{9.4.5a}$$

$$p_f(t) \leq 1, \quad \text{for } t > 1/\lambda. \tag{9.4.5b}$$

This result is independent of any assumption taken to represent the pdf of interarrival time of failures.[16] The associated reliability bound is given by

$$R(t) \geq 1 - \lambda t, \quad \text{for } t \leq 1/\lambda, \tag{9.4.6a}$$

$$R(t) \geq 0, \quad \text{for } t > 1/\lambda. \tag{9.4.6b}$$

Example 9.26. Earthquake damage. Records of earthquakes in Italy over ten centuries indicate that there were 28 earthquakes exceeding a value of 4 of the Mercalli–Canconi–Seeber (SMB) index in Rome (see Problem 1.22). These events produced severe damage to historical buildings, which needed large restoration works. An average interarrival time of $1000/28 \cong 36$ years (that is, an average failure rate of $1/36 = 0.028$ per year) is estimated. If restoration works are completed after the last devastating earthquake, the buildings' risk of failure after 10 years from restoration is

$$p_f(10) \leq 10/36 \cong 0.28,$$

and the associated reliability is

$$R(10) \geq 0.72.$$

[16] The derivation of Eq. (9.4.5) from (9.4.3) can be simplified by considering the distribution function of U instead of its density. Boccotti and Rosso (1984) adopted these relationships, which are interpreted as the upper bound of risk of failure, as equalities when the failure process is unknown. Also, Eq. (9.4.5a) was proposed by Gumbel (1958) to estimate system reliability for values of $t \ll 1/\lambda$.

When the distribution of interarrival times of failure occurrence is known, Eq. (9.4.3) gives the risk of failure, and Eq. (9.4.4) the associated reliability function for the specified design lifetime t. For example, if the failures are exactly displaced in time with a fixed interarrival time of $1/\lambda$, the pdf of U is modeled as a Dirac function. Accordingly, the substitution of $f_U(u) = \delta(u - 1/\lambda)$ in Eq. (9.4.3) gives the pdfs of Eq. (9.4.5). As noted in Subsection 4.1.3, if the failures occur as homogeneous Poisson events with rate λ, the interarrival time is exponentially distributed with parameter λ; thus,

$$p_f(t) = 1 - \exp(-\lambda t), \tag{9.4.7}$$

$$R(t) = \exp(-\lambda t). \tag{9.4.8}$$

Example 9.27. Earthquake damage. Assuming that the interarrival time of earthquake occurrence in Example 9.26 is exponentially distributed with an average failure rate of $1/36 = 0.028$ per year,

$$p_f(10) = 1 - \exp(-10/36) \cong 0.25.$$

The associated reliability is $R(10) = 0.75$, which is higher than the lower bound of 72%.

If the interarrival time of failures follows the Rayleigh distribution,

$$F_U(u) = 1 - \exp(-\pi\lambda^2 u^2/4), \tag{9.4.9}$$

with mean $1/\lambda$, Eq. (9.4.3) gives the risk of failure as

$$p_f(t) = 2\Phi[(\pi/2)^{1/2}\lambda t] - 1. \tag{9.4.10}$$

Also,

$$R(t) = 2 - 2\Phi[(\pi/2)^{1/2}\lambda t] \tag{9.4.11}$$

is the associated reliability function.

Example 9.28. Earthquake damage. Assuming that the interarrival time of earthquake occurrence in Example 9.26 is Rayleigh distributed with an average interarrival time of 36 years, one gets

$$p_f(10) = 2\Phi[(\pi/2)^{1/2} 10/36] - 1 \cong 0.27.$$

The associated reliability is $R(10) = 0.73$, which is higher than the lower bound of 72%, but is lower than that estimated for exponential interarrival time.

The risk of failure and its associated reliability function are plotted in Fig. 9.4.2*a* and 9.4.2*b*, respectively, for the Dirac, exponential, and Rayleigh models of the interarrival time distribution. In many engineering applications, however, the failure mechanism follows the Poisson distribution representing the occurrence of rare events; accordingly, the interarrival time is exponentially distributed, so that the reliability of the system at any time is given by $R(t) = \exp(-\lambda t)$. It is suggested that if there are sufficient data (from similar systems), a goodness-of-fit test (see Sections 5.6 and 5.8) should be made to verify the Poisson assumption.

The expected life of the system can be evaluated from the *mean survival time*, which is defined as the expected value of the time that the system will operate successfully from the beginning of operation,

$$\mu_W = E[W] = \int_0^\infty t f_W(t)\, dt. \tag{9.4.12}$$

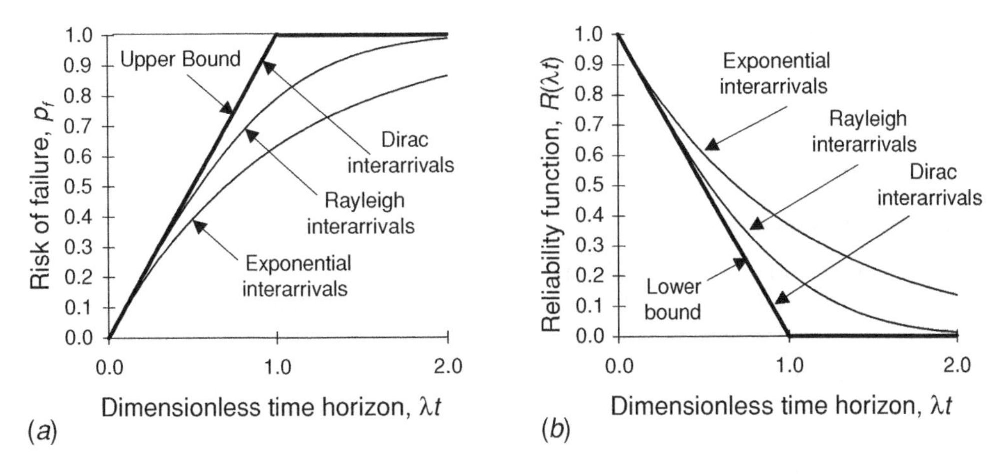

Fig. 9.4.2 Risk of failure (*a*) and reliability function (*b*) for different distributions of interarrival times.

In the case of exponentially distributed interarrival times, substituting $1 - \exp(-\lambda u)$ for $F_U(u)$ in the right side of Eq. (9.4.2) gives

$$f_W(w) = \lambda[1 - F_U(w)] = \lambda \exp(-\lambda w),$$

and from Eq. (9.4.12),

$$\mu_W = \int_0^\infty \lambda w \left[1 - F_U(w)\right] dw = \lambda \int_0^\infty w \exp(-\lambda w)\, dw = 1/\lambda. \tag{9.4.13}$$

As noted in Subsection 4.1.3 for the Poisson process and Sub-subsection 4.2.2.5, in which renewal and point processes are discussed, the interarrival and waiting times are both exponentially distributed variates with parameter λ. The fact that the distributions are identical is attributed to the lack of memory.

Differentiating Eq. (9.4.1) with respect to t gives $f_W(t) = -dR(t)/dt$. From Eq. (9.4.12),

$$\mu_W = -\int_0^\infty t \frac{dR(t)}{dt} dt = -\left[tR(t)\right]_0^\infty + \int_0^\infty R(t)\, dt.$$

Since $R(t)$ tends to zero for moderately large values of t, $[tR(t)]_0^\infty = 0$, and

$$\mu_W = \int_0^\infty R(t)\; dt. \tag{9.4.14}$$

The moments of the waiting times W are obtained from those of the interarrival time U. Thus,

$$E[W^k] = \int_0^\infty k t^{k-1} R(t)\, dt; \tag{9.4.15}$$

for the moments of any kth order about the origin, and

$$\operatorname{Var}[W] = \int_0^\infty 2t R(t)\, dt - \left[\int_0^\infty R(t)\, dt \right]^2. \tag{9.4.16}$$

9.4.2 Hazard function

At any time of system operation prior to first failure, it is of interest to investigate the residual survival probability. Given that a system has survived up to a time t, the probability that it will fail in the next time interval of length Δt is the conditional probability

$$\Pr[t < W \leq t + \Delta t | W > t] = \Pr[t < W \leq t + \Delta t] / \Pr[W > t].$$

Since

$$\Pr[t < W \leq t + \Delta t] = F_W(t + \Delta t) - F_W(t) = R(t) - R(t + \Delta t),$$

and

$$\Pr[W > t] = 1 - F_W(t) = R(t).$$

The hazard function $h(t)$ is defined such that

$$h(t)\Delta t = \Pr[t < W \leq t + \Delta t] / \Pr[W > t].$$

In the limit (as Δt tends to zero),

$$h(t) = \lim_{\Delta t \to 0} \frac{R(t) - R(t + \Delta t)}{R(t)\, \Delta t} = \frac{1}{R(t)} \left[-\frac{dR(t)}{dt} \right] = \frac{f_W(t)}{R(t)}. \tag{9.4.17}$$

Under these stipulations this temporal function can be interpreted as the failure rate of the system. From Eq. (9.4.17),

$$\int_0^t h(t)\, dt = -\int_0^t \frac{dR(\tau)}{R(\tau)} = -\left[\ln R(\tau)\right]_0^t.$$

It is reasonable to take $R(0) = 1$ as the initial condition, meaning that the system is perfect initially. Therefore,

$$R(t) = \exp\left[-\int_0^t h(t)\, dt \right], \tag{9.4.18}$$

which is substituted for $R(t)$ in Eq. (9.4.17) to obtain

$$f_W(t) = h(t) \exp\left[-\int_0^t h(t)\, dt \right]. \tag{9.4.19}$$

Definition: Hazard function. The *hazard function* of a system, defined as $h(t) = f_W(t)/R(t)$, provides the failure rate of the system.

In civil engineering practice, the hazard function characterizing many systems can be modeled as sketched in Fig. 9.4.3, where three different periods are indicated. During the initial period of operation, an expected high quality of management during construction reduces the high initial failure rate, so that the hazard decreases in time. In the second period, corresponding to the useful life of the system, the failures occur randomly, often

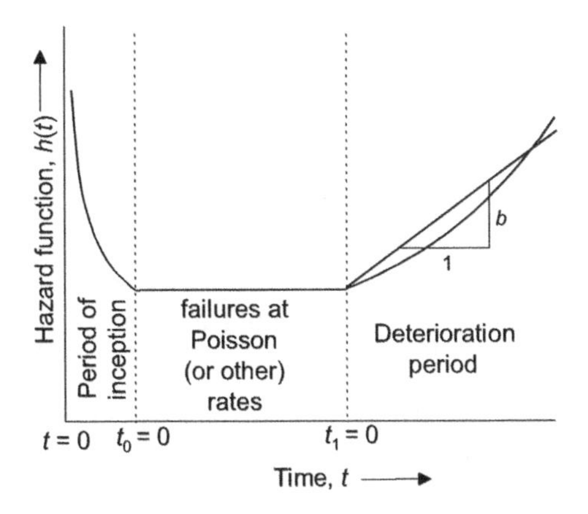

Fig. 9.4.3 Hazard function.

as Poisson or similar rates. In the third period the system is either in a state of deterioration or it runs inefficiently, so that the hazard or failure rate increases accordingly.

Example 9.29. Combined earthquake and deterioration hazard. One must evaluate the reliability of a building to be constructed in an earthquake-prone area, where catastrophic earthquakes are modeled as a sequence of Poisson events with an average annual rate of 0.02. The building is also exposed to hazard owing to deterioration (for example, fatigue). This hazard is negligible at the time $t = 0$ (completion of the construction), but it grows exponentially in time with a rate of 0.002. Let $R_1(t)$ be the reliability function associated with the earthquake hazard. The cdf of the survival time under the earthquake hazard,

$$F_{W1}(t) = 1 - \exp(-0.02t),$$

is used in Eq. (9.4.1) to obtain

$$R_1(t) = \exp(-0.02t).$$

Since

$$f_{W1}(t) = 0.02 \exp(-0.02t),$$

the failure rate $h_1(t)$ due to earthquake occurrence is found using Eq. (9.4.17):

$$h_1(t) = 0.02 \exp(-0.02t)/\exp(-0.02t) = 0.02.$$

Assuming that the failure rate $h_2(t)$ associated with wearout hazard is initially null, the hazard function of the wearout process is modeled as

$$h_2(t) = \exp(0.002t) - 1.$$

Since the wearout process in independent from that of earthquake occurrence, the hazard rates can be added. Thus,

$$h(t) = h_1(t) + h_2(t) = 0.02 + \exp(0.002t) - 1 = \exp(0.002t) - 0.98,$$

which is shown in Fig. 9.4.4a. From Eq. (9.4.18), the reliability function of the building is then found:

$$R(t) = \exp\left\{-\int_0^t [\exp(0.002t) - 0.98]dt\right\} = \exp[0.98t - 500(e^{0.002t} - 1)].$$

This is shown in Fig. 9.4.4b, where the individual reliability functions of the two processes are also displayed. Because $R(0) = 1$, and $R(20) \cong 0.45$, so that the engineer can predict a

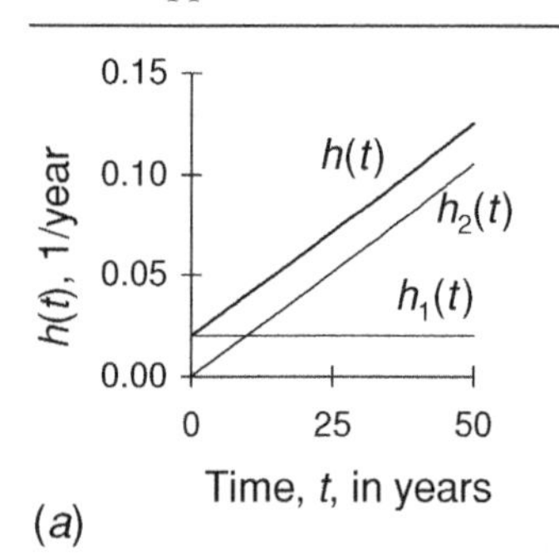

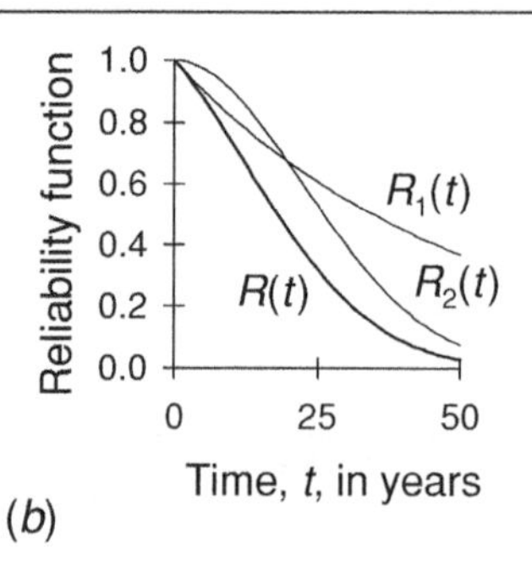

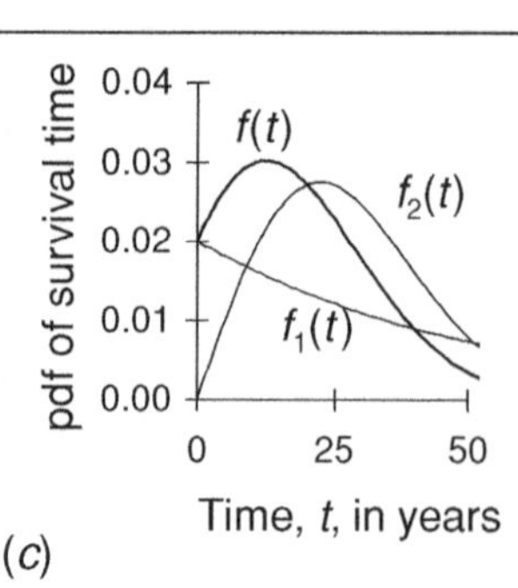

Fig. 9.4.4 Combined earthquake and deterioration illustration: (*a*) hazard function, (*b*) reliability function, and (*c*) pdf of survival time.

reliability of 45% over a 20-year building lifetime. Note that the individual reliabilities are both equal to 67% for $t = 20$ years. The pdf of time to failure,

$$f_W(t) = h(t)R(t) = [\exp(0.002t) - 0.98]\exp[0.98t - 500(e^{0.002t} - 1)],$$

is shown in Fig. 9.4.4*c*. Note that the modal value of the survival time, about 22 years, which is associated with the deterioration hazard, is reduced to about 12 years because of the additional earthquake hazard.

For a constant failure rate of λ, we obtain by substituting λ for $h(t)$ in Eq. (9.4.19)

$$f_W(t) = \lambda \exp(-\lambda t). \tag{9.4.20}$$

This gives an exponential survival time. Then, from Eq. (9.4.18), $R(t) = \exp(-\lambda t)$, and, from Eq. (9.4.12), $\mu_W = 1/\lambda$. Because the expected survival time is proportional to the reciprocal of the hazard rate, lower hazards are associated with longer expected times to failure, and vice versa.

The initial failure state, corresponding to inception time, is generally shorter for civil engineering systems than in the industrial disciplines, where failures of individual parts are quite common. On the other hand, the deterioration of civil engineering systems can be longer than that for other systems, because of costs involved. Generally, to model the deterioration hazard is not an easy task, because of the difficulty of predicting the joint long-term behavior of the system and its environment. Let us assume, for simplicity, that the third period is represented by a linear hazard function $h(t) = bt, b > 0$, and that time begins after the constant hazard phase (see Fig. 9.4.3). Thus, from Eq. (9.4.19),

$$f_W(t) = bt \exp(-bt^2/2), \tag{9.4.21}$$

which gives the *Rayleigh* distribution for survival time. Thus, the expected survival time,

$$\mu_W = \sqrt{\frac{\pi}{2b}}, \tag{9.4.22}$$

is inversely proportional to the square root of the hazard slope. From Eq. (9.4.18),

$$R(t) = \exp(-bt^2/2) \tag{9.4.23}$$

is the corresponding reliability function.[17]

[17] Alternatively, a nonlinear hazard function can be modeled through a Weibull pdf, with two or three parameters. It can be relevant, for example, in planning for irrigation (see, for example, Mukherjee and Kottegoda, 1992).

Example 9.30. Water distribution. A water distribution system is only 95% reliable 1 year after it has begun to deteriorate due to lack of maintenance. Assuming the failure rate to increase linearly with time, the engineer must estimate the reliability of the system for a time horizon of 5 years after the initiation of the deterioration. From Eq. (9.4.23),

$$b = -2 \ln R(t)/t^2 = -2 \ln(0.95)/1 \cong 0.10 \text{ year}^{-2}.$$

Thus, the wearout reliability is estimated from

$$R(t) = \exp(-0.10t^2/2) = \exp(-0.05t^2),$$

which gives $R(5) \cong 0.28$, indicating a residual reliability of about 28% after 5 years. This remaining reliability can be interpreted as the worth of the water distribution system. That is, after 5 years of deterioration without maintenance, the engineer estimates that the system is worth approximately 28% percent of its initial value after the beginning of this phase.

9.4.3 Reliable life

A further measure of temporal reliability is the *reliable life*, denoted by t_r, which is defined as the time required for system reliability to decrease to a specified level, r.

Example 9.31. Combined earthquake and deterioration hazard. Consider again the reliability of the building in Example 9.29 subject to the combined effect of earthquake and deterioration hazards. The engineer wishes to evaluate the reliable life of the building for a selected reliability level of 90%. Since,

$$R(t) = \exp[0.98t - 500(e^{0.002t} - 1)],$$

one must solve

$$0.9 - \exp[0.98t - 500(e^{0.002t} - 1)] = 0$$

for t, thus obtaining $t_{0.9} \cong 4.3$ years.

If the system has a constant failure rate, one can use Eq. (9.4.18) and substitute the specified level r of reliability for R to obtain

$$t_r = -(1/\lambda) \ln r. \tag{9.4.24}$$

For instance, for a reliability level of 10% the historical buildings of Example 9.26 with the constant failure rate of $1/36$ per year have a reliable life of

$$t_{0.1} = -36 \times \ln(0.1) \cong 83 \text{ years}.$$

Example 9.32. Flood control. The records over the last 75 years indicate that there were three floods in the Sansobbia Valley, in northern Italy that produced severe damage to the downstream town of Albisola by overtopping of existing levees. Assuming a constant rate of failure of $\lambda = 3/75 = 0.04 \text{ year}^{-1}$, corresponding to a return period of 25 years, the engineer estimates that the reliable life of the system is very low: $t_{90\%} = -25 \times \ln(0.9) \cong 2.6$ years, for a reliability level of 90%. To improve the performance of the system, the engineer wishes to modify the existing levees in order to achieve a reliable life of 20 years for the specified 90% reliability level. Accordingly, the rate of failure must be reduced to

$$\lambda = -\ln(r)/t_r = -\ln(0.9)/20 \cong 0.0053/\text{year}.$$

Therefore, the new design return period must be of $1/\lambda = 1/0.0053 = 190$ years, which must equal the probability of exceeding the maximum annual peak flow, X; that is,

$$1/\lambda = 1/[1 - F_X(x)],$$

or $\lambda = 1 - F_X(x)$. Since no indications are available for the probability distribution of flood flows, the engineer further assumes that large floods in the area are exponentially distributed. Inverting the cdf of X yields

$$x = -(1/a)\ln\lambda,$$

with a denoting an unknown constant. Thus,

$$x_{\text{new design}}/x_{\text{old design}} = \ln\lambda_{\text{new design}}/\ln\lambda_{\text{old design}} = \ln(0.0053)/\ln(0.04) \cong 1.63.$$

This means that the reconstruction of levees must accommodate 1.63 times the original design flow in order to have a reliable life of 20 years for the specified reliability level of 90%.

9.4.4 Summary of Section 9.4

This section mainly concerns time to failure or survival time. We discuss the survival function and show the use of the hazard function. This is followed by the determination of the reliable life. The next section deals with the important subject of reliability design.

9.5 RELIABILITY-BASED DESIGN

Generally, the design reliability of most civil and environmental engineering systems is in the range from 0.9 to 0.99 or more. The usual design practice is based on recommended factors of safety, which should allow the system to be subject to only a limited number of failures over its lifetime, and to be subsequently repaired or restored. The criteria to determine design tolerable factors of safety are based on the past performances of many systems. These criteria have been established by U.S. and European organizations (for example, the U.S. Army Corps of Engineers and the U.S. Bureau of Reclamation in the United States, the Commission of European Communities in the European Community). They are formulated as recommendations or by-laws in many other countries. For example, the structural safety of a reinforced concrete structure can be taken as $ax^* \geq by_1^* + cy_2^*$, where x^*, y_1^*, and y_2^* indicate the nominal values of resistance, dead load, and maximum live load, respectively; here $a \leq 1$, $b \geq 1$, and $c \geq 1$ are specified coefficients; for example, $a = 0.9$, $b = 1.4$, and $c = 1.7$ as recommended by the American Concrete Institute. Accordingly, the recommended safety factor is $z^* = ax^*/(by_1^* + cy_2^*)$, where the nominal values of resistances and loads in this case are usually higher than the corresponding mean values. However, the accuracy of this approach depends on the actual similarity of the investigated systems; also, the confidence in the results so achieved is related to the number of systems studied. The performance of many systems needs to be analyzed for this purpose.

One must account for the appropriate capacity and demand factors in a reliability-based design. The components and interactions, and the results obtained, should be related to the expected performance of a system during its projected design life. Also, it is necessary to have inputs or parameters or other characterizations that can be estimated within the current state of the art. Indices currently thought to be pertinent, such as the factor of safety or the margin of safety, should be considered because of the relationship with probabilistic aspects.

Reliability-based design usually depends on a maximum probability of failure within a given lifetime l of the system. The failure process is related to the reliability as a decreasing

Table 9.5.1 Suggested minimum design lifetime, in years, for coastal engineering structures

	Required security level		
Type of infrastructure	*a*	*b*	*c*
General use			
General interest and moderate risk of loss of human life or environmental damage in case of failure (for example, work in large ports, outfalls of large cities)	25	50	100
Specific industrial installation			
Works in service of a particular installation or associated with the use of a transitory natural deposit of resources (for example, industry service ports, loading platforms of a mineral deposit, petroleum extraction platforms)	15	25	50

Level a: Local auxiliary interest and small risk of loss of human life or environmental damage in case of failure (for example, defense and coastal regeneration works, works in minor ports and marinas, local outfalls, pavements, buildings).
Level b: General interest and moderate risk of loss of human life or environmental damage in case of failure (for example, works in large ports, outfalls of large cities).
Level c: International interest or protection against flooding, and high risk of loss of human life or environmental damage in case of failure (for example, defense of urban and industrial centers).
Source: By kind courtesy of Leopoldo Franco, University of Roma Tre, Italy.

function of time, $R(t)$, of a given system. If, for example, a system has a constant failure rate during its design life, the probability of failure $p_f(t)$ as given by Eq. (9.4.7) is maximum for $t = l$; that is,

$$\max p_f(t) = p_f(l) = 1 - \exp(-\lambda l). \tag{9.5.1}$$

However, if environmental or climatic changes are foreseen or the original operation of the system is changed, more sophisticated models are required to evaluate the reliability function. There are, of course, no universally recommended values for the lifetime l or the acceptable risk $\max p_f(t)$. The design life varies for different parts of a project and is related to system's time of performance. Also, the choice of the design lifetime depends on the specified level of security as shown, for instance, in Table 9.5.1 for coastal engineering structures. Furthermore, one must consider separately those structural facilities that are related to a particular industrial installation. For assessing risk during the construction period, one can equate the time of construction to the lifetime. The probability of failure during the service period is constrained on the basis of human loss in case of failure or damage and the economic consequences of failure. Also, the circumstances that lead to failure need to be considered together with facilities for the repair of a structure; one considers either economic damages associated with underperformance of the system or the need for its restoration or repair (see Table 9.5.2). There are also damages associated with the loss of life or other hazards to people and other factors that cannot be adequately assessed (such as the reputation of those involved).

Example 9.33. Risk assessment of a dam. The choice of the design lifetime l of a dam depends on the type and size of the dam. From Table 9.5.1, one can argue that 100 years is a correct choice for a large dam, because of its general use and high threat of potential loss of human life and environmental damage in case of failure. The maximum allowable failure

Table 9.5.2 Suggested maximum failure probability during useful life of coastal engineering systems

	Damage initiation		Total destruction	
Economic effects	Unexpected human loss in case of damage or failure	Expected human loss in case of damage or failure	Unexpected human loss in case of damage or failure	Expected human loss in case of damage or failure
Low	0.50	0.30	0.20	0.15
Average	0.30	0.20	0.15	0.10
High	0.25	0.15	0.10	0.05

Economic effects: Low for $c_L/c_I \leq 5$, average for $5 < c_L/c_I \leq 20$, and high for $c_L/c_I > 20$.
c_L: Total costs of direct and indirect losses in case of damage or failure.
c_I: Total investment costs for system construction.
Source: By kind courtesy of Leopoldo Franco, University of Roma Tre, Italy

probability for total destruction and expected human loss in case of damage or failure is 0.05, from Table 9.5.2. From Eq. (9.5.1),

$$\lambda = -l^{-1}\ln[1 - p_f(l)] = -(1/100)\ln(1 - 0.05) = 0.00051,$$

which is also the probability of failure in a year if the annual number of failure events is a Poisson variate. One can compare this value with the observed probabilities of failure and damage reported by Cheng (1993). These are based on statistics of 5450 dam incidents in the United States and 8925 dam incidents that have occurred in 43 ICOLD (International Commission on Large Dams) member countries (see Table 9.5.3). Note that assuming a maximum allowable risk of failure of about 0.05% yields a more conservative design than that of most existing dams.

If overtopping because of a flood exceeding the spillway capacity is the only risk factor in the analysis, the design flood has a return period of $1/\lambda$. Therefore, in the design of the dam one must consider a return period of $1/0.00051 = 1950$ years to prevent dam overtopping. In practice, one must also account for the characteristics of the site where the dam is located. The local factors to be considered are piping and leakage, sliding, and earthquake attack. These factors constitute multiple modes of failure. Also, at least 50% of observed dam failures and accidents have occurred within 5 years after the commencement of dam operation, so that a constant rate of failure is only an approximation to the real situation.

Economic analysis is applied to damages associated with normal operating conditions. One obtains the probability of failure of the system configuration in relation to economic aspects such as project cost, benefit, or benefit-cost ratio. However, the estimation of the costs of damages caused is not straightforward unlike construction and operation costs. The following example is based on the book by Chow et al. (1988).

Example 9.34. Hydroeconomic analysis. The design return period T of a hydraulic structure facing a hydrological hazard can be evaluated by economic analysis. It is assumed that one can estimate the probability distribution of the hydrologic events and costs of damage. The initial cost of a structure increases as the design return period increases. However, there will be a decrease in expected damages because the structure can cope better with larger hazards. One can find a constant failure rate or design return period with minimum total cost by adding the costs of expected annual damage and capital costs. For a variable X, such as

Table 9.5.3 Probability of dam incidents according to causes

Cause	Probability per year per dam	
	Failure	Damage
Overtopping	.0014	.0016
Spillway damage	.0022	.0100
Piping and leakage	.0063	.0180
Sliding	.0017	.0047
Others	.0004	.0026

Source: Adapted from Cheng, 1993.

annual maximum flow, no damage is caused if $x \leq x(T)$, where T is the return period. For a pdf $f_X(x)$ and damage function $d(x)$,

$$d_T = \int_{x(T)}^{+\infty} d(x) f_X(x) dx$$

is the expected annual cost. A finite difference approximation is obtained as

$$d_T = \sum_{i=1}^{+\infty} \frac{d(x_{i-1}) + d(x_i)}{2} \left[F_X(x_i) - F_X(x_{i-1})\right],$$

where $x_0 = x(T)$. One can then add the discounted annual cost c_T to d_T and search for the optimum return period by minimizing the total cost.

Consider, for example, the reliability assessment of an existing urban drainage system. This system has been seen to fail its intended function at least once in 2 years, but no human loss is associated with its failure. For events of various return periods, the annual costs and the annualized capital cost of structures are shown in Table 9.5.4. The engineer must determine the present expected annual damages and evaluate the optimum return period to design a new system. Damage costs and the annualized capital costs are given in Table 9.5.4, where the incremental expected damage is computed as

$$\frac{d(x_{i-1}) + d(x_i)}{2} \left[F_X(x_i) - F_X(x_{i-1})\right] \quad \text{or} \quad \frac{d(T_{i-1}) + d(T_i)}{2} \left(\frac{1}{T_{i-1}} - \frac{1}{T_i}\right),$$

for each increment used to discretize the probability of nonexceedance, $1 - 1/T$.

The expected damage cost is found by summing all the values of incremental expected damage, resulting in 58.8 monetary units. The damage risk cost for each return period is then computed by partial summation of relevant incremental values, and the total costs are obtained by adding to these figures the corresponding capital cost, as given in Table 9.5.4. The results are also shown in Fig. 9.5.1.

The minimum total cost is found for a return period of 6 years. The corresponding risk of failure in a year is 0.167.

Because civil and environmental engineering systems involve multivariate formulations, often with several random variables, one must use analytical methods that provide information for functions of random variables. Probability distributions can be assigned and statements can be made with respect to the reliability of the system. The probability distribution, pertaining to a number of random variables, must be determined. Several probabilistic methods have been developed to give measures of the distribution of

Table 9.5.4 Optimal design return period from hydroeconomic analysis for an urban drainage system

Increment	Return period Return period (years)	Annual probability of nonexceedance	Damage (units)	Incremental expected damage (units/year)	Damage risk cost (units/year)	Capital cost (units/year)	Total cost (units/year)
0	1	0.000	0		58.8	0	58.8
1	2	0.500	40	10.0	58.8	0	58.8
2	3	0.667	45	7.1	48.8	8	57.0
3	4	0.750	50	4.0	41.7	11	52.3
4	5	0.800	55	2.6	37.8	13	50.8
5	6	0.833	60	1.9	35.2	15	50.5
6	7	0.857	70	1.5	33.2	18	50.9
7	8	0.875	80	1.3	31.7	20	51.7
8	9	0.889	85	1.1	30.4	22	52.6
9	10	0.900	120	1.1	29.2	25	53.7
10	15	0.933	140	4.3	28.1	36	63.7
11	20	0.950	190	2.8	23.7	50	73.7
12	30	0.967	270	3.8	21.0	70	91.0
13	40	0.975	380	2.7	17.2	90	107.2
14	50	0.980	500	2.2	14.5	110	124.5
15	100	0.990	800	6.5	12.3	210	222.3
16	200	0.995	1500	5.8	5.8	320	325.8

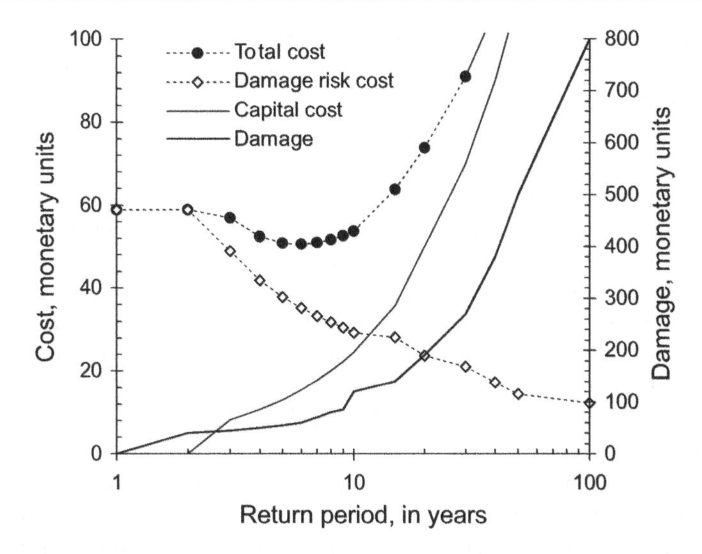

Fig. 9.5.1 Determination of optimal design return period by economic analysis of an existing urban drainage system.

functions of random variables. These range from so-called exact methods that require computer-oriented numerical excursions to approximate procedures that can be accommodated by relatively simple algebraic calculations. Each system has its set of assumptions and its own group of advocates. Briefly, the methodologies can be divided into three categories, or levels, depending on the complexity of the analysis and the amount of required information.

The design of a system is given by a set of values of parameters characterizing the system, and by a set of pertinent safety relations. In this way one identifies the safe and unsafe regions in the space of random variables $X_1, X_2, \ldots$ relevant to the problem as modeled in terms of random load and resistance. A *design point* is usually defined as the most probable point on the failure boundary or limiting state of interest between the safe and unsafe regions, or as some conditioned point on the boundary. Similarly, a *characteristic point* may be defined in the domain of random variables by assigning to each variable a characteristic (average or cautious) value. Quotients between the design and characteristic values (for load, and the opposite for resistance) represent the *partial safety factors*. The probability of failure p_f of the system can be evaluated by integrating over the unsafe region the joint pdf of the involved variates, as indicated by Eq. (9.1.26). This method is conceptually sound and although it is sometimes referred to as the exact method, or *Level 3 reliability assessment*, some approximations are used for integration, with numerical methods in many cases. Because standard numerical integration techniques require unrealistic computational resources as soon as one considers more than five or six variates, the application of Level 3 reliability assessment involves the development of Monte Carlo experiments to simulate system behavior. All variates are simulated according to their known joint statistical properties, and the safe or unsafe behavior of the system is evaluated in a deterministic manner for each simulation. Failure frequency, as evaluated from simulation results gives an approximation of the required failure probability, and the accuracy increases with the length of simulation, as shown in Section 8.1. To achieve a design with a specified reliability, however, one must repeat this exercise with different design parameters within a trial-and-error framework.

Simplified methods to evaluate reliability may be classified in two categories or levels. *Level 2 reliability assessment* is performed by assuming the variates to be jointly normally distributed, and the shape of the failure boundary to be approximated by a linear or circular surface. This method yields the probability of failure, the design point, partial safety factors, and the relative importance of the uncertainty of the single variates in estimating the probability of failure. Its accuracy depends on the combined effect of its assumptions to represent the limiting state of interest and the joint pdf of the variates. *Level 1 reliability assessment*, which is provided by a number of partial safety factors, or safety margins related to characteristic values of variates, is the most popular method for design practice. In most cases these safety factors are not explicitly related to the probability distribution of random variables or to the failure probability, but they reflect a standard variability and tolerance risk. Only if the safety factors are explicitly related to the failure probabilities and to the joint pdf of the involved variates does a Level 1 assessment give explicit information on risk.

The exact methods require that the probability distribution functions of all component variables are known initially. Because of the complexity of the solution process, the unknown component distributions are usually assumed to be normal or lognormal, or even uniform. The lack of data for some random variables makes it difficult to infer their probability distributions. The analytical solution can only be approached for particular cases, and numerical integration of Eq. (9.1.26) is needed. Monte Carlo simulations, for which high-performance computers are often mandatory, are required when the design involves several random variables. The advantage of this methodology is that the complete probability distributions of the dependent random variables are obtained. The disadvantages are that the output may be no better than the (assumed) input and that considerable computer time may be required. Also, each case must be treated separately, so that one may need to assume a large number of inputs or design alternatives to achieve a satisfactory design in terms of system performance.

Level 2 methods assume that the variates are jointly normally distributed, and nonnormal variates are transformed into normal variates. Also, the solution is searched through the use of a tangent hyperplane or hypersurface. This approximation is conservative if the exact nonlinear safe-unsafe boundary is convex toward the origin, but it is nonconservative for a concave boundary. These methods yield the probability of failure, the design point, partial safety factors, and the relative importance of the uncertainty of the single variates that constitutes the probability of failure. Their accuracy depends on the combined effect of the given assumptions to represent the safe-unsafe boundary and the joint pdf of the variates.

Level 3 methods require the evaluation of the probability distribution of the safety index adopted in the analysis. Because of the difficulties in determining this distribution on analytical grounds, the lognormal and the normal distributions are usually assumed to represent the safety factor and the safety margin, respectively, so that one needs only to estimate the mean and standard deviation of these indices. The major advantage of this method is that the solution to the inverse problem of determining reliability-based values of design parameters results in a much more straightforward exercise than that required when using higher levels of reliability assessment.

9.6 SUMMARY FOR CHAPTER 9

Civil and environmental engineers are interested in evaluating the chance that a system is successful over its expected lifetime. Thus, reliability is defined as the probability that, under given operating conditions, a system performs adequately over a specified period of time, and a failure is said to occur when the system is incapable of performing its intended function. Measures of reliability are the factors of safety, the safety margin, and the reliability index, which are determined from the joint distribution of the random capacity X and demand Y of the system. The performance function, which provides a general approach to system reliability, is the random function $g(X, Y)$ of capacity and demand; it describes system performance as related to its possible failure, or limiting state, given by $g(X, Y) = 0$. If $g(\cdot)$ is linear, $g(\cdot) = 0$ can be solved in a straightforward manner for normal capacity and demand, thus giving the required estimate of the reliability index. Nonnormal variates are transformed at the failure point; except for some specific distributions (for example, lognormal capacity and demand), this transformation requires iterative calculations to be performed. In the nonlinear case, the solution always requires iterative calculations.

The concepts of series, redundant, and compound configurations have been applied to multimodal failures of a system. For independent system components, the overall reliability is obtained directly from the individual reliabilities; on the other hand mutually dependent modes of failure can be analyzed using reliability bounds.

The assumption of a beta-distributed random variable for reliability can yield credibility limits for an unknown reliability. The Bayesian approach is also shown to be useful for updating reliability estimates when new information is received on system performance.

The analysis of the failure process has been shown to be necessary for evaluating the temporal reliability of a system, and the definitions of interarrival and survival times have been introduced accordingly. For a stationary failure process, temporal reliability depends only on the distribution of interarrival time, but its upper bound is shown to be distribution free. Representing the failure rate by the hazard function makes it possible to model the useful life of a system and also its breaking-in and wearout phases. Finally, through the concept of reliable life, that is, the time required for the system reliability to decrease to a specified level, one can evaluate the design lifetime on a probabilistic basis.

In the application of reliability concepts to engineering design, one considers the maximum probability of failure in a given life of the system, with incorporated economic issues. Also, reliability is assessed at different levels depending on the required degree of accuracy. Approximation methods such as the FOSM, the Taylor series expansion, and point estimation methods facilitate the computations.

REFERENCES

General. The following references are given for further reading as required:

Ang, A. H. S., and W. H. Tang (1984). *Probability Concepts in Engineering Planning and Design, Vol. 2: Decision, Risk and Reliability*, John Wiley and Sons, New York. Recommended reading on reliability methods, with illustrations, in engineering design.

Ansell, J., and F. Warton, (eds.) (1992). *Risk: Analysis, Assessment and Management*, John Wiley and Sons, New York. An extended review of methodological, industrial, financial, social, governmental, and ethic issues in the engineering approach to risk management.

Ballio, G., and F. M. Mazzolani (1982). *Theory and Design of Steel Structures*, Chapman and Hall, London. Chapter 2 provides a design-oriented introduction to reliability of steel structures.

Ditlevsen, O. D. (1979). *Uncertainty Modeling with Applications to Multidimensional Civil Engineering System,* McGraw-Hill, New York. A comprehensive book on reliability-based engineering design.

Harr, M. E. (1987). *Reliability-Based Design in Civil Engineering*, McGraw-Hill, New York. A comprehensive survey of reliability methods in civil engineering with special emphasis on structural and geotechnical issues.

Galambos, T. V. (ed.) (1998). *Guide to Stability Design Criteria for Metal Structures*, 5th ed., John Wiley and Sons, New York. Provides an outline of structural safety methods for metal structures.

Ghiocel, D., and D. Longu (1975). *Wind, Snow and Temperature Effects on Structures Based on Probability*, Abacus Press, Tunbridge Wells, UK. Reference on reliability-based structural design.

Madsen, M. O., S. Krenk, and N. C. Lind (1986). *Methods of Structural Safety*, Prentice Hall, Englewood Cliffs, NJ. An advanced book on safety in structural design.

Marek, P., M. Gustar, and T. Anagnos (1995). *Simulation-Based Reliability Assessment for Structural Engineers*, CRC Press, Boca Raton, FL. An updated overview of structural reliability assessment that compares simulation-based methods with other concepts, such as partial factor design.

Rethaty, L. (1988). *Probabilistic Solutions in Geotechnics*, Elsevier, Amsterdam. A general overview of statistical methods in geotechnics including reliability and decision analyses.

Tichy, M. (1993). *Applied Methods of Structural Reliability*, Kluwer Academic Publishers, Dordrecht. An advanced book on structural reliability.

Yao, J. T. P. (1985). *Safety and Reliability of Existing Structures*, Pitman Advanced Publications Program, London. An advanced book on structural reliability.

Yen, B. C., and Y. K. Tung (eds.) (1993). *Reliability and Uncertainty Analyses in Hydraulic Design*, American Society of Civil Engineers, New York. An updated survey of current practice in hydrologic and hydraulic design.

Zio, E. (2007). *An introduction to the basics of reliability and risk analysis*, World Scientific, New York. An introduction to reliability for engineers.

Additional references quoted in the text

American Institute of Steel Construction (1986). *Load and Resistance Factor Design Specification for Structural Steel Building.* AISC Publications, Chicago. Design specification standards.

Boccotti, P., and R. Rosso (1984). "Risk analysis of spillway flood design," *Proc. Int. Conf. on Safety of Dams*, Coimbra, Portugal, April 23–28, pp. 85–92. Nonclassical approach to dam safety.

Burcharth, H. F. (1994). "Reliability of a structure at sea," *Proc. Int. Conf. on Wave Barriers in Deepwaters*, Tokyo, January 10–14, pp. 470–517. Risk assessment for coastal structures.

Cheng, S. T. (1993). "Statistics on dam failures," in: *Reliability and Uncertainty Analyses in Hydraulic Design*, edited by B. C. Yen, and Y. K. Tung, American Society of Civil Engineers, New York. Worldwide data on dam failures.

Chow, V. T., D. R. Maidment, and L. W. Mays (1988). *Applied Hydrology*, McGraw-Hill, New York. Hydroeconomic analysis, Section 13.2.

Cox, D. R., and P. A. W. Lewis (1966). *The Statistical Analysis of Series of Events.* Metheun, London. Reference on temporal reliability.

Ditlevsen, O. (1979). "Narrow reliability bounds for structural systems," *J. Struct. Mech.*, Vol. 7, No. 4, pp. 453–472. Bounds of reliability estimates.

Ellingwood, B., J. G. MacGregor, T. V Galambos, and C. A. Cornell (1982). "Probability-based load criteria: Load factors and load combinations," *J. Struct. Engr. Proc. ASCE*, Vol. 108, pp. 978–997. Load evaluation in probability-based design.

Franco, L., A. Lamberti, A. Noli, and U. Tomasicchio (1986). "Evaluation of risk applied to the designed breakwater of Punta Riso at Brindisi, Italy," *Coast. Eng.*, Vol. 10. pp. 169–191. Reliability analysis in simulation.

Gumbel, E. J. (1958). *Statistics of Extremes*, Columbia University Press, New York, NY. Chapman and Hall, New York. Original work on extreme values.

La Barbera, P., R. Rosso, , and F. Siccardi (1983). "Reliability of low-head mini-hydro power plants," *Proc. XX Congr. Int. Assoc. Hydraul. Res.*, Moscow, 5–9, September Vol. 3, pp. 556–566. Reliability assessment in a simulation.

Mukherjee, D., and N. T. Kottegoda (1992). "Stochastic model for soil moisture deficit in irrigated lands," *J. Irrigat. Drain. Eng., Proc. ASCE*, Vol. 118, pp. 527–542. Weibull hazard function for irrigation.

Raiffa, H., and R. Schlaifer (1961). *Applied Statistical Decision Theory*, Harvard University Press, Cambridge. Bayesian decisions and conjugate distributions.

Ravindra, M. K., and T. V. Galambos (1978). "Load and resistance factor design for steel," *J. Struct, Engr., Proc. ASCE*, Vol. 104, pp. 1337–1353. Introducing the LRFD approach.

Rosenblueth, E. (1975). "Point estimates for probability moments," *Proc. Nat. Acad. Sci. USA*, Vol. 72 No. 10, pp. 3812–3814. Introducing the point estimation method.

Serfling, R. J. (1974). "The role of the Poisson distribution in approximating system reliability of k-out-of-n structures," Contract N0014-75-C0551, Office of Naval Research. Special application of the Poisson distribution.

Shinozuka (1983). "Basic analysis of structural safety," *J. Struct. Engr., Proc. ASCE*, Vol. 109 No. 3, pp. 721–740. General approach to nonlinear performance function.

Smith, J. C. (1991). *Structural Steel Design. LRFD Approach.* John Wiley and Sons, New York.

Vannucchi, G. (1985). "Straightforward probabilistic methods in geotechnics," *Ital. J. Geotech.*, Vol. 19 No. 2, pp. 77–87 (in Italian). Comparison between simulation and point estimation.

Worsley, K. J. (1982). "An improved Bonferroni inequality and applications," *Biometrika*, Vol. 69, pp. 297–302. On reliability bounds.

PROBLEMS

9.1. Heuristic reliability predictions. Based on your own knowledge or experience estimate the reliability of (*a*) an automobile, (*b*) a telephone, (*c*) an audio system, (*d*) the operating system of a digital computer, (*e*) a 1-day ahead weather prediction, (*f*) a 1-week ahead weather prediction, and (*g*) the municipal water supply and wastewater removal systems of the place where you live on a scale of 0 to 100%.

9.2. Reliability assessment. Consider the following civil and environmental engineering systems and apply the concepts of capacity and demand to their reliability assessment, also indicating the variables involved:

(*a*) The roof of your house.
(*b*) The water distribution network in a campus or housing subdivision
(*c*) Water quality of the municipal supply system
(*d*) The spillway of a dam
(*e*) A municipal transportation system
(*f*) A retaining wall for highway embankment
(*g*) Air quality in a city

9.3. Structural safety factor. Consider a structure designed with a central safety factor of 2 with a nonrandom load Y. Determine its risk of failure for (*a*) normally, (*b*) lognormally, and (*c*) gamma-distributed load-carrying capacity X with known mean $\mu_X = y$ and coefficient of variation $V_X = 0.5$.

9.4. Pile. The conventional safety factor of a pile is $z^* = 1.2$. Both load-carrying capacity and strength are independent normal variates with coefficients of variation of 30 and 50%, respectively. Find the sigma bound $h_X = h_Y = h$, if the central safety factor is $\zeta = 1.6$.

9.5. Flow meters. The reliability of a standard flow meter used for rating municipal water supply to private buildings is 95%, and it is estimated that a defective meter underestimates flow by 20%. If the tolerable loss is 2% for each supplied building, what is the reliability of the municipal system if 100 buildings are supplied? Calculate also the reliability if 1000 buildings are supplied. Note that for large n the binomial distribution can be approximated by the normal distribution (with the same mean and variance), as shown in Sub-subsection 4.2.6.2.

9.6. Uniform capacity and demand. The joint capacity-demand distribution of a supply system is uniform: $f_{X,Y}(x, y) = (ab)^{-1}$ unit^{-2}, for $0 \le X \le a$ units and $0 \le Y \le b$ units, with $a \ge b$. What is the reliability of the system?

9.7. Pipe flow. The pressure p and water flow q in a circular pipe are measured as $p = 7$ kPa (kN/m^2) and $q = 0.08$ m^3/s, respectively. The pipe is located 2 m above the reference level and its diameter is $d = 20$ cm. The total head h (energy) in the pipe at the point of interest is given by the Bernoulli equation,

$$h = \frac{u^2}{2g} + \frac{p}{\gamma} + z,$$

where $X_1 = u^2/2g$ is called the *kinetic head*, $X_2 = p/\gamma$ the *pressure head*, and $X_3 = z$ the *elevation head*. Assuming that X_1, X_2, and X_3 are normal variates with a coefficient of variation of 0.05, an engineer needs to determine the reliability of system operation for $h > h_0$, with $h_0 = 3$ m. (The flow velocity is defined as the ratio between flow and cross-sectional area of the pipe, say, $u = q/(\pi d^{-2}/4)$, g is the acceleration due to gravity, 9.806 m/s^2, and γ is the specific weight of water, 9.806 kN/m^3.) Assume that all variates are independent of each other.

9.8. Column load. A column of a building is designed with a central safety factor of 1.6. The coefficient of variation of its strength is 25%. The total column load is the sum of several factors: live load, dead load, wind load, and snow load. Assume these factors are independent normal variates.

Factor	Expected value (kN)	Coefficient of variation
Live load	70	0.15
Dead load	90	0.05
Wind load	30	0.30
Snow load	20	0.20

Find the following:

(*a*) The expected value and coefficient of variation of the total column load, Y.

(*b*) The reliability index and the corresponding risk of failure of the column, if the strength is assumed to be a normal variate independent of load.

(*c*) The reliability index and risk of failure, if the strength and load are correlated normal variates with $\rho = 0.6$.

9.9. Earth embankment. For the stability of an earth embankment, the overturning moment eW must not exceed the resisting moment $r(L_A R_A + L_B R_B)$, as shown schematically in Fig. 9.P1. For the given configuration, $L_A = 21$ m, $L_B = 4$ m, $r = 12$ m, $e = 3$ m, and $W = 2000$ kN/m^2. Find the reliability of the system if R_A and R_B are joint normally distributed variates with means 35 and 20 kN/m^2, respectively, coefficient of variation of 20%, and coefficient of correlation of 0.7.

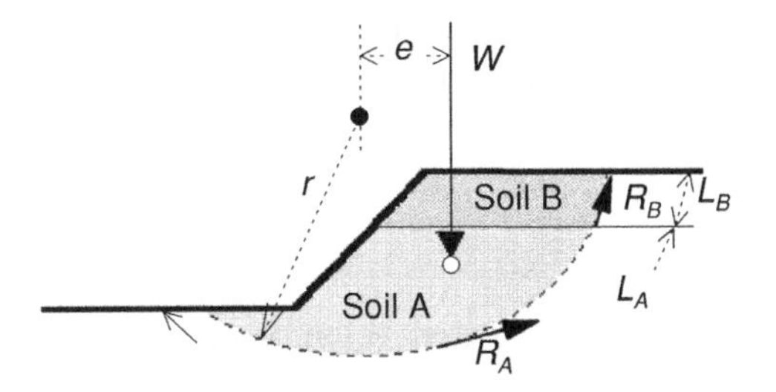

Fig. 9.P1 Sketch for stability analysis of an earth embankment.

9.10. Slope stability. The *wedge method* for analyzing the stability of an earth slope assumes a linear critical surface, such as AB in Fig. 9.P2. The factor of safety is then obtained as

$$z = \frac{2c \sin\theta \cos\varphi}{h\gamma \sin^2[(\theta - \varphi)/2]},$$

where c is the *cohesion parameter*, φ is the internal angle of friction or *friction angle*, θ is the slope angle, γ is specific weight, and h is the slope height. Find the risk of failure for a slope with $h = 10$ m and $\theta = 55°$ if these factors are independent normal variates as follows:

Factor	Expected value	Coefficient of variation
Friction angle	21°	0.12
Cohesion parameter	15 kN/m^2	0.40
Specific weight	20 kN/m^3	0.10

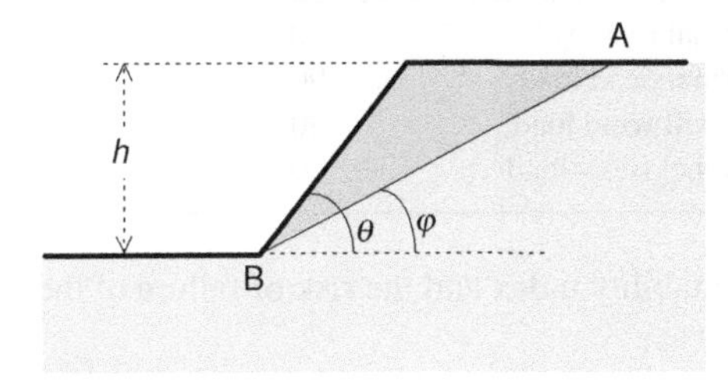

Fig. 9.P2 Wedge method for slope stability.

9.11. Elastic collapse of a steel beam. Consider a simply supported steel beam with normally distributed strength X, with mean of 25 KN/cm^2, and coefficient of variation of 15%. The bending moment Y is also a normal variate with mean 900 kN · cm and coefficient of variation of 20%. Find the reliability of the beam if its section modulus W is normally distributed with mean 20 cm^3 and coefficient of variation of 5%. The limiting state of interest is given by $Y/W - X = 0$. Assume mutually independent X, Y, and W.

9.12. Flexure formula. Consider a timber beam subject to flexure. The stress at the extreme fiber at a distance X_2 from the neutral axis acted upon by a bending moment

X_3 is given by X_2X_3/X_4, where X_4 denotes the moment of inertia of the section. We assume that the factors are normal variates as follows:

Factor	Expected value	Coefficient of variation
Bending moment, X_3	6 kN cm	0.25
Moment of inertia, X_4	90 cm^4	0.10
Distance from neutral axis, X_2	20 cm	0.05

Further assume that X_2, X_3, and X_4 are independent of each other. Find the reliability of the system if the capacity X_1 of the beam is a normal variate with a mean of 4 kN/cm^2 and a coefficient of variation of 30%.

9.13. Surveying using Geosatellite Positioning System. The values of latitude Y_{GPS} and longitude X_{GPS} obtained by GPS readings at a point are affected by a certain random multiplicative error Z. Thus, $X_{\text{GPS}} = Zx$ and $Y_{\text{GPS}} = Zy$, respectively, with x and y denoting longitude and latitude of the point. Find the reliability of measuring the planar distance w between two points with a tolerance of 3% if Z is a lognormally distributed variate with unit mean and coefficient of variation of 0.02, assuming that all (four) readings used in measuring the distance are independent of each other (this reliability can be evaluated as $\Pr[0.97 < W_{\text{GPS}}/w \le 1.03]$).

9.14. Column load. The strength of a building is normally distributed with mean of 336 kN, and coefficient of variation of 25%. The total column load is the sum of several components: live load, dead load, wind load, and snow load. These factors are independent variates as follows:

Factor	Expected value (kN)	Coefficient of variation
Normal live load	70	0.15
Normal dead load	90	0.05
Weibull wind load	30	0.30
Gumbel snow load	20	0.20

Find the reliability index and the risk of failure of the column. Check this result by simulation.

9.15. Stormwater removal. In the Italian method for designing storm sewer systems, the system capacity is estimated by

$$W = qX\left[1 - \exp\left(-\frac{qX}{s}\right)\right]^{-1},$$

where W is the stormwater removal capacity (that is, the volume of stormwater that can be appropriately drained by the system) for a storm with duration X, q is the outlet discharge capacity (that is, the maximum discharge that can be conveyed by the outlet channel under uniform flow conditions), and s is system storage capacity (that is, the volume of water that can be stored in the whole system, including the outlet channel, upstream channel network, and surface detention). System reliability can be evaluated using the concepts of capacity and load by considering the volume of stormwater delivered in a storm event as the storm depth multiplied by drainage area a, that is, aXY, where Y denotes the average intensity of a storm [see R. Rosso and

E. Caroni (1977), "Storm sewer capacity design under risk," *Proc. XVII Congr. Int. Assoc. Hydraul. Res.*, Baden-Baden, 15–19, August, Vol. 4, pp. 537–543]. Consider the stormwater removal system for a drainage area of $205 \times 10^3 m^2$ in a location where the duration X and intensity Y of a severe storm are independent exponentially distributed variates with means of 1.4 hours and 18 mm/h, respectively. The outlet discharge capacity of the system is 4 m^3/s, and its storage capacity is 1500 m^3. Using coherent units, do the following computations:

(*a*) Compute the system reliability for a storm by simulation.

(*b*) Because of gradually varied flow in the system, the system storage during a storm may not achieve storage capacity, and a random variate Z is substituted for s to evaluate the stormwater removal capacity W. Compute the system reliability for a storm if $f_Z(z) = 3z^2/s^3$ for $0 \leq z \leq s$, and 0 elsewhere.

9.16. Dilution requirements. The amount of water into which wastewater can be discharged without creating objectionable conditions is represented by a dilution parameter that is commonly expressed for combined systems as the minimum streamflow Y required. If wastewater with a first-stage BOD of W in newtons per capita is discharged daily into a stream with a permissible loading of Z in newtons per cubic meter, the required streamflow becomes $Y = 0.012W/Z$ in cubic meters per second per 1000 population. Assume that the population is 10,000, that permissible loading Z can vary uniformly from 0.23 to 0.12 N/m^3 and that load W is a normal variate with a mean of 1.2 N per capita and a coefficient of variation of 20%. Find the reliability of the system if streamflow X is a gamma variate with mean and standard deviation of 2 and 0.5 m^3/s, respectively [see G. M. Fair, J. C. Geyer, and D. A. Okun (1968), *Elements of Water Supply and Wastewater Disposal*, 2nd ed., John Wiley and Sons, New York, p. 659].

9.17. Law of diminishing returns. Consider a system with n series components, each constituted by m redundant independent subcomponents with equal probability of failure p.

(*a*) Find the overall reliability of this system.

(*b*) Show that, by determining the rate of increase of system reliability with increasing number m, the advantage of introducing additional redundant subcomponents in each series component rapidly vanishes.

9.18. Pipe network. Consider the portion of a pipeline network for urban water supply of Example 9.18 (see also Fig. 9.2.2). Assuming independent failure modes find the probability that node c remains isolated if there is a common probability of rupture of 1% for all pipes.

9.19. Retaining wall. The retaining wall for road embankment sketched in Fig. 9.P3*a* can fail due to several factors. The failure modes are schematically indicated in the block diagram for system reliability analysis shown in Fig. 9.P3*b* (M. E. Harr (1987), *Reliability-Based Design in Civil Engineering*, McGraw-Hill, New York). Under the assumption of independent modes, compute the reliability of the overall system if the individual probabilities of failure are as indicated in parentheses.

9.20. Rain gage network. The rain gage network for stormwater management in the metropolitan area of Milan, Italy, is constituted by 16 stations. Real-time operation of the urban drainage control system requires telemetered data from at least 12 stations in order to have sufficient information of spatial precipitation. Find the

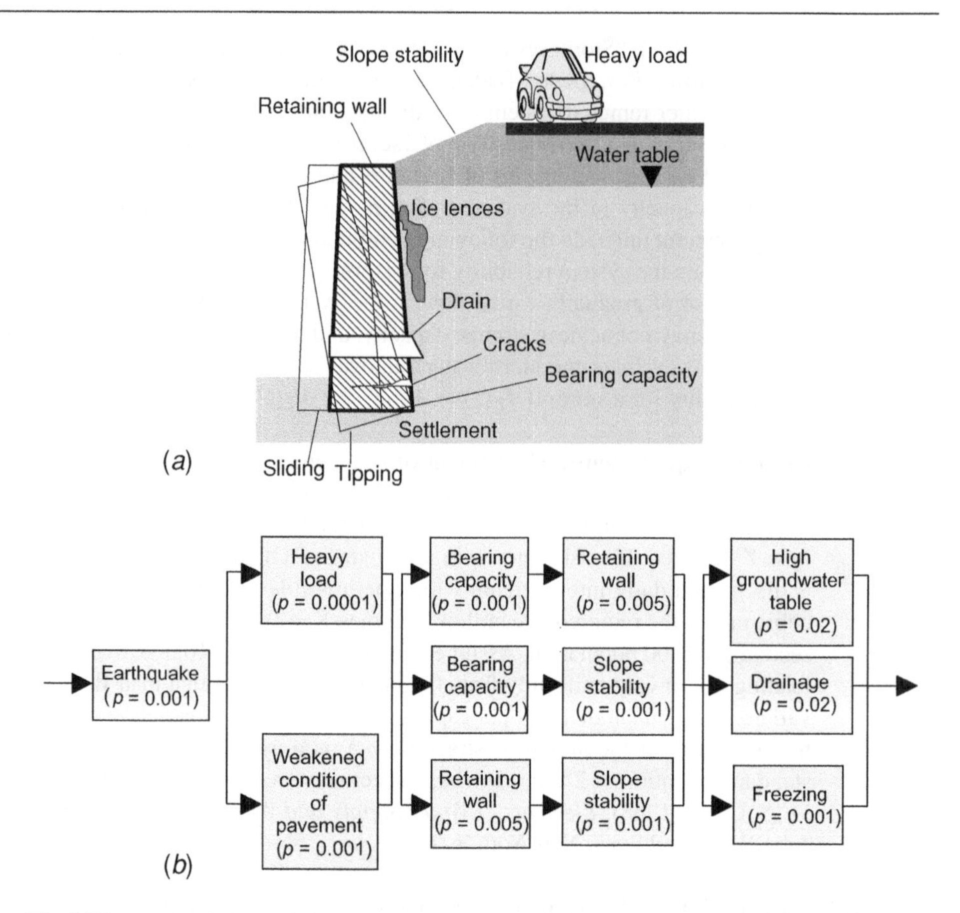

Fig. 9.P3 (*a*) Retaining wall for road embankment and (*b*) block diagram for reliability analysis. (Adapted from Harr, 1987, with permission of the McGraw-Hill.)

reliability of the network if the failures occur independently with a probability of 10%.

9.21. Improved reliability bounds for a *k*-out-of-*m* system. Assuming that failures occur as rare events, one can substitute the Poisson distribution for the binomial distribution to compute the reliability of a k-out-of-m system. Accordingly, R. J. Serfling (1974), "The role of the Poisson distribution in approximating system reliability of k-out-of-n structures," Contract N0014-75-C0551, Office of Naval Research, found

$$F_X(m-k) - l \leq r \leq F_X(m-k) + l,$$

where $F_X(\cdot)$ is the cdf of a Poisson variate X with mean $\sum p_i$, and $2l = \sum p_i^2$, where the summation is made over all m components, each of them having an individual probability of failure of p_i. Compute these bounds for the pumping system shown in Fig. 9.2.3 with $p = 0.03$.

9.22. Levee collapse. The design elevation of a levee built for flood protection at a site in the Po River plain was determined using the estimated 200-year flood, so that overtopping occurs with a risk of 0.5% in a year. However, this structure can fail also due to excessive seepage through the foundation material at high stages of the river, and it is estimated that this can occur with a probability of 10% if the 10-year flood stage is exceeded. Assuming that the two failure modes are normal and correlated with $\rho = 0.7$, find the minimum reliability of the levee and compare with that corresponding to the independent case.

9.23. Redundant and series equally reliable components. A system with a given overall reliability r is composed of l positively correlated components; each of them has an identical probability of failure, say, p. Find the bounds for the reliability of each component if (a) they are series; (b) they are redundant.

9.24. Repeated design. The reliability of a particular design procedure to prevent the collapse of buildings caused by earthquakes was found to be 99% over a long period of time. Since it is planned to construct ten structures using this design, evaluate the probability that none of the ten similar structures fails over the same time span.

9.25. Reservoir sedimentation. When reservoir sedimentation exceeds the dead level in a reservoir, it can affect its efficiency in meeting the target demand. Expensive maintenance work is then necessary to remove sedimentation excess. Assuming that the annual sediment yield trapped by a reservoir with dead capacity c is an exponentially distributed variate with mean μ, find expressions of $(a)R(t)$ and $(b)h(t)$. (c) For $c = 5 \times 10^6$ m^3 and $\mu = 4 \times 10^5$ m^3, determine the design life for a reliability level of 90%. Note that the sum of t exponentially distributed variates having a common scale parameter is a gamma-distributed variate with the same scale parameter and shape parameter equal to t.

9.26. Road pavement. Suppose that a road is made of 1000 pavement sections. The number of surviving pavements n_s after the jth year in service is as follows:

$j =$	1	2	3	4	5	6	7	8	9	10
$n_s =$	865	782	701	362	201	157	86	47	40	36

$j =$	11	12	13	14	15	16	17	18	19	20
$n_s =$	31	27	26	16	10	6	4	3	2	1

Estimate the pdf and cdf of the survival time distribution, $f_W(t)$ and $F_W(t)$; the reliability function $R(t)$; and the hazard function $h(t)$ for a pavement section. Find the reliable life of a pavement section for a specified reliability level of 70%. Note that for the discrete case the hazard function is the ratio between the number of failures in a time interval and the average number of survivors for the period.

9.27. Weibull reliability function. The reliability function of a saltwater conversion unit is taken as $R(t) = \exp[-(t/\tau)^{\gamma}]$, where τ is its characteristic lifetime, t is the test time, and γ is a parameter estimated from observations of several units of this type. For $\gamma = 1$, the reliability is exponentially distributed with constant rate of failure $1/\tau$; and $\gamma = 2$ gives the Rayleigh model associated with a linearly increasing hazard function.

(a) Find the expression of $h(t)$ associated with $R(t)$.

(b) If $\tau = 10$ years, $\gamma = 1.5$, and the unit is observed to be performing properly for 5 years, find its conditional reliability at the end of this period for $t = 10$ years.

9.28. Combined hazards in bridge construction. Evaluate the reliability of a bridge to be constructed in an earthquake-prone area, where the estimated return period of catastrophic earthquakes is 250. The 200-year flood must be taken as a design guidance for that area. The bridge is exposed to the hazard due to obsolescence, which increases in time as a logistic function,

$$h_{\text{obs}}(t) = 0.05/\{1 + \exp[-0.25(t - 25)]\}$$

where t is in years. Assuming that earthquakes and floods are independent sequences of Poisson events find (a) the reliability function of the bridge and (b) its design life for a reliability level of 90%.

9.29. Bearing capacity of soil. Bearing capacity of soil depends on the following three factors:

$$Y = \tan^4(45 + X/2),$$
$$W = e^{\pi \tan(X)} \tan^2(45 + X/2),$$
$$Z = \text{cotan}(X)[e^{\pi \tan(X)} \tan^2(45 + X/2) - 1],$$

which are functions of the friction angle X (see G. Vannucchi (1985). "Straightforward probabilistic methods in geotechnics," *Ital. J. Geotech.*, Vol. 19 No. 2, pp. 77–87 (in Italian).). Using the point estimate method find the mean and coefficient of variation of Y, W, and Z for mean friction angles of 15, 25, and 35°, and coefficients of variations for X of 0.1, 0.2, and 0.3. Compare these estimates with those obtained by simulation.

9.30. Partial load factors for beams. Consider two simply supported beams made of different materials that are designed to carry the same live load, with a nominal value of $q_l = 2$ kN/m. The nominal dead loads are $q_d = 1$ kN/m for beam 1, and $q_d = 3$ kN/m for beam 2, respectively. The mean value of the nominal resistance is $(\gamma_d q_d + \gamma_l q_l)/\phi$ with $\phi = 0.9$, $\gamma_d = 1.33$, and $\gamma_l = 1.5$, and its coefficient of variation is 0.1. Use the FOSM method to compute the values of the reliability index β for the two beams, assuming that both dead and live loads are normally distributed variates with means equal to the nominal values and coefficients of variation of 0.1 for the dead loads and 0.25 for the live load, respectively. Modify the value of the partial factor γ_d for beam 1 in order to obtain the same reliability of beam 2.

Chapter 10

Bayesian Decision Methods and Parameter Uncertainty

The main focus of this book's previous chapters was to find realistic descriptions of the types of random phenomena which an engineer encounters in his or her professional life. These activities should lead to a common goal, that of making decisions. For example, what design should be adopted for a stormwater drainage system with respect to pipe diameter, slope, and other variables when the magnitudes, durations, and frequencies of rainfall events cannot be predicted? Or perhaps an engineer must decide on the sizes of piles to be driven for the foundation of a large building when there is uncertainty regarding the nature of the subsurface. This chapter focuses on decisions that are called for under conditions of unpredictability—that is, when the decision maker is faced with the unknown. Each choice must be logically based with the aim of meeting given objectives, which often have an economic basis. The probability models we adopt are not unique in some practical situations, for it becomes evident when fitting to data from the real world that there is hardly any satisfactory criterion for choosing between two or more competing types. Our philosophy for making decisions should take this problem into account.

Nature is a frequent contributor to the uncertainty, and it has become customary to refer to its aspects of uncertainty as the *states of nature.* For instance, the inflows to a reservoir depend on the rainfall and state of the basin; the settlement of a foundation varies with the characteristics of the soil strata; and the water quality of a river is a function of the flow and other factors. The states of nature are quantified by probabilities, which are often evaluated subjectively when there is no practical alternative. The approach comes under the sphere of Bayesian decision theory, for which we follow Bayes' theorem introduced in Chapter 2 and utilized further in Chapter 9.

Thus the Bayesian viewpoint is an alternative to the classical approach, which is exemplified, for instance, by the estimation of confidence intervals in Section 5.3, wherein population parameters are treated as unknown constants. In Bayesian estimation, on the other hand, the unknown parameter or state of nature is treated as a random variable. For example, the proportion p of defective welds for a steel structure may change from one time period to another and is accordingly described by a prior pdf, $f(p)$. The value of this probability is thus an intrinsic part of the decision process. There is also direct involvement of the engineer's experience in such situations, albeit in a subjective manner.

The next step is the acquisition of new information by the engineer. This can take the form of additional cube tests, soil samples, gaugings, or drill holes, for instance. The Bayes' method, as we have previously seen, then provides a direct means of updating the probabilities. Not surprisingly, it has aroused some controversy, mainly with respect to subjective probabilities and the choice of a prior distribution.

We have already followed the decision-theoretic path in Chapters 5 and 6. Whereas we were previously concerned with the probabilities of making the right or wrong decisions and similar issues, the methodology is now formulated so that monetary loss (or gain) becomes the main criterion when quantifying engineering judgment. Although we have turned the spotlight on economic decision making, there are other factors of importance. For example, environmental benefits are of direct concern; these aspects may, however, entail problems of definition and estimation, as in the case of social and political considerations.

This chapter commences with basic Bayes' rules for action by the engineer followed by decision trees that show the available alternatives such as actions, states of nature, and losses. The associated minimax solution, which minimizes the maximum risk, is then presented. In posterior decision analysis, presented in Section 10.2, we discuss loss and utility functions. Appropriate theoretical distributions are indicated and applied. We return to likelihood ratio tests introduced in Chapter 5 but in a Bayesian context.

There are many situations in which Bayes' theorem is difficult to apply. This happens when we cannot obtain the moments of the posterior distribution except possibly by using some type of cumbersome numerical integration. In such cases one can apply Markov chain Monte Carlo methods. These are now standard in many areas of usage. A particular technique for application is the Gibbs sampler. The subject is discussed here and a case study is presented.

In the final section we demonstrate James-Stein estimators, at an elementary level, as an alternative method of revising forecasts of mean values.

10.1 BASIC DECISION THEORY

The Bayesian decision procedure can be summarized as follows: Let us suppose that the decision is to take an action or set of actions a that belong to a set A, the *action space*, comprising all conceivable courses of action. We specify the states of nature by the values a parameter θ takes. This parameter indexes the probability distribution of a random variable X, the observations of which form the basis of our decisions. The *model* comprises the probability distributions that X can take. We denote the set of all possible values of θ by Θ, the *parameter space*. The viable decision d is viewed as part of a decision space D; it is the link between X and A and maps the sample space of the basic random variable X onto the action space. The actions a that we take may be correct, or partially correct to some extent, or incorrect. In general, we expect that there will be a loss, implying a wastage of resources. This is quantified by a *loss function* $l(\theta, a)$ which specifies the loss, commonly expressed in monetary terms that will be incurred under action a and state of nature θ; if on the contrary such an outcome is negative the loss becomes a gain, say, $g(\theta, a)$.

For the analysis, we need a function to quantify the status or quality of the decision rule. If we adopt the action or set of actions $a = d(X)$ given a state of nature θ, the expected or average loss is defined by the *risk function*

$$R(\theta, d) = E[l\{\theta, d(X)\}]. \tag{10.1.1}$$

In this equation we use either the pmf $p(x \mid \theta)$ or the pdf $f(x \mid \theta)$ in discrete and continuous cases, respectively, as the weighting function to obtain the average loss. The risk is based on the decision and the true state of nature (which will usually be known in retrospect). Our objective is to find a reasonable method of minimizing the risk (or maximizing the gain) in the face of uncertainty.

10.1.1 Bayes' rules

Decision analysis is sometimes based on a *prior distribution* $\pi(\theta)$ to specify the probability distribution of the parameters. Previously, we discussed such distributions in relation to Bayes' theorem by which we can incorporate prior knowledge about a parameter; if we then revise it on receipt of sampling information, it becomes the posterior distribution, which we shall return to in Section 10.2. For making decisions, prior distributions can be used in different ways. Basically, they serve as weighting functions to determine the average risk, as already mentioned. They also give the expected monetary value, or as commonly termed, the *Bayes' risk* of a decision rule $a = d(X)$. This is defined as

$$B(\pi, d) = E[R(\theta, d)], \tag{10.1.2}$$

which is the expected risk, taking account of the prior distribution $\pi(\theta)$ of the state of nature θ. The decision that minimizes the Bayes' risk is called the Bayes' rule (or Bayes' decision). In Examples 10.1 to 10.4 we assume that the prior distribution is known or can be estimated.

Example 10.1. Bayes' risks in road contracts. A road contractor needs to hire or lease machinery and equipment for resurfacing works during a limited period. The first type costs \$120,000 and enables 200 km of roads to be restored. Alternatively, a second type costs \$40,000 but its roadwork is limited to 50 km. These costs include labor and other expenditures. The corresponding decisions that should be made are termed d_1 and d_2, respectively. The contractor expects to be successful in one of two bids. The first contract θ_1 involves 190 km of road and the second θ_2 only 40 km. The rate contracted is \$1900 per km for θ_1 and \$2050 per km for θ_2. Let us assess the Bayes' risks and find the Bayes' rule.

Solution. In the loss function, negative losses signify gains. The expected losses and gains are as specified [so there are no weighting functions and Eq. (10.1.1) follows directly]. The four risk functions are as follows:

$$R(\theta_1, d_1) = -1900 \times 190 + 120{,}000 = -\$241{,}000.$$
$$R(\theta_1, d_2) = -1900 \times 50 + 40{,}000 = -\$55{,}000.$$
$$R(\theta_2, d_1) = -2050 \times 40 + 120{,}000 = \$38{,}000.$$
$$R(\theta_2, d_2) = -2050 \times 40 + 40{,}000 = -\$42{,}000.$$

The probability that the first contract θ_1 is obtained, is denoted by p. Let us assume that the probability that the contractor is unsuccessful in both bids, say, θ_3, is zero. The expected losses or Bayes' risks associated with each decision are as follows:

$$B(\pi, d_1) = -241{,}000p + 38{,}000(1 - p) = \$38{,}000 - \$279{,}000p$$

and

$$B(\pi, d_2) = -55{,}000p - 42{,}000(1 - p) = -\$42{,}000 - \$13{,}000p.$$

We note that $B(\pi, d_1) < B(\pi, d_2)$, if $80{,}000 < 266{,}000p$. This holds if $p > 0.3$ approximately. Hence the Bayes' rule is that if $p > 0.3$—that is if the probability of obtaining the first contract is greater than 0.3—the first decision should be taken. Bayes' risks are plotted against the probability of obtaining the first contract in Fig. 10.1.1.

Example 10.2. Welding. In Problem 1.4 (of Chapter 1), we discussed specifications for welding of some structural components. These concerned lengths of welds. There may be other flaws, such as those measured on a volumetric basis. A welder employed by a civil

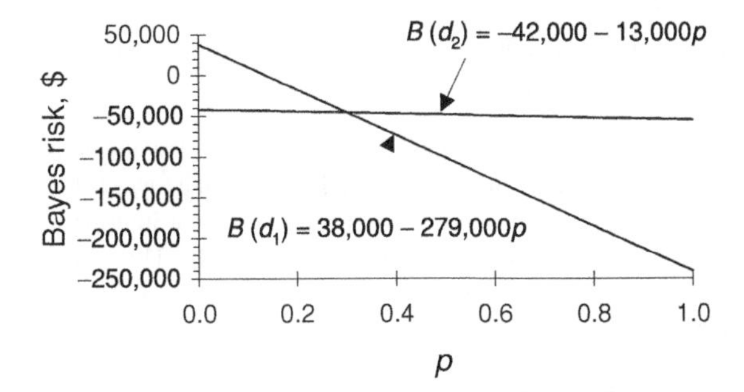

Fig. 10.1.1 Bayes' risk versus probability p of obtaining the first contract.

engineer fabricates 50 welds at a time. Subsequently a random test of a single weld is made. The welder is informed that there are only two possible actions open to him or her after the result of the random test is known. The first action a_1 is to accept payment at the rate of \$10 per weld for those which are satisfactory. However, a penalty of \$25 per weld will be imposed on any one or more welds which are found to be defective; this is to cover the cost of necessary remedial work. The second action a_2 is to accept a payment of \$100 for the lot. Let us assess the Bayes' risks and find the Bayes' rule.

Solution. Let ϕ_1 signify that the initial test is satisfactory and ϕ_2 that it is unsatisfactory. The decision space D can be fully described as follows:

D	ϕ_1	ϕ_2
d_1	a_1	a_2
d_2	a_1	a_1
d_3	a_2	a_2
d_4	a_2	a_1

Let i be the sampling number of bad welds in a lot of 50. This is taken as the state of nature. The probability that the tested weld is unacceptable is $i/50$, so the probability that it is satisfactory is $1 - i/50$. The risk functions are obtained by weighting the losses by these probabilities and are written as follows (with gains indicated by negative losses):

$$R(i, d_1) = [25i - 10(50 - i)](1 - i/50) - 100(i/50) = \$43i - \$0.7i^2 - \$500.$$
$$R(i, d_2) = [25i - 10(50 - i)] = \$35i - \$500.$$
$$R(i, d_3) = -\$100.$$
$$R(i, d_4) = [25i - 10(50 - i)](i/50) - 100(1 - i/50) = -\$8i + \$0.7i^2 - \$100.$$

In order to find the expected risks, we assume that the conditions of the $n = 50$ welds are the outcomes of n Bernoulli trials with parameter p, the probability that a weld is unsatisfactory. This assumption will, however, be violated in practice if the defects in welds are interdependent. If the assumption holds, I is a (n, p) binomial random variable.[1] Hence,

$$E[I] = np = 50p$$

and

$$\begin{aligned} E[I^2] &= \text{Var}[I] + \{E[I]\}^2 = np(1 - p) + (np)^2 \\ &= 50p(1 - p) + 2500p^2 = 50p + 2450p^2. \end{aligned}$$

Hence from Eq. (10.1.2) the Bayes' risks are written with p as a variable as follows:

$$B(\pi, d_1) = 43 \times 50p - 0.7(50p + 2450p^2) - \$500 = \$2115p - \$1715p^2 - \$500.$$
$$B(\pi, d_2) = 35 \times 50p - 500 = \$1750p - \$500.$$
$$B(\pi, d_3) = -\$100.$$
$$\begin{aligned} B(\pi, d_4) &= -8 \times 50p + 0.7 \times (50p + 2450p^2) - \$100 \\ &= -\$365p + \$1715p^2 - \$100. \end{aligned}$$

Bayes' risks are plotted against the probability p that a weld is defective in Figure 10.1.2. It is seen that for small values of p, decision d_2 has the lowest Bayes' risk and for large values of p, decision d_3 gives the lowest risk. There is a narrow intermediate range, in which decision d_1 provides the optimum by a margin of less than \$10. This range, which can be easily found from the given equations for Bayes' risk, is from $p = 0.213$ to $p = 0.233$. Decision d_4 is shown merely for completeness.

[1] Reference may be made to Subsection 4.1.2 and also Eq. (3.2.12) for what follows.

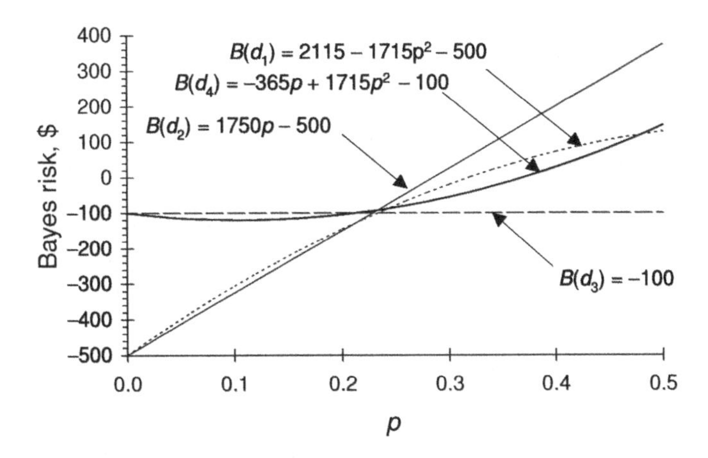

Fig. 10.1.2 Bayes' risk versus probability p of a defective weld.

10.1.2 Decision trees

In Examples 10.1 and 10.2, we considered the best possible courses of action when there is uncertainty regarding the true state of nature. The optimal expected result such as the highest net gain is found by examining each of the decisions $d_1, d_2, d_3, \ldots$, each associated with more than one course of action $a_1, a_2, \ldots$. These possible actions are said to belong to the action space A, as already stated. The engineer may sometimes delete actions that are inferior or illogical, such as decision d_4 in Example 10.2. Nevertheless, there are many options available. Note also that the probability distribution of the state of nature θ did not confine us to a few discrete cases in Examples 10.1 and 10.2; we considered the full range of probabilities. Situations arise, however, when the decision space is limited to a small number of actions or compound actions and the true state of nature can be represented by a few discrete values. In such cases a pictorial representation can be made of a particular problem. A commonly used method is a decision tree.

A decision tree is drawn from left to right. Having identified the various possible decisions and starting from a single node at the left extreme, one provides a fork for each decision. Further subdivisions are made at the ends of these forks to represent alternative actions, such as to test or not to test, to drill or not to drill, and so on, depending on the case studied. Then a complete set of forks is provided, at the ends of the previous forks, to denote the exclusive and true states of nature with estimated probabilities of occurrence. Finally, the losses, benefits, or utilities are given under each decision, action, and state of nature. Examples 10.3 and 10.4 utilize decision trees.

Example 10.3. Pipes for water supply. A manufacturer who supplies pipes for a water project is paid \$95 per pipe. This rate is applicable only for pipes that are found to be satisfactory. Unsatisfactory pipes are used for subnormal work and are paid for at a rate of \$25 per pipe. The overall cost to the manufacturer is \$60 per pipe. Past records show that 92% of the pipes manufactured meet the criteria. Sophisticated tests can be made at a cost of \$15 per pipe in order to exclude those with defects from being delivered. Let us suppose that if a pipe is found to be unsatisfactory by this test, it is brought to a satisfactory condition before delivery at an additional cost \$15. There is a cheaper alternative: doing an initial inspection at a cost of \$2 per pipe, but there is a 4% chance of not detecting an unsatisfactory pipe and a 2% chance of rejecting a good pipe with this test. If it is decided to do the initial inspection on all the pipes, tabulate the decisions available (with or without the tests) and determine the risks and Bayes' rule.

Solution. The decision space D is described in the table below. The *apparent* states of nature as seen from the results of the initial inspection are either ϕ_1 (the pipe is satisfactory) or ϕ_2 (it is unsatisfactory). The two possible actions are a_1 (do the exact test) and a_2 (forego the test).

D	ϕ_1	ϕ_2
d_1	a_1	a_1
d_2	a_1	a_2
d_3	a_2	a_1
d_4	a_2	a_2

The risk functions are calculated as follows. These are written as expected losses per pipe (so negative values are gains) and related to the *true* states of nature: θ_1 (satisfactory pipe) and θ_2 (unsatisfactory pipe) and the decisions given above.

$$R(\theta_1, d_1) = 60 - 95 + 2 + 15 = -\$18.0.$$

$$R(\theta_1, d_2) = (60 - 95 + 2 + 15) \times 0.98 + (60 - 95 + 2) \times 0.02 = -\$18.3.$$

$$R(\theta_1, d_3) = (60 - 95 + 2) \times 0.98 + (60 - 95 + 2 + 15) \times 0.02 = -\$32.7.$$

$$R(\theta_1, d_4) = 60 - 95 + 2 = -\$33.0.$$

$$R(\theta_2, d_1) = 60 - 95 + 2 + 15 + 15 = -\$3.0.$$

$$R(\theta_2, d_2) = (60 - 95 + 2 + 15 + 15) \times 0.04 + (60 - 25 + 2) \times 0.96 = \$35.4.$$

$$R(\theta_2, d_3) = (60 - 25 + 2) \times 0.04 + (60 - 95 + 2 + 15 + 15) \times 0.96 = -\$1.4.$$

$$R(\theta_2, d_4) = 60 - 25 + 2 = \$37.0.$$

As previously stated, the following prior probabilities are estimated from past records: $\pi(\theta_1) = 0.92$ and $\pi(\theta_2) = 0.08$. Bayes' risks are then obtained from Eq. (10.1.2), using the prior probabilities as weighting functions:

$$B(\pi, d_1) = -18 \times 0.92 + -3 \times 0.08 = -\$16.8.$$

$$B(\pi, d_2) = -18.3 \times 0.92 + 35.4 \times 0.08 = -\$14.0.$$

$$B(\pi, d_3) = -32.7 \times 0.92 - 1.4 \times 0.08 = -\$30.2.$$

$$B(\pi, d_4) = -33 \times 0.92 + 37 \times 0.08 = -\$27.4.$$

The minimum Bayes' risk is thus given by the third decision, which becomes the Bayes' rule. If this is adopted, the manufacturer expects to make a profit of above \$30 per pipe on average. The decision tree for pipe testing is given in Fig. 10.1.3.

Example 10.4. Bayes' rule for cofferdam. A small dam is to be built across a stream in a mountainous area where the melting of snow makes a substantial contribution to the runoff and high flows. A cofferdam is planned for protecting the work on foundations during the first year of construction. At the preliminary stage, two designs are submitted for the cofferdam based on the maximum flow expected during the period. The proposed alternative heights for this temporary structure, which are proportional to the costs, are as follows: h_1: 2 m and h_2: 5 m.

The designs are based on the assumption that the true state of nature θ that is related to the magnitude of the maximum flow in the stream during the design period can have one of two values:

θ_1 : head of water is 2 m.

θ_2 : head of water is 5 m.

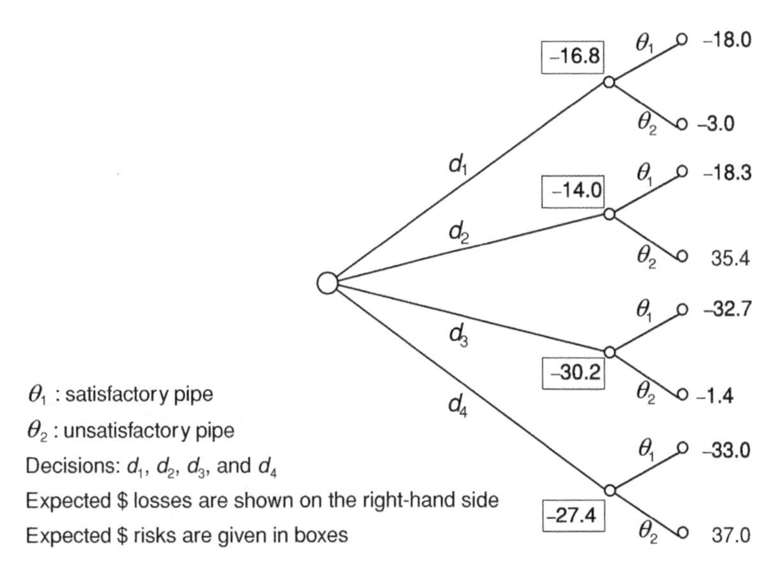

Fig. 10.1.3 Decision tree for pipe testing.

If the shorter cofferdam is built and subsequently found to be inadequate, emergency measures are required to protect the foundations of the main dam. This is estimated to cost $60,000. Conversely, the construction of a larger cofferdam in place of a smaller structure, which would have sufficed under the circumstances, amounts to $15,000 in wasted resources. In this example we deal entirely with losses, and the signs are therefore omitted. We can thus define the loss function as follows:

H	θ_1	θ_2
h_1	$0	$60,000
h_2	$15,000	$0

A forecast is made of the height required for the cofferdam in the coming year. Suppose this gives the probable heights X required, which depend on θ as follows:

X	θ_1	θ_2
2 m	.50	.0
3 m	.30	.10
4 m	.15	.20
5 m	.05	.70

There are five decision rules that give the actions under various forecasts F as specified in the following table:

D	2 m	3 m	4 m	5 m
d_1	h_1	h_1	h_1	h_1
d_2	h_1	h_1	h_1	h_2
d_3	h_1	h_1	h_2	h_2
d_4	h_1	h_2	h_2	h_2
d_5	h_2	h_2	h_2	h_2

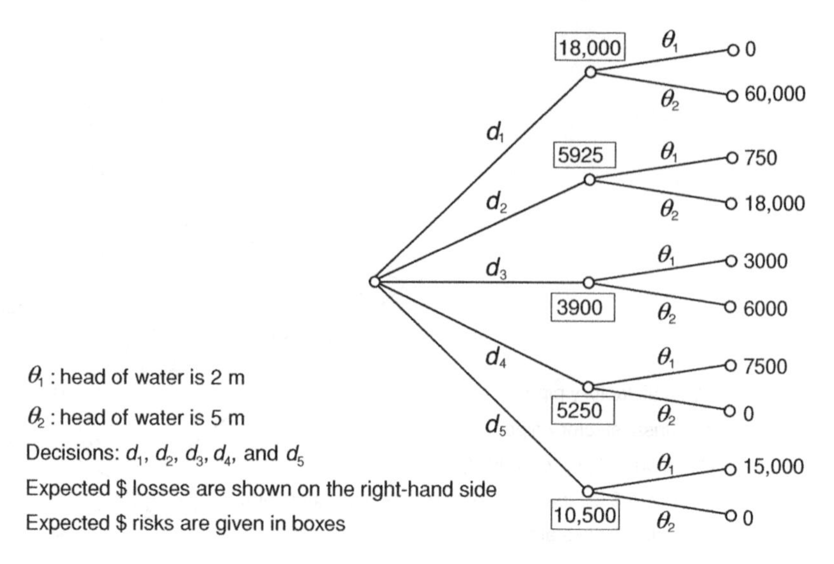

Fig. 10.1.4 Decision tree for cofferdam.

From Eq. (10.1.1) the risk functions for each state of nature and decision are as follows:

$$R(\theta_1, d_1) = 0 \times 0.50 + 0 \times 0.30 + 0 \times 0.15 + 0 \times 0.05 = \$0.$$
$$R(\theta_1, d_2) = 0 \times 0.50 + 0 \times 0.30 + 0 \times 0.15 + 15{,}000 \times 0.05 = \$750.$$
$$R(\theta_1, d_3) = 0 \times 0.80 + 15{,}000 \times 0.20 = \$3000.$$
$$R(\theta_1, d_4) = 0 \times 0.50 + 15{,}000 \times 0.50 = \$7500.$$
$$R(\theta_1, d_5) = \$15{,}000.$$
$$R(\theta_2, d_1) = \$60{,}000.$$
$$R(\theta_2, d_2) = 60{,}000 \times 0.30 + 0 \times 0.70 = \$18{,}000.$$
$$R(\theta_2, d_3) = 60{,}000 \times 0.10 + 0 \times 0.90 = \$6000.$$
$$R(\theta_2, d_4) = 0.$$
$$R(\theta_2, d_5) = 0.$$

From past records, the following prior probabilities are estimated: $\pi(\theta_1) = 0.7$ and $\pi(\theta_2) = 0.3$. Bayes' risks are then obtained from Eq. (10.1.2), using the prior probabilities as weighting functions.

$$B(\pi, d_1) = 0 \times 0.7 + 60{,}000 \times 0.3 = \$18{,}000.$$
$$B(\pi, d_2) = 750 \times 0.7 + 18{,}000 \times 0.3 = \$5925.$$
$$B(\pi, d_3) = 3000 \times 0.7 + 6000 \times 0.3 = \$3900.$$
$$B(\pi, d_4) = 7500 \times 0.7 + 0 \times 0.3 = \$5250.$$
$$B(\pi, d_5) = 15{,}000 \times 0.7 + 0 \times 0.3 = \$10{,}500.$$

The minimum Bayes' risk is thus given by the third decision, which becomes the Bayes' rule. The decision tree for the cofferdam is given in Fig. 10.1.4.

10.1.3 The minimax solution

In Subsection 10.1.2 we used the notion of Bayes' risk to summarize in a unique form the information conveyed by a risk function. The average risk under each decision rule is evaluated in the application of Bayes' risk through the prior distribution of the state of nature. Hence an optimal solution can be found. The *minimax* method provides an alternate summary of the risk function by considering maximum risks and then finding the minimum value of the maximum risks. This procedure is done in the absence of prior information. The risk varies with the state of nature θ, and the optimal decision is found within the decision space D. The procedure seems conservative, but in the absence of prior information it is justifiable if one needs to safeguard against all adverse eventualities. The minimax decision rule thus points to the decision, say, d^*, that minimizes the maximum risk, under each decision, as follows:

$$\max_{\theta \in \Theta}[R(\theta, d^*)] = \min_{d \in D}\left\{\max_{\theta \in \Theta}[R(\theta, d)]\right\}.$$

Example 10.5. Water pipes. Consider the pipes for water supply problem of Example 10.3. The values in dollars of the risk function are reproduced here, for the four decisions and two states of nature, with negative values denoting gains.

	d_1	d_2	d_3	d_4
θ_1	−18	−18.3	−32.7	−33
θ_2	−3	35.4	−1.4	37

The maximum risks are given in the bottom row (for θ_2), so the minimax solution is to take action d_1. As shown in Example 10.3, the Bayes' rule is d_3 based on the prior probabilities $\pi(\theta_1) = 0.92$ and $\pi(\theta_2) = 0.08$. Calculations show that the minimax decision corresponds to the range of prior probabilities in which $0 \le \pi(\theta_1) < 0.1$. Therefore the minimax decision rule seems to be ultraconservative in this case because it is a most uncommon practice to supply pipes 90% of which are defective!

Example 10.6. Cofferdam. Let us reconsider Example 10.4 and find the corresponding minimax solution. Each of the five decisions contain a risk that also varies with the state of nature θ. These risks are shown in the following table. (Given in the last row of the table are Bayes' risks evaluated in Example 10.4 for the prior distribution: $\pi(\theta_1) = 0.7$ and $\pi(\theta_2) = 0.3$.)

	d_1	d_2	d_3	d_4	d_5
θ_1	0	750	3,000	7,500	15,000
θ_2	60,000	18,000	6,000	0	0
$B(\pi, d)$	18,000	5,925	3,900	5,250	10,500

For the minimax solution the maximum risks are \$60,000, \$18,000, \$6,000, \$7,500, and \$15,000 under the respective decisions; hence, the minimax rule is d_3. As shown in Example 10.4 and given here, the Bayes' rule is also d_3. It is interesting to note that for the range of prior probabilities in which $0.6 \le \pi(\theta_1) \le 0.8$, the minimax and Bayes' rules are the same. Hence the minimax solution seems to be reasonable in this case.

Some of the shortcomings of the minimax method can be overcome, at least in part, by adopting the *minimax regret* approach. Suppose we have a net benefit or gain matrix. For each state of nature θ, we find the maximum gain, corresponding to some decision. Then the other benefits for that state of nature are subtracted from the maximum benefit and the differences are termed *regret losses*. We than proceed as before to find the minimax solution.

Example 10.7. Irrigation scheduling. The inflows to a reservoir built for irrigation needs are random variables. For preliminary purposes the annual yield of the reservoir may be represented by variables, say, θ_1, θ_2, and θ_3. Early in the year the engineer needs to take one of three decisions d_1, d_2, and d_3. Each of these are related differently to the following:

(*a*) The preparation of a variable extent of land for farming.
(*b*) Decisions on the types of crops to be grown.
(*c*) Amounts of water to be drawn from external sources or released from the reservoir for other purposes if a surplus is expected.

The annual net benefits in units of $1000 are given as follows:

	d_1	d_2	d_3
θ_1	100	80	60
θ_2	120	180	140
θ_3	150	200	250

We convert the benefit matrix to a regret matrix by subtracting the highest value in each row from the values in that row. The resulting values represent the opportunity loss for each decision and state of nature in units of $1000:

	d_1	d_2	d_3
θ_1	0	20	40
θ_2	60	0	40
θ_3	100	50	0

Under each decision the maximum losses are $100,000, $50,000, and $40,000. The minimax solution is to take the action that minimizes these maximum losses. Hence the decision is d_3.

The regrets approach has its advantages as shown here. It should be noted, however, that differences in regrets may be linearly disproportionate to differences in benefits.

10.1.4 Summary of Section 10.1

In this section we provide a brief introduction to Bayesian decision theory. This involves loss and risk functions, a prior distribution, and the Bayes' risk. Then the Bayes' rule is given for taking the decision that minimizes the risk. When the true state of nature may be represented by a few discrete values, decision trees are helpful, as seen here, in showing the choice of decisions that can be made with respect to expected losses and risks. The alternative minimax rule is adopted when there is no prior information or when the prior distribution is vaguely defined. This method may sometimes give results that are irrational, as in Example 10.5 but not in Example 10.6, which corresponds to a reasonable prior distribution. The minimax regret method (Example 10.7) may overcome some of the problems cited.

10.2 POSTERIOR BAYESIAN DECISION ANALYSIS

In this section we apply Bayes' rule in situations in which a posterior density function $f(\theta \mid x)$ is estimated on the basis of observations or new data x in addition to a known (or sometimes assumed) prior distribution function $\pi(\theta)$. Let $f(x \mid \theta)$ denote the likelihood function of x conditional to a state of nature θ. Then if θ is continuous

$$f(\theta \mid x) = \frac{f(x \mid \theta)\pi(\theta)}{\int_{-\infty}^{\infty} f(x \mid \theta)\pi(\theta)d\theta}. \tag{10.2.1}$$

When discrete values of θ are considered, we should replace the integral in the denominator on the right-hand side of Eq. (10.2.1) by a summation as follows:

$$f(\theta_j \mid x) = \frac{f(x \mid \theta_j)\pi(\theta_j)}{\sum_i f(x \mid \theta_i)\pi(\theta_i)}. \tag{10.2.2}$$

Corresponding to Eq. (10.1.1), the risk function is defined as

$$R(\theta, d) = E[l\{\theta, d(x)\}] = \int l\{\theta, d(x)\} f(\theta \mid x)d\theta, \tag{10.2.3}$$

which depends on the posterior distribution function. The decision d' that minimizes the risk function is called the Bayes' rule.

10.2.1 Subjective probabilities

As previously noted, engineers sometimes face situations that require knowledge of probabilities without the benefit of knowing repeated outcomes under similar conditions, so that the long-run frequency approach is not possible. For example, what is the probability that a pile driven for a foundation will encounter rock at a specified depth? Or what is the probability that traffic flow will exceed a critical rate 10 years from now? In circumstances such as these the frequency approach may not apply because the data are not yet available or it is not possible to carry out the necessary experiment. In such cases it seems justifiable to apply subjective probabilities quantifying personal knowledge or belief, such as the chance of high flows during the coming year that might damage a cofferdam (as in Example 10.4) or the possibility of rain affecting some important construction work.

In Examples 10.3 and 10.4, prior probabilities were used to obtain the Bayesian decision rules. We did not state how these probabilities are obtained. It is sometimes possible to estimate these through the interpretation of frequentists such as Fisher, Neyman, E. S. Pearson, and Wald. In fact, this classical treatment has been the basis of our previous chapters (with the exception of the brief introduction in Chapter 2 and the Bayesian revision of reliability in Chapter 9). However, the approach one follows in situations discussed in this chapter is more often based on the degree of belief a rational individual has, according to information available, on the outcome of an uncertain event. The subjective view of probability, as conveyed by this statement, was expressed by James Bernoulli in the seventeenth century and, also prior to the twentieth century, by De Morgan and Laplace. Subsequently, Savage (1954) and de Finetti (1970) were notable among those advocating this approach.

Prior probabilities can be vague or diffused (that is, noninformative) at times, or more definitive and impersonal on other occasions, in the manner of Jeffreys (1961) who objected to the use of subjective probabilities. In any case, Bayes' theorem provides a method of revising the probabilities on receipt of additional data as demonstrated in the following illustrations. When there is inclusion of the subjective element, the activity is usually called decision making under uncertainty. Decision making with objective probabilities, on the other hand, is called decision making under risk.

10.2.2 Loss and utility functions

In Subsection 10.1.1 we obtained the risk function by taking the expectation of the *loss function*. As previously emphasized, an engineer has to make decisions, in all phases of a design and its implementation. These pertain, for example, to heights of dams, depths of piles, and construction procedures. All are dependent on an unknown state of nature θ. This is a fixed parameter in classical statistical theory, but in Bayesian decision analysis it is a random variable, as already mentioned, and is further revised on receipt of additional information. The objective is to take the optimum decision that minimizes the risk or loss. The choice of a suitable loss function is perhaps the main problem one faces in the Bayesian approach; the other concerns the prior distribution (as discussed in Subsection 10.2.1). It is an accepted fact that accurate estimation of loss functions is not feasible. An incorrect loss function leads to reduced profits, extra costs, losses, damages, compensations, and even disasters in some cases. In most circumstances, however, one can estimate the right type of function with sufficient accuracy. Engineering experience indicates the shape of the function, but it should be amenable to mathematical analysis.

For instance, if errors of estimation are not quite serious we may use the loss function $l(\tilde{\theta}, \theta) = c\,|\,\tilde{\theta} - \theta\,|$, where c is a constant. Thus, our decision is based on the value $\tilde{\theta}$ when the true state of nature is θ. It is more common, however, to choose the stronger alternative function $l(\tilde{\theta}, \theta) = c(\tilde{\theta} - \theta)^2$, based on squared errors. A notable use of this function was made in the 1930s by Neyman, Pearson, and Wald in their classical statistical decision theory for estimating an unknown parameter, having disregarded the use of unbiased estimators. This is also used in James-Stein estimators (Section 10.4). Then from Eq. (10.2.3),

$$R(\tilde{\theta}, d) = E[l(\tilde{\theta}, \theta)] = \int_{-\infty}^{\infty} c(\tilde{\theta} - \theta)^2 f(\theta\,|\,x)d\theta.$$

Hence,

$$\frac{dR}{d\tilde{\theta}} = 2c \int_{-\infty}^{\infty} (\tilde{\theta} - \theta) f(\theta\,|\,x)d\theta.$$

Because $\int_{-\infty}^{\infty} f(\theta\,|\,x)d\theta = 1$, we have for the optimum condition, $dR/d\tilde{\theta} = 0$,

$$\tilde{\theta} = \int_{-\infty}^{\infty} \theta f(\theta\,|\,x)d\theta.$$

It shows that for the squared-error loss function the posterior mean corresponding to our optimal decision is the estimator for the state of nature.

Risk analysis applied through Eq. (10.2.3) is equivalent to defining a so-called *utility function*, $U(\theta\,|\,a)$ for a state of nature θ conditional to an action a. Such a function expresses an expected benefit or loss; utility conveys the same meaning as preferability. The variable can be monetary worth in economic theory or, more importantly for the civil or environmental engineer, the true state of nature such as the actual strength of an engineering material or the magnitude of a flood. These functions are usually nonlinear. When interpreted in a personal sense, a risk seeker's curve is concave

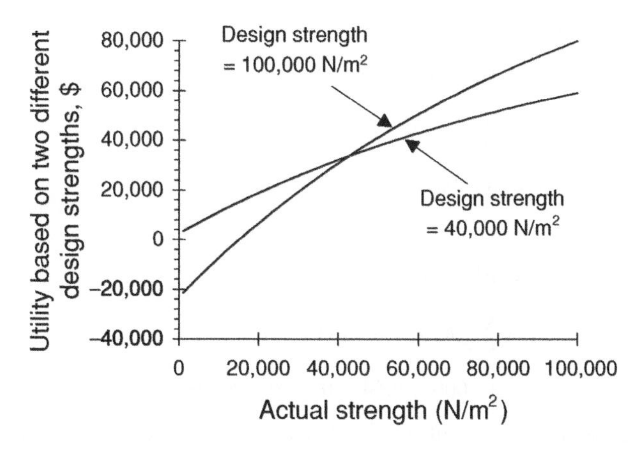

Fig. 10.2.1 Utility functions (with constants specified in the text).

upward.[2] But more commonly they are concave downward signifying an avoidance of risk. For example,

$$U(\theta \mid a) = ba + c[1 - \exp(-k(\theta - a))], \tag{10.2.4}$$

where b, c, and k are constants and action a directly reflects the decision regarding the design value for θ. Two such utility functions are shown in Fig. 10.2.1 for $b = 0.8$, $c = 60{,}000$, and $k = 0.00001$ and design values of 40,000 and 100,000 N/m^2 for θ.

10.2.3 The discrete case

Let us initially consider only discrete values of θ, without the application of a theoretical probability mass function. As in Eq. (10.2.2), a summation takes the place of the integral on the right-hand side of Eq. (10.2.3). The discrete case simply involves the substitution of probabilities and values of risk from data or other means as shown in the following example:

Example 10.8. Cofferdam. We return again to the problem of the cofferdam of Example 10.4. Let us suppose that on the basis of the depth of the snow cover in the upstream basin, long-range weather forecasts, and hydrological calculations, it is predicted that the necessary height of the cofferdam is $x = 4$ m. From the data provided in Example 10.4, $f(4 \mid \theta_1) = 0.15$ and $f(4 \mid \theta_2) = 0.20$; also $\pi(\theta_1) = 0.7$ and $\pi(\theta_2) = 0.3$. Hence, from Eq. (10.2.2) the posterior (conditional) pmf of θ is

$$f(\theta_1 \mid 4) = \frac{f(4 \mid \theta_1)\pi(\theta_1)}{\sum_{i=1}^{2} f(4 \mid \theta_i)\pi(\theta_i)} = \frac{0.15 \times 0.7}{0.15 \times 0.7 + 0.20 \times 0.3} = 0.64.$$

And

$$f(\theta_2 \mid 4) = 1 - (\theta_1 \mid 4) = 0.36.$$

The decision is either to build a cofferdam of height h_1 or a taller one with height h_2. From Eq. (10.2.3) and the loss function defined in Example 10.4, the two actions are associated

[2] See, for example, Ingles (1983).

with posterior risks given by

$$R(\theta, h_1) = E[l(\theta, h_1)]$$
$$= \sum_{i=1}^{2} (\theta_i, h_1) f(\theta_i \mid 4)$$
$$= 0 \times 0.64 + 60{,}000 \times 0.36 = \$21{,}600$$

and

$$R(\theta, h_2) = E[l(\theta, h_2)]$$
$$= \sum_{i=1}^{2} (\theta_i, h_2) f(\theta_i \mid 4)$$
$$= 15{,}000 \times 0.64 + 0 \times 36 = \$9600.$$

The posterior Bayes' rule is to build a cofferdam of height h_2 because this entails a smaller risk. In effect we have reworked the problem of Example 10.4 differently, in a more compact form, using the posterior distributions. We have come to the same conclusion.

10.2.4 Inference with conditional binomial and prior beta

Let the prior distribution of the state of nature θ be beta(α, β), that is,

$$\pi(\theta) = \frac{\Gamma(\alpha+\beta)}{\Gamma(\alpha)\Gamma(\beta)} \theta^{\alpha-1}(1-\theta)^{\beta-1}, \quad 0 < \theta < 1, \alpha > 0, \beta > 0, \qquad (10.2.5)$$
$$= 0, \qquad \text{elsewhere.}$$

This distribution is appropriate if θ represents a probability, for example, $\theta = 1 - F(y)$, where the cdf $F(y) = \Pr[y < Y]$ and Y is a random variable such as soil pressure, concrete strength, or river flow.

If X is the number of exceedances of Y in n independent trials, the sample likelihood function of X given θ is binomial. That is,

$$f(x \mid \theta) = \binom{n}{x} \theta^x (1-\theta)^{n-x}, \quad \text{for } x = 0, 1, 2, \ldots, n. \qquad (10.2.6)$$

Hence, the joint density function of X and θ becomes

$$f(x, \theta) = f(x \mid \theta)\pi(\theta)$$
$$= \binom{n}{x} \frac{\Gamma(\alpha+\beta)}{\Gamma(\alpha)\Gamma(\beta)} \theta^{\alpha+x-1}(1-\theta)^{\beta+n-x-1}. \qquad (10.2.7)$$

The marginal density function of X is obtained by integrating out θ as follows:

$$f_X(x) = \binom{n}{x} \frac{\Gamma(\alpha+\beta)}{\Gamma(\alpha)\Gamma(\beta)} \int_0^1 \theta^{\alpha+x-1}(1-\theta)^{\beta+n-x-1} d\theta$$
$$= \binom{n}{x} \frac{\Gamma(\alpha+\beta)}{\Gamma(\alpha)\Gamma(\beta)} \frac{\Gamma(\alpha+x)\Gamma(\beta+n-x)}{\Gamma(\alpha+\beta+n)}. \qquad (10.2.8)$$

We recognize that Eq. (10.2.8) can be substituted for the denominator of Eq. (10.2.1), and the posterior density function of θ for a given value of X is obtained using Eq. (10.2.7) as follows:

$$f(\theta \mid x) = \frac{\Gamma(\alpha + \beta + n)}{\Gamma(\alpha + x)\Gamma(\beta + n - x)} \theta^{\alpha+x-1}(1-\theta)^{\beta+n-x-1}. \tag{10.2.9}$$

From Eq. (10.2.5) it follows that Eq. (10.2.9) is the pdf of a beta $(\alpha + x, \beta + n - x)$ distribution.

Example 10.9. Protective embankments on the Po. The Po is the largest river basin in Italy. In the lower reaches of the basin, vast areas of agricultural land are protected by embankments along both banks of the river. There is concern about the adequacy of the flood protection scheme. Over a period of 72 years, it is known that the agricultural lands below the city of Piacenza have been extensively flooded five times, that is, in five different years. Assuming quite reasonably that the overtopping of the embankments constitutes a series of independent events, let us determine (a) the probability that the adjacent lands are inundated in the next year and (b) the probability that lands are safe from flood water during the next 10 years.

Solution. Corresponding to Eq. (10.2.8), X is the number of exceedances over a future period of n years. The marginal pdf of X is obtained from Eq. (10.2.8) as follows with $\alpha = t + 1$ and $\beta = m - t + 1$:

$$f(x) = \frac{n!}{x!(n-x)!} \frac{(m+1)!}{t!(m-t)!} \frac{(x+t)!(n+m-t-x)!}{(n+m+1)!},$$

where $m = 72$ and $t = 5$. (Note that $0! = 1$.)

(a) $n = 1; x = 1$

$$f(1) = \frac{1!}{1!0!} \frac{73!}{5!67!} \frac{6!67!}{74!} = \frac{6}{74} = 0.08.$$

(b) $n = 10; x = 0$

$$f(10) = \frac{10!}{0!10!} \frac{73!}{5!67!} \frac{5!77!}{83!} = 0.45.$$

The prior [Eq. (10.2.5), with $\alpha = 6$ and $\beta = 68$)] and posterior [Eq. (10.2.9), with (a) $n = 1$; $x = 1$ and (b) $n = 10; x = 0$] distributions of θ, the probability of inundation of adjacent lands, are shown in Fig. 10.2.2.

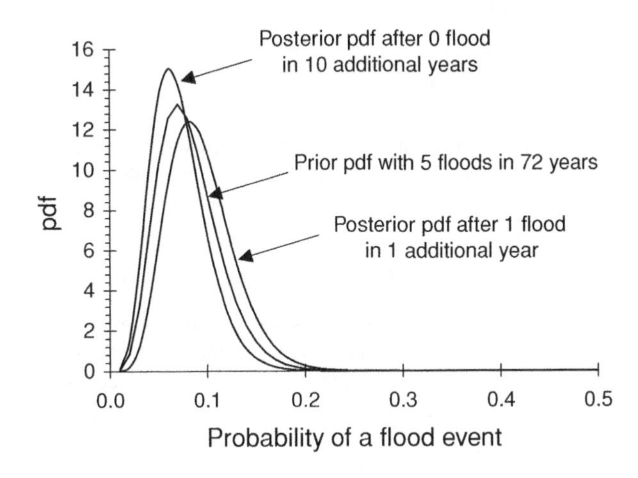

Fig. 10.2.2 Prior and posterior beta distributions of probability of a flood event.

It is noted from the preceding and the following examples that the posterior pdf tends to be taller and hence narrower than the prior pdf. This reflects a reduction in the uncertainty.

10.2.5 Poisson hazards and gamma prior

As discussed in Chapter 4, there are many applications in civil and environmental engineering in which intervals in time or space tend to result from Poisson processes, such as in traffic flow on busy roads and failures at treatment plants. We noted that the waiting time between successive events, X, for a homogeneous Poisson process, with parameter λ, called the hazard rate, is exponentially distributed with pdf

$$\begin{aligned} f_X(x) &= \lambda e^{-\lambda x}, \quad && \text{for } x \geq 0, \lambda > 0, \\ &= 0, && \text{otherwise.} \end{aligned}$$

For n independent waiting times with sum S_x, the joint pdf is as follows:

$$f(x_1, x_2, \ldots, x_n \mid \lambda) = \prod_{i=1}^{n} \lambda e^{-\lambda x_i} = \lambda^n e^{-\lambda S_x}. \tag{10.2.10}$$

In Bayesian estimation we noted that the Poisson rate parameter λ is treated as a random variable. It is reasonable to assume a gamma prior pdf for the hazard rate λ; that is,

$$\begin{aligned} \pi(\lambda \mid \alpha, r) &= \frac{\alpha^r \lambda^{r-1} e^{-\alpha\lambda}}{\Gamma(r)}, \quad && \text{for } \lambda > 0, \\ &= 0, && \text{otherwise.} \end{aligned} \tag{10.2.11}$$

The posterior pdf is proportional to the product of Eqs. (10.2.10) and (10.2.11); that is,

$$f(\lambda \mid S_x) = \frac{1}{k} \frac{\alpha^r \lambda^{r+n-1} e^{-\lambda(\alpha+S_x)}}{\Gamma(r)}, \tag{10.2.12}$$

where k is a constant. We obtain k by integrating out λ as follows [using a substitution such as $z = \lambda(\alpha + S_x)$]:

$$f(S_x) = \int_0^\infty \frac{\alpha^r \lambda^{r+n-1} e^{-\lambda(\alpha+S_x)}}{\Gamma(r)} d\lambda = \frac{\alpha^r \Gamma(r+n)}{\Gamma(r)(\alpha + S_x)^{r+n}}.$$

Thus the posterior pdf of λ is

$$f(\lambda \mid S_x) = \frac{(\alpha + S_x)^{r+n} \lambda^{r+n-1} e^{-\lambda(\alpha+S_x)}}{\Gamma(r+n)}. \tag{10.2.13}$$

Example 10.10. Traffic rates. In Example 4.21 we considered, for a Poisson process with hazard rate λ, the exponentially distributed time intervals between vehicles passing a point of observation on a road. The following are ten time intervals in minutes between vehicles observed at another point:

1.2 0.1 1.5 1.0 2.3 0.2 1.4 0.1 1.1 1.3

As just discussed it is reasonable to assume a gamma prior for the Poisson rate parameter λ. From observations in similar cases, the mean and standard deviation of the rate λ are 2 and 2.828. Hence the parameters for the gamma prior take values $\alpha = 0.25$ and $r = 0.50$. Also, $n = 10$ and $S_x = 10.2$ min. The prior pdf of λ from Eq. (10.2.11) is

$$\pi(\lambda \mid \alpha, r) = \frac{0.25^{0.50} \lambda^{-0.5} e^{-0.25\lambda}}{\Gamma(0.50)}.$$

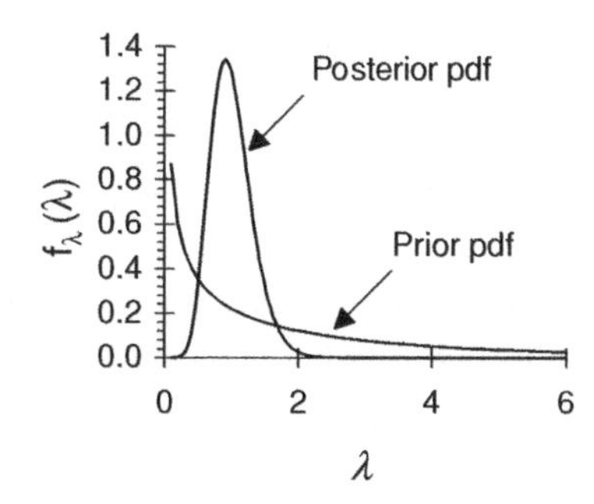

Fig. 10.2.3 Prior and posterior pdfs of the Poisson hazard rate, λ, resulting in exponentially distributed times between successive vehicles.

After the ten observations are made, the posterior pdf becomes

$$f(\lambda \mid S_x) = \frac{(10.45)^{10.5}\lambda^{9.5}e^{-10.45\lambda}}{\Gamma(10.5)}.$$

Figure 10.2.3 shows the prior and posterior pdfs for the Poisson rate.

10.2.6 Inferences with normal distribution

Let $X_1, X_2, \ldots, X_n$ be a random sample taken from a distribution $N(\theta, \sigma^2)$ with known σ^2. Suppose the prior distribution of the mean θ is $N(\mu_0, \sigma_0)$. In the following equations we shall use constants k_1, k_2, k_3, k_4, and k_5 for normalizing purposes. The prior pdf of θ can be written as

$$\pi(\theta) = k_1 \exp\left[-\frac{(\theta - \mu_0)^2}{2\sigma_0^2}\right]. \tag{10.2.14}$$

Also, we can write the pdf of X for a given θ as

$$f(x \mid \theta) = k_2 \exp\left[-\frac{\sum_{i=1}^{n}(x_i - \theta)^2}{2\sigma^2}\right]. \tag{10.2.15}$$

Using the sample mean $\bar{x}$, this can also be written as

$$f(x \mid \theta) = k_2 \exp\left[-\frac{n(\bar{x} - \theta)^2}{2\sigma^2} - \frac{\sum_{i=1}^{n}(x_i - \bar{x})^2}{2\sigma^2}\right].$$

We can treat the second term, which does not include θ, as a constant. Hence the joint pdf of x and θ becomes

$$\begin{aligned} f(x, \theta) &= f(x \mid \theta)\pi(\theta) \\ &= k_3 \exp\left[-\frac{n(\bar{x} - \theta)^2}{2\sigma^2} - \frac{(\theta - \mu_0)^2}{2\sigma_0^2}\right]. \end{aligned}$$

After expanding the above equation, we can treat the two terms that do not involve θ as constants and obtain

$$f(x, \theta) = k_4 \exp\left[-\frac{\theta^2}{2}\left(\frac{n}{\sigma^2} + \frac{1}{\sigma_0^2}\right) + \theta\left(\frac{n\bar{x}}{\sigma^2} + \frac{\mu_0}{\sigma_0^2}\right)\right].$$

Using the procedure known as completing the square (on θ), we can write this equation as

$$f(x,\theta) = k_5 \exp\left[-\frac{1}{2}\left(\theta - \frac{(n\bar{x}/\sigma^2) + \mu_0/\sigma_0^2}{(n/\sigma^2) + 1/\sigma_0^2}\right)^2 \left(\frac{n}{\sigma^2} + \frac{1}{\sigma_0^2}\right)\right].$$

We noted that in Eq. (10.2.1) the denominator is a normalizing constant, so the previous equation can represent the posterior pdf in the form

$$f(\theta \mid x) = \frac{1}{\sigma_1\sqrt{2\pi}} \exp\left[-\frac{1}{2}\frac{(\theta - \mu_1)^2}{\sigma_1^2}\right], \qquad (10.2.16)$$

in which

$$\mu_1 = \frac{\sigma_0^2\bar{x} + \mu_0\sigma^2/n}{\sigma_0^2 + \sigma^2/n} \qquad (10.2.17)$$

is the mean and

$$\sigma_1^2 = \frac{\sigma_0^2\sigma^2/n}{\sigma_0^2 + \sigma^2/n} \qquad (10.2.18)$$

is the variance. Thus the posterior mean μ_1 is a weighted average of the prior mean μ_0 and the sample mean $\bar{x}$ and it approaches the sample mean $\bar{x}$ as n becomes very large. In the case of a diffused or vague prior distribution, σ_0^2 will be high and the posterior mean μ_1 will again approach the sample mean $\bar{x}$, and the posterior variance tends to σ^2/n.

The following example, based on the work of Benjamin and Cornell (1970), is on soil strengths[3]:

Example 10.11. Soil strengths. Buildings and other structures are planned in several localities for a new town. Numerous tests have been made in the region on the load-bearing soils which consist mainly of a reddish-gray clay. It is reported that the standard deviation is in the narrow range of 14,000–16,000 N/m^2, but the mean soil strength S shows a wide variation. The engineer responsible at a particular site decides that the prior distribution of S is N (85,000, 11,000^2) in metric units. Subsequently, three random soil tests are done from which the mean strength of 70,000 N/m^2 is obtained. Assuming that the standard deviation $\sigma = 15{,}000$ N/m^2, the mean and standard deviation of the posterior $N(\mu_1, \sigma_1^2)$ distribution are obtained from Eqs. (10.2.17) and (10.2.18) as follows:

$$\mu_1 = \frac{11{,}000^2 \times 70{,}000 + 85{,}000 \times 15{,}000^2/3}{11{,}000^2 + 15{,}000^2/3} = 75{,}740\ \text{N/m}^2$$

and

$$\sigma_1 = \sqrt{\frac{11{,}000^2 \times 15{,}000^2/3}{11{,}000^2 + 15{,}000^2/3}} = 6804\ \text{N/m}^2.$$

The prior and posterior distribution are shown in Fig. 10.2.4.

Because the mean soil strength is the important design variable for the foundation of structures in this locality, a utility function is used for the actual strength, or state of nature, denoted by μ_s. From previous experience the following function of the type given by Eq. (10.2.4) is used:

$$U(\mu_s \mid a) = ba + c[1 - \exp(-k(\mu_s - a))],$$

[3] See also Bayesian estimation of compressive strengths of concrete by Viola (1983).

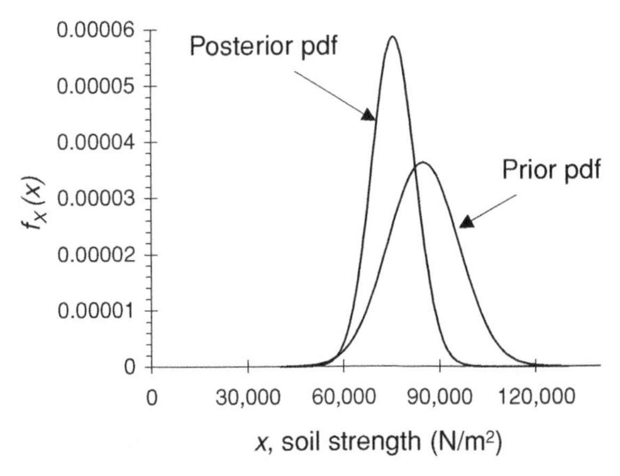

Fig. 10.2.4 Prior and posterior pdfs of soil strengths.

where a is the recommended design strength; we also estimate the constants $b = 0.8$, $c = 90{,}000$, and $k = 0.00001$. The form of this function shows that, if the actual soil strength is greater than the recommended strength, a small increase in benefit is obtained through an increased margin of safety; on the other hand, where the true strength is less than that adopted in the design, the loss increases at a fast rate. One notes also that an increase in the design strength is reflected in a cheaper design. From the adaptations of Eqs. (10.2.4) and (10.2.16), the mean utility for a particular value of a is given as follows after some rearrangement:

$$
\begin{aligned}
E[U \mid a] \\
&= ba + c - c \times \exp(ka)\frac{1}{\sigma_1\sqrt{2\pi}}\int_{-\infty}^{\infty} \exp\frac{-2k\mu_s\sigma_1^2 - \mu_s^2 + 2\mu_1\mu_s - \mu_1^2}{2\sigma_1^2}ds \\
&= ba + c - c \times \exp(ka)\frac{1}{\sigma_1\sqrt{2\pi}}\int_{-\infty}^{\infty} \exp\frac{-\left[\mu_s - (\mu_1 - k\sigma_1^2)\right]^2 - 2\mu_1 k\sigma_1^2 + k^2\sigma_1^4}{2\sigma_1^2}ds \\
&= ba + c - c \times \exp\left(ka - \mu_1 k + \frac{k^2\sigma_1^2}{2}\right)\frac{1}{\sigma_1\sqrt{2\pi}}\int_{-\infty}^{\infty} \exp\frac{-\left[\mu_s - (\mu_1 - k\sigma_1^2)\right]^2}{2\sigma_1^2}ds \\
&= ba + c - c \times \exp\left(ka - \mu_1 k + \frac{k^2\sigma_1^2}{2}\right).
\end{aligned}
$$

Hence the Bayes' solution for the optimum design strength of the soil is obtained by maximizing the expected utility, that is, by differentiating the previous function with respect to a and equating to zero. After simplification and substituting for the constants b, c, and k and the posterior mean and standard deviation,

$$
\begin{aligned}
a &= \mu_1 + \frac{1}{k}\ln\left(\frac{b}{ck}\right) - \frac{1}{2}k\sigma_1^2 \\
&= 75{,}740 + \frac{1}{0.00001}\ln\left(\frac{0.8}{90{,}000 \times 0.00001}\right) - \frac{0.00001 \times 6804^2}{2} \\
&= 63{,}730 \text{ N/m}^2
\end{aligned}
$$

Figure 10.2.5 shows utility curves of the type shown in Fig. 10.2.1, based on design-recommended means of 20,000 and 150,000 N/m^2 and the optimum design value, which is approximately 64,000 N/m^2.

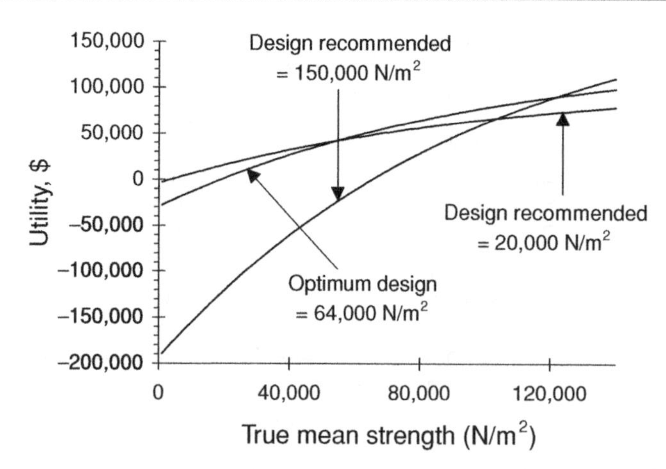

Fig. 10.2.5 Three utility functions for design recommendations of 20,000 and 150,000 N/m^2 and optimum design recommendation of 64,000 N/m^2 for maximum utility, with $c = 90{,}000$, $b = 0.8$, and $a = 0.00001$.

10.2.7 Likelihood ratio testing

We return to the subject of likelihood ratio tests introduced in Subsection 5.4.3. This is applicable in Bayesian theory with respect to an unknown state of nature, θ (which is treated as a random variable). Here we consider only prior probabilities (as in Subsections 10.1.1 and 10.1.2). Suppose a random variable X has conditional pdf $f(x \mid \theta)$. If a random sample of observations $x_1, x_2, \ldots, x_n$, of size n is available, we can formulate the following Bayes' solution for testing hypotheses:

The null hypothesis H_0: $\theta = \theta_0$.
The alternate hypothesis H_1: $\theta = \theta_1$.

The critical region is defined by

$$\frac{L_0}{L_1} = \frac{\prod_{i=1}^{n} f(x_i \mid \theta_0)}{\prod_{i=1}^{n} f(x_i \mid \theta_1)} \leq \frac{p_1}{p_0}, \tag{10.2.19}$$

where p_0 and p_1 are the prior probabilities for the values θ_0 and θ_1 in the null and alternate hypotheses, respectively. If in the critical region the ratio (10.2.19) is low, we thus minimize the Type I error.[4] It will be noted that p_1/p_0 corresponds directly with k_α in the Neyman-Pearson lemma. Thus α and β, the probabilities of a Type I and Type II error, are dependent on p_1/p_0.

Example 10.12. Concrete strengths. Several tests on large samples over a long period have shown that the standard deviation of the 28-day strength of a particular concrete is very nearly 5 N/m^2. There is, however, some uncertainty regarding the mean strength. Compression tests on five test cubes have indicated that the mean 28-day strength is 59 N/m^2. Let us state that the null hypothesis is that the mean strength is 58 N/m^2 and the alternate hypothesis is that it is 62 N/m^2:

H_0: $\mu = \mu_0 = 58$ N/m^2.
H_1: $\mu = \mu_1 (> \mu_0) = 62$ N/m^2.

[4] See also Example 5.19 and Problem 10.10.

Formulate a test of hypothesis with (*a*) the prior probabilities unknown, and (*b*) the prior probabilities (from knowledge of past test results) as .2 and .8 for values of 58 and 62 N/m^2, corresponding to the null and alternate hypotheses, respectively.

Solution. We apply Eq. (10.2.19) and proceed as in Example 5.19 but in a Bayesian framework, considering also the variance σ^2. Within the critical region,

$$\frac{L_0}{L_1} = \frac{\prod_{i=1}^{n} f(x_i \mid \theta_0)}{\prod_{i=1}^{n} f(x_i \mid \theta_1)} = \exp\left[-\left(\sum_{i=1}^{n} x_i\right)\frac{\mu_1 - \mu_0}{\sigma^2} + \frac{n}{2\sigma^2}\left(\mu_1^2 - \mu_0^2\right)\right] \leq \frac{p_1}{p_0}.$$

After taking logarithms and rearranging terms, the critical region is defined by

$$\bar{x} \geq \frac{1}{2}(\mu_1 + \mu_0) - \frac{\sigma^2 \ln(p_1/p_0)}{n(\mu_1 - \mu_0)}.$$

Note that if the alternative hypothesis is such that $\mu_1 < \mu_0$, the critical region becomes

$$\bar{x} \leq \frac{1}{2}(\mu_1 + \mu_0) + \frac{\sigma^2 \ln(p_1/p_0)}{n(\mu_0 - \mu_1)}.$$

(a) If the prior probabilities are unknown, we make $p_0 = p_1$. Then the critical region simplifies to

$$\bar{x} \geq \frac{1}{2}(\mu_1 + \mu_0).$$

That is, $\bar{x} \geq \frac{1}{2}(58 + 62) = 60$ N/m^2. Because $\bar{x} = 59$ N/m^2, the null hypothesis is not rejected.

(b) For the given prior probabilities, the critical region is defined by

$$\bar{x} \geq \frac{1}{2}(58 + 62) - \frac{5^2 \ln(0.8/0.2)}{5 \times 4} = 58.27\,\text{N/m}^2.$$

In this case the ratio of prior probabilities, with $\Pr(\mu_1) > \Pr(\mu_0)$, has caused a negative shift in the boundary of the critical region from 60 (for the condition $p_0 = p_1$) to 58.27 N/m^2. Thus the null hypothesis is rejected. (Of course, it does not mean that we automatically accept the alternative hypothesis.)

If $\Pr(\mu_0) > \Pr(\mu_1)$, the shift will be in the opposite direction. As n increases, however, the influence of the prior probabilities tends to reduce; that is, the information in sample data will gradually swamp the effects of differences in the prior probabilities. Similar results are obtained as $(\mu_1 - \mu_0)$ increases or when the variance σ^2 decreases.

10.2.8 Summary of Section 10.2

Bayesian methods are discussed here on the basis of prior and posterior distributions with utility functions. All these are databased, and we may not always find appropriate functions and suitable constants. More about this follows immediately. How the information collected by engineers is used in a prior distribution varies from one case to another. Almost invariably, specific assumptions are made. These should be subject to careful scrutiny from different perspectives. When no information is available, one uses a noninformative prior.

10.3 MARKOV CHAIN MONTE CARLO METHODS

In Section 10.2 we used the normal, gamma, and beta as prior probabilities and found associated likelihood functions to apply Bayes' theorem. However, there are many situations in which the posterior distribution of Eq. (10.2.1) is not easily obtained. One can use numerical integration techniques in an attempt to solve the problem, but this is usually very cumbersome. The current widely accepted practice in such situations is to use the Markov chain Monte Carlo (MCMC) method. In this way one constructs a Markov chain whose stationary and ergodic distribution is the posterior distribution from Bayes' theorem, that is, intractable by analytical means. The theory and numerous applications from diverse fields are given by Gilks et al. (1996). A commonly used technique of implementation is the Gibbs sampler, which is a special type of MCMC (see, for example, Casella and George, 1992). It is named after the physicist J. W. Gibbs because of the analogy between the sampling algorithm and statistical physics. This multivariate simulation procedure involves the replacement of a component by drawing from its conditional distribution given the current values of all other components, using available data. We present the basic idea of this Bayesian updating method.

For a variable X, a Markov chain with stationary distribution $\pi(x)$ is constructed as follows. Under certain conditions, the sample output from the chain can be used asymptotically to estimate the expected value with respect to $\pi(x)$ of a function, say, $f(x)$ of interest. Let $\pi(x) = \pi(x_1, x_2, \ldots, x_n)$ denote a joint pdf of the components x_i, $i = 1, 2, \ldots, n$. Also, let $\pi(x_i \mid x_{i-1})$ denote the conditional densities of each of the components x_i using values of the other components. We commence at stage 0, say, by taking arbitrary starting values, $\mathbf{x}^0 = (x_1^0, \ldots, x_n^0)$. For the next stage, we draw randomly from conditional distributions $\pi(x_i \mid x_{i-1})$, $i = 1, 2, \ldots, n$, as follows:

$$
\begin{aligned}
&x_1^1 \text{ from } \pi\left(x_1 \mid x_2^0, x_3^0, \ldots, x_n^0\right),\\
&x_2^1 \text{ from } \pi\left(x_2 \mid x_1^1, x_3^0, \ldots, x_n^0\right),\\
&\qquad\vdots\\
&x_n^1 \text{ from } \pi\left(x_n \mid x_1^1, x_2^1, \ldots, x_{n-1}^1\right).
\end{aligned}
$$

Note that the conditionality is the key feature. Thus, after the first iteration we transit from $\mathbf{x}^0 = (x_1^0, \ldots, x_n^0)$ to $\mathbf{x}^1 = (x_1^1, \ldots, x_n^1)$. On completion of t and more such iterations, we have a sequence $\mathbf{x}^0, \mathbf{x}^1, \ldots, \mathbf{x}^t, \ldots$, which is a realization of a Markov chain (see, for example, Gelfand et al., 1990). When all distributions are taken as conditional on the data, as shown in the following example, the marginal equilibrium distributions give the marginal posteriors as in Bayes' theorem. Because multivariate Bayesian methods are usually specified as an ensemble of conditional distributions, the Gibbs sampler can be appropriately and easily adopted.

Repeated simulations are required to reach the desired states of stationarity and ergodicity. The speed of convergence depends on the strength of dependence between the components.

Example 10.13. Climatic trends and periodicities. In this case study we combine initial assumptions about model parameters, such as trend and periodicity with observations to obtain the posterior distribution on the parameter space through a Bayesian framework, after repeated simulations.

Here we consider a monthly time series of mean temperature, as a climatic variable. The series is expressed in the classical additive form by

$$y_i = t_i + p_i + \eta_i,$$

in which at time i, y_i represents the observed value of the variable, t_i and p_i are the parameters of trend and periodicity, respectively, and η_i is a random component.

The trend component is incremented as follows:

$$t_{i+1} = t_i + s_i,$$

using an auxiliary variable s_i. The uncertainty in the trend is modeled as

$$s_{i+1} = s_i + \omega_{1_i},$$

in which ω_{1_i} is a "latent" variable (in Bayesian terminology) that describes the random fluctuations in this component. With regard to the periodicity component, we simplify this component so that it is modeled as a truncated Fourier series, in the usual way, but including only the fundamental frequency, $f = 1/12$ (per month, corresponding to the annual cycle), that is, with only one harmonic. This gives a close representation for monthly mean temperatures and generally provides a good approximation for some other monthly climatic series. More about this follows. Accordingly, the periodicity component is modeled, following West and Harrison (1997), as

$$p_{i+1} = p_i \cos(2\pi f) + q_i \sin(2\pi f)$$

and

$$q_{i+1} = -p_i \sin(2\pi f) + q_i \cos(2\pi f) + \omega_{2_i},$$

using an auxiliary variable q_i, and another "latent" variable ω_{2_i} that describes the random fluctuations in this component. Thus, the trend component is treated as a simple linear process whereas the periodicity component is a linear sum of sine and cosine curves. The advantage in this type of nonstationary modeling is that one can incorporate changes in the mean level from year to year and also differences in the cyclical structure.

The dynamic linear model of the system, also called the state space model, is then written as follows in a form similar to the well-known Kalman filter:

$$\begin{aligned} \mathbf{x}_{i+1} &= \mathbf{F}\mathbf{x}_i + \mathbf{G}\boldsymbol{\omega}_i \\ y_i &= \mathbf{H}\mathbf{x}_i + \eta_i \end{aligned}$$

where y_i represents the input data,

$$\mathbf{x}_i = [t_i, s_i, p_i, q_i]^{\mathrm{T}},$$

and

$$\boldsymbol{\omega}_i = \left[\omega_{1_i}, \omega_{2_i}\right]^{\mathrm{T}}.$$

Also,

$$\mathbf{H} = \begin{bmatrix} 1 & 0 & 1 & 0 \end{bmatrix},$$

$$\mathbf{G} = \begin{bmatrix} 0 & 1 & 0 & 0 \\ 0 & 0 & 0 & 1 \end{bmatrix}^{\mathrm{T}},$$

and

$$\mathbf{F} = \begin{bmatrix} 1 & 1 & 0 & 0 \\ 0 & 1 & 0 & 0 \\ 0 & 0 & \cos(2\pi/12) & \sin(2\pi/12) \\ 0 & 0 & -\sin(2\pi/12) & \cos(2\pi/12) \end{bmatrix}.$$

To facilitate the application of the Gibbs sampler we adopt a generalized approach by modeling the state variables as random walks in which the variances are treated as parameters within a Bayesian estimation framework, in the manner of West and Harrison (1997). Accordingly, we write the stochastic dynamic model in a static form. For the static representation we follow Magni and Bellazi (2004):

$$\mathbf{y} = \mathbf{Az} + \boldsymbol{\eta}$$
$$\mathbf{x} = \mathbf{Bz}$$

Here **y** represents n observations, corresponding to which $\boldsymbol{\eta}$ signifies n measurement errors and **x** denotes the $2n$ values of the trend and periodicity components to be evaluated, and **z** is a variable incorporating the four initial conditions of t, s, p, and q followed by the $2(n-1)$ "latent" variables of the trend and periodicity components.

Thus,

$$\mathbf{x} = [\, t_0 \; p_0 \; \cdots \; t_{n-1} \; p_{n-1} \,]^{\mathrm{T}},$$

$$\mathbf{y} = [\, y_0 \; \cdots \; y_{n-1} \,]^{\mathrm{T}},$$

$$\boldsymbol{\eta} = [\, \eta_0 \; \cdots \; \eta_{n-1} \,]^{\mathrm{T}},$$

and

$$\mathbf{z} = \left[\, t_0 \; s_0 \; p_0 \; q_0 \; \omega_{1_0} \; \omega_{2_0} \; \cdots \; \omega_{1_{n-2}} \; \omega_{2_{n-2}} \,\right]^{\mathrm{T}}.$$

Also, the **A** and **B** matrices, of sizes $n \times (2n+2)$ and $2n \times (2n+2)$, respectively, of constants are functions of the **F**, **G**, and **H** matrixes of the original state space model as follows:

$$\mathbf{A} = \begin{bmatrix} \mathbf{H} & \mathbf{0} & \cdots & \mathbf{0} \\ \mathbf{HF} & \mathbf{HG} & \cdots & \mathbf{0} \\ \vdots & \vdots & & \vdots \\ \mathbf{HF}^{n-1} & \mathbf{HF}^{n-2}\mathbf{G} & \cdots & \mathbf{HG} \end{bmatrix}$$

and

$$\mathbf{B} = \begin{bmatrix} \mathbf{M} & \mathbf{0} & \cdots & \mathbf{0} \\ \mathbf{MF} & \mathbf{MG} & \cdots & \mathbf{0} \\ \vdots & \vdots & & \vdots \\ \mathbf{MF}^{n-1} & \mathbf{MF}^{n-2}\mathbf{G} & \cdots & \mathbf{MG} \end{bmatrix},$$

in which

$$\mathbf{M} = \begin{bmatrix} 1 & 0 & 0 & 0 \\ 0 & 0 & 1 & 0 \end{bmatrix}.$$

We note that the size of the square **F** matrix will increase if one increases the number of harmonics in the periodicity component from 1, toward the maximum of 6 for monthly time intervals; correspondingly, the dimensions of the **H**, **G**, and **M** matrixes and the length of the **z** vector will also increase.

To implement the scheme, one needs to specify the probability distributions of the error terms. This also applies to the initial states. Let us assume that

$$Pr[\omega_{k_i}] = N\left(0, \sigma^2_{\omega_k}\right), \quad \text{for } k = 1, 2, \ldots,$$

$$Pr\,[\eta_i] = N\left(0, \sigma^2_{\eta_i}\right),$$

and

$$Pr\,[x_0] = N\left(0, \mathrm{diag}\left[\, \sigma^2_{x_{0_1}} \; \sigma^2_{x_{0_2}} \; \sigma^2_{x_{0_3}} \; \sigma^2_{x_{0_4}} \,\right]\right),$$

where $N(.,.)$ denotes the multinormal distribution and diag $[\cdot\cdot]$ signifies a diagonal matrix.

Unlike in the Kalman filter, we treat $\sigma^2_{\omega_1}$ and $\sigma^2_{\omega_2}$ as parameters to be estimated from the data within the Bayesian framework through repeated simulations. We assume that these unknown variances pertaining to trend and periodicity, respectively, have inverse gamma distributions, which is a reasonable assumption, as in Gilks et al. (1996). For the prior state these distributions are vaguely defined; that is, they are taken to be somewhat like uniform so that the final estimates do not have a large latent variable. Since the initial values of the t, s, p, and q variables of the trend and periodicity components are assumed arbitrarily, as stated, their variances are taken sufficiently large so that the data will dominate the posterior distribution.

In order to update trend and periodicity through Bayes' theorem, we need to compute the first two moments of the joint posterior probability

$$Pr\left[\mathbf{t}, \mathbf{p}, \sigma_{\omega 1}^2, \sigma_{\omega 2}^2 \,|\mathbf{y}\right],$$

where for n values of data $\mathbf{y}$, $\mathbf{t} = [\,t_0 \cdots t_{n-1}\,]^{\mathrm{T}}$ and $\mathbf{p} = [\,p_0 \cdots p_{n-1}\,]^{\mathrm{T}}$ are the trend and periodicity components, respectively. To solve this problem, we implement the MCMC procedure through Gibbs sampling. We commence by partitioning the random parameters into three groups or subsets: $\sigma_{\omega_1}^2$, $\sigma_{\omega_2}^2$, and $\mathbf{z}$. It means that, the algorithm should accommodate a scheme that extracts a sample in each iteration from each of the following three conditional distributions (that is, the distribution of a variable conditioned on the other variables and sample data, $\mathbf{y}$) based on the previously stated strategy for Gibbs sampling:

$$Pr\left[\frac{1}{\sigma_{\omega_1}^2}\middle|\sigma_{\omega_2}^2, \mathbf{z}, \mathbf{y}\right] = \Gamma\left(\frac{n}{2} + \gamma_1, \frac{(\boldsymbol{\omega}_1^{\mathrm{T}}\boldsymbol{\omega}_1)}{2} + \gamma_2\right),$$

$$Pr\left[\frac{1}{\sigma_{\omega_2}^2}\middle|\sigma_{\omega_1}^2, \mathbf{z}, \mathbf{y}\right] = \Gamma\left(\frac{n}{2} + \gamma_3, \frac{(\boldsymbol{\omega}_2^{\mathrm{T}}\boldsymbol{\omega}_2)}{2} + \gamma_4\right),$$

and

$$Pr\left[\mathbf{z}\,|\sigma_{\omega_1}^2, \sigma_{\omega_2}^2, \mathbf{y}\right] = N\left(\mathbf{D}^{-1}\mathbf{A}^{\mathrm{T}}\sum_{\eta-1}^{-1}\mathbf{y}, \mathbf{D}^{-1}\right),$$

with $\mathbf{D} = \mathbf{A}^{\mathrm{T}}\sum_{\eta}^{-1}\mathbf{A} + \sum_{z}^{-1}$,

in which $\Gamma\,(.,.)$ is the gamma distribution. $N\,(.,.)$ is the multinormal distribution, and $\boldsymbol{\omega}_i = [\omega_{i_0} \cdots \omega_{i_{n-2}}]^{\mathrm{T}}$ for $i = 1, 2$, and $(\gamma_1, \ldots, \gamma_4)$ pertain to the (initially prior) parameters of the gamma distributions of $1/\sigma_{\omega_1}^2$, $1/\sigma_{\omega_2}^2$. The sample data are given by $\mathbf{y}$ as defined previously. Also,

$$\sum_{\eta} = \mathrm{diag}\left(\left[\,\sigma_{\eta_0}^2 \cdots \sigma_{\eta_{n-1}}^2\,\right]\right) \text{ and } \sum_{z} = \mathrm{diag}\left(\left[\,\sigma_{x0_1}^2 \;\sigma_{x0_2}^2 \cdots \sigma_{x0_4}^2 \;\boldsymbol{\sigma}_{\omega}^2\,\right]\right),$$

where $\boldsymbol{\sigma}_{\omega}^2$ is a vector containing the sequence $\{\sigma_{\omega_1}^2, \sigma_{\omega_2}^2\}$ repeated $(n - 1)$ times.

The algorithm can be summarized as follows:

(1) We commence with prior values of the vectors $\boldsymbol{\omega}_1$ and $\boldsymbol{\omega}_2$, on which the current distributions of the unknown variances $\sigma_{\omega_1}^2$ and $\sigma_{\omega_2}^2$ in the random components of trend and periodicity, respectively, depend as previously defined. Hence, the dependent parameters of the respective inverse gamma distributions, as specified above, are taken to the next step; we tried values of 1, 10, 0.1, and 10, respectively, for γ_1, γ_2, γ_3, and γ_4 as prior values, but note that γ_1 and γ_3 are invariable in our scheme.

(2) The samples of $Pr[\mathbf{z}\,|\,\sigma_{\omega_1}^2, \sigma_{\omega_2}^2, \mathbf{y}]$ are drawn iteratively by the Bayesian estimator and Gibbs sampler. This is in the main loop of the computer programme for simulation.

    The vector $\mathbf{z}$ is simulated through the multinormal distribution conditioned on sample data $\mathbf{y}$ and the two variances of the $\boldsymbol{\omega}_1$ and $\boldsymbol{\omega}_2$ vectors. Recall that the variable $\mathbf{z}$ contains, except for the four initial conditions of t, s, p, and q at the beginning, alternating values of ω_{1_i} and ω_{2_i}, for each successive value of i. The current vector $\mathbf{z}$ forms a new row in the matrix $\mathbf{Z}$, at each iteration.

(3) New values of the $\boldsymbol{\omega}_1$ and $\boldsymbol{\omega}_2$ vectors are abstracted from $\mathbf{Z}$.

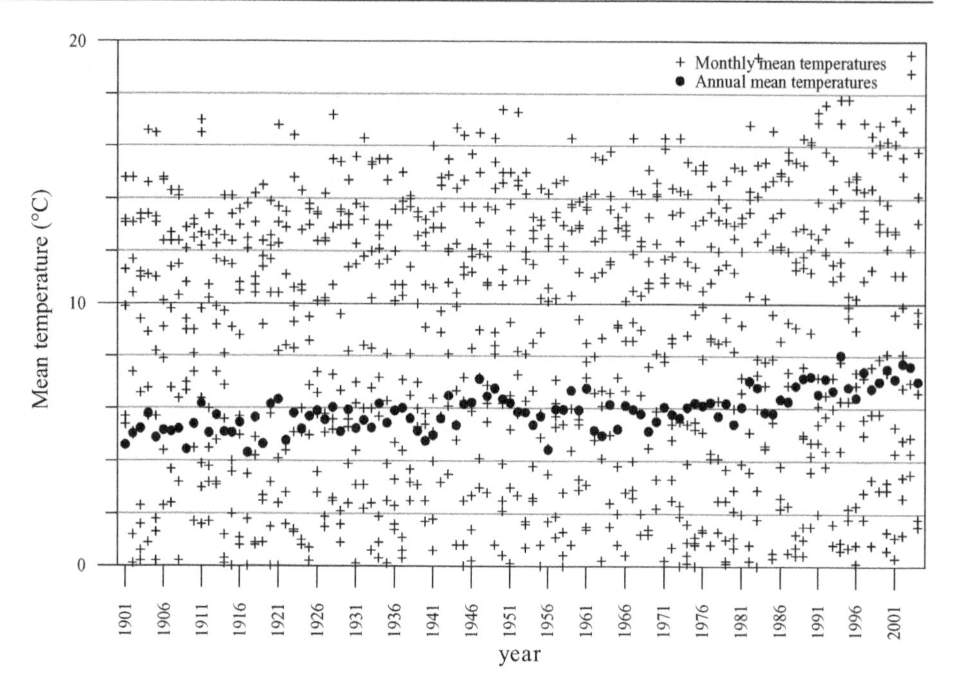

Fig. 10.3.1 Monthly mean temperature in Chateau-d'Oex, in the Lake Geneva region of southwestern Switzerland, for the period 1901–2004 with the data grouped by years.

(4) The new values of the ω_1 and ω_2 vectors are taken to step (1) each time in the stated loop, considering current (post prior) values of the gamma parameters. Their respective updated variances are simulated using the specified gamma simulators.
(5) From these iterations k series (runs or vectors $\mathbf{z}$) are obtained for $\mathbf{Z}$**:** step (2).
(6) Then the initial si series are discarded and the mean values of a $\mathbf{z}$ row vector are formed by averaging the $sr = k - si$ remaining series for each point in time. Desirable values of si and sr are subject to experimental investigation in order to stabilize the final series.
(7) We draw the sampling posterior distribution of the trend and periodicity components by a direct transformation of (the mean of) $\mathbf{z}$ to $\mathbf{x}$ using the equation $\mathbf{x} = \mathbf{Bz}$ of our model.

Figure 10.3.1 gives the monthly mean temperature at Chateaux-D'oex, in the Geneva region of southwestern Switzerland, for the period 1901–2001. The data are listed in Table E.10.1 in degrees centigrade. Also shown in Fig. 10.3.1 are annual mean temperatures. The variability in the annual mean temperatures is of course much smaller than in the monthly data. The annual temperature data does not show a significant trend except for a rise after the third quarter of the past century, or thereabouts.

The Bayesian estimation and Gibbs sampling scheme is then applied to the monthly temperatures in Chateaux-D'oex from 1901 to 2001. Figure 10.3.2 shows the stochastic trend component. This does not indicate any general movement, except that there is a late rise during the fourth quarter of the last century as seen in Fig. 10.3.1. However, it shows a similar rise in the first quarter followed by a steady decline in the second and third quarters. Figure 10.3.3 shows the stochastic periodicity component for the some period. This has sharp irregular movements within periods of 12 months. The differences in amplitudes between years indicate the nonstationarity of the periodicity component.

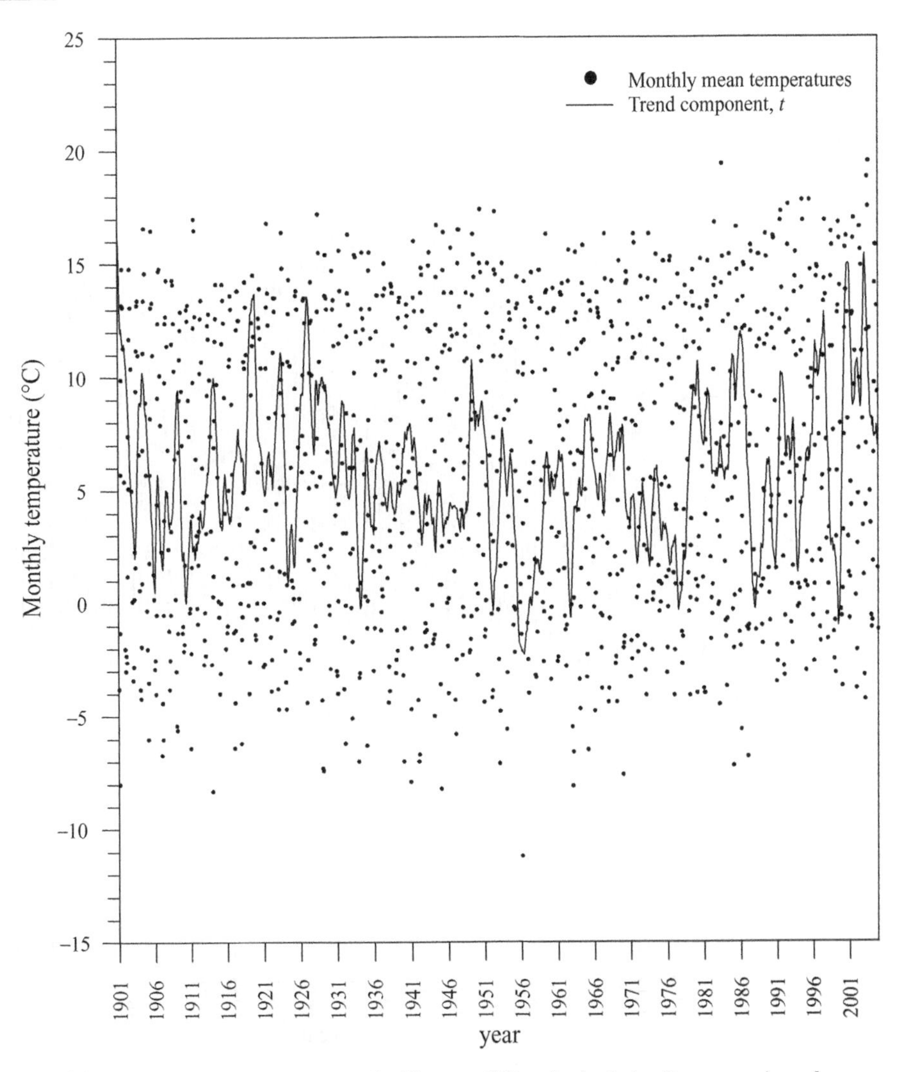

Fig. 10.3.2 Monthly mean temperature in Chateau-d'Oex, in the Lake Geneva region of southwestern Switzerland, for the period 1901–2004 with the trend component from Gibbs sampling.

Conditional distributions that are not of a standard form that is easy to apply, or the fact that the model is not tractable enough, warrant the use of the original Metropolis-Hastings algorithm for the individual components. In this wide approach, a powerful Markov chain method is used to simulate multivariate distributions by implementing a random walk procedure and an acceptance-rejection sampling scheme (see, for example, Chib and Greenberg, 1995). Details of the Gibbs sampler and the Metropolis-Hastings algorithm are also provided by Smith and Roberts (1993), among others. Applications of the Gibbs sampler have been made in diverse fields, for example, in medicine as discussed by Gilks et al. (1993).[5] The Bugs manual (Spiegelhalter et al., 2003) originally written for medical research is an aid for users (website: http://www.mrc-bsu.com.ac.uk/bugs).

[5] See also Kottegoda et al. (2007).

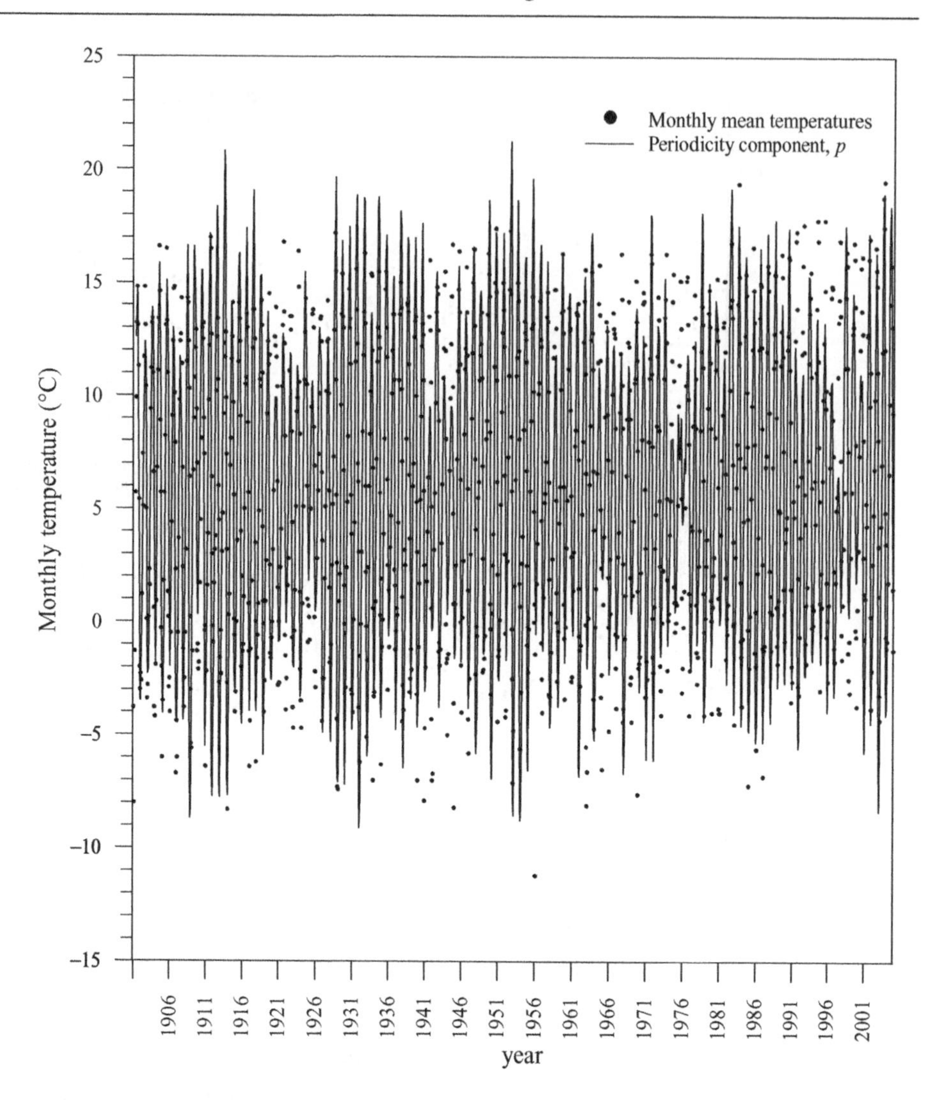

Fig. 10.3.3 Monthly mean temperature in Chateau-d'Oex, in the Lake Geneva region of southwestern Switzerland, for the period 1901–2004 with the periodicity component from Gibbs sampling.

10.4 JAMES-STEIN ESTIMATORS

In this section we consider briefly a class of estimators called *James-Stein estimators* that have proved to be efficient as data analytic tools. The procedure concerns the simultaneous estimation of several means $\mu_i, i = 1, 2, \ldots, k$, where $k > 2$, coming from mutually independent normal populations that are not identically distributed.[6] These estimators have the advantage that information on prior distributions is not required. The subject is included in this chapter because of similarities with Bayesian methods. (We noted that the

[6] See James and Stein (1981), Lehmann and Casella (1998), and Casella and Berger (2002, pp. 574–576).

uncertainty in the estimation of the prior distribution had provided grounds for criticism of the Bayesian approach.)

Efron and Morris (1977) give a nonmathematical introduction. They show how the James-Stein results sometimes contradict a basic law in classical statistics, going back to Gauss, Legendre, and others, which states that the arithmetic mean is uniformly superior to other comparable estimators. Originally, Stein (1956) had shown that in terms of a form of mean squared error, this estimator was suboptimal, in a multivariate context; an improvement can be obtained by deliberately introducing some bias. Efron and Morris (1977) demonstrate the apparent paradox by finding estimators X_i primarily for the batting averages of 18 baseball players who performed in the 1970 season. Using James-Stein procedures for the estimate of a player's future batting ability is more likely to be closer to a particular average obtained from the grand average

$$\bar{X} = \frac{\sum_{i=1}^{k} \bar{X}_i}{k}$$

of all $k = 18$ players than an individual player's own past average. It is shown that such averages shrink toward the overall average.

A simple version of the James-Stein estimator gives the shrunken mean for each player as

$$Z(X_i) = \bar{X} + c(X_i - \bar{X}), \qquad \text{for } i = 1, 2, \ldots, k, \tag{10.4.1a}$$

in which c is a constant called the *shrinking factor*. This factor is obtained from

$$c = 1 - \frac{(k-3)\sigma^2}{\sum_{i=1}^{k} (X_i - \bar{X})^2}, \tag{10.4.1b}$$

where σ^2 is the variance. In the application cited, observed batting averages and the James-Stein estimates from the early part of the 1970 season were compared with the averages from the players' subsequent performances during the same year, in which there were nine times as much data. The sum of the squared errors of the observed arithmetic averages was found to be 3.5 times that of the James-Stein estimates. This result was expected because the risk function of the James-Stein estimator is less than that for the sample average. However, this estimator is inadmissible if the constant c becomes negative. Then one simply takes the grand average, that is, $Z(X_i) = \bar{X}$.

There are alternatives to Eq. (10.4.1), such as the following James-Stein estimator:

$$Z^*(\bar{X}_i) = \left(1 - \frac{(k-2)\sigma^2}{n \sum_{j=1}^{k} \bar{X}_j^2}\right) \bar{X}_i, \qquad \text{for } i = 1, 2, \ldots, k, \tag{10.4.2}$$

in which $X_{ij} \approx N(\mu_i, \sigma^2)$ for $i = 1, 2, \ldots, k$ and $j = 1, 2, \ldots, n$. That is, the series are of length n and have a common variance σ^2 but their means are different as stated before. Also, they are all independent.

Example 10.14. Annual rainfall. There are several data series in civil and environmental engineering in which the James-Stein estimators may be applicable. However the somewhat stringent conditions preclude some types. One application that seems to be appropriate is annual precipitation. This is true because by the central limit theorem it tends to be normally distributed and the series are not usually correlated between different series and within them. In this example we use ten series from the Puyango basin, Ecuador. They span 23 years from 1963 to 1985 and are listed in Table E.10.2 in millimeters. The means and standard deviations are also shown. The means vary over a narrow range with six values in the 1300s.

Table 10.4.1 Correlation matrix of annual rainfall at ten stations in Ecuador

Station	1	2	3	4	5	6	7	8	9	10
1	1.000	0.009	0.036	0.210	0.176	−0.002	0.111	0.029	−0.153	0.200
2	0.009	1.000	0.590	0.665	0.078	0.872	0.233	0.676	0.467	0.329
3	0.036	0.590	1.000	0.858	0.250	0.780	0.166	0.735	0.441	0.107
4	0.210	0.665	0.858	1.000	0.342	0.850	0.368	0.725	0.332	0.368
5	0.176	0.078	0.250	0.342	1.000	0.165	0.047	0.097	0.023	0.512
6	0.002	0.872	0.780	0.850	0.165	1.000	0.351	0.792	0.425	0.237
7	0.111	0.233	0.166	0.368	0.047	0.351	1.000	0.257	0.096	0.180
8	0.029	0.676	0.735	0.725	0.097	0.792	0.257	1.000	0.610	0.321
9	−0.153	0.467	0.441	0.332	0.023	0.425	0.096	0.610	1.000	0.011
10	0.200	0.329	0.107	0.368	0.512	0.237	0.180	0.321	0.011	1.000

The standards deviation of the first series is significantly lower than the others. Many of the other values are not very different from each other. Furthermore, it is seen in the correlation matrix, Table 10.4.1, that some series are correlated.

Many of the correlations are not significant. Although the model assumptions (constant variance and independence) are violated in some cases, we proceeded with the application of the James-Stein estimator. Initially we estimated the averages from the first n years, where $n = 8, 9, 10, \ldots, 20$. We find the reduction factors and compare the averages of the remaining data with (a) the averages of the first n years and (b) the James-Stein averages. As shown in the last two columns of Table 10.4.2 from the sum of least squares, the James-Stein estimator performed marginally better for values of n ranging from 10 to 18 years. Because of the inadequate length of the series, strong conclusions cannot be made.

This procedure can be extended, for instance, to estimate the variance from replicates.

Table 10.4.2 James-Stein estimation of Ecuador annual rainfalls

				Forecast for next $(23 - n)$ years[a]	
Estimated from first n years	Mean (mm)	Standard deviation (mm)	Reduction coefficient	SSEOBM (mm^2)	SSEJSE (mm^2)
$n = 8$	1099.9	547.2	0.9768	751,876	791,459
9	1109.6	518.6	0.9816	662,611	694,286
10	1187.6	676.1	0.9754	498,937	468,100
11	1218.1	669.7	0.979	452,658	399,002
12	1203.6	648.9	0.9815	502,320	464,007
13	1224.8	635.5	0.9841	569,913	515,726
14	1245.8	628.6	0.9859	778,903	708,549
15	1228.9	613.7	0.9871	695,300	639,621
16	1207.8	607.1	0.9877	868,762	831,572
17	1190.4	599.1	0.9884	706,874	685,292
18	1173.5	592.4	0.989	785,884	782,017
19	1158.1	587.7	0.9895	825,835	842,967
20	1156.8	580.9	0.9903	1,059,582	1,080,825

[a] SSEOBM: sum of squared errors from observed mean; SSEJSE: sum of squared errors from James-Stein estimator.

10.5 SUMMARY AND DISCUSSION OF CHAPTER 10

In this chapter we have introduced some aspects of Bayesian decision theory that should serve as useful tools for civil and environmental engineers. We have also provided details of the associated subjects of likelihood ratio testing and James-Stein method of estimation with practical applications.

Depending on the subject investigated and data available, one may decide whether a Bayesian or classical approach or a combination of both is suitable. (It is noted, incidentally, that if sample sizes are small, more statisticians tend to follow Bayesian methods.) Previously, Bayesian methods seem to have found favor in the economic and business communities, partly because loss functions seem more straightforward to them. In the toolkits of scientists and engineers, on the other hand, there have been some signs of rust. The main reasons that practicing statisticians have tended to avoid Bayesian methods are the problem of prior probabilities, which are generally unknown or vaguely defined, and hence the use of subjective probabilities. Another reason has been inadequate computing facilities, but this has progressively become less of an issue. Difficulties have also arisen because of the lack of experience, which is assumed. However, a paper presented by Racine et al. (1986) to the Royal Statistical Society on the practical experiences in the pharmaceutical industry (with published data, which is a rarity, and numerous high-level discussions) provided insights, albeit at a sophisticated level.

During the past decade the Bayesian approach has been revolutionized. The recent popularity is due to the advent of Markov chain Monte Carlo methods. The Gibbs sampler and the Metropolis-Hastings algorithm are used increasingly by a wide spectrum of applicants in common situations where direct application of Bayesian methods is not possible.

In spite of the initial problems and uncertainties involved, revision of probabilities on receipt of additional data and consequent decision making is being implemented in one way or another. (This is in line with the Laplacian dictum of treating probability as calculations based on common sense.) There is widespread recognition of the iterative nature of scientific investigations. One learns from the past, then reformulates the analysis and design, makes additional experiments, and gathers more information. Further fitting and verification follow. The problem of prior probabilities will then fade into the past. This procedure should make the task of the decision maker less and less subject to error.

REFERENCES

General. The following references are given for further reading as required:

Aitchison, J. (1970). *Choice Against Chance*, Addison-Wesley, Reading, MA. For preliminary reading.

Ang, A. H.-S., and W. H. Tang (1984). *Probability Concepts in Engineering Planning and Design, Vol. 2: Decision, Risk and Reliability*, John Wiley and Sons, New York. See Chapter 2 on decision analysis (11 pages). Utility functions and multiple objectives are also treated. There are 40 examples including numerous practical applications from quoted sources. There are also 34 problems.

Arrow, K. J. (1970). *Essays in the Theory of Risk-Bearing*, North-Holland, Amsterdam. On utility theory.

Aykaç, A., and C. Brumat (eds.) (1977). *New Developments in the Applications of Bayesian Methods*, North-Holland, Amsterdam. Bayesian methods in economics.

Benjamin, J. R., and C. A. Cornell (1970). *Probability, Statistics and Decision for Civil Engineers*, McGraw-Hill, New York. See Chapters 5 (Elementary Bayesian Decision Theory) and 6 (Decision Analysis of Independent Random Processes).

Box, G. E. P., and G. C. Tiao (1973). *Bayesian Inference in Statistical Analysis*, John Wiley and Sons, New York. Published in 1992 as a Wiley Classic Library Edition.

Bradley, J. V. (1976). *Probability; Decision; Statistics*, Prentice Hall, Englewood Cliffs, NJ. Includes subjective probabilities for decision making under uncertainty.

Broadbent, D. E. (1973). *In Defence of Empirical Psychology*, Methuen, London. Use of subjective probabilities in psychology.

DeGroot, M. H. (1970). *Optimal Statistical Decisions*, McGraw-Hill, New York. Includes conjugate distributions.

de Finetti, B. (1970). *Theory of Probability*, Vols. 1 and 2, John Wiley and Sons, New York. Advocates subjective probabilities, a classic with manageable math.

Erickson, G. J., and C. R. Smith (eds.) (1988). *Maximum-Entropy and Bayesian Methods in Science and Engineering, Vol. 2: Applications*, Kluwer Academic Press, Dordrecht. Some useful articles.

Gilks, W. R., S. Richardson, and D. J. Spiegelhalter (1996). *Markov Chain Monte Carlo in Practice*, Chapman and Hall, London. Widely used in applications.

Halter, A., and C. Dean (1971). *Decisions under Uncertainty with Research Applications*, South-Western, Cincinnati, OH. Applications of Bayesian methods in meteorology.

Jeffreys, H. (1961). *The Theory of Probability*, 3rd ed., Oxford University Press, Oxford. A masterpiece of original thought with a blend of theory and practice from a versatile scientist. In Bayesian applications he adopted an impersonal view of probability, unlike de Finetti and Savage.

Kyberg, H. E., Jr., and H. E. Smokler (eds.) (1964). *Studies in Subjective Probability*, John Wiley and Sons, New York. Work by de Finetti (degree of belief and betting odds placed on a chance event) and others.

Lee, P. M. (1989). *Bayesian Statistics: An Introduction*, Oxford Press, New York. A mathematical introduction.

Lee, S. M., and L. J. Moore (1975). *Introduction to Decision Science*, Petrocelli-Charter, New York. Utility curves.

Lindley, D. V. (1985). *Making Decisions*, 2nd ed., John Wiley and Sons, New York. Manageable math.

Maritz, J. S., and T. Lwin (1989). *Empirical Bayes Methods*, 2nd ed., Chapman and Hall, London. Provides a sophisticated approach to parameter estimation.

O'Hagan, A. (1994). *Kendal's Advanced Theory of Statistics, Vol. 2B: Bayesian Inference*, Chapman and Hall, London. An advanced introduction.

Raiffa, H., and R. Schlaifer (1961). *Applied Statistical Decision Theory*, Harvard University Press, Cambridge, MA. Pioneering work in conjugate distributions and Bayesian decision theory.

Riggs, J. L. (1968). *Economic Decision Models for Engineers and Managers*, McGraw-Hill, New York. Discusses present worth, future worth, utility functions, and decision trees at a comprehensible level.

Rose, L. M. (1976). *Engineering Investment Decisions: Planning under Uncertainty*, Elsevier, Amsterdam. Utility curves.

Savage, L. J. (1954). *The Foundations of Statistics*, John Wiley and Sons, New York. Includes prior and subjective probabilities.

Schmitt, S. A. (1969). *Measuring Uncertainty, An Elementary Introduction to Bayesian Statistics*, Addison-Wesley, Reading, MA. Helpful for the first-time reader.

Siddall, J. N. (1972). *Analytical Decision-Making in Engineering Design*, Prentice Hall, Englewood Cliffs, NJ. Good introduction to decision theory.

West, M., and J. Harrison (1997). *Bayesian Forecasting and Dynamic Models*, 2nd ed., Springer, New York. A classic; provides simulation-based methods.

White, D. J. (1976). *Fundamentals of Decision Theory*, North-Holland, New York. Instructive though sophisticated text on subjective probabilities.

Winkler, R. A. (1972). *An Introduction to Bayesian Inference and Decision*, Holt, Rinehart and Winston, New York. Unsophisticated mathematical inference.

Zellner, A. (1971). *An Introduction to Bayesian Inference in Econometrics*, John Wiley and Sons, New York. Includes data-based and non-data-based prior probabilities.

Additional references quoted in text

Casella, G., and R. L. Berger (2002). *Statistical Inference*, 2nd ed., Duxbury, Pacific Grove, CA. Discussion on loss and risk functions in relation to the Bayesian approach and the work of James and Stein (1961); see pp. 574–576.

Casella, G., and E. I. George (1992). "Explaining the Gibbs sampler," *Am. Stat.*, Vol. 46, pp. 167–174. Clarification of the Gibbs sampler.

Chib, S., and E. Greenberg (1995). "Understanding the Metropolis-Hastings algorithm," *Am. Stat.*, Vol. 49, pp. 327–335. Clarification of the MCMC algorithm.

Efron, B., and C. Morris (1977). "Stein's paradox in statistics," *Sci. Am.*, Vol. 236, No. 5, pp. 119–127. Shows in simple terms how in some circumstances there are better estimators of the future average than the sample mean.

Gelfand, A. E., S. E. Hills, A. Racine-Poon, and A. F. M. Smith (1990). "Illustration of Bayesian inference in normal data models using Gibbs sampling," *J. Am. Stat. Assoc.*, Vol. 85, pp. 972–985. Shows the attainment of the Markov chain from Gibbs sampling and the stationary distribution from which is the posterior joint distribution.

Gilks, W. R., D. G. Clayton, D. J. Spiegelhalter, N. G. Best, A. J. Neil, L. D. Sharples, and A. J. Kirby (1993). "Model complexity: Applications of Gibbs sampling in medicine," *J. R. Stat. Soc.*, B, Vol. 55, pp. 39–52. With additional discussions including diverse applications.

Ingles, O. G. (1983). "Measurement of risk and rationality in civil engineering," in *Procs. Appl. of Stats. and Prob. in Soil and Struct. Eng.*, 4th Int. Conf. 13–17 June 1983, edited by G. Augusti, A. Borri, and G. Vannucchi, pp. 357–373, Pitagora Editrice, Bologna. Some examples of risk aversion and engineering judgment from Australia.

James, W., and C. Stein (1981). "Estimation with quadratic loss," in *Proceedings of the Fourth Berkeley Symposium on Probability*, Vol. 1, pp. 361–379. University of California Press, Berkeley. Reference on James-Stein estimators.

Kottegoda, N. T., L. Natale, and E. Raiteri (2007). "Gibbs sampling of climatic trends and periodicities," *J. Hydrol.*, Vol. 339, pp. 54–64. Practical use of the Gibbs sampler.

Lehmann, E. L., and G. Casella (1998). *Theory of Point Estimation*, 2nd ed., Springer, New York. Includes an explanation of Stein's paradox.

Magni, P., and R. Bellazi (2004). "Analysing Italian voluntary abortion data using a Bayesian approach to the time series decomposition," *J. Stat. Med.*, Vol. 23, pp. 105–123. Gibbs sampler applied to time series.

Racine, A., A. P. Grieve, H. Flühler, and A. F. M. Smith (1986). "Bayesian methods in practice: Experiences in the pharmaceutical industry," *Appl. Stat.*, Vol. 35, pp. 93–150. Of practical interest albeit at a sophisticated level: includes an in-depth discussion.

Smith, A. F. M., and G. O. Roberts (1993). "Bayesian computation via the Gibbs and related Markov chain Monte Carlo methods," *J. R. Stat. Soc.*, B, Vol. 55, pp. 3–23. Provides details of MCMC methods.

Spiegelhalter, D. J., A. Thomas, N. G. Best, and W. R. Wilks (2003). "BUGS: Bayesian inference using Gibbs sampling," Version 1.4, Cambridge Medical Research Council Biostatistics Unit, England. Practical guidance for Gibbs sampling (MCMC) methods.

Stein, C. (1956). "Inadmissibility of the usual estimator for the mean of a multinormal distribution," in *Proceedings of the Third Berkeley Symposium on Probability*, Vol. 1, pp. 197–206. University of California Press, Berkeley. The Stein paradox which took decades to fully understand.

Viola, E. (1983). "On the Bayesian approach to estimate compressive strengths of concrete in situ," in *Proc. Appl. of Stats. and Prob. in Soil and Struct. Eng.*, 4th Int. Conf. 13–17 June 1983 edited by G. Augusti, A. Borri, and G. Vannucchi, pp. 1155–1168, Pitagora Editrice, Bologna. Optimal sample size and the value of statistical information.

PROBLEMS

10.1. Treatment plant design. Design I of a wastewater plant for a new community is based on the expectation that some heavy industries will be established in the area. This eventuality has an estimated probability of 0.9. Some alternative designs are considered. The cost of implementing Design I(a) is $300,000 and Design I(b) is $400,000. These designs are based on effluents of two types (depending on types of possible heavy industries) that have equal probabilities, given a positive decision on citing the industries here. If these industries are not established in the area, a loss $150,000 will be incurred because of modifications to the plant in Design I(a). Furthermore, if Design I(a) is implemented, whereas subsequently Design I(b) is found to be necessary, an additional cost of $150,000 will be incurred. (Design I(b) is versatile in all these aspects.)

Design II, on the other hand, costs $170,000 to implement and does not take account of the extra industrial effluent; however, if the industries are cited in the area, it is deemed that extensions costing an estimated $330,000 will be necessary to meet the increased demand.

Sketch a tree diagram and show the expected risks. What decision should be taken?

10.2. Decision tree and utility curve. A structural engineer has to choose an action from three alternatives a_1, a_2, and a_3. The state of nature to cope with is the modulus of elasticity used in the design. Suppose the unknown states of nature are approximated discretely by θ_1, θ_2, and θ_3. The monetary values of action a_1 consequent to the three states of nature are estimated as $500,000, $150,000, and −$300,000, respectively. The corresponding values are $250,000, $200,000, and −$100,000 for action a_2 and $200,000, $50,000, and −$50,000 for action a_3. Sketch the decision tree. Assigning a utility value of +1 unit to the highest of these monetary values and −1 unit to the lowest value, draw three utility curves to typically represent (*a*) a risk seeker, (*b*) a risk avoider, and (*c*) a large organization with a balanced view on risk taking. In case (*c*) what utility should be assigned for the outcome of the second action and the third state of nature.

10.3. Rural water supply. A contractor has the job of providing water to communities in a rural area by drilling boreholes. There is uncertainty regarding the depth of the groundwater level. Experience elsewhere suggests an assumption of either 10 m or 20 m. It is necessary to acquire well-casing, pumps, and other equipment in advance because of time factors. If an incorrect choice of depth is made, a loss will be incurred in monetary units as follows:

	Water depth	
Purchase equipment for depth	10 m	20 m
10 m	No loss	150 units
20 m	50 units	No loss

From data of other wells in the region the following prior probabilities are assumed:

Pr(depth 10 m) = 0.7.

Pr(depth 20 m) = 0.3.

Draw the decision tree and show the expected risks. Determine the Bayes' rule. Compare with the minimax solution.
A hydrogeologist is consulted on the optimum depth. The following likelihoods are assigned to the predictions:

	Actual depth	
Indicated depth	10 m	20 m
10 m	0.7	0.1
20 m	0.3	0.9

Obtain the posterior probabilities of the two states of nature conditional to the predictions. Determine the expected risks for each prediction.

10.4. Pipes and cofferdam. Find solutions to the pipes-for-water-supply (Example 10.3) and cofferdam (Example 10.4) problems using the following criteria:
(*a*) Bayesian theory with uniform prior.
(*b*) Maximax (maximize maximum profit or minimize minimum loss).

10.5. Water projects. Water supply schemes are planned for three new towns: X, Y, and Z. Designs are based on projected populations 10 years hence. Future populations with approximated probabilities are as follows:

	Probability		
Town	0.25	0.50	0.25
X	90,000	100,000	120,000
Y	125,000	150,000	175,000
Z	160,000	190,000	250,000

Assume that the water demand is 100 L per day per head of population. The cost C in dollars per million liters per day varies with size S of water supply scheme in liters per day as follows:

$$C = -\frac{S}{10} + 100,000.$$

Assume additional supply is sold to local industries at \$70,000 per million liters per day and any shortfall is met from alternate sources at \$130,000 per million liters per day. Determine the optimum sizes of plants at X, Y, and Z on the basis of expected minimum costs and sizes given by the above forecasts of population.

10.6. Contractor's utility function. Suppose the utility function of the contractor in Example 10.1 is defined by the following pairs of utilities (in the range 0–100 units) and gains in units of \$100,000:

Utility	Gain
100	2.30
80	1.25
60	0.90
40	0.60
20	0.50
0	0.25

What decision should be taken if the expected utility were to be maximized and the probabilities of winning either contract are equal.

10.7. Earth dam. Two designs are submitted for an earth dam. Design I is based on locally available materials, and its implementation is estimated to cost \$1,000,000. For Design II 5000 $m^3$of a particular type of clay is required. The engineer's estimated pdf of the availability X of the clayey material in the vicinity of the dam is uniform (0, 7000 m^3). The estimated cost of implementing Design II is \$650,000; however, the average cost of hauling any extra material from outside the area at \$100 per m^3should be added. Which design should be accepted on the basis of expected least cost? What decision should be taken if the engineer's pdf for X is

$$f_X(x) = \frac{1}{3500} \exp\left(-\frac{x}{3500}\right)?$$

10.8. Soil strengths with uniform prior. In Example 10.11, the engineer decided that the prior distribution of the soil strength is $N(85{,}000,\ 11{,}000^2)$. Determine the posterior distribution and the optimum design strength assuming that the prior distribution is uniform (75,000, 95,000).

10.9. Loss functions. It was shown that in the case of squared loss function, the Bayes' estimator is the mean value of the state of nature θ. What estimator is obtained if the loss function is (*a*) constant and (*b*) linear with respect to θ?

10.10. Traffic rates. In Example 10.10 ten exponentially distributed waiting times between successive vehicles are given. Using this data, formulate and apply a likelihood ratio test in which the null hypothesis is that the parameter is 1 minute and the alternative hypothesis is that it is 0.9 minute, if (*a*) the prior probabilities are 0.4 and 0.6, respectively, (*b*) the prior probabilities are unknown. Show how the Type I and Type II errors of the test, α and β, respectively, can be calculated.

10.11. Traffic rates. In Example 10.10, estimate the Poisson parameter λ using the posterior mean. Compare with the moments or ML estimator. What is the significance of the difference?

10.12. Ecuador rainfalls. From the data used in Example 10.14 choose some series which meet the model requirements more closely and repeat the exercise of comparing the past averages with the James-Stein estimators for predicting the future averages (see Table E.10.2). The lag-l serial correlation (of variables that are l units apart in time) may be estimated, say, for $l = 1, 2$, and 3, as follows:

$$r_l = \frac{\sum_{t=1}^{n-l} (x_t - \bar{x})(x_{t+l} - \bar{x})}{\sum_{t=1}^{n} (x_t - \bar{x})^2}.$$

In an independent time series the $r_l,\ l \neq 0$ have an approximate $N(0, 1/n)$ distribution. Are the conclusions from the reduced data set substantially different?

Appendix A
Further Mathematics

A.1 CHEBYSHEV INEQUALITY

Consider a nonnegative function $h(X)$ of a continuous random variable X with pdf $f_X(x)$

$$E[h(X)] = \int_{-\infty}^{\infty} h(x) f_X(x) dx = \int_{x, h(x)<m} h(x) f_X(x) dx + \int_{x, h(x) \geq m} h(x) f_X(x) dx,$$

where m is a positive constant. Because the two terms on the right-hand side are positive,

$$E[h(X)] \geq \int_{x, h(x) \geq m} h(x) f_X(x) dx \geq \int_{x, h(x) \geq m} m f_X(x) dx = m \Pr[h(X) \geq m].$$

Hence,

$$\Pr[h(X) \geq m] \leq m^{-1} E[h(X)].$$

A.2 CONVEX FUNCTION AND JENSEN INEQUALITY

A continuous function $h(x)$ is by definition *convex* if for every x on the real line, there is a straight line which passes through the point $\{x, h(x)\}$ and is situated on or under the curve representing $h(x)$. Hence the saying that $h(x)$ can "hold water."

Consider a continuous random variable X with mean $E[X]$ and a straight line $f(x) = \beta_0 + \beta_1 x$ that passes through the point $[E[X], h\{E[X]\}]$. Because

$$E[f(X)] = \beta_1 E[X] + \beta_0 = f\{E[X]\},$$
$$h\{E[X]\} = f\{E[X]\} = E[f(X)].$$

Also, in the case of two functions $h(x)$ and $f(x)$ where $h(x)$ is equal to or greater than $f(x)$ for all values of x, $E[h(X)] \geq E[f(X)]$. Thus,

$$E[h(X)] \geq h\{E[X]\}.$$

A.3 DERIVATION OF POISSON DISTRIBUTION

Let the term $o(\Delta t)$ denote a function such that

$$\lim_{\Delta t \to 0} \frac{o(\Delta t)}{\Delta t} = 0.$$

Let $P_x(t + \Delta t)$ denote the probability that there are exactly x arrivals in the interval $(0, t + \Delta t]$. Similarly, let $P_x(t)$ and $P_x(\Delta t)$ denote the probabilities that there are x arrivals in the intervals $(0, t]$ and $(t, t + \Delta t]$, respectively. Because of the Poisson postulate 3 of independence,

$$P_0(t + \Delta t) = P_0(t) P_0(\Delta t).$$

Considering also that the mean rate of occurrence of a Poisson arrival is λ,

$$P_0(\Delta t) = 1 - P_{x>0}(\Delta t) = (1 - \lambda\Delta t - o(\Delta t)),$$

in which the last term is introduced because, as already discussed, the probability $P_{x>0}(\Delta t) = P_1(\Delta t) = \lambda\Delta t$ is only an approximation. From the previous equations,

$$P_0(t + \Delta t) = P_0(t)(1 - \lambda\Delta t - o(\Delta t)).$$

By rearranging terms and in the limit as $\Delta t \to 0$,

$$\lim \Delta t \to 0 \left[\frac{P_0(t + \Delta t) - P_0(t)}{\Delta t}\right] = P_0'(t) = -\lambda P_0(t).$$

Because $P_0(0) = 1$, it follows that

$$P_0(t) = e^{-\lambda t}.$$

Also, it follows from the theorem of total probability that

$$\begin{aligned} P_1(t + \Delta t) &= P_1(t)P_0(\Delta t) + P_0(t)P_1(\Delta t) \\ &= P_1(t)[1 - \lambda\Delta t - o(\Delta t)] + P_0(t)[\lambda\Delta t + o(\Delta t)]. \end{aligned}$$

As $\Delta t \to 0$,

$$P_1'(t) = \frac{P_1\,(t + \Delta t) - P_1\,(t)}{\Delta t} = -\lambda P_1(t) + \lambda P_0(t).$$

Similarly, by induction

$$P_x'(t) = -\lambda P_x(t) + \lambda P_{x-1}(t).$$

We thus have a system of differential equations, and it is easy to verify that the solution is

$$P_x(t) = \frac{(\lambda t)^x e^{-\lambda t}}{x!}.$$

We let $v = \lambda t$ and hence obtain the Poisson pmf as defined by Eq. (4.1.7),

$$p_X(x) \equiv \Pr(X = x|v) = \frac{v^x e^{-v}}{x!}, \quad \text{for } x = 0, 1, 2, \ldots, \quad \text{and } v > 0.$$

A.4 DERIVATION OF THE NORMAL DISTRIBUTION

The normal curve was originated in the eighteenth century by De Moivre and developed as a mathematical tool notably by Gauss, whose work in astronomy led to the normal law of errors (which denote differences between estimated values and observations), and also by Laplace. The representative bell-shaped curve takes the form

$$g(x) = e^{-z^2/2}, \quad \text{for } -\infty < z < \infty,$$

which is proportional to the pdf of a large number of independent errors.

To put this in the form of a pdf, one must divide the right-hand side by its complete integral which can be equated to a constant, say, c. This is, of course, an essential requirement for a pdf. Thus,

$$c = \int_{-\infty}^{\infty} \exp\left(-\frac{y^2}{2}\right) dy.$$

We can also write

$$c^2 = \int_{-\infty}^{\infty} \exp\left(-\frac{u^2}{2}\right)du \int_{-\infty}^{\infty} \exp\left(-\frac{v^2}{2}\right)dv = \int_{-\infty}^{\infty}\int_{-\infty}^{\infty} \exp\left(-\frac{u^2+v^2}{2}\right)du\,dv,$$

where the double integral replaces the product of two integrals. Let us change the variables to polar coordinates by the transformations $u = r\sin\theta$ and $v = r\cos\theta$. Hence $u^2 + v^2 = r^2$. The double integral given in the preceding equation represents the volume under a surface; in the polar system of coordinates an elementary area is $r\delta\theta\delta r$. Also, because $0 \leq \theta \leq 2\pi$ and $0 \leq r \leq \infty$,

$$c^2 = \int_0^{\infty}\int_0^{2\pi} \exp\left(-\frac{r^2}{2}\right)rd\theta dr = 2\pi \int_0^{\infty} \exp\left(-\frac{r^2}{2}\right)rdr$$

$$= 2\pi\left[-\left(\exp\left(-\frac{r^2}{2}\right)\right)\right]_0^{\infty} = 2\pi.$$

Hence the pdf of a $N(0, 1)$ variable is

$$\phi(z) = \frac{1}{\sqrt{2\pi}} \exp\left(-\frac{z^2}{2}\right).$$

If we use the transformation $x = \sigma z + \mu$,

$$f_X(x) = \frac{1}{\sigma\sqrt{2\pi}} \exp\left[-\frac{1}{2}\left(\frac{x-\mu}{\sigma}\right)^2\right], \quad \text{for } -\infty \leq x \leq \infty,$$

is the pdf of an $N(\mu, \sigma^2)$ distribution; the first σ term is a consequence of Eq. (3.4.5) and the relationship $dz/dx = 1/\sigma$.

A.5 MGF OF NORMAL DISTRIBUTION

For a variate $X \sim N(\mu, \sigma^2)$,

$$M_X(t) = E[e^{tX}] = e^{t\mu}E\left[e^{t(X-\mu)}\right]$$

$$= \frac{e^{t\mu}}{\sigma\sqrt{2\pi}} \int_{-\infty}^{\infty} e^{t(x-\mu)}e^{-(x-\mu)^2/2\sigma^2}dx = \frac{e^{t\mu}}{\sigma\sqrt{2\pi}} \int_{-\infty}^{\infty} e^{-[(x-\mu)^2-2\sigma^2 t(x-\mu)]/2\sigma^2}dx.$$

This can be put in the form

$$\frac{e^{t\mu}}{\sigma\sqrt{2\pi}} \int_{-\infty}^{\infty} e^{-[(x-\mu-\sigma^2 t)^2-\sigma^4 t^2]/2\sigma^2}dx = e^{t\mu}e^{\sigma^2 t^2/2}\left[\frac{1}{\sigma\sqrt{2\pi}} \int_{-\infty}^{\infty} e^{-(x-\mu-\sigma^2 t)^2/2\sigma^2}dx\right].$$

The term within square brackets is the area under the pdf of a $N(\mu + \sigma^2 t, \sigma^2)$ variate. Hence,

$$M_X(t) = e^{(2\mu t+\sigma^2 t^2)/2}.$$

It follows that for a variate $Z \sim N(0, 1)$, $M_Z(t) = e^{t^2/2}$.

A.6 CENTRAL LIMIT THEOREM

Let $X_1, X_2, \ldots, X_n$ be a random sample of identically distributed *independent* random variables (of unspecified distribution), with sample mean $\bar{X}$, from a population with mean μ and *finite* variance σ^2. Assume that the mgf $M_X(t)$ exists for $|t| < \delta$, where $\delta > 0$.

Let $Z_i = (X_i - \mu)/\sigma$ and the random variable Y_n be defined as

$$Y_n = \frac{\bar{X} - \mu}{\sigma/\sqrt{n}}.$$

Then,

$$\begin{aligned} M_{Y_n}(t) = E[\exp(tY_n)] &= E\left[\exp\left(\frac{t}{n}\sum_{i=1}^{n}\sqrt{n}Z_i\right)\right] \\ &= \prod_{i=1}^{n} E\left[\exp\left(\frac{t}{\sqrt{n}}Z_i\right)\right] = \left[M_Z\frac{t}{\sqrt{n}}\right]^n. \end{aligned}$$

From the expansion of the exponential term, as given by Eq. (3.2.17),

$$M_{Y_n}(t) = \left\{1 + \frac{t}{\sqrt{n}}E[Z] + \frac{1}{2!}\left(\frac{t}{\sqrt{n}}\right)^2 E[Z^2] + \frac{1}{3!}\left(\frac{t}{\sqrt{n}}\right)^3 E[Z^3] + \cdots\right\}^n.$$

Equating $E[Z]$ to zero and taking logarithms,

$$\ln\{M_{Y_n}(t)\} = n\ln\left\{1 + \frac{1}{n}\left(\frac{1}{2}t^2E[Z^2] + \frac{1}{6}\frac{t^3}{n^{1/2}}E[Z^3] + \cdots\right)\right\}.$$

From Eq. (4.1.6*b*) or a Taylor series expansion,

$$\ln(1+a) = a - \frac{a^2}{2} + \frac{a^3}{3} - \frac{a^4}{4} + \cdots \quad \text{if} -1 < a < +1.$$

Let

$$a = \frac{1}{n}\left(\frac{1}{2}t^2E[Z^2] + \frac{1}{6}\frac{t^3}{n^{1/2}}E[Z^3] + \cdots\right).$$

Then

$$\ln\{M_{Yn}(t)\} = na - \frac{na^2}{2} + \frac{na^3}{3} - \frac{na^4}{4} + \cdots.$$

Since $E[Z^2] = 1$,

$$\lim_{n\to\infty} M_{Y_n}(t) = e^{t^2/2}.$$

From Section A.5, this is the mgf of an $N(0, 1)$ variate.

If the random sample $X_1, X_2, \ldots, X_n$ is from a normal population,

$$\begin{aligned} M_{\bar{X}}(t) &= E\left[\exp\frac{t\sum_{i=1}^{n}X_i}{n}\right] = \prod_{i=1}^{n} E\left[\exp\frac{tX_i}{n}\right] = \prod_{i=1}^{n} M_{X_i}\left(\frac{t}{n}\right) \\ &= \prod_{i=1}^{n}\exp\left[\mu\left(\frac{t}{n}\right) + \frac{1}{2}\sigma^2\left(\frac{t}{n}\right)^2\right] = \exp\left[\mu t + \frac{1}{2}\left(\frac{\sigma^2}{n}\right)t^2\right], \end{aligned}$$

using the result from A.5. This is the mgf of a normal distribution with mean μ and variance σ^2/n (from A.5). Therefore, for a normal population, the result of the central limit theorem holds regardless of sample size.

A.7 PDF OF STUDENT'S *T* DISTRIBUTION

Let the random variables $Z \sim N(0, 1)$ and $Y \sim \chi^2(v)$ be independent (see Subsections 4.2.3 and 5.3.5). Then the random variable

$$T = \frac{Z}{\sqrt{Y/v}}$$

has the Student's t distribution with v degrees of freedom. From Eqs. (4.2.12*d*) and (4.2.21) the joint pdf of Z and Y, which are mutually independent, is given by

$$f_{Z,Y}(z, y) = \frac{1}{\sqrt{2\pi}} \frac{1}{\Gamma(v/2)} \left(\frac{1}{2}\right)^{v/2} y^{(v/2)-1} \exp\left[-\frac{(z^2 + y)}{2}\right],$$

$$\text{for } -\infty < z < \infty \quad \text{and } 0 < y < \infty.$$

In order to determine the pdf of the t distribution, we use the method of transformations of Subsection 3.4.2, defining the random variable $X = Y$ to solve for

$$t = \frac{z}{\sqrt{y/v}}.$$

The inverse relationships are $z = t\sqrt{x/v}$ and $y = x$.

We first need to define the bivariate pdf of T and X as in Eq. (3.4.22), for which we follow closely the steps outlined in Subsection 3.4.2 to obtain the Jacobian:

$$J = \begin{vmatrix} \dfrac{\partial z}{\partial t} & \dfrac{\partial z}{\partial x} \\ \dfrac{\partial y}{\partial t} & \dfrac{\partial y}{\partial x} \end{vmatrix} = \begin{vmatrix} \sqrt{\dfrac{x}{v}} & \dfrac{t}{2\sqrt{xv}} \\ 0 & 1 \end{vmatrix} = \sqrt{\frac{x}{v}}.$$

Hence from Eq. (3.4.22),

$$f_{T,X}(t, x) = \sqrt{\frac{x}{v}} \frac{1}{\sqrt{2\pi}\Gamma(v/2)2^{v/2}} x^{(v/2)-1} e^{-(x/2)(1+t^2/v)}$$

for $-\infty < t < \infty$ and $0 < x < \infty$. We then integrate out the X variable to obtain

$$f_T(t) = \frac{1}{\sqrt{2\pi v}\Gamma(v/2)2^{v/2}} \int_0^{\infty} x^{(v-1)/2} e^{-(x/2)(1+t^2/v)} dx.$$

We note from Eq. (4.2.8) that

$$\int_0^{\infty} z^{\frac{v+1}{2}-1} e^{-z} dz = \Gamma\left(\frac{v+1}{2}\right)$$

and we let $z = (x/2)(1 + t^2/v)$. Hence,

$$f_T(t) = \frac{\Gamma[(v+1)/2]}{\sqrt{\pi v}\Gamma(v/2)} \frac{1}{[(t^2/v) + 1]^{(v+1)/2}}, \quad \text{for } -\infty < t < \infty.$$

A.8 PDF OF THE F DISTRIBUTION

Let U and V be independent chi-squared random variables with m and n degrees of freedom, respectively. Taking account of the independence between the two variables, their joint density function is given by

$$f_{U,V}(u, v) = \frac{u^{(m/2)-1} v^{(n/2)-1}}{\Gamma(m/2)\Gamma(n/2)2^{(m+n)/2}} e^{-(u+v)/2}, \quad \text{for } 0 < u, v < \infty.$$

The ratio of the two random variables scaled by the respective degrees of freedom,

$$F = \frac{U/m}{V/n},$$

which is sometimes called the *variance ratio*, has the F distribution.

Following the methods of Subsection 3.4.2 as applied in Section A7, we can obtain the distribution of F. First define a new variable $Y = V$. The inverse solutions of $f = (u/m)/(v/n)$ and $y = v$ are $u = myf/n$ and $v = y$. Hence the Jacobian is

$$J = \begin{vmatrix} \dfrac{\partial u}{\partial f} & \dfrac{\partial u}{\partial y} \\ \dfrac{\partial v}{\partial f} & \dfrac{\partial v}{\partial y} \end{vmatrix} = \begin{vmatrix} \dfrac{my}{n} & \dfrac{mf}{n} \\ 0 & 1 \end{vmatrix} = \frac{my}{n}.$$

Thus,

$$g_{F,Y}(f, y) = \frac{my}{n} \frac{(mfy/n)^{(m/2)-1} y^{(n/2)-1}}{2^{(m+n)/2}\Gamma(m/2)\Gamma(n/2)} e^{-[(m/n)fy+y]/2}.$$

We then integrate out the Y variable. From the properties of the gamma integral as given by Eq. (4.2.8) and used in Section A.7,

$$h_F(f) = \int_0^\infty g_{F,Y}(f, y)dy$$

$$= \frac{\Gamma[(m+n)/2]}{\Gamma(m/2)\Gamma(n/2)} \left(\frac{m}{n}\right)^{m/2} \frac{f^{(m/2)-1}}{[1 + fm/n]^{(m+n)/2}} \quad \text{for } 0 < f < \infty.$$

Corollaries: It can be shown if $X \sim t_v$, $X^2 \sim F_{1v}$.
Also, if $X \sim F_{m,n}$,

$$\frac{mX/n}{1 + mX/n} \sim \text{beta}(m/2, n/2).$$

A.9 WILCOXON SIGNED-RANK TEST: MEAN AND VARIANCE OF THE TEST STATISTIC

The test statistic is the sum of the ranks of the positive ranks T. Under the null hypothesis, this is the same as the sum of the ranks of the negative ranks. Furthermore, the probability that a rank will be assigned to either a positive or negative difference is 1/2. On the assumption that the variable investigated is random, we can therefore model the random process by which the n ranks are divided into positive and negative differences by a Bernoulli distribution as defined by Eq. (4.1.1).

Suppose that $X_i, i = 1, 2, \ldots, n \sim$ Bernoulli (1/2) and thus the variable can take values of only 0 and 1. The mean and variance of the X_i are 1/2 and 1/4 from Eq. (4.1.2*a*) and (4.1.2*b*), respectively. Accordingly, we can express T as

$$T = 1 \times X_1 + 2 \times X_2 + \cdots + n \times X_n = \sum_{i=1}^{n} i \times X_i.$$

The mean of T then becomes

$$\begin{aligned}\mu_T \equiv E[T] &= E[1 \times X_1 + 2 \times X_2 + \cdots + n \times X_n] \\ &= \sum_{i=1}^{n} i \times E[X_i] = \frac{1}{2}\sum_{i=1}^{n} i = \frac{n(n+1)}{4},\end{aligned}$$

because the summation of the i values is the sum of an arithmetic series and is equal to $n(n+1)/2$. The variance of T is

$$\mathrm{Var}[T] = \sum_{i=1}^{n} i^2\, \mathrm{Var}[X_i] = \frac{1}{4}\sum_{i=1}^{n} i^2.$$

One method of obtaining the summation of the first n values of i^2 is by equating the right-hand side of

$$\sum_{i=1}^{n} [i(i+1)(i+2) - (i+1)i(i-1)] = \sum_{i=1}^{n} 3(i^2 + i)$$

to the net sum, $n(n+1)(n+2)$, obtained by inserting $i = 1, 2, \ldots, n$ on the left-hand side. Hence,

$$\mathrm{Var}[T] = \frac{n(n+1)(2n+1)}{24}.$$

A.10 SPEARMAN'S RANK CORRELATION COEFFICIENT

Let x_i denote the rank i of a set of n objects placed in order. The sum of the ranks is the sum of the first n integers,

$$\sum_{i=1}^{n} x_i = \frac{1}{2}n(n+1),$$

as in Section A.9. Hence the mean rank is

$$\bar{x} = \frac{1}{2}(n+1).$$

We also note from Section A.9 that the sum of squares of the first n ranks is

$$\sum_{i=1}^{n} x_i^2 = \frac{1}{6}n(n+1)(2n+1).$$

Hence,

$$\sum_{i=1}^{n} (x_i - \bar{x})^2 = \sum_{i=1}^{n} x_i^2 - n\bar{x}^2 = \frac{1}{6}n(n+1)(2n+1) - \frac{1}{4}n(n+1)^2 = \frac{n(n^2-1)}{12}.$$

Similarly, for another set of n objects in which y_i denotes the rank i,

$$\sum_{i=1}^{n} (y_i - \bar{y})^2 = \frac{n(n^2 - 1)}{12}.$$

We can use the product-moment correlation coefficient as defined by Eq. (1.4.3) and substitute these relationships to obtain an equation for Spearman's rank correlation coefficient:

$$r_s = \frac{\sum_{i=1}^{n} (x_i - \bar{x})(y_i - \bar{y})}{\sqrt{\sum_{i=1}^{n} (x_i - \bar{x})^2 \sum_{i=1}^{n} (y_i - \bar{y})^2}} = \frac{\sum_{i=1}^{n} x_i y_i - n\bar{x}\bar{y}}{\frac{1}{12}n(n^2 - 1)}$$

$$= \frac{\sum_{i=1}^{n} x_i y_i - \frac{1}{4}n(n+1)^2}{\frac{1}{12}n(n^2 - 1)}.$$

It now remains to substitute for the first term in the numerator. We note that the difference in ranks $d_i = x_i - y_i$. Hence,

$$\sum_{i=1}^{n} d_i^2 = \sum_{i=1}^{n} (x_i - y_i)^2 = \sum_{i=1}^{n} x_i^2 - 2\sum_{i=1}^{n} x_i y_i + \sum_{i=1}^{n} y_i^2$$

$$= \frac{1}{3}n(n+1)(2n+1) - 2\sum_{i=1}^{n} x_i y_i$$

and

$$r_s = \frac{\frac{1}{12}n(n^2 - 1) - \frac{1}{2}\sum_{i=1}^{n} d_i^2}{\frac{1}{12}n(n^2 - 1)} = 1 - \frac{6\sum_{i=1}^{n} d_i^2}{n(n^2 - 1)}.$$

Appendix B
Glossary of Symbols

$\mathbf{A}$	Matrix of coefficients in principal component analysis
A	Event space
A_i	Atkinson's modification to Cook's distance
A^c	Complement of event A
$A \mid B$	A conditional to B
$A \subset B$	Event A contained in event B
$A \in \mathrm{A}$	The event space A is a special set containing A
A^2, A^*	Test statistic in Anderson-Darling test, modified version
ANOVA, ANCOVA	Analysis of variance, covariance
$\mathbf{a}_j$	Column vector j of variables
a	Correlation distance or radius of influence; action in decision theory
a_{ij}	Coefficient in principal components analysis
$\hat{a}$	Moments estimator of parameter a (typical)
$\tilde{a}$	Maximum likelihood estimate of parameter a (typical)
a, b	Parameters of uniform, beta distributions; constants in Gutenberg-Richter relationship for earthquakes
B	Studentized deviate; Bayes' risk
BLUE	*Best* linear unbiased estimator
b	Location parameter for Gumbel distribution; width in simulation
b_{ij}	Factor loading of variables i and j
$\mathbf{C}$	Variance-covariance matrix, matrix of correlation coefficients
C	Copula function
C_l, C_u; C_i, C_α	Lower and upper confidence limits; Cook's distance and critical value in detecting outliers in regression
c_{ij}	Correlation coefficients of i and j variables
cdf	Cumulative distribution function
Cor, Cov	Correlation, covariance
$\mathbf{D}$	Diagonal matrix of eigenvalues
DFFITS	Modified Cook's distance
D_n, $D_{m,n}$; D_n^+, D_n^-	Difference statistics in Kolgomorov-Smirnov test, for sample sizes n and m; cumulative positive and negative departures from the mean for sample size n
d	Mean absolute deviation; sediment diameter in simulation; decision
d_i	Difference between ranks for rank correlation test
d_n	Sample difference in Kolgomorov-Smirnov test, for sample size n
$E[\cdot]$	Expectation

E	Modulus of elasticity
E_i	Expected number in class i in chi-squared test
EV1, EV2, EV3	Three types of extreme value distributions
e,exp	Exponential, base of natural logarithms $= 2.71828\ldots$
$\mathbf{F}$	Matrix of factor scores
$F_N(n)$	Empirical or sample distribution function or step function, for sample size n
$F_X(x)$	Cumulative distribution function X at x
F	F distribution, variance ratio
F_j	jth uncorrelated common factor
$F_{m,n}$	F variate with numerator m and denominator n degrees of freedom
$F_{m,n,\alpha}$	F variate with numerator m and denominator n degrees of freedom and probability of exceedance α
$f_X(x)$	Probability density function of X at x
$f_{X,Y}(x, y)$	Joint probability density function of X, Y at x, y
$f_{X\|Y}(x \mid y)$	Probability density function of X at x conditional to Y at y
GEV	General extreme value distribution
g	Gini's mean difference; gain; performance function; acceleration due to gravity
g_1	Sample coefficient of skewness
g_2	Sample coefficient of kurtosis
$\mathbf{H}$	Matrix of coefficients in factor analysis; leverage or "hat" matrix
H	Kruskal-Wallis test statistic; depth in simulation illustration; Hurst exponent
$H(x)$	Entropy function of X
H_0, H_1	Null and alternative hypotheses
$\mathbf{h}$	Distance vector
h_i, h_i'	ith diagonal element of leverage matrix, leverage measure
$h_F(f)$	pdf of F distribution at f
$\mathbf{I}$	Identity matrix (with 1s in the leading diagonal and 0s elsewhere)
I	Cross-sectional moment of inertia
iqr	Interquartile range
J	Jackknife estimator; Jacobian
K	Frequency factor
k	Variable used in binomial distribution; shape parameter in general extreme value distribution
$L(\theta)$	Likelihood function of θ
L_0, L_1	Likelihoods for ratio test
$L_1, L_2, \ldots$	L moments
l	Number of classes for chi-squared test; loss function
ln	Natural logarithm (base e)
$M(\cdot)$	Moment generation function
M_{ijk}	Probability weighted moments with indices i, j, and k
ML	Maximum likelihood

m	Number of successes in Bernoulli trials
max[·], min[·]	Maximum and minimum of a function
mgf	Moment generation function
mL	Milliliters
$N(a, b)$	Normal distribution with parameters a and b
N, n	Sample size; number of Bernoulli trials; scaling exponent
n_c;n_e	Number of classes in histogram; Euler's constant = 0.5772 (approximately)
O_i	Observed number in class i in chi-squared test
P	Matrix for rotation of principal components
Pr[·]	Probability
p	Probability of a success in Bernoulli trials
p_f	Probability of failure
p_i	ith plotting position
p_0; p_A; p_B	Probability mass at origin of cdf of low-flow index; strike probability; speed probability
$p_X(x)$	Probability mass function of X at x
$p_{X,Y}(x,y)$	Joint probability mass function of X, Y at x, y
$p_{X\|Y}(x \| y)$	Probability mass function of X conditional to Y at x, y
pdf, pmf	Probability density and mass functions
pwm	Probability weighted moment
Q_1, Q_2, Q_3	Quantiles; Q_2 is the median
R	Matrix of correlation coefficients between X variables and principal components (or factors)
R^*	Sum of ranks in Kruskal-Wallis test; risk function; correlation function
R'; R_n^*, R_n^{**}	Design reliability; adjusted range and adjusted rescaled range for sample size n
$R(f)$	Reliability function
R_n	Range of sample of size n
R^2	Coefficient of determination or multiple correlation
r	Range; shape parameter in gamma distribution; correlation coefficient (product moment); reliability
r_s; r_i; r_U, r_L	Spearman's rank sample correlation coefficient; internally Studentized residual; upper and lower credibility limits of reliability
$r_{X,Y}$	Coefficient of correlation between X and Y
S; S_n	Safety margin; standard deviation for sample size n
S^2, s^2; $\hat{S}^2, \hat{s}^2$	Variance; unbiased variance
S_p	Pooled variance in hypothesis testing
S_{xx}, SS	Sum of squares
S_{xy}	Sum of cross-products
Sup[·]	Upper bound of a function
T	Transpose of a matrix
T	Student's t variate; test statistic in Wilcoxon signed-rank test
$T_.$	Grand total in ANOVA
T_j	Test statistic in test for exponentiality

t	Time variable; variable in MGF; t distribution
t_i	Externally Studentized residual
$t_{n,\alpha}$	Student's t variate with n degrees of freedom and probability of exceedance α
U, u	Uniform random variates; utility function
V	Velocity of flow
v, V_Z	Coefficient of variation, of variable Z
$\text{Var}[\cdot]$	Variance
W	Test statistic in Shapiro-Wilk's for normality; width or sediment gradation
WE_j	Test statistic in Shapiro-Wilk's test for exponentiality
$\mathbf{X}$	Matrix of (explanatory) variables
X, Y, Z	Random variables; depths in simulation illustration
X^*	Smallest-value variate having extreme value distribution
X^2	Statistic in chi-squared test, with an asymptotic chi-squared distribution
$X_{(1)}, X_{(2)}, \ldots X_{(n)}$	Series of sample size n, ranked in increasing order
$\mathbf{x}$	Vector of explanatory variables in multiple regression
$\bar{x}; \bar{x}_.$	Sample mean; mean value in ANOVA
$\bar{x}_g$	Geometric mean
$\bar{x}_h$	Harmonic mean
$x_i^{(r)} = x_i(x_i - 1)(x_i - 2)\cdots \times (x_i - r + 1)$	Partial factorial
x_0	Lower bound in Pareto distribution; scale parameter in Fréchet distribution
x^*	Nominal value of capacity
x, y, z	Values taken by random variables
$\hat{Y}$	Mean response in fitted regression model
$\mathbf{y}$	Vector of response variables
y^*	Nominal value of demand; reduced (Gumbel) smallest-value variate
y	Reduced (Gumbel) variate; observed response variable
$\mathbf{Z}$	Matrix of principal components
Z	Safety factor
$\mathbf{z}_j$	jth vector of principal components
z^*	Nominal factor of safety
z_α	Standard normal variate with exceedance probability α
α	Shape parameter for Type I extreme value (Gumbel) distribution; level of significance in hypothesis testing
α, β	Type I and II errors in hypothesis testing; parameters of beta distribution
α_i, β_j	Treatment effect at level i and block effect at level j, respectively
α_n, β_n	Parameters used in general extreme value distribution
$\alpha\beta$	Interaction effect
$\boldsymbol{\beta}$	Vector of regression coefficients
β	Reliability index; shape parameter in Weibull distribution
$\beta_0, \beta_1, \ldots, \beta_p$	Parameters in multiple linear regression

$\Gamma(p)$	Gamma function for argument p
$\Gamma(h)$, $\Gamma(\mathbf{h})$	Semivariogram at distance h, **h** as a vector
γ	Limiting constant in extreme value theory
$\hat{\gamma}(h)$	Empirical semivariogram at h
γ_1	Population coefficient of skewness
γ_2	Population coefficient of kurtosis
ε	Vector of error terms
δ	Dependence parameter of bivariate copula
$\varepsilon; \hat{\varepsilon}$	Residual in ANOVA, location parameter of a continuous distribution, error term in regression; residual in regression
ζ	Central safety factor of a system
η	Coefficient used for plotting position and in simulation parameter space; scale parameter in modeling Weibull parameter λ
Θ	Parameter space for state of nature, θ
θ	Shape parameter in Fréchet and Pareto distributions; state of nature
$\hat{\theta}$	Unbiased estimator of the population parameter θ
κ	Constant in rainfall depth-duration; normalizing factor
Λ	Variate used in modeling Weibull parameter λ
$\lambda; \lambda_1$	Poisson mean-rate parameter (ν/t), scale parameter in exponential, gamma, and Weibull distributions; Lagrange multiplier
λ_i	Eigenvalue, weight in Kriging
μ_r	Moment of order r about the mean, also called central moment
μ'_r	Moment of order r about the origin, that is, zero, also called absolute, crude, or raw moment
μ^*_r	Moment of order r about a point other than the origin
$\mu_{(r)}$	Factorial moment of order r
μ_w	Mean survival time
$\hat{\mu}_{Y\,\vert\,\mathbf{x}=\mathbf{a}}$	Estimate of conditional mean response in multiple regression at $\mathbf{x} = \mathbf{a}$
ν	Poisson mean parameter; degrees of freedom for t and chi-squared distributions; ratio of coefficients of variation
Ξ	Parameter vector used in regional flood estimation
ξ_q	The qth quantile of a random variable
$\pi = 3.14159\ldots$	Circumference of a circle with unit diameter or area of a circle of unit radius
π_k	Weight or reaction in point estimation method of reliability analysis at kth point
$\pi(\theta)$	Prior distribution of state of nature θ
ρ	correlation parameter; physical growth rate
σ	standard deviation of population
σ^2	Variance of population
τ	Kendall's rank correlation coefficient
$\phi(x)$	Normal pdf at x
$\Phi(x)$	Normal cdf at x

$\chi^2(p)$	Chi-squared distribution with parameter p
$\chi^2_{n,\alpha}$	Chi-squared variate with n degrees of freedom and probability of exceedance α
Ω	Sample space, multidimensional volume in simulation
ω	Outcome; state; upper bound for EV3 distribution
$\emptyset$	Null set
$\cup$	Union
$\cap$	Intersection
Σ	Addition
Π	Multiplication
$\langle \cdot \rangle$	Arithmetic mean over the n sample points
$(a, b]$	b is within interval, a outside interval
$\binom{n}{x}$	Number of possible combinations selecting x objects at a time over a total of n objects
$\|J\|$	Absolute value of J, determinant
$n!$	Factorial of n $(=n(n-1)(n-2)\cdots 1)$
$\sim$	Distributed as
$\approx$	Approximated as
$\mathbf{X}^{-1}$	Inverse of matrix $\mathbf{X}$
$\stackrel{d}{=}$	Equality in distribution
$\forall_x$	For all x
$\Rightarrow$	Implies
$x \to x_0^+$	$x \to x_0$ and $x > x_0$
$x \to x_0^-$	$x \to x_0$ and $x < x_0$

Appendix C
Tables of Selected Distributions

Table C.1 Cumulative standard normal distribution: $\Phi(z) = \int_{-\infty}^{z} \frac{1}{\sqrt{2\pi}} \exp(-\frac{t^2}{2})dt$

z	0.00	0.01	0.02	0.03	0.04	0.05	0.06	0.07	0.08	0.09
0.0	0.5000	0.5039	0.5079	0.5119	0.5159	0.5199	0.5239	0.5279	0.5318	0.5358
0.1	0.5398	0.5438	0.5477	0.5517	0.5556	0.5596	0.5635	0.5674	0.5714	0.5753
0.2	0.5792	0.5831	0.5870	0.5909	0.5948	0.5987	0.6025	0.6064	0.6102	0.6140
0.3	0.6179	0.6217	0.6255	0.6293	0.6330	0.6368	0.6405	0.6443	0.6480	0.6517
0.4	0.6554	0.6591	0.6627	0.6664	0.6700	0.6736	0.6772	0.6808	0.6843	0.6879
0.5	0.6914	0.6949	0.6984	0.7019	0.7054	0.7088	0.7122	0.7156	0.7190	0.7224
0.6	0.7257	0.7290	0.7323	0.7356	0.7389	0.7421	0.7453	0.7485	0.7517	0.7549
0.7	0.7580	0.7611	0.7642	0.7673	0.7703	0.7733	0.7763	0.7793	0.7823	0.7852
0.8	0.7881	0.7910	0.7938	0.7967	0.7995	0.8023	0.8051	0.8078	0.8105	0.8132
0.9	0.8159	0.8185	0.8212	0.8238	0.8263	0.8289	0.8314	0.8339	0.8364	0.8389
1.0	0.8413	0.8437	0.8461	0.8484	0.8508	0.8531	0.8554	0.8576	0.8599	0.8621
1.1	0.8643	0.8665	0.8686	0.8707	0.8728	0.8749	0.8769	0.8790	0.8810	0.8829
1.2	0.8849	0.8868	0.8887	0.8906	0.8925	0.8943	0.8961	0.8979	0.8997	0.9014
1.3	0.9032	0.9049	0.9065	0.9082	0.9098	0.9114	0.9130	0.9146	0.9162	0.9177
1.4	0.9192	0.9207	0.9222	0.9236	0.9250	0.9264	0.9278	0.9292	0.9305	0.9318
1.5	0.9331	0.9344	0.9357	0.9369	0.9382	0.9394	0.9406	0.9417	0.9429	0.9440
1.6	0.9452	0.9463	0.9473	0.9484	0.9495	0.9505	0.9515	0.9525	0.9535	0.9544
1.7	0.9554	0.9563	0.9572	0.9581	0.9590	0.9599	0.9608	0.9616	0.9624	0.9632
1.8	0.9640	0.9648	0.9656	0.9663	0.9671	0.9678	0.9685	0.9692	0.9699	0.9706
1.9	0.9712	0.9719	0.9725	0.9732	0.9738	0.9744	0.9750	0.9755	0.9761	0.9767
2.0	0.9772	0.9777	0.9783	0.9788	0.9793	0.9798	0.9803	0.9807	0.9812	0.9816
2.1	0.9821	0.9825	0.9830	0.9834	0.9838	0.9842	0.9846	0.9850	0.9853	0.9857
2.2	0.9861	0.9864	0.9867	0.9871	0.9874	0.9877	0.9880	0.9884	0.9887	0.9889
2.3	0.9892	0.9895	0.9898	0.9901	0.9903	0.9906	0.9908	0.9911	0.9913	0.9915
2.4	0.9918	0.9920	0.9922	0.9924	0.9926	0.9928	0.9930	0.9932	0.9934	0.9936
2.5	0.9937	0.9939	0.9941	0.9943	0.9944	0.9946	0.9947	0.9949	0.9950	0.9952
2.6	0.9953	0.9954	0.9956	0.9957	0.9958	0.9959	0.9960	0.9962	0.9963	0.9964
2.7	0.9965	0.9966	0.9967	0.9968	0.9969	0.9970	0.9971	0.9972	0.9972	0.9973
2.8	0.9974	0.9975	0.9976	0.9976	0.9977	0.9978	0.9978	0.9979	0.9980	0.9980
2.9	0.9981	0.9981	0.9982	0.9983	0.9983	0.9984	0.9984	0.9985	0.9985	0.9986
3.0	0.9986	0.9986	0.9987	0.9987	0.9988	0.9988	0.9988	0.9989	0.9989	0.9990
3.1	0.9990	0.9990	0.9991	0.9991	0.9991	0.9991	0.9992	0.9992	0.9992	0.9992
3.2	0.9993	0.9993	0.9993	0.9993	0.9994	0.9994	0.9994	0.9994	0.9994	0.9995
3.3	0.9995	0.9995	0.9995	0.9995	0.9995	0.9996	0.9996	0.9996	0.9996	0.9996
3.4	0.9996	0.9996	0.9996	0.9997	0.9997	0.9997	0.9997	0.9997	0.9997	0.9997
3.5	0.9997	0.9997	0.9997	0.9997	0.9998	0.9998	0.9998	0.9998	0.9998	0.9998
3.6	0.9998	0.9998	0.9998	0.9998	0.9998	0.9998	0.9998	0.9998	0.9998	0.9998
3.7	0.9998	0.9999	0.9999	0.9999	0.9999	0.9999	0.9999	0.9999	0.9999	0.9999
3.8	0.9999	0.9999	0.9999	0.9999	0.9999	0.9999	0.9999	0.9999	0.9999	0.9999
3.9	0.9999	0.9999	0.9999	0.9999	0.9999	0.9999	0.9999	0.9999	0.9999	0.9999
4.0	0.9999	0.9999	0.9999	0.9999	0.9999	0.9999	0.9999	0.9999	0.9999	0.9999
4.1	0.9999	0.9999	0.9999	0.9999	0.9999	0.9999	0.9999	0.9999	0.9999	0.9999
4.2	0.9999	0.9999	0.9999	0.9999	0.9999	0.9999	0.9999	0.9999	0.9999	0.9999
4.3	0.9999	0.9999	0.9999	0.9999	0.9999	0.9999	0.9999	0.9999	0.9999	0.9999
4.4	0.9999	0.9999	1.0000	1.0000	1.0000	1.0000	1.0000	1.0000	1.0000	1.0000
4.5	1.0000	1.0000	1.0000	1.0000	1.0000	1.0000	1.0000	1.0000	1.0000	1.0000

Table C.2 Cumulative Student's t Distribution with v degrees of freedom: $F(t) = \int_{-\infty}^{t} \frac{\Gamma((v+1)/2)}{\sqrt{v\pi}\Gamma(v/2)(z^2/v+1)^{(v+1)/2}} dz$ **(entry in table is t)**

$F =$	0.75	0.80	0.85	0.90	0.95	0.975	0.99	0.995
$v = 1$	1	1.376	1.963	3.078	6.314	12.71	31.82	63.66
2	0.816	1.061	1.386	1.886	2.920	4.303	6.965	9.925
3	0.765	0.978	1.250	1.638	2.353	3.182	4.541	5.841
4	0.741	0.941	1.190	1.533	2.132	2.776	3.747	4.604
5	0.727	0.920	1.156	1.476	2.015	2.571	3.365	4.032
6	0.718	0.906	1.134	1.440	1.943	2.447	3.143	3.707
7	0.711	0.896	1.119	1.415	1.895	2.365	2.998	3.499
8	0.706	0.889	1.108	1.397	1.860	2.306	2.896	3.355
9	0.703	0.883	1.100	1.383	1.833	2.262	2.821	3.250
10	0.700	0.879	1.093	1.372	1.812	2.228	2.764	3.169
11	0.697	0.876	1.088	1.363	1.796	2.201	2.718	3.106
12	0.695	0.873	1.083	1.356	1.782	2.179	2.681	3.055
13	0.694	0.870	1.079	1.350	1.771	2.160	2.650	3.012
14	0.692	0.868	1.076	1.345	1.761	2.145	2.624	2.977
15	0.691	0.866	1.074	1.341	1.753	2.131	2.602	2.947
16	0.690	0.865	1.071	1.337	1.746	2.120	2.583	2.921
17	0.689	0.863	1.069	1.333	1.740	2.110	2.567	2.898
18	0.688	0.862	1.067	1.330	1.734	2.101	2.552	2.878
19	0.688	0.861	1.066	1.328	1.729	2.093	2.539	2.861
20	0.687	0.860	1.064	1.325	1.725	2.086	2.528	2.845
21	0.686	0.859	1.063	1.323	1.721	2.080	2.518	2.831
22	0.686	0.858	1.061	1.321	1.717	2.074	2.508	2.819
23	0.685	0.858	1.060	1.319	1.714	2.069	2.500	2.807
24	0.685	0.857	1.059	1.318	1.711	2.064	2.492	2.797
25	0.684	0.856	1.058	1.316	1.708	2.060	2.485	2.787
26	0.684	0.856	1.058	1.315	1.706	2.056	2.479	2.779
27	0.684	0.855	1.057	1.314	1.703	2.052	2.473	2.771
28	0.683	0.855	1.056	1.313	1.701	2.048	2.467	2.763
29	0.683	0.854	1.055	1.311	1.699	2.045	2.462	2.756
30	0.683	0.854	1.055	1.310	1.697	2.042	2.457	2.750
40	0.681	0.851	1.050	1.303	1.684	2.021	2.423	2.704
50	0.679	0.849	1.047	1.299	1.676	2.009	2.403	2.678
60	0.679	0.848	1.045	1.296	1.671	2.000	2.390	2.660
70	0.678	0.847	1.044	1.294	1.667	1.994	2.381	2.648
80	0.678	0.846	1.043	1.292	1.664	1.990	2.374	2.639
90	0.677	0.846	1.042	1.291	1.662	1.987	2.368	2.632
100	0.677	0.845	1.042	1.290	1.660	1.984	2.364	2.626
∞	0.674	0.842	1.036	1.282	1.645	1.960	2.326	2.576

Table C.3 Cumulative Chi-square Distribution with ν degrees of freedom: $F(\chi^2) = \frac{1}{2}\int_0^{\chi^2} \frac{(t/2)^{(\nu/2)-1}\exp(-t/2)}{\Gamma(\nu/2)}\,dt$ (entry in table is χ^2)

$F =$	0.001	0.005	0.010	0.025	0.050	0.100	0.250	0.500	0.750	0.900	0.950	0.975	0.990	0.995
$\nu = 1$	0.000002	0.00004	0.0002	0.0010	0.0039	0.016	0.102	0.455	1.32	2.71	3.84	5.02	6.63	7.88
2	0.0020	0.0100	0.0201	0.0506	0.103	0.211	0.575	1.386	2.77	4.61	5.99	7.38	9.21	10.6
3	0.0243	0.0717	0.115	0.216	0.352	0.584	1.213	2.366	4.11	6.25	7.81	9.35	11.3	12.8
4	0.0908	0.207	0.297	0.484	0.711	1.06	1.92	3.36	5.39	7.78	9.49	11.1	13.3	14.9
5	0.210	0.412	0.554	0.831	1.15	1.61	2.67	4.35	6.63	9.24	11.1	12.8	15.1	16.7
6	0.381	0.676	0.872	1.24	1.64	2.20	3.45	5.35	7.84	10.6	12.6	14.4	16.8	18.5
7	0.599	0.989	1.24	1.69	2.17	2.83	4.25	6.35	9.04	12.0	14.1	16.0	18.5	20.3
8	0.857	1.34	1.65	2.18	2.73	3.49	5.07	7.34	10.2	13.4	15.5	17.5	20.1	22.0
9	1.15	1.73	2.09	2.70	3.33	4.17	5.90	8.34	11.4	14.7	16.9	19.0	21.7	23.6
10	1.48	2.16	2.56	3.25	3.94	4.87	6.74	9.34	12.5	16.0	18.3	20.5	23.2	25.2
11	1.83	2.60	3.05	3.82	4.57	5.58	7.58	10.3	13.7	17.3	19.7	21.9	24.7	26.8
12	2.21	3.07	3.57	4.40	5.23	6.30	8.44	11.3	14.8	18.5	21.0	23.3	26.2	28.3
13	2.62	3.57	4.11	5.01	5.89	7.04	9.30	12.3	16.0	19.8	22.4	24.7	27.7	29.8
14	3.04	4.07	4.66	5.63	6.57	7.79	10.2	13.3	17.1	21.1	23.7	26.1	29.1	31.3
15	3.48	4.60	5.23	6.26	7.26	8.55	11.0	14.3	18.2	22.3	25.0	27.5	30.6	32.8
16	3.94	5.14	5.81	6.91	7.96	9.31	11.9	15.3	19.4	23.5	26.3	28.8	32.0	34.3
17	4.42	5.70	6.41	7.56	8.67	10.1	12.8	16.3	20.5	24.8	27.6	30.2	33.4	35.7
18	4.90	6.26	7.01	8.23	9.39	10.9	13.7	17.3	21.6	26.0	28.9	31.5	34.8	37.2
19	5.41	6.84	7.63	8.91	10.1	11.7	14.6	18.3	22.7	27.2	30.1	32.9	36.2	38.6
20	5.92	7.43	8.26	9.59	10.9	12.4	15.5	19.3	23.8	28.4	31.4	34.2	37.6	40.0

(continued)

Table C.3 *(continued)*

F =	0.001	0.005	0.010	0.025	0.050	0.100	0.250	0.500	0.750	0.900	0.950	0.975	0.990	0.995
21	6.45	8.03	8.90	10.3	11.6	13.2	16.3	20.3	24.9	29.6	32.7	35.5	38.9	41.4
22	6.98	8.64	9.54	11.0	12.3	14.0	17.2	21.3	26.0	30.8	33.9	36.8	40.3	42.8
23	7.53	9.26	10.2	11.7	13.1	14.8	18.1	22.3	27.1	32.0	35.2	38.1	41.6	44.2
24	8.08	9.89	10.9	12.4	13.8	15.7	19.0	23.3	28.2	33.2	36.4	39.4	43.0	45.6
25	8.65	10.5	11.5	13.1	14.6	16.5	19.9	24.3	29.3	34.4	37.7	40.6	44.3	46.9
26	9.22	11.2	12.2	13.8	15.4	17.3	20.8	25.3	30.4	35.6	38.9	41.9	45.6	48.3
27	9.80	11.8	12.9	14.6	16.2	18.1	21.7	26.3	31.5	36.7	40.1	43.2	47.0	49.6
28	10.4	12.5	13.6	15.3	16.9	18.9	22.7	27.3	32.6	37.9	41.3	44.5	48.3	51.0
29	11.0	13.1	14.3	16.0	17.7	19.8	23.6	28.3	33.7	39.1	42.6	45.7	49.6	52.3
30	11.6	13.8	15.0	16.8	18.5	20.6	24.5	29.3	34.8	40.3	43.8	47.0	50.9	53.7
31	12.2	14.5	15.7	17.5	19.3	21.4	25.4	30.3	35.9	41.4	45.0	48.2	52.2	55.0
32	12.8	15.1	16.4	18.3	20.1	22.3	26.3	31.3	37.0	42.6	46.2	49.5	53.5	56.3
33	13.4	15.8	17.1	19.0	20.9	23.1	27.2	32.3	38.1	43.7	47.4	50.7	54.8	57.6
34	14.1	16.5	17.8	19.8	21.7	24.0	28.1	33.3	39.1	44.9	48.6	52.0	56.1	59.0
35	14.7	17.2	18.5	20.6	22.5	24.8	29.1	34.3	40.2	46.1	49.8	53.2	57.3	60.3
40	17.9	20.7	22.2	24.4	26.5	29.1	33.7	39.3	45.6	51.8	55.8	59.3	63.7	66.8
50	24.7	28.0	29.7	32.4	34.8	37.7	42.9	49.3	56.3	63.2	67.5	71.4	76.2	79.5

Table C.4 Cumulative F Distribution with m and n degrees of freedom in numerator and denominator, respectively:
$G(F) = \int_0^F \frac{\Gamma(\frac{m+n}{2}) m^{m/2} n^{n/2} x^{(m-2)/2} (n+mx)^{-(m+n)/2}}{\Gamma(m/2)\Gamma(n/2)} dx$ (entry in table is F).

G	n	$m = 1$	2	3	4	5	6	7	8	9	10	25	50	100	∞
0.9	1	39.9	49.5	53.6	55.8	57.2	58.2	58.9	59.4	59.9	60.2	62.1	62.7	63.0	63.3
0.95	1	161	199	216	225	230	234	237	239	241	242	249	252	253	254
0.975	1	648	799	864	900	922	937	948	957	963	969	998	1,008	1,013	1,018
0.99	1	4,052	4,999	5,404	5,624	5,764	5,859	5,928	5,981	6,022	6,056	6,240	6,302	6,334	6,363
0.995	1	16,212	19,997	21,614	22,501	23,056	23,440	23,715	23,924	24,091	24,222	24,959	25,213	25,339	25,451
0.9	2	8.53	9.00	9.16	9.24	9.29	9.33	9.35	9.37	9.38	9.39	9.45	9.47	9.48	9.49
0.95	2	18.5	19.0	19.2	19.2	19.3	19.3	19.4	19.4	19.4	19.4	19.5	19.5	19.5	19.5
0.975	2	38.5	39.0	39.2	39.2	39.3	39.3	39.4	39.4	39.4	39.4	39.5	39.5	39.5	39.5
0.99	2	98.5	99.0	99.2	99.3	99.3	99.3	99.4	99.4	99.4	99.4	99.5	99.5	99.5	99.5
0.995	2	199	199	199	199	199	199	199	199	199	199	199	199	199	200
0.9	3	5.54	5.46	5.39	5.34	5.31	5.28	5.27	5.25	5.24	5.23	5.17	5.15	5.14	5.13
0.95	3	10.1	9.55	9.28	9.12	9.01	8.94	8.89	8.85	8.81	8.79	8.63	8.58	8.55	8.53
0.975	3	17.4	16.0	15.4	15.1	14.9	14.7	14.6	14.5	14.5	14.4	14.1	14.0	14.0	13.9
0.99	3	34.1	30.8	29.5	28.7	28.2	27.9	27.7	27.5	27.3	27.2	26.6	26.4	26.2	26.1
0.995	3	55.6	49.8	47.5	46.2	45.4	44.8	44.4	44.1	43.9	43.7	42.6	42.2	42.0	41.8
0.9	4	4.54	4.32	4.19	4.11	4.05	4.01	3.98	3.95	3.94	3.92	3.83	3.80	3.78	3.76
0.95	4	7.71	6.94	6.59	6.39	6.26	6.16	6.09	6.04	6.00	5.96	5.77	5.70	5.66	5.63
0.975	4	12.2	10.6	9.98	9.60	9.36	9.20	9.07	8.98	8.90	8.84	8.50	8.38	8.32	8.26
0.99	4	21.2	18.0	16.7	16.0	15.5	15.2	15.0	14.8	14.7	14.5	13.9	13.7	13.6	13.5
0.995	4	31.3	26.3	24.3	23.2	22.5	22.0	21.6	21.4	21.1	21.0	20.0	19.7	19.5	19.3
0.9	5	4.06	3.78	3.62	3.52	3.45	3.40	3.37	3.34	3.32	3.30	3.19	3.15	3.13	3.11
0.95	5	6.61	5.79	5.41	5.19	5.05	4.95	4.88	4.82	4.77	4.74	4.52	4.44	4.41	4.37
0.975	5	10.0	8.43	7.76	7.39	7.15	6.98	6.85	6.76	6.68	6.62	6.27	6.14	6.08	6.02
0.99	5	16.3	13.3	12.1	11.4	11.0	10.7	10.5	10.3	10.2	10.1	9.45	9.24	9.13	9.03
0.995	5	22.8	18.3	16.5	15.6	14.9	14.5	14.2	14.0	13.8	13.6	12.8	12.5	12.3	12.2

(continued)

Table C.4 *(continued)*

G	n	$m = 1$	2	3	4	5	6	7	8	9	10	25	50	100	∞
0.9	6	3.78	3.46	3.29	3.18	3.11	3.05	3.01	2.98	2.96	2.94	2.81	2.77	2.75	2.72
0.95	6	5.99	5.14	4.76	4.53	4.39	4.28	4.21	4.15	4.10	4.06	3.83	3.75	3.71	3.67
0.975	6	8.81	7.26	6.60	6.23	5.99	5.82	5.70	5.60	5.52	5.46	5.11	4.98	4.92	4.86
0.99	6	13.7	10.9	9.78	9.15	8.75	8.47	8.26	8.10	7.98	7.87	7.30	7.09	6.99	6.89
0.995	6	18.6	14.5	12.9	12.0	11.5	11.1	10.8	10.6	10.4	10.3	9.45	9.17	9.03	8.89
0.9	7	3.59	3.26	3.07	2.96	2.88	2.83	2.78	2.75	2.72	2.70	2.57	2.52	2.50	2.47
0.95	7	5.59	4.74	4.35	4.12	3.97	3.87	3.79	3.73	3.68	3.64	3.40	3.32	3.27	3.23
0.975	7	8.07	6.54	5.89	5.52	5.29	5.12	4.99	4.90	4.82	4.76	4.40	4.28	4.21	4.15
0.99	7	12.2	9.55	8.45	7.85	7.46	7.19	6.99	6.84	6.72	6.62	6.06	5.86	5.75	5.66
0.995	7	16.2	12.4	10.9	10.1	9.52	9.16	8.89	8.68	8.51	8.38	7.62	7.35	7.22	7.09
0.9	8	3.46	3.11	2.92	2.81	2.73	2.67	2.62	2.59	2.56	2.54	2.40	2.35	2.32	2.30
0.95	8	5.32	4.46	4.07	3.84	3.69	3.58	3.50	3.44	3.39	3.35	3.11	3.02	2.97	2.93
0.975	8	7.57	6.06	5.42	5.05	4.82	4.65	4.53	4.43	4.36	4.30	3.94	3.81	3.74	3.68
0.99	8	11.3	8.65	7.59	7.01	6.63	6.37	6.18	6.03	5.91	5.81	5.26	5.07	4.96	4.87
0.995	8	14.7	11.0	9.60	8.81	8.30	7.95	7.69	7.50	7.34	7.21	6.48	6.22	6.09	5.96
0.9	9	3.36	3.01	2.81	2.69	2.61	2.55	2.51	2.47	2.44	2.42	2.27	2.22	2.19	2.16
0.95	9	5.12	4.26	3.86	3.63	3.48	3.37	3.29	3.23	3.18	3.14	2.89	2.80	2.76	2.71
0.975	9	7.21	5.71	5.08	4.72	4.48	4.32	4.20	4.10	4.03	3.96	3.60	3.47	3.40	3.34
0.99	9	10.6	8.02	6.99	6.42	6.06	5.80	5.61	5.47	5.35	5.26	4.71	4.52	4.41	4.32
0.995	9	13.6	10.1	8.72	7.96	7.47	7.13	6.88	6.69	6.54	6.42	5.71	5.45	5.32	5.20
0.9	10	3.29	2.92	2.73	2.61	2.52	2.46	2.41	2.38	2.35	2.32	2.17	2.12	2.09	2.06
0.95	10	4.96	4.10	3.71	3.48	3.33	3.22	3.14	3.07	3.02	2.98	2.73	2.64	2.59	2.54
0.975	10	6.94	5.46	4.83	4.47	4.24	4.07	3.95	3.85	3.78	3.72	3.35	3.22	3.15	3.09
0.99	10	10.0	7.56	6.55	5.99	5.64	5.39	5.20	5.06	4.94	4.85	4.31	4.12	4.01	3.92
0.995	10	12.8	9.43	8.08	7.34	6.87	6.54	6.30	6.12	5.97	5.85	5.15	4.90	4.77	4.65

0.9	25	2.92	2.53	2.32	2.18	2.09	2.02	1.97	1.93	1.89	1.87	1.68	1.61	1.56	1.52
0.95	25	4.24	3.39	2.99	2.76	2.60	2.49	2.40	2.34	2.28	2.24	1.96	1.84	1.78	1.72
0.975	25	5.69	4.29	3.69	3.35	3.13	2.97	2.85	2.75	2.68	2.61	2.23	2.08	2.00	1.91
0.99	25	7.77	5.57	4.68	4.18	3.85	3.63	3.46	3.32	3.22	3.13	2.60	2.40	2.29	2.18
0.995	25	9.48	6.60	5.46	4.84	4.43	4.15	3.94	3.78	3.64	3.54	2.90	2.65	2.52	2.39
0.9	50	2.81	2.41	2.20	2.06	1.97	1.90	1.84	1.80	1.76	1.73	1.53	1.44	1.39	1.33
0.95	50	4.03	3.18	2.79	2.56	2.40	2.29	2.20	2.13	2.07	2.03	1.73	1.60	1.52	1.45
0.975	50	5.34	3.97	3.39	3.05	2.83	2.67	2.55	2.46	2.38	2.32	1.92	1.75	1.66	1.56
0.99	50	7.17	5.06	4.20	3.72	3.41	3.19	3.02	2.89	2.78	2.70	2.17	1.95	1.82	1.70
0.995	50	8.63	5.90	4.83	4.23	3.85	3.58	3.38	3.22	3.09	2.99	2.35	2.10	1.95	1.80
0.9	100	2.76	2.36	2.14	2.00	1.91	1.83	1.78	1.73	1.69	1.66	1.45	1.35	1.29	1.22
0.95	100	3.94	3.09	2.70	2.46	2.31	2.19	2.10	2.03	1.97	1.93	1.62	1.48	1.39	1.30
0.975	100	5.18	3.83	3.25	2.92	2.70	2.54	2.42	2.32	2.24	2.18	1.77	1.59	1.48	1.36
0.99	100	6.90	4.82	3.98	3.51	3.21	2.99	2.82	2.69	2.59	2.50	1.97	1.74	1.60	1.45
0.995	100	8.24	5.59	4.54	3.96	3.59	3.33	3.13	2.97	2.85	2.74	2.11	1.84	1.68	1.51
0.9	∞	2.71	2.31	2.09	1.95	1.85	1.78	1.72	1.68	1.64	1.61	1.38	1.27	1.20	1.00
0.95	∞	3.85	3.00	2.61	2.38	2.22	2.11	2.02	1.95	1.89	1.84	1.52	1.36	1.26	1.00
0.975	∞	5.04	3.70	3.13	2.80	2.58	2.42	2.30	2.20	2.13	2.06	1.64	1.45	1.32	1.00
0.99	∞	6.66	4.63	3.80	3.34	3.04	2.82	2.66	2.53	2.43	2.34	1.79	1.54	1.38	1.00
0.995	∞	7.91	5.33	4.30	3.74	3.37	3.11	2.92	2.77	2.64	2.54	1.90	1.61	1.43	1.00

Table C.5 Values of the gamma function, $\Gamma(k)$

k	0.00	0.01	0.02	0.03	0.04	0.05	0.06	0.07	0.08	0.09
0.0		99.433	49.442	32.785	24.461	19.470	16.146	13.774	11.997	10.616
0.1	9.5135	8.6127	7.8633	7.2302	6.6887	6.2203	5.8113	5.4512	5.1318	4.8468
0.2	4.5908	4.3599	4.1505	3.9598	3.7855	3.6256	3.4785	3.3426	3.2169	3.1001
0.3	2.9916	2.8903	2.7958	2.7072	2.6242	2.5461	2.4727	2.4036	2.3383	2.2765
0.4	2.2182	2.1628	2.1104	2.0605	2.0132	1.9681	1.9252	1.8843	1.8453	1.8081
0.5	1.7725	1.7384	1.7058	1.6747	1.6448	1.6161	1.5886	1.5623	1.5369	1.5126
0.6	1.4892	1.4667	1.4450	1.4242	1.4041	1.3848	1.3662	1.3482	1.3309	1.3142
0.7	1.2981	1.2825	1.2675	1.2530	1.2390	1.2254	1.2123	1.1997	1.1875	1.1757
0.8	1.1642	1.1532	1.1425	1.1322	1.1222	1.1125	1.1031	1.0941	1.0853	1.0768
0.9	1.0686	1.0607	1.0530	1.0456	1.0384	1.0315	1.0247	1.0182	1.0119	1.0059

Table C.6 Values of $\Gamma^2(1 + k)/\Gamma(1 + 2k)$

k	0.00	0.01	0.02	0.03	0.04	0.05	0.06	0.07	0.08	00.09
0.0	1.000	1.000	0.999	0.999	0.998	0.996	0.995	0.993	0.991	0.988
0.1	0.986	0.983	0.980	0.977	0.974	0.970	0.966	0.963	0.959	0.954
0.2	0.950	0.946	0.941	0.937	0.932	0.927	0.922	0.917	0.912	0.907
0.3	0.901	0.896	0.891	0.885	0.880	0.874	0.868	0.863	0.857	0.851
0.4	0.845	0.839	0.833	0.828	0.822	0.816	0.810	0.804	0.798	0.791
0.5	0.785	0.779	0.773	0.767	0.761	0.755	0.749	0.743	0.737	0.731
0.6	0.725	0.719	0.713	0.706	0.700	0.694	0.688	0.682	0.677	0.671
0.7	0.665	0.659	0.653	0.647	0.641	0.635	0.630	0.624	0.618	0.612
0.8	0.607	0.601	0.596	0.590	0.584	0.539	0.573	0.568	0.562	0.557
0.9	0.552	0.546	0.541	0.536	0.531	0.525	0.520	0.515	0.510	0.505

Weibull shape parameter $\beta = 1/k$.

Table C.7 Values of $D_{n,\alpha}$ for the Kolmogorov-Smirnov goodness-of-fit test

n	$D_{n,0.10}$	$D_{n,0.05}$	$D_{n,0.02}$	$D_{n,0.01}$
10	0.369	0.409	0.457	0.489
11	0.352	0.391	0.437	0.468
12	0.338	0.375	0.419	0.449
13	0.325	0.361	0.404	0.432
14	0.314	0.349	0.390	0.418
15	0.304	0.338	0.377	0.404
16	0.295	0.327	0.366	0.392
17	0.286	0.318	0.355	0.381
18	0.279	0.309	0.346	0.371
19	0.271	0.301	0.337	0.361
20	0.265	0.294	0.329	0.352
21	0.259	0.287	0.321	0.344
22	0.253	0.281	0.314	0.337
23	0.247	0.275	0.307	0.330
24	0.242	0.269	0.301	0.323
25	0.238	0.264	0.295	0.317
26	0.233	0.259	0.290	0.311
27	0.229	0.254	0.284	0.305
28	0.225	0.250	0.279	0.300
29	0.221	0.246	0.275	0.295
30	0.218	0.242	0.270	0.290
31	0.214	0.238	0.266	0.285
32	0.211	0.234	0.262	0.281
33	0.208	0.231	0.258	0.277
34	0.205	0.227	0.254	0.273
35	0.202	0.224	0.251	0.269
36	0.199	0.221	0.247	0.265
37	0.196	0.218	0.244	0.262
38	0.194	0.215	0.241	0.258
39	0.191	0.213	0.238	0.255
40	0.189	0.210	0.235	0.252
>40	$1.22/\sqrt{n}$	$1.36/\sqrt{n}$	$1.52/\sqrt{n}$	$1.63/\sqrt{n}$

Entries in the last row are approximations for large samples. For $n = 40$ the closeness of this approximation is seen from the entries in the penultimate row.

Source: From Miller, L. H. (1956), "Table of percentage points of Kolgomorov statistics," *J. Am. Stat. Assoc.*, Vol. 51, p. 111–121 (with permission of the publishers).

Table C.8 Lilliefors's test for normality[a]

n	$D_{n,0.20}$	$D_{n,0.10}$	$D_{n,0.05}$	$D_{n,0.01}$	$D_{n,0.001}$
10	0.217	0.241	0.262	0.304	0.352
11	0.208	0.231	0.251	0.291	0.338
12	0.200	0.222	0.242	0.281	0.325
13	0.193	0.215	0.234	0.271	0.314
14	0.187	0.208	0.226	0.262	0.305
15	0.181	0.201	0.219	0.254	0.296
16	0.176	0.195	0.213	0.247	0.287
17	0.171	0.190	0.207	0.240	0.279
18	0.167	0.185	0.202	0.234	0.273
19	0.163	0.181	0.197	0.228	0.266
20	0.159	0.176	0.192	0.223	0.260
25	0.143	0.159	0.173	0.201	0.236
30	0.131	0.146	0.159	0.185	0.217
40	0.115	0.128	0.139	0.162	0.189
100	0.074	0.082	0.089	0.104	0.122
400	0.037	0.041	0.045	0.052	0.061
900	0.025	0.028	0.030	0.035	0.042

[a]Corrected values of $D_{n,\alpha}$ for the Kolmogorov-Smirnov goodness-of-fit test.
This is applicable when parameters are estimated from the same sample used in the test.
Source: Dallal, G. E., and L. Wilkinson (1986), "An analytic approximation to the distribution of Lilliefors's test statistic for Normality," *Am. Stat.*, Vol. 40, p. 294–296. (Reprinted with permission from the *American Statistician*.

Table C.9 Tests for multiple outliers: Critical values for 5 and 1% significance levels of k Studentized deviates B_j; samples of size n from a normal population

	$\alpha = 0.05$					$\alpha = 0.01$				
sample size	$k = 1$	$k = 2$	$k = 3$	$k = 4$	$k = 5$	$k = 1$	$k = 2$	$k = 3$	$k = 4$	$k = 5$
$n = 20$	2.56	2.83	2.88	2.95	2.97	2.88	3.09	3.13	3.20	3.18
		2.52	2.60	2.63	2.65		2.76	2.83	2.83	2.89
			2.45	2.49	2.51			2.68	2.68	2.69
				2.39	2.42				2.58	2.61
					2.37					2.57
$n = 30$	2.74	3.05	3.12	3.16	3.19	3.10	3.35	3.41	3.48	3.48
		2.67	2.73	2.77	2.78		2.92	3.01	3.02	3.03
			2.56	2.59	2.60			2.75	2.79	2.80
				2.49	2.51				2.70	2.74
					2.45					2.62
$n = 40$	2.87	3.17	3.22	3.32	3.31	3.24	3.52	3.58	3.64	3.63
		2.77	2.81	2.86	2.88		2.98	3.03	3.10	3.13
			2.62	2.67	2.69			2.82	2.87	2.89
				2.55	2.55				2.74	2.74
					2.47					2.65
$n = 50$	2.96	3.27	3.34	3.40	3.45	3.34	3.61	3.68	3.74	3.77
		2.85	2.89	2.93	2.96		3.08	3.15	3.18	3.21
			2.68	2.72	2.74			2.89	2.92	2.94
				2.59	2.61				2.78	2.79
					2.52					2.70
$n = 60$	3.03	3.34	3.42	3.48	3.51	3.41	3.70	3.75	3.82	3.81
		2.90	2.95	2.98	3.01		3.17	3.20	3.20	3.24
			2.73	2.77	2.77			2.95	2.97	2.96
				2.63	2.65				2.82	2.83
					2.56					2.72
$n = 80$	3.14	3.45	3.49	3.57	3.61	3.53	3.80	3.85	3.91	3.93
		2.97	3.03	3.05	3.11		3.23	3.27	3.31	3.36
			2.81	2.84	2.86			3.01	3.04	3.08
				2.69	2.72				2.87	2.89
					2.62					2.76
$n = 100$	3.21	2.52	3.60	3.64	3.70	3.27	3.87	3.97	3.96	4.01
		3.03	3.10	3.13	3.16		3.28	3.34	3.34	3.42
			2.86	2.89	2.91			3.06	3.06	3.10
				2.74	2.77				2.90	2.93
					2.67					2.84

Source: See references in Chapter 6. Adapted partly from Jain (1981). Reprinted with permission from *Technometrics*. and from Barnett and Lewis (1994).

Appendix D

Brief Answers to Selected Problems

CHAPTER 1

1.1 RDEN: 61.0, 17.39, 0.63; ACCEL: 0.327, 0.142, 0.70; $r = 0.28$. **1.2** 5408, 1735; $p = 0.59$. **1.3** 2837, 1301; 1980, 2410, 3245; larger v, g_1; 3 outliers. **1.5** frequency reversal in midwinter; 2.92, 0.23. **1.6** load: 918, 86, 58, −2.1; strength: 40, 4.2, 3.2, −1.5; $r = 0.85$. **1.7** all: 11.66, 7.55, 10.89, 9.92; no zero: 11.47, 7.36, 10.54, 9.46; 1.63, 2.48, 3.20, 4.63%. **1.8** 985,000. **1.10** 88.9. **1.12** large skew, 5 outliers. **1.15** 4.9, 10.7% $r = 0.027$. **1.16** trimmed 3%: 7.94, 0.14, 2.48; all: 9.92, 0.15, 4.46. **1.17** $r = 0.069$. **1.19** 1.2, 3.0%; $r = -0.15$. **1.20** increase with duration. **1.21** 1.65 per year. **1.24** x: 6.29, 0.291, 4.6%; y: 49.6, 0.166, 0.3%.

CHAPTER 2

2.3 $p = 1/6, 1/3, 1/3, 1/6$. **2.4** $\Pr[AB] = 0.2$, $\Pr[A^c + B^c] = 0.8$, $\Pr[A|B] = 1/3$, $\Pr[B|A] = 2/5$. **2.5** $\Pr[B] = 0.03$, $\Pr[BC] = 0.01$. **2.6** $p = 4/9$. **2.9** 13/14, 3/14, 0.5. **2.10** $n = 55$, $p = 0.70, 0.85$. **2.11** $p = 0.9$. **2.12** $p = 0.68256, 0.65152$. **2.13** (a) \$141,000, (d) rate a. **2.14** $p = 0.1875$. **2.15** (b) 2/9. **2.16** $p = 0.942$. **2.17** (a) $p_s = (1-q)^n\, e^{-\beta n(n+1)/2}$, ($b$) $q_n = (1-q)^{n-1}\, e^{-\beta n(n-1)/2}(1 - e^{-\beta n} + qe^{-\beta n})$, ($c$) $p_s = 0.4061$, $q_n = 0.02513$. **2.18** $p = 0.81, 0.75$, through Chambery. **2.19** $p = 0.402$. **2.20** $p = 0.8$, 2/329. **2.21** 0.75. **2.22** $p = 0.04, 0.50, 0.45, 0.01$. **2.23** $p = 0.055, 0.65$. **2.24** $p = 0.95, 0.3$. **2.25** (a) 54, (b) 19, (c) yes. **2.26** (b) 1/3, (c) 0.32 kg/m^2, (d) two subcatchments.

CHAPTER 3

3.1 1.88, 3.13. **3.2** 3/800, $p = 0.875$. **3.3** 0.00182, 0.138; 12.857, 0.298; 0.298. **3.4** 0.982. **3.5** $\geq$0.82. **3.7** 0.135, 7.81. **3.8** 9/275. **3.10** 1.28, 1.39. **3.11** 1.58, 1.40, 1.30; 1.61. **3.12** (a) 2.7, 5.3. (c) 0.140. **3.13** 81,200, 313,000,000. **3.15** 0.988. **3.16** $k = (4/297)10^{-10}$; $(2x/99)10^{-8}$, $10{,}000 < x < 100{,}000$; $y/150$, $10 < y < 20$. **3.17** 1/16. **3.18** 0. **3.19** (1) $10^{-9}(p - 15{,}000)/8.3655$, $28{,}000 \leq p \leq 145{,}000$. **3.20** ($a$) 0.91. **3.21** 2/9, 3/9; 4/9; 3/12, 4/12, 5/12; 11/9, 14/12. **3.24** 0.435, 1.4 −0.8y, −1/11. **3.25** $44a/(125b^4)$. **3.26** $\theta = \beta$; $x_0 = av^{1/\theta}$.

CHAPTER 4

4.1 0.74. **4.2** 975. **4.3** 46. **4.4** 0.366, 0.00904. **4.5** 0.0729. **4.6** 0.678. **4.8** 0.000019996. **4.9** 0.0000200, after 10 years 0.0000198. **4.10** 0.472. **4.11** between 1 and 2. **4.12** 5. **4.13** 5 days. **4.14** 0.471, 0.533. **4.15** 3. **4.16** 0.691. **4.17** $f_X(x) = x^3 e^{-x/3}/486$. **4.18** 5/12. **4.19** 0.999. **4.20** 0.15. **4.21** 7.9. **4.22** 29 min. **4.23** 0.844. **4.24** 0.378. **4.25** 0.0000, 0.0067. **4.26** $N(500, 80^2)$; 0.03. **4.27** $N(350, 150)$, 1, 0.08. **4.28** 0.866. **4.29** 0.988.

CHAPTER 5

5.1 (a) Bernoulli (0, 1), (c) 18/500, (d) 0.008331; (0.020, 0.052). **5.2** (2436, 2449), (2435, 2449). **5.3** 59; 84%. **5.4** 1910. **5.5** (102.3, 377.8). **5.6** 5 years. **5.7** reject H_0. **5.8** do not reject H_0. **5.9** do not reject H_0. **5.10** reject, reject, reject do not reject H_0.

5.11 5 and 11, 0.0455; 0.752; 0.642; to reduce (*a*) Type I error, take 3 standard errors and (*b*) Type II error, increase n. **5.12** 0.937, 0.628, 0.382, 0.175, 0.014. **5.13** $X^2 = 5.93$: do not reject H_0. **5.14** do not reject H_0. **5.15** do not reject H_0. **5.16** reject H_0. **5.17** reject H_0. **5.18** $r_s = -0.934$. **5.19** $X^2 = 7.81$: do not reject H_0, 0.2. **5.20** $\alpha = 0.01$, reject H_0; do not reject H_0. **5.21** reject H_0. **5.22** reject H_0. **5.23** $X^2 = 7.29$: do not reject H_0. **5.24** $X^2 = 14.87$: do not reject H_0. **5.25** do not reject H_0; do not reject H_0. **5.26** reject H_0. **5.27** do not reject H_0. **5.28** do not reject H_0. **5.29** $X^2 = 3.31$: do not reject H_0. **5.32** do not reject H_0. **5.33** do not reject H_0; do not reject H_0. **5.36** 0.50. **5.39** mean 3224, std. dev. 1042; 6225.

CHAPTER 6

6.1 0.394, −0.0656; 0.985. **6.2** 0.994; 240, 260. **6.4** (*a*) 0.137, 0.924; (*b*) 0.352; (*c*) 0.862; (*d*) 0.570, 0.961. **6.5** (*a*) −0.1212, 0.1062; (*b*) 0.095, 0.118; −0.633, 0.391; (*c*) reject H_0. **6.6** 475, 0.943; reject H_0; subtract evapotranspiration, include rainfall from previous years. **6.7** 2.188, 1.712; 16.74; 18.84, 14.64. **6.11** reject H_0: slope is not significant. **6.12** (*b*) 3.09, 9.92, 25.51; 80.67; 0.50. **6.15** eigenvalues: 98.4572752, 4.7230437, 0.5996811; vectors: 0.05262043, −0.99809260, 0.03228406; 0.82395523, 0.06165978, 0.56329020; 0.564206404, 0.003039957, −0.825628180. **6.20** 11.54.

CHAPTER 7

7.1 (*b*) 0.983, (*c*) 0.991. **7.2** 0.63 (rainfall), 0.72 (runoff). **7.3** $E[X_{\min}] = x_0 + 1/\lambda n$; $\mathrm{Var}[X_{\min}] = E^2[X_{\min}] - 2E[X_{\min}](x_0 + \frac{1}{\lambda}) + (x_0^2 + 2\frac{x_0}{\lambda} + \frac{1}{\lambda^2})$. **7.5** (*a*) $\alpha = 0.192$, $b = 0.889$; (*b*) L: $k = 0.052$; $\alpha = 0.201$, $\varepsilon = 0.894$; moments: $k = 0.033$; $\alpha = 0.199$, $\varepsilon = 0.891$; (*c*) $x_{\max}(100, t) = 23.01t^{0.46}$. **7.7** $F_{X_{\max}}(x) = \int_{\lambda_0}^{+\infty} \alpha\, e^{-\alpha(\lambda-\lambda_0)-\nu \exp[-(x/\lambda)^2/2]}\, d\lambda$. **7.8** $F_{Y_{\max}}(y) = \exp[-e^{-\lambda(y-\lambda^{-1}\ln \nu)}]$; $y_{\max}(100) = 23.8 \times 10^6\ \mathrm{m}^3$. **7.9** 6.545; 5.569. **7.10** (*a*) $\alpha = 1.46$, $b = 16.97$, $x_{\max}(50) = 22.68$ (Cagliari); (*b*) $\alpha = 3.18$, $b = 24.98$, $x_{\max}(50) = 37.38$ (Pantelleria). **7.11** (*a*) $\alpha = 2.36$, $b = 13.76$; (*b*) $\alpha = 2.22$, $b = 12.93$; (*c*) $x_{max,B}(50) = 19.47$. **7.12** (*a*) $X_{\max} \sim$ Gumbel (2.70, 18.42), $x_{\max}(50) = 28.96$; (*b*) $X_{\max} \sim$ Frechèt(18.69, 8.47), $x_{\max}(50) = 29.62$. **7.14** (*a*) $T = 7.89 \times 10^6$; (*b*) $p = 1.27 \times 10^{-7}$. **7.20** $F_{Z_{\max}}(z) = \exp(-[\frac{1}{z}(\frac{1230}{(l+25)^2})[\exp(\lambda x_0 + \ln \nu]^{0.8/\lambda}]^{\lambda/0.8})$. **7.21** (*a*) $W \sim \exp(0.32)$; (*b*) $a = 2.55 \times 10^4$, $b = 0.91$, $T = 92$. **7.22.** (*a*) $T = 69$; (*b*) $T = 97$.

CHAPTER 8

8.1 $N(0.32, 0.02)$; $N(7.91, 1.63)$. **8.4** 0.7388; 361. **8.5** Gumbel (19.06, 0.10). **8.6** Gumbel (7, 0.74). **8.9** 288, 0.22. **8.10** lognormal (14,200, 9915^2). **8.11** 0.340.

CHAPTER 9

9.4 0.370. **9.5** 98.9, 100. **9.6** $r = \Pr[X \geq Y] = 1 - \Pr[X < Y] = 1 - b/(2a)$. **9.7** 0.661. **9.8** (*a*) 210, 0.072; (*b*) 1.476, 0.07; (*c*) 1.660, 0.048. **9.9** 0.976. **9.10** 0.023. **9.11** 0.979. **9.12** 98.3%. **9.13** 0.867. **9.14** 1.451, 0.073. **9.15** 0.983, 0.982. **9.17** (*a*) $r = (1 - p^m)^n$; (*b*) $dr/dm = n \ln(1/p)(1 - p^m)^{n-1} p^m$. **9.18** 0.000102. **9.19** 0.9441. **9.20** 0.983. **9.21** 0.9916, 0.9952. **9.22** 0.9888, 0.9851. **9.23** (*a*) r to $r^{1/l}$; (*b*) $1 - (1 - r)^{1/l}$ to r. **9.24** 0.9044. **9.25** (*a*) $R(t) = \int_0^c \frac{1}{\mu\Gamma(t)}(\frac{x}{\mu})^{t-1}\exp(-\frac{x}{\mu})dx$; (*b*) $h(t) = \frac{1/\mu\Gamma(t)(c/\mu)^{t-1}\exp(-c/\mu)}{\int_0^c 1/\mu\Gamma(t)(x/\mu)^{t-1}\exp(-x/\mu)dx}$; (*c*) $t_{0.9} \approx 8.5$. **9.26** 3. **9.27** (*a*) $(\gamma/\tau)(t/\tau)^{\gamma-1}$; (*b*) 52.4. **9.28** (*a*) $R(t) = \exp\{-0.009t - \int_0^t \frac{0.05}{1+\exp[-0.25(u-25)]}du\}$; (*b*) ≈ 11. **9.30** 2.395 (beam 1), 2.867 (beam 2), 1.75 (beam 1).

CHAPTER 10

10.1 −382,500; −400,000; −467,000. **10.2** −0.5. **10.3** 45, 35; Bayes 20 m; minimax 20 m; 0.94, 0.06; 0.44, 0.56; 8.7, 21.9. **10.4** $d_3, d_3; d_3, d_1$ or d_4 or d_5. **10.5** X: 10,000,000; Y: 15,000,000; Z: 19,000,000. **10.6** d_1. **10.7** Design II, Design II. **10.8** $N(70{,}000, 15{,}000^2/3)$; 57,847. **10.9** mode, median. **10.10** reject, do not reject H_0. **10.11** 1.005, 0.98. **10.12** no change.

Appendix E

Data Lists

Table E.1.1 Appearance quality grade 5[a]

48.78	32.02	45.54	32.40	48.37	50.98	35.58	40.53	29.11	65.35
41.64	39.34	34.12	33.06	29.93	40.71	28.97	47.25	65.61	45.19
39.77	46.33	45.92	33.47	36.38	34.63	34.56	32.68	37.78	70.22
35.89	46.99	36.47	35.67	46.86	24.84	28.69	43.26	43.33	41.75
54.04	22.67	28.98	28.46	36.00	28.83	38.64	47.61	53.63	37.51
35.43	39.62	40.85	23.16	23.19	42.31	24.25	28.13	41.85	31.60
22.75	44.78	56.60	44.51	36.88	39.33	44.54	32.48	33.19	37.65
44.78	26.63	28.76	42.47	44.30	39.93	40.85	36.81	39.15	28.00
43.99	43.48	47.42	48.39	44.59	39.60	39.97	35.88	54.71	46.01
47.74	30.05	33.61	38.05	44.00	38.16	37.69	33.92	43.64	43.48
25.39	30.33	44.36	35.03	40.39	43.33	41.78	57.99	56.80	40.27
38.00	39.21	35.30	31.33	41.72	69.07	33.14	49.57	43.07	39.05
25.98	51.39	33.18	27.31	29.90	51.90	55.23	40.20	43.12	32.76
36.84	50.91	36.85	53.99	35.17	33.71	36.53	49.59	30.02	45.97
34.49	49.65	17.98	43.41	34.44	46.50	22.74	32.03	38.81	23.14
38.71	47.83	27.90	28.71	27.93	36.92	34.40	39.20	24.09	53.00
30.53	44.07	44.36	58.34	0.00					

[a]Modulus of rupture data from 50 mm × 150 mm Swedish redwood and whitewood timber in neutons per square millimeter.

Source: By kind courtesy of the Building Research Establishment, Timber Division, Garston; Watford, England.

Table E.1.2 Density and compressive strength at 28 days from examination of 40 concrete cube test records during the period 8 July 1991 to 21 September 1992, and arranged in reverse chronological order

Date	Density (kg/m^3)	Compressive strength (N/mm^2)
21 September 1992	2437	60.5
29 June 1992	2437	60.9
26 June 1992	2425	59.8
14 April 1992	2427	53.4
31 March 1992	2428	56.9
19 March 1992	2448	67.3
9 March 1992	2456	68.9
7 February 1992	2436	49.9
28 January 1992	2435	57.8
18 December 1991	2446	60.9
6 December 1991	2441	61.9
6 December 1991	2456	67.2
6 December 1991	2444	64.9
5 December 1991	2447	63.4
4 December 1991	2433	60.5
3 December 1991	2429	68.1
2 December 1991	2435	68.3
22 October 1991	2471	65.7
18 October 1991	2472	61.5
14 October 1991	2445	60.0
9 October 1991	2436	59.6
7 October 1991	2450	60.5
3 October 1991	2454	59.8
2 October 1991	2449	56.7
30 September 1991	2441	57.9
27 September 1991	2457	60.2
23 September 1991	2447	55.8
20 September 1991	2436	53.2
17 September 1991	2458	61.1
13 September 1991	2415	50.7
10 September 1991	2448	59.0
9 September 1991	2445	63.3
6 September 1991	2436	52.5
3 September 1991	2469	54.6
2 September 1991	2455	56.3
29 August 1991	2473	64.9
23 August 1991	2488	69.5
12 July 1991	2454	58.9
9 July 1991	2427	54.4
8 July 1991	2411	58.8

Source: By kind courtesy of Mr. L. K. Moore, Technical Manager, Douglas Concrete & Aggregates, Barton-on-Trent, England.

Table E.1.3 Dissolved oxygen (DO) and biochemical oxygen demand (BOD) at 38 stations along the Blackwater River in units of milligrams per liter

DO	8.15	5.45	6.05	6.49	6.11	6.46	6.22	6.05
BOD	2.27	4.41	4.03	3.75	3.37	3.23	3.18	4.08
DO	6.3	6.53	6.74	6.9	7.05	7.19	7.55	6.92
BOD	4	3.92	3.83	3.74	3.66	3.58	3.16	3.43
DO	7.11	7.28	7.44	7.6	7.28	7.44	7.59	7.73
BOD	3.36	3.3	3.24	3.19	3.22	3.17	3.13	3.08
DO	7.85	7.97	8.09	8.19	8.29	8.38	8.46	8.54
BOD	3.04	3	2.96	2.93	2.89	2.86	2.82	2.79
DO	8.62	8.69	8.76	9.26	9.31	9.35		
BOD	2.76	2.73	2.7	2.51	2.49	2.46		

Source: By kind courtesy of the Severn Trent Authority, Birmingham, England.

Table E.5.1 Road rutting measurements at Section 1, Site 6 in millimeters; base thickness 305 mm; base material, dense bituminous macadam

Date	Location 1	Location 2	Location 3	Location 4	Location 5
7 May 1960	0.01	2.9	0.9	0.01	1.5
14 November 1960	0.6	3.7	2	3	0.9
14 May 1961	0.2	3.7	1.8	3	1.1
28 August 1961	0.6	3.8	2.3	3.5	0.8
5 March 1962	1.8	4.1	1.8	3.2	0.6
19 September 1962	0.7	4.1	2.3	3.5	0.8
24 April 1963	0.7	4	1.9	2.9	0.8
10 October 1963	0.7	4.1	2.3	4.1	0.5
7 February 1964		4.1	2.1	3.4	0.6
3 October 1964	1.3	3.5	2.1	3.4	0.8
12 March 1965	1.8	4.3	2.3	3.4	0.8
26 September 1965	1.1	4	2	4	1.1
26 March 1966	1.1	3.6	1.9	5.3	0.01
21 May 1967	1.8	3.2	2.7	5.5	0.5
10 October 1969	3.1	3.8	3.8	4.1	0.01
3 October 1970	1.9	4.7	4	5	0.01
3 April 1971	1.7	5.3	2.4	4	0.01
10 October 1971	3.7	4.4	2.3	4.1	1.4
10 April 1972	1.5	3.5	3	5.5	1.3
3 October 1972	1.7	5.8	2.1	5.3	0.3
14 May 1973	1.8	5.2	3.2	4	0.01
31 July 1974	1.8	4.5	3.2	6.2	0.01
12 September 1975	3.7	4	4.6	5.3	0.01
17 October 1976	1.8	4	2.9	5.8	1.1
17 October 1977	2.4	4	3.7	5.8	0.3
17 October 1978	1.8	4.1	2.9	5.9	0.01
19 September 1979	2.9	5.2	2.5	5.3	0.01
14 May 1980	2.8	3.6	3.2	6	0.2

Note: The original data are in inches; measurements of 0.01 mm in this and in the following five tables denote zero rutting.

Source: Tables E.5.1 to E.5.6 are used by kind courtesy of the Transport Research Laboratory, Old Wokingham Road, Crowthorne, Berkshire, England.

Table E.5.2 Road rutting measurements at Section 2, Site 6 in millimeters; base thickness 229 mm; base material, dense bituminous macadam

Date	Location 1	Location 2	Location 3	Location 4	Location 5
7 May 1960	1.7	1.4	0.01	0.9	2.1
14 November 1960	0.01	4.9	4.6	1.5	2.3
14 May 1961	0.01	4.9	3	1.2	2
28 August 1961	0.01	5.3	4	1.7	2.3
5 March 1962	0.01	5.1	4	1.8	2.4
19 September 1962	0.01	5.8	4.1	1.5	2.4
24 April 1963	0.01	7.6	3	1.6	2.9
10 October 1963	0.01	5.3	4.5	2	2
7 February 1964	0.01	5.8	4	2	2.5
3 October 1964	0.01	5.6	3.7	1.4	2.6
12 March 1965	0.6	7.7	4.6	2.1	2.6
26 September 1965	0.5	6.1	4.3	2	2.7
26 March 1966	0.3	5.6	5.9	4.1	2
21 May 1967	0.4	5.8	4.6	1.7	2
10 October 1969	0.7	6.2	5.6	2.3	2.9
3 October 1970	1	6.4	5.6	4.4	1.4
3 April 1971	0.01	6.7	5.6	3.8	2.3
10 October 1971	1	7.2	4.9	2.7	3
10 April 1972	1.6	7	6.1	4.1	3.4
3 October 1972	0.9	8.5	5.6	2.4	2.4
14 May 1973	0.01	7.3	4.6	3.3	3
31 July 1974	0.01	8.5	4.8	3.7	2.7
12 September 1975	1.3	8.2	5.6	3.8	4
17 October 1976	1.2	7.6	5.9	1.5	2.8
17 October 1977	1.2	8.7	6.4	3.7	3.8
17 October 1978	0.9	6	5.2	3.8	3.4
19 September 1979	0.01	5.2	6.4	3.1	2.8
14 May 1980	0.5	5.8	5.2	4.1	3.2

Table E.5.3 Road rutting measurements at Section 3, Site 6, in millimeters; base thickness 152 mm; base material, dense bituminous macadam

Date	Location 1	Location 2	Location 3	Location 4	Location 5
7 May 1960	0.8	2.4	2.7	3	1.1
14 November 1960	2.9	2.7	1.8	3.8	1.7
14 May 1961	2.5	3	0.8	3.2	1.7
28 August 1961	2.9	2.9	2.1	2.9	1.7
5 March 1962	2.9	3	2	3.3	1.7
19 September 1962	2.6	3	2.3	2.9	1.5
24 April 1963	2.9	2.9	2.9	4.2	2.4
10 October 1963	3.4	3.4	2.3	3.5	2.3
7 February 1964	3.4	3.4	3.2	2.9	1.8
3 October 1964	3	3	2	3.5	2.1
12 March 1965	3.1	2.7	2.9	3.2	2.6
26 September 1965	3.8	3	1.8	4.4	2.3
26 March 1966	4.7	3.3	3.7	4	2.7
21 May 1967	3.5	5.6	2.4	5.4	2.4
10 October 1969	4.8	4.5	2.3	4.1	2.8
3 October 1970	5.7	4	2.4	4.8	4.5
3 April 1971	6.2	4.3	3.6	4.8	4.3
10 October 1971	3.4	4.4	3.4	5.5	3.1
10 April 1972	4.9	5.4	3.2	5.9	3
3 October 1972	6.1	5.1	3.2	3.3	3.8
14 May 1973	7.2	3.8	3.5	3.9	2.9
31 July 1974	7.3	4.5	3	4.1	3.7
12 September 1975	5.5	5.6	2.8	4.9	4
17 October 1976	5.2	4.5	2.9	4.4	3.5
17 October 1977	5.4	5.9	3.2	4.9	4.6
17 October 1978	5.8	5.6	3.8	4.3	4
19 September 1979	5.1	6	4	4.8	4.1
14 May 1980	5.9	6.6	3.9	5.6	4.1

Table E.5.4 Road rutting measurements at Section 4, Site 6 in millimeters; base thickness 152 mm; base material, hot-rolled asphalt

Date	Location 1	Location 2	Location 3	Location 4	Location 5
7 May 1960	2.9	0.01	0.01		1.8
14 November 1960	4	1.4	0.5	1.4	3.2
14 May 1961	3.6	1.4	0.01	2	3.4
28 August 1961	4.1	1.6	0.01	2	3.4
5 March 1962	3.9	1.3	0.8	2.3	3.7
19 September 1962	4.2	1.3	0.01	2.1	3.5
24 April 1963	4.1	1.8	0.6	1.5	3.7
10 October 1963	4.7	1.3	0.01	1.8	4.1
7 February 1964	4.8	1.5	1	2	5.2
3 October 1964	4.7	1.3	0.7	1.8	4.3
12 March 1965	3.7	1.3	0.01	2	3.8
26 September 1965	4.5	1.6	0.9	2.4	4.4
26 March 1966	4.8		1.5	2.6	4.9
21 May 1967	5.3	1.4	0.6	1.5	3.8
10 October 1969	5.4	1.8	0.3	2.6	4.1
3 October 1970	6.1	1.1	1.1	2.1	5.1
3 April 1971			1.4	2.7	3.7
10 October 1971	6.1	2.1	0.8	2.7	3.4
3 April 1972	6.3	1.4	1	2.1	6.2
3 October 1972	5.7	2.6	0.8	1.1	5.2
30 April 1973	6.5	1.6	1.4	2.3	
28 August 1975	6.2	2.8	1.4	1.3	5.5
3 October 1976	6.8	1.5		2.6	4.5
3 October 1977	6.9		1.4	2.5	5.4
17 October 1978	7.9	2.3	1.3	2.1	5.9
28 August 1979	6.9	3.7	1.7	2	5.3
30 April 1980	7.5	2.4	0.6	2.3	6.1

Table E.5.5 Road rutting measurements at Section 5, Site 6 in millimeters; base thickness 229 mm; base material, hot-rolled asphalt

Date	Location 1	Location 2	Location 3	Location 4	Location 5
7 May 1960	1.2		0.01	0.6	1.1
14 November 1960	2.3	4.1	1.5	2.9	
14 May 1961	2.6	4.1	1.2	3.2	2.6
28 August 1961	2.9	4.4	1.2	3	2.9
5 March1962	3.2	4.3	1.1	2.9	3
19 September 1962	2.9	4	1.4	3.5	3.3
24 April 1963	3.1	4.7	1.5	3.1	3.2
10 October 1963	3.2	4.7	0.8	3.5	3.1
7 February 1964	3.2	5.3	1.5	3.2	3.3
3 October 1964	3	4.7	1.3	3.4	2.9
12 March 1965	2.5	4.9	2	3.4	4.6
26 September 1965	3	4	1.5	3.6	3.3
26 March 1966	3.3	4.7	3	3.8	4.6
21 May 1967	4.4	4.9	1.8	4	3.7
10 October 1969	3.2	4.1	2	4.9	3.5
3 October 1970	3.8	3.7	3.7	3.4	3
3 April 1971	3.2	2.9	3.8	4	3.6
10 October 1971	3	5		3.5	3.8
3 April 1972	3.4	4.7	1.6	3.9	3.4
3 October 1972		3.7	2.9	4.2	4.3
30 April 1973	3	3.5	1.8	5.1	3.9
28 August 1975	3	5	2.4	3.2	
3 October 1976	4.4	3.7	3.2	4.8	4.8
3 October 1978	3.5	5.8	1.4	3.7	5.5
28 August 1979	2.8	3.2	1.4	3.7	4.1
30 April 1980	4.3	4.1	1.2	4.1	4.1

Table E.5.6 Road rutting measurements at Section 6, Site 6 in millimeters; base thickness 305 mm; base material, hot-rolled asphalt

Date	Location 1	Location 2	Location 3	Location 4	Location 5
7 May 1960		0.8	4.4	2.7	0.01
14 November 1960	3.5	2.8	3.9	4.2	2.9
14 May 1961	3.7	3.2	6.2	3.9	3
28 August 1961	3.6	3.4	6.4	4.2	3
5 March 1962	3.2	3.5	6.1	4	3.5
19 September 1962	4.2	3.3	6.4	4	3.2
24 April 1963	4.5	4.6	6.4	4.2	3.9
10 October 1963	4.5	4.5	6.8	4.4	3.8
7 February 1964	4.6	4.5	6.9	4.5	3.6
3 October 1964	5	3.8	6.2	4.4	3.6
12 March 1965	4.4	4.8	6.4	5.6	4.3
26 September 1965	5.2	4.9	6.5	4.5	4.2
26 March 1966	4.3	4.7		6.6	
21 May 1967	5.3	5.1		4.6	4.2
10 October 1969	5.1	5.1	6.2	4.8	5.5
3 October 1970	4.1	5.9	7.9	5.6	3.9
3 April 1971	6.2	7.9			3.8
10 October 1971	3.8	6.8		5.9	3.6
3 April 1972	4.6	5.4	7.6	5.2	4
3 October 1972	4.2	4.2	7.5	5	
30 April 1973	5.2	6.2	7.8	5	6.3
28 August 1975	3.8	6.1	6.8	5.6	6.3
3 October 1976	4.6	7.3	6.5	5.9	5.9
3 October 1978	4	7.4	8.5	6.5	5.9
19 September 1979	4.6	6.6	9.6	6.2	6.2
30 April 1980	4.6	7.7	9	5.7	6.2

Table E.5.7 Annual maximum flows in cubic feet per second from North America with suspected outliers

1. N. Fork Sun River near Augusta, Montana
06078500; 1911; 25; (1911 2 1946 23); 1968

2,390	2,280	2,190	3,520	4,840	2,720	3,340	3,170	2,140	3,990
4,580	2,900	4,170	3,330	2,830	3,960	2,660	3,400	2,830	1,680
51,100	3,370	2,810	3,370	2,600					

2. Two Medicine River near Browning, Montana
06092000; 1907; 42; (1907 1 1909 4 1914 11 1951 26); 1976

7,950	7,600	1,790	4,140	2,900	1,640	1,840	4,300	5,160	5,160
2,030	2,150	2,320	3,180	2,340	2,890	4,620	2,280	6,520	5,860
2,590	5,020	3,960	2,720	3,940	2,100	3,350	1,790	1,540	100,000
4,000	1,970	5,920	2,700	3,500	8,310	3,880	4,020	1,910	2720
74,500	3,390								

3. Twin Creek near Germantown, Ohio
03272000; 1913; 60; (1913 1 1915 9 1927 50); 1976

66,000	9,390	7,880	6,950	6,510	7,640	8,480	5,630	6,070	4,270
5,400	6,410	7,640	6,800	2,470	5,520	7,350	3,370	4,370	4,790
7,890	6,150	5,850	5,250	1,720	4,460	7,200	5,370	6,360	5,250
7,040	5,870	7,520	6,740	6,690	8,790	2,610	1,420	3,970	7,340
7,860	7,010	8,590	2,970	6,300	5,790	8,400	7,420	4,900	4,430
5,980	7,760	6,010	6,280	5,260	4,210	5,300	5,800	7,460	5,140

4. Waterton River near international boundary
05011500; 1948; 17; ; 1964

2,510	1,890	2,560	2,190	1,520	2,520	2,710	2,260	2,360	1,870
1,850	2,160	2,020	2,520	1,440	1,840	12,400			

5. Belly River near Mountain View, Alberta
05011000; 1912; 65; ; 1976

1,030	2,070	1,370	1,372	2,729	3,100	1,643	1,934	1,925	1,730
2,040	2,230	1,500	2,030	644	2,670	2,040	1,400	1,360	1,300
1,800	1,920	2,900	1,640	1,300	3,960	2,070	1,008	1,080	936
2,510	1,910	1,050	1,680	1,540	1,420	3,220	1,330	2,210	3,020
1,020	4,500	2,470	2,140	2,180	2,000	1,900	1,960	1,550	1,950
1,130	1,810	16,400	2,540	1,890	2,330	1,550	2,450	3,600	1,910
2,710	1,590	2,510	14,700	1,730					

6. Little River at Buffumville, Massachussets
01124500; 1940; 38; 1977

516	168	502	264	337	331	334	255	448	181
208	520	356	518	1220	8340	518	280	396	429
382	259	278	255	224	137	145	294	340	288
268	182	321	330	360	284	330	294		

7. Belly River at international boundary
05010000; 1948; 17; ; 1964

1,810	972	1,770	1,570	751	2,450	1,720	1,540	1,740	1,320
1,240	1,570	1,270	1,630	790	1,320	12,000			

(*continued*)

Table E.5.7 *(continued)*

8. Yakdin River at Wilkesboro, North Carolina 02112000; 1904; 63; (1904 6 1916 1 1921 56); 1976									
9,250	16,000	26,800	24,200	14,700	22,600	11,6000	10,700	11,300	13,600
13,700	12,300	12,000	4,020	22,000	10,300	29,000	6,030	7,500	19,200
10,600	11,000	10,800	17,100	19,300	11,600	160,000	10,200	13,200	9,530
11,800	23,200	9,350	14,200	9,250	12,800	5,770	12,300	12,800	11,300
11,800	10,200	7,650	15,500	6,640	15,200	11,600	8,600	6,800	6,100
5,570	8,120	9,300	5,180	5,800	7,800	12,700	4,620	9,080	8,570
9,740	7,660	10,400							

9. Great Miami River at Taylorsville, Ohio 03263000; 1913; 60; (1913 5 1922 55); 1976									
12,7000	11,400	11,400	26,400	14,700	17,300	10,600	21,600	5,610	13,800
21,600	14,000	23,700	20,600	4,750	9,430	25,500	5,500	11,400	15,600
25,500	16,600	13,600	14,600	4,980	12,800	21,900	17,700	17,200	8,410
20,200	16,200	17,200	20,300	18,300	21,500	9,210	7,520	9,000	14,000
17,200	21,400	31,400	6,310	16,100	12,400	24,300	18,800	11,500	8,570
12,200	10,800	12,700	13,900	13,100	11,600	14,100	15,700	22,600	11,400

10. Quinebaug River at Quinebaug, Connecticut 01124000; 1932; 45; (1932 42 1975 3); 1977									
1,260	1,800	2,310	2,140	10,500	2,280	19,000	1,700	2,550	1,210
2,110	1,340	1,460	1,560	1,490	1,180	2,370	1,020	1,020	2,940
1,990	2,240	5,990	49,300	2,830	1,410	1,970	3,160	2,170	1,490
1,730	1,570	1,350	985	838	1,480	2,760	2,100	2,500	907
2,070	2,440	1,850	2,620	2,170					

11. North Fork New River at Crumpler, North Carolina 03162500; 1878; 43; (1878 1 1901 1 1909 8 1928 32 1966 1); 1966									
44,300	23,500	6,660	2,700	4,190	4,320	7,960	2,590	5,290	39,000
5,970	6,250	6,670	2,980	4,570	3,670	4,220	7,500	5,910	3,450
3,900	3,670	73,000	3,450	3,850	7,780	6,620	11,400	8,830	7,290
6,380	11,300	4,870	17,100	4,380	5,730	7,920	7,900	9,500	10,000
3,930	7,950	10,600							

Key to line beneath station listings: station index; starting year; number of years of data; (excluding missing years: starting year and number of years); last year.

Source: By kind courtesy of U.S. Geological Survey.

Table E.5.8 Annual maximum flow of Tevere (Tiber) River observed at Ripetta, a guaging station in Rome, central Italy, from 1921 to 1974

Year	x (m^3/s)	Year	x (m^3/s)	Year	x (m^3/s)	Year	x (m^3/s)	Year	x (m^3/s)	Year	x (m^3/s)
1921	1092	1930	775	1939	985	1948	1600	1957	612	1966	1325
1922	1099	1931	1166	1940	1346	1949	714	1958	822	1967	528
1923	1440	1932	843	1941	1553	1950	794	1959	1370	1968	622
1924	1083	1933	1508	1942	1370	1951	1460	1960	1380	1969	355
1925	1621	1934	1876	1943	743	1952	1240	1961	510	1970	468
1926	1132	1935	1696	1944	1340	1953	1230	1962	810	1971	472
1927	935	1936	1690	1945	896	1954	1270	1963	735	1972	664
1928	1540	1937	2730	1946	1600	1955	861	1964	259	1973	717
1929	1966	1938	1440	1947	2190	1956	1355	1965	1290	1974	950

Table E.6.1 Some characteristics of stream basins on the left bank of the Po basin in northern Italy

Index	Stream basin outlet	Annual runoff (mm)	Annual rainfall (mm)	Mean elevation (m)	Longest flow path (km)	Area of basin (sq km)
1	Pian di Nambron	1654	1350	2329	5.59	20.42
2	Saone	1374	1621	1593	35.38	506.19
3	Nago	910	1263	1479	73.82	937.63
4	Capo di Ponte	1189	1293	1857	50.42	781.11
5	Ponte Cene	1453	1666	1335	43.12	426.86
6	Ponte Briolo	1278	1593	1140	54.03	763.11
7	Tirano	818	932	2136	55.86	616.83
8	Fuentes	1047	1121	1844	138.60	2323.09
9	Colombaio	589	1398	144	12.98	38.90
10	Ponte Gurone	769	1615	472	13.69	84.41
11	Santino	1730	2113	1230	17.43	62.80
12	Caderese	1571	1457	2146	25.93	185.45
13	Candoglia	1382	1519	1641	74.66	1461.86
14	Ponte Folle	1600	1936	1350	23.56	149.41
15	Campertognio	1295	1427	2120	22.12	171.37
16	Ponte Aranco	1428	1735	1480	49.39	697.75
17	Passobreve	1461	1803	1495	13.63	73.78
18	D'Ejola	1733	1280	3112	3.08	28.51
19	Gressoney St. Jean	1357	1191	2615	12.17	89.79
20	Saint Oyen	1023	1283	2206	10.44	68.85

Table E.6.2 Biochemical oxygen demand (BOD), nitrates (NO_3-N) and ammonia (NH_3-N) at 38 stations along the Blackwater river in units of milligrams per liter

Station	BOD	NO_3-N	NH_3-N
1	2.27	1.97	0.11
2	4.41	12.83	0.61
3	4.03	11.11	0.53
4	3.75	9.86	0.47
5	3.37	9.54	0.62
6	3.23	8.85	0.56
7	3.18	8.02	0.64
8	4.08	8.94	1.14
9	4	8.76	1.11
10	3.92	8.59	1.07
11	3.83	8.43	1.04
12	3.74	8.27	1
13	3.66	8.13	0.97
14	3.58	7.99	0.94
15	3.16	6.72	0.83
16	3.43	9.23	0.94
17	3.36	9.1	0.93
18	3.3	8.97	0.91
19	3.24	8.85	0.89
20	3.19	8.74	0.88
21	3.22	9.8	0.95
22	3.17	9.64	0.93
23	3.13	9.49	0.9
24	3.08	9.34	0.88
25	3.04	9.2	0.86
26	3	9.06	0.84
27	2.96	8.03	0.82
28	2.93	8.81	0.8
29	2.89	8.69	0.78
30	2.86	8.57	0.76
31	2.82	8.45	0.74
32	2.79	8.35	0.73
33	2.76	8.24	0.71
34	2.73	8.14	0.7
35	2.7	8.04	0.68
36	2.51	6.54	0.48
37	2.49	6.51	0.47
38	2.46	6.46	0.46
Mean	3.218	8.533	0.781
Standard deviation	0.496	1.580	0.213
Coefficient of variation	15.6%	18.5%	27.3%

Source: By kind courtesy of the Severn Trent Authority, Birmingham, England.

Table E.7.1 Annual maximum hourly storm depth in millimeters at Genoa University, Italy, from 1931 to 1988

Year	x (mm)	Year	x (mm)	Year	x (mm)	Year	x (mm)	Year	x (mm)	Year	x (mm)
1931	38.6	1941	40.2	1951	76.2	1961	66.5	1971	50.4	1981	89.4
1932	33.7	1942	53.8	1952	27.4	1962	24.5	1972	43.2	1982	27.2
1933	33.8	1943	26.9	1953	69.4	1963	64.1	1973	39.6	1983	32.7
1934	79.2	1944	34.7	1954	22.8	1964	53.9	1974	38.7	1984	105.7
1935	58.6	1945	72.6	1955	34.8	1965	66.5	1975	40.2	1985	25.3
1936	39.3	1946	30.2	1956	38.8	1966	32.9	1976	55.7	1986	27.6
1937	33.2	1947	42.7	1957	39.8	1967	52.4	1977	118.9	1987	128.5
1938	29.2	1948	54.5	1958	29.3	1968	27.8	1978	25.0	1988	24.7
1939	46.7	1949	30.0	1959	58.1	1969	23.3	1979	55.6		
1940	80.0	1950	30.0	1960	48.5	1970	80.0	1980	40.1		

Table E.7.2 Hydrologic data of the Po basin at Pontelagoscuro, northern Italy, from 1918 to 1978

Year	Annual maximum flow (m^3/s)	Annual minimum flow (m^3/s)	Annual minimum 7-day flow (m^3/s)	Mean annual flow (m^3/s)	Rainfall (mm)	Runoff (mm)
1918	5390	910	944	2010	1133	904
1919	4240	826	856	1440	999	648
1920	7220	969	1005	2400	1501	1080
1921	3000	569	590	1220	807	549
1922	2590	380	394	1070	1051	481
1923	2980	563	580	1280	969	576
1924	3920	749	813	1400	997	630
1925	3460	696	713	1530	1090	688
1926	8850	740	757	2040	1356	918
1927	3760	825	861	1630	1133	733
1928	8600	538	563	1800	1171	810
1929	2220	428	463	1090	876	490
1930	5400	607	618	1660	1159	747
1931	3700	508	527	1180	993	531
1932	4150	555	571	1420	1112	639
1933	4690	437	463	1310	1128	589
1934	6810	855	875	2050	1345	922
1935	6620	529	565	1750	1290	787
1936	6620	787	797	2310	1259	1039
1937	7700	668	675	2130	1529	958
1938	4380	287	305	1150	940	517
1939	3900	745	800	1780	1196	801
1940	5420	424	447	1350	1046	607
1941	6870	720	749	1860	1218	837
1942	4600	366	383	1160	948	522
1943	3270	310	311	987	896	444
1944	3660	306	307	905	950	407
1945	6830	304	306	916	846	412
1946	5130	655	669	1510	1011	679
1947	5460	588	594	1300	1096	585
1948	6630	711	735	1610	1100	724
1949	7220	275	278	967	922	435
1950	3260	400	411	1020	978	459
1951	8940	830	837	2200	1496	990
1952	4200	450	510	1110	913	499
1953	7400	520	550	1370	1046	616
1954	4450	440	456	1500	1100	675
1955	2400	423	441	1060	886	477
1956	5090	426	445	1220	1028	549
1957	6990	540	544	1390	1215	625
1958	5680	425	466	1320	1142	594
1959	7730	470	524	1900	1422	855
1960	6510	939	981	2620	1654	1179
1961	4880	424	450	1330	987	598
1962	4540	354	361	1070	909	481
1963	6430	732	737	1980	1362	891
1964	5630	444	461	1370	1026	616
1965	6110	321	363	1300	1015	585

Table E.7.2 *(continued)*

Year	Annual maximum flow (m^3/s)	Annual minimum flow (m^3/s)	Annual minimum 7-day flow (m^3/s)	Mean annual flow (m^3/s)	Rainfall (mm)	Runoff (mm)
1966	7240	359	425	1570	1228	706
1967	2470	414	454	1120	885	504
1968	7830	653	711	1650	1264	742
1969	6080	577	593	1410	995	634
1970	3170	376	394	1070	986	481
1971	5270	433	469	1380	1017	621
1972	5940	664	688	1900	1349	855
1973	4030	474	486	1270	1029	571
1974	5590	421	442	1422	959	640
1975	5360	488	508	1730	1323	778
1976	8030	323	340	1750	1318	787
1977	7800	1030	1103	2630	1564	1183
1978	5540	650	668	1920	1197	864

Table E.7.3 Earthquake catalog for California north of latitude 32° and south of latitude 36°

Date and time (day/month/year hour minute)	Magnitude	Zone type[a]	Date and time (day/month/year hour minute)	Magnitude	Zone type[a]
29/11/1852 20.00	6.5	C	08/06/1934 4.47	6.1	B
11/7/1855 4.15	6.0	B	25/02/1937 16.49	6.0	A
9/1/1857 16.00	7.8	B	19/05/1940 4.36	6.9	B
16/12/1858 10.00	6.0	C	21/10/1942 16.22	6.6	A
27/5/1862 20.00	6.0	B	15/03/1946 13.49	6.0	C
15/11/1875 22.30	6.2	B	10/04/1947 15.58	6.6	C
5/9/1883 12.30	6.2	C	04/12/1948 23.43	6.0	A
9/2/1890 12.06	6.5	A	21/07/1952 11.52	7.5	B
24/2/1892 7.20	7.0	B	21/07/1952 12.05	6.4	B
28/5/1892 11.15	6.5	A	29/07/1952 7.03	6.3	C
30/7/1894 5.12	6.0	A	22/11/1952 7.46	6.0	C
25/12/1899 12.25	6.4	A	19/03/1954 9.54	6.4	A
03/03/1901 7.45	6.4	B	28/06/1966 4.26	6.0	B
19/04/1906 0.30	6.2	B	09/04/1968 2.28	6.5	A
04/11/1908 8.37	6.0	C	09/02/1971 14.00	6.7	B
23/06/1915 3.59	6.0	B	15/10/1979 23.16	6.4	B
10/11/1916 9.11	6.1	C	02/05/1983 23.42	6.4	C
21/04/1918 22.32	6.8	A	08/07/1986 9.20	6.2	A
10/03/1922 11.21	6.1	B	24/11/1987 13.16	6.5	A
23/07/1923 7.30	6.0	A	23/04/1992 4.50	6.1	C
29/06/1925 14.42	6.9	C	28/06/1992 11.57	7.3	B
04/11/1927 13.50	7.3	B	28/06/1992 15.05	6.2	C
11/03/1933 1.54	6.2	B	17/01/1994 11.18	6.7	B

[a]Type A zones contain faults for which paleoseismic data suffice to estimate conditional probabilities. Type B zones contain faults with insufficient data for conditional probability analysis. Type C zones contain diverse or hidden faults.

Source: Data from Working Group on California Earthquake Probabilities (1995).

Table E.7.4 Annual maximum 10-minute wind velocity in meters per second at three stations in Italy, from 1951 to 1973

	Station		
Year	Milan	Cagliari	Pantelleria
1951	13.89	20.06	26.24
1952	10.29	20.58	27.78
1953	16.46	15.43	30.87
1954	11.83	16.98	26.75
1955	11.83	15.43	27.78
1956	10.29	18.52	36.01
1957	9.77	15.43	29.84
1958	11.32	15.95	30.87
1959	18.01	19.03	25.72
1960	14.40	19.03	26.75
1961	19.55	18.52	33.44
1962	14.92	16.98	31.90
1963	10.29	14.40	25.21
1964	13.89	18.01	25.72
1965	11.32	19.03	28.29
1966	13.89	20.58	26.75
1967	13.38	19.55	21.61
1968	13.89	19.55	22.12
1969	12.35	16.46	20.58
1970	12.86	18.52	20.58
1971	15.43	18.52	21.61
1972	13.38	15.43	24.69
1973	20.06	—	25.72

Table E.10.1*a* Monthly mean temperatures at Chateaux-D'oex, Switzerland: Part 1

1901	3.8	8.0	1.3	5.7	9.9	13.2	14.8	13.1	11.3	5.4	2.0	3.0
1902	2.3	2.6	1.2	7.4	5.1	11.7	14.8	13.1	10.4	5.0	0.1	3.4
1903	2.8	0.2	2.3	1.6	9.4	11.2	13.2	13.4	11.0	6.6	0.6	3.8
1904	4.2	1.9	0.9	6.8	11.1	13.4	16.6	14.6	8.9	5.7	0.3	2.0
1905	6.0	3.5	1.8	5.7	8.2	13.1	16.5	13.3	11.0	1.3	0.2	2.9
1906	2.5	4.0	0.5	4.4	9.1	12.4	14.7	14.8	10.1	7.9	2.3	6.7
1907	4.4	6.0	0.5	3.7	9.8	12.4	12.7	14.3	11.4	6.8	2.4	1.2
1908	3.8	2.5	0.5	3.2	11.5	14.3	14.1	12.4	10.3	6.4	0.2	3.0
1909	5.4	5.6	1.3	6.7	9.0	10.8	12.1	12.9	9.4	7.0	1.3	1.0
1910	2.1	1.8	1.7	4.5	8.1	13.0	12.5	13.2	9.0	7.4	0.4	0.2
1911	6.4	2.2	1.6	3.9	9.8	12.2	17.0	16.5	12.7	6.4	3.0	0.2
1912	0.9	1.7	3.2	3.8	10.2	12.6	13.4	10.7	6.0	4.5	2.7	1.6
1913	0.4	2.3	3.1	4.8	9.2	12.3	11.7	12.8	9.9	7.4	3.2	2.7
1914	8.3	0.3	1.2	6.9	8.1	11.6	12.6	14.1	9.7	5.6	0.1	0.6
1915	4.0	3.1	0.0	3.6	11.5	14.1	13.4	12.4	9.1	4.0	2.0	2.0
1916	1.3	1.0	1.1	5.0	10.5	10.8	13.6	13.0	8.8	5.7	0.8	1.3
1917	6.4	4.4	1.2	1.8	12.1	13.8	13.8	12.5	12.5	3.5	0.0	6.2
1918	1.6	0.6	0.8	4.9	10.7	10.4	14.2	13.1	11.0	4.2	0.9	0.1
1919	4.0	2.6	0.9	2.7	9.2	12.4	11.4	14.5	11.8	2.5	1.4	1.6
1920	0.7	0.0	3.2	5.8	12.6	12.2	13.9	11.7	10.4	6.2	1.5	2.8
1921	0.0	2.7	2.4	4.1	10.4	13.1	16.8	13.7	12.3	8.2	0.6	1.5
1922	3.9	2.5	1.6	2.8	11.1	13.5	12.9	13.5	8.4	4.4	0.7	3.8
1923	4.7	0.4	1.4	5.1	9.3	9.9	16.4	14.8	10.6	8.3	1.3	2.1
1924	3.5	4.7	0.8	5.1	10.7	12.8	14.3	10.5	10.5	6.0	1.0	0.9
1925	0.7	0.2	0.8	5.0	9.5	13.6	13.8	13.0	8.6	6.9	0.2	2.2
1926	2.9	2.8	2.8	5.8	7.4	10.1	13.4	13.5	12.4	6.6	3.6	4.4
1927	2.5	2.5	1.9	5.1	10.2	12.4	14.2	12.5	10.1	5.8	1.5	1.8
1928	1.6	1.0	2.5	5.2	7.3	12.9	17.2	15.5	10.7	5.6	2.6	4.3
1929	7.3	7.4	0.9	2.1	9.6	13.6	15.4	13.5	13.0	6.7	1.6	0.1
1930	0.5	2.9	2.4	5.3	8.2	14.7	13.0	13.4	11.4	5.6	3.9	3.0
1931	3.2	4.0	0.1	4.4	11.5	15.6	13.8	12.3	7.0	6.2	3.1	3.8
1932	1.5	6.2	0.1	3.1	8.4	11.8	13.7	16.3	13.2	6.0	2.2	0.2
1933	5.1	1.7	2.4	6.0	8.3	10.2	15.4	15.3	12.0	6.8	0.6	7.0
1934	3.3	3.1	0.9	7.2	11.5	13.0	15.5	12.6	12.1	5.7	0.3	1.9
1935	6.3	1.1	0.1	3.9	8.1	14.7	15.5	13.0	11.7	6.3	2.5	3.0
1936	1.4	1.1	3.3	4.7	10.1	12.0	13.6	13.6	10.7	2.3	1.6	1.2
1937	0.4	0.6	0.3	4.4	10.7	13.6	15.0	13.9	10.3	7.1	1.1	4.4
1938	2.8	3.9	3.2	2.5	8.1	14.1	14.0	13.7	11.4	6.5	3.7	3.1
1939	2.5	1.2	2.1	6.0	7.0	12.6	13.3	13.5	10.0	5.3	3.1	3.2
1940	7.0	1.1	1.7	5.4	9.1	12.2	13.2	12.7	10.7	5.8	2.5	7.9
1941	4.7	2.0	1.8	4.0	6.4	13.5	16.0	12.9	10.6	5.1	0.6	4.3
1942	7.0	6.7	3.2	5.7	9.7	13.7	14.8	14.5	12.9	8.9	1.3	0.9
1943	1.9	1.2	3.5	6.1	10.8	12.1	14.9	15.5	12.0	8.1	0.1	1.8
1944	1.6	5.0	1.4	6.7	9.9	12.3	14.4	16.7	10.4	4.8	0.8	3.6
1945	8.2	0.8	2.5	7.2	11.1	14.7	16.4	13.7	11.4	6.3	0.0	1.9
1946	4.0	0.2	2.7	8.3	11.2	12.1	15.7	13.7	11.8	5.9	1.4	4.3
1947	5.8	2.5	3.0	9.0	11.9	15.0	16.5	16.5	13.1	7.2	4.1	2.3
1948	0.2	0.6	5.5	6.2	10.7	11.8	12.3	13.7	10.8	6.9	2.8	2.2
1949	2.1	1.6	0.6	8.1	8.9	13.6	16.3	15.4	14.4	8.4	0.8	0.3
1950	3.2	0.4	2.5	3.9	11.3	15.0	17.4	14.6	10.7	6.2	2.1	4.4
1951	1.5	1.2	0.1	5.2	9.3	12.8	15.0	13.8	12.2	6.5	3.0	0.5
1952	4.2	3.9	2.7	7.3	10.9	14.7	17.3	14.5	8.2	5.8	0.3	2.5

Table E.10.1*b* Monthly mean temperatures at Chateaux-D'oex, Switzerland: Part 2

1953	7.1	4.8	1.6	6.3	10.9	11.7	15.0	14.2	12.0	8.1	1.7	0.7
1954	5.6	3.1	2.6	3.8	8.5	13.3	12.8	12.5	11.5	6.7	2.5	1.0
1955	0.2	2.2	0.3	5.8	8.9	13.0	14.4	13.2	10.2	4.9	0.0	0.8
1956	1.4	11.2	1.7	3.5	10.1	10.6	14.2	12.5	12.2	4.5	0.9	2.5
1957	4.0	1.1	5.3	5.7	7.1	13.5	14.6	13.3	10.2	6.2	1.9	3.2
1958	3.3	0.4	1.3	2.8	11.7	12.2	14.7	14.7	13.0	5.4	1.3	0.0
1959	2.6	0.4	4.4	6.0	10.3	13.7	16.3	13.8	12.8	6.0	0.8	0.7
1960	3.4	0.2	3.3	5.4	11.2	13.9	12.9	13.0	9.4	5.6	2.9	2.6
1961	2.4	1.4	3.1	7.8	8.5	13.7	13.6	13.6	14.1	7.2	1.5	0.6
1962	1.2	2.8	1.0	4.8	8.0	12.4	14.2	15.6	11.0	6.6	0.1	5.5
1963	8.1	6.6	0.2	6.1	8.7	12.5	15.5	12.8	11.2	6.7	4.1	3.4
1964	4.7	0.2	1.5	6.6	11.5	14.1	15.8	13.6	11.6	5.0	2.4	3.3
1965	2.3	6.5	0.8	3.8	9.1	13.3	13.3	12.9	9.2	7.2	2.0	0.2
1966	4.8	2.8	0.4	6.7	10.1	13.7	13.0	12.8	12.6	8.6	0.5	2.0
1967	3.8	0.6	2.9	4.0	9.3	11.9	16.3	14.3	10.5	8.6	2.6	4.4
1968	3.8	0.7	1.2	6.5	9.0	12.4	14.2	12.2	10.3	9.0	2.0	2.7
1969	2.4	4.4	1.2	4.5	10.6	10.8	15.1	13.3	11.6	7.6	1.4	7.6
1970	2.0	1.8	1.5	2.2	8.1	14.1	14.6	14.2	12.3	5.9	3.3	3.3
1971	3.3	1.7	2.5	8.0	10.8	11.7	16.3	15.9	11.0	7.8	0.2	1.5
1972	2.2	0.7	3.3	4.8	8.7	11.8	14.4	13.4	8.5	5.4	2.5	1.9
1973	3.0	4.1	0.0	2.3	10.8	13.5	14.3	16.3	12.4	5.5	1.9	2.1
1974	0.1	0.4	3.9	5.4	8.8	11.6	14.2	15.4	10.7	1.6	0.9	0.4
1975	0.9	0.0	0.8	5.2	9.6	11.4	15.1	15.1	13.0	5.5	1.1	3.1
1976	2.0	0.2	1.4	5.1	10.5	15.1	15.3	13.2	9.9	7.9	1.3	4.1
1977	2.1	0.8	4.3	4.1	8.7	12.2	14.5	13.5	10.8	8.6	0.8	1.3
1978	2.7	1.5	2.5	4.1	8.5	12.0	14.0	13.2	11.3	6.3	1.3	0.4
1979	4.1	0.2	2.5	3.7	9.5	13.7	14.7	13.1	11.5	8.6	1.0	0.1
1980	4.0	0.7	2.1	2.9	8.2	11.2	13.0	15.2	12.5	6.0	1.1	3.8
1981	4.0	4.0	3.9	7.2	9.2	13.1	13.3	15.1	12.0	6.6	2.0	1.6
1982	0.2	0.5	0.8	5.1	10.3	14.0	16.8	14.3	13.5	7.0	3.6	0.1
1983	0.0	4.5	2.9	6.0	7.9	14.1	19.4	15.3	12.4	7.3	1.8	0.8
1984	1.5	3.7	0.7	4.5	6.9	12.7	15.4	14.5	10.2	7.8	4.6	0.3
1985	7.2	2.0	0.3	5.5	9.6	12.1	16.6	14.7	13.3	7.9	1.3	0.5
1986	1.9	5.6	2.2	3.8	12.1	13.5	15.0	14.9	12.2	8.6	2.6	1.1
1987	6.8	1.2	1.0	6.9	7.4	12.0	15.8	15.5	14.7	9.3	2.3	0.5
1988	1.2	1.2	0.4	6.9	11.3	12.8	15.4	15.4	11.6	9.1	0.7	1.0
1989	1.8	1.0	5.0	4.9	11.4	12.9	16.3	15.3	11.8	7.6	1.5	0.4
1990	0.8	3.6	4.2	4.7	12.1	12.7	16.2	16.1	11.2	8.9	1.5	3.5
1991	1.9	2.6	5.4	4.7	7.2	12.5	16.9	17.3	13.9	6.1	2.0	2.8
1992	3.2	0.6	3.8	6.5	11.9	12.9	15.8	17.6	12.0	5.7	4.4	0.9
1993	0.8	1.8	1.6	7.2	11.5	13.7	14.4	15.5	10.9	6.0	0.2	0.8
1994	0.1	0.6	6.3	4.4	11.1	13.9	17.8	16.9	11.6	7.8	5.4	0.9
1995	3.0	2.2	0.7	6.3	9.9	12.4	17.8	14.7	9.5	10.3	1.8	0.6
1996	0.1	2.6	0.8	6.9	10.2	14.4	14.9	14.8	9.0	7.4	2.3	1.1
1997	1.7	1.8	5.0	5.6	10.9	13.3	14.3	16.9	13.6	7.2	2.8	0.5
1998	1.6	0.8	3.3	5.8	11.3	14.4	16.4	15.8	11.3	7.6	1.3	2.5
1999	0.8	3.1	2.9	5.8	12.8	13.1	16.8	16.1	14.0	7.9	0.3	0.6
2000	3.7	0.6	3.2	7.4	12.1	15.7	13.8	16.2	12.8	8.1	2.9	1.3
2001	0.7	1.1	5.3	4.3	12.7	12.8	16.1	17.0	9.6	11.1	0.3	3.8
2002	2.1	2.6	4.8	6.8	9.9	16.6	15.6	14.9	11.1	8.1	3.4	1.2
2003	3.2	4.3	4.3	7.0	12.0	18.8	17.5	19.5	12.1	4.9	3.5	0.6
2004	1.1	0.8	1.8	6.6	9.7	14.1	15.8	15.8	13.1	9.3	1.5	1.2

Table E.10.2 Annual precipitation in mm from ten stations in the Puyango Basin, Ecuador

	Station									
Year	Saraguro	Celica	Zaruma	Marcabeli	San Lucas	Abamar	Pindo	Catocha	Cigne	Saracay
1963	865.5	1258.5	1305.6	1434.9	1031.6	1276	893.7	869.4	1933.5	1097.4
1964	637.1	1159.7	1629.7	1139.5	1245.7	989.5	1077.1	723.3	1768	71
1965	581.6	2444.7	1848.1	1881.8	1080.6	2581.4	1156	1494.8	3964.7	1110.1
1966	462.3	812.1	1207.1	1015	878	1143.5	843.4	895.7	1167.4	576.2
1967	570.1	992.4	1008.1	883.2	1024	862.3	1593.3	820.7	1476.4	705
1968	630.5	573.5	837.7	404.8	867.7	296.6	860.6	580.9	758.6	192.6
1969	793.8	1196.1	1499.7	1511.5	1046.7	1497.2	1200	1236.1	1825.4	1130.7
1970	971.3	703.1	1120.3	1068.1	1454	594.3	1057.3	708	980.7	906
1971	960	1130.3	1370.9	1338.3	1291.7	1224.2	1284.4	972.2	1187.3	1111.8
1972	833.2	1779.1	1374.5	1951	1095.5	2277.9	5477	1043.6	1570.7	1500
1973	726.6	1374.9	1739.8	2234.2	1938.7	1678.1	1250.3	897.4	1110.2	2272.9
1974	1041.7	934.1	1052.4	1297	1019.4	830.8	1139.7	729.7	1769.5	636.1
1975	880.3	1376.8	1715.4	2139.1	1356.8	1868	1500.6	1167.3	1141.5	1645.5
1976	920.9	2127.9	1574.7	1693.3	1063.3	1841.7	1222.9	1263.4	1141.5	2327.2
1977	634.5	1350.4	1179.3	1214.9	1023.5	1160.8	1081.8	568.5	873.7	846
1978	810.7	1257	769.3	719.1	1053.9	747.4	602.8	486.8	642.3	1818
1979	687.3	1167.8	1304.6	1366.6	834.4	1073.6	1167.1	580.5	745.7	201.3
1980	878.8	701.4	1279.3	1263.3	1168.3	508.8	948	622.3	1297.7	191.5
1981	761.5	1124	1246.9	1296.8	75.2	899.1	1151.4	898.3	1147.9	194.7
1982	885.3	1147.1	1772.7	1444.5	779	1293	1539.1	787.4	1420.9	252.2
1983	809.4	2508.1	1968.4	2335.9	1070.7	3618.2	1156	1231.5	1394.5	272.4
1984	866.9	1382.6	1452	1425.2	1126	1491.8	1723.7	1009.1	1256.5	221.3
1985	820.5	1490.3	772.8	723.3	1035.6	549.2	741.6	517.5	1195.7	810.8
Mean	783.9	1304	1349.1	1381.8	1067.8	1317.5	1333.4	874.1	1381.3	873.5
Standard deviation	146.1	503.9	338.3	492.3	323	755.9	942.8	272	662.7	674.8

Index

Printed and bound in the UK by
CPI Antony Rowe, Eastbourne

Printed and bound by CPI Group (UK) Ltd, Croydon, CR0 4YY

16/04/2025

14658504-0002